Lecture Notes in Computer Science 8485

Commenced Publication in 1973
Founding and Former Series Editors:
Gerhard Goos, Juris Hartmanis, and Jan van Leeuwen

Feifei Li   Guoliang Li   Seung-won Hwang
Bin Yao   Zhenjie Zhang (Eds.)

Web-Age Information Management

15th International Conference, WAIM 2014
Macau, China, June 16-18, 2014
Proceedings

Springer

Volume Editors

Feifei Li
University of Utah, Salt Lake City, UT 84112, USA
E-mail: lifeifei@cs.utah.edu

Guoliang Li
Tsinghua University, Beijing 100084, China
E-mail: liguoliang@tsinghua.edu.cn

Seung-won Hwang
POSTECH, Republic of Korea
E-mail: swhwang@postech.edu

Bin Yao
Shanghai Jiao Tong University, Shanghai, 200240, China
E-mail: yaobin@cs.sjtu.edu.cn

Zhenjie Zhang
Advanced Digital Sciences Center (ADSC), Singapore 138632
E-mail: zhenjie@adsc.com.sg

ISSN 0302-9743   e-ISSN 1611-3349
ISBN 978-3-319-08009-3   e-ISBN 978-3-319-08010-9
DOI 10.1007/978-3-319-08010-9
Springer Cham Heidelberg New York Dordrecht London

Library of Congress Control Number: Applied for

LNCS Sublibrary: SL 3 – Information Systems and Application, incl. Internet/Web
and HCI

Typesetting: Camera-ready by author, data conversion by Scientific Publishing Services, Chennai, India

Printed on acid-free paper

Springer is part of Springer Science+Business Media (www.springer.com)

Foreword

Welcome to the proceedings of the 15th International Conference on Web-Age Information Management (WAIM 2014). WAIM is a leading international conference for researchers, practitioners, developers, and users to share and exchange ideas, results, experience, techniques, and tools in connection with all aspects of Web data management. The rapid development and prevalent use of the Web highlight the need for new technologies for the design, implementation, and management of Web-based information systems. It is evident that high-volume, distributed, heterogeneous data increasingly found on the Web introduces new challenges beyond the capability of traditional database systems. It calls for new Web-based information systems to provide effective and efficient techniques for specifying, retrieving, integrating, exchanging, querying, archiving, managing, and cleaning the data on the Web. The study of Web-based information systems requires seamless integration of technologies from database systems and Internet management, and is essential to the development of Web services, e-commerce and e-science, among other things.

The WAIM conference provides a forum for Web data management researchers to exchange new results. The call for papers attracted submissions from many research institutions in the USA, UK, Australia, Germany, Japan, India, and China. The Program Committee accepted 48 high-quality full papers and 35 short papers covering theoretical research as well as empirical studies. We hope that the proceedings serve as a valuable reference for researchers and developers interested in Web data management techniques.

We would like to thank all authors for their contributions. We would like to express our gratitude to the Program Committee members who worked hard in reviewing papers and providing suggestions for improvements. Special thanks to our keynote speakers, Prof. Jiawei Han, Prof. Amr El Abbadi, Prof. Yufei Tao, Dr. Wen-syan Li, and Dr. Ricky Sun. We would also like to express our gratitude to Prof. Weiyi Meng for leading the Steering Committee of the workshop.

We hope that WAIM helps expand the community of Web data management researchers and practitioners all over the world, and that it provides a ground for new ideas in this important field.

April 2014

Feifei Li
Guoliang Li
Seung-won Hwang

Organization

Honorary Chair

Wei Zhao   University of Macau, Macau

General Co-chairs

Zhiguo Gong   University of Macau, Macau
Jian Pei   Simon Fraser University, Canada

Program Co-chairs

Feifei Li   University of Utah, USA
Guoliang Li   Tsinghua University, China
Seung-Won Hwang   POSTECH, Korea

Workshop Co-chairs

Yueguo Chen   Renmin University, China
Wolf-Tilo Balke   Technische Universität Braunschweig,
     Germany
Jianliang Xu   Hong Kong Baptist University, Hong Kong

Demo Co-chairs

Tingjian Ge   University of Massachusetts at Lowell, USA
Hongzhi Wang   Harbin Institute of Technology, China

Panel Co-chairs

Lizhu Zhou   Tsinghua University, China
Xiaofang Zhou   University of Queensland, Australia

DYL Series Co-chairs

Xuemin Lin   University of New South Wales, Australia
Ke Yi   Hong Kong University of Science and
     Technology, Hong Kong
Aoying Zhou   East China Normal University, China

Industry Co-chairs

Shimin Chen   Chinese Academy of Sciences, China
Cong Yu   Google Co., USA

Publication Co-chairs

Bin Yao   Shanghai Jiao Tong University, China
Zhenjie Zhang   Advanced Digital Sciences Center, Singapore

Best Paper Award Co-chairs

Jianzhong Li   Harbin Institute of Technology, China
Kyuseok Shim   Seoul National University, Korea
Ge Yu   Northeast University, China

Publicity Co-chairs

Xiangfeng Luo   Shanghai University, China
Hua Wang   University of Southern Queensland, Australia
Wenjun Zhou   University of Tennessee at Knoxville, USA

Local Organization Co-chairs

Leong Hou U   University of Macau, Macau
Lawrence Si   University of Macau, Macau

Sponsorship Chair

Wong Fai   University of Macau, Macau

Registration Co-chairs

Shirley Siu   University of Macau, Macau
Zhuang Yan   University of Macau, Macau

Finance Chair

Leong Hou U   University of Macau, Macau

Web Co-chairs

William Sio   University of Macau, Macau
Junjie Zhang   University of Macau, Macau

Steering Committee Liaison

Weiyi Meng   SUNY Binghamton, USA

CCF DBS Liaison

Xiaofeng Meng   Renmin University, China

SAP Summer School

Wen-Syan Li   SAP Co., China

APWeb Liaison

Xuemin Lin   University of New South Wales, Australia

WISE Liaison

Yanchun Zhang   Victoria University, Australia

Program Committee

Toshiyuki Amagasa   University of Tsukuba, Japan
Ning An   Hefei University of Technology, China
Zhifeng Bao   National University of Singapore, Singapore
Sourav Bhowmick   Nanyang Technological University, Singapore
Gang Chen   Zhejiang University, China
Jinchuan Chen   Renmin University, China
Yueguo Chen   Renmin University, China
Enhong Chen   University of Science and Technology of China,
                             China
Lei Chen   Hong Kong University of Science and
                             Technology, Hong Kong
Hong Chen   Renmin University, China
Ling Chen   University of Technology, Sydney, Australia
Reynold Cheng   The University of Hong Kong, Hong Kong
James Cheng   The Chinese University of Hong Kong,
                             Hong Kong
David Cheung   The University of Hong Kong, Hong Kong
Gao Cong   Nanyang Technological University, Singapore
Bin Cui   Peking University, China

Alfredo Cuzzocrea	University of Calabria, Italy
Ting Deng	Beihang University, China
Zhiming Ding	Institute of Software, Chinese Academy of Sciences, China
Dejing Dou	University of Oregon, USA
Ju Fan	National University of Singapore, Singapore
Yaokai Feng	Kyushu University, Japan
Aimin Feng	Nanjing University of Aeronautics and Astronautics, China
Jun Gao	Peking University, China
Yunjun Gao	Zhejiang University, China
Hong Gao	Harbin Institute of Technology, China
Yong Ge	Rutgers University, USA
Xueqing Gong	East China Normal University, China
Jingfeng Guo	Yanshan University, China
Zhenyin He	Fudan University, China
Xiaofeng He	East China Normal University, China
Luke Huan	University of Kansas, USA
Jimmy Huang	York University, Canada
Jianbin Huang	Xidian University, China
Xuanjing Huang	Fudan University, China
Yan Jia	National University of Defense Technology, China
Lili Jiang	Max Planck Institute, Germany
Peiquan Jin	University of Science and Technology of China, China
Wu Kui	Victoria University, Canada
Wookey Lee	Inha University, Korea
Carson K. Leung	University of Manitoba, Canada
Guohui Li	Huazhong University of Science and Technology, China
Chengkai Li	University of Texas at Arlington, USA
Zhoujun Li	Beihang University, China
Chuan Li	Sichuan University, China
Tao Li	Florida International University, USA
Qingzhong Li	Shandong University, China
Zhanhuai Li	Northwestern Polytechnical University, China
Jian Li	Tsinghua University, China
Cuiping Li	Renmin University, China
Xiang Lian	Hong Kong University of Science and Technology, Hong Kong
Hongyan Liu	Tsinghua University, China
Qi Liu	University of Science and Technology of China, China
Yiqun Liu	Tsinghua University, China

Jiaheng Lu	Renmin University, China
Ping Luo	HP Research Labs, China
Jizhou Luo	Harbin Institue of Technology, China
Shuai Ma	Beihang University, China
Yang-Sae Moon	Kangwon National University, Korea
Akiyo Nadamoto	Konan University, Japan
Shinsuke Nakajima	Kyoto Sangyo University, Japan
Weiwei Ni	Southeast University, China
Baoning Niu	Taiyuan University of Technology, China
Hiroaki Ohshima	Kyoto University, Japan
Zhiyong Peng	Wuhan University, China
Tieyun Qian	Wuhan University, China
Xiaolin Qin	Nanhang University, China
Jianbin Qin	University of New South Wales, Australia
Ning Ruan	Kent State University, USA
Leong Hou U	University of Macau, Macau
Jie Shao	National University of Singapore, Singapore
Lidan Shou	Zhejiang University, China
Weiwei Sun	Fudan University, China
Chih-Hua Tai	National Taiwan University, China
Nan Tang	Qatar Computing Research Institute, Qatar
Taketoshi Ushiama	Kyushu University, Japan
Changxuan Wan	Jiangxi University, China
Jianmin Wang	Tsinghua University, China
Hua Wang	University of Southern Queensland, Australia
Wei Wang	University of New South Wales, Australia
Yijie Wang	National University of Defense Technology, China
Wei Wang	Fudan University, China
Jiannan Wang	University of California, Berkerley, USA
Peng Wang	Fudan University, China
Hongzhi Wang	Harbin Institue of Technology, China
Qiang Wei	Tsinghua University, China
Junjie Wu	Beihang University, China
Yinghui Wu	University of California, Santa Barbara, USA
Shengli Wu	Jiangsu University, China
Yanghua Xiao	Fudan University, China
Reynold Xin	University of California, Berkerley, USA
Jianliang Xu	Hong Kong Baptist University, Hong Kong
Ning Yang	Sichuan University, China
Lianghuai Yang	Zhejiang University of Technology, China
Xiaochun Yang	Northeast University, China
Bin Yao	Shanghai Jiao Tong University, China
Ke Yi	Hong Kong University of Science and Technology, Hong Kong

Jian Yin	Zhongshan University, China
Ge Yu	Northeast University, China
Jeffrey Yu	The Chinese University of Hong Kong, Hong Kong
Xiaohui Yu	Shandong University, China
Hua Yuan	University of Electronic Science and Technology of China, China
Yong Zhang	Tsinghua University, China
Zhenjie Zhang	Advanced Digital Sciences Center, Singapore
Ming Zhang	Peking University, China
Dongxiang Zhang	National University of Singapore, Singapore
Xiangliang Zhang	King Abdullah University of Science and Technology, Saudi Arabia
Ying Zhao	Tsinghua University, China
Wenjun Zhou	University of Tennessee at Knoxville, USA
Aoying Zhou	East China Normal University, China
Junfeng Zhou	Yanshan University, China
Xuan Zhou	Renmin University, China
Shuigeng Zhou	Fudan University, China
Feida Zhu	Singapore Management University, Singapore
Xingquan Zhu	University of Technology, Sydney, Australia
Fuzhen Zhuang	ICT, Chinese Academy of Sciences, China
Yi Zhuang	Zhejiang Gongshang University, China
Lei Zou	Peking University, China
Quan Zou	Xiamen University, China
Zhaonian Zou	Harbin Institute of Technology, China

Industrial Program Committee

Zhao Cao	IBM Research, China
Peng Jiang	Alibaba Group, China
Bin Liu	Google Co., USA
Shin-Ming Liu	Intel Labs, China
Tantan Liu	Google Co., USA
Yue Lu	Twitter Co., USA
Ping Luo	HP Research Labs, China
Junfeng Pan	Facebook Co., USA

Table of Contents

Data Mining I

Data and Information Quality

Data Mining II

Information Extraction

XML, Semi-structured and Uncertain Data

Session 9: Mobile and Pervasive Computing

Data Mining III

Stream, Time-Series

Security and Privacy I

Semantic Web

Cloud Computing

New Hardware

Crowdsourcing

Data Mining IV

Social Computing

Information Retrieval II

Information Retrieval III

Data Mining V

Information Retrieval IV

Data Mining VI

Privacy and Security II

Information Retrieval V

Faster MaxScore Document Retrieval
with Aggressive Processing

Kun Jiang, Xingshen Song, and Yuexiang Yang

College of Computer
National University of Defense Technology
Changsha, Hunan Province, China
{jiangkun,songxsh,yyx}@nudt.edu.cn

Abstract. Large search engines process thousands of queries per second over billions of documents, making a huge performance gap between disjunctive and conjunctive queries. An important class of optimization techniques called top-k processing is therefore used to narrow the gap. In this paper, we propose an aggressive algorithm based on the document-at-a-time (DAAT) MaxScore, aiming at further reducing the query latency of disjunctive queries. Essentially, our approach, named Aggressive MaxScore (AMaxScore), can speed up quickly by fine-tuning the initial top-k threshold, which allows a first aggressive process and then a supplementary process if not enough results are returned. Experiments with TREC GOV2 collection show that our approach reduces disjunctive query processing time by almost 15.4% on average over the state-of-the-art MaxScore baseline, while still returns the same results as the disjunctive evaluation.

Keywords: Query processing, Top-k processing, DAAT, Aggressive MaxScore.

1 Introduction

Given that search engines need to answer queries within fractions of a second, naively traversing the huge amount of basic index structure, which could take hundreds of milliseconds, is not acceptable [1]. This efficiency problem has been largely addressed by top-k processing methods [2-4, 7, 8], which in a nutshell means returning the best k results without an exhaustive traversal of the relevant posting lists. However, most top-k processing algorithms start with a threshold of zero, and then update the value while results are discovered. Thus, the algorithms can only speed up until there are k results obtained, and too small initial threshold can make lots of documents be promising candidates temporarily.

The slow startup problem in top-k processing has motivated a lot of work on optimization techniques. Fontoura et al. [5] propose a new technique to speed up MaxScore [2-4], the most famous top-k processing method, by splitting the query into a short query and a long query. The results (or threshold) obtained by quickly evaluated short query is used to speed up the evaluation of the long query. The term bounded MaxScore method [3] improves upon the DAAT MaxScore to set a better initial threshold by using toplist index structures [6]. Recent work in [7] builds a two-tiered index upon the

F. Li et al. (Eds.): WAIM 2014, LNCS 8485, pp. 1–4, 2014.

blockmax index, resulting in considerable performance gains. Different from previous work, we propose a novel improvement of MaxScore that requires no modifications to the index structure. Our method allows a first aggressive process with one posting list's max score as the top-k threshold and then a supplementary process if not enough results are returned, thus the name Aggressive MaxScore (AMaxScore). In the following sections, we present and evaluate our AMaxScore on TREC GOV2 collection.

2   Aggressive MaxScore

As mentioned, a proper threshold can exactly avoid processing documents that are unable to be in the top results. Due to the fact that the final threshold θ can only be obtained by a complete query evaluation, one possible approach is to estimate the threshold as θ' with a guarantee that $\theta' \leq \theta$, in order to ignore any documents with scores lower than θ'. The threshold θ' is commonly automatically set to the kth score of the documents in the results heap during the query processing. It is obvious that the closer θ' is to θ, the faster the algorithm runs, because more documents can be ignored.

If a proper threshold that satisfies the condition of $\theta' \leq \theta$ can be estimated at the initial stage of the query, the whole query latency will be reduced. The threshold estimation methods are very popular for most optimization techniques, but most of these methods require modification to the underlying index structures for storing high quality documents. Also, it is hard to exactly estimate threshold that satisfies the condition of $\theta' \leq \theta$, and provide extra memory for processing more indexes. In AMaxScore, we estimate an approximate threshold θ' that don't need to be strictly guaranteed that $\theta' \leq \theta$, but somehow closer to θ. When complete the query processing, we check the number of the results. If the number of the results is not enough, we can conclude that the initial threshold θ' is a little bigger, i.e., $\theta' \geq \theta$, and we should iteratively conduct supplementary passes with smaller initial thresholds until the results is enough. The threshold θ' can be set to various accumulative score of term upper bounds, i.e., the minimum or maximum score of term upper bounds, the sum of all term upper bounds etc. Suppose the threshold is set to the sum of some or all of the term upper bounds, it means a document that can make it into the top result heap only if it contains term not just appear in the terms set above. Also, if the threshold is set to one of the term upper bounds, it means a document that only contains that term, but no other terms, can never make it into the top results heap. Thus, there is no need to score any documents that contain only that term. Performance gains are achieved when the processing does not run a supplementary pass.

This kind of method tends to be less studied for top-k processing methods. Though the idea of the Two-pass method by Broder et al. [8] is similar to our approach, it focuses on WAND approach and is just used to transfer normal queries into AND-queries. Thus, we can consider it as a special case of our method. Note that in our method, the partial results by the aggressive pass can be used in the supplementary pass to reduce a large amount of heap sort operations.

3 Experiments

We use the TREC GOV2 collection containing about 25.2 million documents and about 32.8 million terms with an uncompressed size of 426GB. We build inverted index with 128 docIDs per block, using PForDelta compression, removing stopwords, and applying Porter's stemmer. The final compressed index size is 7.57GB. We use 10000 queries randomly selected from the TREC2005 Efficiency track queries using distinct amounts of terms with $|q| \geq 2$. Our experiments were performed on an Intel(r) Xeon(r) E5620 processor running at 2.40 GHz with 8GB of RAM and 12,288KB of cache. All methods were implemented in JAVA on Terrier IR platform [9] with Okapi BM25 as the ranking function. In every experiment, the index was preloaded into memory and the numbers are averaged over 5 independent runs.

3.1 Results

In this section, we compare AMaxScore with the state-of-the-art MaxScore [4] by fine tuning the threshold θ'. The threshold θ' is set to the minimum value, maximum value and average value of the query term upper bounds respectively in the aggressive pass. In extreme cases, too large threshold will lead to completely ineffectiveness of the aggressive pass, resulting in some of the top-k results obtained in the supplementary pass. Thus, we run an experiment of AMaxScore with a threshold set to the sum of all term upper bounds. In additional, we also test the effect of resetting the heap after an aggressive pass for query processing performance. Table 1 shows the query processing time using different algorithms with different number of documents returned.

Table 1. Average processing times in ms of the AMaxScore series of algorithms with different number of results based on TREC GOV2

Algorithm	avg	k=10	k=50	k=100	k=500	k=1000
MaxScore_Baseline	55.7	34.8	44.2	50.3	69.4	79.6
AMaxScore_Min(θ')	52.3	34.3	42.7	47.8	63.5	73.2
AMaxScore_Max(θ')	47.1	32.4	40.1	44.2	55.7	63.0
AMaxScore_Avg(θ')	50.5	33.4	42.0	46.6	61.2	69.5
AMaxScore_Max(θ')_HR	48.6	33.7	41.6	45.6	58.2	63.8
AMaxScore_TooLarge(θ')	55.5	34.2	43.4	48.8	68.7	82.7

From Table 1, we can see that all optimized techniques improve greatly over the MaxScore baseline. For different k, AMaxScore_Max(θ') always performs the best, which achieves an average improvement of 15.4% over the MaxScore baseline. We also observe a little performance degradation in AMaxScore_TooLarge(θ') over the baseline. This is mainly because too large threshold value will cause ineffectiveness of the aggressive pass. Most results are returned in the supplementary pass, and the aggressive processing becomes an extra cost for the whole query processing. In addition, the improvement achieved by AMaxScore_Max(θ')_HR is not as much as that achieved by AMaxScore_Max(θ'). This suggests that keeping the query results stored in the heap is helpful for enhancing the performance. Although the results in the first pass can be obtained by a supplementary pass again, a couple of heap sort operations are reduced when the same results are inserted into the heap. As we increase k, though query processing time increases for all methods, our AMaxScore achieves an increasingly improvement over the baseline. The explanation is that the estimated threshold θ' is closer to the final top-k threshold θ with a larger k. The best case occurred in AMaxScore_Max(θ') with k equals 1000, being about 20% faster. Thus, we can conclude that fine-tuning the top-k threshold can result in significant performance improvement for different corpus.

4   Conclusions

In this paper we have proposed a new method to speed up MaxScore by fine-tuning the top-k threshold, which can trigger an aggressive processing with supplementary original MaxScore processing if not enough results returned. Experimental results with TREC GOV2 showed that our AMaxScore significantly outperforms the previous methods with an average improvement of 15.4%. The best case produced almost 20% performance gains without sacrificing result quality.

References

1. Dean, J.: Challenges in building large-scale information retrieval systems: invited talk. In: Proceedings of the Second ACM International Conference on Web Search and Data Mining, p. 1. ACM (2009)
2. Turtle, H., Flood, J.: Query evaluation: strategies and optimizations. Information Processing & Management 31(6), 831–850 (1995)
3. Strohman, T., Turtle, H., Croft, W.B.: Optimization strategies for complex queries. In: Proceedings of the 28th Annual International ACM SIGIR Conference on Research and Development in Information Retrieval, pp. 219–225. ACM (2005)
4. Jonassen, S., Bratsberg, S.E.: Efficient compressed inverted index skipping for disjunctive text-queries. In: Clough, P., Foley, C., Gurrin, C., Jones, G.J.F., Kraaij, W., Lee, H., Mudoch, V. (eds.) ECIR 2011. LNCS, vol. 6611, pp. 530–542. Springer, Heidelberg (2011)
5. Fontoura, M., Josifovski, V., Liu, J., Venkatesan, S., Zhu, X., Zien, J.: Evaluation strategies for top-k queries over memory-resident inverted indexes. In: Proceedings of the VLDB Endowment, vol. 4(12), pp. 1213–1224 (2011)
6. Brown, E.W.: Fast evaluation of structured queries for information retrieval. In: Proceedings of the 18th Annual International ACM SIGIR Conference on Research and Development in Information Retrieval, pp. 30–38. ACM (1995)
7. Rossi, C., de Moura, E.S., Carvalho, A.L., da Silva, A.S.: Fast document-at-a-time query processing using two-tier indexes. In: Proceedings of the 36th International ACM SIGIR Conference on Research and Development in Information Retrieval, pp. 183–192. ACM (2013)
8. Broder, A.Z., Carmel, D., Herscovici, M., Soffer, A., Zien, J.: Efficient query evaluation using a two-level retrieval process. In: Proceedings of the Twelfth International Conference on Information and Knowledge Management, pp. 426–434. ACM (2003)
9. Terrier, http://www.terrier.org/

Finding Novel Patents Based on Patent Association

Ling Feng[1], Zhiyong Peng[1], Bin Liu[1,*] and Dunren Che[2]

[1] Computer School, Wuhan University, Wuhan 430072, P.R. China
{fengling,Peng,binliu}@whu.edu.cn
[2] Dept. of Computer Science, Southern Illinois University, Carbondale, IL62901, USA
dche@cs.siu.edu

Abstract. Patent retrieval is critical for technological researches, inventions, and innovative entrepreneurship. There exist a lot of methods for patent retrieval, which usually merely find relevant patents. As a result, outdated patents are frequently found and even ranked ahead of more interesting ones in the result list. However, in most cases, enterprisers and researchers only concern cutting-edge techniques and research results. Novelty-based patent retrieval thus becomes extremely important nowadays. In this paper, we propose an innovative patent finding method that exploits the broad associations between patents. More specifically in this paper, a new concept of patent novelty is first introduced and a novel ranking algorithm is then proposed for proper ranking of patents. Following that, in order to handle rank variations caused by the arrival of new patents, an efficient rank-update algorithm is designed. We have done extensive experimental study that well shows the effectiveness and the efficiency of our proposed method.

Keywords: patent novelty, patent association, comparative novelty rate.

1 Introduction

Patents nowadays represent one of the largest technical information sources and are major carriers of scientific and technological knowledge. According to the statistical data from World Intellectual Property Organization (WIPO), 90% to 95% of the inventions and research results in the world could be found in the patents, in which about 70% of the inventions have never been published in other literatures [1]. Due to the importance of patent, the information retrieval technique for patent documents is in urgent need.

Generally, researches on patent retrieval include prior-art retrieval [2-4], invalidity retrieval [5-7] and cross-language retrieval [8-10]. These methods first find patents that match the queries, and then rank them by semantic similarity. Since there is no novelty criterion in the process of retrieval, some (even large amounts of) old-fashioned and invalid patents may be retrieved and ranked ahead in the result list. However, these patents may be meaningless to searchers in certain scenario [11], e.g., enterprisers seeking novel techniques for product quality enhancement.

In [11], Mohammad *et al.* presented a COA method to find novel patents in the patent dataset. The method evaluates the novelty of a patent by the novelty of the

---

* Corresponding author.

F. Li et al. (Eds.): WAIM 2014, LNCS 8485, pp. 5–17, 2014.
© Springer International Publishing Switzerland 2014

keywords contained in the patent. The novelty of keywords is measured by a time-dependent function, which decreases as the time is passing. In other words, the novelty of a patent in the COA method relates and only relates to the time factor. However, in practice the novelty of a patent is also heavily influenced by the newer, related patents. That is, with a specific technical problem, when an improved solution is proposed via a new patent, the existing related patents become less novel. Therefore, in order to more effectively find genuine novel patents, a better approach shall consider not only the time factor but also the influence of other related patents, which forms the main theme of this paper.

In this paper, we provide a new method to find novel patents based on the technical associations among the patents, besides the time factor. The main contributions of our work can be highlighted as follows:

- A patent network structured is proposed to model the technical associations among patents. Based on the network, a new concept of patent novelty is defined to take both the time factor time and the technical associations into consideration.
- An original novelty ranking algorithm, called *Novelty-Rank*, is proposed that can effectively compute patents' novelty and rank them.
- In order to efficiently handle the rank variations caused by the arrival of new and related patents, a companion algorithm, i.e., *Update-Rank*, is proposed.
- The effectiveness and efficiency of our method (and algorithms) is verified by a set of comprehensive experiments.

The rest of this paper is organized as follows. Section 2 reviews related works. Section 3 introduces our definitions w.r.t. patent novelty and presents our *Novelty-Rank* algorithm. Section 4 expounds our *Update-Rank* algorithm. Section 5 discusses experimental results. Finally, Section 6 gives the conclusions.

2   Related Work

As patent contains a large number of latest technical information, patent retrieval is paid more attention in recent years. Some famous international conferences or organizations, such as SIGIR[12], ACL[13] and NTCIR[14], have held the corresponding workshops on patent retrieval, and many fruitful research results are discussed. Divided by purpose, patent retrieval includes prior-art retrieval, invalidity retrieval, and cross-language retrieval at present. Prior-art retrieval is intended to access prior state of a technology or domain, invalidity retrieval aims to verify that whether a patent is invalid or not, and cross-language patent retrieval enables us to retrieve information from other languages using a query written in the language we are familiar with. In prior-art retrieval, as the query is always too long, it generally transforms the query into a group of relevant search queries. Because of the presence of ambiguous terms, the selection to search queries is a challenge task. Parvaz *et al.* [2] propose a query modeling approach utilizing patent characteristics and semantic disambiguation to generate more precise queries. Shariq *et al.* [3] not only transform the query into several search queries, but also expand queries with pseudo relevance feedback to increase the retrieval effectiveness. Walid *et al.* [4] make a comparison between two search approaches in patent prior-art search, and find less time and effort can be exerted by applying simple IR approaches when initial citations are provided. In invalidity retrieval, Hisao *et al.* [5] propose a two-stage patent retrieval method to find the relevant patents. In the first stage, general text analysis and retrieval methods are applied to improve recall. In the second stage, the top N documents retrieved in the first stage are rearranged to improve precision using the claim structure. As the query usually includes more than one topic, Toru *et al.* [6] provide a method to divide the query as several sub-queries, and the relevance scores in each sub-query are integrated to determine the final relevant documents. To further improve the retrieval effectiveness, Atsushi *et al.* [7] propose a method to combine text-based and citation-based retrieval methods in the invalidity patent search. In cross-language patent retrieval, Chen *et al.* [8] apply Latent Semantic Indexing to extract concepts from each set of patent documents and utilizes the IPC codes to construct a cross-language mediator that express patent documents in different languages. And Li *et al.* [9,10] propose a method of cross-language patent retrieval using Kernel Canonical Correlation Analysis. All of the above methods focus on how to find the patents which are semantically relevant with the query. Specially, Mohammad *et al.* [11] propose a COA method to find novel patents, and patent novelty is measured by the novelty of keywords a patent contains.

3   Association-Based Patent Novelty

We first define patent novelty before we calculate it and rank the patents by it. In this section, we first construct a novelty-association network through measuring the relations between patent documents, and then define the concept of patent novelty based on the network. Finally, we give the novelty-rank algorithm to rank patents.

3.1   Novelty-Association Network

Generally, there are two kinds of relations among patents, i.e., citation and semantic similarity. Citation implies that there is technical association between two patents, and the citing patent generally has certain technique improvement to the cited patent on a same topic. Semantic similarity is often used to measure the strength of association between pairs of patents. The greater semantic similarity between two patents is, the stronger association they have. In this paper, we consider both the relations when constructing our novelty-association network, and we take the assumption that the semantic similarity between each pair of patents has been calculated using the method presented in [17]. Thus, we define *Technical Association* and *Strength of Technical Overlap* respectively as follows.

Definition 1. (Technical Association) Given a patent dataset $D = \{d_1, d_2, \ldots, d_n\}$, $\forall d_i \in D$, there is a time element T_i indicating its publishing time. Then, $\forall d_j \in D$, $T_j > T_i$, the technical association (*TA*) between d_i and d_j is defined as follows:

$$TA(d_j, d_i) = \begin{cases} 1 & if\ d_j\ cites\ d_i \\ 0 & otherwise \end{cases} \tag{1}$$

If $TA(d_j, d_i)=1$, we say d_i and d_j are technique-associated patents.

Definition 2. (Strength of Technical Overlap) $\forall d_i, d_j \in D$, if $TA(d_j, d_i) = 1$, then the strength of technical overlap *(STO)* between d_i and d_j is defined as the semantic similarity between them as follows:

$$STO(d_j, d_i) = \begin{cases} sim(d_i, d_j) & if\ TA(d_j, d_i) = 1 \\ 0 & otherwise \end{cases} \quad (2)$$

where $sim(d_i, d_j)$ is similarity between the two patents and can be calculated by [17].

Next, we use the concept of *Comparative Novelty Rate* to model the influence of a newly published patent on a previous, technique-related patent. Intuitively, for d_i, $d_j \in D$, if $TA(d_j, d_i) = 1$, a *low* $STO(d_j, d_i)$ value indicates that the newer patent d_j does not adopt a similar technique as d_i, but introduces quite new ideas or makes significant improvement on the technique of d_i, which naturally causes the technique d_i losing its novelty and attractiveness. Namely, d_i is *less novel* compared to d_j. Thus, *Comparative Novelty Rate* of d_i to d_j is positively related to the strength of technical overlap between them, and is defined as follows.

Definition 3. (Comparative Novelty Rate) $\forall d_i, d_j \in D$, if d_j cites d_i, then the comparative novelty rate of d_i to d_j is defined as

$$CNR(d_i, d_j) = \begin{cases} \frac{STO(d_j, d_i)}{2 - STO(d_j, d_i)} & if\ TA(d_j, d_i) = 1 \\ 0 & otherwise \end{cases} \quad (3)$$

where *CNR* (d_i, d_j) denotes the comparative novelty rate of d_i to d_j.

Using *Technical Association* and *Comparative Novelty Rate*, we define novelty-association network.

Definition 4. (Novelty-Association Network) Given a patent dataset $D = \{d_1, d_2, \ldots, d_n\}$, the novelty-association network is defined as a directed graph $G = \{V, E, T\}$, in which each node respects each patent in D, and $T = \{T_1, T_2, \ldots, T_n\}$ denotes the publishing time of each patent. For $v_i, v_j \in V$, if d_j cites d_i, then there exist an edge $<v_j, v_i>$ from v_j to v_i and the weight on $<v_j, v_i>$ is $CNR(d_i, d_j)$.

Example 1. In Fig. 1, there exists a novelty-association network $G = \{V, E, T\}$ of a patent dataset D. v_1, v_2, v_3, v_4, v_5 represent the patents documents d_1, d_2, d_3, d_4, d_5 in D respectively. The lateral axis value of each node denotes the publishing time of each patent. The edge denotes the citation between patents, and the weight on each edge is the comparative novelty rate between two nodes. For example, node v_1 and v_2 respect patent d_1 and d_2, which are published at 2010 and 2011 respectively. In addition, an edge $<v_2, v_1>$ denotes that d_2 cites d_1, and comparative novelty rate of d_1 to d_2 is 0.5.

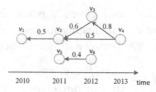

Fig. 1. An Example of Novelty-Association Network

Based on above definitions, we establish novelty-association network via the following procedure (Algorithm 1).

Algorithm 1. Establishing Novelty-Association Network

Input: $D=\{d_1,d_2,\ldots,d_n\}$, $T=\{T_1,T_2,\ldots,T_n\}$
Output: $G=(V,E,T)$
1. *foreach* d_i in D
2.   add node v_i in G;
3. *endfor*
4. *for* $i=1$ to N
5.   *for* $j=1$ to N
6.   *if* d_j cites d_i *then*
7.   add an edge $<v_j,v_i>$;
8.   calculate $CNR(v_i, v_j)$ by formula (1) (2)(3);
9.   *endfor*
10. *endfor*
11. *Output* $G=(V,E,T)$;

3.2   Patent Novelty and Novelty-Rank Algorithm

After the novelty-association network is established, we define patent novelty on it. As we know, the novelty of a patent is strongly related with its publishing time and the current state of the research problem. If a patent has been published for a long time, the novelty of the patent will be low. Moreover, if a patent has many latter-presented technique-associated patents of the same topic, the patent novelty of the patent will be low too. Thus, we define patent novelty based on both time factor and current state of the patent.

Definition 5. (Patent Novelty) $\forall v_i \in V$, V_i is the patent dataset which cites v_i in D, then the patent novelty of v_i is defined as

$$PNovelty(v_i) = \begin{cases} min_{(v_j,v_i)\in E} \{PNovelty(v_j) \times CNR(v_i, v_j) \times e^{-\beta \Delta t_{ji}}\} & if \; \exists (v_j, v_i) \in E \\ 1 \times e^{-\beta(T_{current}-T_i)} & otherwise \end{cases} \tag{4}$$

where $PNovelty(v_i)$ is the patent novelty of v_i; Δt_{ji} is the interval of the publishing time between v_j and v_i and computed as T_j-T_i, in which T_j and T_i are the publishing time of v_j and v_j respectively; $T_{current}$ is the current time; β is a damp factor and $\beta \in [0,+\infty)$.

As in Definition 5, for $v_i \in V$, if v_i has no newer and technique-associated patents, then its patent novelty $PNovelty(v_i)$ only depends on its publishing time. The longer one patent has been published, the lower its patent novelty will become (we therefore assume patent novelty obeys the law of exponential decay for the parameter that is the elapsed time since the patent was published). Otherwise, its patent novelty $PNovelty(v_i)$ is calculated based on the patents which cite v_i. The less novel the citing patents are, the less novel the cited one is. And $e^{-\beta \Delta t_{ji}}$ takes the gap of publishing time between the two patents considered. The longer this gap is, the lower the novelty of patent v_i is.

In Fig. 1, assume the current time is 2013 and β is set to be 0.5. Thus, We can compute the patent novelty of v_6 as $PNovelty(v_6)=e^{-0.5(2013-2012)}=e^{-0.5}$. The patent novelty of v_2 is calculated by the patents which cite v_2. Namely, $PNovelty(v_2)$

$=min\{$ $PNovelty(v_3)\times CNR(v_2,v_3)\times e^{-\beta(2012-2011)}$, $PNovelty(v_4)\times CNR(v_2,v_4)\times e^{-\beta(2013-2011)}\}$
$=min\{0.6e^{-0.5}\times PNovelty(v_3), 0.5e^{-1}\times PNovelty(v_4)\}$. As it can be seen, the computing of patent novelty of v_2 can be performed only after $PNovelty(v_3)$ and $PNovelty(v_4)$ are obtained. Thus, we must find the processing sequence before computing and ranking the patent novelty of each patent.

Based on Definition 5, we can easily get the following lemma (Lemma 1).

Lemma 1. The patent novelty of a patent which is published later is independent with the patents which are published earlier.

According to Lemma 1, we never need to compute the novelty of each patent iteratively. Instead, we first rank the nodes by the patents' publishing time, and then compute the novelty of each patent *in sequence* and rank them (as shown in Algorithm 2).

Algorithm 2. Novelty-Rank

Input: $G=(V,E,T)$
Output: Patent ranking list
1. Rank the nodes in V by their publishing time;
2. *for* i=1 to N
3.   compute $PNovelty(v_i)$ by formula (4);
4. *endfor*
5. Rank the patents by their patent novelty with Merge Sort
6. Output the patent ranking list;

For *example* 1, we first rank the nodes in G by their publishing time, and get the rank $\{v_4,v_3,v_6,v_2,v_5,v_1\}$. Then we compute patent novelty of each node as $PNovelty(v_4)=1$, $PNovelty(v_3)=0.8\times e^{-0.5}=0.49$, $PNovelty(v_6)=e^{-0.5}=0.61$, $PNovelty(v_2)=0.48e^{-1}=0.18$, $PNovelty(v_5)=0.4e^{-1}=0.15$, $PNovelty(v_1)=0.24e^{-1.5}=0.05$. Therefore, the ranking list of patents is $\{v_4,v_6,v_3,v_2,v_5,v_1\}$.

4   Rank-Update Algorithm

Novelty-Rank algorithm could rank the patents by their patent novelty on a specific time t. However, when new patents are published at time $t+1$, namely the new nodes are added into the novelty-association network, the rank will change. (In Fig.2, when v_7 is added to the network in time $t+1$, and patent novelty of each node varies.)

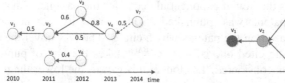

Fig. 2. The Novelty-Association Network at time t+1

Fig. 3. The Level of Node

Although we can compute and re-rank the patent novelty of each patent at time $t+1$ with Novelty-Rank algorithm, but it must be time-wasting as the ranking list at time t is not used. As it can be known, if most of patents keep the same sequences at time t and $t+1$, then the running time of ranking the patents at $t+1$ can be cut down.

Firstly, the level of node is defined to reflect the computing sequence of patent novelty for further measuring the relation of the ranks at t and $t+1$.

Definition 6. (Level of Node) Given novelty-association network $G=(V,E,T)$, $v_i \in V$, $V_i=\{v_j|v_j \in V$ and $<v_j,v_i> \in E\}$. If $|V_i|=0$, then the v_i is on level 1, and denoted as $level(v_i)=1$. Otherwise, v_i is on level k $(k>1)$ which is denoted as $level(v_i)=k$, if and only if $max_{(v_j \in V_i)}\{level(v_j)\} = k - 1$. Namely, the level of v_i is denoted as

$$level(v_i) = \begin{cases} 1 & if \ |V_i| = 0 \\ k & if \ |V_i| \neq 0 \ and \ max_{(v_j \in V_i)}\{level(v_j)\} = k - 1 \end{cases} \quad (5)$$

In Fig. 3, $level(v_6)=level(v_7)=level(v_5)=1$, $level(v_3)=level(v_4)=2$, $level(v_2)=3$, $level(v_1)=4$.

Based on the level of node, we prove the following lemmas.

Lemma 2. Given $G=(V,E,T)$ at $t+1$, V_{add} is the node set in which each node is added to G at time $t+1$. $V_{unreach}$ is the node set in which there are no paths from any node in V_{add} to a node in $V_{unreach}$. If $v_i \in V_{unreach}$, and $PNovelty^{(t)}(v_i)$ and $PNovelty^{(t+1)}(v_i)$ are the patent novelty of v_i at time t and $t+1$ respectively, then $PNovelty^{(t+1)}(v_i)=PNovelty^{(t)}(v_i) \times e^{-\beta}$.

Proof. Suppose $v_i \in \{v_j|level(v_j)=k\}$, and V_i and V_i' are sets of the adjacent nodes of v_i in which there exists an edge to v_i at t and $t+1$ respectively. As $v_i \in V_{unreach}$, $V_i=V_i'$. If $k=1$, as v_i is not cited by any patent, $PNovelty^{(t+1)}(v_i)=PNovelty^{(t)}(v_i) \times e^{-\beta}$.

$k=2$, $PNovelty^{(t+1)}(v_i)=min_{(v_j \in V_i' and \ level(v_j)<2)}\{PNovelty^{(t+1)}(v_j) \times CNR(v_i,v_j) \times e^{-\beta \Delta t_{ji}}\}$

$=min_{(v_j \in V_i \ and \ level(v_j)<2)}\{e^{-\beta} \times PNovelty^{(t)}(v_j) \times CNR(v_i,v_j) \times e^{-\beta \Delta t_{ji}}\} = e^{-\beta} \times PNovelty^{(t)}(v_i)$.

...

$k=n$, $PNovelty^{(t+1)}(v_i)= min_{(v_j \in V_i' \ and \ level(v_j)<n)}\{PNovelty^{(t+1)}(v_j) \times CNR(v_i,v_j) \times e^{-\beta \Delta t_{ji}}\} = min_{(v_j \in V_i \ and \ level(v_j)<n)}\{e^{-\beta} \times PNovelty^{(t)}(v_j) \times CNR(v_i,v_j) \times e^{-\beta \Delta t_{ji}}\} = e^{-\beta} \times PNovelty^{(t)}(v_i)$. Therefore, the lemma is proved.

As shown in Lemma 2, if $v_i \in V_{unreach}$, then

$$PNovelty^{(t+1)}(v_i) = e^{-\beta} \times PNovelty^{(t)}(v_i) \quad (6)$$

Lemma 3. Given $G=(V,E,T)$ at $t+1$, V_{reach} is the node set in which there is at least one path from any node in V_{add} to a node in V_{reach}. If $v_i \in V_{reach}$, then $PNovelty^{(t+1)}(v_i) \leq e^{-\beta} \times PNovelty^{(t)}(v_i)$.

Proof. Suppose $v_i \in \{v_j|level(v_j)=k\}$, as $v_i \in V_{reach}$, then $k \geq 2$. V_i and V_i' are sets of the adjacent nodes in which there exists an edge to v_i at t and $t+1$ respectively, then $V_i \subseteq V_i'$. If $k=2$, $PNovelty^{(t+1)}(v_i)= min_{(v_j \in V_i' and \ level(v_j)<2)}\{PNovelty^{(t+1)}(v_j) \times CNR(v_i,v_j) \times e^{-\beta \Delta t_{ji}}\}$. For $v_j \notin V_{add}$, as $level(v_j)=1$, $PNovelty^{(t+1)}(v_j)= PNovelty^{(t)}(v_j) \times e^{-\beta}$. As $V_i \subseteq V_i'$, $PNovelty^{(t+1)}(v_i) \leq min_{(v_j \in V_i \ and \ level(v_j)<2)}\{e^{-\beta} \times PNovelty^{(t)}(v_j) \times CNR(v_i,v_j) \times e^{-\beta \Delta t_{ji}}\}$. Namely, $PNovelty^{(t+1)}(v_i) \leq e^{-\beta} \times PNovelty^{(t)}(v_i)$.

...

$k=n$, $PNovelty^{(t+1)}(v_i)= min_{(v_j \in V_i' \text{ and } level(v_j)<n)} \{PNovelty^{(t+1)}(v_j) \times CNR(v_i,v_j) \times e^{-\beta \Delta t_{ji}}\}$

$\leq e^{-\beta} \times min_{(v_j \in V_i \text{ and } level(v_j)<n)} \{PNovelty^{(t)}(v_j) \times CNR(v_i,v_j) \times e^{-\beta \Delta t_{ji}}\} = e^{-\beta} \times PNovelty^{(t)}(v_i)$.

Therefore, the lemma is proved.

Theorem 1. Given $G=(V,E,T)$ at time $t+1$. For $v_i \notin V_{add}$, $v_j \in V_{unreach}$, if $PNovelty^{(t)}(v_i)<PNovelty^{(t)}(v_j)$, then $PNovelty^{(t+1)}(v_i)< PNovelty^{(t+1)}(v_j)$.

Proof. Based on Lemma 2 and 3, if $v_i \notin V_{add}$, $v_j \in V_{unreach}$, then $PNovelty^{(t+1)}(v_j)$ $=PNovelty^{(t)}(v_j) \times e^{-\beta}$, $PNovelty^{(t+1)}(v_i) \leq PNovelty^{(t)}(v_i) \times e^{-\beta}$. As $PNovelty^{(t)}(v_i) <PNovelty^{(t)}(v_j)$, $PNovelty^{(t+1)}(v_i)< PNovelty^{(t+1)}(v_j)$。 The theorem is proved.

Theorem 2. Given $G=(V,E)$ at a specific time. For $v_i,v_j \in V$, the patent novelty of v_i and v_j are $PNovelty(v_i)$ and $PNovelty(v_j)$ respectively. If $<v_j,v_i> \in E$, then $PNovelty(v_i)$ $\leq PNovelty(v_j)$.

Proof. If $<v_j,v_i> \in E$, then $PNovelty(v_i)=min_{(v_k \in V_i)}(PNovelty(v_k) \times CNR(v_i,v_k) \times e^{-\beta \Delta t_{ki}})$. As $v_j \in V_i$ and $CNR(v_i,v_j) \times e^{-\beta \Delta t_{ji}} \leq 1$, $PNovelty(v_i) \leq PNovelty(v_j)$. The theorem is proved.

Based on Theorem 1 and 2, quite a few of nodes in the ranking lists at time $t+1$ keep the same orders with the time t. Therefore, we consider updating the rank at time $t+1$ with Insertion Sort to improve the efficiency. The update algorithm is as showed in Algorithm 3.

As shown in Algorithm 3, the main running time lies in comparison when each node inserts into the list2 (*step* 20). In Inference 1, we proof the comparison count is at a low level with insertion sort when a node inserts into the list.

Inference 1. Suppose list1 and list2 are the ranking lists at time t and $t+1$ respectively. For $v_i,v_j,v_k \in list1$, v_j and v_k are inserted into list2 before v_i. v_j is the nearest node before v_i in list1 which satisfies $v_j \in V_{unreach}$, and v_k is the nearest node before v_i in list1 which satisfies $v_k \in V_i$ ($V_i=\{v_m|v_m \in V$ and $<v_m,v_i> \in E\}$). Suppose $Pos^{(t+1)}(v_i)$, $Pos^{(t+1)}(v_j)$, $Pos^{(t+1)}(v_k)$ are the final position of v_i, v_j and v_k respectively in list2. When v_i is inserted into list2, the comparison count is $Min\{Pos^{(t+1)}(v_i)-Pos^{(t+1)}(v_j), Pos^{(t+1)}(v_i)-Pos^{(t+1)}(v_k)\}$ at most.

Proof. As v_j is inserted into list2 before v_i, then $PNovelty^{(t)}(v_j) \geq PNovelty^{(t)}(v_i)$. Based on Theorem 1, $PNovelty^{(t+1)}(v_j) \geq PNovelty^{(t+1)}(v_i)$. Based on Theorem 2, $PNovelty^{(t+1)}(v_k) \geq PNovelty^{(t+1)}(v_i)$. As a result, the position of v_i in list2 must be after v_j and v_k. Therefore, when v_i is inserted into list2, v_i must compare with the nodes which are inserted into list2 before v_i. When v_i meet v_j or v_k, the swap operation will not be proceeded, and the loop terminates. Thus, the comparing count is $Min\{Pos^{(t+1)}(v_i)-Pos^{(t+1)}(v_j), Pos^{(t+1)}(v_i)-Pos^{(t+1)}(v_k)\}$ at most. The inference thus holds.

Algorithm 3. Update-Rank

Input: $G=(V,E,T)$ at time $t+1$, V_{add}, Ranking List list1 at time t

Output: Ranking List list2 at time $t+1$

1. *foreach* node v_i in V_{add}
2.   compute $PNovelty(v_i)=1$ by fomula (4);
3.   add v_i to list2;
4.   *foreach* node v_j cited by v_i
5.   add a link $<v_j,v_i>$ in G;
6.   compute $CNR(v_i,v_j)$ by formula (1)(2)(3);
7.   *endfor*
8. *endfor*
9. Find the node set V_{reach} which is related with the nodes in V_{add};
10. $V_{unreach} =V-V_{add}-V_{reach}$;
11. *foreach* $v_j \in V_{reach}$
12.   compute $PNovelty(v_j)$ by formula (4);
13. *endfor*
14. *foreach* $v_k \in V_{unreach}$
15.   compute $PNovelty(v_k)$ by formula (6) ;
16. *endfor*
17. *for* $m=1$ to list1.*length*
18.   put list1[m] to the end of list2;
19. n=list2.length;
20.   *while*($n>0$&&($PNovelty$(list2[n])>$PNovelty$(list2[n-1])))
21.   *swap*(list2[n-1],list2[n]);
22. n--;
23.   *endwhile*
24. *endfor*
25. Output list2;

Fig. 4 is the process of v_3 inserting into list2 on Fig. 2. As $v_6 \in V_{unreach}$ and $PNovelty^{(t)}(v_6) \geq PNovelty^{(t)}(v_3)$, $PNovelty^{(t+1)}(v_6) \geq PNovelty^{(t+1)}(v_3)$. As $<v_4,v_3> \in E$, $PNovelty^{(t+1)}(v_4) \geq PNovelty^{(t+1)}(v_3)$. Therefore, the comparison count is 1 at most. Actually $PNovelty^{(t+1)}(v_3)=0.15$, $PNovelty^{(t+1)}(v_4)=0.31$, $PNovelty^{(t+1)}(v_6)=0.37$. Therefore, the comparison count is 1 too.

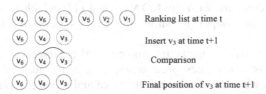

Fig. 4. The process of v_3 inserting to list2

Complexity Analysis. As can be seen in Fig.4, when a node is inserted to a new list at time $t+1$, the general cost of comparison is $O(1)$. Therefore the total cost of sort is $O(N)$ in which N is the number of nodes in the network at time $t+1$. However, if we use general sort method as Algorithm 2 does, the cost of sort will be $O(N\log N)$. As the time complexity of computing patent novelty of each patent is $O(N)$ after the network is established, the cost of finding novel patents mainly lies on sort. Therefore, the cost could be cut down by Update-Rank algorithm compared to Novelty-Rank algorithm.

5   Experiments

In this section, we use three patent datasets crawled from USPTO [15] to evaluate the effectiveness and efficiency of our proposed method. Among them, one dataset is used as a training set to evaluate the damp factor β and the others are used to show the effectiveness and efficiency. The utilized datasets are described in Table 1.

Table 1. The Description of Used Datasets

Dataset	Number of patents	Effect	Year
Semiconductor	21236	Training Set	2001-2008
Phone	25245	Testing Set	2001-2008
Image Processing	23628	Testing Set	2001-2008

5.1   Text processing

Before using the method in this paper to rank the patents by patent novelty, we firstly process the collected patent documents containing stop-words removal, word stemming, keywords extraction and similarity computing. Since there is no generally acknowledged list of stop-words in use, we use a self-defined list of stop-words. After the removal, words can be stemmed based on Porter's algorithm [16] to better match concepts among terms in the patent documents. Then a TF-IDF [17] method is employed to compute the weight of a word in a patent, and top m words are identified as the keywords of each patent(In this paper, m is set to be 20 empirically). Finally, a Jaccard Coefficient [17] method is applied to measure the similarity of two patent documents.

5.2   Experimental Results

After text processing, we evaluate the effectiveness and efficiency of proposed methods. First, we compare with the COA method [11] which is most relevant our work presented in this paper to show the effectiveness of our Novelty-Rank algorithm.

Patent is a type of paid property intelligence, and patentees keep their patents in effect through paying a few fees on a yearly basis. Because of the high cost of patent renewal fees, when there are new improved and novel patents of current ones, then the old ones will become out-fashioned and are considered to be useless. As a result, the renewal will be abandoned and the old-fashioned patents will become invalid. Therefore, we use $VRate$, the rate of valid patents within top k ones in the ranking list, to evaluate the effectiveness of our method which is denoted as:

$$VRate = \frac{N_{vaild,k}}{k} \tag{7}$$

where $N_{valid,k}$ is the valid patents within the top k patents of the ranking list.

In Fig 5, we show the effectiveness on "Phone" and "Image Processing" datasets respectively (The damp factor β is set to be 0.1 based on the training set of "Semiconductor"). As it can be seen, in the "Phone" dataset the *VRates* of Novelty-Rank are greater than COA method when k=20, 40, 60, 80, 100 and 120. Similarly, the *VRates* of novelty-rank is greater than COA on most selected k in "Image Processing" dataset. Namely, Novelty-Rank is more effective than COA on novelty evaluation and rank.

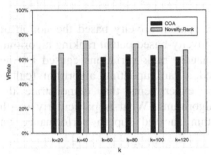 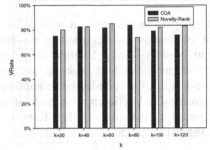

Fig.5(a) The Effectiveness on Phone   **Fig.5(b)**The   Effectiveness   on   Image Processing

Fig. 5. The Effectiveness of Novelty-Rank Algorithm

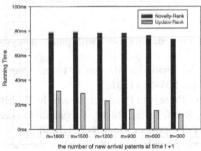 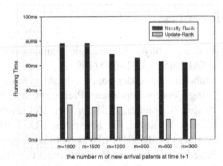

Fig.6(a) The Efficiency on Phone   **Fig.6(b)** The Efficiency on Image Processing

Fig. 6. The Efficiency of Update-Rank Algorithm

As the COA method does not provide corresponding update algorithm to re-rank the patents, we do experiments and comparisons only between our own algorithms, Novelty-Rank and Update-Rank. The two test datasets are both first divided into two parts: the patents published from 2001 to 2007 and the other ones published in 2008. The former patents are preprocessed with novelty-rank algorithm and the latter ones are taken as the added ones at the next time slice. The Novelty-Rank and Update-Rank algorithms are then invoked to re-rank the patents respectively.

In Fig.6, we compare the running time of novelty computing and ranking between Novelty-Rank algorithm and Update-Rank algorithm. From Fig 6(a), we can see that, the running time of both of Novelty-Rank and Update-Rank on the "Phone" dataset decrease as the number of new added patents decreases, and Update-Rank is obviously more efficient than Novelty-Rank. We draw the same conclusion from the test results with the "Image Processing" dataset (as shown in Fig 6(b)). That is, our Update-Rank algorithm is significantly more efficient than Novelty-Rank algorithm on updating the rank of patents when new patents arrive.

6   Conclusions

In this paper, we proposed a new definition of patent novelty based the notion of patent association and designed an innovative novelty-based patent ranking algorithm (named Novelty-Rank). Further, in order to handle rank variations caused by the arrival of new (related) patents, we designed a rank-updating algorithm (called Update-Rank). We conducted comprehensive experiments that demonstrate the effectiveness and efficiency of both of our algorithms. What is particularly worth mentioning is that our Update-Rank algorithm showed superb performance on updating patent ranks.

Acknowledgements. This work is supported by the Key Program of National Natural Science Foundation of China (61232002) and the National "863" High-tech Research Development Plan Foundation (2012AA011004).

References

1. Liu, S.-J., Shyu, J.: Strategic planning for technology development with patent analysis. IJTM 13(5/6), 661–680 (1997)
2. Mahdabi, P., Andersson, L., Keikha, M., et al.: Automatic Refinement of Patent Queries using Concept Importance Predictors. In: SIGIR, pp. 505–514 (2012)
3. Bashir, S., Rauber, A.: Improving Retrievability of Patents in Prior-Art Search. In: Gurrin, C., He, Y., Kazai, G., Kruschwitz, U., Little, S., Roelleke, T., Rüger, S., van Rijsbergen, K. (eds.) ECIR 2010. LNCS, vol. 5993, pp. 457–470. Springer, Heidelberg (2010)
4. Magdy, W., Lopez, P., Jones, G.J.F.: Simple vs. Sophisticated Approaches for Patent Prior-Art Search. In: Clough, P., Foley, C., Gurrin, C., Jones, G.J.F., Kraaij, W., Lee, H., Mudoch, V. (eds.) ECIR 2011. LNCS, vol. 6611, pp. 725–728. Springer, Heidelberg (2011)
5. Mase, H., Matsubayashi, T., Ogawa, Y., et al.: Proposal of Two-Stage Patent Retrieval Method. TALIP 4(2), 190–206 (2005)
6. Takaki, T., Fujii, A., Ishikawa, T.: Associative Document Retrieval by Query Subtopic Analysis and its Application to Invalidity Patent Search. In: CIKM, pp. 399–405 (2004)
7. Fujii, A.: Enhancing Patent Retrieval by Citation Analysis. In: SIGIR, pp. 793–794 (2007)
8. Chen, Y.-L., Chiu, Y.-T.: Cross-language patent matching via an international patent classification-based concept bridge. JIS 39(4), 1–17 (2013)
9. Li, Y., Shawe-Taylor, J.: Advanced Learning Algorithms for Cross-Language Patent Retrieval and Classification. IP&M 43(5), 1183–1199 (2007)
10. Li, Y., Shawe-Taylor, J.: Using KCCA for Japanese–English cross-language information retrieval and document classification. JIIS 27(2), 117–133 (2006)

11. Hasan, M.A., Spangler, W.S., Griffin, T., et al.: COA-Finding Novel Patents through Text Analysis. In: KDD, pp. 1175–1184 (2009)
12. ACM SIGIR 2000 workshop on patent retrieval (2000), http://www.aueb.gr/conferences/sigir-2000/English/workshop.htm
13. ACL-2003 Workshop on Patent Corpus Processing (2003), http://acl.ldc.upenn.edu/acl2003/
14. NTCIR, http://research.nii.ac.jp/ntcir/index-en.html
15. USPTO, http://www.uspto.gov/
16. Porter, M.F.: An algorithm for suffix stripping. Program: Electronic Library and Information Systems 14(3), 130–137 (1980)
17. Han, J., Kamber, M.: Data Mining: Concepts and Techniques (2006)

Optimizing Top-k Retrieval: Submodularity Analysis and Search Strategies

Chaofeng Sha[1], Keqiang Wang[2], Dell Zhang[3], Xiaoling Wang[2], and Aoying Zhou[2]

[1] School of Computer Science,
Shanghai Key Laboratory of Intelligent Information Processing,
Fudan University, Shanghai 200433, China
[2] Shanghai Key Laboratory of Trustworthy Computing,
East China Normal University, Shanghai 200062, China
[3] Department of Computer Science and Information Systems (DCSIS),
University of London, London, United Kingdom
cfsha@fudan.edu.cn

Abstract. The key issue in top-k retrieval — finding a set of k documents (from a large document collection) that can best answer a user's query — is to strike the optimal balance between relevance and diversity.

In this paper, we study the top-k retrieval problem in the framework of facility location analysis and prove the submodularity of that objective function which provides a theoretical approximation guarantee of factor $1 - \frac{1}{e}$ for the (best-first) greedy search algorithm. Furthermore, we propose a two-stage hybrid search strategy which first obtains a high-quality initial set of top-k documents via greedy search, and then refines that result set iteratively via local search.

Experiments on two large TREC benchmark datasets show that our two-stage hybrid search strategy approach outperforms the existing ones.

1 Introduction

For top-k document search, users are usually only interested in a few documents at the top of the returned results list. The traditional approaches to this top-k retrieval problem focus on the *relevance* of results only: they make use of a standard retrieval model [9] that computes the relevance score of each document with regard to the query individually and then just return the k documents with the highest relevance scores. However, many recent studies have shown theoretically and empirically that it is beneficial to take the *diversity* of results into account as well [4], particularly when different users are interested in different meanings or aspects of the query. Retrieving a diverse set of the most representative relevant documents is more likely to ensure that all possible information needs of a given query are satisfied. So finding the best set of top-k documents for a given query is of crucial importance to IR systems.

In order to resolve the query ambiguity and avoid the information redundancy, a number of diversification techniques [1,14,13,15,2,12,16] have been designed

F. Li et al. (Eds.): WAIM 2014, LNCS 8485, pp. 18–29, 2014.

to optimize the top-k documents collectively, in terms of both relevance and diversity. Most these former work, for example MMR, define some objective functions to trade off between maximizing relevance and maximizing diversity. [16] is the latest work which addressed this problem within the framework of *facility location analysis* [6] – a branch of operations research concerning optimal placement of facilities. Objective functions based on *Desirable Facility Placement* and *Obnoxious Facility Placement* are designed and optimization of the objective functions is carried out by using the *local search* [11] algorithm. However the local search is expensive, so [16] can't obtain the top-k documents efficiently and fast approximations are desired.

In this paper, we revisit this problem through a *submodularity* analysis of the *desirable-facility-placement* objective function, which provides a theoretical approximation guarantee of factor $1 - \frac{1}{e}$ with the (best-first) greedy search algorithm. Furthermore, in order to improve effectiveness, we propose a two-stage search strategy which first obtains a high-quality initial set of top-k documents via greedy search, and then refines that result set iteratively via local search. Compared to the previous approach where the local search algorithm starts from the set of k documents with highest relevance scores, the proposed approach significantly improves not only the effectiveness (i.e., the quality of final result set) but also the efficiency (i.e, the speed of convergence). The advantages of our new approach over the existing ones are confirmed by extensive experiments on two large TREC benchmark datasets.

The rest of this paper is organized as follows. The framework of *facility location analysis* is discussed and the *submodularity* of the objective functions is analyzed in Section 2. A two-stage hybrid search strategy is employed to optimize the objective function in Section 3. The experiments and the analysis of the results is presented in Section 4. We conclude our work in Section 5.

2 Top-k Retrieval and Submodular Functions

2.1 Top-k Retrieval Problem

Suppose that for a given query q, the set of relevant documents found by the retrieval system is $D = \{d_1, d_2, \ldots, d_n\}$.

The relevance score of a document with respect to q is calculated by a function $r : D \to R$. Without loss of generality, we assume that $r(d_1) \geq r(d_2) \geq \ldots \geq r(d_n)$. Furthermore, the distance or dissimilarity between any two documents in D is calculated by a function $w : D \times D \to R$.

The task of top-k retrieval is to pick a subset $S \subseteq D$ of k documents that is both relevant and diverse simultaneously.

Recently facility location analysis from operations research has been introduced as a unified framework for top-k retrieval [16]. To find the optimal subset, we formulate it as a *facility location problem* [6] in operations research — given a set of customer "locations" D, we would like to find a subset $S \subseteq D$ to open k "facilities" there so as to optimize a graph-theoretic objective that is dependent on the cost of opening a facility at each location and also the distance between each pair of locations. The facility opening cost at location d_i is set to $-r(d_i)$ which reflects our preference for high relevance. Making use of different optimization objectives, facility location analysis would lead to different retrieval techniques for search result diversification.

Several state-of-the-art techniques for diversified retrieval, including MMR [1], QPRP [15], and MPT [13], treat the top-k documents as *obnoxious* facilities to be dispersed as far as possible from each other, but a much better performance could be obtained by considering the top-k documents as *desirable* facilities to be placed as close as possible to their "customers" (other relevant documents). In the following sub-sections, we firstly give the sub-modularity analysis for optimization function in top-k retrieval problem, and then prove the sub-modularity of both Obnoxious Facility Dispersion(OBN) function and Desirable Facility Placement(DES) function, where OBN and DES are two kinds of approaches in the framework of Facility Location analysis.

2.2   Submodular Function Maximization

A function $f : 2^X \to R$ is submodular if for any $S \subseteq T \subseteq X$ and $i \notin T$, $f(S \cup \{i\}) - f(S) \geq f(T \cup \{i\}) - f(T)$. In other words, submodular functions are characterized by the diminishing return property: the marginal gain of adding an element to a smaller subset of X is higher than that of adding it to a larger subset of X. We are often interested in finding the subset with cardinality constraint $|S| \leq k$ that maximizes a submodular function, i.e., $\arg\max_S f(S)$ [7]. However this problem is in general NP-hard, even for some simple submodular functions. In recent years, submodular functions have been receiving much attention from the Artificial Intelligence (AI) community because of the following insight about their approximate optimization. The greedy algorithm was employed to optimize the normalized, monotone, and submodular function with approximation ratio $1 - \frac{1}{e}$ [10].

2.3   Obnoxious Facility Dispersion (OBN)

The optimization objective of obnoxious-facility-dispersion is two-fold: (1) to minimize the total cost of opening k facilities and (2) to maximize the spread of those facilities. ALL the standard forms of MMR [1], QPRP [15], and MPT [13] could be expressed as the dispersion of the top-k documents.

The Maximum Marginal Relevance (MMR) [1] is the first diversification method which balances the relevance and diversity of the result list. The marginal gain of MMR, i.e., $f_{MMR}(S \cup \{d\}) - f_{MMR}(S)$, is defined as $\lambda r(d) - (1 - \lambda) \max_{d' \in S} s(d, d')$, where $r(d)$ is the relevance between the document d and the query, S is the set of select documents and $s(d, d')$ is the similarity between the document d and d'. [8] has proved that f_{MMR} is submodular, but it has also pointed out that f_{MMR} is not monotone.

The Quantum Probability Ranking Principle (QPRP) [15] states that documents should be ranked according to $d_i = \arg\max \left(P(d) + \sum_{d' \in S} I_{d, d'} \right)$. With the substitution $I_{d, d'} = -\sqrt{P(d)}\sqrt{P(d')}s(d, d')$, we see that the marginal gain $f_{QPRP}(S \cup \{d\}) - f_{QPRP}(S)$ can be rewritten as: $\frac{1}{2}P(d) - \frac{1}{2}\sum_{d' \in S}\sqrt{P(d)}\sqrt{P(d')}s(d,d')$, where $P(d)$ is the probability of d being relevant to the query. The diminishing return property obviously holds due to the fact that $\sum_{d' \in S}\sqrt{P(d)}\sqrt{P(d')}s(d,d') \leq \sum_{d' \in T}\sqrt{P(d)}\sqrt{P(d')}s(d,d')$ for all $S \subseteq T$ and $d \notin T$, and therefore f_{QPRP} is submodular. However, it is easy to show that f_{QPRP} is monotone nondecreasing.

In the Modern Portfolio Theory (MPT) for IR   [13], the marginal gain $f_{MPT}(S \cup \{d\}) - f_{MPT}(S)$ is given by $w_d E[r_d] - bw_d^2\sigma_d^2 - 2b\sum_{d' \in S}w_d w_{d'}\sigma_d\sigma_{d'}\rho_{d,d'}$

where w_d is the importance weight of d's rank position using nDCG, $E[r_d]$ is the expected relevance scores, b is a specified risk preference parameter for users, and σ_d is the standard deviation of d. This means that f_{MPT} would not be submodular since $\rho_{d,d'}$ could be negative. Nevertheless, if we replace the Pearson's coefficient with a non-negative similarity function $s(d,d')$ as employed in the experiments of [16], then we get a submodular function due to fact that
$$\sum_{d' \in S}w_d w_{d'}\sigma_d\sigma_{d'}s(d,d') \leq \sum_{d' \in T}w_d w_{d'}\sigma_d\sigma_{d'}s(d,d')$$
for all $S \subseteq T$ and $d \notin T$, However, once again, it is easy to show that f_{MPT} is not monotone.

2.4   Desirable Facility Placement (DES)

The optimization objective of desirable-facility-placement is also two-fold: (1) to minimize the total cost of opening k facilities; and (2) to minimize the distances from the customer locations to their closest facilities. In the context of top-k retrieval, by treating the top-k documents like desirable facilities, we are in fact selecting the best representatives of the relevant documents so that the top-k result set consists of a concise summary of all the relevant information, and as such novelty and diversity naturally arise.

In order to facilitate submodularity analysis, we re-write the original minimization problem of desirable-facility-placement into its equivalent maximization problem as follows. The optimal set of top-k documents, considered as desirable facilities, is given by $S^* = \arg\max_{S \subseteq D, |S|=k} f(S)$, where the objective function f is defined as follows:

$$f(S) = \lambda \sum_{d \in S} r(d) - (1-\lambda) \sum_{d \in D \setminus S} \min_{d' \in S} w(d,d'). \tag{1}$$

where $w(d,d')$ could be the dissimilarity between the document d and d'. Essentially it can be regarded as a mixture of relevance-based ranking and k-medoids clustering.

The above mentioned diversified retrieval models — MMR, QPRP, and MPT — can all be transformed to use the above desirable-facility-placement objective function instead, as pointed out by [16].

Theorem 1. *The desirable-facility-placement objective function $f(S)$ is submodular and also monotone nondecreasing.*

Proof. Let $S \subseteq T \subseteq D$ and $i \in D \setminus T$. For any $d \in D \setminus (S \cup \{i\})$, the nearest neighbor is i or in $D \setminus S$. Consequently, we have that

$$
\begin{aligned}
&f(S \cup \{i\}) - f(S) \\
&= \lambda \sum_{d \in S \cup \{i\}} r(d) - (1 - \lambda) \sum_{d \in D \setminus (S \cup \{i\})} \min_{d' \in S \cup \{i\}} w(d, d') \\
&\quad - \lambda \sum_{d \in S} r(d) + (1 - \lambda) \sum_{d \in D \setminus S} \min_{d' \in S} w(d, d') \\
&= \lambda r(i) + (1 - \lambda) \min_{d' \in S} w(i, d') + (1 - \lambda) \sum_{d \in D \setminus (S \cup \{i\})} \max\{0, \min_{d' \in S} w(d, d') - w(d, i)\}
\end{aligned}
$$

The marginal gain is nonnegative, i.e. $f(S)$ is nondecreasing set function. We know that $\min_{d' \in S} w(i, d') \geq \min_{d' \in T} w(i, d')$ and $\max\{0, \min_{d' \in S} w(d, d') - w(d, i)\} \geq \max\{0, \min_{d' \in T} w(d, d') - w(d, i)\}$ thanks to $S \subseteq T$. Combined with $D \setminus (T \cup \{i\}) \subseteq D \setminus (S \cup \{i\})$, we have that $f(S \cup \{i\}) - f(S) \geq f(T \cup \{i\}) - f(T)$.

It is obvious that f is monotone nondecreasing.

3   Search Strategies

Both the obnoxious-facility-dispersion (OBN) problem and the desirable-facility-placement (DES) problem are in general NP-hard [16], therefore we are only able to solve them through an approximation algorithm. The greedy search algorithm is $(1 - \frac{1}{e})$-approximation for DES problem thanks to the monotonicity and submodularity of the objective function. Both greedy search and local search, can be straightforwardly applied to top-k retrieval. In greedy search, the most relevant document d_1 is initialized to S, and then the document $d_{D/S} \in D/S$ with maximum marginal gain $f(S \cup \{d\}) - f(S)$ is added into S, until the size of result set S is k. In local search, the set S is initialized with a random documents subset $\{d_1, d_2, ..., d_k\}$. Then if there is a document $d_{D/S} \in D/S$ improve the objective, i.e for some $f(S \cup \{d_{D/S}\} \setminus \{d_S\}) > f(S)$ for some $d_S \in S$, then d_S is replaced with $d_{D/S}$, until the result set does not change.

Generally, greedy search is simpler and thus more efficient, but local search is often more effective at the expense of running time. In [16], it has been demonstrated that using the local search (LS) algorithm to optimize the desirable-facility-placement objective function works much better than using the greedy search (GS) algorithm to optimize the obnoxious-facility-dispersion objective function. Particularly, the DES variant of the MPT retrieval model optimized by local search achieves the highest performance scores. However, on large document collections, local search is very slow. Is it possible to accelerate local search without hurting its excellent performance?

The effectiveness and efficiency of local search actually depend on its starting point, i.e., the initial result set S in our context. If the starting point is within the vicinity of the global optimal solution, local search is not only more likely to end up with the best possible result set, but also more likely to reach that

Algorithm 1. GSLS algorithm for top-k retrieval

Input: A ranked list of relevant documents D, k, f
Output: $S \subseteq D$ with $|S| = k$

1: $S \leftarrow \{d_1\}$
2: **while** $|S| < k$ **do**
3:    $d^* \leftarrow \arg\max_{d \in D \setminus S} f(S \cup \{d\}) - f(S)$
4:    $S \leftarrow S \cup \{d^*\}$
5: **end while**

6: **repeat**
7:   **for** $d \in S$ **do**
8:     **for** $d' \in D \setminus S$ **do**
9:       $S' \leftarrow (S \setminus \{d\}) \cup \{d'\}$
10:      **if** $f(S') > f(S)$ **then**
11:        $S \leftarrow S'$
12:      **end if**
13:     **end for**
14:   **end for**
15: **until** S does not change
16: **return** S

result set quickly. It is non-trivial to find a good starting point for local search: quite often people just run the local search algorithm for a number of times with randomly generated "seeds", and then pick the best one from all those result sets produced. Fortunately, for top-k retrieval, Theorem 1 tells us that a near-optimal result set could be obtained quickly through greedy search (which itself is quite fast). Inspired by this fact, we propose a two-stage hybrid search strategy named GSLS (outlined in Algorithm 1): it first obtains a high-quality initial set of top-k documents via greedy search (line 1-5), and then refines that result set iteratively via local search (line 6-15).

4 Experiments

4.1 Datasets

The document collection used for our experiments was the ClueWeb09 corpus[1] part B (the first 50 million documents). It was indexed by employing the toolkit Lemur/Indri[2], where standard stop-word removal and Porter stemming were applied. There are two (query) topic sets from the TREC-2009 and TREC-2010 Web Tracks[3] that have been widely adopted as the benchmarks for diversified retrieval: the former contains 50 topics and the latter contains 48 topics. The title of each topic was used as a query over the document collection.

---

[1] http://boston.lti.cs.cmu.edu/Data/clueweb09/
[2] http://www.lemurproject.org/indri.php
[3] http://trec.nist.gov/data/webmain.html

4.2   Settings

We conducted a series of experiments to evaluate and compare the performances of the following different approaches to top-k retrieval: 1. LM: standard language modeling. 2. OBN-GS: obnoxious-facility-dispersion, using the greedy search algorithm. 3. DES-LS: desirable-facility-placement, using the local search algorithm. 4. DES-GS: desirable-facility-placement, using the greedy search algorithm. 5. DES-GSLS: desirable-facility-placement, using the two-stage hybrid search algorithm.

Among them, LM is the baseline without diversification; OBN-GS and DES-LS are described in [16]; DES-GS and DES-GSLS are proposed in this paper.

The ranking provided by standard language modeling was used as the basis by all the diversity based top-k retrieval techniques.

We computed the relevance score of a document d with respect to a query q as $r(d) = \log Pr(q|\hat{M}(d))/|q|$ where $\hat{M}(d_i)$ was the unigram language model estimated from document d_i with Dirichlet smoothing (the Dirichlet prior $\mu = 2000$), and $|q|$ was the number of terms in the query. The scaling factor $1/|q|$ was introduced to remove the influence of query length and facilitated setting the parameter λ across queries. We then calculated the distance between d and d' using Kullback-Leibler divergence.

$s(d, d') = -D_{KL}(\bar{M}(d)||\hat{M}(d'))$ where $\bar{M}(d)$ was the unigram language model estimated from document d using the maximum likelihood estimation, and $\hat{M}(d')$ was the unigram language model estimated from document d' with Dirichlet smoothing. To better compare the retrieval techniques independently from the distance function, we used the KL divergence based similarity $s(d, d')$ for replacing the Pearson's correlation in MPT and for approximating the quantum interference term in QPRP.

For each query q in the test set Q, we retrieved $n = 100$ documents with the highest relevance scores to form the set D. Then the retrieval techniques — MMR, MPT, QPRP (as OBN variants), and their DES variants — were employed to select the top $k = 20$ documents as the set S to be returned to users.

4.3   Evaluation Metrics of Effectiveness

The effectiveness of top-k retrieval was measured at rank positions 5, 10 and 20 by three metrics: ERR-IA [3], α-nDCG [5] and subtopic-recall (S-recall) [14], where both ERR-IA and α-nDCG were following the standard configurations in the TREC-2011 Web Track (i.e. in ERR-IA all intents were given the same probability, and in α-nDCG, α was set to 0.5).

Expected Reciprocal Rank   [3]. This metric is defined as the expected reciprocal length of time that the user will take to find a relevant document, which overcomes implicitly discounts documents which are shown below very relevant documents. The criterion for the PRP model(Probability Ranking Principle) is defined as $ERR := \sum_{r=1}^{n} \frac{1}{r} \prod_{i=1}^{r-1}(1 - R_i)R_r$, where r is the rank position, n is the number of selected documents and R_i is the probability of satisfying the user which is always set to 0.5. To measure the diversity of document result set, the ERR can be extended to $ERR := \sum_{r=1}^{n} \frac{1}{r} \sum_t P(t|q) \prod_{i=1}^{r-1}(1 - R_i^t)R_r^t$.

Normalized Discounted Cumulative Gain [5]. The premise of DCG is that highly relevant documents appearing lower in a search result list should be penalized as the graded relevance value is reduced logarithmically proportional to the position of the result. The discounted CG accumulated is defined as $DCG := \sum_{i=1}^{n} rel_i / (\log_2(1 + i))$.

The DCG for the diversity task is designed as $DCG := \sum_{i=1}^{n} \left(\sum_t \frac{rel_{i_t}(1-\alpha)^{c_{i,t}}}{d \log_2(1+i)} \right)$ where t is the subtopic of the document i, α reflects the possibility of assessor error with $0 < \alpha \leq 1$ and c is the count of subtopic t appeared.

Subtopic Recall [14]. This metric is to measure the number of different subtopics covered as a function of rank. More precisely, consider a topic T with n_A subtopics $A_1..., A_{n_A}$, and a ranking $d_1..., d_m$ of m documents. Let subtopics(d_i) be the set of subtopics to which d_i is relevant. So the objective function of the first K documents is follows: $S - recall_K := \frac{\bigcup_{i=1}^{K} subtopics(d_i)}{n_A}$.

Obviously, it makes the retrieval to cover as many different subtopics as possible quickly.

4.4   Results

The experimental results for the MMR and MPT retrieval models (under both OBN and DES formulations) are reported here, while the experimental results for the QPRP retrieval model are omitted due to the space limit. As observed in [16], the performance of QPRP was consistently lower than that of MPT.

Effectiveness. The effectiveness of top-k retrieval was measured at rank positions 5, 10 and 20 by three metrics described in the part 4.3.

Table 1 show the results on TREC-2009 and TREC-2010. For each measure, the best result is reported in bold for MMR or MPT. As expected, all approaches with diversification OBN-GS, DES-GS, DES-GS, and DES-GSLS beat the baseline without diversification LM easily.

While both using greedy search for optimization, DES-GS achieved much higher performance than OBN-GS, which confirms the main conclusion of [16]: the desirable-facility-placement formulation is indeed better than the the the obnoxious-facility-dispersion formulation.

While both optimizing the desirable-facility-placement objective function, DES-LS achieved much higher performance than DES-GS, indicating that local search is in general more effective than greedy search for top-k retrieval.

More importantly, it can be seen that the proposed two-stage hybrid search strategy DES-GSLS significantly outperformed all other approaches including DES-LS (which is the best performing one in [16]). This is probably because

Table 1. The top-k retrieval performances measured by ERR-IA, alpha-nDCG and s-recall on the TREC 2009 and TREC 2010 topic sets

TREC 2009										
		ERR-IA			alpha-DCG			s-rec		
top-k		@5	@10	@20	@5	@10	@20	@5	@10	@20
LM		0.068	0.086	0.095	0.112	0.145	0.181	0.171	0.246	0.349
MMR	OBN+GS	0.071 3.55%	0.091 6.04%	0.098 2.54%	0.111 -0.46%	0.157 8.13%	0.182 0.55%	0.171 -0.23%	0.283 14.88%	0.343 -1.63%
	DES+LS	0.109 59.83%	0.128 48.18%	0.135 41.57%	0.161 43.98%	0.194 33.72%	0.222 22.54%	0.223 30.17%	0.311 26.26%	0.390 11.85%
	DES+GS	0.099 45.04%	0.118 37.34%	0.125 30.65%	0.150 33.93%	0.186 28.31%	0.210 16.07%	0.218 27.25%	0.311 26.26%	0.378 8.50%
	DES+GSLS	**0.131** 90.86%	**0.151** 75.58%	**0.156** 63.98%	**0.187** 67.31%	**0.226** 55.94%	**0.245** 35.11%	**0.233** 36.41%	**0.352** 43.06%	**0.397** 13.76%
MPT	OBN+GS	0.092 34.42%	0.115 33.94%	0.120 26.18%	0.149 33.10%	0.197 35.62%	0.213 17.83%	0.196 14.78%	0.329 33.85%	0.364 4.30%
	DES+LS	0.267 290.00%	0.277 221.71%	0.279 192.49%	0.381 240.59%	0.354 144.00%	0.355 96.30%	0.353 106.66%	0.411 67.04%	0.419 20.26%
	DES+GS	0.254 271.72%	0.265 207.11%	0.266 178.49%	0.343 207.32%	0.344 136.93%	0.342 89.03%	0.376 120.01%	0.403 63.92%	0.403 15.67%
	DES+GSLS	**0.286** 318.38%	**0.298** 245.50%	**0.301** 215.86%	**0.386** 245.57%	**0.387** 166.96%	**0.395** 118.18%	**0.423** 147.49%	**0.460** 86.95%	**0.481** 37.85%

TREC 2010										
		ERR-IA			alpha-DCG			s-rec		
top-k		@5	@10	@20	@5	@10	@20	@5	@10	@20
LM		0.087	0.103	0.120	0.105	0.137	0.196	0.173	0.281	0.441
MMR	OBN+GS	0.102 16.98%	0.121 17.63%	0.131 9.24%	0.132 25.50%	0.170 23.43%	0.204 4.32%	0.233 34.81%	0.348 23.86%	0.458 3.70%
	DES+LS	0.131 50.39%	0.154 50.09%	0.165 36.87%	0.159 50.52%	0.205 49.44%	0.238 21.88%	0.252 46.08%	0.418 48.70%	**0.494** 11.96%
	DES+GS	0.123 41.53%	0.118 15.01%	0.157 30.96%	0.153 45.42%	0.200 45.51%	0.234 19.84%	0.252 46.08%	0.410 45.86%	**0.494** 11.96%
	DES+GSLS	**0.160** 83.70%	**0.181** 76.33%	**0.189** 57.06%	**0.185** 75.08%	**0.226** 64.49%	**0.250** 27.63%	**0.289** 67.40%	**0.424** 50.93%	0.492 11.49%
MPT	OBN+GS	0.147 68.60%	0.172 67.39%	0.180 49.91%	0.198 87.57%	0.245 78.04%	0.269 37.56%	0.299 73.24%	0.430 53.03%	0.502 13.85%
	DES+LS	0.343 294.69%	0.361 251.06%	0.365 203.46%	0.385 265.49%	0.408 196.85%	0.415 112.08%	0.468 171.23%	0.538 91.47%	0.551 24.94%
	DES+GS	0.337 287.95%	0.356 245.76%	0.359 198.47%	0.380 261.01%	0.405 194.56%	0.410 109.57%	0.469 171.83%	0.543 93.33%	0.551 24.94%
	DES+GSLS	**0.384** 341.14%	**0.400** 288.26%	**0.404** 236.24%	**0.440** 317.82%	**0.454** 230.92%	**0.466** 138.03%	**0.555** 221.53%	**0.623** 121.63%	**0.634** 43.59%

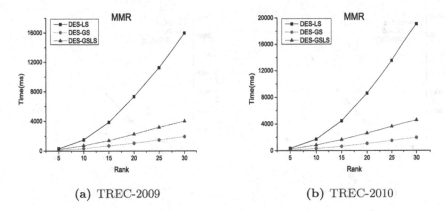

(a) TREC-2009   (b) TREC-2010

Fig. 1. Running time of MMR on TREC-2009 and TREC-2010

greedy search in its first stage could find a near-optimal initial result set (which is close to the global optimum as guaranteed by the submodularity, see Theorem 1) as the starting point for local search in its second stage, and thus combining them sequentially could greatly improve the overall effectiveness compared to just starting from the set of k most relevant documents.

Efficiency. Figure 1 and 2 show the average time costs of different search strategies for optimizing the desirable-facility-placement objective function: DES-LS, DES-GS and DES-GSLS. Not surprisingly, greedy search DES-GS is the fastest while local search DES-LS is the slowest. More importantly, it can be seen that the proposed two-stage hybrid search strategy DES-GSLS runs much faster than pure local search DES-LS (which is the best performing one in [16]). The larger k, the bigger time gap between DES-GSLS and DES-LS. Furthermore, running DES-GSLS does not require much extra time compared to DES-GS, which suggests that starting from the high-quality initial result set obtained via greedy search in the first stage, local-search in the second stage of DES-GSLS can converge to the optimal solution within a small number of iterations. The fast speed of greedy search and the quick convergence of local search (as a consequence of good starting point) make DES-GSLS efficient.

Comparison of Different Similarity Measures. In the above experiments, we used the KL divergence based similarity $s(d, d_0)$ for these retrieval models. To assess the influence of different document similarity measures to our approaches, we have also done experiments to compare the KL divergence with the cosine similarity and Pearson's correlation coefficient. Since the experimental results with the cosine similarity are very similar to results with Pearson's correlation coefficient, we omit the former here. Figure 3 shows the effectiveness scores at rank position 10 on the TREC-2009 dataset. As expected, these top-k document

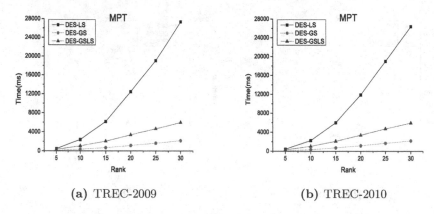

(a) TREC-2009   (b) TREC-2010

Fig. 2. Running time of MPT on TREC-2009 and TREC-2010

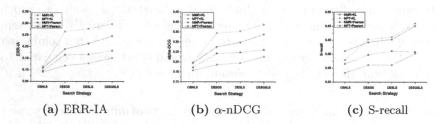

(a) ERR-IA   (b) α-nDCG   (c) S-recall

Fig. 3. Performance with different Distance Measures on TREC-2009

retrieval methods using different document similarity measures exhibit the same pattern. The four top-k document retrieval methods, in the ascending order of effectiveness, are OBN-GS, DES-GS, DES-LS, DES-GSLS. More importantly, we see that the MPT retrieval model with KL divergence generally performs the best.

5   Conclusions

In this paper, we formalized the top-k retrieval problem as a submodular function maximization problem to search a subset of documents that had nice accuracy-diversity trade-off.

It would be very promising to apply the same idea of integrating greedy search and local search to solve other submodular function optimization problems in a variety of domains.

Acknowledgments. This work was supported by the 973 project (No. 2010CB328106), NSFC grant (No. 61170085 and 61033007), Program for New Century Excellent Talents in China (No.NCET-10-0388) and Shanghai Knowledge Service Platform Project (No. ZF1213).

References

1. Carbonell, J.G., Goldstein, J.: The use of MMR, diversity-based reranking for reordering documents and producing summaries. In: SIGIR, pp. 335–336 (1998)
2. Chandar, P., Carterette, B.: Diversification of search results using webgraphs. In: SIGIR, pp. 869–870 (2010)
3. Chapelle, O., Metlzer, D., Zhang, Y., Grinspan, P.: Expected reciprocal rank for graded relevance. In: CIKM, pp. 621–630 (2009)
4. Chen, H., Karger, D.R.: Less is more: Probabilistic models for retrieving fewer relevant documents. In: SIGIR, pp. 429–436 (2006)
5. Clarke, C.L.A., Kolla, M., Cormack, G.V., Vechtomova, O., Ashkan, A., Buttcher, S., MacKinnon, I.: Novelty and diversity in information retrieval evaluation. In: SIGIR, pp. 659–666 (2008)
6. Gonzalez, T.F. (ed.): Handbook of Approximation Algorithms and Metaheuristics. Chapman and Hall (2007)
7. Krause, A., Golovin, D.: Submodular Function Maximization. In: Tractability: Practical Approaches to Hard Problems. Cambridge University Press (2012)
8. Lin, H., Bilmes, J.: A class of submodular functions for document summarization. In: ACL, pp. 510–520 (2011)
9. Manning, C.D., Raghavan, P., Schütze, H.: Introduction to Information Retrieval. Cambridge University Press (2008)
10. Nemhauser, G., Wolsey, L., Fisher, M.: An analysis of approximations for maximizing submodular set functions – i. Mathematical Programming 14(1), 265–294 (1978)
11. Russell, S., Norvig, P.: Artificial Intelligence: A Modern Approach, 3rd edn. Prentice Hall (2009)
12. Santos, R.L.T., Macdonald, C., Ounis, I.: Intent-aware search result diversification. In: SIGIR, pp. 595–604 (2011)
13. Wang, J., Zhu, J.: Portfolio theory of information retrieval. In: SIGIR, pp. 115–122 (2009)
14. Zhai, C., Cohen, W.W., Lafferty, J.D.: Beyond independent relevance: Methods and evaluation metrics for subtopic retrieval. In: SIGIR, pp. 10–17 (2003)
15. Zuccon, G., Azzopardi, L.: Using the quantum probability ranking principle to rank interdependent documents. In: Gurrin, C., He, Y., Kazai, G., Kruschwitz, U., Little, S., Roelleke, T., Rüger, S., van Rijsbergen, K. (eds.) ECIR 2010. LNCS, vol. 5993, pp. 357–369. Springer, Heidelberg (2010)
16. Zuccon, G., Azzopardi, L., Zhang, D., Wang, J.: Top-k retrieval using facility location analysis. In: Baeza-Yates, R., de Vries, A.P., Zaragoza, H., Cambazoglu, B.B., Murdock, V., Lempel, R., Silvestri, F. (eds.) ECIR 2012. LNCS, vol. 7224, pp. 305–316. Springer, Heidelberg (2012)

Improving Recommendations with Collaborative Factors

Penghua Yu[1], Lanfen Lin[1], Feng Wang[1,2], Jing Wang[1], and Meng Wang[1]

[1] Zhejiang University, Hangzhou, China, 310027
[2] Hangzhou Dianzi University, Hangzhou, China, 310018
{yph719,llf,wangfeng,cswangjing,wangmeng}@zju.edu.cn

Abstract. Collaborative filtering has become the most popular technique in the field of recommender system to deal with the information overload problem. Most collaborative filtering approaches either based on the intuitive nearest neighbor methods or the scalable latent factor models. In order to benefit from the advantages of these two paradigms, some hybrid strategies are proposed by taking weighted averages on near neighbors' ratings as effects, or factorizing neighborhood to model interactions and relationships directly. However, these methods usually assume that the latent factors of users/items are independent of each other. Yet in fact, there are relationships among latent factors would affect the performance of recommendations. Motivated by this, in this paper, we introduce the collaborative factors, which are smoothed by near neighbors' factors, to better capture the intrinsic features for users and items. We further propose a novel collaborative matrix factorization (CoMF) model in order to elaborately incorporate these collaborative factors into latent factor models. Finally, experimental results on two datasets show that our CoMF significantly outperforms some state-of-the-art methods in prediction accuracy.

Keywords: Recommender Systems, Collaborative Filtering, Collaborative Factors, Matrix Factorization.

1   Introduction

Recommender system is an indispensable tool used to produce item recommendations based on users' preferences. In many applications, degrees of users' preferences are presented by the explicit scores provided by users, or these implicit feedbacks inferred from users' behaviors [3]. We refer to all types of interactions as ratings. The ratings expressed by users on items are stored in a rating matrix, which is usually extremely sparse [1]. Existing recommendation algorithms are mainly based on two strategies: content based filtering versus collaborative filtering (CF). The former methods require gathering content information that might not be easy to collect. The latter only rely on users' historical records [10]. The CF methods are further adopted into two directions: the nearest neighbor based method and the latent factor model. The latter gains much more attention since the success of the Netflix competition, and many variants [2], [3], [8], [11] have been proposed to improve the prediction accuracy.

In order to take full advantages of these two strategies, the task on combining these CF approaches has been studied recently. Existing research can be classified into the

F. Li et al. (Eds.): WAIM 2014, LNCS 8485, pp. 30–33, 2014.
© Springer International Publishing Switzerland 2014

following categories: (1) incorporating the neighbors' preferences into latent factor approaches[2], [5]; (2) factorizing the neighborhood to directly model neighbors, e.g., a factorized neighborhood method [4] is presented; (3) making use of external data to model these relationships more realistically and precisely [6], [7]. However, both users and items are not independent of each other in the real world. The latent factors of similar entities are not directly utilized into latent factor models in previous studies. In this paper, we aim at improving the accuracy by injecting the dependent latent factors rather than the individual user and item factor.

The remainder of this paper is structured as follows. In Section 2, we introduce the collaborative factors and then elaborate our proposed model. The performances of our approach is reported in Section 3. Finally, we draw some conclusions and discuss the future work in Section 4.

2   Collaborative Matrix Factorization

Given an active user and a target item, we refer to factors of user' neighbors as the collaborative user factors, so as for collaborative item factors. The collaborative user factors and collaborative item factors are named together as collaborative factors:

$$p_{uc} = \alpha p_u + (1 - \alpha) \sum_{v \in N(u)} s_{uv} p_v \tag{1}$$

$$q_{ic} = \beta q_i + (1 - \beta) \sum_{j \in N(i)} s_{ij} q_j \tag{2}$$

where p_{uc} represents the factor of user u smoothed by the user' neighbors $N(u)$, q_{ic} is the factor for item i smoothed by its neighboring items $N(i)$. s_{uv} is similarity between user u and v, and s_{ij} is the similarity between item i and j. Then α and β are constants to determine weights between individual factor and collaborative factors.

Model Representation. We denote b_{ui} as the bias term effect, which makes up of global effect, user effect and item effect. Then we replace the factor of user and item in traditional latent factor model with collaborative factors of user and item, thus leading to a novel formulation represented as:

$$\hat{r}_{ui} = b_{ui} + \left(\alpha p_u + (1 - \alpha) \sum_{v \in N(u)} s_{uv} p_v \right)^T \left(\beta q_i + (1 - \beta) \sum_{j \in N(i)} s_{ij} q_j \right) \tag{3}$$

This model is referred to as the Collaborative Matrix Factorization (CoMF), and it is utilized to better infer the latent features of users and items.

Learning Algorithm. To find the optimization of the CoMF, we perform the minimization problem on the sum of squared error between the actual observed ratings and their predicted values. The objective function minimizing the regularized squared error on the set (denoted as S) of known ratings is then defined as:

$$L = \min_{P,Q} \sum_{(u,i) \in S} \left(r_{ui} - \hat{r}_{ui} \right)^2 + \lambda_1 \left(\sum_u \| p_u \|^2 + \sum_i \| q_i \|^2 \right) + \lambda_2 \left(\sum_u \| b_u \|^2 + \sum_i \| b_i \|^2 \right) \tag{4}$$

where $\hat{r}_{ui}$ obeys the rule formulated in Eq. (3). To learn optimization (b_*, p_* and q_*), we apply the stochastic gradient descent algorithm, which loops through all known ratings. The derivations are omitted as they are easy for readers to perform.

3 Experiments

We evaluate the methods on two data sets: the Movielens dataset[1] and the Epinions dataset [9]. In addition, we adopt the RMSE [1] on 5-fold cross validation to measure the result. We further compare our proposed CoMF with the following methods:

- **UserMean.** It utilizes the mean value of other users to predict the missing values.
- **ItemMean.** It applies the mean value of every item to predict the missing values.
- **Biased SVD.** The regularization term is set to 0.05, and the learning rate is 0.005.
- **SVD++.** It is proposed in [4] and regularization is 0.055, and learning rate 0.07.
- **RSTE.** It is proposed in [6] and α is set to 0.6 with the regularization term 0.04 and the learning rate is 0.01.

On the Movielens dataset, the regularization coefficients are set to 0.005and 0.001 with learning rate set to 0.005. In the Epinions dataset, all of them are all set to 0.01.

In the CoMF model, α and β play very important roles. To determine the sensitivity of α and β, we carried out an experiment where we varied the value of α and β from 0.0 to 1.0 in an increment of 0.2. We finally selected 0.6 for both α and β on both Movielens dataset and Epinions dataset as an optimum value.

Table 1. Performance comparisons of the CoMF with other approaches

Datasets	D	Metric	UserMean	ItemMean	BSVD	SVD++	RSTE	CoMF
Epinions	5	RMSE	1.1988	1.0942	1.0380	1.0408	1.0480	1.0094
		Improve	15.80%	7.75%	2.76%	3.02%	3.68%	
	10	RMSE	1.1988	1.0942	1.0378	1.0411	1.0406	1.0021
		Improve	16.41%	8.42%	3.44%	3.75%	3.70%	
Movielens	5	RMSE	1.036	0.983	0.9001	0.8932	0.8875	0.8423
		Improve	18.70%	14.31%	6.42%	5.70%	5.09%	
	10	RMSE	1.0360	0.9830	0.8980	0.8831	0.8702	0.8291
		Improve	19.97%	15.66%	7.67%	6.11%	4.72%	

Table 1 reports the results of our CoMF method compared with several state-of-the-art methods. "D" means dimensionality of factors. It can be observed that our CoMF method significantly outperforms the other methods both on the Movielens and the Epinions dataset.

---

[1] http://www.grouplens.org/datasets/movielens/

4   Conclusions

In this paper, we focus on improving recommendation accuracy of the latent factor models. Motivated by the fact that both users and items are not independent of each other in the real world, we introduce the collaborative factors to capture factors of the neighbors of users and items, respectively. We further investigate integrating these collaborative factors into the matrix factorization models to improve the prediction accuracy, and propose a novel collaborative matrix factorization (CoMF) model.

With the explosive increase of information in the Web, the modalities and types of information evolve at the same time. Such heterogonous information is useful for recommender systems. Therefore, it is important to consider the heterogeneous data to improve performance of recommendations. Meanwhile, scalable methods to address other recommendation tasks, such as the Top-N task, and other performance measures should be considered under this circumstance.

Acknowledgments. The work described in this paper is supported by grants from Programs under Grant No: 2012BAD35B01-3, 20110101110065, 2013BAF02B10. In addition, the authors thank the anonymous reviewers for their helpful comments.

References

1. Adomavicius, G., Tuzhilin, A.: Towards the Next Generation of Recommender Systems: A Survey of the State-of-the-Art and Possible Extensions. IEEE TKDE 17(6), 734–749 (2005)
2. Bell, R., Koren, Y.: Scalable Collaborative Filtering with Jointly Derived Neighborhood Interpolation Weights. In: IEEE ICDM, pp. 43–52 (2007)
3. Hu, Y., Koren, Y., Volinsky, C.: Collaborative Filtering for Implicit Feedback Datasets. In: IEEE ICDM, pp. 263–272 (2008)
4. Koren, Y.: Factorization Meets the Neighborhood: a Multifaceted Collaborative Filtering Model. In: ACM SIGKDD, pp. 426–434 (2008)
5. Koren, Y.: Factor in the Neighbors: Scalable and Accurate Collaborative Filtering. TKDD 4(1), 1 (2010)
6. Ma, H., King, I., Lyu, M.R.: Learning to Recommend with Social Trust Ensemble. In: ACM SIGIR, pp. 203–210 (2009)
7. Ma, H., King, I., Lyu, M.R.: Learning to Recommend with Explicit and Implicit Social Relations. ACM TIST 2(3), 29 (2011)
8. Paterek, A.: Improving regularized singular value decomposition for collaborative filtering. In: Proceedings of KDD Cup and Workshop, pp. 5–8 (2007)
9. Paolo, M., Avesani, P.: Trust-aware Bootstrapping of Recommender Systems. In: Workshop on Recommender Systems, pp. 29–33 (2006)
10. Sarwar, B., Karypis, G., Konstan, J., Riedl, J.: Item-based Collaborative Filtering Recommendation Algorithms. In: ACM WWW, pp. 285–295 (2001)
11. Takács, G., Pilászy, I., Németh, B., Tikk, D.: A unified approach of factor models and neighbor based methods for large recommender systems. In: IEEE Applications of Digital Information and Web Technologies (ICADIWT), pp. 186–191 (2008)

Evolutionary Personalized Hashtag Recommendation

Jianjun Yu[1,*] and Yi Shen[1,2]

[1] Computer Network Information Center, Chinese Academy of Sciences
[2] University of the Chinese Academy of Sciences
{yujj,shenyi}@cnic.ac.cn

Abstract. Hashtags, starting with a symbol "#" ahead of terms, are widely used and inserted anywhere within posts as they present rich sentiment information on topics that people are really interested in. In this paper, we focus on the problem of hashtag recommendation considering its personalized and evolutionary aspects. We introduce three features to model personal user interest and its evolution, including (1) hashtag popularity; (2) hashtag textual information; and (3) hashtag time factor. We construct a hybrid model combining these features to learn user preference and recommend personalized hashtags consequently.

Keywords: recommendation, micro-blogging system, hashtag, popularity.

1 Introduction

A hashtag is a set of words prefixed by a # (or #xxx# for Chinese hashtag) symbol in a post on micro-blogging systems. Hashtag is always generated to present a topic that aims to make posts easily understandable by other relevant users and facilitate conversations among the users with its rich sentiment information. Hashtag recommendation in nature is to predict users' interests and provide possible preferred hashtags to users that they have not yet considered.

Current work introduces simple yet incomplete one or two features to rank items for recommendation, while some other important factors are neglected, i.e., item popularity and user interest evolution [2] [3] [5]. Correspondingly, people other get a recommended list only with breaking events, or only with personal user interest when applied with traditional methods. While our proposed approach would provide fascinating results. Our contributions in this paper can be summarized as follows.

(1) We introduce the first work on hashtag recommendation with three features: hashtag popularity, hashtag textual information, and hashtag time factor. All the proposed features are theoretically and experimentally demonstrated important for hashtag recommendation.

(2) We scheme a novel hashtag recommendation approach considering personalized and evolutionary aspects. We model user preference for recommendation based on the three features, and a hybrid recommendation approach is introduced to extract the top N hashtags.

* This research was supported by NSFC Grant No.61202408, CNIC Grant CNIC-PY-1407, and CAS 125 Informatization Project XXH12503.

F. Li et al. (Eds.): WAIM 2014, LNCS 8485, pp. 34–37, 2014.

2 Proposed Method

In our hashtag recommendation method, we balance three features: hashtag popularity, hashtag textual information, hashtag time factor to model user preference and its evolution, and take several steps to get the final recommendation results.

2.1 Hashtag Popularity

Intuitively, if a hashtag is a hot one on a micro-blogging system, it will attract more users involved in, which also brings higher popularity of the hashtag. According to this observation, we use the following equations to measure the hashtag popularity and score the final recommendation.

First, we define hashtag hotness as follows:

$$Hot_{hashpop} = \frac{Count_{users} - MIN_{users}}{MAX_{users} - MIN_{users}} \tag{1}$$

where $Count_{users}$ is the count of users that use the current hashtag, MAX_{users} is the maximum count of users involved in a hashtag, and MIN_{users} is the minimum count of users involved in a hashtag.

The hashtag hotness works well if all hashtags are generated to reflect user interest. Whereas, as we observed, there exist parts of unfavorable hashtags from advertisers or with sale information. These hashtags are with massive referred posts, while with few people involved in, which are valueless for recommendation and should be ignored. We borrow the concept "authorship entropy" [1] to measure the hashtag authorship.

$$Auth_{hashpop} = -\sum_{i=1}^{l} \frac{c_i}{n} \cdot \log(\frac{c_i}{n}) \tag{2}$$

where l is the count of users that involve in the current hashtag, n is the number of posts associated with the hashtag, and c_i is the count of posts that the ith user creates.

Thus we get the final score of hashtag popularity for recommendation:

$$Score_{hashpop} = Hot_{hashpop} \times Auth_{hashpop} \tag{3}$$

After hashtag popularity computation, the hashtags would be ranked with their hotness degrees. Correspondingly, hashtags expressing breaking events would come out at the top of the ranking list.

2.2 Hashtag Content Similarity

To predict personal user preference on hashtags, we model the relationship between personal user interest and visited hashtags as text similarity. That is, user interest (labeled as D_{user}) can be presented with the textual documents of all posts related with the current user, and each post associated with the current hashtag (labeled as $D_{hashtag}$) can be viewed as another document. Thus personal user preference on hashtags can be calculated with a similarity function. We take two steps to calculate text similarity between D_{user} and $D_{hashtag}$.

1) Topic probability distribution generation. We adopt Latent Dirichlet Allocation (LDA) to calculate the probability distribution. LDA is one of the increasingly popular tools for summarization and discovery with the capability of automatically extracting the topical structure of large document collections. We first calculate keywords w distribution on topics to, and topics to distribution on posts pd with LDA computation equations.

We train the LDA parameters to generate a set of probability distributions: P_{to}^{pd} that is topics to_n probability distribution on post pd, and P_w^{to} is keywords w probability distribution on topic to_n. Then we use P_w^{to} to determine D_{user} and $D_{hashtag}$ probability distributions on each topic labeled as P_{user} and $P_{hashtag}$ respectively. $P_{user} = \{P_U(to_1), P_U(to_2), ..., P_U(to_k)\}$ and $P_{hashtag} = \{P_H(to_1), P_H(to_2), ..., P_H(to_k)\}$. Where k is the total number of topics, and we set $k = 200$.

2) Similarity calculation. We use negative symmetric Kullback-Leibler divergence function to measure the similarity between D_{user} and $D_{hashtag}$.

$$Sim_c = \frac{-1}{2[KL(P_{user}||P_{hashtag}) + KL(P_{hashtag}||P_{user})]} \qquad (4)$$

where $KL(P_{user}||P_{hashtag}) = \sum P_{user}(to_i) \log(P_{user}(to_i)/P_{hashtag}(to_i))$. KL is a non-symmetric measure of the difference between two probability distributions P_{user} and $P_{hashtag}$.

With this step, most of hashtags reflecting personal user interest would be promoted in the ranking list. Still the ranking list rarely provides time-insensitive hashtags presenting long-term user interest. We then add hashtag time feature to capture user interest evolution, and promote the importance of time-insensitive hashtags.

2.3   Hashtag Time Factor

As described above, hashtags can be categorized into two classifications: time-sensitive hashtags and time-insensitive hashtags. User interest would evolve when time changes, e.g. football fans would like to attend football match events for a long time, whereas they may watch live webcast on #World Championships in Athletics in Moscow during the competitions. We take two steps to calculate the contribution of hashtag time adapting to short-term and long-term user interest.

1) Hashtag classification. We first address the classification strategy according to hashtag time. Hashtag, especially time-sensitive hashtag exists the phenomenon of micro-meme. As we observed, most of time-sensitive hashtags have a hot period within a week. On the other hand, time-insensitive hashtags represent long-term user interest, which are almost time independence. We view this problem as a sequence classification problem and simplify the approach introduced in [4]. For each hashtag H, we rank all associated posts within a time period Δt ($\Delta t = 1$ day in this paper), and then get a ranked set of posts: $S(H) = \{N_1, N_2, ..., N_k\}$. Where N_k is the count of posts at k^{th} day after the hashtag is created. We set $k = 7$ since time-sensitive hashtags have a hot period within a week. If a hashtag is created less than 7 days, we just set $N_i = 0$ for the days before its creation. We then transform $S(H)$ to a 7-dimensional vector: $V_{S_H} = \{1, ||N_2 - N_1||, ||N_3 - N_2||, ..., ||N_7 - N_6||\}$, and construct a SVM (Support Vector Machine) using RBF(Radial Basis Function) kernel with feature V_{S_H} based on

the libsvm toolkit. We label 428 hashtags manually for the training set (253 for positive classification, and 175 negative one), and verify with 5-cross-validation approach, which gets 82.3% accuracy.

2) The contribution of hashtag time. We introduce a decay factor $TF_{hashtag}$ to calculate the contribution of hashtag time, which aims to reduce the importance of time-sensitive hashtags.

$$TF_{hashtag} = \begin{cases} e^{-\mu(t-t_{max})}, & H\ is\ time-sensitive \\ 1, & otherwise \end{cases} \tag{5}$$

where μ is a constant attenuation coeffcient. We set parameter $\mu = 0.3$ through heuristic learning approach. t is current day, t_{max} is the day with the maximum count of posts, and the unit of t is one day. When we calculate $TF_{hashtag}$ at the day of hashtag creation, we just set $TF_{hashtag} = 1$ for simplicity. Also we make $TF_{hashtag} = 1$ if the hashtags are time-insensitive.

2.4 Hashtag Recommendation

Hashtag recommendation gives top N candidates for users with a combination function utilizing three features, including hashtag content, hashtag popularity and hashtag time factor:

$$RecH(U, H) = (\gamma \times Sim_c + (1 - \gamma) \times Score_{hashpop}) \times TF_{hashtag} \tag{6}$$

where Sim_c and $Score_{hashpop}$ would be normalized into $[0, 1]$. For each user, top N candidate hashtags would be ranked with $RecH(U, H)$ as final recommendation results.

After this step, the recommendation list would include those hot hashtags, personal short-term interest and long-term interest hashtags. That means the recommendation results appeal to a broader constituency.

References

1. Cui, A., Zhang, M., Liu, Y., et al.: Discover breaking events with popular hashtags in twitter. In: CIKM 2012 Conference Proceedings, pp. 1794–1798 (2012)
2. Godin, F., Slavkovikj, V., Neve, W.D.: Using topic models for twitter hashtag recommendation. In: WWW 2013 Conference Proceedings, pp. 593–596 (2013)
3. Kywe, S.M., Hoang, T.-A., Lim, E.-P., et al.: On recommending hashtags in twitter networks. In: SocInfo 2012 Conference Proceedings, pp. 337–350 (2012)
4. Lehmann, J., Goncalves, B., Ramasco, J.J., et al.: Dynamical classes of collective attention in twitter. In: WWW 2012 Conference Proceedings, pp. 251–260 (2012)
5. Yang, L., Sun, T., Zhang, M., et al.: We know what you #tag: does the dual role affect hashtag adoption? In: WWW 2012 Conference Proceedings, pp. 261–270 (2012)

PathSimExt: Revisiting PathSim in Heterogeneous Information Networks

Leong Hou U.*, Kun Yao, and Hoi Fong Mak

Department of Computer and Information Science, University of Macau, Macau
{ryanlhu,leomak}@umac.mo, yaokun527@qq.com

Abstract. Similarity queries in graph databases have been studied over the past few decades. Typically, the similarity queries are used in homogeneous networks, where random walk based approaches (e.g., Personalized PageRank and Sim-Rank) are the representative methods. However, these approaches do not well suit for heterogeneous networks that consist of multi-typed and interconnected objects, such as bibliographic information, social media networks, crowdsourcing data, etc. Intuitively, two objects are similar in heterogeneous networks if they have strong connections among the heterogeneous relationships. PathSim is the first work to address this problem which captures the similarity of two objects based on their connectivity along a semantic path. However, PathSim only considers the information in the semantic path but simply omit other supportive information (e.g., number of citations in bibliographic data). Thus we revisit the definition of PathSim by introducing external support to enrich the result of PathSim.

1 Introduction

A heterogeneous network is a logical network that usually consists of a large amount of multi-typed and interconnected components. The inter-connections in the heterogeneous networks often indicate different kind of relations, such as bibliographic networks, epidemic network, and social media network [1]. As an example in bibliographic networks, users may be interested in querying similar authors to the author of a paper that they just read. There are variant ways to measure the similarity of two authors in a bibliographic network. For instance, two authors is similar if their papers co-appear in the same venue frequently or they are the co-authors in many publications.

Similarity queries in homogenous networks have been extensively studied over the past few decades, where random walk based approaches (e.g., P-PageRank [2] and Sim-Rank [3]) are the representative methods in this category. However, the random walk solutions cannot be used in heterogenous networks since the walks over different relationships have different meanings behind. To address the similarity queries in heterogenous networks, Yizhou Sun et al. [4] proposed PathSim that computes the similarity of two objects based on the affiliation of a semantic path. For instance, the affiliations in bibliographic networks may include *authors-venues* (**AV**), *authors-papers* (AP), and

---

* This work was partially supported by MYRG109(Y1-L3)-FST12-ULH from UMAC Research Committee.

F. Li et al. (Eds.): WAIM 2014, LNCS 8485, pp. 38–42, 2014.

papers-terms (**PT**) relationships. A possible semantic path is *authors-venues-authors* (**AVA**) which indicates the similarity of the authors based on their co-appearance in the same venue. Given a semantic path (e.g., **AVA**) and a query object (e.g., an author), PathSim returns the most similar objects based on the affiliations in the semantic path.

Table 1. Similarity search result under meta-path 'VAV' of a query 'PKDD' on DBLP dataset

Rank	P-PageRank	SimRank	PathSim	PathSimExt
q	PKDD	PKDD	PKDD	PKDD
1	KDD	Local Pattern Detection	ICDM	PAKDD

In this work we enrich PathSim by introducing *external support* into the similarity computation. We observe that some factor can be used as an external support for improving the similarity search. More specifically, the similarity of venues is not only reflected by the semantic path *venues-authors-venues* (**VAV**) but also the reputation similarity of the venues (e.g., average citations per paper). Table 1 compares this idea with other similarity methods using a DBLP example that lists the most similar venues to 'PKDD' based on a semantic path *venues-authors-venues* (**VAV**). 'PKDD' is a European conference on machine learning and knowledge discovery in databases. As shown in the table, P-PageRank [2] and PathSim [4] return 'KDD' and 'ICDM', respectively; however, these results may be not the best answer for the query since the reputation of 'KDD' and 'ICDM' is quite different from 'PKDD'. SimRank returns a less related conference 'Local Pattern Detection' as the top-1 result which manifests the problem of random walk based solutions in heterogenous networks. Thus, we revise the definition of PathSim, named as PathSimExt, that additionally introduces *external support* into the similarity measures.

2 PathsimExt

A heterogenous information network is a special type of information networks that either contains multiple types of objects or multiple types of relationships. More specifically, a heterogenous information network is a directed graph G, where each vertex belongs to one particular object type T and each edge belongs to one particular semantic relationship R.

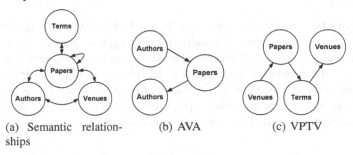

(a) Semantic relationships   (b) AVA   (c) VPTV

Fig. 1. Bibliographic semantic relationships and 2 semantic paths

Semantic paths, Δ. Based on the semantic relationships, we can construct a semantic paths Δ that typically represents some semantic meaning behind. For instance, a semantic path *authors-papers-authors* (**APA**) (shown in Figure 1(b)), that uses two semantic relationships **AP** and **PA**, indicates the co-authorship. Note that the semantic path is not necessary symmetric, where a asymmetric semantic path **VPTV** can be found in Figure 1(c). To facilitate reasonable search queries, we assume that the begin and the end object type in a semantic path are always identical in this work.

PathSim. As studied in [4], PathSim is shown to capture better semantics of peer similarity in heterogenous network than other methods proposed for homogenous information networks, such as P-PageRank and SimRank. Formally, PathSim is defined as follows.

$$S_\Delta(x,y) = \frac{2 \times w(x,y)}{w(x,x) + w(y,y)} \tag{1}$$

where $w(x,y)$ represents the number of path instances between x and y under the semantic path Δ. The denominators, $w(x,x)$ and $w(y,y)$, can be viewed as the normalization factor in the definition.

Table 2. Search result under **APA** on a query 'Mike'

(a) Co-authorship and author information

	#collaborations $w(x,y)$	#publications $w(x,x)$	#cititations $ext(\mathbf{A},x)$
Mike	-	100	300
Jim	20	20	50
Tom	19	100	300

(b) Similarity value between Mike and others

	Jim	Tom
PathSim	0.33	0.19
PathSimExt	0.028	0.095
PathSimExt (norm.)	0.057	0.19

Table 2 shows an example of PathSim that finds the most similar authors of 'Mike' under the semantic path **APA**. According to the definition of PathSim (Equation 1), 'Mike' is similar to 'Jim' than 'Tom' since the number of path instances (i.e., the number of collaborations) between 'Mike' and 'Jim' is higher than 'Mike' and 'Tom'. However, if we take the fame of the authors (i.e., number of citations) into consideration, 'Tom' should be a more similar result to 'Mike' since both of them have similar number of citations and publish certain amount of papers together. This example shows that external support (i.e., number of citations) can enrich the result of PathSim but is not considered in [4].

External Supports. An *external support* can be any factor that reflects the importance of the objects from global views. Typically, for each type of object $\alpha \in T$, we can facilitate some support (e.g., well accepted knowledge and common sense) to rank the objects $o_i \in O$ in a reasonable way, which is defined as $ext(\alpha, o_i)$ in this work. For instance, in the bibliographic network, we can use the citation numbers as the external support for every author a_i in the author type $\mathbf{A}$, which is denoted as $ext(\mathbf{A}, a_i)$. Given the external support, we revise PathSim to PathSimExt as follows.

$$S_\Delta^E(x,y) = \frac{|w(x,y)| \times sim_T(x,y)}{ext(T,x) + ext(T,y)} \tag{2}$$

where $sim_T(x,y) = \frac{min(ext(T,x),ext(T,y))}{max(ext(T,x),ext(T,y))}$ can be viewed as the similarity of two objects in the object type T. The denominators in Equation 2, $ext(T,x)$ and $ext(T,y)$, can be viewed as the normalization factor.

3   Experiments

In this paper, we used a DBLP citation network dataset downloaded from arnetminer.org that contains 7.7K venues, 1M authors, 1.6M papers, 34K terms, and 2.3M citations.

Matrix	Data size	Density
AV	2,988,422	0.0374%
AP	4,227,433	0.00025%
VP	1,632,442	0.013%
VT	3,741,075	1.42%
VAV	7,422,651	12.5%

Rank	AVA	APA	VAVTV	VAV
q	Jiawei Han	Jiawei Han	ACM Trans. Graph.	ACM Trans. Graph.
1	Philip S. Yu	Philip S. Yu	SIGGRAPH	SIGGRAPH
2	Christos Faloutsos	Jian Pei	IEEE Visualization	Trans. Vis. Comput. Graph.
3	Divesh Srivastava	Charu C. Aggarwal	Trans.Vis.Comput.Graph.	Journal of Computer Vision
4	H. V. Jagadish	ChengXiang Zhai	ICCV	Symp. on Comput. Geometry
5	Surajit Chaudhuri	Laks V. S. Lakshmanan	CVPR	Computer Aided Geometric Design

We show the result of different semantic paths in Table 3, including author-to-author relationships (**AVA** and **APA**) and venue-to-venue relationships (**VAVTV** and **VAV**). The result of **AVA** shows that 'Philip S. Yu' and 'Christos Faloutsos' are the two most similar authors to 'Jiawei Han'. The result is meaningful as both 'Philip S. Yu' and 'Christos Faloutsos' publish lot of papers in the common venues of 'Jiawei Han' and their citation records (i.e., external support) are strong. In another set of experiments, we evaluate the path **VAVTV** and **VAV** for a journal 'ACM Transaction on Graphics'. The results using the semantic path **VAVTV** are more relevant to the area of computer graphics and computer vision research since the returned venues should have similar topics according to **VTV** (i.e., the terms). In addition, 'ICCV' and 'CVPR' are ranked in the top-5 result since these venues have similar reputation (i.e., by external support) to 'ACM Transaction on Graphics'. For clarity, the results using the semantic path **VAV** are more diverse since the path only take author-venue relationship into consideration.

4   Conclusions

In this work, we revise the definition of PathSim that enriches the result of PathSim by introducing the external support into the similarity measure. Our result demonstrates that the external support is effective in bibliography data. In the future, we attempt to further improve the query performance in terms of both efficiency and effectiveness.

References

1. Han, J.: Mining heterogeneous information networks by exploring the power of links. In: Gavaldà, R., Lugosi, G., Zeugmann, T., Zilles, S. (eds.) ALT 2009. LNCS, vol. 5809, p. 3. Springer, Heidelberg (2009)
2. Jeh, G., Widom, J.: Scaling personalized web search. In: WWW, pp. 271–279 (2003)
3. Jeh, G., Widom, J.: Simrank: a measure of structural-context similarity. In: KDD, pp. 538–543 (2002)
4. Sun, Y., Han, J., Yan, X., Yu, P.S., Wu, T.: Pathsim: Meta path-based top-k similarity search in heterogeneous information networks. In: PVLDB, vol. 4(11), pp. 992–1003 (2011)

AdaMF:Adaptive Boosting Matrix Factorization for Recommender System

Yanghao Wang, Hailong Sun, and Richong Zhang

School of Computer Science and Engineering, Beihang University
Beijing, 100191, China
{wangyh,sunhl,zhangrc}@act.buaa.edu.cn

Abstract. Matrix Factorization (MF) is one of the most popular approaches for recommender systems. Existing MF-based recommendation approaches mainly focus on the prediction of the users' ratings on unknown items. The performance is usually evaluated by the metric Root Mean Square Error (RMSE). However, achieving good performances in terms of RMSE does not guarantee a good performance in the top-N recommendation. Therefore, we advocate that treating the recommendation as a ranking problem. In this study, we present a ranking-oriented recommender algorithm AdaMF, which combines the MF model with AdaRank. Specifically, we propose an algorithm by adaptively combining component MF recommenders with boosting methods. The combination shows superiority in both ranking accuracy and model generalization. Normalized Discounted Cumulative Gain (NDCG) is chosen as the parameter of the coefficient function for each MF recommenders. In addition, we compare the proposed approach with the traditional MF approach and the state-of-the-art recommendation algorithms. The experimental results confirm that our proposed approach outperforms the state-of-the-art approaches.

Keywords: Recommender System, Matrix Factorization, Learning to Rank.

1 Introduction

Recommender systems are widely adopted by online websites to discover the implicit needs of consumers. In recent years, existing approaches have focused on the rating prediction for users on unobserved items, in which Root Mean Square Error (RMSE) is usually chosen as the performance evaluation metric. We refer these approaches as rating-oriented recommender systems. One of the popular approaches for solving this problem is Matrix Factorization (MF) [1] which can also be called Latent Factor Model. However in real recommender systems, instead of rating prediction performance, top-N recommendation precision may attract more attention as it generates a ranked list which is directly presented to users. The uncertainty of the prediction in rating-oriented recommender system brings instability in the ranking performance. Moreover, other factors such as

F. Li et al. (Eds.): WAIM 2014, LNCS 8485, pp. 43–54, 2014.

the position information that measure the ranking performance are not taken in to consideration in rating-oriented recommender systems. Paolo et.al show that lower RMSE does not always translate into a better ranking result in [2]. Thus, instead of achieving better rating prediction, directly optimizing the relevance of the ranking list can offer user better experience. In recent years, Learning to Rank (LTR) [3] has becoming a powerful technique to solve ranking problem by using machine learning. For existing LTR-based recommender systems [4,5], in practice, the single model cannot be adjusted adaptively to the shifty real world data and the ranking performance evaluation metrics, such as NDCG, are not usually taken into the account when building the objective function.

In this paper, we propose AdaMF, an adaptive boosting approach for ranking-oriented recommender systems, by combining a state-of-the-art ranking model AdaRank [6] with MF. AdaRank creates a *strong ranker* by composing *weak rankers* linearly and the coefficient of each weak ranker is calculated from the weight function of a certain performance metric. Different with the traditional LTR problem, recommendation problem usually does not have explicit features in dataset to build model-based weak rankers. However Matrix Factorization can allocate latent factors for each user and item only based on user history data. Therefore, we choose Matrix Factorization to build the *component recommender* and component recommenders are combined to the *ensemble recommender* linearly according to the recommendation performance. In our approach we choose Normalized Discounted Cumulative Gain (NDCG) as the parameter for the coefficient calculation, which can measure the effect of the overall relevance of the recommendation lists.

In summary, the contributions of this work are three folded:

- We introduce a ranking-oriented approach for recommender system AdaMF, which is an adaptive boosting algorithm that combines component recommenders built by simple Matrix Factorization Models. The combination insures the ranking performance and decreases the generalization error.
- We propose a gradient descend algorithm for training component recommenders. The algorithm is ease in implementation and efficient in choosing parameters.
- We conduct a validation towards the efficiency and performance of AdaMF. The experimental results confirm the effectiveness and show an outstanding increase from the state-of-the-art approaches.

This paper is organized as follows: In Section 2, we introduce some related works for our work. In Section 3 we present our AdaMF algorithm. Some experiment and comparison are described in Section 4. We conclude our study of AdaMF at Section 5.

2   Related Work

2.1   Matrix Factorization

Matrix Factorization model uses low dimensional vectors to represent the feature for users and items. The MF model has high scalability and accuracy, and gets high success in Netflix challenge [1]. During the competition, more elements are considered to enhance the factor vector. For example, traditional neighbor based collaborative filtering is combined with MF model [7]. Temporal dynamic of users' taste and items' timeliness is also important in modeling. In temporal approaches of MF model, function based temporal dynamic factors have proven its effectiveness by presenting the latent factor vector as linear function of the time stamp [8]. MF can be viewed as graphical models in some probabilistic approaches of recommender system [9]. Latent Dirichlet Allocation [10] has been used to build hidden topic feature variables of users and items, thus Probabilistic Matrix Factorization model can be combined to improve performance and solve cold start problem [11].

2.2   Learning to Rank

Traditional ranking algorithm generates a list by sorting the relevance function of query and document. With the development of linguistics model and more accumulation of labeled data from search engine, researchers try to make machines 'learn' the ranking model.

LTR is a series of supervised learning algorithm to learn better ranking model from the labeled documents and feedbacks of users. LTR methods are categorized as Point-wise, Pair-wise and List-wise [3] models. Point-wise approaches predict the score of a document to the query by regression methods in machine learning. Pair-wise ranking models use classification to learn the relative preference of each item pair, classification methods like SVM [12] can be adapted into solving the problem. In recent years List-wise algorithms have drawing much attentions. The key point of List-wise LTR is to define loss function based on the whole result list for each query. The simple intuition is to use performance metrics in information retrieval such as MAP, NDCG, and so on, but these metrics are associated with the rank of documents thus they are discontinuous with model's parameters, therefore different smoothing technology for metrics lead to various approaches [13].

Based on the direct optimization of the performance metric, AdaRank [6] uses boosting technique to solve the problem. The basic idea of AdaRank is to employ an exponential loss function of performance metrics. It generates a strong ranker which linearly combines weak rankers, where the coefficients are computed by the performance of each ranker. So that different weak rankers are correlated according to each one's performance. To create different weak rankers, AdaRank maintains a distribution of weights over the queries in the training data. The distribution is determined according to the performance of strong ranker on each query. AdaRank increases the weights on the queries that are not ranked well by the strong ranker. For each round in the training, AdaRank trains a new weak ranker according to the weight distribution, and the newly trained weak ranker is appended to the strong ranker.

2.3   Ranking-Oriented Recommender System

Learning to Rank has made contribution to a wide variety of applications. One of the mainly used areas is in recommender system [4,5]. In Ranking-Oriented recommender systems, several ranking-based model have been built. In [4], Balakrishnan et.al. propose to take the trained latent factor of each user and item as feature vector. The extra model parameters are added for building point-wise and pair-wise model. They denote as Collaborative Ranking. List-wise LTR algorithm was also used to build MF model. Shi et.al. propose ListRank-MF [5], which optimize the loss function which is conducted based on the top one probability of item when given a predicted score. However, Collaborative Ranking proposes an EM-like training process, which cost more time on model training. Ranking based loss function in ListRank-MF also increases the complexity in gradient descent algorithm.

3   Adaptive Boosting Matrix Factorization

In this section we present our model AdaMF which combines Matrix Factorization with AdaRank, that provides a boosting technique for conducting the ranking-oriented recommender system. Firstly, we present our problem definition. Secondly, we introduce Matrix Factorization for recommender system and the framework of our AdaMF model. Finally, we discuss the advantages of combining Matrix Factorization with AdaRank.

3.1   Problem Definition

The objective of the recommender system is to learn the preference model from users' rating history to generate a top-N recommendation list. Suppose that there are N ratings given by U users to I items. For each user u, R_u is the set of rated items $\{R_{ui}\}_{i=1}^{N_u}$ where N_u is the size of R_u. The average of N_u is simplified to n. The rating given by user u to item i is denoted as R_{ui}. The goal of our algorithm is to create a recommender $f(u, i)$ to rank the unrated items for users.

3.2   Matrix Factorization

In the MF model [1], the prediction of the unobserved rating of item i given by user u is formulated as:

$$\hat{r}_{ui} = \sum_{k=0}^{K} P_{uk} Q_{ik} \tag{1}$$

where P_u is the latent factor vector of user u to reflect the interests of user u to each latent factor, similarly Q_i shows the relevance of item i with the latent factors, and K is the dimension of the latent factor.

The traditional approach of MF model is built for the prediction of missing ratings. To measure the performance of the accuracy of prediction, RMSE is one of the most commonly used evaluation metric. The RMSE-based loss function is defined as:

$$L = \sum_{u=1}^{U} \sum_{i \in R_u}^{N_u} (R_{ui} - \hat{r}_{ui})^2 + \frac{\lambda}{2}(||P||_F + ||Q||_F) \tag{2}$$

where $\hat{r}_{ui}$ denotes the predicted rating given by (1). In the loss function, $||\cdot||_F$ is Frobenius norm, which is used as the normalized term to generalize the model. Once the latent vectors have been learned, the unobserved ratings of each user can be estimated by (1). Intuitively, the recommendation list is generated by sorting $\hat{r}_{ui}$ for unrated items of each user.

3.3   AdaMF

Similar to AdaBoost, the AdaMF maintains a weight distribution on each user in the training set. The training process contains T rounds in all. In the tth round of training, the algorithm builds a component recommender P_u^t, Q_u^t by MF according to the weight distribution on users. The distribution reflects the focal point of the component recommender. During the learning, weights on the users with low training performance are increased so that the component recommender of next round would be forced to focus on users with bad fitting. A coefficient α_t of the component recommender is calculated according to the performance metric. Then the newly trained recommender is integrated to the ensemble recommender $f(u, i)$. New distribution for each user is updated after testing the performance of the ensemble recommender $f(u, i)$ on the training set.

In recommender system, the user's rating on items can be labeled by different level, such as one to five star scale. A metric that measures both multi-level relevance and position information is required to measure the ranking performance. For the reason of better representing the ranking performance, we use NDCG [14] measurement to measure the performance of the recommenders. Specifically, NDCG is defined as:

$$NDCG(u)@m = Z_u \sum_{j=1}^{m} \frac{2^{r_{u,j}} - 1}{log(1 + j)} \tag{3}$$

where $r_{u,j}$ is the real rating data given by user u to the item at the jth position of recommendation list. Z_u is the normalized parameter for the user u, so that the perfect ranker will get NDCG of 1. In the information retrieval domain, $r_{u,j}$ stands for the relevance of document on j^{th} position to a certain query. For our recommender problem, the rating of user given to the item can be viewed as the relevance. From the definition of NDCG, we can summarize that both various relevance level and position information is taken into consideration. Thus, we use NDCG for the calculation of the coefficient α_t. The AdaMF algorithm is shown in Algorithm 1.

AdaMF builds a strong recommender $f(u, i) = \sum_{t=1}^{T} \alpha_t P_u^{tT} Q_i^t$. The recommender list can be created by sorting $f(u, i)$ of the unrated items for each

---

Algorithm 1. AdaMF

Input: Rating history $\{r_{ui}\}$
Output: MF ranker $f(u, i)$
1. Initialize $D_1(u) = 1/U, f^0(u, i) = 0$
2. **for** $t = 1, t <= T, t + +$ **do**
3. $P_u^t, Q_u^t = ComponentRecommender(D_t(u), \{r_{ui}\});$
4.   Sort the observed items for each user by the component recommender $\hat{r}_{ui} = \sum_{k=0}^{K} P_{uk}^t Q_{ik}^t;$
5.   Calculate the current component recommender performance $NDCG(u)@m_c^t$
6.   Choose α_t:

$$\alpha_t = \frac{1}{2} * ln \frac{\sum_{u=1}^{U} D_t(u)(1 + NDCG(u)@m_c^t)}{\sum_{u=1}^{U} D_t(u)(1 - NDCG(u)@m_c^t)} \tag{4}$$

7.   Create ensemble recommender $f^t(u, i)$:

$$f^t(u, i) = f^{t-1}(u, i) + \alpha_t P_u^{tT} Q_i^t \tag{5}$$

8.   Sort the observed items for each user by the ensemble recommender $f(u, i)$
9.   Calculate the ensemble recommender performance $NDCG(u)@m_e^t$
10. Update $\mathbf{D_{t+1}}$ with $NDCG(u)@m_e^t$ of the ensemble recommender:

$$D_{t+1}(u) = \frac{exp\{-NDCG(u)@m_e^t\}}{\sum_{j=1}^{U} exp\{-NDCG(u)@m_e^t\}} \tag{6}$$

11. **end for**
12. $f(u, i) = f^T(u, i)$
13. **return** $f(u, i);$

---

user. Theoretical analysis shows that the adaptive boosting approaches can continuously improve the performance on the training set until reach the upper bound [6].

3.4   Component Recommender

In the original AdaRank, similar to AdaBoost, the weak ranker is conducted with a simple approach, only one feature is chosen to build the weak ranker. This approach can build the weak models which generate relatively low performances. However as mentioned by Li et.al. in [15], AdaBoost can get better generalization result with correlated *strong classifiers*. They present AdaBoostSVM by taken Support Vector Machine as the component classifier. We note that SVM is viewed as a strong classifier. In collaborative filtering, we utilize Matrix Factorization as the component recommender model. Matrix Factorization can easily achieve a good performance in rating prediction and recommender ranking. We can view MF as a relatively *strong recommender*.

One of the problem when building recommender model is the generalization problem. For matrix factorization, the single model would meet overfitting easily when the training gets deeper, as in usual, the latent factor vector's dimension would be close to the real user's rating number. The number of model parameter is larger than traditional model-based approaches. The common solution for solving generalization problem is to add the normalized term on the loss function, as illustrated in (2). This would solve the generalization problem to some extent. However, user's multi-faceted interest cannot be easily modeled by single model with normalization. AdaMF provides an adaptive voting mechanism for different component recommenders to determine the final result together. Different faceted models are correlated. The generalization is maintained by the combination of different component recommenders.

Similar with AdaBoost's robustness toward overfitting, AdaMF can rarely meet dilemma in generalization when more component recommenders are combined. Moreover, if the component recommender is generalized by the additional normalized term, the component recommender cannot reflect the preference over the adjusted user weight distribution. Therefore the component recommenders need to prune the normalized term.

AdaMF gives weight distribution on users which depends on the performance of current component recommender, the next component recommender model must fit the distribution. The current model should fit users with higher weights, i.e. minimize the errors on higher weighted users.

In summary we define the loss function of component recommender as:

$$L(P,Q) = U \sum_{u-1}^{U} D(u) \sum_{i \in R_u}^{N_u} (R_{ui} - P_u^T Q_i)^2 \qquad (7)$$

From this loss function, we can generate derivatives of each P_u and Q_i, then the parameters can be learned by Stochastic Gradient Descend (SGD). The training algorithm of component recommender is shown in Algorithm 2. η stands for the learning rate during the training, we set $\eta = 0.01$ in our study. Notice that if D(u) equals to $\frac{1}{U}$ for every user u, the component recommender equals to the basic MF model.

Algorithm 2. ComponentRecommender

Input: Distribution $\mathbf{D_t}$, Rating history $\{r_{ui}\}$
Output: Latent Factor Vector P_u, Q_i
  1. Initialize P_u^0, Q_i^0 randomly, initialize learning rate η, iteration rounds I
  2. **for** $n = 1, n <= I, n++$ **do**
  3.   **for** each r_{ui} **do**
  4. $err = r_{ui} - P_u^{n-1T} Q_i^{n-1}$;
  5. $P_u^n := P_u^{n-1} + \eta(UD(u)err * Q_i^{n-1})$;
  6. $Q_i^n := Q_i^{n-1} + \eta(UD(u)err * P_u^{n-1})$;
  7.   **end for**
  8. **end for**
  9. **return** P_u^I, Q_i^I;

3.5   Discussion

AdaMF is easy to be implemented and can be trained fast with high accuracy. Since the complexity of training component recommender is $O(K \cdot T \cdot (I \cdot N + U \cdot m \cdot nlogn))$. The algorithm's cost linearly increases with the scale of case. One of the advantage of AdaMF is that it combines component recommenders with various focal point to prevent the overfitting problem. We think that in real world data with more rating history, different aspects of users' preference are covered, in this occasion our model may show its effectiveness.

4   Experiment

In this section, we conduct experiments to evaluate the effectiveness of our method and make comparison with two baseline methods.

4.1   Experimental Setup and Datasets

We use MovieLens-100K[1] dataset to conduct our experiment. MovieLens-100K dataset contains 100K ratings(scale 1-5) given by 943 users to 1682 movies. In our conduction, 10, 20, 50 ratings are randomly chosen from each user to build a training set, and the remaining ratings are gathered for the testing set. For each condition, the chosen user must have 20, 30, 60 ratings so that we can test NDCG@10 on the testing set. In addition, we build a mixed length dataset. In reality, users' rating number varies in a large range, and only a few active users would contribute most of the rating feedbacks. Therefore, to involve the different activeness from different user, we select 50, 20, 10 ratings for users with more than 60, 30, 20 ratings then put the remain ratings into testing sets. In this way we can evaluate the adaptiveness of our method to various users' activeness and stickiness. In the following sections, we use 10-set, 20-set, 50-set, mix-set to denote these datasets representatively.

Table 1. Data sets statistic

	10-set	20-set	50-set	mix-set
Training Rating number	9430	14480	24850	31780
Test Rating number	90570	85520	75150	68220
User number	943	724	493	943

---

[1] http://www.grouplens.org/node/73

Table 2. Statistic on test set

	10-set	20-set	50-set	mix-set
Users' rating number in $[0,20]$	213	106	51	356
Users' rating number in $[20,50]$	236	192	85	226
Users' rating number in $[50,100]$	172	144	135	135
Users' rating number more than 100	322	302	226	226

4.2   Improvement by Iteration

Since AdaMF combines different weak recommenders, the overall performance would get improved with the increasing rounds of combination. In this experiment we test our AdaMF on four training sets to validate the improvement of NDCG@10 when increasing the iteration rounds number T. In specially, we choose T in 1, 2, 3, 4, 5, 7, 10, 15, 20. For each weak recommender we set the latent factor dimension K to be 10, learning rate as 0.01 and iteration upper bound T_w to be 5 rounds. For each condition, we run the algorithm for 10 times and compare the mean and variance of NDCG@10.

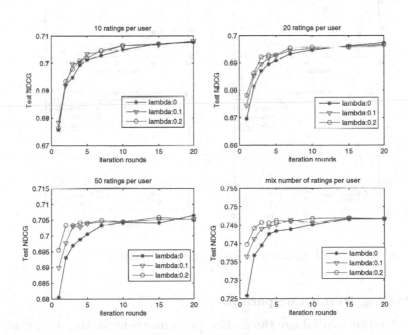

Fig. 1. The performance curve of AdaMF

Fig.1 shows the performance of AdaMF. It can be observed that after the combination of weak recommenders, the original MF model is enhanced in terms of accuracy. It is obvious from the curve that AdaMF reaches its best performance after about 10 rounds of iteration.

In traditional MF model, the regularization coefficient λ controls the degree of overfitting on the training set. For the single MF model, good performance could be achieved by selecting parameters from Cross-Validation, and this may take a great quantity of works. However as expressed in Fig.1, when setting different parameters in a certain range,, AdaMF would converge to the best performance, thus we do not need to pay too much attentions to the parameter selection.

Not only does AdaMF improve the accuracy of recommendation, it also decreases the uncertainty caused by randomly initialized factor value. We use the variance of the results in 10 time running to express the uncertainty. From Fig.2, it is evident to observe that the variance of NDCGs decreases with the increasing of iteration number T. The result confirms the expectation that AdaMF reduces the uncertainty from the initial value of the model when increasing the round of iterations.

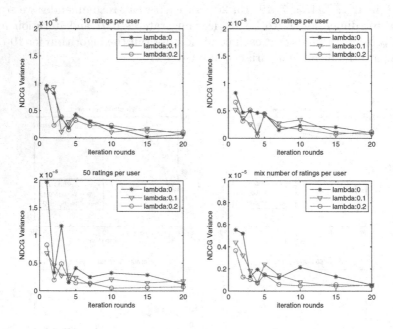

Fig. 2. The variance curve of AdaMF

4.3   Ranking Result Comparison

In this subsection we evaluate the ranking performance by making a comparison for AdaMF with two baseline methods. We choose ListRank-MF [5] and the basic MF model to be the baseline. ListRank-MF is a state-of-the-art ranking-oriented recommendation model based on List-wise learning to rank approach, which uses top one probability for the items to build the loss function. We choose the parameters of baselines by reach their best performance. For each algorithm, we run for 10 times on different dataset to compare the average NDCG@10 metric.

From the results presented in Table 3, we observe that AdaMF performs better on all of four datasets. Especially for the mix-set, AdaMF shows a gratifying improvement in comparison to the ListRank-MF model and the basic MF model.

Table 3. The NDCG@10 comparison between ListRank-MF, Basic MF and AdaMF

	10-set	20-set	50-set	mix-set
ListRank-MF	0.6968	0.6861	0.6875	0.7309
Basic MF	0.6784	0.6745	0.6899	0.7365
AdaMF	0.7065	0.6961	0.7043	0.7464

It can be manifested from Table 3 that ListRank-MF model performs better on 10-set and 20-set. MF shows its disadvantage when meeting a sparse rating matrix, thus the model would easily get overfitted. However AdaMF delivers a 1.43% increase to ListRank-MF. Such result reflects the superiority of AdaMF when facing the sparsity. The composition of weak MF models can prevent overfitting on the training set.

With the grow of matrix density, higher RMSE can somehow transform into better ranking result, therefore basic MF presents a similar result with ListRank-MF on the 50-set and mix-set. AdaMF can take the advantage of different MF models which fit different users better, so that AdaMF can enhance the performance of single MF model. For the 50-set and the mix-set, AdaMF improves the basic MF model by 1.73%.

5   Conclusions and Future Work

In this paper, we have introduced the AdaMF, a boosting algorithm for ranking-oriented recommender system. Optimization for NDCG is done by the boosting of the component recommenders which trained by different weight distribution for each user. The model offers several advantages: ease in implementation, low complexity in training and high performance in ranking. Experimental results demonstrate the effectiveness and accuracy of our model. In comparison with ListRank-MF and basic MF model, we show the superiority of AdaMF. In the future, we are going to focus on the the distributed training algorithm by adaptively composing component recommenders that are trained on different processers in parallel, so that the training process can be easily extended to the multi-processer environment.

Acknowledgments. This work was supported partly by National Natural Science Foundation of China (No. 61300070, No. 61103031), partly by China 863 program (No.2013AA01A213), China 973 program (No.2014CB340305), partly by the State Key Lab for Software Development Environment (SKLSDE-2013ZX-16), partly by A Foundation for the Author of National Excellent Doctoral Dissertation of PR China(No. 201159), partly by Beijing Nova Program(2011022) and partly by Program for New Century Excellent Talents in University.

References

1. Koren, Y., Bell, R., Volinsky, C.: Matrix factorization techniques for recommender systems. Computer 42(8), 30–37 (2009)
2. Cremonesi, P., Koren, Y., Turrin, R.: Performance of recommender algorithms on top-n recommendation tasks. In: Proceedings of the Fourth ACM Conference on Recommender Systems, RecSys 2010, pp. 39–46. ACM, New York (2010)
3. Liu, T.Y.: Learning to rank for information retrieval. Found. Trends Inf. Retr. 3(3), 225–331 (2009)
4. Balakrishnan, S., Chopra, S.: Collaborative ranking. In: Proceedings of the Fifth ACM International Conference on Web Search and Data Mining, WSDM 2012, pp. 143–152. ACM, New York (2012)
5. Shi, Y., Larson, M., Hanjalic, A.: List-wise learning to rank with matrix factorization for collaborative filtering. In: Proceedings of the Fourth ACM Conference on Recommender Systems, RecSys 2010, pp. 269–272. ACM, New York (2010)
6. Xu, J., Li, H.: Adarank: a boosting algorithm for information retrieval. In: Proceedings of the 30th Annual International ACM SIGIR Conference on Research and Development in Information Retrieval, SIGIR 2007, pp. 391–398. ACM, New York (2007)
7. Koren, Y.: Factorization meets the neighborhood: a multifaceted collaborative filtering model. In: Proceedings of the 14th ACM SIGKDD International Conference on Knowledge Discovery and Data Mining, KDD 2008, pp. 426–434. ACM, New York (2008)
8. Koren, Y.: Collaborative filtering with temporal dynamics. Commun. ACM 53(4), 89–97 (2010)
9. Salakhutdinov, R., Mnih, A.: Probabilistic matrix factorization. In: NIPS (2007)
10. Blei, D.M., Ng, A.Y., Jordan, M.I.: Latent dirichlet allocation. J. Mach. Learn. Res. 3, 993–1022 (2003)
11. Agarwal, D., Chen, B.C.: flda: matrix factorization through latent dirichlet allocation. In: Proceedings of the Third ACM International Conference on Web Search and Data Mining, WSDM 2010, pp. 91–100. ACM, New York (2010)
12. Herbrich, R., Graepel, T., Obermayer, K.: Large margin rank boundaries for ordinal regression. MIT Press, Cambridge (2000)
13. Chapelle, O., Wu, M.: Gradient descent optimization of smoothed information retrieval metrics. Inf. Retr. 13(3), 216–235 (2010)
14. Järvelin, K., Kekäläinen, J.: Ir evaluation methods for retrieving highly relevant documents. In: Proceedings of the 23rd Annual International ACM SIGIR Conference on Research and Development in Information Retrieval, SIGIR 2000, pp. 41–48. ACM, New York (2000)
15. Li, X., Wang, L., Sung, E.: AdaBoost with SVM-based component classifiers. Engineering Applications of Artificial Intelligence 21(5) (2008)

A Generic Approach for Bulk Loading Trie-Based Index Structures on External Storage

Dongzhe Ma and Jianhua Feng

Department of Computer Science and Technology
Tsinghua University, Beijing 100084, P.R. China
mdzfirst@gmail.com, fengjh@tsinghua.edu.cn

Abstract. A wide range of applications require efficient management of sorted data on external storage. Recently, trie-based data structures have attracted much attention from the academia as a competitive alternative for the ubiquitous B-tree. In this paper, we present a novel approach for bulk loading disk-based trie structures (a.k.a. B-trie). Our algorithm sorts raw data at first and then builds the B-trie directly from the sorted data. Data in the output data structure are compacted and physically ordered, and thus efficient sequential access can be obtained. We test the proposed algorithm with both real-world and synthetic datasets. Experimental results show that our algorithm outperforms the baseline insertion method dramatically when the dataset is large enough and is almost always superior to the basic sort-and-insert algorithm.

1 Introduction

Recently, trie-based data structures [1–3] have attracted much attention from the academia as a competitive alternative for hash table [4–7] and B-tree [8, 9]. A trie is a multi-way tree. Unlike B-tree, a trie does not store any string explicitly in its nodes, but represents a string by the path from the root to a node associated with the string. To search a string, we start at the root, examine each character of the string, and follow the corresponding edges. When the string is completely consumed, we reach the node that represents the string and get the information associated with it. If no matching path is found during the retrieval, we are sure that the string does not exist.

In this paper, we propose a novel approach for bulk loading trie-based data structures especially on external storage such as magnetic disks and flash memories. In particular, we implement a B-trie [3] and apply the proposed bulk loading algorithm. Like most B-tree bulk loading algorithms, our algorithm consists of two steps. First, the data are sorted. Then a B-trie is built directly from the sorted data. Data in the bulk loaded B-trie are compacted and physically sorted, so that range queries (e.g., prefix matching) can be performed very efficiently. To help understand the performance of the proposed algorithm, we test our implementation with six real-world datasets and two synthetic datasets. Experimental results show that our algorithm outperforms the baseline insertion method dramatically if the dataset is large enough and is almost always superior to the basic sort-and-insert algorithm.

F. Li et al. (Eds.): WAIM 2014, LNCS 8485, pp. 55–66, 2014.

2 Background

The trie access method is first proposed in [10, 11]. In standard trie structure, a node maintains a pointer for each character in the alphabet and is therefore memory-hungry. One possible optimization is to compact the nodes by removing null pointers. Another is to reduce the total number of trie nodes, such as compressing chain paths [12, 13], adjacent levels [14], or both [15]. [16] proposes compressed trie, or C-trie, which is only feasible to store static data. And [17] studies the complexity of trie construction.

Burst trie [2] employs a different method to reduce the space requirement of a trie. Taking advantage of the fact that nodes close to the root tend to be dense and vice versa, burst trie stores the upper-level nodes as a conventional trie called *access trie*, and compacts others in small data structures, called *containers*. To search a string, we first follow the path corresponding to a prefix of the string in the access trie, reach a container, and then search in the container using the remaining suffix of the string. When a container contains too many strings to search, it is *burst*-ed according to the first characters of the stored suffixes.

B-trie [3] is a variant of burst trie which is designed for external storage. In order to reduce the overhead of split operations, [3] suggests that only one bucket[1] should be created if a bucket overflows. Obviously, this goal requires that a bucket should be able to accommodate suffixes with different leading characters, and therefore the concepts of *pure* and *hybrid* buckets are introduced.

One potential drawback of B-trie and other trie structures is that the position of a string is uniquely determined by its prefix. (This is also the reason why trie can be more efficient than binary search tree and B-tree.) To build a B-tree, we can split the sorted data into leaves at arbitrary position as long as the leaves do not exceed the size limit. But in a trie, we always need to search from the root to find the right place of a string, and therefore efficient bulk loading is not applicable for trie structures, as asserted in [3]. We are going to show, in this paper, that this assertion is not accurate. By carefully choosing potential bucket boundaries, we can build a B-trie directly from the sorted data.

3 Basic Idea

Like other bulk loading algorithms, the proposed algorithm first sorts the data in desired order and then builds the B-trie directly from the sorted data. The fundamental question during this process is how to split the data into buckets.

Before going into details, we first introduce the basic idea of the proposed bulk loading algorithm by a top-down approach. In Fig. 1, we build a B-trie step by step. Initially, all strings are put in a single bucket (Fig. 1(a)). It is obvious that the total size exceeds the space limit of a bucket (10 bytes in our example). Therefore, the bucket is split according to the first characters (Fig. 1(b)). This procedure continues until all buckets meet the size limitation.

---

[1] Buckets in a B-trie are similar to containers in a burst trie. Unlike containers, a bucket is split when its size (in bytes) exceeds a predefined threshold.

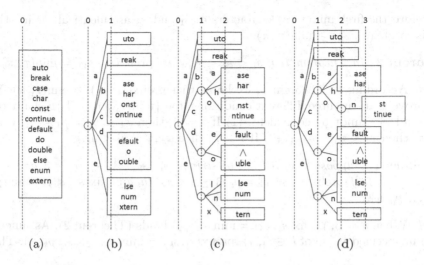

Fig. 1. Building a B-trie in a top-down approach. The B-trie contains some of the reserved keywords defined in the C programming language. Each bucket can store 10 characters at most, including the terminal *null*.

It should be noted that in Fig. 1(c), the second characters of *case* and *char* are not consumed by the path after the split, because they are in a hybrid bucket, while *const* and *continue* have the same third character and the bucket in Fig. 1(d) is pure, and therefore the character is discarded in the bucket.

From this example, we can see that strings in a bucket share the same prefix (not necessarily the longest one) as indicated by the path[2]. And for a group of strings to be packed in the same bucket, the path that leads to the bucket should be able to uniquely distinguish those strings, or in other words, no string in other buckets may share the same prefix[3]. Accordingly, the proposed algorithm works by examining the length of the common prefix between adjacent strings and choosing potential bucket boundaries.

First of all, we need to know how to calculate the common prefix length between any arbitrary pair of strings in a sorted list. We start with two theorems.

Theorem 1. *Suppose s_1, s_2 and s_3 are three strings, and we have* prefix(s_1, s_2) *= τ and* prefix(s_2, s_3) *= π, where* prefix() *returns the common prefix length of the two input strings, then* prefix(s_1, s_3) $\geq \min(\tau, \pi)$.

Proof. Since prefix(s_1, s_2) = τ, we can represent s_1 and s_2 as $a_1 a_2 \ldots a_\tau x_1 x_2 \ldots$ and $a_1 a_2 \ldots a_\tau y_1 y_2 \ldots$, respectively. Similarly, s_2 and s_3 can also be represented as $b_1 b_2 \ldots b_\pi p_1 p_2 \ldots$ and $b_1 b_2 \ldots b_\pi q_1 q_2 \ldots$, respectively. The two representations of s_2 must be equivalent, which means that $a_i = b_i$ for all $i \in [1, \min(\tau, \pi)]$.

---

[2] In hybrid buckets, the last character may vary and can be represented as a range.
[3] For hybrid buckets that share the same immediate parent, the last character ranges should not intersect.

Therefore the first $\min(\tau, \pi)$ characters of s_1 and s_3 are identical, or in other words, $\text{prefix}(s_1, s_3) \geq \min(\tau, \pi)$.

Theorem 2. *In Theorem 1, if $s_1 < s_2 < s_3$, then $\text{prefix}(s_1, s_3) = \min(\tau, \pi)$.*

Proof. According to Theorem 1, we know that $\text{prefix}(s_1, s_3) \geq \min(\tau, \pi)$. We will prove that, in three different conditions, the $\{\min(\tau, \pi) + 1\}$-th characters of s_1 and s_3 cannot be equivalent. (1) If $\tau = \pi$, then $x_1 < y_1 = p_1 < q_1$. (2) If $\tau < \pi$, then $x_1 < y_1 = b_{\tau+1}$. (3) If $\tau > \pi$, then $a_{\pi+1} = p_1 < q_1$.

Corollary 1. *Suppose a sorted list of strings, $s_1, s_2, \ldots, s_n$ ($n \geq 3$ and $s_1 < s_2 < \ldots < s_n$). If $\text{prefix}(s_i, s_{i+1}) = p_i$ ($1 \leq i < n$), then $\text{prefix}(s_1, s_n) = \min(p_1, p_2, \ldots, p_{n-1})$.*

Proof. When $n = 3$, $\text{prefix}(s_1, s_3) = \min(p_1, p_2)$ holds (Theorem 2). Assume for some unspecified value of $k \in [3, n)$, $\text{prefix}(s_1, s_k) = \min(p_1, p_2, \ldots, p_{k-1})$. Then

$$\begin{aligned}
\text{prefix}(s_1, s_{k+1}) &= \min(\text{prefix}(s_1, s_k), \text{prefix}(s_k, s_{k+1})) \\
&= \min(\min(p_1, p_2, \ldots, p_{k-1}), p_k) \\
&= \min(p_1, p_2, \ldots, p_k)
\end{aligned}$$

thereby showing that the above equation holds for $k+1$. It has now been proved by mathematical induction that $\text{prefix}(s_1, s_n) = \min(p_1, p_2, \ldots, p_{n-1})$.

Since all data are sorted in the first step, we could calculate the common prefix length of any two strings in the dataset according to Corollary 1.

In the rest of this section, we assume a sorted list of strings, $s_1, s_2, \ldots, s_n$ ($s_1 < s_2 < \ldots < s_n$) and $\text{prefix}(s_i, s_{i+1}) = p_i$ ($1 \leq i < n$). A bucket candidate that consists of strings $s_i, s_{i+1}, \ldots, s_j$ ($1 \leq i \leq j \leq n$) is represented by $\mathcal{B}_{ij}$.

To determine the maximal length of the common prefix shared by a continuous segment of the string list, we define bucket depth as follows.

Definition 1. *The depth of bucket candidate $\mathcal{B}_{ij}$ (abbreviated as $\mathcal{D}_{ij}$) is defined as the length of the longest common prefix shared by all strings in $\mathcal{B}_{ij}$. $\mathcal{D}_{ij}$ is the maximal possible depth in the B-trie if $\mathcal{B}_{ij}$ is a pure bucket. Therefore let $\widetilde{\mathcal{D}_{ij}}$ be the actual depth of $\mathcal{B}_{ij}$ in the B-trie, and we have*

$$\widetilde{\mathcal{D}_{ij}} \leq \begin{cases} \mathcal{D}_{ij} & (i = j), \\ \mathcal{D}_{ij} + 1 & (i < j). \end{cases} \tag{1}$$

A bucket can be hybrid only when it contains at least two strings. It should also be noted that in order to discuss feasibility and correctness, we use the maximal value of $\widetilde{\mathcal{D}_{ij}}$ most of the time in this paper.

Theorem 3. *Let $\mathcal{D}_{ij}$ be the depth of bucket candidate $\mathcal{B}_{ij}$, then*

$$\mathcal{D}_{ij} = \begin{cases} s_i.length & (i = j), \\ \min(p_i, p_{i+1}, \ldots, p_{j-1}) & (i < j). \end{cases} \tag{2}$$

Proof. If $i = j$, the only string, s_i, is of course the longest common prefix. If $i < j$, for any $k \in [i + 1, j]$, the first $\min(p_i, p_{i+1}, \ldots, p_{k-1})$ characters in s_i and s_k are identical according to Corollary 1. Since $\min(p_i, p_{i+1}, \ldots, p_{k-1}) \geq \min(p_i, p_{i+1}, \ldots, p_{k-1}, \ldots, p_{j-1})$, all strings in the bucket should share the first $\min(p_i, p_{i+1}, \ldots, p_{j-1})$ characters as s_i. Therefore (2) has been proved according to Corollary 1, because the $\{\min(p_i, p_{i+1}, \ldots, p_{j-1}) + 1\}$-th characters of s_i and s_j cannot match.

In order to make sure that two buckets do not conflict with each other, we define the repellency of two buckets.

Definition 2. *The repellency of two buckets, $\mathcal{B}_{ij}$ and $\mathcal{B}_{kl}$, is defined as*

$$\text{repel}(\mathcal{B}_{ij}, \mathcal{B}_{kl}) = \max_{\forall s_1 \in \mathcal{B}_{ij}, s_2 \in \mathcal{B}_{kl}} \{\text{prefix}(s_1, s_2)\}. \tag{3}$$

We further define the repellency between bucket $\mathcal{B}_{ij}$ and its left (right) neighbor as $\mathcal{B}_{ij}$'s left (right) repellency and abbreviate it as $\overleftarrow{\mathcal{R}}_{ij}$ ($\overrightarrow{\mathcal{R}}_{ij}$). Especially, $\overleftarrow{\mathcal{R}}_{ij}$ ($\overrightarrow{\mathcal{R}}_{ij}$) is 0 if $\mathcal{B}_{ij}$ has no left (right) neighbor.

Repellency measures the minimal length (exclusive) of path that is needed to separate two buckets. Given two strings, s_1 and s_2, they cannot be distinguished by the first $\text{prefix}(s_1, s_2)$ characters. Therefore, if two strings, s_1 and s_2, come from two different buckets, the paths that lead to the two buckets in the access trie should be longer than $\text{prefix}(s_1, s_2)$.

Theorem 4. *Assume a bucket generation plan splits the sorted string list, s_1, $s_2, \ldots, s_n$, into k buckets, $\mathcal{B}_1, \mathcal{B}_2, \ldots, \mathcal{B}_k$. The plan is legal iff*

$$\forall i, j \in [1, k], \ i \neq j: \quad \widetilde{\mathcal{D}}_i > \text{repel}(\mathcal{B}_i, \mathcal{B}_j). \tag{4}$$

Proof. **Necessity:** A split plan is legal only when the path to any bucket can distinguish the strings in that bucket with those in others. Suppose the paths to $\mathcal{B}_i$ and $\mathcal{B}_j$ are $a_1, a_2, \ldots, a_m$ and $b_1, b_2, \ldots, b_n$, respectively (a_m and b_n can be ranges). There must be some difference between the two paths in the first $\min(m, n)$ characters. Therefore, $\text{repel}(\mathcal{B}_i, \mathcal{B}_j) < \min(m, n) \leq m = \widetilde{\mathcal{D}}_i$ holds. **Sufficiency:** If (4) holds, the path to a bucket will never be a prefix of another path. As a result, inserting a bucket in the B-trie will always create a new path in the access trie and no conflict will occur. So the plan is legal.

Theorem 4 provides a method to check the validity of a bucket generation plan. However, it is still impossible to design a practical algorithm accordingly, since Theorem 4 requires the existence of all buckets. To solve this problem, we introduce another theorem.

Theorem 5. *Assume $\mathcal{B}_i$, $\mathcal{B}_j$ and $\mathcal{B}_k$ are three different buckets from a sorted bucket list and $i < j < k$. Then we have*

$$\text{repel}(\mathcal{B}_i, \mathcal{B}_j) \geq \text{repel}(\mathcal{B}_i, \mathcal{B}_k) \text{ and } \text{repel}(\mathcal{B}_j, \mathcal{B}_k) \geq \text{repel}(\mathcal{B}_i, \mathcal{B}_k). \tag{5}$$

Proof. For any $s_x \in \mathcal{B}_i$, $s_y \in \mathcal{B}_j$, and $s_z \in \mathcal{B}_k$, we have

$$\text{prefix}(s_x, s_y) = \min(p_x, p_{x+1}, \ldots, p_{y-1}),$$
$$\text{prefix}(s_y, s_z) = \min(p_y, p_{y+1}, \ldots, p_{z-1}),$$
$$\text{prefix}(s_x, s_z) = \min(p_x, p_{x+1}, \ldots, p_{y-1}, p_y, p_{y+1}, \ldots, p_{z-1}),$$

according to Theorem 1. Therefore $\text{prefix}(s_x, s_y) \geq \text{prefix}(s_x, s_z)$ and $\text{prefix}(s_y, s_z)$ $\geq \text{prefix}(s_x, s_z)$. Based on Definition 2, it can be easily seen that (5) holds.

Therefore it is unnecessary to check the repellency with all other buckets when building a bucket. Instead, we only have to pay attention to its adjacent neighbors.

Corollary 2. *Bucket $\mathcal{B}_{ij}$ is legal when*

$$\mathcal{D}_{ij} \geq \overleftarrow{\mathcal{R}_{ij}} \text{ and } \mathcal{D}_{ij} \begin{cases} > \overrightarrow{\mathcal{R}_{ij}} & (i = j), \\ \geq \overrightarrow{\mathcal{R}_{ij}} & (i < j). \end{cases} \tag{6}$$

Proof. According to Theorem 5, $\mathcal{B}_{i-1}$ has the largest repellency with $\mathcal{B}_i$ among all buckets on $\mathcal{B}_i$'s left, while $\mathcal{B}_{i+1}$ has the largest repellency with $\mathcal{B}_i$ among all buckets on $\mathcal{B}_i$'s right. Therefore, bucket $\mathcal{B}_i$ is legal when $\widetilde{\mathcal{D}_{ij}} > \overleftarrow{\mathcal{R}_{ij}}$ and $\widetilde{\mathcal{D}_{ij}} > \overrightarrow{\mathcal{R}_{ij}}$ according to Theorem 4. The maximal possible value of the actual depth can be easily figured out by the depth of a bucket (Definition 1). Therefore, we have proved (6). Note that we do not distinguish the left part according to whether $i < j$, because duplication is not allowed in a B-trie and for a single string as a bucket, it is impossible for its left neighbor string to share a prefix as long as its length. (See Theorem 6 for more information.)

Now, the last question lies in how to calculate the repellency of two adjacent buckets, especially when the right one has not been generated. We introduce a much easier method in Theorem 6, leveraging the fact that all strings are sorted.

Theorem 6. *Repellency of two adjacent buckets equals to the common prefix length of the two nearest strings from the two buckets, or in other words,*

$$\text{repel}(\mathcal{B}_{ij}, \mathcal{B}_{j+1,k}) = \text{prefix}(s_j, s_{j+1}) = p_j. \tag{7}$$

Proof. First, we choose two strings, $s_j \in \mathcal{B}_{ij}$ and $s_{j+1} \in \mathcal{B}_{j+1,k}$. According to Definition 2, $\text{repel}(\mathcal{B}_{ij}, \mathcal{B}_{j+1,k}) \geq \text{prefix}(s_j, s_{j+1}) = p_j$ holds. Meanwhile, since for any $s_s \in \mathcal{B}_{ij}$ and $s_t \in \mathcal{B}_{j+1,k}$, $\text{prefix}(s_s, s_t) = \min(p_s, \ldots, p_j, \ldots, p_{t-1}) \leq p_j$ according to Corollary 1, we get $\text{repel}(\mathcal{B}_{ij}, \mathcal{B}_{j+1,k}) \leq p_j$. Therefore, (7) holds.

Finally, we provide a general description of the proposed algorithm and leave the details to the next section. First, all strings are sorted in the desired order and the common prefix length between adjacent strings are calculated. Then, we split the sorted data into buckets and build the access trie directly. The buckets are generated by scanning the sorted data and looking for the largest continuous segment of the string list that fits for a bucket and meets Corollary 2. After deciding a bucket, we obtain the prefix shared by its content according to the bucket depth (Definition 1) and then build the path in the access trie.

Table 1. Global Variables

Name	Description
input	A queue that contains all input strings.
internal	List of internal strings that will end up in the access trie.
cnt	Number of strings that have been scanned.
prefix	Common prefix length of the scanned strings, or depth of the bucket.
left	Common prefix length with former bucket, or left repellency.
right	Common prefix length with latter bucket, or right repellency.
sum	Total length of the scanned strings.

Algorithm 1. bucket_size() - Calculate the candidate bucket size

1. $sum' \leftarrow sum$
2. $cnt' \leftarrow cnt$

3. **if** $prefix = input[0].length$ **then**
4.   **if** $cnt = 1$ **then**
5.   **return** 0
6. $sum' \leftarrow sum' - input[0].length$
7. $cnt' \leftarrow cnt' - 1$
8. $a \leftarrow input[1][prefix]$
9. **else**
10. $a \leftarrow input[0][prefix]$
11. $b \leftarrow input[cnt - 1][prefix]$

12. **if** $a = b$ **then**
13. $data \leftarrow sum' - (prefix + 1) * cnt'$
14. **else**
15. $data \leftarrow sum' - prefix * cnt'$

16. **return** $6 + 5 * cnt' + data$

---

4   A Bulk Loading Algorithm

In this section, we explain the details of the proposed algorithm. For the sake of simplicity, we maintain a few global variables that indicate the current state of the algorithm. They are summarized in Table 1.

Algorithm 1 calculates the amount of bytes that the current strings will take if they are going to be packed together. First, we calculate the space that the suffixes will take (Line 3-15). This can be done by removing the shared prefix from the total length. In case the first string is of the same length as the common prefix, we may either return 0 directly[4] if there is only one string in the current set or treat it as an internal string otherwise (Line 4-8). It should be noted that if all the suffixes have the same leading character, which indicates a pure bucket, we may remove it from the suffixes (Line 13). Otherwise, the characters have to be stored in the bucket (Line 15). Finally, we obtain the total amount of bytes by the expression in Line 16. (Layout of a bucket will be explained in Section 5.1.)

---

[4] A single string can always fit in a bucket as long as the bucket is deep enough.

Algorithm 2. build_bucket() - Build the current bucket

1. **if** $prefix = input[0].length$ **then**
2. $internal.add(input[0])$
3.   remove the first string from $input$
4. $cnt \leftarrow cnt - 1$

5. $a \leftarrow input[0][prefix]$
6. $b \leftarrow input[cnt - 1][prefix]$
7. $bucket \leftarrow$ new_bucket(a, b)

8. **for** $i \leftarrow 0$ **to** $cnt - 1$ **do**
9. $len \leftarrow input[i].length$
10.   **if** $a = b$ **then**
11.   insert $input[i][(prefix + 1)..len]$ into $bucket$
12.   **else**
13.   insert $input[i][prefix..len]$ into $bucket$

14. $leaf \leftarrow$ build_path$(input[0][0..(prefix - 1)])$
15. **for** $i \leftarrow a$ **to** b **do**
16. $leaf.pointer[i] \leftarrow bucket$

17. **for** each $string \in internal$ **do**
18.   insert $string$ into the access trie
19. $internal \leftarrow \phi$

If the current group of strings can fit in a bucket, we may build the bucket and insert it in the B-trie (Algorithm 2). First, we create a bucket with proper pointer range (Line 5-7), and then insert all the suffixes in the bucket (Line 8-13). Line 14 builds the path that leads to the bucket. It should be noted that all pointers in the range should point to the bucket (Line 15-16), including those that do not appear in the bucket. Finally, we insert the internal strings in the access trie (Line 17-19).

With Algorithm 1 and 2 as building blocks, we can now introduce the proposed bulk loading algorithm. The pseudo code is provided in Algorithm 3. As described in the previous section, the proposed algorithm works on sorted data (Line 1). It scans the sorted data and builds the buckets one by one. Therefore, the body of Algorithm 3 is mainly a single while-loop (Line 3-29). In each loop, we build a single bucket. The algorithm runs until all strings are consumed. To build a bucket, we scan the remaining strings, find the largest set of strings that fit in a bucket (Line 6-26), build the bucket (Line 27), and remove them from the dateset (Line 28). Line 2, 4-5, 7-13, and 29 are used to maintain the global variables, and we will focus on the other three if-conditions in the for-loop.

Since we are trying to find the largest set that can fit in a bucket, we will continue scanning until the size limitation is exceeded. And then we will use the largest possible group that is ever found (Line 21-22). Otherwise, the first remaining string should be put in the internal set (Line 16-19). The reason lies in that the first string should always be able to make a potential bucket, unless it is a prefix of the second one.

Algorithm 3. bulk_load() - Bulk load a B-trie

1. sort *input*

2. *left* ← 0
3. **while** *input.size* > 0 **do**
4.   *cnt* ← 0
5.   *sum* ← 0

6.   **for each** *str* ∈ *input* **do**
7.   *cnt* ← *cnt* + 1
8.   **if** *cnt* = 1 **then**
9.   *prefix* ← *str.length*
10.   **else**
11.   *prefix* ← min(*prefix*, *right*)
12.   *sum* ← *sum* + *str.length*
13.   *right* ← common prefix with the next string or 0 if *str* is the last one

14.   **if** bucket_size() > BUCKET_SIZE **then**
15.   **if** there is no saved bucket **then**
16.   *internal.add*(*input*[0])
17.   *cnt* ← 1
18.   *right* ← common prefix of the first two strings
19.   **goto** line 28
20.   **else**
21.   restore < *cnt*, *prefix*, *right*, *sum* > from < *cnt'*, *prefix'*, *right'*, *sum'* >
22.   **goto** line 27

23.   **if** *right* < *left* **then**
24.   **goto** line 27

25.   **if** *prefix* > *right* **or** (*prefix* = *right* **and** *cnt* > 1) **then**
26.   save < *cnt*, *prefix*, *right*, *sum* > in < *cnt'*, *prefix'*, *right'*, *sum'* >

27.   build_bucket()

28.   remove the first *cnt* elements of *input*
29.   *left* ← *right*

During the search procedure, if we should ever find a position where the right repellency is smaller than the left one, we can stop here and build the bucket (Line 23-24), because beyond this position, the bucket depth will always be under the left repellency and there can be no valid bucket candidate any more according to Corollary 2.

The last issue that is worth mentioning is that we should remember the last valid bucket candidate during the search process (Line 25-26). Acute readers may have noticed that we only check right part of the condition (Corollary 2). Actually, the left part (*prefix* ≥ *left*) always holds during the search process. Initially, *prefix* is set to the length of the first string (Line 9), which is, of course, at least as long as the common prefix with another string. After that, *prefix* can only be lowered by the right repellency *right* (Line 11), and when *right* is getting too low, we build the bucket immediately (Line 23-24).

(a) A hybrid bucket   (b) A pure bucket

Fig. 2. Bucket Layout

Table 2. Dataset Summary

Dataset	# Keys	Size	Avg Len	Max Len	Description
aol[5]	1,607,028	42 MB	28	151	URLs from web queries
dblp[6]	2,169,411	148 MB	72	1016	Titles from DBLP
enron[7]	517,424	36 MB	73	141	Email subjects
enwiki[8]	12,791,923	290 MB	24	260	Titles of Wikipedia articles
trec[9]	291,655	23 MB	84	402	Titles of medical documents
uniref[10]	1,000,000	285 MB	299	999	Protein sequences
random	50,000,000	954 MB	20	20	Synthetic, random distrib.
zipf	50,000,000	954 MB	20	20	Synthetic, zipf distrib.

5 Performance Evaluation

5.1 Implementation Details

We implement a B-trie as well as the proposed algorithm. The standard ASCII code is employed as the alphabet. We assume that the access trie is small enough to reside in main memory entirely.

The buckets are limited to 4 KB. Figure 2 shows the layout of our bucket implementation. The first two bytes contain the character (pointer) range of the bucket. The next four bytes contain the number of strings stored in the bucket. Then follows an array of exactly the same number of pointers, each of which points to a suffix. Buckets are stored in external files.

The experiments are done on a Lenovo T430s laptop running Ubuntu 12.04.2 LTS. We allocate 64 MB as the memory pool to generate initial runs or as the bucket cache. If the data are sorted before loading, we will flush the cache each time it is filled up.

Six publicly available real-world datasets and two synthetic datasets are used. We believe that these datasets are representative and can cover a wide range of applications. Table 2 summarizes the major characteristics of the datasets. Besides, duplicate strings and non-ASCII characters are removed from the datasets. Before execution, we always drop the system cache to minimize its effects. All results reported in this paper are the average of at least five independent runs.

---

[5] http://www.gregsadetsky.com/aol-data/
[6] http://www.informatik.uni-trier.de/~ley/db/
[7] http://www.cs.cmu.edu/~enron/
[8] http://en.wikipedia.org/wiki/Main_Page
[9] http://trec.nist.gov/data/t9_filtering.html
[10] http://www.uniprot.org/

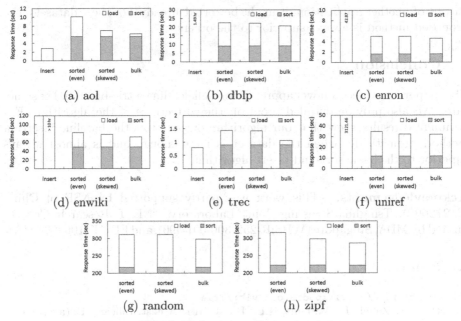

Fig. 3. Performance Comparison

5.2   Results and Analysis

We implement four different algorithms, the naïve insert method, two sort-and-insert methods, and the proposed bulk loading algorithm. The first sort-and-insert method always tries to split full buckets evenly, and therefore may result in many half-filled buckets, while the second always chooses the last possible split point, which is a common practice for bulk loading tree-structured indexes.

Results of the experiments are shown in Fig. 3. As the figures indicate, performance of the naïve insert method[11] is very well when the dataset can be kept in memory (*aol* and *trec*). Sometimes, it even outperforms the proposed bulk loading algorithm, thanks to its linear complexity. All the other three methods have to sort the data first and may not be very competitive. However, when the dataset exceeds the memory capacity, performance of the naïve insert method drops rapidly due to its random access pattern. On the contrary, the other three sort-based algorithms can always respond in reasonable time, and for all datasets except *uniref*, the proposed bulk loading algorithm outperforms the other sort-based algorithms, which insert the sorted data in the normal way.

The synthetic datasets (*random* and *zipf*) aim to simulate large-scale applications. As can be seen from Fig. 3(g)(h), the external sort procedure causes long response time. Fortunately, this step can be easily parallelized and there exist many implementations and tools (e.g., Hadoop). Besides, if the dataset is the

---

[11] We did not test the naïve insert method against the synthetic datasets due to the unacceptable response time.

output of another application, it might be sorted already. For these datasets, the proposed method is always superior to the others.

6   Conclusion

This paper introduces a novel approach for bulk loading a trie-based index structure. We also provide detailed proof on the correctness of the algorithm. Experimental results show that our algorithm outperforms the baseline insertion method dramatically when the dataset is large enough and is almost always superior to the basic sort-and-insert algorithm.

Acknowledgements.   This work was partly supported by NSF of China (61272090), Tsinghua-Samsung Joint Laboratory, "NExT Research Center" funded by MDA, Singapore (WBS:R-252-300-001-490), and FDCT/106/2012/A3.

References

1. Trie, https://en.wikipedia.org/wiki/Trie
2. Heinz, S., Zobel, J., Williams, H.E.: Burst Tries: A Fast, Efficient Data Structure for String Keys. ACM Transactions on Information Systems 20(2), 192–223 (2002)
3. Askitis, N., Zobel, J.: B-tries for Disk-based String Management. VLDB Journal 18(1), 157–179 (2009)
4. Litwin, W.: Linear Hashing: A New Tool for File and Table Addressing. In: VLDB (1980)
5. Larson, P.A.: Dynamic Hash Tables. Communications of the ACM 31(4), 446–457 (1988)
6. Fagin, R., Nievergelt, J., Pippenger, N.: Extendible Hashing - A Fast Access Method for Dynamic Files. ACM Transactions on Database Systems 4(3), 315–344 (1979)
7. Enbody, R.J., Du, H.C.: Dynamic Hashing Schemes. ACM Computing Surveys 20(2), 85–113 (1988)
8. Bayer, R., Mcreight, E.: Organization and Maintenance of Large Ordered Indices. In: SIGFIDET (now SIGMOD) (1970)
9. Comer, D.: The Ubiquitous B-Tree. ACM Computing Surveys 11(2), 121–137 (1979)
10. Briandais, R.D.L.: File Searching Using Variable Length Keys. In: Proceedings of the Western Joint Computer Conference (1959)
11. Fredkin, E.: Trie Memory. Communications of the ACM 3(9), 490–499 (1960)
12. Edward, H., Sussenguth, J.: Use of Tree Structures for Processing Files. Communications of the ACM 6(5), 272–279 (1963)
13. Sedgewick, R.: Algorithms. Addison-Wesley (1984)
14. Andersson, A., Nilsson, S.: Improved Behaviour of Tries by Adaptive Branching. Information Processing Letters 46(6), 295–300 (1993)
15. Nilsson, S., Tikkanen, M.: Implementing a Dynamic Compressed Trie. In: WAE (1998)
16. Maly, K.: Compressed Trie. Communications of the ACM 19(7), 409–415 (1976)
17. Comer, D., Sethi, R.: The Complexity of Trie Index Construction. Journal of the ACM 24(3), 428–440 (1977)

Density-Based Local Outlier Detection
on Uncertain Data[*]

Keyan Cao[1], Lingxu Shi[3], Guoren Wang[1,2], Donghong Han[1], and Mei Bai[1]

[1] College of Information Science & Engineering, Northeastern University, China
[2] Key Laboratory of Medical Image Computing (NEU), Ministry of Education
[3] Logistic Engineering University of People's Liberation Army, China
caokeyan@gmail.com

Abstract. Outlier detection is one of the key problems in the data mining area which can reveal rare phenomena and behaviors. In this paper, we will examine the problem of density-based local outlier detection on uncertain data sets described by some discrete instances. We propose a new density-based local outlier concept based on uncertain data. In order to quickly detect outliers, an algorithm is proposed that does not require the unfolding of all possible worlds. The performance of our method is verified through a number of simulation experiments. The experimental results show that our method is an effective way to solve the problem of density-based local outlier detection on uncertain data.

1 Introduction

Uncertainty is inherent to many important applications, such as location-based services (LBS), sensor monitoring and radio-frequency identification (RFID) [1,9]. In these applications, outlier detection often is essential when analyzing uncertain data. In many real-world applications, determining whether the object is an outlier, not only its distance to neighbors is considered, but also the density of surrounding neighbors should be considered. As such density-based outlier detection can consider local information [4].

Uncertain objects as referred to in this paper, by default, is in a d-dimensional numerical tuple. Given two uncertain instances $\widetilde{u}_i$ and $\widetilde{u}_j$, the distance between these is denoted by $d(\widetilde{u}_i, \widetilde{u}_j)$. In this paper, we employ the Euclidian distance metric, but the developed techniques can be easily extended to other distance metrics.

Uncertain Objects. The uncertain object set U consists of $\{u_1, u_2, \cdots, u_i, \cdots, u_\mu\}$. Each uncertain object u_i has d dimensions. In this paper, we focus on the discrete case, an uncertain object u_i consists of a set of m instances $\{u_i^1, u_i^2, \cdots, u_i^j, \cdots, u_i^m\}$ where $1 \leq j \leq m$, $p(u_i^j)$ $(0 \leq p(u_i^j) \leq 1)$ denotes the probability of instance u_i^j occurring.

[*] This research are supported by the NSFC (Grant No. 61025007, 61328202, 61173029, 61100024, 61332006, 61073063), National High Technology Research and Development 863 Program of China (GrantNo.2012AA011004), National Basic Research Program of China (973, Grant No. 2011CB302200-G).

F. Li et al. (Eds.): WAIM 2014, LNCS 8485, pp. 67–71, 2014.

Definition 1. *Distance sum of k-neighbors:* *We assume that instance $\widetilde{u}_j$ is k-neighbor of instance $\widetilde{u}_i$ in a possible world W, and $n_k(\widetilde{u}_i)$ is a k-neighbor set of $\widetilde{u}_i$ in W. Let $dis_k(\widetilde{u}_i)$ denote the distance sum of k-neighbors of instance $\widetilde{u}_i$ in possible world W, then*

$$dis_k(\widetilde{u}_i) = \sum_{\widetilde{u_j} \in n_k(\widetilde{u}_i)} dis(\widetilde{u}_j, \widetilde{u}_i)$$

Definition 2. *Density of an instance:* *Let instances $\widetilde{u}_j$ and $\widetilde{u}_i$ be entities from different uncertain objects, i.e. $j \neq i$. If $\widetilde{u}_j$ is a k-neighbor of $\widetilde{u}_i$ in possible world W, the density of instance $\widetilde{u}_i$ is defined as*

$$den(\widetilde{u}_i) = \frac{k}{\sum_{W \in \mathbb{W}} dis_k(\widetilde{u}_i)P(W)}$$

Definition 3. *k-neighbor set of an instance:* *Let $\widetilde{u}_i$ denote an instance of uncertain object u_i, $(u_i \in U)$. $n_k(\widetilde{u}_i)$ denotes the k-neighbor set of instance $\widetilde{u}_i$ in possible world W. Let $N_k(\widetilde{u}_i)$ denote the k-neighbor set of instance $\widetilde{u}_i$ in all possible worlds, then*

$$N_k(\widetilde{u}_i) = \bigcup_{W \in \mathbb{W}} n_k(\widetilde{u}_i)$$

Definition 4. *Local Outlier Factor of an instance:* *Given that instance $\widetilde{u}_j$ is a k-neighbor of instance $\widetilde{u}_i$, $den(\widetilde{u}_j)$ denotes the density of $\widetilde{u}_j$, $LOF(\widetilde{u}_i)$ denotes the LOF of instance $\widetilde{u}_i$.*

$$LOF(\widetilde{u}_i) = \frac{\sum_{\widetilde{u_j} \in n_k(\widetilde{u}_i)} den(\widetilde{u}_j)P(W)}{k \times den(\widetilde{u}_i)}$$

Definition 5. *Local Outlier Factor of an uncertain object:* *Let $\widetilde{u}_i$ denote any one instance of object u_i, $P(\widetilde{u}_i)$ denotes the probability of $\widetilde{u}_i$, and $LOF(\widetilde{u}_i)$ denotes the LOF of $\widetilde{u}_i$. $LOF(u_i)$ is then defined as*

$$LOF(u_i) = \sum_{\widetilde{u}_i \in u_i} LOF(\widetilde{u}_i)P(\widetilde{u}_i)$$

Definition 6. *Density-based local outlier:* *When the uncertain objects are sorted in descending order based on their Local Outlier Factor, then the top-n uncertain objects are the density-based local outliers of the uncertain data set.*

2 Algorithms

In this section, we present an algorithm for Density-based Local Outlier detection on Uncertain data ($UDLO$), which transforms the outlier definition into a probability problem. Density based outliers can be calculated based on the definition in a naive way by finding all k-neighbors in all possible worlds. This solution however is impractical as the number of possible worlds grows exponentially with the number of instances. To overcome this, we propose an exact algorithm to compute the density of an instance.

Theorem 1. *Let $N_k(\widetilde{u}_i)$ denote the k-neighbor set of instance $\widetilde{u}_i$. If all instances of an uncertain object are all in the k-neighbor set of instance $\widetilde{u}_i$, then object is a complete k-neighbor object of instance $\widetilde{u}_i$. The number of complete k-neighbor objects of instance $\widetilde{u}_i$ equals k. The maximal distance from instance to its complete k-neighbor object is denoted by $dis_c(\widetilde{u}_i)$. The k-neighbor set of instance $\widetilde{u}_i$ consists of the instances whose distance to instance $\widetilde{u}_i$ is not larger than $dis_c(\widetilde{u}_i)$.*

Theorem 2. *In the basic case, for $1 \le i,j \le |N_k(\widetilde{u}_j)|$, the order of instance $\widetilde{u}_j$ is t_i in list L. $P(S_{t_i}, \kappa)$ denotes the probability that there are κ instances existing in S_{t_i}. $P(S_{t_i},0) = P(S_{t_{i-1}},0)(1 - P(t_i)) = \prod_{j=1}^{i}(1 - P(t_j))$, and $P(S_{t_i}, \kappa) = P(S_{t_{i-1}}, \kappa - 1)P(t_i) + P(S_{t_{i-1}}, \kappa)(1 - P(t_i))$*

Theorem 3. *We assume that the instances $\widetilde{u}_j$ and $\widetilde{u}_i$ are from different uncertain objects, $\widetilde{u}_j$ is a k-neighbor of $\widetilde{u}_i$, and $P_k(\widetilde{u}_j)$ denotes the probability, then the density of instance $\widetilde{u}_i$ is calculated as follows:*

$$den(\widetilde{u}_i) = \frac{k}{\sum_{\widetilde{u}_j \in N_k(\widetilde{u}_i)} dis(\widetilde{u}_j, \widetilde{u}_i)P_k(\widetilde{u}_j)}$$

Theorem 4. *If we assume that instance $\widetilde{u}_j$ is a k-neighbor of $\widetilde{u}_i$, then $P_k(\widetilde{u}_j)$ is its k-neighbor probability. $den(\widetilde{u}_i)$ and $den(\widetilde{u}_j)$ are the densities of $\widetilde{u}_i$ and $\widetilde{u}_j$, then the LOF is given by:*

$$LOF(\widetilde{u}_i) = \frac{\sum_{\widetilde{u}_j \in N_k(\widetilde{u}_i)} den(\widetilde{u}_j)P_k(\widetilde{u}_j)}{k \times den(\widetilde{u}_i)}$$

3 Experiments

We conducted several experiments on two real data sets and a synthetic data set to examine the efficiency and accuracy. In the remainder of this paper, these algorithms will be referred to as follows: outlier detection algorithm (denoted by $UDLO$). For comparison, we implemented the outlier detection algorithms (denoted by $BULOF$ and $ULOF$) which are proposed in [7].

3.1   Efficiency

Efficiency is an important term frequently used in outlier detection studies. Figure 1 shows the results on the two different datasets. As expected $UDLO$ performs better than $ULOF$. Parameter n varies from 20 to 100, the running time increases as n increases. The running time of $ULOF$ is much higher than $UDLO$ algorithm.

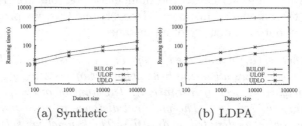

(a) Synthetic   (b) LDPA

Fig. 1. Running time vs. Data size

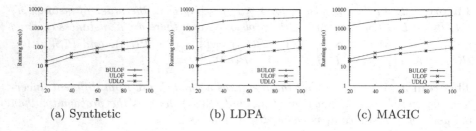

(a) Synthetic   (b) LDPA   (c) MAGIC

Fig. 2. Running time vs. n

3.2   Accuracy

In this section, we give the experimental results on accuracy, as shown in Figure 3. Since of the outliers lie in the center of a cluster, it is hard for the $ULOF$ algorithm to pick out this kind of objects from the entire dataset. The $UDLO$ algorithm adheres more strictly to the outlier definition, and therefore the accuracy of $UDLO$ algorithm is higher than that of $ULOF$. As expected, $UDLO$ can deliver the best results on all datasets.

4   Related Work

Aggarwal, C.C. *et al.* [2] were the first to investigate the problem of outlier detection on uncertain data. Wang *et al.*[8] focused on distance-based outlier detection on uncertain data, in which each data is affiliated with a confidence value. Jiang *et al.* [6] started with a comprehensive model considering both uncertain objects and their instances. In Ref.[3] they attempted to find outliers by building a global classifier. However, it is difficult to build a clear boundary

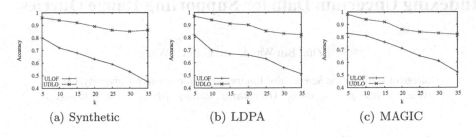

(a) Synthetic   (b) LDPA   (c) MAGIC

Fig. 3. Accuracy vs. k

between normal data and abnormal data. Fan [5] proposed density-based top-k outlier detection algorithm on uncertain objects. In their work, due to the distance between two objects is approximation, the density of object can not be accurate, so that affect the detection results. In Liu *et al.* [7], the authors proposed a signed outlier detection algorithm based on local information (local density and local uncertainty level) on uncertain data.

5   Conclusions

There are many important applications that require outlier detection on uncertain data. These applications always require that outlier are identified based on local information. We first derived an algorithm, which can effectively detect outliers without unfolding all possible worlds.

References

1. Aggarwal, C.: On density based transforms for uncertain data mining. In: ICDE, pp. 866–875 (2007)
2. Aggarwal, C., Yu, P.: Outlier detection with uncertain data. In: SDM, pp. 483–493 (2008)
3. Bo, L., Jie, Y., Shan, X.Y., Longbing, C., Philip, Y.: Exploiting local data uncertainty to boost global outlier detection. In: ICDM, pp. 304–303 (2010)
4. Breunig, M.M., Kriegel, H.-P., Ng, R.T., Sander, J.: Lof:identifying density-based local outliers. Sigmod 29(2), 93–104 (2000)
5. Gaofeng, F., Hongmei, C., Zhiping, O.Y., Lizhen, W.: Density-based top-k outlier detection on uncertain objects. In: ICCSNT, vol. 4, pp. 2469–2472 (2011)
6. Jiang, B., Pei, J.: Outlier detection on uncertain data: objects, instances, and inferences. In: ICDE, pp. 422–433 (2011)
7. Liu, J., Deng, H.: Outlier detection on uncertain data based on local information. KBS (2013)
8. Wang, B., Xiao, G., Yu, H., Yang, X.: Distance-based outlier detection on uncertain data. CIT 1, 293–298 (2009)
9. Zhan, L., Zhang, Y., Zhang, W., Lin, X.: Finding top-k most influential spatial facilities over uncertain objects. In: CIKM, pp. 922–931 (2012)

Indexing Uncertain Data for Supporting Range Queries*

Rui Zhu, Bin Wang, and Guoren Wang

College of Information Science and Engineering, Northeastern University, China
neuruizhu@gmail.com, {binwang,wanggr}@ise.neu.edu.cn

Abstract. Probabilistic range query is a typical and a fundamental problem in probabilistic DBMS. Although the existing solutions provide a good performance, there are some shortages that are needed to be overcomed. In this paper, we firstly propose a novel structure called MRST to approximately capture the probability density function of uncertain object. Through considering the gradient of the probability density function, MRST could provide uncertain object with strong pruning power and consume fewer space cost. Based on characters of MRST, we also design an efficient algorithm to access MRST. We propose a novel index named R-MRST to efficiently support range query on multidimensional uncertain data. Its has a strong pruning power. At the same time, it has a lower cost both in space and dynamic update. Theoretical analysis and extensive experimental results demonstrate the effectiveness of the proposed algorithms.

1 Introduction

Recently, many emerging applications over uncertain data are attracting a wide attention of researchers [3]. The causes of the uncertainty are greatly different in various applications. For example, in a habitat monitoring system, due to the impreciseness of sensing devices [5], the data obtained are often noisy. As another example, in moving objects tracking [6], the location information of objects collected by a GPS system may not be exact due to the delay on data updating. Among a large number of queries, range query is the most fundamental and important operation in managing uncertain data [1].

A probabilistic range query is to find out the objects that appear in the query region with the probability at least θ(the probabilistic threshold of the query). Since such a computation involves the expensive and complex integral [4] [2], the *filter-refinement* is preferable. In the filtering phase, the probabilistic objects, which must(or not) be the query results, are quickly filtered without proceeding the complex integral. For the objects that cannot be filtered, in the refinement phase, the integral has to be done to verify the answers. Thus, the key of optimizing a prob-range query is to provide, as tight as possible, a bound for flittering with a small cost.

Several indexes have been proposed to answer the queries on uncertain data. The key idea is pre-computing the summary [8] of each object's PDF (short for probability density function), augmenting existing index techniques to organize summary, and

* The work is partially supported by the National Basic Research Program of China (973 Program) (No. 2012CB316201,2011CB302200-G), the National Natural Science Foundation of China (Nos. 61322208, 61272178, 61129002), the Doctoral Fund of Ministry of Education of China (No. 201100042110028), and National High Technology Research and Development 863 Program of China (GrantNo.2012AA011004).

F. Li et al. (Eds.): WAIM 2014, LNCS 8485, pp. 72–83, 2014.

then using the summary for filtering.[8]. One of the most popular index named U-tree employs the PCR(short for probabilistically constrained region) technique to summary the PDF of the uncertain object. However, PCR could not provide the uncertain object with a strong pruning/validating power, and the dynamic update cost of U-Tree is high (detailed in Section2).

Another two popular indexes UI-tree [7] and UD-tree [8] employ the partition technique to summary the PDF of uncertain object. Using the partition technique, the summary of an object could provide it with a stronger pruning/validating ability than the PCR-based [4] summary. However, it still has room for improving. The partition does not fully consider the gradient of PDF. And they both consume too much space cost (e.g., given a 62K data, the index size is 20M).

Contributions: In this paper we study the problem of answering prob-range queries on uncertain data. The contributions are as follows:

Firstly, we propose a novel summary called *MRST* (multi-resolution summary tree) to approximately capture the PDF of uncertain object. The MRST fully considers the gradient of PDF and more effectively captures an object's PDF. It has a more powerful filtering ability and consumes lower space cost. We propose a novel algorithm to access the MRST. Through using the key idea of greedy algorithm, this algorithm could reduce the computational cost as much as possible.

Secondly, we propose a new index called R-MRST to organize the summary of objects. R-MRST augments the R-tree technique. The filtering ability of R-MRST's node is as strong as that of U-Tree, but it has the lower cost both in space and dynamic update.

The rest of this paper is organized as follows: Section 2 gives related work and the problem definition. Section 3 proposes the MRST. Section 4 proposes R-MRST that is used to effectively indexing uncertain data. Section 5 evaluates the proposed methods with extensive experiments. Section 6 is the conclusion and the future work.

2   Related Work and Problem Definition

In Section 2.1, we review the existing indexing approaches. In Section 2.2 we formally define the problem of probabilistic range query on uncertain data. Table 1 summaries the mathematical notations used in the paper.

2.1   Related Work

In recent years, many effective indexes have been proposed to answer prob-range query on the uncertain data. The PCR-based index named U-Tree (and U-catalog-Tree) is proposed by Tao et al [4]. The problem of U-Tree is that the filter ability of PCR is not strong, and the dynamic update cost is high. Given a set of objects O, U-Tree constructs a group of PCRs for every object, and employs the R-tree technique for organizing them. For simplicity, we introduce U-Tree in the 2-dimension space. As is depicted in Fig 1(a), given an object o and a probability threshold $\theta(0 < \theta < 0.5)$(eg.0.2), $o.$PCR(θ) is constructed as follows: 2 lines in each dimension are computed. In the horizontal dimension, o has the probability θ to occur on the left(right) side of line $l_1(l_2)$. In the vertical dimension, $l_3(l_4)$ is computed in the similar way with $l_1(l_2)$. $o.$PCR(0.2) is the

Table 1. The Summary of Notations

Notation	definition
o	probabilistic object
$o.pdf(x)$	probability density function of o
o_r	probability region of o
q_r	the search region of the query
q_p	the probability threshold of the query
θ	probability threshold
$o(i)$	the subregion i of o
$PBD(o, i)$	probability bound difference of $o(i)$
$app(o, i)$	the likelihood of o falling in $o(i)$
$app(o, q)$	the likelihood of o falling in q_r
$app(q, i)$	the likelihood of o falling in $q_r \cap o(i)$
$lb(o, i)(ub(o, i))$	the maximal(minimal) probability density in $o(i)$
$S(o, i)$	the area of $o(i)$
$ZS(o, i)$	the blank(and zero-pdf) area of $o(i).MBR$
$o(i).MBR$	the MBR bounding $o(i)$
$S(q, i)$	the area of $q_r \cap o(i).MBR$

rectangle bound by these four lines. Given a prob-query q with $q_p \leq \theta(q_p$ denotes the threshold of q), $o.PCR(0.2)$ is used for pruning/validating if $q_p \geq \theta$. As is depicted in Fig 1(a), q_1, q_2, q_3 and q_4, we assume that their query threshold are all 0.2. Given q_1, o could be pruned because $o.PCR(0.2)$ does not intersect with q_r(short for the query region). On the other hand, given q_2, o could be validated because q_r completely contains the part of the left, upper, down border of $o.MBR$ and l_3. However, the pruning/validating ability of PCR is not powerful if q_r overlaps with an objects but can not contain d-1 dimension planes of an object in a d-dimension space. For example, o obeys uniform distribution. Obviously, o is the query result of q_3. o is not the query result of q_4. However, they can not prune(or validate) o because no filter pruning can be used to prune/validate them. As another problem, the dynamic update cost of U-tree is high. In Fig 1(b), because every object uses a group of PCR to summary its PDF, the node of U-tree also has to use a group of MBRs for bounding these PCRs. Obviously, the cost of maintaining these boundaries is much higher than that of R-Tree once the dynamic update happens.

Zhang et al proposed UI-Tree(and UD-tree) for indexing uncertain objects [7]. The filtering ability of them are stronger than that of U-Tree. However, the space cost of them are all high. To construct the summary of each object's PDF, the key idea of UI-Tree is partitioning the uncertain region of every object, pre-computing the appearance probability of the partitioned sub-region, and using R-tree technique to organize these sub-regions. Given a prob-range query, UI-tree retrieves the sub-regions that overlap with the query region, finds the corresponding objects, and then computes the lower and upper bounds of $app(o, q)$(short for the appearance probability that o lies in the query region). Specifically, given an object, if a subregion $o(i)$ is contained in q_r, $app(o, i)$ (short for the appearance probability that o lies in $o(i)$) contributes to both lower and upper bound of $app(o, q)$. Similarly, if a subregion $o(j)$ overlaps with q_r, $app(o, j)$

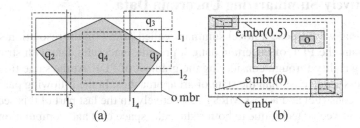

Fig. 1. Answering Prob-range Queries Using U-tree

contributes to the upper bound of $app(o,q)$. Then o may be validated (pruned) based on the lower (upper) bound of $app(o,q)$. Although UI-Tree has the stronger pruning ability than U-Tree, its space cost is too high. In the other hand, the partition do not reflect the PDF's gradient, and the filtering algorithm does not consider the intersection area between the query region and the subregions.

2.2 Problem Definition

Given a multidimensional probabilistic object o in the d dimension space, it is described either continuously or discretely. In the continuous case, an object has two attributes: o_r and $o.pdf(x)$. The o_r is a d-dimension uncertainty region, where o may appear at any locations with certain probabilities. The $o.pdf(x)$ is the probability of o appearing at location x. In the discrete case, o is represented by a set of sampled points $x_1, x_2, \ldots, x_m$, and o occurs at location x_i with probability $x_i.p$. Given a query region q_r, we use $app(o,q)$ to represent the likelihood of o falling in the query region q_r. $app(o,q)$ is also calculated by two cases. In the continuous case:

$$app(o,q) = \int_{o_r \cap q_r} o.pdf(x)dx \tag{1}$$

where $o_r \cap q_r$ denotes the intersection of o_r and q_r, and o is a result if $p_{app}(o,q) \geq \theta$ (query probability threshold). In the discrete case:

$$app(o,q) = \sum_{i=1}^{n2} o.pdf(x_i) / \sum_{i=1}^{n1} o.pdf(x_i) \tag{2}$$

where n_1 is amount of the sampled points in o_r, and n_2 is the amount of the sampled points falling into $o_r \cap q_r$.

Definition 1. *(**Probabilistic Range Query**). Given a set of probabilistic objects O and a range query q, the probabilistic range query retrieves all probabilistic objects $o \in O$ with $app(o,q) \geq \theta$, where θ is the probabilistic threshold and $0 \leq \theta \leq 1$.*

3   Effectively Summarizing Uncertain Data

In this section, we propose a novel summary called *MRST*(multi-resolution summary tree) to capture the PDF of uncertain data. It provides uncertain data with strong pruning/valiating ability through considering the gradient of PDF. At the same time, *MRST* consumes less space cost than the state of art approaches. In the following part, we discuss how to construct and access MRST respectively. In the last part of this section, we employ the bit-vector technique to both reduce the space cost and computational cost.

3.1   A Tight Probabilistic Bound For Filtering

In this section, we introduce how to provide the object with a tight bound. It is the guide of the summary construction.

We firstly discuss how to provide each sub-region $o(i)$ with a tight bound. Given an object o, a sub-region $o(i)$ and a query q, if q_r overlaps with $o(i).MBR$, Equation 3 and Equation 4 show the probabilistic lower-bound and upper-bound of o lying in $o_r \cap q_r$ respectively. Obviously, by fully considering the intersection area between $o(i).MBR$ and q_r, even if our partition is as the same as that of UD-Tree and UI-Tree, the probabilistic bound proposed in this paper is tighter.

$$lb_{app}(q,i) = lb(o,i) \times (max(0, S(q,i) - ZS(o,i))) \qquad (3)$$

$$ub_{app}(q,i) = min(ub(o,i) \times S(q,i), app(o,i)) \qquad (4)$$

where $app(o,i)$ represents the likelihood of o falling in $o(i)$. The $lb(o,i)(ub(o,i))$ denotes the maximal(minimal) probability density in $o(i)$. $ZS(o,i)$ represents the blank (and zero-pdf) area of $o(i)$. $lb_{app}(o,i)$ $(ub_{app}(o,i))$ denotes the lower-bound (or upper-bound) of the probability o lying in $q_r \cap o(i)$.

Property 1. Given an object o and a query q, when q_r overlaps with o's subregion $\bigcup_{i=1}^{i=n_1} o(i)$, the $lb_{app}(o,q) = \sum_{i=1}^{i=n_1} lb_{app}(q,i)$ and $ub_{app}(o,q)$ is $\sum_{i=1}^{i=n_1} ub_{app}(q,i)$.

$lb_{app}(o,q)(ub_{app}(o,q))$ denotes the lower-bound(upper-bound) of the probability o lying in q_r. For each object o, $ub_{app}(o,q)$-$lb_{app}(o,q)$ is to evaluate whether the bound is tight enough. According to Equation 3, Equation 4 and Property 1, the following conditions should be satisfied for the tighter bound: (i) $ub(q,i) - lb(q,i)$ should be relatively small; (ii) the amount of subregions should be relatively small.

3.2   Effective Summary Construction Using Multi-Resolution Technique

In this section, we employ the multi-resolution technique to construct the summary (called MRST). The MRST could provide the uncertain object with a more effective partition and a tighter probabilistic bound. Now, we formally define the PBD(short for probability bound difference) which is used as the criterion of construction.

Definition 2. *(PBD). Given a sub-region $o(i)$ of an object o, $PBD(o,i) = (ub(o,i) - lb(o,i)) \times S(o,i)$.*

Given an object o, its corresponding MRST is constructed in the following two steps: they are *spilt* and *shrink*. The split is to partition the subregions where the probability density changes dramatically. The procedure is that we recursively partition the object region o_r until the PBD of each sub-region is less than λ. And then, we use a quad-tree to temporarily organize this split result. After spilt, the probability density in each sub-region $o(i)$ changes smoothly, and $ub(o, i) - lb(o, i)$ may be small enough. Obviously, the bound provided by MRST is tighter. For example, in Fig 2(a), the shadow region is the object region o_r bounded by a MBR, and the blank region may be seem as the sub-region of o_r with a zero-pdf. Fig 2(g) is designed to show the PBD of each subregion. According to Fig 2(g), because the $PBD(o, A)$ and $PBD(o, C)$ are less than $\lambda(=0.1$ in this section), we stop splitting them. Because $PBD(o, B)$ and $PBD(o, D)$ are more than λ, we subdivide them into four parts respectively. The Fig 2(b) is the result of spilt, and Fig 2(c) shows the corresponding quad-tree.

After the spilt, the shrink is done to merge the subregions where the probability density of them are roughly the same. Given two subregions $o(i)$, $o(j)$ of o, they are merged if $PBD(o, i + j) \leq \lambda$. We access the quad-tree in the post-order. We firstly merge the leaf nodes within the same subtree. Then, we merge the leaf nodes among different subtrees. Specifically, in each subtree, the leaf node with the minimal $app(i, o)$ is selected as the candidate node(eg,. d_1, b_1, A and C). Given two candidate node u and v from different nodes, if $PBD(o, u + v) \leq \lambda$, they are merged. According to Fig 2(g), b_1, because $PBD(o, b_1+b_2+b_4) \leq \lambda, b_1$, b_2 and b_4 can be merged. The Fig 2(d) shows the result of merging the nodes from the same subtree, where b_1 and C are merged. The Fig 2(f) is the finally $MRST$.

After constructing the MRST of an object, an interesting result is that if the probability density of a sub-region is dramatically changing, it has a fine partition; otherwise, it has a coarse partition. By this property, it guarantees that the MRST could more effectively reflect the gradients of the PDF, and the amount of subregions is relatively small(shown in experiment). In addition, because the $PBD(o, i)$ of each subregion $o(i)$ is also relatively small, MRST could provide the object with a tight bound. We could build a cost model to find the optimal λ that need to consider both the filtering ability and I/O cost. A similar method was proposed in [8]. Due to the limitation of space, we do not discuss it.

3.3   Accessing the Summary of Uncertain Data

In this section, we propose Algorithm 1 to efficiently access the summary of uncertain data. Algorithm 1 employs the key idea of greedy algorithm. The Algorithm 1 uses a field called $d(q, i)$ to determine the accessing order of the nodes in MRST so as to early terminating the accessing of MRST as much as possible.

$$d(i, q) = min(u(i, o) \times S(q, i), app(o, i)) - lb(i, o) \times (max(0, S(q, i) - ZS(o, i))) \tag{5}$$

Given a query q, an object o and a subregion $o(i)$, if $q.r$ overlaps with $o.MBR$, we access the MRST of o to check whether o is a result of q. The $d(q, i)$ is computed through Equation 5. Obviously, the larger the $d(q, i)$ is, the greater it contributes

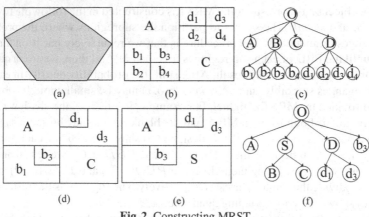

Fig. 2. Constructing MRST

to $ub_{app}(o)$-$lb_{app}(o)$, and the corresponding $o(i)$ should be prior accessed. Compared with the traditional accessing method such as preorder traversal and inorder traversal, introducing this field to control the nodes accessing order is more efficiently to compute the bound.

As shown in Algorithm 1, we firstly access the root of MRST, compute $lb_{app}(o)$ and $ub_{app}(o)$ according to Equation 3 and Equation 4. If o can not be pruned(or validated), we initialize the array L(line 2-6). After initializing L, the following things are repeatedly done to compute the probabilistic bound. Firstly, we find the node e whose corresponding $d(i, q)$ is maximal in L. Secondly, based on e, we tighten the bound: (i) eliminate the contribution of e_i to $ub_{app}(o)$ and $lb_{app}(o)$ (line 8 to 10); (ii) access every children of e_i to compute the new bound according to by Equation 3 and Equation 4,and property 1. Thirdly, if o is not still filtered, we update L: we insert the children e_{ij} of e_i into L, when e_{ij} satisfies the conditions that (i) e_{ij} also has children; (ii) the corresponding subregion of e_{ij} overlaps with q_r.

After accessing the MRST of an object, o is validated if the lower-bound of $app(o, q)$ is more than q_p. Also, o is pruned if the upper-bound of $app(o, q)$ is less than q_p. If o can not be pruned/validated, we have to use the integral to check whether o is the result.

3.4   Efficient Summary Storage

Now, we discuss how to efficiently store the MRST. The MRST stores three types of information to capture the PDF of a given object. Given an object o and a subregion $o(i)$, they are the probabilistic information (eg,.$app(o, i)$, $lb(o, i)$ and $ub(o, i)$), location information, blank area information, and the hierarchical relationship between parent and its children. Because too many information has to be stored, we employ the bit vector to compress MRST as much as possible.

Firstly, we use a m-bits vector to represent the probabilistic information, and its domain is 2^m. As the tradeoff between the degree of accuracy and the space, given an object o and a sub-region $o(i)$, we use 6 bits to express $app(o, i)$, where the domain is 0 to 63. $app(o, i) = 0.2$, it is expressed by $\lfloor 0.2 \times 63 \rfloor$ =12(001100). We use 4bit to express $lb(o, i)$(also $ub(o, i)$), where the domain is 0 to 15. Secondly, we use a n-bits bit vector to express $o(i)$'s location information.

Specifically, given an object o, we use a MBR to bound it. Next, we could use a "virtual grid" with a $2^n \times 2^n$ resolution to partition the MBR. Lastly, the "virtual coordinate" expressed by bit vector is used to express $o(i)$'s location information. For example, using a 7-bits vector, the resolution of the grid is 128×128. The left-bottom(right-upper) coordinates are described by the cell Id. As shown in Fig 2, the "virtual coordinate" of node d_1 is expressed by (64,111) and (80,127). The area information depends on the resolution of the "virtual grid".

Algorithm 1. Accessing MRST

Input: MRST, o, probabilistic range query, q, Node e
1 ; **Output**: lower-bound, lb; upper-bound, ub
2 ; $ub_{app}(o) \leftarrow min(1, ub(o) \times S(o))$;
3 $lb_{app}(o) \leftarrow max(0, S(o) - ZS(o)) \times lb(o)$;
4 **if** $ub_{app}(o) < p_q \vee lb_{app}(o) \geq p_q$ **then**
5 $\quad\lfloor$ *return*;

6 *Insert(L,o, d(q,o),$ub_{app}(o),lb_{app}(o)$)*;
7 **while** *Empty(L)$\neq$ true* **do**
8 $\quad$ *Node e=PopFront(L)*;
9 $\quad$ $ub_{app}(o) \leftarrow ub_{app}(o)-e.ub_{app}(o,i)$;
10 $\quad$ $lb_{app}(o) \leftarrow lb_{app}(o)-e.lb_{app}(o,i)$;
11 $\quad$ **for** i *from* 0 *to e.Len* **do**
12 $\quad\quad$ $ub_{app}(o) \leftarrow ub_{app}(o) + min(1, ub(o,i) \times S(o,q))$;
13 $\quad\quad\lfloor$ $lb_{app}(o) \leftarrow lb_{app}(o) + max(0, S(o,i) - ZS(o,i)) \times lb(o,i)$;
14 $\quad$ **if** $ub_{app}(o) < p_q \vee lb_{app}(o) \geq p_q$ **then**
15 $\quad\quad\lfloor$ *return*;
16 $\quad$ **else**
17 $\quad\quad$ **for** i *from* 0 *to R.Len* **do**
18 $\quad\quad\quad$ **if** $q_r \cap o(i).r \neq \emptyset \wedge q_r \cap o(i) \neq o(i).r$ **then**
19 $\quad\quad\quad\quad\lfloor$ *Insert(L,R(i,o), d(q,i),$ub_{app}(o,i),lb_{app}(o,i)$)*;

20 return ;

For example, as shown in Fig 2, base on the "virtual grid", because the area of d_1 is $32 \times 32 = 1024$ and half on d_1 is blank, the blank area of d_1 is 512(10000000). Finally, we use a static array to organize the nodes in MRST. We use k-bits vector to express "offset+len" so as to describe the hierarchical relationship between the parent and its children. As shown in Fig 2, D is a interval node that has two children d_1 and d_3, where the offset is 3(11),and len=2(10).

Another advantage of data compression is that we could use the bit-operations to do the above operations shown in algorithm 1. Due to the limitation of space, we do not discuss how to store the node in MRST, and how to access MRST using bit-operations.

4   Indexing uncertain data

In this section, we propose an index called *R-MRST* to organize the MRST of uncertain data. Its pruning ability is roughly the same with the other indexes such as U-Tree, but cost of dynamic update and space are much lower than them.

As is discussed in Section 2.1, it is unworthy to store too much probabilistic information in each node(leaf or interval). For example, given a leaf node based on U-Tree, although using a group of MBRs to bound its children's PCR could obtain a tighter boundary, as shown in Fig 1(b), the shrunken degree of the boundary is relatively small, and it causes both a high space cost and high dynamic update cost. The other indexes such as UI-Tree and UD-Tree also have the similar problem.

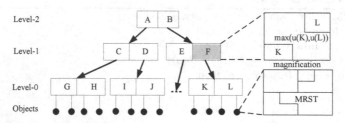

Fig. 3. The Framework of R-MRST

Based on the above analysis, we propose the *R-MRST*. As shown in Fig 3, it is the framework of *R-MRST*. It is similar with R-Tree. Due to the limitation space, we mainly discuss how to maintain the probabilistic information in each node of R-MRST, and how to use it for pruning according to Property 2.

Property 2. Given a prob-range query q and a node e of R-MRST, the intersection area between $e.MBR$ and q_r is S. If $S \times ub(e) < q_p$, e can be pruned.

For each node e in R-MRST, we maintain the maximal probability density called $ub(e)$ among all the objects in the subtree of e. Given a query q, if q_r overlaps with the MBR of e, we employ Property 2 for pruning. Although the pruning method seems simple, as shown in Fig 1, it also could prune the node whose MBR's margin overlaps with the query region. Thus, it is suitable for processing range query over uncertain data, and both the space cost and update cost are low.

Query on R-MRST: Given a prob-range query q, the search starts from the root of R-MRST, and eliminates its entries according to Property 2. For each remaining entry, we retrieve its child node, and perform the above process recursively until a leaf node is reached. For an object o encountered, we attempt to filter it through accessing its MRST. For the object o which can not be filtered, in the refinement phase, we use the integral to check whether o is the result of q.

Dynamic Update Algorithm: Compared with R-tree, the update method of our index is roughly the same. The difference is the maintenance of $ub(e)$. Specifically, given a leaf node e, when a newly arrived object o inserts into e, the following cases cause the updating of $ub(e)$. (i) $ub(o) \geq ub(e)$; (ii) the number of objects in e exceeds to the bucket size, and causes e spilt. In the first case, we set $ub(e)=ub(o)$. In the second case, if e is split into e_1 and e_2, we compute $ub(e_1)$ and $ub(e_2)$. When a object o leaves e, the following cases cause the $ub(e)$ updating. (i) $ub(o)$ is maximal probability density among all the object in e, in this case, we select the new $ub(e)$ from e. (ii) if two node e_1 and e_2 are merged into e, the $ub(e)$ is max($ub(e_1)$, $ub(e_2)$). After updating $ub(e)$, we access the parent e' of e to check whether the $ub(e')$ need to be updated. If so, we update $ub(e')$ and continuous access the upper-level node until no interval node should be updated.

5 Experimental Evaluation

This section experimentally evaluates the efficiency of the proposed techniques. The R-MRST will be compared with U-Tree (a classic technique) and UD-Tree, where U-Tree is a classic index and UD-Tree is the most advanced index technology presently.

Two real spatial data sets LB and CA are employed to represent the center of probabilistic regions, which has been used as the test data set such as [8] [4]. They contain 53k and 62k two-dimension points representing locations in Long Beach and Los Angeles respectively. In addition, three synthetic data sets containing 128k/256k/512k two-dimension points are employed. In our experiments, the region of probabilistic data is a rectangular with side-length varying from 100 to 500 and the default value of the side-length is 200. In this paper, we call the half of side-length as radius. Because it is unfair to select uniform distribution as the $o.pdf(x)$, we use two other common distributions: poisson distribution and normal distribution. In the default case, all dimensions are normalized to domain [0,10000] and LB with constrained normal distribution is employed as the default data set. A workload contains 100 queries in our experiment. The region of the queries are a rectangular with r_q varying from 500 to 1500. In our experiments, we randomly choose the probabilistic threshold $\theta \in (0,1]$ for each query. R-MRST was implemented in C++. Experiments are run on a PC with i3-core and 4 GB memory.

5.1 Index Construction

Firstly, we compare the index size among R-MRST, U-Tree and UD-Tree. Secondly, we compare the constructing time among these three indexes. Thirdly, we compare the space cost of summary based on these three indexes. The experiment is employed in different data sets. One of them is based on the CA. Another one is a synthetic data set with 200k 2-dimension data. Since UD-Tree can not work when the PDF is in the continual case, we use the sampled points to simulate the PDF of an object.

The Fig. 4(a) to Fig. 4(b) uses the synthetic data set. In the Fig. 4(a), the storage cost of R-MRST is less than that of both U-Tree and UD-Tree. Fig. 4(b) shows the space cost of MRST. As we see, ours performs best of all.

5.2 Query Performance

In this section, we evaluate the query performance. In the first group of experiments, we evaluate the performance of the R-MRST, UD-Tree [8] and U-Tree against different r_q.

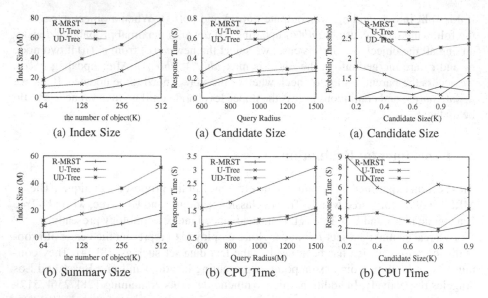

(a) Index Size   (a) Candidate Size   (a) Candidate Size

(b) Summary Size   (b) CPU Time   (b) CPU Time

Fig. 4. Index size Based on Dif-   **Fig. 5.** Cost vs.diff R_u **Fig. 6.** Cost vs. diff θ
ferent Data Sets

The parameters of the experiments are same as the previous one. Firstly, we evaluate the ability of pruning/validating. In the Fig. 5(a), the candidate size of UD-Tree and R-MRST are roughly the same which both perform better than U-Tree. Secondly, we evaluate the response time. In the Fig. 5(b), R-MRST performs best of all.

In the second group of experiments, we evaluate the performance of the R-MRST, UD-Tree and U-Tree against different threshold θ. The θ varies from 0.1 to 0.9, and the other parameters are default. The experiment content are the same as the first group. The Fig. 6(a)-Fig. 6(a) are the results of the experiments. In the Fig. 6(a), R-MRST performs best of all.

The third group of experiments evaluate the filtering ability and the computational cost of the node in R-MRST. All of parameters are default. Firstly, we count the number of the entry nodes needed to be accessed. In Fig. 7(a), the filtered ability of R-MRST and U-Tree are are roughly the same. Secondly, we record the response time. In the Fig. 7(a), the computational cost of them are roughly the same.

In the forth group of experiments, we study the probability filtering ability of MRST. We count the amount of probabilistic data that should be checked and then calculate the recall $ratio(rr)$ and the response time. The result are reported in the Fig. 8(a)-Fig. 8(a). As expected, MRST has a stronger filtering ability. In the Fig. 8(a), although the computational cost based on MRST is higher than of PCR, the difference of their response time can be accepted.

In the last experiments, we compare the performance of R-MRST, UD-Tree, and U-Tree by different data sets. Five data sets (LB,CA and three synthesize) are employed. The number of data points of each data set is 53k, 62k, 128k, 256k and 512k. We use default parameters in these experiments. As expected, R-MRST performs best of all.

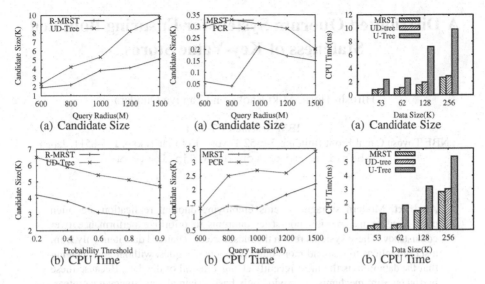

(a) Candidate Size	(a) Candidate Size	(a) Candidate Size
(b) CPU Time	(b) CPU Time	(b) CPU Time

Fig. 7. S-node vs. U-Tree   **Fig. 8.** PCR vs.MRST   **Fig. 9.** Cost vs. diff dataset

6   Conclusions

In this paper, we studied the problem of range query on probabilistic data. Through deep analysis, we proposed an effective indexing technique named R-MRST to manage uncertain data. R-MRST could provided a very tight bound for pruning/validating the objects that overlap(or non-overlap) with the query region in a lower cost. Our experiments convincingly demonstrated the efficiency of our indexing techniques. In the future, we will further study other indexes which are suitable for high-dimensional uncertain data and support probabilistic data update frequently.

References

1. Agarwal, P.K., Cheng, S.W., Tao, Y., Yi, K.: Indexing uncertain data. In: PODS, pp. 137–146 (2009)
2. Kalashnikov, D.V., Ma, Y., Mehrotra, S., Hariharan, R.: Index for fast retrieval of uncertain spatial point data. In: GIS, pp. 195–202 (2006)
3. Lian, X., Chen, L.: Set similarity join on probabilistic data. PVLDB 3(1), 650–659 (2010)
4. Tao, Y., Cheng, R., Xiao, X., Ngai, W.K., Kao, B., Prabhakar, S.: Indexing multi-dimensional uncertain data with arbitrary probability density functions. In: VLDB, pp. 922–933 (2005)
5. Tran, T.T.L., Sutton, C.A., Cocci, R., Nie, Y., Diao, Y., Shenoy, P.J.: Probabilistic inference over rfid streams in mobile environments. In: ICDE, pp. 1096–1107 (2009)
6. Zhang, M., Chen, S., Jensen, C.S., Ooi, B.C., Zhang, Z.: Effectively indexing uncertain moving objects for predictive queries. In: PVLDB, vol. 2(1), pp. 1198–1209 (2009)
7. Zhang, Y., Lin, X., Zhang, W., Wang, J., Lin, Q.: Effectively indexing the uncertain space. IEEE Trans. Knowl. Data Eng. 22(9), 1247–1261 (2010)
8. Zhang, Y., Zhang, W., Lin, Q., Lin, X.: Effectively indexing the multi-dimensional uncertain objects for range searching. In: EDBT, pp. 504–515 (2012)

A Distributed Quorum System for Ensuring Bounded Staleness of Key-Value Stores

Hiroshi Horii, Miki Enoki, and Tamiya Onodera

IBM Research – Tokyo
NBF Toyosu Canal Front Building 5-6-52 Toyosu, Koto-ku Tokyo, 135-8511, Japan
{horii,enomiki,tonodera}@jp.ibm.com

Abstract. Modern storage systems employing quorum replication are often configured to use partial, non-strict quorums to prioritize performance over consistency. These systems return the most recently changed data item only from a set of replicas to respond more quickly to a read request without guaranteeing that the data item is the most recently changed for all of the data. Because these partial quorum mechanisms provide only basic eventual consistency guarantees, with no limit on the freshness of the data returned, sometimes these configurations are not acceptable for certain applications. In this work, we have devised a new key-value store with partial quorums while ensuring bounded staleness. Our store reports the expected bounds on staleness with respect to wall clock. We evaluated our new key-value store with Yahoo! Cloud Service Benchmarks and show its performance.

Keywords: Key-Value Store, Quorum System, Bounded Staleness.

1 Introduction

Good scalability, high availability and low latency are required by modern distributed storage systems [11][15][16]. These systems typically replicate data across different machines to improve availability and performance by duplicating and parallelizing services among multiple replicas. To provide predictably low latency and throughput for reads and writes, systems often sacrifice consistency guarantees for data items [1]. However, weak-consistency systems make no general guarantees on the staleness of data items returned except that the system will eventually return the most recent version in the absence of new writes [2].

Distributed quorum systems [3][4] can be used to ensure strict consistency across multiple replicas of a data item by overlapping the read and write replica sets: given N replicas and read and write quorum sizes R and W, $R+W>N$ is guaranteed. However, if R replicas have the same data items, the quorum system has duplicated overheads when responding to requests for the same data item. By employing *partial* quorums where $R+W \leq N$, the duplicated overheads are reduced in quorum replication. However, the staleness of the data item is potentially *unbounded* and sometimes enterprise customers cannot accept such unbounded staleness, even though these quorum systems have a high probability of returning the freshest value [1].

F. Li et al. (Eds.): WAIM 2014, LNCS 8485, pp. 84–95, 2014.

We have devised a key-value store with a new algorithm for partial quorum replication that guarantees bounded staleness. In our partial quorum replication, we use t-visibility [1], which means a data item must be available for reads within t seconds after it is written. With this constraint, read data items retuned to clients are the most recently changed t seconds before or written within t seconds.

To guarantee t-visibility, we propose a new protocol to share the replication statuses among replicas. With this mechanism, each replica is aware of the existence of a data item written t seconds before in the other replicas. We evaluated overhead of this mechanism in our key-value store.

We make the two major contributions in this paper:

- Protocol to guarantee t-visibility: We developed a new protocol to share replication statuses using wall clock information among the replicas of quorum systems. The shared replication status is used to guarantee the t-visibility of the returned values.
- Cadidas: We implemented a new key-value store, Cadidas, which guarantees t-visibility by using our new protocol. Cadidas clients can use partial quorums to improve their performance by specifying t- visibility in their SLAs.

This paper is structured as follows: Section 2 describes quorum systems in theory and in practice. Section 3 describes the Cadidas implementation. Section 4 reports empirical performance results for Cadidas. Section 5 surveys the related work.

2   Background

In this section, we provide background regarding quorum systems both in the theoretical academic literature and in practice. We begin by introducing prior work on traditional quorum systems. We next discuss practical quorum systems, focusing on the most widely deployed protocol for storage systems employing quorum replication. Finally, we survey consistency to ensure bounded staleness in quorum systems.

2.1   Quorum Systems in Theory

A quorum system is one of the popular strategies to replicate data in distributed systems [3][4]. With quorum replication, a data item is shared among multiple replicas. When a new version of the data item is written, copies of the data item are sent to a set of replicas, its write quorum. To read the data item, the copies are fetched from a possibly different set of replicas, its read quorum, and the most recent data item is selected by comparing the versions of the fetched copies. For each operation, the read and write quorums are chosen from all of the replicas in the quorum system. Usually, the sizes of the read and write quorums are fixed, R and W respectively, that are less than the total number of the replicas N.

If the total size of R and W is larger than N ($R+W>N$), which means at least one overlapping replica in the read and write quorums always exists, consistency is ensured by the quorum system (strict quorum system). After a data item is written in a write quorum, this data item is available by reading the copies in its read quorum, because some of the copies are in the overlapping replicas and at least one of them will be included in the read quorum. When multiple overlapping replicas exist in the two quorums, there are redundant operations for reads and writes.

If the total size of R and W is equal or smaller than N ($R+W \leq N$), which means overlapping replicas in the read and write quorums may not exist, there are fewer redundant operations because the number of overlapped replicas in the two quorums is decreased. Though the performance of these partial quorum systems is better than for strict quorum systems, they cannot ensure consistency because there may be no overlapping replica that knows the latest value of a written data item. Usually, the replicas in quorums are randomly selected and the gaps between the versions of data items in the replicas probabilistically small after the data item is written multiple times. However, the staleness of data items in replicas is potentially unbounded and the clients may read too obsolete data items.

2.2   Quorum Systems in Practice

In practice, many of the key-value store systems use quorums as a replication mechanism for eventual consistency, including Dynamo [11], Cassandra [16], and Voldemort [15]. These systems use one quorum system for each key-value pair and the replicas for that pair are deterministically configured with their mapping algorithms. Their clients select a subset of the replicas to access the value of the key.

Though practical quorum systems are based on the well-studied theoretical quorum systems, there are some differences. First, in practical quorum systems, clients write copies to all of the replicas for each write and W configures the quorum systems differently: the number of replica responses required for a successful write (configurations of $R+W>N$ ensures consistency).

Second, theoretical quorum systems assume that all of the writes for a data item can be ordered only by the replicas holding that data, unlike some of the implemented quorum systems [12][15]. In practical system, the clients must assist in setting the order of the writes. In Dynamo [12] and Voldemort [14], replicas generate vector clocks to order the writes for a value of a key and the clients send a value with a new vector clock generated on a replica for each write. The vector clocks are partially ordered and some of them may be in conflict. All of the values of the conflicting versions are kept in the replicas and the clients select or generate a value from them when they read a value of the key.

In spite of these differences, the way partial quorums are used is the same: The sizes for R and W are set low so there may be no overlapping nodes exist in the read and write quorums ($R+W \leq N$). In practical quorum systems, the sizes of R and W are configurable in the clients or replicas. With a small size for R, the throughput of the system is improved while reducing the redundant processing in the replicas. However, the staleness of the values returned in partial quorums is potentially unbounded, which means their clients may read obsolete values. These inconsistencies are not acceptable for many applications, such as banking applications.

2.3   Bounded Staleness

Staleness is measured by two axes in the literature: versions [1][11][8] and wall clock time [1][6][8][9][10][11]. Using versions k, reading one of the last k versions of a data item is guaranteed. Using a wall clock, reading a version t seconds after it is written is guaranteed. These metrics are called k-staleness and t-visibility respectively in [1]. In the work described here, we focus only on t-visibility because financial systems have many time-based constraints (For example, Japanese FX systems must guarantee slippage within 10 seconds).

The t-visibility idea is defined in [1][1].In this definition, a committed version means a value written in W replicas.

Definition. A quorum system guarantees t-visibility consistency if any read quorum started at least t units of time after the latest version committed returns at least one value that is at least as recent as the last committed version.

Fig. 1 shows two writes and two reads in a partial quorum system that consists of three replicas, r_0, r_1 and r_2, configured with $R=1$, $W=2$, and $N=3$. vc_1 and vc_2 are vector clocks and vc_2 follows vc_1. The k_1's value of version vc_1 is written as $wvc_1(k_1)$ and read as $r(k_1){:}vc_1$. The two values of vc_1 and vc_2 are committed at clocks c_2 and c_5 respectively because two replicas ($W=2$) must write them.

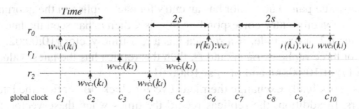

Fig. 1. An example of t-visibility consistency in a partial quorum system ($R=1$, $W=2$, $N=3$)

In Fig. 1, $2s$-visibility is guaranteed in the read at c_6. This is because c_4 is the clock $2s$ before c_6, and at c_4, vc_1 is the latest vector clock of the committed value. On the other hand, $2s$-visibility is not guaranteed in the read at c_9. This is because a value of vc_2 has been already committed at c_5 that is more than $2s$ before c_9.

Though constraints for staleness have been studied in database replication [13][14], they assumes writes are totally ordered. There are few studies for practical quorum systems [1] and no implementation to support t-visibility in our survey.

3   Guaranteeing Bounded Staleness in Quorum Systems

We have designed Cadidas (Continuously Available DIstributed DAta Store), a new key-value store that ensures t-visibility with replicated values. In Cadidas, replicas exchange key and versions of stored values and recognize each others replication

---

[1]   Our definition is slightly modified from [1] omitting the probability extension.

statuses. With these shared statuses, the replicas can check the freshness of their stored values and respond to their read requests from clients while ensuring t-visibility. If the replica can ensures t-visibility, then a client only needs to access a few replicas in the partial read quorum.

3.1   Architecture

Cadidas consists of a number of machines that can store large number of key-value pairs. The key-value pairs are partitioned among the shards by using hashcodes for the keys, where N servers (replicas) are assigned to each shard. A quorum system is constructed for the reading and writing of a key-value pair with R and W replicas, and vector clocks are used to order all of the writes for each key-value pair. The value of N, R, and W are set when the replica machines are configured. For each read or write operation, a client identifies the N replicas for a key from its hashcode and selects the R or W replicas from the N replicas to construct the read or write quorums.

All of the clients and the replicas are implemented in Java. Key-value pairs are stored in memory, and they are recovered when a replica restarts due to a failure.

3.2   Synchronizing the Replication Statuses

To bound the staleness of a replica, each replica maintains a real time vector [9][10] for each key-value pair. The vector has an entry for each replica in the quorum system. If r_0's real time vector entry corresponding to r_1 is t, then r_0 has seen the latest value in r_1 at real time t. For example, if a replica has a real time vector [10:00:02, 10:00:00, 10:00:01] for three replicas, r_0, r_1, and r_2, then this replica has the latest values of r_0 at 10:00:02, of r_1 at 10:00:00, and of r_2 at 10:00:01.

Replicas periodically exchange their latest vector clocks to confirm the latest values in the others. In addition, the replicas record the times when these vector clocks are exchanged. With these exchanged vector clocks and their times, each replica can maintain a real time vector for each key-value pair.

In a distributed environment, there may be no global clock. Therefore, each replica uses its local clock to record when the vector clocks are exchanged. All of the communication is pull-based and the time is recorded when the communication starts. Fig. 2 shows an example of the message exchanges between two replicas, r_0 and r_1 to share the r_1 vector clocks. In this example, c_i is the local clock for r_0 and the two conflicting vector clocks are vc_2 and $vc_{2'}$ following vc_1. At c_1, r_0 requests the latest vector clock from r_1, and at c_3, r_0 receives vc_1 from r_1. Since vc_1 was received in response to the communication started at c_1, r_0 determines vc_1 was the latest r_1 at c_1. Similarly, from the second and third communication exchanges started at c_4 and c_7, r_0 can determine that both vc_2 and $vc_{2'}$ were the latest values for r_1 at c_4 and c_7. In the third communication exchange, to reduce the communication overheads, r_1 sends an empty message because r_1 knows[2] vc_2 and $vc_{2'}$ were received by r_1.

---

[2] The message r_0 reports the success of the previous communication in its new request and then r_1 can determine which versions were successfully exchanged.

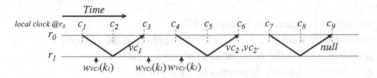

Fig. 2. An example of t-visibility consistency in a partial quorum system ($R=1$, $W=2$, $N=3$)

While sharing the latest vector clocks among the replicas, each replica maintains a real time vector for each key-value pair. Fig. 3 shows an example of the change history of the real time vectors for the same read and write requests shown in Fig. 1. The local clock for r_0 is c_i. r_0 receives the latest vector clocks from r_1 at c_3, c_5 and c_8. A real time vector for a key-value pair is expressed in the form $[c_i, c_j, -]$, that means this pair was the latest at c_i and c_j in r_0 and r_1, respectively, but was not stored at r_2.

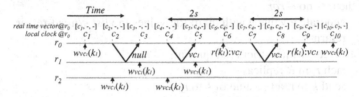

Fig. 3. An example of t-visibility consistency with real time vectors ($R=2$, $W=2$, $N=3$)

In the beginning of the exchanges, the entries of r_1 and r_2 of the real time vectors at r_0 were null (-) because r_0 did not know any of the vector clocks for the other replicas. After receiving vc_1 from r_1 at c_5, r_0 knew that vc_1 was fresh at c_4, and the r_1 entry becomes c_4 while vc_1 was still fresh at r_0. When r_0 writes the value of vc_2 at c_{10}, the r_1 entry becomes c_7 since r_0 knew vc_2 was fresh at r_1 based on the exchange from c_7.

Since each real time vector indicates when the value of each replica was fresh in each of the other replicas, t-visibility can be guaranteed by using the real time vectors. For the example of Fig. 3, $[c_6, c_4, -]$ at c_6 shows that the value was fresh at R replicas at time c_4. Because there are two seconds between c_4 and c_6 and two replicas ($R=2$) have the value, r_0 can guarantee the 2s-visibility for the returned value at c_6.

Though there is overhead in sharing the vector clocks to maintain the real time vectors for all of the key-value pairs, the overhead can be reduced by using certain implementation techniques. First, the requests and responses of versions for multiple key-value pairs are combined when a replica is requesting one replica's versions. Second, if there are no updates for some key-value pairs since the last exchange, the responses don't need to include those vector clocks. Third, each replica only needs to receive the vector clocks from (R - 1) replicas. Some other optimizations are also possible, such as sending only the differences in vector clocks, avoiding sending vector clocks for write-only values. Because the numbers of replicas in practice are limited to three or four, the communication overheads is small after these optimization are applied.

write (k, v)
    identify N replicas for a quorum of k
    vc=generate a new vector clock on a replica in the N replicas
    send v and vc to the N replicas to write a value of k
    wait for W acknowledgements from replicas

read (k, t)
    identify N replicas for a quorum of k
    //start read with t-visibility
    send k and t to read a value of k to a replica r in the N replicas
    receive three results from r
        $vlist$: a list of values of k,
        $vclist$: a list of vector clocks for $vlist$, and
        $error$: a Boolean describes the existence of an error in this read
    if there is no $error$
        generate a value v with all of the values in $vlist$ //read-repair
        return v
    //start read with a strict read quorum
    for each r_i in R replicas
        send k to read a value of k to r_i
        receive two results from r_i
            $vlist_i$: a list of values of k, and
            $vclist_i$: a list of vector clocks of $vlist_i$
    if all of $vlist_i$ have no entry
        return null
    generate a value v with all of the values in all of $vlist_i$ //read-repair
    if there are differences in all of $vlist_i$
        generate a vector clock vc
            with maximum numbers of each entry in vector clocks of $vclist_i$
        send v and vc to R replicas
        wait for R acknowledgements from replicas
    return v

Fig. 4. Read and write implementation of clients

3.3   Implementation

Replica. The replicas store lists of values and their vector clocks for each key-value pair. When the pair of a new value and a new vector clock is received to write the value of a key, each replica adds them into the lists of values and vector clocks, and then returns an acknowledgement to the sender. These lists store only the latest values. If a written vector clock is later than the others in the list of vector clocks, then the older vector clocks and its corresponding values are removed from the lists.

    When a key arrives for reading a key-value pair, the replica returns to the sender the key's lists of values and vector clocks. When time t is received with the key to specify

t-visibility, each replica checks the real time vector of the key-value pair, and returns an error if t-visibility is not guaranteed.

Client. Fig. 4 shows the client algorithm to read and write key-value pairs. Mostly, it behaves the same as the clients of Voldemort except for reading the staleness values.

Each client identifies all of the replicas of a quorum for a key-value pair and sends requests to a subset of the replicas to read and write the values of the keys. To write a value, a client selects a replica in the identified replicas and generates a new vector clock on the replica. With this new vector clock, the client sends the new value to all of the replicas and waits for W acknowledgements.

To read a value, a client selects a replica from the identified replicas and tries to read a value that guarantees t-visibility from that replica (partial quorum). If the selected replica cannot guarantee t- visibility, the client reads values of R replicas among the N replicas (strict quorum). If they return multiple values, then the client repairs them and reads the repaired value (read-repair). The logic to repair values is configured by application of Cadidas when the application initializes the client.

The repaired value is sent only to a strict quorum and is not sent to a partial quorum. This is because the returned values from a partial quorum must be shared among at least R replicas (an error is returned if the values are not shared in R replicas). On the other hand, the returned values from a strict quorum may not be shared among R replicas, and the client needs to ensure the repaired value is available for reads in the future.

4   Evaluation

In this section, we evaluate how much t-visibility improves the performance. First, we compare the throughputs of the original Cadidas and an incomplete Cadidas that disables the version exchange function and evaluate the overhead to support bounded staleness. Second, we assess the performance by changing the t of t-visibility. For both of the evaluations, we used the YCSB (Yahoo! Cloud Serving Benchmark).

4.1   Configurations

We configured one shard in the system to focus on the maximum throughput the system can provide. We set N to three because some practical systems use three replicas for each quorum. The values of R and W were set to two.

To evaluate the overhead, we implemented a generic quorum system by disabling the function for version exchanges in Cadidas. By setting R to one and then two, we created partial and strict quorums.

We then used YCSB to evaluate the performance and memory usage of the key-value stores. YCSB is intended to benchmark data storage systems for cloud systems by providing configurable workloads that simulate Web applications. There are six predefined core workloads with YCSB: update-heavy, read-mostly, read-only, read-latest, short-ranges and read-modify-write. We used update-heavy, read-mostly and read-only with the respective read-write ratios of 50:50, 95:5, and 100:0. In these workloads, the clients ask to access the values of the keys with a zipfian distribution. The number of the records was configured as 100,000. All of the servers in the key-value store share all of the data to evaluate the replication performance. That is, there is no partitioning of the key-value stores.

The machine configuration for this evaluation consists of four machines (each a 64-bit 2-core POWER6 4.0-GHz x2 with 12 GB of RAM, Red Hat Enterprise Linux Server release 6.4) for the four replicas and one machine (a 64-bit 8-core Xeon E5-2680 2.7-GHz x2 with 32 GB of RAM, Red Hat Enterprise Linux Server release 6.3) for the client. All of the machines are connected via 1-Gbps Ethernet cables.

4.2   Results

First, we evaluated the overhead to guarantee bounded staleness. We configured Cadidas to allow any staleness of the values, so the staleness of the returned values was ignored. Because the Cadidas always uses partial quorums to read values, the overhead of the version-exchange mechanism described in 3.2 can be evaluated by comparing that mechanism with the partial quorum. We used several intervals between the versions exchanged in the replicas and measured throughputs with YCSB. We used sixty threads of the test driver and measured five times the average of five throughputs of 60-seconds runs.

Fig. 5 shows the throughputs of the strict quorum, partial quorums, and Cadidas with different frequencies for the version exchanges. To evaluate only overheads of the version exchange, we configured t-visibility with infinite value. Though some throughputs of Cadidas with zero intervals showed worse performance than the strict quorum, the throughputs of Cadidas with the non-zero intervals were better than the strict quorum and mostly similar to the partial quorum. In addition, the frequency of version exchanges did not affect the throughputs when the interval was set to at least 50 milliseconds. These results showed the overhead for version exchange in Cadidas was small.

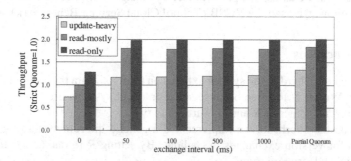

Fig. 5. Overhead of the version exchanges in Cadidas

Second, we evaluated the performance improvement by allowing reads of stale data. The frequency of version exchanges was set to 50 ms. Fig. 6 shows the throughput of the strict quorum, the partial quorum and Cadidas configured with different t-visibilities. When the read ratio in workloads increased, Cadidas performed better for all t-visibility settings. Also, with 50-ms-visibility and larger visibility, there was little performance improvement.

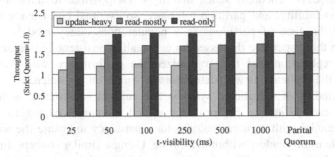

Fig. 6. Performance with various t-visibilities and a 50-ms frequency for version exchanges

Fig. 7 shows the throughputs of Cadidas with 1-s-visibility, a strict quorum, and a partial quorum with different numbers of clients. The performance of Cadidas was degraded when more than fifty clients were running. This is because the Cadidas replicas became overloaded and the Cadidas clients used their strict quorums to read the values. When 128 clients were running, 84.7% values in update-heavy, 12.8% values in read-mostly, and 0.2% values in read-only were read from strict quorums.

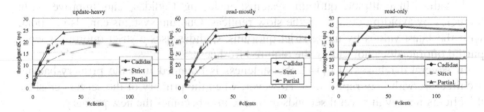

Fig. 7. Scalability of the number of clients with 1-s-visibility

Fig. 8 shows the guaranteed staleness with the replicas of Cadidas in the evaluation of Fig. 7. When the ratio of writes increased, the guaranteed staleness became worse as the replicas were overloaded. This is because all of the replicas handled each write request and the version exchanges were stalled among the replicas.

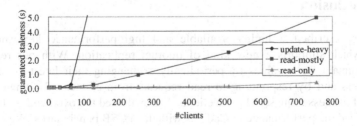

Fig. 8. Guaranteed staleness by replicas of Cadidas

5   Related Work

Recall that the CAP Theorem states that it is not possible to have all three of consistency, availability, and partition tolerance, so eventual consistency to provide availability in the face of partitions is a popular consistency model in practical systems [1]. In this paper, we discussed an eventually consistent key-value store that uses quorum replication and their optimized the performance with reducing the consistency of the replications with deterministic bounds on the staleness.

There are studies of distributed systems that support consistency with bounded staleness for reads. Beehive [6] provides shared memory systems to a parallel programming model with a guarantee of delta consistency to ensure the written data becomes visible to all readers within *delta* time. Using a similar concept, timed serial consistency and timed causal consistency were proposed based on sequential and causal consistency [7]. FRACS [8] provides optimized read performance with replicas that buffer the updates to be replicated within a configured time window and responds to read requests with stale data that has been already replicated. In addition to staleness, TACT [9][10] uses the numbers of numerical errors and order errors as criteria for the data to be returned by exchanging replication statuses and maintaining real time vectors within the replicas. AQuA [11] provides optimization for each client to select a replica that responds to a read request with specified bound on the staleness while predicting response time. These systems do not manage the numbers of replicas to write and read each value. In contract, quorum systems, including Cadidas, can improve write latencies by avoiding delays for the slow replicas. Quorum systems can also respond with the latest values by using strict quorums. Cadidas uses a strict quorum when a partial quorum does not satisfy the required freshness.

In several databases [13][14], the staleness is bounded in a the primary-backup architecture with lazy replication. However, when the primary of the servers crashes, the clients must wait several seconds until a replica becomes the new primary.

As discussed in Section 2.2, Dynamo [12] and Voldemort [15] use quorum systems to read and write data. Cassandra [16], behaves in the same way except for the customized read-repair function. Though they support strict and partial quorums, bounded staleness is not guaranteed. Cadidas provides a mechanism to support bounded staleness for read values.

6   Conclusion

This paper described a highly available and high-performance key-value store, Cadidas, which supports a new kind of quorum replication. With our replication, replicas maintain fresh values by periodically exchanging their latest versions with other replicas and by responding to read requests from clients only when they can ensure the freshness requested by the clients. We evaluated the overhead of the version exchanges and the performance of Cadidas with the YCSB benchmarks. We confirmed our replication outperforms the existing replication approach that supports strict consistency, while Cadidas has only small overhead for the version exchanges.

References

[1] Bailis, P., Venkataraman, S., Franklin, M.J., Hellerstein, J.M., Stoica, I.: Probabilistically Bounded Staleness for Practical Partial Quorums. In: Proceedings of the VLDB Endowment, vol. 5(8), pp. 776–787 (2012)

[2] Vogels, W.: Eventually consistent. Communications of the ACM 52(1), 40–44 (2009)

[3] Thomas, R.H.: A majority consensus approach to concurrency control for multiple copy databases. ACM Transactions on Database Systems 4(2), 180–209 (1979)

[4] Gifford, D.K.: Weighted voting for replicated data. In: Proceedings of the 7th ACM Symposium on Operating Systems, pp. 150–162. ACM, New York (1979)

[5] Gilbert, S., Lynch, N.: Brewer's conjecture and the feasibility of consistent, available, partition-tolerant web services. ACM SIGACT News 33(2), 51–59 (2002)

[6] Singla, A.: Temporal notions of synchronization and consistency in Beehive. In: Proceedings of the 9th Annual ACM Symposium on Parallel Algorithms and Architectures, pp. 211–220 (1997)

[7] Torres-Rojas, F.J., Ahamad, M., Raynal, M.: Timed consistency for shared distributed objects. In: Proceedings of the 18th ACM Symposium on Principles of Distributed Computing, pp. 163–172 (1999)

[8] Zhang, C., Zhang, Z.: Trading replication consistency for performance and availability: an adaptive approach. In: Proceedings of the IEEE International Conference on Distributed Computing Systems, pp. 687–695 (2003)

[9] Yu, H., Vahdat, A.: Design and evaluation of a conit-based continuous consistency model for replicated services. ACM Transactions on Computer Systems 20(3), 239–282 (2002)

[10] Yu, H., Vahdat, A.: The costs and limits of availability for replicated services. ACM Transactions on Computer Systems 24(1), 70–113 (2006)

[11] Krishnamurthy, S., Sanders, W.H., Cukier, M.: An adaptive quality of service aware middleware for replicated services. IEEE Transactions on Parallel and Distributed Systems 14(11), 1112–1125 (2003)

[12] DeCandia, G., Hastorun, D., Jampani, M., Kakulapati, G., Lakshman, A., Pilchin, A., Sivasubramanian, S., Vosshall, P., Vogels, W.: Dynamo: Amazon's highly available key-value store. In: Proceedings of the 26th ACM Symposium on Operating Systems Principles, pp. 205–220 (2007)

[13] Rohm, U., Bohm, K., Schek, H.-J., Schuldt, H.: FAS - a freshness-sensitive coordination middleware for a cluster of OLAP components. In: Proceedings of the 28th Very Large Data Bases, pp. 754–765 (2002)

[14] Guo, H., Larson, P.-A., Ramakrishnan, R., Goldstein, J.: Relaxed currency and consistency: how to say "good enough" in SQL. In: Proceedings of the ACM SIGMOD International Conference on Management of Data, pp. 815–826 (2004)

[15] Project Voldemort, http://project-voldemort.com/

[16] Apache Cassandra, http://cassandra.apache.org/

Multimodal Data Fusion in Text-Image Heterogeneous Graph for Social Media Recommendation[*]

Yu Xiong[1], Daling Wang[1,2], Yifei Zhang[1,2], Shi Feng[1,2], and Guoren Wang[1,2]

[1] School of Information Science and Engineering, Northeastern University
[2] Key Laboratory of Medical Image Computing (Northeastern University),
Ministry of Education, Shenyang 110819, P.R. China
xiongyu@research.neu.edu.cn,
{wangdaling,zhangyifei,fengshi,wangguoren}@ise.neu.edu.cn

Abstract. Every day, millions of texts, images, audios, videos, and other information with different modalities are posted on social media. These multimodal data provide abundant resources for information recommendation. In this paper, a new method based on multimodal data fusion is proposed for more effective recommendation on social media. Firstly, a heterogeneous graph on texts and images is created effectively to represent the relationship of multimodal data. Then the relationship of multimodal data is fused based on graph clustering to improve the quality of social media recommendation. Finally, the multimodal social media information recommendation is performed as a process of walk on the proposed heterogeneous graph. The experiment on texts and images of microblogs shows social media recommendation using multimodal data fusion is better than that on single modality.

Keywords: multimodal data, data fusion, heterogeneous graph, graph clustering, recommendation.

1 Introduction

The explosion of the web data has witnessed multi-modality. Every day, millions of people over the world discuss hot topics on Twitter, share travel photos on Flickr or life videos on YouTube, add new friends on Facebook, etc. These multimodal data are potential good resources for valuable information and provide new applications for us, one of which is the recommendation. For multimodal recommendation, if a user browsed a flower image, the recommended results to him (her) should contain both similar flower images and texts about the flower, such as species, habitat, habit of the flower. However, in traditional single modal recommendation, such task implements hardly when there is not explicit links between above images and texts.

It is useful to fuse multimodal data. By fusing multimodal data, we can obtain a global and vivid view of events or a better recommended result. Besides, the fusion makes the similar query of the recommendation more effectively and efficiently.

[*] Project supported by the State Key Development Program for Basic Research of China (Grant No. 2011CB302200-G), the National Natural Science Foundation of China under Grant No. 61370074, 61100026.

F. Li et al. (Eds.): WAIM 2014, LNCS 8485, pp. 96–99, 2014.

However, it is difficult to fuse multimodal data. Firstly, computing the similarity of objects from different modalities is very hard. Different modalities have different features, and most feature spaces are not compatible. Secondly, multimodal data is often in big size, and finding an efficient fusion method is also tough.

In this paper, we propose an approach of multimodal data fusion, which (1) creates a text-image heterogeneous graph to represent the semantic relationship of texts and images, (2) fuses the graph based on graph clustering, and (3) applies the fusing results to a new social media recommendation strategy. Fig.1 shows our framework.

The rest of the paper is organized as follows. Section 2 describes the heterogeneous graph for multimodal data. The method of multimodal data fusion is proposed in Section 3. Section 4 shows the experiments. We conclude our work in Section 5.

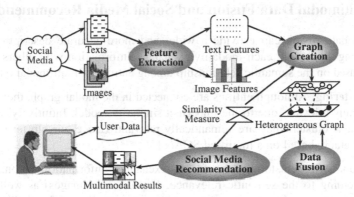

Fig. 1. Social media recommendation based on multimodal data fusion in text-image

2   Heterogeneous Graph Construction

A heterogeneous graph $G=(V, E, W)$ is an undirected weighted graph. Each vertex $v \in V$ is an entity of multimodal data. Each link $e \in E$ is a relationship between entities, denoted by its weight $w \in W$. Our heterogeneous graph consists of two modalities: text (white) and image (black), is shown in Fig.1. Although the heterogeneous graph can effectively model the relationship of texts and images, it is very difficult. First, it is a tough job to calculate the similarity of entities in a large data set. Second, due to the incompatible feature spaces of texts and images, it is hard to calculate their semantic relevance. The following details the creating of each part of our heterogeneous graph.

Text Modal-Graph. Text modal-graph keeps the semantic relationship of texts. For the text of a microblog tells stories by words, each text is represented as a bag of words detected from it. Different datasets have different characteristics and require different similarity measurements. As previous work [1] confirmed the effectiveness of Boolean model for microblog texts, words in the bag only have boolean value. Thus, the text similarity is the Jaccard coefficient of word bags. To speed up, an inverted index of words is created to point out which texts share same words.

Image Modal-Graph. Image modal-graph keeps the content-based relationship of images. First of all, we extract the SIFT (Scale-Invariant Feature Transform) features [2] as image features. To compute image similarity, many works use bag of visual words, but obtaining discriminative visual words in a large data set is very hard [3]. Here, the image similarity is the Jaccard coefficient of SIFT features. To speed up, images with similar color histograms are grouped together, using p-stable LSH [4].

Text-Image Similarity Graph. Text-Image similarity graph keeps the semantic relationship between text and image entities. Generally, if entities belong to the same document, they must be semantically relevant. Due to the incompatible feature spaces of different modalities, the cross-modality similarity is based on the Boolean model of co-occurrence. Thus, if two entities from different modalities are semantically relevant, their weight w is 1; otherwise w is 0.

3   Multimodal Data Fusion and Social Media Recommendation

Multimodal data fusion can provide a wider and more accurate description of the object. During the fusion, each modality exchanges information with others. The exchange is based on the semantic relationship among entities of multimodal data.

Graph Clustering. Although entities are connected in the modal-graph, they may not be semantically relevant, due to the imperfect similarity model. Intuitively, if entities are in the same cluster, they are semantically relevant. Hence, we cluster entities in each modal-graph, based on a modified SCAN [5].

Data Fusion. After clustering, the fusion exchanges information among modal-graphs, according to the semantic relevance. If two texts (images) as well as their corresponding images (texts) are semantically relevant, their relationship will be enhanced. If two texts (images) as well as their corresponding images (texts) are semantically irrelevant, their relationship will be weakened.

Social Media Recommendation. Based on the clustering and fusing result, the recommendation is treated as a process of walking on the heterogeneous graph. Generally, if there are many paths between a target and the query, the target is likely semantically relevant to the query. To obtain candidate targets, the method starts with the query entities, and walks along the link in the graph. The candidate targets with large weights are returned as the recommended results for the query.

4   Experiment

We download microblogs from Tencent website (t.qq.com). The microblogs contains 1.2 million texts and 0.4 million images, over April 7~13, 2012. As people prefer up-to-date news, we create the heterogeneous graph on a per day basis.

First, we evaluate the efficiency of heterogeneous graph construction and multimodal data fusion. The heterogeneous graph consists of text modal-graph, image modal-graph and text-image similarity graph. To create modal-graphs, it takes 6h on 170k texts and 8h on 57k images. Creating text-image similarity graph takes only 40s. Though it is time-consuming to create the graph, it is an offline job. Multimodal data fusion includes graph clustering and data fusion. It costs 1.5s to clustering images and 200s to clustering texts. The time cost of data fusion is 150s.

Second, we evaluate the effectiveness of the recommendation for texts and images, based on AP (average precision). The recommendation based on text modal-graph, image modal-graph, and heterogeneous graph are denoted as T-Re, I-Re and Fu-Re respectively. The baseline for text recommendation is the Lucene and for image recommendation is the SMH (Sim-Min-Hash) [6]. Fig.2 shows that the Fu-Re is suited for the big size of text recommendations, and the T-Re is suited for the middle size. Fig.3 shows that the Fu-Re is good at image recommendation. We also compare the Fu-Re with the QRCs (Query-Relative Classifiers) on five groups in [7], based on MAP. The advantage on ALL groups in Fig.4 shows the effectiveness of the Fu-Re.

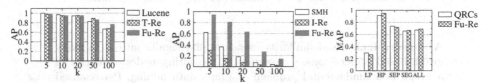

Fig. 2. Text recommendation   **Fig. 3.** Image recommendation   **Fig. 4.** QRCs vs Fu-Re

5   Conclusion and Future Works

To provide more effective social media recommendation results, in this paper we represent the semantic relationship among entities of multimodal data by a heterogeneous graph, and propose a multimodal data fusion algorithm to refine the recommendation result based on graph clustering. As the heterogeneous graph is based on entities of different modalities, our algorithm is insensitive to the modality. The proposed multimodal data fusion based on graph clustering in the heterogeneous graph can improve the recommendation efficiently and effectively. The experimental results validate the advantages of the proposed approach over the baseline.

References

1. Massoudi, K., Tsagkias, M., de Rijke, M., Weerkamp, W.: Incorporating query expansion and quality indicators in searching microblog posts. In: Clough, P., Foley, C., Gurrin, C., Jones, G.J.F., Kraaij, W., Lee, H., Mudoch, V. (eds.) ECIR 2011. LNCS, vol. 6611, pp. 362–367. Springer, Heidelberg (2011)
2. Lowe, D.G.: Distinctive image features from scale-invariant keypoints. IJCV 60, 91–110 (2004)
3. Ling, H., Yan, L., Zou, F., et al.: Fast image copy detection approach based on local fingerprint defined visual words. Signal Process 93, 2328–2338 (2013)
4. Datar, M., Indyk, P., Immorlica, N., et al.: Locality-sensitive hashing scheme based on p-stable distributions. In: 20th SCG, pp. 253–262. ACM, NY (2004)
5. Xu, X., Yuruk, N., Feng, Z., et al.: SCAN: A structural clustering algorithm for networks. In: 13th SIGKDD, pp. 824–833. ACM, NY (2007)
6. Zhao, W.-L., Jegou, H., Gravier, G.: Sim-Min-Hash: An efficient matching technique for linking large image collections. In: 21st MM, pp. 577–580. ACM, NY (2013)
7. Krapac, J., Allan, M., Verbeek, J., et al.: Improving web image search results using query-relative classifiers. In: CVPR, pp. 1094–1101. IEEE (2010)

Accuracy Estimation of Link-Based Similarity Measures and Its Application*

Yinglong Zhang[1,2], Cuiping Li[1], Chengwang Xie[2], and Hong Chen[1]

[1] Key Lab of Data Engineering and Knowledge Engineering of MOE,
and Department of Computer Science, Renmin University of China, China
`zhang_yinglong@126.com`, `cuiping_li@263.net`
[2] School of Software, East China Jiaotong University, China

Abstract. Link-based similarity measures play significant role in many graph based applications. Consequently, measuring nodes similarity in a graph is a fundamental problem of graph data mining. Personalized PageRank (PPR) and SimRank (SR) have emerged as the most popular and influential link-based similarity measures. In practice, PPR and SR scores are achieved by iterative computing. With increasing of iterations, the computations incur heavy overhead. The ideal solution is that computing similarity within the minimum number of iterations is sufficient to guarantee a desired accuracy. However, the existing upper bounds are too coarse to be useful in general. Therefore, we focus on designing accurate and tight upper bounds of PPR and SR in the paper. Our upper bounds are designed based on following human intuition: "the smaller the difference between the two consecutive iteration step results is, the smaller the difference between iterative similarity scores and theoretical ones is". Furthermore, we demonstrate effectiveness of our novel upper bounds in the scenario of top-k similar nodes query, where our upper bounds accelerate speed of the query. At last, we run a comprehensive set of experiments on real data sets to verify effectiveness and efficiency of our upper bounds

Keywords: Personalized PageRank, SimRank, Upper bound, Difference.

1 Introduction

In Web Age, graphs are ubiquitous, such as the web, social networks, bibliographic graphs and entity-relationship graphs, calling for solutions to measure similarity between nodes on graphs. Measures of similarity between objects play significant role in many graph based applications: recommendation systems [3], link prediction [9], fraud detection [7], and collaborative filtering [1].

* This work was supported by National Basic Research Program of China (973 Program)(No. 2012CB316205), NSFC under the grant No.61272137, 61033010, 61202114, 61165004 and NSSFC (No: 12&ZD220). It was partially done when the authors worked in SA Center for Big Data Research in RUC. This Center is funded by a Chinese National 111 Project Attracting.

F. Li et al. (Eds.): WAIM 2014, LNCS 8485, pp. 100–112, 2014.

Many link-based similarity measures have been proposed, such as Personalized PageRank (PPR) [5], SimRank (SR) [4], Hitting time [13] and Commute time [12]. Among them, both PPR and SR have emerged as the most popular and influential link-based similarity measures due to their effectiveness and solid theoretical foundation.

Although iterative similarity scores of SR and PPR are convergent [4][5], in practice the corresponding computations naturally involve performing a finite number of iterations. SR and PPR computations are time-consuming. With increasing of iterations, the computations incur significant overhead, especially on a large graph. In [10], given a graph which consists of 10,000 nodes, Lizorkin et al run the original iterative SR on a 2.1GHz Intel Pentium processor with 1Gb RAM. After 5 iterations, it took 46 hours and 5 minutes for the algorithm to obtain all node pairs similarities.

However, the existing upper bounds are too coarse to be useful in general. It is advised that choosing the decay factor value c = 0.8 and the total number of iterations K = 5 to compute iterative SR similarity in research [4], and, according to proposition 1 in [10], the corresponding difference between theoretical and computed similarity scores is 0.26 . Based on the lemma 2 in [14], the difference between theoretical and iterative PPR scores is also 0.26 when c = 0.8 and the total number of iterations K = 5. Obviously, the existing upper bounds of PPR and SR are relatively large because the interval of the theoretical scores is [0, 1].

Accordingly, an accurate difference between iterative similarity scores and theoretical ones remains an open question. The ideal solution is that computing similarity within the minimum number of iterations is sufficient to guarantee a desired accuracy.

At ith iteration, if the iterative score is P_i and the difference between theoretical and computed similarity score is $P - P_i \leq \delta_i$, then the corresponding upper bound is $P_i + \delta_i$, and vice versa. Based on this relationship, for convenience we abuse terminology: the difference and upper bound are considered to be same in the paper. The intended meaning should be clear from the context.

In summary, it is significance that designing accurate and tight upper bounds in theory and practice. Given a desired accuracy, we terminate the iteration as soon as possible to save overhead by leveraging tight upper bound. Furthermore we can accelerate graph-based query such as link-based similarity join [14] and top-k similarity search by utilizing tight upper bound.

We say that difference between iterative similarity scores and theoretical ones is good if the following properties hold: 1. Accurate: the difference is very close to the true difference and is not coarse. 2. Fast: we can efficiently obtain the difference. Our differences are accurate and can be efficiently obtained.

In the paper we focus on designing accurate and tight upper bounds of PPR and SR (sect. 3). Our upper bounds are designed based on following human intuition: "the smaller the difference between the two consecutive iteration step results is, the smaller the difference between iterative similarity scores and theoretical ones is". We can efficiently obtain the bounds without spending extra overhead. Furthermore, we tailor our upper bounds to accelerate the top-k similar nodes query (sect. 4). At the experiments, we show that our upper bounds significantly outperform the state-of-the-art upper bounds (sect. 5).

2   Preliminary

Given a directed graph $G = (V, E)$ where nodes in V represent objects and edges in E represent relationships between objects. For any $v \in V$, Set $I(v)$ and $O(v)$ denote in-neighbors and out-neighbors of v, respectively. $I_i(v)$ or $O_j(v)$ is an individual member of $I(v)$, for $1 \leq i \leq |I(v)|$, or of $O(v)$, for $1 \leq j \leq |O(v)|$.

Like PageRank, PPR is the steady-state probabilities of random walks; at each step, a surfer random walk along out link with probability c, and with probability 1-c jump back to the random node of the set of preferred nodes. If the preferred set contains only one node, PPR actually is RWR. RWR is a special case of PPR. In the paper we only consider the situation that the preferred set contains one node (query node).

According to [5][14], the equation of PPR is

$$r(q, v) = (1 - c) \sum_{\tau : q \to v} P(\tau) c^{l(\tau)} \tag{1}$$

where τ is the unidirectional path from q to v: $(q, w_1, \ldots, w_n, v)$, $l(\tau)$ is the length of the path τ, $P(\tau) = \frac{1}{|O(q)|} \prod_{i=1}^{n} \frac{1}{|O(w_i)|}$ is the probability of traversing the τ. $r(q, v)$ is the similarity between q and v from the q's personalized view. In practice

$$r_k(q, v) = (1 - c) \sum_{\substack{\tau : q \sim v \\ l(\tau) \leq k}} P(\tau) c^{l(\tau)} \tag{2}$$

is used to estimate $r(q, v)$.

SR measures similarity of nodes is based on following human intuition: "two objects are similar if they are related to similar objects" [4]. So SR score (a, b) actually is the average SR score between in-neighbors of a and in-neighbors of b:

$$s(a, b) = \begin{cases} 1, & \text{if } a = b \\ \frac{c \sum_i^{|I(a)|} \sum_j^{|I(b)|} s(I_i(a), I_j(b))}{|I(a)||I(b)|}, & I(a) \text{ and } I(b) \neq \emptyset \\ 0 & \text{otherwise} \end{cases} \tag{3}$$

Correspondingly, as showed in [4] the iterative formula is:

$$s_{k+1}(a, b) = \frac{c}{|I(a)||I(b)|} \sum_i^{|I(a)|} \sum_j^{|I(b)|} s_k(I_i(a), I_j(b)) \tag{4}$$

$(k = 0, 1, 2, \ldots)$.

From the view of random surfer model, the SR score measures how soon two random surfers are expected to meet at the same node if they started at nodes a and b and randomly walked the graph backwards [4]. According to [4], the formula of SR can be written as follows:

$$s(a, b) = \sum_{\tau : (a, b) \to (x, x)} P(\tau) c^{l(\tau)} \tag{5}$$

where τ is a tour (paths may have cycles) along which two random suffers walk backwards starting at nodes a and b respectively until they first and only first meet at any node x , $l(\tau)$ is the length of tour τ.

Based on [16], the corresponding iterative formula is:

$$s_k(a,b) = \sum_{\substack{\tau:(a,b)\to(x,x)\\ l(\tau)\leq k}} P(\tau)c^{l(\tau)} \quad .$$
(6)

3   Bounds of PPR and SR

The existing upper bounds of PPR and SR are too coarse to be useful in general. In this section we introduce novel upper bounds of PPR and SR.

3.1   Bounding of PPR

By walking one step beforehand from node q, the formula (1) can be transformed as:

$$r(q,v) = (1-c) \sum_{\tau:q\to v} P(\tau)c^{l(\tau)} = \frac{(1-c)c}{|O(q)|} \sum_{i=1}^{|O(q)|} \sum_{\tau':O_i(q)\to v} P(\tau')c^{l(\tau')}$$

$$= \frac{c}{|O(q)|} \sum_{i=1}^{|O(q)|} r(O_i(q),v).$$

Similar, we have:

$$r_{k+1}(q,v) = \frac{c}{|O(q)|} \sum_{i=1}^{|O(q)|} r_k(O_i(q),v) \quad .$$
(7)

At mth iteration, let $\delta_m = MAX_{\forall a\in V, b\in V}\{(r_m(a,b) - r_{m-1}(a,b))\}$, we have following theorem:

Theorem 1. *the difference between theoretical and iterative PPR scores is*

$$r(q,v) - r_m(q,v) \leq \delta_m \frac{c}{1-c} \quad .$$
(8)

Proof. According to (2), $r_{m+1}(q,v) - r_m(q,v) = (1-c)\sum_{\substack{t:q\sim v\\ l(t)=m+1}} P(\tau)c^{l(\tau)}$.
Thus

$$r(q,v) - r_m(q,v) = (1-c) \sum_{\substack{t:q\sim v\\ l(t)=m+1}}^{\infty} P(\tau)c^{l(\tau)} = \sum_{i=1}^{\infty}(r_{m+i}(q,v) - r_{m+i-1}(q,v)) \quad .$$

Based on (7) and $\delta_m = MAX_{a\in V, b\in V}\{(r_m(a,b) - r_{m-1}(a,b))\}, \forall a,b \in V$

$$r_{m+1}(a,b) - r_m(a,b) = \frac{c}{|O(a)|} \sum_{i=1}^{|O(a)|} (r_m(O_i(a),v) - r_{m-1}(O_i(a),v)) \leq \frac{c}{|O(a)|} \sum_{i=1}^{|O(a)|} \delta_m$$

$$= c\delta_m \quad .$$

Likewise, $r_{m+k}(a,b) - r_{m+k-1}(a,b) \leq c^k \delta_m$. Therefore

$$r(q,v) - r_m(q,v) = \sum_{i=1}^{\infty}(r_{m+i}(q,v) - r_{m+i-1}(q,v)) \leq \sum_{i=1}^{\infty} c^i \delta_m = \delta_m \frac{c - c^{\infty}}{1 - c} = \delta_m \frac{c}{1-c} \quad .$$

□

Theorem 1 gives the lower and upper bounds of PPR at mth iteration: $r_m(q,v) \leq r(q,v) \leq r_m(q,v) + \delta_m \frac{c}{1-c}$. At $(m+k)$th iteration, we update δ_{m+k} as follows $\delta_{m+k} = MIN(MAX_{a\in V, b\in V}\{(r_{m+k}(a,b) - r_{m+k-1}(a,b))\}, \delta_{m+k-1})$.

The lemma 2 in [14] gives an upper bound of PPR, c^{m+1}, at mth iteration. The following proposition states our upper bound is better than that in [14].

Proposition 1. *At mth iteration, $\delta_m \frac{c}{1-c} \leq c^{m+1}$.*

Proof.

$$\delta_m \leq MAX_{a\in V, b\in V}\{(r_m(a,b) - r_{m-1}(a,b))\} = MAX_{a\in V, b\in V}\{(1-c)\sum_{\substack{t:a\sim b \\ l(t)=m}} P(\tau)c^{l(t)}\}$$

$$\leq (1-c)c^m$$

because $\sum_{\substack{t:a\sim b \\ l(t)=m}} P(\tau) \leq 1$.   □

Proposition 1 guarantees that our upper bound is superior to that in [14] in theory. Obviously, our upper bound decreases drastically with increasing of iterations due to $\delta_m \frac{c}{1-c} \leq c^{m+1}$.

3.2   Bounding of SR

At mth iteration, let $\delta_m = MAX_{\forall a\in V, b\in V}\{s_m(a,b) - s_{m-1}(a,b)\}$, we have following theorem:

Theorem 2. *the difference between theoretical and iterative SR scores is*

$$s(a,b) - s_m(a,b) \leq \delta_m \frac{c}{1-c} \quad . \tag{9}$$

Proof. Based on e.q.(4),

$$S_{m+1}(a,b) - S_m(a,b) = \frac{c}{|I(a)||I(b)|}\sum_{i=1}^{|I(a)|}\sum_{j=1}^{|I(b)|}(S_m(I_i(a), I_j(b)) - S_{m-1}(I_i(a), I_j(b)))$$

$$\leq \frac{c}{|I(a)||I(b)|}\sum_{i=1}^{|I(a)|}\sum_{j=1}^{|I(b)|}\delta_m = c\delta_m$$

as $\delta_m = max\{s_m(a,b) - s_{m-1}(a,b)\}$ $(\forall a,b \in V)$. Accordingly,

$$S_{m+2}(a,b) - S_{m+1}(a,b) = \frac{c}{|I(a)||I(b)|}\sum_{i=1}^{|I(a)|}\sum_{j=1}^{|I(b)|}(S_{m+1}(I_i(a), I_j(b)) - S_m(I_i(a), I_j(b)))$$

$$\leq \frac{c}{|I(a)||I(b)|}\sum_{i=1}^{|I(a)|}\sum_{j=1}^{|I(b)|}c\delta_m = c^2\delta_m$$

likewise: $s_{m+k}(a,b) - s_{m+k-1}(a,b) \leq c^k \delta_m$

On the other hand, according to e.q.(6)(5), $s_{m+1}(a,b) - s_m(a,b)$
$= \sum\limits_{\substack{\tau:(a,b)\to(x,x)\\ l(\tau)=m+1}} P(\tau)c^{l(\tau)}$, and,

$$s(a,b) - s_m(a,b) = \sum_{\substack{\tau:(a,b)\to(x,x)\\ l(\tau)=m+1}}^{\infty} P(\tau)c^{l(\tau)} = \sum_{k=1}^{\infty} s_{m+k}(a,b) - s_{m+k-1}(a,b)$$

$$= \sum_{k=1}^{\infty} c^k \delta_m = \delta_m \sum_{k=1}^{\infty} \frac{c - c^{\infty}}{1-c} = \delta_m \frac{c}{1-c}$$

$\square$

Theorem 2 gives the lower and upper bounds of SR at mth iteration. As soon as the difference $\delta_m \frac{c}{1-c}$ satisfies the given precision, we can terminate iteration. When $m \geq 3$, the difference is largely less than the difference, c^{m+1}, in [10] although we can not prove it in theory. In practice, $MIN\{\delta_m \frac{c}{1-c}, c^{m+1}\}$ is SR difference in order to obtain better result.

3.3 Obtain Upper Bounds

We do not need to spend extra overhead to obtain our upper bounds by incremental updating similarity scores.

E.q. (2) can be written as

$$r_{k+1}(q,v) = (1-c) \sum_{\substack{\tau:q\sim v\\ l(\tau)\leq k+1}} P(\tau)c^{l(\tau)} = r_k(q,v) + (1-c) \sum_{\substack{\tau:q\sim v\\ l(\tau)=k+1}} P(\tau)c^{l(\tau)} \quad . \quad (10)$$

Likewise, e.q. (6) can be transformed as

$$s_{k+1}(a,b) = s_k(a,b) + \sum_{\substack{\tau:(a,b)\to(x,x)\\ l(\tau)=k+1}} P(\tau)c^{l(\tau)} \quad . \quad (11)$$

The above two equations say that the current similarity score is the sum of previous score and the increment. At each iteration we actually compute the increment to obtain similarity score. And the δ_m in e.q.(8) (or e.q.(9)) is the maximal increment. Optimization computation of PPR in [18] and SR in [16] are based on e.q. (10) and e.q. (11) respectively. Consequently, we do not need to spend extra overhead to obtain upper bounds.

4   Top-k Similar Nodes Query

By leveraging tight upper bound, we terminate the iteration as soon as possible to reduce overhead. Furthermore, we also can accelerate graph-based query by utilizing more tight upper bound. In this section, we demonstrate effectiveness of our novel upper bounds in the scenario of top-k similar nodes query.

In the paper, P denote as PPR or SR similarity score. Observe from e.q.(2) and (5) that P(a,b) involves an infinite number of random walks. Consequently, it is infeasible to achieve accurate P(a,b). It is effective to compute $P_w(a,b)$ instead:

$$|P(a,b) - P_w(a,b)| \leq \epsilon \qquad (12)$$

where ϵ controls the accuracy of $P_w(a,b)$ in estimating $P(a,b)$, and w is the minimum value that satisfies the inequation.

Algorithm 1. Top-k similar nodes query

---

    input : Graph $g,c,q,\ k,\ \varepsilon$
    output: top-k lists of q

1   Set $pathProb \leftarrow 1.0$;
2   push pair $(v, pathProb)$ into queue que;
3   **while** $!obtained$ **do**
4      $(currentNode, pathProb) \leftarrow que.front()$;
5      $que.pop()$;
6   **if** $currentNode\ \ != -1$ **then**
7   **foreach** a neighbors j of $currentNode$ **do**
8             $queTemp[j] \leftarrow queTemp[j] + \frac{pathProb}{OutDegree}$;   // walk one step
9   **else**
10        $i \leftarrow i + 1$;
11   **foreach** element j of $queTemp$ **do**
            // e.q.(10):
12             $rwrScore[j] \leftarrow rwrScore[j] + queTemp[j] \times (1-c) \times c^i$;
13             $\delta \leftarrow \delta + queTemp[j] \times (1-c) \times c^i$;
14   push $(j, queTemp[j])$ into que;
15        $\varepsilon' \leftarrow \delta/queTemp.size()$;   // average local difference
16        $\varepsilon' \leftarrow \varepsilon' \times c/(1-c)$;   // Upper bound e.q.(8)
17   sort $rwrScore$ to obtain top k nodes;
         // $\widehat{P(q,v)}(\forall v \in V(G(V))/T_k(q))$:
18   **if** $T_k > \widehat{P(q,v)}$ or $\varepsilon' < \varepsilon$ **then**
19          $obtained \leftarrow$ True;
20   push $(-1, pathProb)$ into que;
21   clear $queTemp$;
22        $\delta \leftarrow 0$;
23   continue;

24   return $result$;

Problem statement (Top-k similar nodes query) Given a query node q, a number k and ϵ, the result of query, *the top-k similarity nodes of q* , is $T_k(q) = \{t_1, \ldots, t_k\}$ iif similarity score $P_w(q, t_i) \geq P_w(q, t)$ $(\forall t \in V(G(V))/T_k(q))$ on the graph $G(V)$.

The general framework of top-k similar nodes query. Starting at a query node q, we do a breadth-first traverse to visit remaining nodes. At mth iteration, when a node v is visited, we compute $P_m(q, v)$. After mth iteration, we obtain its upper bound value $\widehat{P_m(q, v)}$ and ϵ_m (ϵ_m is the difference between theoretical and iterative similarity scores at current iteration), then find a set of k nodes with the highest scores of lower bounds. Let T_k be the kth largest score. We terminate the query and obtain the final result of the top-k query if one of following conditions is true:

- $\epsilon_m \leq \epsilon$
- $T_k \geq \widehat{P_m(q, t)}$ $(\forall t \in V(G(V))/T_k(q))$.

With help of upper bound, we can obtain top-k nodes via local expansion around the query node. The local top-k similarity search method avoids accessing the whole graph.

The local top-k similar nodes search method effectively handle similarity search because it does not need to access the whole graph, especially, when the graph is very large. While, the upper bounds, discussed in section 3, are obtained based on the whole global information on graphs. Therefore, the upper bounds can not directly be applied to the top-k query when the only local information of the query node is available. Furthermore, we tailor upper bounds based on theory mentioned in section 3.

We customize upper bounds based on the local information. At each iteration, we obtain the similarity scores between the query node and accessed nodes, which belong to the neighborhood of the query node. The average difference between the two consecutive iteration step results can be used to estimate δ_m in e.q.(8)(9). As the result, we obtain variants of the upper bounds based on the local information. Although the variants of the upper bounds can not be proved to be accurate in theory, they are reasonable: the average difference is very close to the true difference due to principle of locality.

Algorithm 1 is top-k similar nodes query method based on PPR. Top-k method based on SR is similar to algorithm 1 and is not listed in the paper.

It is worth to mention that top-k query is exploited to demonstrate effectiveness of the upper bounds. Top-k query algorithm is not our target in the paper, although our algorithm is concise and efficient. From above analysis, the upper bounds play a key role in the algorithm. The upper bounds accelerate the query speed and avoid accessing the whole graph.

5   Experiments

We implemented all experiments on a PC with $i3 - 550$ CPU, 4G main memory, running windows 7 operating system. All codes are written in C++.

The data sets used in the experiments are showed in table 1. Cora[1] is a citation graph. Graphs FaceBook, Hamster (social graph) and Subelj (E-road network)

---

[1] http://www.cs.umd.edu/projects/linqs/projects/lbc/index.html

Table 1. Data Sets

	Cora	FaceBook	Hamster	Subelj	P2P-Gnutella06	ego-Twitter	web-Stanford
Nodes	2,485	2,887	1,999	1,039	8,717	81,306	281,903
Edges	5,209	5,388	31,676	2,484	31,525	1,768,149	2,312,497

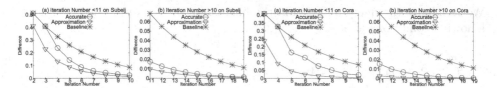

Fig. 1. Our SR difference vs. Baseline difference

can be visited at KONECT[2]. The remaining data can be visited at SNAP[3]. For the first 5 data sets in table 1, we use their maximum graph components instead of the original graphs (The corresponding informations in table 1 are that of their maximum components).

The state-of-the-art upper bounds, PPR upper bound in [14] and SR upper bound in [10], are used to be **baselines** to compare with our upper bounds in experiments.

When δ_m is estimated by the average of top 100 largest difference between the two consecutive iteration step results, the corresponding upper bounds are denoted as approximate and achieve a very high precision ($\geq 99\%$).

Figures 1-7 show the results of our accurate and approximate upper bounds compared with the baselines on 5 real data sets. When $m \geq 3$, our SR difference is largely less than the baseline although the rate of SR convergence is different on these different real data sets. Our approximate PPR upper bound is superior to the baseline. And our accurate PPR difference is less than the baseline in some data sets. In worst case, our accurate PPR difference equals the baseline.

Then we test efficiency of our bounds in the scenario of top-k node query on two large real data (figures 8-9). All the top-k queries are repeated 200 times and the reported values are the average values. Our SR bound is 1.5-2.3 times faster than the baseline while it achieves a high precision ($\geq 96\%$). Our PPR bound is 200-400 times faster than the baseline while it achieves a very high precision ($\geq 97\%$).

In a nut shell, our bounds significantly outperform the state-of-the-art upper bounds.

6   Related Works

Recently, the link-based similarity measures attract the attention of many researchers. The related works are outlined as follows:

---

[2] http://konect.uni-koblenz.de/networks/
[3] http://snap.stanford.edu/data/index.html

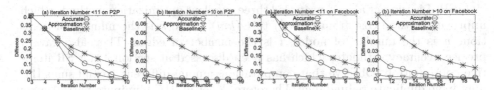

Fig. 2. Our SR difference vs. Baseline difference

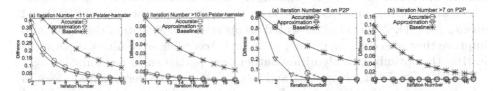

Fig. 3. Our SR difference vs. Baseline difference

Fig. 4. Our SR difference vs. Baseline difference

Fig. 5. Our PPR difference vs. Baseline difference

Fig. 6. Our PPR difference vs. Baseline difference

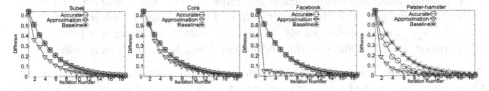

Fig. 7. Our PPR difference vs. Baseline difference

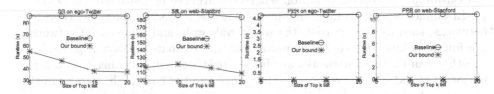

Fig. 8. Runtime of Top-k Query

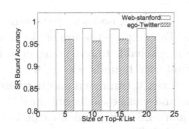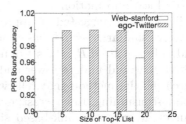

Fig. 9. Accuracy of Top-k Query

Optimization computation. The PPR and SR involve an infinite number of random walks. The nature incurs heavy overhead of computation.

1. SR. Lizorkin et al. [11] proposed three excellent optimization methods which improve the time cost from $O(kn^4)$ to $O(knl)$ where k is the number of iteration, n is the number of nodes, l is the number of edges. They also gave a precise accuracy estimate, which has been discussed at section 1, for SR iterative computation. Observe that computations among different partial sums [11] may have duplicate redundancy. Therefore, Yu et al. [15] eliminate partial sums redundancy by using an adaptive clustering strategy. Yu et al. also proposed a variant of SR and give a corresponding difference between theoretical and iterative scores [15]. In contrast, we focus on the upper bound of original SR. Based on e.q. (6) (11), Zhang et al presented an optimization algorithm [16] which also improve the time cost from $O(kn^4)$ to $O(knl)$. According to results of experiment in [16], the optimization algorithm outperforms partial sums method.

2. PPR. For PPR, computing and storing all possible personalized views in advance is impractical [5]. Jeh et al. suggested a scalable solution for PPR based on the observation that PPR vectors actually is a linear combination of basis vectors and considering hub-pivoted paths that pass through some important "hub" nodes [5]. Based on e.q. (10), Zhu et al proposed FastPPV, an approximate PPV computation algorithm that is incremental [18]. They proposed L1 error to control accuracy of PPR at query time. The L1 error is defined as follows $\phi^k = 1 - \sum_{p \in V} r_k(q, p)$. Obviously, L1 error measures overall error, while we give PPR upper bound of any specific node pairs. They both are totally different.

Application of link-based Measure. By utilizing upper/lower relevance estimations, the speed of the query, computing top-k relevant nodes w.r.t a query node, can be accelerated [2,8].

Considering a general situation where the average in/out-degree is D (D $\geq$ 1), Li defined a new average SR upper bound as a function of D [8]. However, networks, such as the Internet, the world wide web, and some social networks, are found to have degree distributions that approximately follow a power law. In other words, these networks are highly right-skewed, meaning that a large majority of nodes have low degree but a small number have high degree. On the other hand, the top-k similarity search method only access the local neighborhood of the query node. Therefore, the average SR upper bound does not reflect the true local information.

Fujiwara et al suggested an approach to find the top-k nodes so as to support interactive similarity search based on PPR [2]. To compute the upper similarity bound, They utilized R_i, the set of nodes that is reachable to any nodes in S_i for which they would update lower and upper similarity bounds. However, due to research [6], the method that tells whether a vertex u can reach another vertex v is time-consuming. In contrast, our upper bound can be easily obtained.

Sun et al proposed link-based similarity join (LS-join), which extends the similarity join operator to link-based measures [14]. They accelerated the speed of the join query by utilizing upper bounds of PPR and SR. The upper bounds in [14] are used to be baseline to compare with our upper bounds .

Zheng et al proposed an estimated shortest-path distance based upper bound for SR [17]. However, it is also expensive to compute the shortest path between two vertices on the fly. Furthermore, as with the upper bound in [10], the shortest-path distance based upper bound is also coarse.

7 Conclusion

We proposed upper bounds of PPR and SR, which are based on following human intuition: "the smaller the difference between the two consecutive iteration step results is, the smaller the difference between iterative similarity scores and theoretical ones is". Our upper bounds are accurate and can easily be achieved. Furthermore, we customized our bounds to accelerate top-k similar nodes query. At last, we showed that our upper bounds significantly outperform the state-of-the-art upper bounds.

References

1. Antonellis, I., Molina, H.G., Chang, C.C.: Simrank++: query rewriting through link analysis of the click graph. In: Proc. VLDB Endow., vol. 1(1), pp. 408–421 (2008)
2. Fujiwara, Y., Nakatsuji, M., Shiokawa, H., Mishima, T., Onizuka, M.: Efficient ad-hoc search for personalized pagerank. In: SIGMOD 2013, pp. 445–456 (2013)
3. Gupta, P., Goel, A., Lin, J., Sharma, A., Wang, D., Zadeh, R.: Wtf: the who to follow service at twitter. In: WWW 2013, pp. 505–514 (2013)
4. Jeh, G., Widom, J.: Simrank: a measure of structural-context similarity. In: KDD, pp. 538 543 (2002)
5. Jeh, G., Widom, J.: Scaling personalized web search. In: WWW, pp. 271–279 (2003)
6. Jin, R., Ruan, N., Xiang, Y., Wang, H.: Path-tree: An efficient reachability indexing scheme for large directed graphs. ACM Trans. Database Syst. 7, 1–7 (2011)
7. Joshi, A., Kumar, R., Reed, B., Tomkins, A.: Anchor-based proximity measures. In: WWW 2007, pp. 1131–1132 (2007)
8. Lee, P., Lakshmanan, L.V.S., Yu, J.X.: On top-k structural similarity search. In: ICDE, pp. 774–785 (2012)
9. Liben-Nowell, D., Kleinberg, J.: The link-prediction problem for social networks. J. Am. Soc. Inf. Sci. Technol. 58(7), 1019–1031 (2007)
10. Lizorkin, D., Velikhov, P., Grinev, M.N., Turdakov, D.: Accuracy estimate and optimization techniques for simrank computation. In: PVLDB, vol. 1(1), pp. 422–433 (2008)
11. Lizorkin, D., Velikhov, P., Grinev, M.N., Turdakov, D.: Accuracy estimate and optimization techniques for simrank computation. VLDB J. 19(1), 45–66 (2010)
12. Sarkar, P., Moore, A.W.: A tractable approach to finding closest truncated-commute-time neighbors in large graphs. In: UAI, pp. 335–343 (2007)
13. Sarkar, P., Moore, A.W., Prakash, A.: Fast incremental proximity search in large graphs. In: ICML, pp. 896–903 (2008)
14. Sun, L., Cheng, R., Li, X., Cheung, D.W., Han, J.: On link-based similarity join. In: PVLDB, vol. 4(11), pp. 714–725 (2011)
15. Yu, W., Lin, X., Zhang, W.: Towards efficient simrank computation on large networks. In: ICDE 2013, pp. 601–612 (2013)

16. Zhang, Y., Li, C., Chen, H., Sheng, L.: Fast simRank computation over disk-resident graphs. In: Meng, W., Feng, L., Bressan, S., Winiwarter, W., Song, W. (eds.) DASFAA 2013, Part II. LNCS, vol. 7826, pp. 16–30. Springer, Heidelberg (2013)
17. Zheng, W., Zou, L., Feng, Y., Chen, L., Zhao, D.: Efficient simrank-based similarity join over large graphs. In: PVLDB, vol. 6(7), pp. 493–504 (2013)
18. Zhu, F., Fang, Y., Chang, K.C.-C., Ying, J.: Incremental and accuracy-aware personalized pagerank through scheduled approximation. In: PVLDB, vol. 6(6), pp. 481–492 (2013)

An Efficient Influence Maximization Algorithm
to Discover Influential Users in Micro-blog

Qian Ma and Jun Ma

School of Computer Science & Technology, Shandong University, Jinan, 250101, China
maqiansdu@gmail.com, majun@sdu.edu.cn

Abstract. Micro-blog, as an emerging social network platform, provides good opportunities for viral marketing. For advertisers, an important issue is how to find a small subset of influential users which is called seed set (SS) in social network that can maximize the spread of influence. This problem is considered as "influence maximization". For advertisers with a limited budget, finding SS quickly and making the spread maximization are both important. To achieve these two goals, we propose the Candidates-Based Greedy (CBG) algorithm. Our approach is composed of two parts: a) for a given size of SS k, all the users are ranked by heuristic methods and the top-N ($N >= k$) of them who have good spread ability are selected as candidates; b) select SS from the candidates with a greedy algorithm to maximize the influence. In this way the nodes participating in the seed selection of the greedy algorithm are reduced obviously and only the important nodes are reserved, so that the running time is greatly reduced without affecting the accuracy. Our experimental results demonstrate that, comparing with StaticGreedyCELF which is a very efficient greedy algorithm, our algorithm achieves much better running time in micro-blog, almost 70% less, and does not lose any accuracy.

Keywords: influence maximization, viral marketing, micro-blog, independent cascade model.

1 Introduction

Viral marketing or word-of-mouth has been acknowledged as an effective marketing strategy. The development of online social networks, such as Facebook, Twitter, micro-blog, provide new opportunities for companies who want to popularize their products or brands through the powerful word-of-mouth effect online. However, the key issue behind the applications is that how to find the seed set that can make the influence maximization in the social network eventually?

There are a lot of works in influence maximization on social networks [1-8]. One direction is proposing greedy algorithms which have guaranteed accuracy but can't handle the large-scale networks. The other direction is finding heuristic methods. Heuristic methods are efficient enough to handle the large-scale networks but they can't guarantee the accuracy.

F. Li et al. (Eds.): WAIM 2014, LNCS 8485, pp. 113–124, 2014.
© Springer International Publishing Switzerland 2014

Incorporating the advantages of the greedy algorithms and the heuristic methods, we propose an algorithm for influence maximization in micro-blog network named Candidates-Based Greedy algorithm. Micro-blog is a scale-free network and only a few users have many followers [9]. We can infer only a small part of the users in the network have good spread ability, and most of them will not help much in influence maximization. So there is no need to get all the nodes involved in the seed selection of the greedy algorithm. We choose the seeds from some important nodes which are selected firstly. In this way, a lot of time will be saved. These important nodes, called candidates here, are firstly selected with low complexity and effective heuristic methods, such as degree, closeness, betweenness, k-shell. Considering some important nodes may have intimate relationships with each other, and selecting some separate nodes as candidates may have better influence spread result, we also try community-based candidate selection method. In the process of seed selection from candidates, we choose StaticGreedy as the greedy algorithm. The contributions of this paper are summarized as follows:

- We integrate the characteristics of micro-blog into the diffusion model to make the diffusion model more practical.
- We propose a two-stage algorithm called Candidates-Based Greedy algorithm which select top-N influential nodes as candidates with heuristic methods firstly and then choose the seeds from the candidates with a greedy algorithm. In this way we improve the efficiency of the greedy algorithm by reducing the number of nodes in each iteration.
- We compare degree centrality, betweenness centrality, closeness centrality, k-shell centrality and community-based method in candidate selection stage to find out which one has the best effect for the micro-blog network.
- We optimize CBG with CELF strategy to form CBGCELF. Our experiments show that the CBGCELF can reduce the running time observably without affecting the accuracy comparing with StaticGreedyCELF which is a very efficient greedy algorithm. For a micro-blog network with tens of thousands of nodes and millions of edges, CBGCELF only needs less than 1/3 of the StaticGreedyCELF's time to find the same number of seeds and they almost have the same spread effect.

The remainder of this paper is organized as follows. Section 2 reviews related work. Section 3 presents preliminaries. Section 4 details the Candidates-Based Greedy algorithm. Section 5 shows our experimental results. We conclude the paper in Section 6.

2   Related Work

Influence maximization is first studied as an algorithmic problem by Domingos and Richardson [1]. Kempe et al. [2] formulate the problem as discrete optimization problem and prove the optimal solution is NP-hard for most models. They present a greedy algorithm and prove that the optimal solution for influence maximization can be approximated within a factor of ($1-1/e-\varepsilon$). However, their algorithm has a serious drawback: its efficiency. After that, CELF [3], NewGreedy [4], StaticGreedy [5] have been proposed to improve the greedy algorithm's efficiency so that it can deal with large-scale network. Leskovec et al. [3] improve the greedy algorithm using the submodularity property of influence maximization to form a new algorithm called CELF which achieves as much as 700 times speedup in seed selection. Chen et al. [4] propose NewGreedy which removes the edges that will not contribute to the propagation from the original graph firstly to form a small graph. Recently Cheng et al. [5] propose the StaticGreedy algorithm which can reduce the number of Monte Carlo simulations by two orders of magnitude compared with the previous greedy algorithms without affecting the guaranteed accuracy. But there is still room for improvements for large-scale networks. There are also some works in reducing the nodes in the seed selection of the greedy algorithm. Luo et al. [6] propose to conduct the greedy algorithm on the top nodes ranked by PageRank on social networks. Chen et al. [7] and Wang et al. [8] propose community-based algorithms to reduce the nodes in seed selection and get better running time. Comparing with them, our method is simpler and more effective for large-scale networks.

Otherwise, Micro-blog is also a hot topic in recent years. Fan et al. [9] show the network has apparent small-world effect and scale-free characteristic. By analyzing the node centrality of micro-blog networks, the importance of networks' nodes and the impact on the propagation of information are discussed in [10].

3   Preliminaries

3.1   Micro-blog Network Modeling

Micro-blog network is directed and more complicated compared with other social networks. We extract the following relation from Tencent Weibo dataset and get a directed graph: a node corresponds to a user, a directed edge exits from node u to node v if v follows u, and v is called a follower, u is called a followee. We denote the graph as $G=(V, E, W)$ which represent nodes, edges and weights respectively. The "following" relation in micro-blog can be considered as showing interest in someone, and the weight represents the intensity of the interest. In this paper the weight is consist of three parts: content similarity, act intensity and mutual followees.

- **Content Similarity:** The study in [11] indicates that the "following" relations among users are related to their topical similarity. In this paper we regard the similarity on all the topics as the content similarity among users. If the tweets, retweets, comments or tags published by two users contain a lot of same keywords, we think they are like-minded. The like-mined users are more likely to influence each other. The content similarity (cs) is calculated as showed in (1). $k(v)$ denotes the keywords appeared in v's tweets, retweets, comments and tags.

$$cs(u,v) = \frac{|k(u) \cap k(v)|}{\sqrt{|k(u)| \cdot |k(v)|}} \tag{1}$$

- **Act Intensity:** Akshay et al. [12] find that people use micro-blog to seek or share information. If a user finds something interesting, he will share with his friends or make some responses through the actions provided by the micro-blog platform. If user v makes more actions on u which is followed by v than on other users v follows, we can assume that v is more interested in u and is more likely to accept the advices from u. The act intensity (ai) of u and v is defined as (2). N_v is the users that v follows. $a(v,u)$ denotes the actions (retweet, mention, @) that v performs on u.

$$ai(u,v) = \frac{|a(v,u)|}{\sum_{w \in N_v} |a(v,w)|}$$ (2)

- **Mutual Followees:** As mentioned before, the "following" relation in micro-blog can be considered as showing interest in someone. Let $f(v)$ denotes the followees of v (the users that v follows in micro-blog). The more common followees v and u have, the more likely they have common interest and they are more likely to influence each other. The mutual followees (mf) is defined in (3):

$$mf(u,v) = \frac{|f(u) \cap f(v)|}{|f(u) \cup f(v)|}$$ (3)

So the weight of edge (u, v) is defined as (4):

$$w(u,v) = cs(u,v) + ai(u,v) + mf(u,v)$$ (4)

3.2   The Diffusion Model

We adopt the Independent Cascade (IC) model. Given the network $G=(V, E, W)$, an initial set of active nodes A_0 and the propagation probability p associated with each edge in G, the independent cascade model can be described as follows: When an inactive node u becomes active in step t, it has one and only one chance to independently activate each of its inactive neighbors v with success probability $p_{u,v}$. If it succeeds, then v will be active in step $t+1$. The diffusion process continues until no further activation happens.

The model has an important parameter, propagation probability $p_{u,v}$, which denotes the probability that u influences v successfully. Most works adopt the two commonly used IC model: the UIC model with a uniform propagation probability such as $p_{u,v}= 0.01$ and the WIC model with $p_{u,v} = 1/d_v$, d_v is the indegree of v [2]. In this paper we extend the independent cascade model to accommodate the edge weight as the model adopted in [8], the propagation probability is set as:

$$p_{u,v} = w_{u,v} \cdot \rho$$ (5)

ρ is the diffusion spread of the whole network. If v has a closer relation with u, it is reasonable that v is more likely to be influenced by u.

4   Candidates-Based Algorithm

4.1   Selecting the Candidates

To select the candidates who are more likely to be the seeds, firstly we can rank the nodes with the measures such as centralities [13], PageRank [14], Hits [15], Leader-Rank [16] etc. As what we want is to reduce the number of nodes participating in the greedy algorithm, the method's efficiency is more important than its accuracy in this stage, so the network-based centralities including degree, betweenness, closeness and k-shell which are simple and efficient are adopted here. Then the top-N ($N \geq k$) influential nodes according to the ranking measure are selected as candidates. The community structure is also considered to reduce the spreading overlap when multiple seeds are selected.

- Degree Centrality:

$$C_D(v) = d_v \tag{6}$$

d_v is the number of the users who follow v.

- Betweenness Centrality: v's betweenness counts the number of shortest paths in a network that will pass v. Those nodes with high betweenness play key roles in the communication within the network [17].

$$C_B(v) = \sum_{s \neq v \neq t \in V} \frac{\sigma_{st}(v)}{\sigma_{st}} \tag{7}$$

$\sigma_{st}(v)$ is the number of the shortest paths that pass v between s and t. σ_{st} is all the number of shortest paths between s and t.

- Closeness Centrality: closeness centrality measures how close a node is to all the other nodes [17]. Those nodes with high closeness will spread information more quickly. We adopt the general form of closeness called residual closeness in [18].

$$C_c(v) = \sum_{t \in V / v} 2^{-d_G(v,t)} \tag{8}$$

$d_G(v,t)$ is the geodesic distance between v and t.

- K-shell Centrality: the k-shell centrality indentifies a node's location in the network by k-shell decomposition. The most influential spreaders are located in the core of the network [19]. In k-shell decomposition, nodes are assigned to k shells according to their remaining degree, which is obtained by successively pruning of nodes with degree smaller than the k.

- Community-Based Heuristic: When multiple seeds are selected, considering the community structure will help reduce the overlap in the spread. For example, there are two communities in Fig.1, and node 1 is selected as a seed firstly. When we select the second seed, node 5, node 10 are all good candidates according to their degree. However, node 1 and node 5 belong to the same community, most of their influence is limited in community 1. In this case, node 10 will be a better choice so that more nodes (especially nodes in community 2) will have chance to be activated.

A fast, efficient and natural-partition community detection algorithm is needed here as the micro-blog network is large and the seeds number k is not fixed. The agglomerative community detection method based on node similarity in [20] is adopted. If the community number in the network is larger than k, we reserve the largest k communities and merge the rest of them to the k large ones. Then sorting the nodes by their degree in each community and selecting the candidates with large degree from each community proportionally.

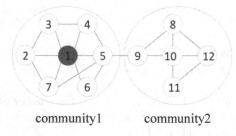

community1   community2

Fig. 1. A network with two communities

The number of the candidates is also important. It will affect the efficiency if it is too large and the algorithm will lose accuracy if it is too small. Here the number of the candidates N is defined as:

$$N = k \cdot \lambda \tag{9}$$

λ is a parameter to adjust the efficiency and accuracy.

4.2   Candidates-Based Greedy Algorithm

Algorithm 1. CBG Algorithm

Input: Graph $G = (V, E, W)$, number of total seeds k, parameter λ, global influence probability ρ, iterations number r.
Output: Seed set S
01:    $N = k \cdot \lambda$ //N is the number of the candidates
02:    Ranking the nodes according the measures in 4.1
03:    Selecting the top-N nodes as the candidate set C

```
04:     Initialize S =∅, R(S)=∅ // R(S) is the spread re-
        sult of S
05:     for i =1 to r do
06:        produce G'=(V',E') by removing each edge from G
           with probability 1−w•ρ // w is the weight of the
           edge
07:     end for
08:     for i =1 to k do
09:        set sᵥ = 0 for all v∈ C\S // sᵥ is the number of
           nodes activated by v
10:        for j =1 to r do
              for all v∈ C\S do
11:                sᵥ+=|R(G_j,S∪{v})|
              end for
12:        end for
13:        S = S∪{arg max{sᵥ / r}}
                    v∈C\S
14:     end for
15:     return S
```

The Candidates-Based Greedy algorithm can be further optimized by CELF strategy to improve its efficiency, we call it CBGCELF.

4.3   Complexity

We analyze the time complexity of CBG algorithm here. The CBG algorithm consists of two parts, lines 01-03 are the candidate selection part and lines 04-15 are the seed selection part. As the methods in candidate selection part are all efficient and just need to be computed once, it will not take much time for us to compute them in advance. So we pay more attention to the seed selection part.

Like the complexity of StaticGreedy in [5], the time complexity of our seed selection part also includes two parts. The first part which includes lines 05-07 in the algorithm is generating r subgraphs. The algorithm needs to traverse all the edges in G to produce a subgraph G', so the time complexity of this part is $O(rm)$, m is the number of edges in G. The second part includes lines 08-14. k is the number of the seeds. For each seed, the algorithm needs to visit all the subgraphs. In each subgraph G', for each node v in C, it takes $O(m')$ time to compute v' s influence s_v by a linear scan of G'. So the time complexity of the second part is $O(kNrm')$, m' is the average number of edges in G', N is the number of nodes in C. The time complexity of CBG is $O(rm+kNrm')$. [5] shows the time complexity of StaticGreedy is $O(rm+knrm')$, n is the number of nodes in G. As N is far more less than n (for example, n is tens of thousands while N is several hundred in our experiment), the time complexity of CBG is much less than StaticGreedy.

5   Experiments

5.1   Dataset

The experiments are to evaluate the performance of the proposed CBG algorithm on the micro-blog network we extract from Tencent Weibo dataset in KDD CUP 2012. We summarize some basic statistics of the network as showed in Table 1. Some other information, such as user profiles, keywords, acts, is not listed here.

Table 1. Dataset Statistics

Number of nodes	82025
Number of edges	1298473
Average out degree	15.8306
Maximal out degree	21145

In order to verify the scale-free characteristic of the micro-blog network, we show the distribution of the nodes' degree in Fig.2. It follows a power law distribution that few users have many followers and many users have few followers, which means the network in our dataset has the scale-free characteristic. Fig.3. shows the distribution of the edges' weight. Most edges' weights are very small (less than 0.1) and we normalize all the weights to 0-1. This will be helpful in the spread of influence to avoid that the propagation probability is too large to affect the seed selection.

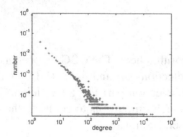

Fig. 2. Distribution of degree   **Fig. 3.** Distribution of weight

5.2   Algorithms

As the micro-blog network is large and dense, we employ the CELF optimization on CBG and StaticGreedy to ensure the running time in a reasonable range. We call them CBGCELF and StaticGreedyCELF respectively. We compare CBGCELF with the greedy algorithm—StaticGreedyCELF and the simple heuristic method—degree. The parameter r in 4.2 is set to be 100. In the Independent Cascade model the number of simulations is set as 10000.

5.3   Varying λ

We first compare the spread results of different methods in the candidate selection part of CBGCELF. We set the spread result of StaticGreedyCELF as baseline.

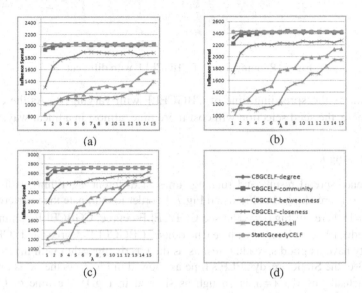

(a)   (b)

(c)   (d)

Fig. 4. Influence spreads of different methods with different λ when $k=10$(a), $k=20$(b) and $k=30$(c)

Fig.4 shows the influence spread results of different methods with different λ when k=10, k=20, k=30. We can get: (1) the CBGCELF-degree and CBGCELF-community perform much better than other centralities. They spread as much as the seeds selected by StaicGreedyCELF with a small λ ($\lambda = 5$, or 6) which indicates less seed selection time. This indicates that the number of followers is an important symbol of influence in micro-blog network. (2) Actually CBGCELF-community has a bit advantage in spreading number over CBGCELF-degree, but it's not so obvious and stable, because we are not selecting the seeds from all the nodes. Moreover, considering the community detection is more time-consuming than the degree method, CBGCELF-degree is the best choice. (3) K-shell centrality is the worst candidate selection method. It works well in single-seed selection but is not suitable for multi-seeds selection. (4) The spread results of CBGCELF-betweenness, CBGCELF-closeness and CBGCELF-kshell become closer to the baseline as λ increases. If λ continues to increase, their spreading results will become as good as the greedy algorithm, but this will cost time. (5) The convergence in Fig.4. becomes better as k increases. As we set the candidate number $N = k \cdot \lambda$, there are more nodes participating in the seed selection as k increases, so the spreading results will become closer to the baseline for all the centralities(especially CBG-betweenness, CBG-closeness, CBG-kshell) in the same range of λ.

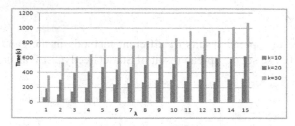

Fig. 5. Running times of CBGCELF with different λ

Fig.5 shows the spreading time of CBGCELF with varying λ. As the candidates become more, the seed selection will cost more time, almost in a linear way.

5.4   Varying k

The influence spread results and running times of different algorithms with varying k when $\rho = 0.1$ are showed in Fig.6 and Fig.7. In order to eliminate the interference of λ we set $\lambda=15$ here. In Fig.6 we can see CGBCELF-degree, CGBCELF-community and StaticGreedyCELF coincide with each other. CBGCELF-degree and CBGCELF-community have as good spreading results as the greedy algorithm with much less time, almost 1/3 of the StaticGreedyCELF's time as showed in Fig.7. As the λ is much larger than we actually need ($\lambda=6$ is enough as showed in Fig.4), the time of CBGCELF should be less. The degree heuristic needs very little time, but its accuracy is poor.

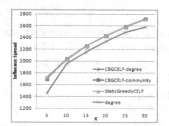

Fig. 6. Influence spread of different k **Fig. 7.** Running times of different k

5.5   Varying ρ

Fig.8 shows the influence spread results of different methods with different ρ when $k =10$ and $\lambda=15$. CBGCELF has good influence spread as StaticGreedyCELF. Degree heuristic is worse than them. Comparing the running times in Fig.9, degree heuristic is the best unsurprisingly. CBGCELF is much better than StaticGreedy-CELF, especially when ρ becomes large. StaticGreedyCELF will cost days while CBGCELF only needs several hours without losing accuracy when $\rho = 0.2$. CBG-CELF-degree costs less time than CBGCELF-community while they almost have the same spread result.

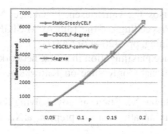
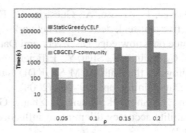

Fig. 8. Influence spreads of different ρ **Fig. 9.** Running times of different ρ

6   Conclusions and Future Work

In this paper we integrate the characteristics of micro-blog into the process of network modeling to make the spread model more practical. Then we empirically propose Candidates-Based Greedy (CBG) algorithm, which firstly select the candidates with heuristic methods then choose the seeds from the candidates with a greedy algorithm, as an efficient influence maximization algorithm in micro-blog network. Through comparing several centralities in the candidate selection part, we find that degree-based CBG algorithm is the best choice in terms of less time (less than 1/3 of the greedy algorithm), low complexity and good spread result as the greedy algorithm for the micro-blog network. Furthermore this algorithm is scalable to large-scale social networks.

Theoretically CBG algorithm can apply to most social networks because most social networks are scale-free, however it still requires further verification. We will validate it on more datasets in the future. We will also try to find whether the best candidate selection method will change with different networks.

Acknowledgement. We would like to thank the anonymous reviewers for their helpful comments in reviewing this paper. This work is supported by Natural Science Foundation of China (61272240, 61103151), the Doctoral Fund of Ministry of Education of China (20110131110028) and the Natural Science foundation of Shandong province (ZR2012FM037).

References

1. Domingos, P., Richardson, M.: Mining the network value of customers. In: Proceedings of the 7th ACM SIGKDD Conference on Knowledge Discovery and Data Mining, pp. 57–66 (2001)
2. Kempe, D., Kleinberg, J.M., Tardos, E.: Maximizing the spread of influence through a social network. In: Proceedings of the 9th ACM SIGKDD Conference on Knowledge Discovery and Data Mining, pp. 137–146 (2003)

3. Leskovec, J., Krause, A., Guestrin, C., Faloutsos, C.: vanBriesen, J., Glance, N.S.: Cost-effective outbreak detection in networks. In: Proceedings of the 13th ACM SIGKDD Conference on Knowledge Discovery and Data Mining, pp. 420–429 (2007)
4. Wei, C., Yajun, W., Siyu, Y.: Efficient influence maximization in social networks. In: Proceedings of the 15th ACM SIGKDD International Conference on Knowledge Discovery and Data Mining, pp. 199–208 (2009)
5. Cheng, S., Shen, H., Huang, J., Zhang, G., Cheng, X.: Static greedy: solving the apparent scalability-accuracy dilemma in influence maximization. arXiv preprint arXiv:1212.4779 (2012)
6. Luo, Z.-L., Cai, W.-D., Li, Y.-J., Peng, D.: A pageRank-based heuristic algorithm for influence maximization in the social network. In: Gaol, F.L. (ed.) Recent Progress in DEIT, Vol. 2. LNEE, vol. 157, pp. 485–490. Springer, Heidelberg (2012)
7. Chen, Y.C., Chang, S.H., Chou, C.L., et al.: Exploring Community Structures for Influence Maximization in Social Networks. In: The 6th SNA-KDD Workshop on Social Network Mining and Analysis Held in Conjunction with KDD, vol. 12, pp. 1–6 (2012)
8. Wang, Y., Cong, G., Song, G., Xie, K.: Community-based greedy algorithm for mining top-k influential nodes in mobile social networks. In: Proceedings of the 16th ACM SIGKDD International Conference on Knowledge Discovery and Data Mining, pp. 1039–1048 (2010)
9. Pengyi, F., Hui, W., Zhihong, J., Pei, L.: Measurement of Microblogging Network. Journal of Computer Research and Development 49(4), 691–699 (2012)
10. Yang, K., Zhang, N.: The analysis of the structure of micro-blog users relation networks. In: International Conference on Computer, Networks and Communication Engineering (ICCNCE 2013). Atlantis Press (2013)
11. Weng, J., Lim, E.-P., Jiang, J., He, Q.: Twitterrank: finding topic-sensitive influential twitterers. In: WSDM, pp. 261–270 (2010)
12. Akshay, J., Song, X.D., Tim, F., Belle, T.: Why We Twitter: Understanding Microblogging Usage and Communities. In: Proceedings of Joint 9th WEBKDD and 1st SNA-KDD Workshop (2007)
13. Hou, B., Yao, Y., Liao, D.: Identifying all-around nodes for spreading dynamics in complex networks. Physica A: Statistical Mechanics and its Applications 391(15), 4012–4017 (2012)
14. Page, L., Brin, S., Motwani, R., Winograd, T.: The pagerank citation ranking: Bringing order to the web. Technical Report 1999-66, Stanford InfoLab (1999)
15. Kleinberg, J.M.: Authoritative sources in a hyperlinked environment. Journal of the ACM (JACM) 46(5), 604–632 (1999)
16. Lü, L., Zhang, Y.C., Yeung, C.H., Zhou, T.: Leaders in social networks, the delicious case. PloS One 6(6), e21202 (2011)
17. Tang, L., Liu, H.: Community detection and mining in social media. Synthesis Lectures on Data Mining and Knowledge Discovery 2(1), 1–137 (2010)
18. Dangalchev, C.: Residual closeness in networks. Physica A: Statistical Mechanics and its Applications 365(2), 556–564 (2006)
19. Kitsak, M., Gallos, L.K., Havlin, S., Liljeros, F., Muchnik, L., Stanley, H.E., Makse, H.A.: Identifying influential spreaders in complex networks. arXiv, Tech. Rep., physics and Society, 1001.5285 (2010)
20. Pan, Y., Li, D.H., Liu, J.G., Liang, J.Z.: Detecting community structure in complex networks via node similarity. Physica A: Statistical Mechanics and its Applications 389(14), 2849–2857 (2010)

Detecting Errors in Numeric Attributes

Grace Fan[1], Wenfei Fan[2], and Floris Geerts[3]

[1] Conestoga High School, USA
[2] University of Edinburgh, UK
[3] University of Antwerp, Belgium

Abstract. To detect errors in numeric data, this paper proposes *numeric functional dependencies* (NFDs), a class of dependencies that allow us to specify arithmetic relationships among numeric attributes. We show that NFDs subsume conditional functional dependencies (CFDs); hence, we can catch data inconsistencies, numeric or not, in a uniform logic framework by using NFDs as data quality rules. Better still, NFDs do not increase the complexity of reasoning about data quality rules. We show that the satisfiability and implication problems for NFDs remain NP-complete and coNP-complete, respectively, the same as their counterparts for CFDs. Moreover, NFDs can be implemented in SQL and hence, error detection can be readily supported by DBMS. In addition, we show that NFDs and CFDs can be extended across multiple tables, without increasing the complexity of static analyses and error detection.

1 Introduction

One of the central problems with real-life data is data consistency. Indeed, data in the real world is often dirty, with errors, conflicts and discrepancies. Inconsistent data inflicts a daunting cost. For example, it costs US businesses 600 billion dollars annually [5], and errors in medical data may have disastrous consequences such as death [18]. The scale of the problem is even worse in big data, and it poses one of the most pressing challenges to big data management.

To detect data inconsistencies, a number of dependency formalisms have been studied, such as denial constraints [3], conditional functional dependencies (CFDs) [7] and conditional inclusion dependencies (CINDs) [14]. Using these dependencies as data quality rules, several systems have been developed for detecting errors and repairing real-life data (see [6] for a recent survey).

These dependencies, however, fall short of effectively catching errors in numeric data, *e.g.*, integer and real numbers. Numeric values are widely found in, *e.g.*, medical, scientific and financial data, and should by no means be overlooked.

Example 1. (1) A relation r_1 is shown in Fig. 1(a), in which each tuple specifies a composer with his name, year of birth (YoB), year of death (YoD) and origin (country, town). The data in r_1 is, however, inconsistent: (a) Bonn is a city in Germany, not in Belgium (tuple t_2), and (b) it is normally expected that no one lives more than 120 years, *i.e.*, YoD − YoB ≤ 120, in contrast to tuple t_1.

The error in tuple t_2 can be detected by a CFD ψ_1: town = "Bonn" → country = "Germany", which asserts that if the town is Bonn then the country is Germany. When ψ_1 is used as a data quality rule, the inconsistency in t_2 emerges

F. Li et al. (Eds.): WAIM 2014, LNCS 8485, pp. 125–137, 2014.
© Springer International Publishing Switzerland 2014

	name	YoB	YoD	town	country
t_1:	Wolfgang Amadeus Mozart	1756	1891	Salzburg	Austria
t_2:	Ludwig van Beethoven	1770	1827	Bonn	Belgium

(a) A composer relation r_1

	SS#	name	cno	hw	tests	lab	proj
t_3:	14311	Joe Lee	C_1	5%	35%	40%	35%
t_4:	14311	Joe Lee	C_2	15%	55%	0%	30%

(b) A report relation r_2

	cno	start	end	day
s_1:	C_1	9am	11am	Tue
s_2:	C_2	10am	11am	Tue

(c) Course relation r_2'

	CC#	name	street	city	zip	when	where	amnt
t_5:	610000253253775	Mark Smith	Main St.	Edi	EH8 9LE	21:00/7/7/2013	EDI	£350
t_6:	610000253253775	Mark Smith	Main St.	Edi	EH8 9LE	22:00/7/7/2013	NYC	£500

(d) A transaction relation r_3

	Hid	Pid	relationship	sex	age	status
t_7:	237654	1	reference	M	35	married
t_8:	237654	2	child	F	24	single

(e) A census relation r_4

Fig. 1. Example relations

as a violation of ψ_1. However, to detect the inconsistency in t_1, it requires an arithmetic operation YoD − YoB on the numeric attributes YoB and YoD, which is, unfortunately, not supported by denial constraints, CFDs or CINDs.

(2) Relation r_2 of Fig. 1(b) shows the academic report of a high-school student, one tuple for each course taken, with the distribution of the score into homework (hw), tests, lab and projects (proj) (ignore Fig. 1(c) for now). Obviously the sum of these elements should be 100%. However, for tuple t_3, its total percentage is 115%. To detect this error, arithmetic operations are again needed.

(3) Relation r_3 of Fig. 1(d) is a sample of data from a bank. A tuple in r_3 specifies a transaction record of a credit card: the card number (CC#), card holder in the UK (name, street, city, zip), when and where the card was physically used, and the amount charged to the card (amnt). Note that tuples t_5 and t_6 indicate a possible *fraud*. Indeed, Edinburgh (Edi) is 5 hours ahead of New York city (NYC), and flights from Edi to NYC take 7 hours. Hence, we have the constraint $t_6[\text{when}] - t_5[\text{when}] \geq 2$, where $t_5[\text{when}]$ and $t_6[\text{when}]$ are the local time in Edi and NYC, respectively. That is, there is no way for one to use a card in Edinburgh at 21:00, and use the same card again in New York at 22:00 on the same day. The fraud, however, cannot be detected by the dependencies mentioned earlier.

(4) Relation r_4 of Fig. 1(e) is a piece of census data from [10]. Each household is identified by an Hid, and within each household, Pid is a key for persons. There is a *reference* person for each Hid, and every other person in the household specifies a relationship with the reference, *e.g.*, child, spouse, along with age, sex and marital status. One rule for census data is that a parent must be at least 12 years older than a child [10]. However, $t_7[\text{age}] - t_8[\text{age}] = 11$, where t_8 is a child of t_7. Unfortunately, the dependencies mentioned earlier are unable to detect this. □

This example tells us that to clean real-life data, we need new dependencies to detect errors in numeric attributes. Moreover, as the dependencies will be incorporated into existing data cleaning systems that support, *e.g.*, CFDs, the extension with the new dependencies should not incur substantial extra costs.

Contributions. This paper studies new dependencies in response to the need.

(1) We propose *numeric functional dependencies* (NFDs, Section 2). NFDs are defined in a QBE-like syntax [16], and support arithmetic operations. We show that CFDs are a special form of NFDs. Hence, NFDs can serve as data quality rules for detecting inconsistencies commonly found in practice, numeric or not.

(2) We show that NFDs do not increase the complexity of the static analyses of data quality rules (Section 3). Indeed, the classical problems – the satisfiability and implication problems – are NP-complete and coNP-complete for NFDs, respectively, the same as their counterparts for CFDs [7].

(3) We show that error detection with NFDs can be built on top of relational DBMS without requiring any additional functionality (Section 4). Indeed, for each NFD φ, an SQL query Q_φ can be automatically generated such that when Q_φ is evaluated on a dataset D, $Q_\varphi(D)$ returns all and only those tuples in D that violate φ, *i.e.*, data inconsistencies, in low polynomial time (PTIME).

(4) Finally, we present an extension of NFDs and CFDs (Section 5). While NFDs and CFDs are defined on a single table, we show that they can be naturally extended to span across multiple tables, in a QBE-like syntax. Better still, the extended NFDs do not make our lives harder: they retain the same complexity bounds of the static analyses and error detection as their NFD counterparts.

We contend that NFDs are a natural extension of CFDs. They can be readily employed by data cleaning systems that already support CFDs [6], and extend the capabilities of those system to detect numeric errors in real-life data.

Related work. A variety of dependencies have been studied as data quality rules for detecting data inconsistencies, from traditional functional and inclusion dependencies [2] to CFDs, CINDs and denial constraints [6]. Several data quality systems based on CFDs are already in place, and have proven effective in various applications. However, such dependencies cannot express arithmetic relationships and hence, do not effectively catch inconsistencies in numeric data.

The need for detecting numeric errors has long been recognized [4, 9–11, 13, 17]. Metric functional dependencies [13] and sequential dependencies [11] extend functional dependencies by supporting (numeric) metrics and intervals on ordered data, respectively. However, they do not support arithmetic operations and cannot capture the inconsistencies of Example 1. A class of powerful aggregation constraints was proposed in [17], defined in terms of aggregate functions (*e.g.*, max, min, sum, avg, count). Using these constraints as data quality rules, data repairing and consistent query answering were studied in [4]. It is, however, too expensive to use aggregation constraints: it is undecidable even to decide whether a given set of aggregation constraints is satisfiable. There has also been work on repairing numeric data using constraints defined in terms of aggregate functions [9] and disjunctive logic programming [10]. These constraints are far more complicated than data quality rules that are already employed by data cleaning systems such as CFDs. The complexity of their static analyses (satisfiability and implication) is not yet known and is suspected high. In contrast to the previous work, NFDs aim to strike a balance between the complexity of their reasoning and the expressive power needed for detecting numeric inconsistencies commonly found in practice. We want NFDs to be seamlessly incorporated into data quality systems that already use CFDs, without incurring much extra cost.

It is known that the satisfiability and implication analyses of CFDs are NP-complete and coNP-complete, respectively [7]. Moreover, for each CFD ψ, two SQL queries can be automatically generated to detect all violations of ψ in a dataset [7]. We will show that NFDs retain the same complexity and property.

2　Numeric Functional Dependencies

Below we first define NFDs. We then show that CFDs are a special case of NFDs.

Numeric Functional Dependencies. NFDs are defined on instances of a single relation schema $R(A_1, \ldots, A_n)$. Each A_i is an attribute, with domain $\mathsf{dom}(A_i)$.

A *numeric functional dependency* φ (NFD) defined on R is a pair of tables:

(1) a *pattern table* T_p of schema R has two tuples p_1 and p_2; for $j \in [1, 2]$ and $i \in [1, n]$, $p_j[A_i]$ is a constant in $\mathsf{dom}(A_i)$, a variable $x_{(i,j)}$ or wildcard '_'; and

(2) a *condition table* T_c with a single *condition tuple* of the form e op z, where
   - e is either a variable or a *linear* arithmetic expression;
   - op is one of the built-in predicates $=, \neq, <, \leq, >, \geq$; and
   - z is either a constant c or a variable y in T_p.

Here an *arithmetic expression* is built up from terms of numeric constants c or variables x in T_p with a numeric domain, by closing them under arithmetic operators $+, -, \times, \div$ and $|\cdot|$ (for absolute value). Note that variables $x_{(i,j)}$ and $x_{(l,s)}$ that appear in pattern tuples p_1 and p_2 of T_p may be identical, asserting condition $p_j[A_i] = p_s[A_l]$. When $p_j[A_i]$ is a constant c, it denotes condition $p_j[A_i] = c$. If p_1 and p_2 are identical, T_p consists of a single tuple p_1 only.

Example 2. (1) An NFD $\varphi_1 = (T_{P1}, T_{C1})$ is defined on the composer relation of Fig. 1(a). As shown in Figures 2(a) and 2(b), T_{P1} consists of a single pattern tuple t_{p1} with $t_{p1}[\mathsf{YoB}] = x$ and $t_{p1}[\mathsf{YoD}] = y$, and T_{P1} consists of a single condition. It is to ensure that for any composer tuple t, $t[\mathsf{YoD}] - t[\mathsf{YoB}] \leq 120$.

(2) Figures 2(c) and 2(d) define an NFD $\varphi_2 = (T_{P2}, T_{C2})$ on the academic report relation of Fig. 1(b). It states that for any report tuple t, the sum $t[\mathsf{hw}] + t[\mathsf{tests}] + t[\mathsf{lab}] + t[\mathsf{proj}]$ of the distribution of the marks should be equal to 100%.

(3) Another NFD $\varphi_3 = (T_{P3}, T_{C3})$ is given in Figures 2(e) and 2(f), defined on the transaction data of Fig. 1(d). Note that pattern table T_{P3} consists of two tuples. The NFD states that for any two transaction records of the same credit card

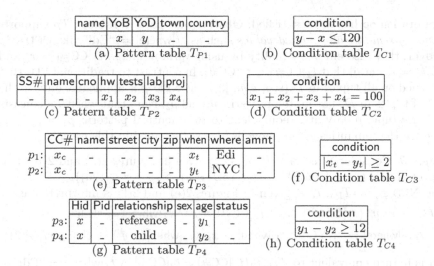

name	YoB	YoD	town	country
-	x	y	-	-

(a) Pattern table T_{P1}

condition
$y - x \leq 120$

(b) Condition table T_{C1}

SS#	name	cno	hw	tests	lab	proj
-	-	-	x_1	x_2	x_3	x_4

(c) Pattern table T_{P2}

condition
$x_1 + x_2 + x_3 + x_4 = 100$

(d) Condition table T_{C2}

	CC#	name	street	city	zip	when	where	amnt
p_1:	x_c	-	-	-	-	x_t	Edi	-
p_2:	x_c	-	-	-	-	y_t	NYC	-

(e) Pattern table T_{P3}

condition
$|x_t - y_t| \geq 2$

(f) Condition table T_{C3}

	Hid	Pid	relationship	sex	age	status
p_3:	x	-	reference	-	y_1	-
p_4:	x	-	child	-	y_2	-

(g) Pattern table T_{P4}

condition
$y_1 - y_2 \geq 12$

(h) Condition table T_{C4}

Fig. 2. Example NFDs

(specified by $p_1[\text{CC\#}] = x_c$ and $p_2[\text{CC\#}] = x_c$), if it was used in Edi and NYC, then the two transactions had to be at least two-hour apart. Note that constants "Edi" and "NYC" are used to specify a pattern, along the same lines as CFDs [7]. In addition, variable x_c is used to enforce condition $p_1[\text{CC\#}] = p_2[\text{CC\#}]$.

(4) Finally, an NFDs $\varphi_4 = (T_{P4}, T_{C4})$ is given in Figures 2(g) and 2(h), defined on the census relation of Fig. 1(e). It assures that for any two persons in the same household $(p_3[\text{hid}]) = x$ and $p_4[\text{hid}] = x$), if one is the "reference" and the other is his/her "child", then the parent must be at least 12-years older than the child. Again, constants "reference" and "child" specify patterns.   □

Semantics. To give a formal semantics of NFDs, we use an operator $\trianglerighteq$ defined on constants, variables and '_': $\eta_1 \trianglerighteq \eta_2$ is interpreted as the truth value true if η_2 is '_', and it is an equality predicate $\eta_1 = \eta_2$ otherwise. The operator $\trianglerighteq$ naturally extends to tuples and produces true or a *conjunction of equality atoms*, e.g., (Main St, EDI, NYC) $\trianglerighteq$ (_, x, y) yields 'Edi' $= x \wedge$ 'NYC' $= y$.

Given $\trianglerighteq$, an NFD $\varphi = (T_P, T_C)$ in the QBE-like syntax [16] given above can be rewritten into an equivalent first-order logic (FO) sentence as follows. Assume that T_p has two pattern tuples p_1 and p_2, T_c is e op z, and that all the variables appearing in T_P are $x_1, \ldots, x_m$. Then φ can be written as the FO sentence:

$$\varphi = \forall t_1, t_2, x_1, \ldots, x_m ((t_1 \trianglerighteq p_1 \wedge t_2 \trianglerighteq p_2) \rightarrow e \text{ op } z).$$

As remarked earlier, $t_1 \trianglerighteq p_1$ and $t_2 \trianglerighteq p_2$ yield two conjunctions ω_1 and ω_2, possibly equal to true. The variables x_i are specified in $t_j \trianglerighteq p_j$ and used in e and z.

An instance D of relation schema R *satisfies* the NFD φ, denoted by $D \models \varphi$, if for *all* tuples t_1 and t_2, if t_1 and t_2 satisfy $\omega_1 \wedge \omega_2$ following the standard semantics of first-order logic, then the condition e op z is also satisfied. We say that D *satisfies* a set Σ of NFDs, denoted by $D \models \Sigma$, if for all $\varphi \in \Sigma$, $D \models \varphi$.

Intuitively, if tuples t_1 and t_2 match pattern tuples p_1 and p_2, respectively, then the predicate e op z defined with arithmetic operations in e and the comparison operation op has to be satisfied. Observe that pattern tableau T_P supports *patterns of semantically related values* such as "t[where] = "Edi" like CFDs [7]. Moreover, they also enforce equality by using variables, *e.g.*, $t_1[\text{CC}\#] = x_c$ and $t_2[\text{CC}\#] = x_c$ entail that $t_1[\text{CC}\#] = t_2[\text{CC}\#]$. In contrast to traditional FDs that are defined on all tuples in D (see, *e.g.*, [2]), the NFD φ is applicable only to the *subset* of tuples in D that match patterns p_1 and p_2. Note that when T_P consists of a single tuple p only, it is equivalent to two identical patterns $p_1 = p_2 = p$, and hence the semantics given above is also well-defined in this case.

Example 3. Consider NFD $\varphi_1 = (T_{P1}, T_{C1})$ given in Figures 2(a) and 2(b). It is interpreted as $\forall\, t, x, y\ \big(t[\text{YoB}] = x \wedge t[\text{YoD}] = y\big) \rightarrow (y - x \leq 120)\big)$.

The NFD $\varphi_3 = (T_{P3}, T_{C3})$ given in Figures 2(e) and 2(f) is interpreted as

$$\forall\, t_1, t_2, x_c, x_t, y_t\ \big(t_1[\text{CC}\#] = x_c \wedge t_2[\text{CC}\#] = x_c \wedge t_1[\text{where}] = \text{"Edi"}$$
$$\wedge\, t_2[\text{where}] = \text{"NYC"} \wedge t_1[\text{when}] = x_t \wedge t_2[\text{when}] = y_t) \rightarrow (|y_t - x_t| \geq 2)\big).$$

which is in turn equivalent to $\forall\, t_1, t_2 \big(t_1[\text{CC}\#] = t_2[\text{CC}\#] \wedge t_1[\text{where}] = \text{"Edi"} \wedge t_2[\text{where}] = \text{"NYC"} \rightarrow (|t_2[\text{when}] - t_1[\text{when}]| \geq 2)\big)$. The semantics of φ_2 and φ_3 given in Example 2 can be similarly interpreted in first-order logic.   □

Taking NFDs and CFDs Together. Recall that CFDs on schema R can be written in a normal form $\psi = (X \rightarrow A,\ t_p)$ [6], where (a) X is a set of attributes of R, A is a single attribute of R, and (b) t_p is a *pattern tuple* with attributes in X and A such that for each $B \in X \cup \{A\}$, $t_p[B]$ is either a constant in $\text{dom}(B)$ or a wildcard '_'. An instance D of R satisfies the CFD ψ, denoted by $D \models \psi$, if for any two tuples $t_1, t_2 \in D$, when $t_1[X] \unrhd t_p[X]$ and $t_2[X] \unrhd t_p[X]$, then $t_1[A] = t_2[A] \unrhd t_p[A]$ (here CFDs restrict "$\unrhd$" to constants and '_' only).

One can readily verify that ψ is equivalent to an NFD φ of the following form.

(1) If $t_p[A]$ is a constant c, then the NFD $\varphi = (T_p, T_c)$, where T_p consists of a single pattern tuple p such that (a) $p[A]$ is a distinct variable, and (b) for all the other attributes B of R, $p[B] = t_p[B]$ if $B \in X$, and $p[B] = \text{'_'}$ otherwise; and (c) T_c is $x_A = c$. Such a CFD is referred to as a *constant* CFD in [6].

(2) If $t_p[A]$ is wildcard '_', then $\varphi = (T_p, T_c)$, where T_p consists of two pattern tuples p_1 and p_2, such that (a) $p_1[A]$ and $p_2[A]$ are distinct variables x_1 and x_2, respectively; (b) for all the other attributes B of R, $p_1[B]$ and $p_2[B]$ are defined in the same way as for constant CFDs; and (c) T_c is $x_1 = x_2$. Such CFDs are called *variable* CFDs [6], which subsume traditional functional dependencies [2].

Example 4. The constraint ψ_1 given in Example 1 is a constant CFD, and can be expressed as an equivalent NFD $\varphi_0 = (T_{P0}, T_{C0})$, where

$$T_{P0} = \begin{array}{|c|c|c|c|c|} \hline \text{name} & \text{YoB} & \text{YoD} & \text{town} & \text{country} \\ \hline \_ & \_ & \_ & \text{Bonn} & \text{x} \\ \hline \end{array} \qquad T_{C0} = \begin{array}{|c|} \hline \text{condition} \\ \hline x = \text{Germany} \\ \hline \end{array} \qquad □$$

Based on the discussions above, one can readily verify the following.

Proposition 1: CFDs of [7] are a special cases of NFDs.   □

This tells us that NFDs provide us with a uniform logic formalism to express data quality rules for detecting data inconsistencies, numeric or not. In other words, NFDs are capable of capturing all errors that CFDs can catch, and moreover, errors in numeric attributes that CFDs are not able to detect.

3   Reasoning about NFDs

NFDs are strictly more expressive than CFDs. Indeed, Proposition 1 tells us that NFDs subsume CFDs; moreover, NFDs can specify arithmetic relationships among numeric attributes such as those in φ_1–φ_4, which CFDs cannot express. Despite the increase in expressive power, we next show that NFDs do not increase the complexity of reasoning about data quality rules. More specifically, we study two classical problems that are associated with any dependency class C.

Consider a set Σ of dependencies in C defined on a relation schema R. To use Σ as data quality rules, we have to answer the following questions.

- The *satisfiability problem* for C is to decide, given a set Σ of dependencies in C, whether there exists a nonempty instance D of R such that $D \models \Sigma$.
- The *implication problem* for C is to determine, given Σ and another dependency φ in C that is also defined on R, whether Σ *implies* φ, denoted by $\Sigma \models \varphi$, i.e., for each instance D of R, if $D \models \Sigma$, then $D \models \varphi$.

The satisfiability analysis checks whether Σ has conflicts, *i.e.,* whether the data quality rules in Σ are "dirty" themselves. It help us find out what goes wrong in the rules. The implication analysis allows us to optimize our rules by eliminating redundant ones: if $\Sigma \models \varphi$, then it suffices to use Σ instead of $\Sigma \cup \{\varphi\}$.

When C is the class of traditional functional dependencies, any finite subset Σ of C is satisfiable, and its implication problem is in linear-time [2]. When it comes to NFDs, however, these problems are no longer simple.

To understand where the complication arises, we consider a special form of NFDs, denoted by AFDs, in which conditions have the form of e op c, where e is a linear arithmetic expressions defined on numeric attributes, and c is a constant. One can show that AFDs cannot express CFDs (not even constant CFDs $(X \to A, t_p)$ when A is a non-numeric attribute), and that CFDs cannot express AFDs. We show below that there exists a set of AFDs that is not satisfiable, even when all the attributes in R have an infinite domain. In contrast, for a set of CFDs to be unsatisfiable, the CFDs must be defined on some attributes with a finite domain (see [7] for more details).

Example 5. Consider a relation schema $R(A, B)$, where A and B have an infinite integer domain. Let Σ be a set consisting of two AFDs $\phi_1 = (T_P, T'_{C1})$ and $\phi_2 = (T_P, T'_{C2})$, where T_P consists of a single tuple (x, y), the condition of T'_{C1} is $x - y = 0$, and T'_{C2} is $y - x > 0$. Then there exists no nonempty instance D of R such that $D \models \Sigma$. Indeed, for any instance D of R, if there exists a tuple $t \in D$, then ϕ_1 requires $t[A] = t[B]$ whereas ϕ_2 asks for $t[A] \neq t[B]$.   □

The good news is that the extra expressive power introduced by NFDs over CFDs does not make our lives harder: their static analyses have the same complexity as their counterparts for CFDs [7], and as well as for AFDs.

Theorem 2. *For NFDs, (1) the satisfiability problem is* NP-*complete, and (2) the implication problem is* coNP-*complete. For AFDs, (3) the satisfiability and implication problems remain* NP-*complete and* coNP-*complete, respectively.*   □

Proof Sketch. (1) NFDs. The lower bounds follow from their counterparts for CFDs [7], since CFDs are a special case of NFDs. The upper-bound proofs are far more involved. To see the complications, observe that if arbitrary arithmetic expressions were allowed in NFDs, the satisfiability problem would be undecidable, which could be verified by reduction from undecidable Diophantine equations [12]. That is why we restrict arithmetic expressions in NFDs to be linear.

For the satisfiability problem, we first show that NFDs have a small model property: given a set Σ of NFDs defined on a schema R, if there exists a nonempty instance of R that satisfies Σ, then there exists an instance D_0 of R such that $D_0 \models \Sigma$, D_0 consists of a single tuple t_0, and moreover, it suffices to consider the values of the attributes of t_0 from a finite domain determined by the domains of the attributes and those constants in Σ. The proof of the small model property is nontrivial and needs to distinguish integers and non-integers. Based on the small model property, one can develop an NP algorithm for satisfiability checking as follows: (a) guess a tuple t_0 with values drawn from the finite domain, and (b) check whether $\{t_0\} \models \Sigma$. The algorithm is in NP since step (b) is in PTIME.

Similarly, the implication problem for NFDs is shown in coNP by establishing a small model property, where D_0 consists of two tuples.

(b) AFDs. The upper bounds follow from their counterparts for NFDs, since AFDs are a special case of NFDs. The lower bound for the satisfiability problem is verified by reduction from the linear integer programming problem (LIP), which is NP-complete (cf. [15]). The latter problem is to determine whether a set of linear inequalities has an integer solution. Similarly, the implication problem for AFDs is verified coNP-complete by reduction from the complement of LIP.   □

4   Validating NFDs in SQL

Recall that we introduce NFDs to detect data inconsistencies. We next present an SQL technique for relational DBMS to detect inconsistencies as violation of NFDs.

Consider a set Σ of NFDs defined on a relation schema R, an instance D of R, and an NFD $\varphi = (T_p, T_c)$ in Σ, where T_p consists of two pattern tuples p_1 and p_2, and T_c is a condition e op z (see Section 2 for the definition of NFDs). We say that a tuple $t \in D$ is a *violation* of φ if there exists a tuple $t' \in D$ such that $\{t, t'\} \not\models \varphi$, i.e., either $t \trianglerighteq p_1$ and $t' \trianglerighteq p_2$ or $t' \trianglerighteq p_1$ and $t \trianglerighteq p_2$, but $\neg(e$ op $z)$. Here we use $\neg(e$ op $z)$ to denote that the condition e op z is not satisfied. For instance, if op is '$=$', then $\neg(e$ op $z)$ is $e \neq z$, and if op is '$\leq$', then $\neg(e$ op $z)$ is $e > z$. Note that violations of φ are defined on single tuples.

The *error detection problem* can then be stated as follows. Given Σ and D, it is to find the set of all tuples in D that are violations of some NFD in Σ, denoted by $\mathsf{vio}(\Sigma, D)$; i.e., $\mathsf{vio}(\Sigma, D) = \{t \in D \mid \exists \varphi \in \Sigma, t$ is a violation of $\varphi\}$.

The main result of this section is as follows.

Proposition 3: (1) For any NFD φ defined on a schema R, there exists an SQL query Q_φ such that for any instance D of R, $Q_\varphi(D)$ returns all violations of φ in D. (2) For any set Σ of NFDs defined on R and any instance D of R, vio(Σ, D) is computable in at most $O(\|\Sigma\|\,|D|^2)$ time, where $\|\Sigma\|$ is the cardinality of Σ (*i.e.*, the number of NFDs in Σ), and $|D|$ is the size of D. $\square$

To prove Proposition 3, we show how Q_φ is (automatically) generated from φ. Assume that $\varphi = (T_p, T_c)$, $T_p = \{p_1, p_2\}$, and T_c is condition e op z. Then Q_φ is:

> **select** t
> **from** $R\ t,\ R\ t'$
> **where** $(\Omega_1(t, t', p_1, p_2)$ **and** $\mathsf{C}_1(t, t', e, z))$ **or** $(\Omega_2(t, t', p_1, p_2)$ **and** $\mathsf{C}_2(t, t', e, z))$

Here $\Omega_1(t, t', p_1, p_2)$ encodes $t \trianglerighteq p_1$ and $t' \trianglerighteq p_2$ in SQL, as a conjunction of *terms* such that for each attribute A of R, (1) $t[A] = c$ is a term if $p_1[A]$ is a constant c; similarly for $t'[A] = c$; and (2) $t[A] = t'[A]$ is a term if $p_1[A]$ and $p_2[A]$ are the same variable x_A. The conjunct $\mathsf{C}_1(t, t', e, z)$ encodes $\neg(e$ op $z)$ as described above, by substituting attributes of t and t' for variables in e or z; more specifically, each variable x occurring in e or z is replaced with $t[A]$ if $p_1[A] = x$ (resp. with $t'[A]$ if $p_2[A] = x$). This is possible since SQL supports arithmetic operations (see, *e.g.*, [1]). Along the same lines, $\Omega_2(t, t', p_1, p_2)$ encodes $t \trianglerighteq p_2$ and $t' \trianglerighteq p_1$ in SQL, and $\mathsf{C}_2(t, t', e, z)$ encodes $\neg(e$ op $z)$ accordingly.

Example 6. The NFD φ_1 given in Example 2 can be implemented in SQL as Q_1:

> **select** t
> **from**   composer t
> **where** $t[\mathsf{YoD}] - t[\mathsf{YoB}] > 120$

Here we need a single variable t to range over composer tuples in this simple SQL query, since T_{P1} has a single pattern tuple. Similarly, φ_4 can be validated by Q_4:

select t
from   census t, census t',
where $t[\mathsf{Hid}] = t'[\mathsf{Hid}]$ **and**
  ($t[\mathsf{relationship}] =$ "reference" **and** $t'[\mathsf{relationship}] =$ "child" **and** $t[\mathsf{age}] - t'[\mathsf{age}] < 12$) **or**
  ($t[\mathsf{relationship}] =$ "child" **and** $t'[\mathsf{relationship}] =$ "reference" **and** $t'[\mathsf{age}] - t[\mathsf{age}] < 12$)

In contrast to Q_1, the SQL query Q_4 uses two variables t and t' to range over census tuples, since T_{P4} consists of two pattern tuples. Similarly, one can get SQL queries to validate the NFDs φ_2 and φ_3 given in Example 2.

As another example, recall the NFD φ_0 of Example 4 for expressing the CFD ψ_1; this NFD can be validated by using the following SQL query Q_0:

> **select** t
> **from**   composer t
> **where** $t[\mathsf{town}] =$ "Bonn" **and** $t[\mathsf{country}] \neq$ "Germany" $\square$

One can readily verify that given any instance D of R, it takes $O(|D|^2)$ time to evaluate $Q_\varphi(D)$ in the worst case, to compute all violations of the NFD φ in D (note that $|\varphi|$ is determined by the arity of R). From this it follows that to validate a set Σ of NFDs, it takes at most $O(\|\Sigma\|\|D|^2)$ time. The cost can be substantially reduced via, *e.g.*, indexing. This completes the proof of Proposition 3.

Remark. Proposition 3 tells us that one can support inconsistency detection based on NFDs directly on top of commercial DBMS, without requiring any additional functionality. We conclude this section with the following remarks.

(1) To detect violations of a CFD ψ in a database, an SQL method has been presented in [7], which requires two SQL queries in general. In contrast, we show that a single SQL query Q_φ suffices to validate an NFD φ, although NFDs are more expressive than CFDs. Nonetheless, the SQL queries in [7] are in $O(|\psi\|D|)$ time, where $|\psi|$ denotes the size of ψ, whereas the SQL query Q_φ takes $O(|D|^2)$ time (although the cost can be reduced as mentioned above).

(2) After we have seen the static analyses and validation of NFDs, we now justify the definition of NFDs (Section 2). One might be tempted to extend NFDs by allowing (a) non-linear arithmetic expressions in T_c, or (b) an unbounded number of pattern tuples in T_p instead of at most two. However, either extension would lead to substantial increase in the complexity. Indeed, (a) non-linear arithmetic expressions in conditions would make the satisfiability and implication analyses of NFDs undecidable, as shown in the proof of Theorem 2; and (b) an unbounded number of pattern tuples would require exponential time to validate an NFD φ; more specifically, it would take $O(|D|^n)$ time to validate φ in the worst case, where n is the number of pattern tuples of φ. As remarked earlier, we want to strike a balance between the expressive power and the complexity of NFDs. The current definition of NFDs, on one hand, suffices to catch (numeric) errors commonly found in the real world, and on the other hand, extends CFDs without increasing the complexity of the static analyses of data quality rules.

5   Extending NFDs and CFDs across Multiple Tables

Both NFDs and CFDs are defined on a single relation schema R. We next show that they can be readily extended to detect data inconsistencies across multiple tables, without increasing the complexity of their validation and reasoning.

To illustrate the need for such an extension, consider the following example.

Example 7. Consider the report relation r_2 of Fig. 1(b) and course relation r_2' of Fig. 1(c). Suppose that the inconsistency in tuple t_1 of r_2 is fixed by the NFD φ_2 of Example 2. Then each of the relations r_2 and r_2', when taken separately, seems consistent. However, when r_2 and r_2' are put together, inconsistencies emerge: course C_1 specified by tuple s_1 of r_2' and course C_2 given by tuple s_2 overlap with each other for an hour; this accounts for *a conflict* when some student takes both courses, which is witnessed by tuples t_3 and t_4 in relation r_2.   □

Such a conflict cannot be directly captured by NFDs or CFDs that are defined on a single table. Moreover, it cannot be detected by CINDs [14] although CINDs are defined on two tables. To catch this, one may want to compute the natural join r_2'' of r_2 and r_2' on attribute cno and then define an NFD on r_2''. This is doable, but costly: to detect such inconsistencies one has to compute a number of joins.

Extended NFDs. This motivates us to extend NFDs across multiple tables, so that we can directly catch data inconsistencies between these tables.

Consider a relational schema $\mathcal{R}$, which is a collection $(R_1, \ldots, R_m)$ of relation schemas. Let $k \geq 2$ be a predefined natural number (a constant). We define an *extended* NFD φ on relational schema $\mathcal{R}$ to be a pair of $(\overline{T_p}, T_c)$, where

(1) $\overline{T_p}$ is a set of k pattern tables defined on k relation schemas of $\mathcal{R}$; each T_p of $\overline{T_p}$ is a pattern table of a schema R of $\mathcal{R}$, consisting of two pattern tuples p_1 and p_2 defined with constants, variables and wildcard '_' as before; and

(2) T_c is the *condition table* of φ with a tuple that is either (a) e op z as before, or (b) a *Boolean expression* e defined with terms of comparison predicates $(=, \neq, \leq, <, \geq, >)$ on constants and variables, by closing them under $\wedge, \vee$ and $\neg$. Here e and z may include variables from any pattern table T_p of $\overline{T_p}$.

The semantics of extended NFDs is defined along the same lines as NFDs.

Example 8. We define an extended NFDs φ_e across report and course relations:

SS#	name	cno	hw	tests	lab	proj
x	$-$	y_1	$-$	$-$	$-$	$-$
x	$-$	y_2	$-$	$-$	$-$	$-$

T_{Pr}

cno	start	end	day
y_1	z_1	w_1	d
y_2	z_2	w_2	d

T_{Pc}

condition
$(z_1 > w_2) \vee (z_2 > w_1)$

T_C

It asserts that for any two courses y_1 and y_2, (a) if there exists a student x taking both (pattern table T_{Pr}), and (b) if the two courses are on the same day d (table T_{Pc}), then they do not overlap (condition T_C). To ensure this, we also use an NFD to assert that for any course tuple s, $s[\text{start}] < s[\text{end}]$ (omitted). □

Validating and reasoning about extended NFDs. Extended NFDs do not increase the complexity of validation and static analyses of data quality rules. Along the same lines as the argument of Section 4, one can verify the following.

Corollary 4: (1) For any extended NFD φ defined on a relational schema $\mathcal{R}$, there exists an SQL query Q_φ such that for any instance $\mathcal{D}$ of $\mathcal{R}$, $Q_\varphi(\mathcal{D})$ finds all violations of φ in $\mathcal{D}$. (2) For any set Σ of extended NFDs on $\mathcal{R}$ and instance $\mathcal{D}$ of $\mathcal{R}$, $\text{vio}(\Sigma, \mathcal{D})$ can be computed in $O(|\Sigma||\mathcal{D}|^2)$ time, where $|\Sigma|$ is the size of Σ. □

We should remark that Corollary 4 would no longer hold if one would allow either arbitrary number of pattern tables (*i.e.*, without the constant bound k) or unbounded number of tuples in a pattern table in extended NFDs.

Extending the proof of Theorem 2, one can verify the following corollary, in which *extended* AFDs refer to the subclass of extended NFDs in which the conditions are linear arithmetic expressions e op c for some constant c.

Corollary 5: For extended NFDs and for extended AFDs, (1) the satisfiability problem is NP-complete, and (2) the implication problem is coNP-complete. □

6 Conclusion

We have proposed NFDs and shown the following. (1) NFDs extend CFDs [7] and are capable of detecting inconsistencies in numeric attributes. (2) Despite the increased expressive power, NFDs do not increase the complexity of the satisfiability and implication analyses of data quality rules. (3) Better still, NFDs allow us to detect errors in low PTIME by using existing relational DBMS, by means of automatically generated SQL queries. (4) In addition, NFDs (and hence CFDs) can be extended to catch inconsistencies across different tables, without incurring substantial extra overhead. In light of these, we suggest to use NFDs to detect errors, numeric or not, in a uniform logic framework.

Several topics are targeted for future work. (1) It is known that CFDs are finitely axiomatizable [7]: there exists a finite set of axioms for the implication analysis of CFDs. The finite axiomatizability of NFDs remains to be investigated. (2) To make practical use of NFDs, algorithms need to be developed for automatically discovering NFDs from (possibly dirty) data, along the same lines as discovery algorithms for CFDs (*e.g.*, [8]). (3) To simplify the discussion, we prove Proposition 3 by using a separate SQL query for each NFD in a given set Σ of NFDs (Section 4). It is possible to find a fixed number of SQL queries for error detection, regardless of the cardinality of Σ, along the same lines as CFDs [7]. (4) Finally, data repairing algorithms based on NFDs should be in place to fix errors detected by NFDs (see [6] for data repairing based on CFDs).

Acknowledgments. Wenfei Fan is supported in part by NSFC 61133002, 973 Program 2012CB316200, Guangdong Innovative Research Team Program 2011D005 and Shenzhen Peacock Program 1105100030834361, China.

References

1. MySQL, http://dev.mysql.com/doc/refman/5.0/en/func-op-summary-ref.html
2. Abiteboul, S., Hull, R., Vianu, V.: Foundations of Databases. Addison-Wesley (1995)
3. Arenas, M., Bertossi, L.E., Chomicki, J.: Consistent query answers in inconsistent databases. TPLP 3(4-5) (2003)
4. Bertossi, L.E., Bravo, L., Franconi, E., Lopatenko, A.: Complexity and approximation of fixing numerical attributes in databases under integrity constraints. Inf. Syst. 33(4-5), 407–434 (2008)
5. Eckerson, W.W.: Data Quality and the Bottom Line: Achieving Business Success through a Commitment to High Quality Data. In: The Data Warehousing Institute (2002)
6. Fan, W., Geerts, F.: Foundations of Data Quality Management. Morgan & Claypool Publishers (2012)
7. Fan, W., Geerts, F., Jia, X., Kementsietsidis, A.: Conditional functional dependencies for capturing data inconsistencies. TODS 33(1) (2008)
8. Fan, W., Geerts, F., Li, J., Xiong, M.: Discovering conditional functional dependencies. TKDE 23(5), 683–698 (2011)
9. Flesca, S., Furfaro, F., Parisi, F.: Querying and repairing inconsistent numerical databases. TODS 35(2) (2010)

10. Franconi, E., Palma, A.L., Leone, N., Perri, S., Scarcello, F.: Census data repair: A challenging application of disjunctive logic programming. In: Nieuwenhuis, R., Voronkov, A. (eds.) LPAR 2001. LNCS (LNAI), vol. 2250, pp. 561–578. Springer, Heidelberg (2001)
11. Golab, L., Karloff, H.J., Korn, F., Saha, A., Srivastava, D.: Sequential dependencies. In: PVLDB, vol. 21(1) (2009)
12. Jones, J.P.: Undecidable Diophantine equations. Bull. Amer. Math. Soc. 3(2), 859–862 (1980)
13. Koudas, N., Saha, A., Srivastava, D., Venkatasubramanian, S.: Metric functional dependencies. In: ICDE (2009)
14. Ma, S., Fan, W., Bravo, L.: Extending inclusion dependencies with conditions. TCS 515, 64–95 (2014)
15. Papadimitriou, C.H.: Computational Complexity. Addison Wesley (1994)
16. Ramakrishnan, R., Gehrke, J.: Database Management Systems. McGraw-Hill Higher Education (2000)
17. Rossa, K.A., Srivastava, D., Stuckeyc, P.J., Sudarshan, S.: Foundations of aggregation constraints. TCS 193(1-2), 149–179 (1998)
18. The New York Times. Articles about Jesica Santillan, http://topics.nytimes.com/topics/reference/timestopics/people/s/jesic_santillan/index.html

Distributed Entity Resolution Based on Similarity Join
for Large-Scale Data Clustering

Tiezheng Nie[1], Wang-chien Lee[2], Derong Shen[1], Ge Yu[1], and Yue Kou[1]

[1] College of Information Science and Engineering, Northeastern University
Shenyang 110819, P.R. China
{nietiezheng,shenderong,yuge,kouyue}@ise.neu.edu.cn
[2] Department of Computer Science and Engineering, The Pennsylvania State University
University Park, PA 16802, USA
wlee@cse.psu.edu

Abstract. Entity resolution has been widely used in data mining applications to find similar records. However, the increasing scale and complexity of data has restricted the performance of entity resolution. In this paper, we propose a novel entity resolution framework that clusters large-scale data with distributed entity resolution method. We model the clustering problem as finding similarity sub connected graphs from records. Firstly, our approach finds pairs of records whose similarities are above a given threshold based on *appjoin* algorithm which extends the *ppjoin* algorithm and are executed on MapReduce framework. Then, we propose a cache-based algorithm which cluster entities with similar pairs based on the Disjoin Set algorithm and are also designed for MapReduce framework. Experimental results on real dataset show that our algorithms can achieve more efficiency than previous algorithms on the entity resolution and clustering.

Keywords: Entity Resolution, similarity join, MapReduce, clustering, disjoin set.

1 Introduction

In recent years, *Entity Resolution (ER)*[1] [1] has become a crucial technique for several application domains, such as data mining, data integration, and data cleansing. The task of Entity Resolution is to find similar record pairs based on a specified similarity function, where two records are considered as similar, if the value returned by the similarity function for these two records is above a given threshold. Similarity join [2, 3, 4] is one of effective approaches to implement *ER* over one or two datasets of interest. Generally, similarity joins take as input two datasets, and identifies all pairs of records they are considered as similar from the two datasets. If we apply *ER* on one dataset, i.e., employ (self) similarity join on the same dataset, we can perform a clustering on the dataset. Figure 1 shows the process of clustering using *ER* based on similarity join. Many mining applications can benefit from *ER* based on similarity join. Web document clustering [5] and replicated web collections [6] can be achieved

---

[1] Also known as record linkage, object identification, duplicate detection, or field matching.

F. Li et al. (Eds.): WAIM 2014, LNCS 8485, pp. 138–149, 2014.
© Springer International Publishing Switzerland 2014

by clustering web documents based on their similarity, community discovery in a social network site can be achieved by mining graph data based on the similarity between nodes[7], and large dense graphs can be obtained by similarities[8].

With the rapid development of data mining applications and ever-growing scale of data used, a great research challenge faced in ER is the performance issue. Take the social networking services as an example. The data increment of Facebook has reached up to 500TB per day, producing 27 million times Like Button clicks. Moreover DNA matching in biology also incurs significant computational cost over a huge amount of data. Therefore, the similarity joins for *Entity Resolution* need to be performed in an efficient and scalable fashion. If we compute every pair of records, the computation bears a prohibitive $O(n^2)$ time complexity. With a high computational cost, traditional centralize approaches have the obvious limitation on the performance and thus resort to an approximate solution, i.e., not finding all similar pairs. In this paper, we aim to address the performance issue by exploiting the power of distributed computing.

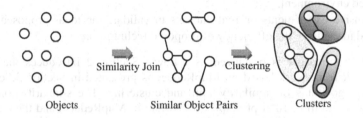

Objects   Similar Object Pairs   Clusters

Fig. 1. The process of clustering using ER based on similarity join

In recent years, the works of similarity join for *ER* have received growing research momentum. These works typically adopt a two-phase approach: (i) generate candidate pairs and (ii) verify candidate pairs [4]. In the candidate generation phase, the goal is to filter record pairs that can't be matched, in order to minimize the amount of candidate record pairs in calculating the similarities. Therefore, current works rely on effectively filtering the candidate pairs to improve the performance of similarity join. On the other hand, the verification phase calculates the similarity of candidate pair to decide whether the pair shall be maintained based on a given threshold. With large-scale datasets, as mentioned earlier, efficiency remains a major problem in these conventional approaches. Recently, several techniques have been proposed for the similarity join based on the MapReduce framework which provides the ability of processing massive data in a distributed and parallel fashion. They mostly focus on the phase of filtering record pairs by separating candidate pairs into different blocks for distributed processing. However, the performance surfers from the data skew which caused deteriorated performance in distributed processing.

In this paper, we propose novel similarity join techniques to build a distributed processing framework for *ER*, which provides high performance on computing similarity of objects pairs and clustering objects. Many similarity functions have been proposed to quantify similarity between objects, such as Jaccard similarity, cosine similarity, and edit distance similarity. Note that each of these similarity functions has been specially designed with some specific kinds of data in mind. This paper aims to work on the similarity function that matches records with the textual similarity of token-based algorithms, e.g., Jaccard algorithm. To facilitate parallel similarity joins, we use MapReduce as the framework of distributed processing. In our *ER* framework, we propose similarity join algorithms based on multiple filtering techniques to reduce the candidate size of compare pairs. Regarding the resulted objects as a graph in which edges connect similar objects, we cluster objects based on the connection between them. The contributions of this paper are as follows:

- We propose an *appjoin* algorithm for similarity join which can efficiently filter candidate pairs for efficient *Entity Resolution* in distributed processing.
- For all similar object pairs, we use the algorithm of *Disjoint Set* to cluster them based on the connection in their similarity graph.
- We propose an *ER* framework based on MapReduce, which integrates the algorithm of similarity join and *Disjoint Set* to support efficiently clustering objects in a distributed environment.
- Our extensive experiments on real datasets to validate the ideas proposed in this paper and to evaluate the efficiency of proposed techniques.

The rest of the paper is organized as follows: Section 2 introduces the related works. The *ER* framework based on MapReduce is presented in Section 3.Section 4 presents the algorithms on similarity join and clustering. The similarity join algorithms integrate multiple filter principles to execute in MapReduce, and the clustering algorithm is based on the *Disjoin Set*. We present our experimental results in Section 5. Section 6 concludes the paper.

2   Related Works

Entity Resolution has been widely studied for data mining applications [6, 9, 10]. Existing research works on *ER* mainly focus on algorithms for similarity join. Similarity Join is defined as follows: Given two datasets R and S, for a entity-pair *(r, s)* where *r* in R and *s* in S, if similarity measure *Sim(r, s)* is higher than the given threshold, then the pair *(r, s)* is returned as the results. Several similarity join techniques [3, 11] have been proposed for processing in-memory and external memory data. There are also techniques [4] which make use of database operators to address similarity join.

The similarity measure is defined by similarity functions. For textual data, token-based functions are widely used for measuring the similarity by considering a document of a string as a bag of words (tokens). Thus, some existing works on the similarity join over text data use the overlap of two token sets, i.s., Jaccard similarity, as the similarity function. There are two popular methods in solving the problem of filtering candidate pairs: Local Sensitive Hash and Prefix Filtering. LSH is appropriate for the joins with lower threshold [12], while the prefix filtering is superior in a higher threshold for the exact case. The technique has been proposed in [14] to reduce the amount of the candidate pairs and thus improve the overall computational cost. The length filter and prefix hash algorithm have also been proposed in [15] for the same purpose.

In addition, the position information of prefix introduced in [13] filters the unqualified candidate pairs that have passed the length-filter. The algorithm proposed in [16] generates characters for different types of strings based on the adaptive selection of prefix length. However, it took much time in selecting the length of prefix.

To achieve fast *Entity Resolution*, several research works have started to study distributed processing. MapReduce framework [17] has been applied to improve the efficiency of finding all similar pairs from terabytes of data. The basic idea is to partition data into subsets match records in the same subset. The work in [18] has studied the problem of set-similarity join that runs multiple MapReduce steps to find similar pairs on set-based data. MRSimJoin [19], implemented on Hadoop, focuses on distributing the data based on window-pair partition in which the subsets are small enough to be processed in a single node. However, MRSimJoin requires its dataset to lie in a metric space. The work in [20] focuses on analyzing cost to pick the optimal algorithm when using MapReduce. A MapReduce implementation of set similarity proposed in [21] includes three stages, computing data statistics, outputting RID pairs of similar records, and generating actual pairs of joined records. Dealing with large documents, [22] investigated the virtual node method.

3   The Entity Resolution Framework for Clustering

In this section, we first describe the proposed *Entity Resolution* framework for clustering, based on the MapReduce model, to improve the efficiency by exploiting distributed processing. In our framework, the process of *ER* and clustering consists of two steps: 1) matching similar objects based on similarity join, 2) clustering objects based on the graph of similarities, i.e., each connected sub-graph is regarded as a cluster. In each step, we propose a set of MapReduce based algorithms to partition large datasets and distribute them into multiple nodes. The architecture of our framework for *ER* is shown in Figure 2.

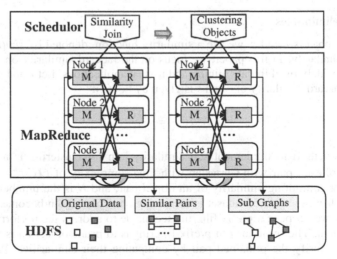

Fig. 2. The architecture of Entity Resolution framework

In the architecture, we use the Hadoop distributed file system (HDFS) to store data that includes original datasets, similar object pairs generated with similarity joins, and even clustering results. To achieve high performance on massive data, we make use of the MapReduce model as the execution engine to process diverse *ER* jobs. Each *ER* job is divided into multiple Map tasks and Reduce tasks to execute on multiple nodes. We carefully design map tasks so that data can be accurately distributed into appropriate nodes. In this architecture, we design two functional components to handle the jobs of (i) Similarity Join and (ii) Object Graph Cluster. The similarity join component executes a self similarity join on the given dataset, and returns all matched pairs of objects whose similarities are above the given threshold. Then the Object Graph Cluster component cluster objects into clusters based on similar pairs such that members of a cluster are in similar to each other. In summary, we regard a connected sub graph as a basic cluster of objects.

In both phases of similarity join and clustering, algorithms require multiple MapReduce job loops to run the task. For our similarity join algorithm, to reduce candidate pairs, it first uses one MapReduce loop to sort elements of objects based on their frequency for selecting objects' features. Then, in an additional MapReduce loop, we map each object into multiple nodes based on elements in its feature, and verify similarities between objects in each node. The clustering algorithm used in Object Graph Cluster is based on the *Disjoin Set* algorithm which requires merge groups with iterative function. Thus, we design an algorithm with multiple MapReduce loop for clustering in next section.

4   Algorithms for Similarity Join and Clustering

In this section, we first introduce the preliminaries for filter-based similarity join in Section 4.1. Then in Section 4.2, we propose the improved prefix filter algorithms for similarity join on MapReduce Model. Finally Section 4.3 presents the algorithm for clustering objects with similarities.

4.1   Preliminaries

Given two objects r and s, we use a similarity function, denoted by *sim(r, s)*, to measure their similarity. In this paper, we focus on the Jaccard similarity on *sets* or *strings* which are widely used in data mining and textual applications. Let r and s be two token sets, the Jaccard similarity of r and s is given by Eq. (1):

$$SimJ(r,s) = \frac{|r \cap s|}{|r \cup s|} \tag{1}$$

The basic method to implement self similarity join for clustering is to compute the similarity of each pair with Cartesian with time complexity of $O(C \cdot N^2)$, in which C is the cost for computing similarity of an object pair, and N is the number of objects in dataset. When the scale of dataset is increasing, the cost of join becomes quite expensive. Thus, we propose a prefix filtering technique to address the performance issue in similarity join. The basic idea of prefix filtering is to prune object pairs that cannot be similar and verify the remained pair by computing their similarities. When the threshold is high, the cost of similarity join with prefix filtering will reduce significantly.

Let L_r, and L_s be the size of object r and s, respectively, θ be the threshold of the similarity, and *prefix(r)* denotes the elements prefix of object r. According to prefix filter principle, it first sorts the elements of all objects based on frequency to get a global ordering. Then it sorts the elements of each object with the global ordering, and gets top t elements as the prefix of object. The t of object r is obtained by $\lfloor (1-\theta)L_r \rfloor +1$. With this value of t, if an object s is similar with r, the prefix of s and r must have at least one common element in their prefix sets, $prefix(r) \cap prefix(s) \neq \varnothing$. In the MapReduce model, we map objects into distributed nodes based on their prefix elements. This can ensure s and r are mapped into at least one common bucket.

For example, given three objects r_1, r_2 and r_3, where $r_1 =(EACBD)$, $r_2=(ACBD)$ and, $r_2=(CBDF)$, where their elements are ordered based on global frequency in the dataset, and the threshold is specified as $\theta = 0.7$. The prefix size of each object is computed by $t(r)= \lfloor (1-\theta)L_r \rfloor +1$. Thus, we get $prefix(r_1)=(E, A)$, $prefix(r_2)=(A)$ and $prefix(r_3)=(C)$. Accordingly r_1 is mapped into *two* buckets (E, A), r_2 is mapped into bucket A, and r_3 is mapped into bucket C. Next, we verify the similarity of object r_1 and r_2 in the bucket A and get the similar pair (r_1, r_2). Note that we don't need to verify the similarity of (r_1, r_3) and (r_2, r_3) since they had no common prefix elements.

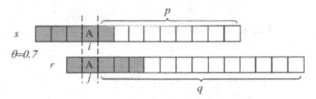

Fig. 3. An example of *PPjoin* and *PPjoin+* filter

When an object r is mapped into a bucket based on its element e, we use the tuple $(e, r(e), L_r)$ to denote the position information of e in r, where $r(e)$ is the position of e in the *prefix(r)*. Thus, the position information is further used to improve filtering performance in *PPjoin* algorithm. An example is shown in Figure 3. Objects r and s are mapped into the bucket A with their position information. Suppose it has $r(A)=j$ and $s(A)=i$, then the suffix length of r and s in A is $q= L_r-j$ and $p= L_s-i$. For the suffix of r and s, the maximum number of their common elements is $min(p, q)$. Thus, the size of element intersection is at most $|i-j|+min(p, q)$, and the size of elements union is at least $max(L_s, L_r)$. We use $SimP(r, s)$ to denote the maximum value of estimate similarity for *PPjoin*. If $SimP(r, s)$ is not higher than θ, the pair of (r,s) should be filtered in the verification. So the value of $SimP(r, s)$ must satisfy Eq. (2).

$$SimP(r,s) = \frac{|i-j|+min(p,q)}{max(L_s,L_r)} \geq \theta \qquad (2)$$

Let us consider the above facts that transforming the inequality $SimJ(r, s) \geq \theta$ is effective to prune the number of the candidate pairs. Therefore we can propose a proper way to find an estimated expression that is closer to the real value, and get a better filter performance.

4.2   The All Prefix Filter for Similarity Join

Considering the cases where two objects may have more than one common prefix, an efficient way is to compute them in one bucket based on one common element and filter other candidate pairs. As shown in Figure 4, *Prefix(r)* and *Prefix(s)* has two common elements *A* and *B*, the positions of *B* are ahead of the positions of *A* in *r* and *s*. As the prefix *B* must have been dealt in the *B* bucket, so the computation in the bucket *A* is redundant and thus can be avoid.

To filter the candidate pairs, existing works have proposed the position filter principle that using the positions information being indexed can prune the candidate pairs to be verified. However, it doesn't know whether it is the first common prefix and whether the estimated size of the intersections (r, s) is close to the real length. As a result, a larger number of candidate pairs may still need to be verified.

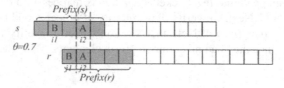

Fig. 4. Multiple common elements in prefixes

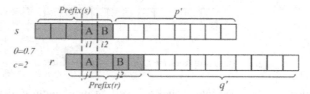

Fig. 5. An example of *appjoin* filter

In this paper, we use the information of all prefix elements to filter candidate pairs. We name this filtering principle as *appjoin*. Figure 5 shows an example of *appjoin* filter, where *r* and *s* has a common element *A* in prefixes with position *i* in *s* and position *j* in *r*. If the positions *(i, j)* are not the first common key in prefix, we will filter it out. It's obvious that the more accurate we estimate the ratio of the intersections and the union of elements of two objects, the better the filter's performance is. Different from the position prefix filter, we make a full scan in the prefixes of each candidate pair to compute the number *c* of the common elements in the two prefixes, *prefix(r)* and *prefix(s)*. Suppose *p'* is the suffix of *r* and *q'* is the suffix of *s*. The maximum size of intersection is $c+min(p',q')$ and the minimum size of union is $L_s+L_r-min(p', q')-c$. Therefore, we propose the *Estimated Similarity* of *appjion* denoted as *SimE(r, s)*. The *SimE(r, s)* is defined by Eq. (3) below:

$$SimE(r,s) = \frac{c+Min(p',q')}{L_s+L_r-Min(p',q')-c} \tag{3}$$

Consider the value of *SimE(s,r)*, we can get $SimE(r, s) \geq SimJ(r, s)$ since it has $c+min(p',q') > |r \cap s|$. Therefore, given a pair of objects *(r, s)*, if there exists *SimE(r,*

$s) < \theta$, $SimJ(r, s)$ must be lower than θ and the pair (r, s) *is filter out*. Then we compare the value of $SimE(r, s)$ with $SimP(r, s)$. For the value of $c+min(p',q')$, it has filtered elements in prefixes that do not exist in both objects, so we can get $c+min(p',q')</i-j/+min(p,q)$. Therefore, we have $SimE(r,s){\leq}SimP(r,s)$, and thus $SimJ(r,s){\leq}SimE(r,s){\leq}SimP(r,s)$. As the value of $SimE(s,r)$ is more close to the value of $SimJ(s,r)$, the *appjoin* filter can more efficient filter candidate pairs for verification.

4.3   Clustering Objects with Similarities

After verifying all candidate pairs of objects, we get all similar pairs with their similarities. We regard the similar relationship between objects as a graph data, and each connected sub graph is a cluster. Therefore, the problem of finding clusters of objects can be mapped to construct connected sub graphs with similarities of pairs. It can be defined as $SubG(R)=\{sg/ \forall x, y{\in}sg \rightarrow (x,y{\in} R) \land path(x, y), \forall x{\in}sg, z{\notin}sg \rightarrow path(x, z)=false\}$, where R is the set of objects, and $path(x, y)$ denotes there is a similarity path between x and y.

Algorithm 1. The MapReduce-based *Disjoint Sets* algorithm

Input：Similar pairs of objects

Output：Clusters of objects

1. **MAP** (key=*unused*,value=*text*)
2.   key=*r;* value=*(min(HashNO,r,s),<r,s,sim>)*; output(key,value);
3. **REDUCE**(key=*item*,values=*iterator*)
4.   **while** *iterator*.hasnext()
5.   HashNo=*listMerger*(results, it.next()); *count++;*
6.   **if** *count==*1
7.   *counter*.increment(1)
8.   **while** *resultList*
9.   output(*record, HashNO*);
10. **MAP** (key=*unused*, value=*text*)
11.   output(value+key, value);
12. **REDUCE** (key=*item*, values=*iterator*)
13.   **while**
14.   minHashNo=min(minHashNo,values.next);
15.   output(key,*HashNo*);

Since we use MapReduce model to execute the similarity join with our *appjoin* algorithm, all similar pairs are stored in distributed nodes. We need combine objects of sub graphs as a cluster. To achieve this problem, we use the algorithm based on *Disjoint Sets*. We first annotate each object in dataset with a unique ID. Then in each MapReduce node, all similar pairs are sorted based on the smaller object IDs in them. Therefore, the similarity graph is transformed into a directed graph in which each edge always directs from an object node with larger ID to an object node with smaller ID. Finally, we execute multiple MapReduce tasks to merge connected graphs with algorithm based on *Disjoint Sets*. The algorithm of clustering for MapRecude model is shown in Algorithm 1. The final result of this algorithm is all connected sub graphs that are mapping to clustering groups.

However, the *Disjoint Set* algorithm may include too many MapReduce loops if the graph contains long paths between nodes. To reduce the number of MapReduce loops, we further propose a cache-based algorithm. The algorithm caches the records of Map tasks into two hash tables, *rtable* and *ntable*. Where *rtable* holds similar pairs and their cluster ID, and *ntable* holds the smaller object ID of each similar pair. Then a merge operation is executed on two hash tables. When the cache is full or there is no new similar pair, it output similar pairs with their cluster ID. The detail of algorithm is shown in Algorithm 2.

Algorithm 2. The Cache Merge Algorithm

---
1.   Intiliztion (rtable, ntable)
2.   Map(key=*unused*, value=*text*)
3.   *record*=parser(text);
4.   **if** rtable.size$\geq c$
5.   output(rtable);
6.   *rtable*.clean(); *ntable*.clean();
7.   *rtable*.add(value);
8.   *ntable*.update(record.*s*); *ntable*.update(record.*r*);
9.   Close()
10.   output(*table*);

---

5   Experimental Results

5.1   Experimental Setup

In this paper, we implement the Entity Resolution framework based on the proposed techniques, and conduct an experimental evaluation. The framework is deployed on a Hadoop cluster consists of a master node and 20 slaver nodes. Each node has 8GB memory and 3.1GHz four Core i3-2100 Processor with Gigabit Ethernet. The operating system of each node is Red Hat Enterprise Linux 6.1(Santiago).

In our experiments, we use the real world datasets collected from DBLP and CiteseerX. The DBLP dataset consists of almost 1.2 million records which include information about conference papers, conferences, journals, authors and so on. Our experiments mainly use conference papers. The CiteseerX data includes more than 1.5 million records of papers with citation information. To evaluate the performance of our algorithms over large-scale data, we also generate records based on existing real data. This allows us to increase the scale of datasets to evaluate the similarity join. In our experiments, we execute similarity join algorithms to achieve entity evaluation task.

5.2   Experimental Results

1)   The performance on different data size

In this experiment, we focus on evaluating the time cost of our similarity join algorithms with different scale of data. We use various sizes of experimental data to compare the performances of our *appjoin* method with *ppjoin* method (which serves as the baseline). We set the value of similarity threshold as 0.8, and the size of data has a range from 5 to 25 million objects.

Figure 6 shows the results of time cost on different filter methods. It compares the *ppjoin* algorithm and the *appjoin* algorithm. Note that we do not compare with the Cartesian method in the evaluation due to its unacceptable cost in processing large scale data. From the results, we find that our *appjoin* method has better performance on a large data set. With the scale of data increasing, *appjoin* method can achieve a better performance than the *ppjoin* method. In our experiments, we also measure the number of filtered record pairs before the verification stage of entity resolution. As shown the large data set generates much more pairs and our filter method is more efficient to filter unqualified candidate pairs as data size increases.

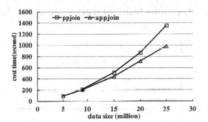

Fig. 6. The result of time cost on *ppjoin* and *appjoin* algorithms

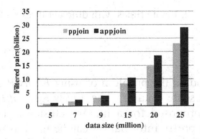

Fig. 7. Size of filtered record pairs on *ppjoin* and *appjoin* algorithms

2)   The performance on different threshold

We also evaluate the performances of different methods under different values of similarity thresholds, varied from 0.65 to 0.95. The evaluation is performed using a dataset of 7 million records. Figure 8 shows the cost time of the *appjoin* and *ppjoin* methods under different thresholds. These experimental results prove that our *appjoin* method outperforms the *ppjoin* method, especially when the threshold has a lower value.

As shown in Figure 8, the differences between the compared two methods are significant as the threshold decrease because the size of candidate record pairs becomes larger. Our *appjoin* method performs well in low threshold, because our *appjoin* method can filter more candidate pairs than *ppjoin*.

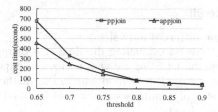

Fig. 8. The cost time on different thresholds

3) The performance on cache-based algorithm

We further evaluate the performances of our cache-based algorithm for clustering under different values of similarity thresholds, varied from 0.65 to 0.95. The evaluation use a setting with 100 million cached similar pairs. Figure 9 shows the number of MapReduce loops of *no-cache* and *cache-based* method for the Disjoint Set under different thresholds. This experimental result proves that the *cache-based* algorithm achieves a higher performance on reduce MapReduce loops, especially when the threshold has a lower value. This is because connected sub graphs hold a larger scale when the threshold is low.

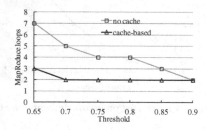

Fig. 9. Iterate times with different thresholds

6   Conclusion

In this paper, we have studied the problem of using *Entity Resolution* to support efficient clustering on large-scale dataset. The *Entity Resolution* framework based on MapReduce model was proposed for achieving clustering tasks with the similarity join. We argue that more prefix information can be used to filter candidate pairs before verify stage and propose a new *appjoin* algorithm for similarity join to improve filtering. We also discuss the algorithm for clustering objects with similar pairs. We have implemented our method and compared with a state-of-the-art method, i.e., *ppjoin*. Experimental results on real-world datasets show that our method achieves high performance for similarity join, especially when a lower value of similarity threshold is setup for large-scale data clustering.

Acknowledgement. Project supported by the National Basic Research Program of China (973 Program) under grant No. 2012CB316201, the National Natural Science Foundation of China (Grant No. 61003060), the Fundamental Research Funds for the Central Universities (Grant No. N110404010), and the 863 High Technology Foundation of China (Grant No. 2012AA010704).

References

1. Benjelloun, O., Garcia-Molina, H., Menestrina, D., Su, Q., Whang, S.E., Widom, J.: Swoosh: a generic approach to entity resolution. VLDB J., 255–276 (2009)
2. Arasu, A., Ganti, V., Kaushik, R.: Efficient Exact Set-Similarity Joins. In: VLDB, pp. 918–929 (2006)
3. Dohnal, V., Gennaro, C., Zezula, P.: Similarity Join in Metric Spaces Using eD-Index. In: Mařík, V., Štěpánková, O., Retschitzegger, W. (eds.) DEXA 2003. LNCS, vol. 2736, pp. 484–493. Springer, Heidelberg (2003)
4. Chaudhuri, S., Ganti, V., Kaushik, R.: A Primitive Operator for Similarity Joins in Data Cleaning. In: ICDE (2006)
5. Broder, A.Z., Glassman, S.C., Manasse, M.S., Zweig, G.: Syntactic Clustering of the Web. Computer Networks, 1157–1166 (1997)
6. Cho, J., Shivakumar, N., Garcia-Molina, H.: Finding Replicated Web Collections. In: SIGMOD Conference, pp. 355–366 (2000)
7. Spertus, E., Sahami, M., Buyukkokten, O.: Evaluating similarity measures: a large-scale study in the orkut social network. In: KDD, pp. 678–684 (2005)
8. Gibson, D., Kumar, R., Tomkins, A.: Discovering Large Dense Subgraphs in Massive Graphs. In: VLDB, pp. 721–732 (2005)
9. On, B., Elmacioglu, E., Lee, D., Kang, J., Pei, J.: Improving Grouped-Entity Resolution Using Quasi-Cliques. In: ICDM, pp. 1008–1015 (2006)
10. Chaudhuri, S., Ganti, V., Xin, D.: Mining Document Collections to Facilitate Accurate Approximate Entity Matching. In: PVLDB, pp. 395–406 (2009)
11. Jacox, E.H., Samet, H.: Metric space similarity joins. ACM Trans. Database Syst. 33(7), 1–38 (2008)
12. Lee, H., Ng, R.T., Shim, K.: Similarity Join Size Estimation using Locality Sensitive Hashing. In: PVLDB, pp. 338–349 (2011)
13. Xiao, C., Wang, W., Lin, X., Yu, J.X., Wang, G.: Efficient similarity joins for near-duplicate detection. ACM Trans. Database Syst., 15 (2011)
14. Ribeiro, L.A., Härder, T.: Efficient set similarity joins using min-prefixes. In: Grundspenkis, J., Morzy, T., Vossen, G. (eds.) ADBIS 2009. LNCS, vol. 5739, pp. 88–102. Springer, Heidelberg (2009)
15. Bayardo, R.J., Ma, Y., Srikant, R.: Scaling up all pairs similarity search. In: WWW, pp. 131–140 (2007)
16. Wang, J., Li, G., Feng, J.: Can we beat the prefix filtering?: an adaptive framework for similarity join and search. In: SIGMOD Conference, pp. 85–96 (2012)
17. Dean, J., Ghemawat, S.: MapReduce: Simplified Data Processing on Large Clusters. In: OSDI, pp. 137–150 (2004)
18. Vernica, R., Carey, M.J., Li, C.: Efficient parallel set-similarity joins using MapReduce. In: SIGMOD Conference, pp. 495–506 (2010)
19. Silva, Y.N., Reed, J.M.: Exploiting MapReduce-based similarity joins. In: SIGMOD Conference, pp. 693–696 (2012)
20. Afrati, F.N., Sarma, A.D., Menestrina, D., Parameswaran, A.G., Ullman, J.D.: Fuzzy Joins Using MapReduce. In: ICDE, pp. 498–509 (2012)
21. Vernica, R., Carey, M.J., Li, C.: Efficient parallel set-similarity joins using MapReduce. In: SIGMOD Conference, pp. 495–506 (2010)
22. Wang, C., Wang, J., Lin, X., Wang, W., Wang, H., Li, H., Tian, W., Xu, J., Li, R.: Map-DupReducer: detecting near duplicates over massive datasets. In: SIGMOD Conference, pp. 1119–1122 (2010)

Exploring the Intervention Problem with the Networked Poisson Process in a Real Heterogeneous Social Network*

Wang Yue

Department of Computer Science, School of Information
Central University of Finance and Economics. 100081 Beijing, P.R. China
wangyuecs@cufe.edu.cn

Abstract. We model the microblog as a heterogeneous network with individuals in different roles, and introduce "intervention" to describe the phenomenon that some tiny variations about a small subset of the individuals change the whole network's status. Our main contributions are: (1) proposing the Networked Poisson Process (NPP) to model the dynamic tweeting patterns for the microblog; (2) formalizing a NP-hard problem: the intervention impacts maximization (IIM); (3) proposing heuristic algorithms to solve the IIM; (4) providing sufficient experiments to test our methods. The experimental results show that NPP captured the real interaction patterns for the users in the microblog.

Keywords: intervention, heterogeneous network, networked system modeling, data mining, graph mining.

1 Introduction

Intervention is pervasive in many fields: some insignificant events may cause a great disturbance on the internet by a small group of intended online promoters; some tiny mutations of an single gene may cause the critical illnesses for human; several news events may result the financial crisis for the whole world.

To study the rules of the intervention phenomenon in the microblog from a heterogeneous network view, our main contributions of this paper are: (1) proposing the networked Poisson process (NPP) to model the patterns for the dynamic tweeting process in the social network by considering the interaction between the latent features of people, and proposing a parameters learning algorithm, NPP-learning, to generate the NPP model; (2) formalizing the concepts about the intervention and defining the problem of the intervening impact maximization(IIM); proving that the problems IIM is NP-hard;(3) proposing algorithms greedy-IIM to find the solutions for the problems IIM; (4) providing the experiments on the real microblog data to verify the effectiveness of our methods.

---

* This research was supported by National Nature Science Foundation of China (Grant No. 61272398), Nature Science Foundation of Beijing (Grant No. 4112053), and Graduate Student Education Innovation Project of Central University of Finance and Economics (CUFE).

F. Li et al. (Eds.): WAIM 2014, LNCS 8485, pp. 150–154, 2014.

2   Problem Definition

We extend the methods in [1] to define the heterogeneous social network.

Definition 1. *Heterogeneous Social Network.* *Let V be a set of users from R types $X = \{X_r\}_{r=1}^R$, then a heterogeneous social network is a weighted graph $G = \langle V, E, W \rangle$, where $R \geq 2$; edge set E is the relations between the users in V, and W is the set of the trust degrees among the users v_1, v_2 ($\forall \langle v_1, v_2 \rangle \in E$).*

Definition 2. *Networked Poisson Process (NPP) model.* *Let G be a heterogeneous social network with 2 types of users (X_{r_1} = intended, X_{r_2} = ordinary), then $\{N(t)_v\}_{\forall v \in V}$ is a networked poisson process, where $N(t)_v$ is a Poisson-distributed random variable with $\lambda_{v,t}$ representing the tweeting rate of v at time t; for $\forall v$ in type X_{r_1}, $\lambda_{v,t} \equiv \lambda_v(t = 0, 1, 2, 3...)$; for $\forall v \in V$ and v in type X_{r_2}, the initial $\lambda_{v,0}$ is the initial tweeting rate of v, and $\lambda_{v,t}$ for $t > 0$ can be calculated by equation $\lambda_{v,t} = \alpha\lambda_{v,t-1} + (1 - \alpha)\frac{\sum_{\forall u \in NBS(v)}[W_{u,v} \cdot N(t)_u]}{T}$.*

In definition 2, $NBS(v)$ is the neighbors' set of v in G, $W_{u,v}$ is u's trust degree toward v, $\alpha(\alpha \in [0,1])$ is a priori parameter to adjust the weight of the social relationships; T is the average time period for the observation; λ_v is a constant, and it measures the "tweeting times per hour" of v.

Definition 3. *Intervening Action and Impact.* *Given a heterogeneous social network $G = \langle V, E, W \rangle$ with 2 types of users (X_{r_1} = intended, X_{r_2} = ordinary), an **intervening action** of v ($v \in V$), $A(v)_{r_2,r_1}$, happens when v turns his role from X_{r_2} to X_{r_1}; If $A(v)_{r_2,r_1}$ takes place by setting v's tweeting rate to λ_v, then the **intensity** of the $A(v)_{r_2,r_1}$ is λ_v; denote this intensity as $I(A(v)_{r_2,r_1}) = \lambda_v$; The **intervening impact** of S ($S \subseteq V$) on G, $Y(S|G)_{T,T'}$, can be calculated by the equation: $Y(S|G)_{T,T'} = E(\sum_{\forall v \in V}|N(v)_T - N(v)_{T'}|)$.*

Where T and T' are two non-intersected time periods; $N(v)_T$ and $N(v)_{T'}$ are the simulated tweeting numbers for v by the NPP model within T and T'. In the rest sections, when the context is fixed, we use $A(v)$ to instead $A(v)_{r_2,r_1}$ and simplify $Y(S|G)_{T,T'}$ to $Y(S|G)$ or $Y(S)$.

Problem 1. **Intervening Impact Maximization, IIM.** Given a heterogeneous social network $G = \langle V, E, W \rangle$ and its corresponding NPP in the time period T_0, suppose $\forall v \in V$ are in type X_{r_1} initially; the **problem IIM** is to find out the best set S ($S \subseteq V$) with k users to make the maximum $Y(S|G)_{T_0,T}$ by taking the intervening action $A(v)$ for $\forall v$ in S with the fixed λ^* in a time period T.

Theorem 1. *The problem IIM is NP-hard.*

Proof. By the definition 2 and 3, for a set S_1 of intended users, we have:
$\lambda_{v,t} = \alpha\lambda_{v,t-1} + (1 - \alpha)\frac{\sum_{\forall u \in (NBS(v) - S_1)} W_{u,v} N(t)_u}{T} + C$; **Where** $C = (1 - \alpha)\sum_{\forall u \in NBS(v) \cap S_1} W_{u,v}\lambda_v$ is a positive value. As not every new intended user (in type X_{r_2}) will be in the set of $NBS(v)$, the increasing speed of $NBS(v) \cap S_1$ is slower than S_1. Then with the two sets: S_1, S_2 ($S_1 \subseteq S_2$) and $v'(v' \notin S_1)$, the

equation $Y(S_1 \bigcup \{v'\}|G) - Y(S_1|G) \geq Y(S_2 \bigcup \{v'\}|G) - Y(S_2|G)$ holds. Therefore, $Y(S|G)$ is a non-negative monotone submodular function, by [2], to search a set S of k users to maximize $Y(S|G)$ is NP-hard, and the greedy algorithms can be applied on it to make a $(1-\frac{1}{e})$-approximation [3].

3   Algorithms for NPP and Intervention Problem

Algorithm for the NPP Learning. We list the detail steps of the algorithm NPP-learning in the algorithm 1. Where "Poisson-random$(u, \lambda_u)_T$" is a Poisson random number in time period $T(D_{test})$; D_{train} and D_{test} are the training and testing set, and the Λ is a vector of the λ_v for $\forall v$ in V.

Algorithm for IIM. The pesudocode of the algorithm for IIM is listed in the algorithm 2. In algorithm 2, the function "$\arg\max_{v^* \in V}(*)$" contains the detail steps to search the best node to make the maximum intervening impact.

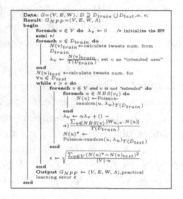

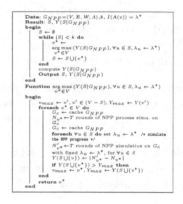

Fig. 1. Algorithm 1. NPP-Learning   **Fig. 2.** Algorithm 2. greedy-IIM

4   Experiments and Discusses

Data Set and Configs. We extracted the tweets and user relations related to 3 hot topics on Twitter. The details of this data are listed in the table 1. We set the trust degrees to 1 for all the pairs of users to simplify the computation.

Parameters Learning for the NPP. Figure 3 records the results of the experiments about the parameters learning for NPP. One can see from figure 3 that the learning error (RMSE) for NPP is converging with the progress. We did another 2 experiments on the topics "#GalaxyS4" and "#NBAplayoffs", and both results show the similar converging rules. The more, with the decreasing of α, the learning error is getting smaller. This proves that the tweeting behaviors of the users in Twitter are really affected by their social relations.

Table 1. A Summary of Data Sets

	#GalaxyS4	#BostonBomb	#NBAplayoffs
tweets	34,310	204,130	39,026
seeds	24,963	121,886	30,863
nodes	2,138,594	465,863	1,853,264
edges	3,746,768	930,587	2,927,329
start	2013-5-3	2013-4-20	2013-5-4
end	2013-5-7	2013-4-24	2013-5-7

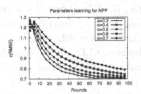

Fig. 3. Parameters learning for NPP

Comparison for NPP, IC, LT. We implemented the algorithms greedy-IC and greedy-LT by following [4][5]. We compare the intervening impacts and the information spreading range for the top users discovered by the algorithms in the table 2. As it is shown in the table 2, the greedy-IIM found out the set S with the biggest intervening impacts and the information spreading range.

Table 2. Comparison of the greedy-IIM, greedy-LT and greedy-IC

	Results of greedy-IIM			Results of greedy-LT			Results of greedy-IC		
rank of v	tweets	followers	$Y(\{v\}\|G^*)$	tweets	followers	$Y(\{v\}\|G^*)$	tweets	followers	$Y(\{v\}\|G^*)$
1	84	5,001	0.44	173	445	0.17	83	2,185	0.55
2	0	0	N/A	0	0	N/A	128	441	0.12
3	51	2,904	0.1	83	5,001	0.69	0	0	N/A
4	10	192	0.99	0	0	N/A	259	5,226	0.71
5	80	2,185	0.8	0	0	N/A	0	0	N/A
6	0	0	N/A	0	0	N/A	0	0	N/A
7	0	0	N/A	216	5,226	0.01	0	0	N/A
8	369	5,226	0.64	0	0	N/A	75	5,001	0.83
9	0	0	N/A	108	727	0.53	0	0	N/A
10	0	0	N/A	0	0	N/A	0	0	N/A
sum	594	15,138	2.97	580	11,114	1 4	545	12,482	2.21

5   Conclusion

In this paper, we model the microblog as a heterogeneous network, propose the NPP model to capture the tweeting behaviors for the users in the microblog, and address a NP-hard problem: the intervening impact maximization (IIM). Then, we propose the algorithm NPP-learning to learn the NPP model, and propose the algorithm greedy-IIM to solve the IIM. Our experiments show that NPP captured the real interaction patterns for the users in the microblog, and greedy-IIM can discover the users to make more intervening impact than the results of greedy-IC and greedy-LT. Our future work will focus on the knowledge discovering in a heterogeneous network with more than two types of individuals. The more, how to predict the best time or the users to take actions and how to apply our methods to solve the intervention problems in other fields are also the interesting and challenging topics for our future research.

References

1. Sun, Y., Han, J., Aggarwal, C.C., Chawla, N.V.: When will it happen?: relationship prediction in heterogeneous information networks. In: WSDM 2012, pp. 663–672 (2012)

2. Kempe, D., Kleinberg, J., Tardos, E.: Maximizing the spread of influence through a social network. In: Proceedings of the Ninth ACM SIGKDD International Conference on Knowledge Discovery and Data Mining, pp. 137–146. ACM (2003)
3. Nemhauser, G.L., Wolsey, L.A., Fisher, M.L.: An analysis of approximations for maximizing submodular set functions-I. Mathematical Programming 14(1), 265–294 (1978)
4. Chen, W., Wang, Y., Yang, S.: Efficient influence maximization in social networks. In: Proceedings of the 15th ACM SIGKDD International Conference on Knowledge Discovery and Data Mining, pp. 199–208. ACM (2009)
5. Wang, C., Chen, W., Wang, Y.: Scalable influence maximization for independent cascade model in large-scale social networks. Data Mining and Knowledge Discovery 25(3), 545–576 (2012)

An Infinite Latent Generalized Linear Model

Jianbo Luo and Jiangtao Ren

Sun Yat-sen University, Guangzhou 510006, China
issrjt@mail.sysu.edu.cn

Abstract. We propose an Infinite Latent Generalized Linear Model (ILGLM), a Dirichlet process mixture of generalized linear model in latent space for classification problem. In ILGLM, we assume latent variable z_n is generated from a low-dimensional DPM model in latent space, and the corresponding observed feature x_n and class label y_n are generated from some latent probability model and local linear classification model independently conditioned on z_n. Then in ILGLM, we will jointly learn the latent variable model and multiple local generalized linear model under the framework of Dirichlet process mixture. On one hand, ILGLM can model the multiple local linearity of data distribution adaptively according to data complexity; on the other hand, it avoid the curse of dimensionality problem. ILGLM can be extended to semi-supervised setting, training the model using both labeled and unlabeled data. Because ILGLM is a general model framework, it can incorporate any kind of latent variable models and linear classification models. Then we implement ILGLM based on Factor Analysis and MultiNomial Logit model, which results in the Infinite Latent MultiNomial Logit (ILMNL) model as an example of ILGLM. We also develop an approximate posterior inference algorithm for ILMNL using Gibbs sampling. Experiments on several real-world datasets demonstrate the advantages of ILMNL in dealing with high-dimensional data classification problems compared with competitive models.

Keywords: nonparametric Bayesian, Dirichlet process, generalized linear model.

1 Introduction

Dirichlet process mixture (DPM) models introduced by [2] is a promising tool to intelligently determine the optimal number of components according to the data distribution, which is a hard problem for finite mixture model like finite mixture-of-Gaussian or finite mixture-of-experts. Although applications of Dirichlet process mixture are mainly focused on clustering models, quite a few excellent classification models build on Dirichlet process prior have been proposed recently. They include infinite mixtures of Gaussian process models [4], Dirichlet process mixtures of MultiNomial Logit models (DPMNL) [5], Dirichlet process mixtures of generalized linear models (DPGLM) [6] and Dirichlet process mixture of large-margin kernel machines (iSVM) [7], etc. These models can be considered as special cases of mixture-of-experts models, since they all split the input space into many components, learn a linear classifier within each component, then form globally nonlinear classifiers. But instead of being specified by users, the number of components can be learned automatically from data. However,

F. Li et al. (Eds.): WAIM 2014, LNCS 8485, pp. 155–166, 2014.

since these models train classifiers in the input space, if the data dimension is very high, they will suffer from the curse of dimensionality problem.

Fortunately, many high-dimensional data, such as images, biological data, and social-network data often reside in a low-dimensional subspace or latent space. This fact draws great interest on the research of latent variable models, which focus on generating high-dimensional observed data from low-dimensional latent spaces. Most latent variable models [8], such as Probabilistic Principal Component Analysis (PPCA), Factor Analysis (FA) and Independent Component Analysis (ICA) are unsupervised, they don't incorporate class label information in the learning process. In order to solve this problem, [3] proposes Supervised Probabilistic PCA (SPPCA) model, and demonstrates that the latent variables learned from SPPCA are more discriminative than PPCA. [9] extends SPPCA by using a weighted objective likelihood function to adjust the relative importance weight on predicting covariates and associated response variables.

Inspired by unsupervised and supervised latent variable models, we try to extend DPGLM model [6] with them, which can train DPGLM model in low-dimensional latent space. In this paper, we propose an Infinite Latent Generalized Linear Model (IL-GLM). In ILGLM, we assume latent variable z_n is generated from a low-dimensional DPM model in latent space, and the corresponding observed feature x_n and class label y_n are generated from some latent probability model and local linear classification model independently conditioned on z_n, which are $P(x_n|z_n)$ and $P(y_n|z_n)$. Then in ILGLM, we will jointly learn the latent variable model and multiple local generalized linear model under the framework of Dirichlet process mixture. On one hand, ILGLM can model the multiple local linearity of data distribution adaptively according to data complexity; on the other hand, it avoid the curse of dimensionality problem.

More interestingly, ILGLM can be extended to semi-supervised setting, training the model using both labeled and unlabeled data. However, ILGLM is a general model framework, it can incorporate any kind of latent variable models and linear classification models. We realize it based on Factor Analysis and MultiNomial Logit model, which results in the Infinite Latent MultiNomial Logit (ILMNL) model as an example of ILGLM. An approximate inference algorithm based on Gibbs sampling is also proposed. We compare the performance of ILMNL with SVM, MNL and DPMNL. Experimental results on several real-world datasets demonstrate the good performance of our proposed model in dealing with high-dimensional data classification problems.

The remainder of this paper is organized as follows. After reviewing DPM and DPGLM in Section 2, we introduce ILGLM model in Section 3, then realize it with ILMNL model in Section 4. Section 5 illustrates experimental results on real-world datasets. Finally, we conclude this paper in Section 6.

2 Dirichlet Process Mixtures and DPGLM

2.1 Dirichlet Process Mixtures

Dirichlet process mixture can be understood as a special mixture model with the number of components being infinite large[10]. Suppose exchangeable random samples $\{x_n\}$ are drawn independently from a mixture model with $p(x_n) = \sum_{k=1}^{K} \pi_k f(x|\theta_k)$ where $f(x|\theta_k)$ denotes the density function for component k with parameter θ_k.

If we further assume that all parameters $\theta_{1:K}$ be drawn from base distribution G_0. With a symmetric Dirichlet prior for π_k and $c_n = k$ indicating x_n generated by component k, the mixture model can be expressed hierarchically as follows:

$$
\begin{aligned}
x_n|c_n, \Theta &\sim f(x|\theta_{c_n}) \\
c_n|\pi_{1:K} &\sim Discrete(\pi_1, ..., \pi_K) \\
\theta_k &\sim G_0 \\
\pi_1, ..., \pi_K &\sim Dir(\tfrac{\alpha}{K}, ..., \tfrac{\alpha}{K})
\end{aligned}
\tag{1}
$$

2.2 DPGLM

DPGLM builds on Dirichlet process mixtures and generalized linear models(GLM). It assumes covariate x_n is generated from a mixture of exponential-family distributions, and the response variable y_n is modeled by GLM conditioned on x_n. With Dirichlet process mixtures, DPGLM can automatically learn the number of components according to data distribution, and achieve a global nonlinear classification or regression model if it learns more than one component. However, DPGLM may suffer from curse of dimension since it is trained in original input space. Its generative process is summarized as follows:

$$
\begin{aligned}
y_n|x_n, c_n, \theta_y &\sim GLM_y(y|x_n, \theta_{c_n, y}) \\
x_n|c_n, \theta_x &\sim p_x(x|\theta_{c_n, x}) \\
c_n|\pi_{1:K} &\sim Discrete(\pi_1, ..., \pi_K) \\
\theta_k &\sim G_0(\phi) \\
\pi_1, ..., \pi_K &\sim Dir(\tfrac{\alpha}{K}, ..., \tfrac{\alpha}{K})
\end{aligned}
\tag{2}
$$

where c_n indicates which component (x_n, y_n) belongs to, and $\theta_k = (\theta_{k,x}, \theta_{k,y})$ is the parameter of component k; p_x describe the covariate distribution with parameter $\theta_{k,x}$ for component k, and the form of GLM_y depends on problems. For example, Gaussian for regression or logistic for classification.

3   ILGLM and Its Semi-supervised Extension

Now we turn to our Infinite Latent Generalized Linear Model (ILGLM), an extension of DPGLM in latent space that tries to avoid the curse of dimensionality problem.

3.1   Notation

The covariates matrix after normalized is $\mathbf{X} = [x_1, ..., x_N] \in \mathbf{R}^{D \times N}$ and response variables vector $\mathbf{Y} = [y_1, ..., y_N]$. Latent variables matrix is $\mathbf{Z} = [z_1, ..., z_N] \in \mathbf{R}^{d \times N}$, and model parameters $\theta = [\theta_1, ..., \theta_K]$. $\mathbf{C} = [c_1, ..., c_N]$ is the indicator vector, and $\mathbf{X_k}, \mathbf{Y_k}, \mathbf{Z_k}$ are data generated from component k parameterized by $\theta_k = (\theta_{k,x}, \theta_{k,y}, \theta_{k,z})$. N, D, d and K represent data number, data dimension number, latent dimension number and component number respectively.

3.2 Graphical Model

In ILGLM, we assume covariate x_n and response variable y_n are independent conditioned on low-dimensional latent variable z_n. (This is different from DPGLM, because y_n is directly dependent on x_n as indicated in (2)) We also assume latent variable z_n is generated from a Dirichlet process mixture of exponential-family distributions $p_z(z|\theta_{c_n,z})$, covariate x_n is generated by z_n through latent variable model $p_x(x|z_n, \theta_{c_n,x})$, and response variate y_n conditioned on z_n is modeled with any suitable generalized linear model $p_y(y|z_n, \theta_{c_n,y})$. Here c_n indicates the index of component that sample (z_n, x_n, y_n) belongs to.

With above assumptions, our DP mixtures model is trained in low-dimensional latent space other than original high-dimensional input space. This strategy not only speeds up the training but also makes the model more robust. The latent variable model used in ILMNL is supervised because the posterior distribution of parameters θ_z is affected by $\mathbf{X}$ and $\mathbf{Y}$ jointly; besides, it will enjoy the advantages of local dimension reduction shared by all mixture latent variable models if ILMNL has more than one component.

The generative process of ILGLM can be summarized as follows, with the number of components $K \to \infty$:

$$
\begin{aligned}
y_n|z_n, c_n, \theta_y &\sim p_y(y|z_n, \theta_{c_n,y}) \\
x_n|z_n, c_n, \theta_x &\sim p_x(x|z_n, \theta_{c_n,x}) \\
z_n|c_n, \theta_z &\sim p_z(z|\theta_{c_n,z}) \\
c_n|\pi_{1:K} &\sim Discrete(\pi_1, ..., \pi_K) \\
\theta_k &\sim G_0(\phi) \\
\pi_1, ..., \pi_K &\sim Dir(\tfrac{\alpha}{K}, ..., \tfrac{\alpha}{K})
\end{aligned}
\tag{3}
$$

The graphical model of ILGLM is demonstrated in Figure 1, and its joint probability distribution is:

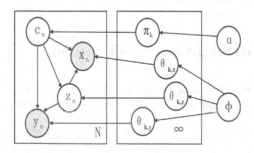

Fig. 1. Graphical model of ILGLM

$$p(\mathbf{X}, \mathbf{Y}, \mathbf{Z}, \mathbf{C}, \theta)$$
$$= \prod_{n=1}^{N} p_z(z_n|c_n, \theta_z) p_x(x_n|c_n, z_n, \theta_x) p_y(y_n|c_n, z_n, \theta_y) \tag{4}$$
$$\prod_{k=1}^{K} p_\theta(\theta_{k,x}, \theta_{k,y}, \theta_{k,z})$$

3.3 Semi-supervised Extension

Many datasets may be short of labeled data, and unlabeled data can be used to enhance the training of model parameter θ_x and θ_z, so we extend ILGLM to its semi-supervised form. Unlabeled covariates matrix after normalized is denoted as $\mathbf{U} = [u_1, ..., u_{N_1}] \in \mathbf{R}^{D \times N_1}$, latent matrix as $\mathbf{H} = [h_1, ..., h_{N_1}] \in \mathbf{R}^{d \times N_1}$, and indicator vector as $\mathbf{V} = [v_1, ..., v_{N_1}]$.

Both $\mathbf{H}$ and $\mathbf{U}$ are assumed to be generated exactly the same as $\mathbf{Z}$ and $\mathbf{X}$ in ILGLM. After incorporating unlabeled data, joint probability distribution becomes:

$$p(\mathbf{X}, \mathbf{Y}, \mathbf{Z}, \mathbf{C}, \mathbf{H}, \mathbf{V}, \theta)$$
$$= \prod_{n-1}^{N} p_z(z_n|c_n, \theta_z) p_x(x_n|c_n, z_n, \theta_x) p_y(y_n|c_n, z_n, \theta_y)$$
$$\prod_{n=1}^{N_1} p_z(h_n|v_n, \theta_z) p_x(u_n|v_n, h_n, \theta_x) \tag{5}$$
$$\prod_{k=1}^{K} p_\theta(\theta_{k,x}, \theta_{k,y}, \theta_{k,z})$$

3.4 Predictive Distribution

According to Bayes' theorem, predictive distribution for new data x is:

$$p(y|x, \mathbf{X}, \mathbf{Y}, \mathbf{U}) = \int p(y, z|x, \theta) p(\theta|\mathbf{X}, \mathbf{Y}, \mathbf{U}) dz d\theta \tag{6}$$

Unfortunately, it is infeasible to precisely evaluate $p(y|x, \mathbf{X}, \mathbf{Y}, \mathbf{U})$ since $p(\theta|\mathbf{X}, \mathbf{Y}, \mathbf{U}) = \int \frac{p(\mathbf{X}, \mathbf{Y}, \mathbf{Z}, \mathbf{U}, \mathbf{H}, \theta)}{p(\mathbf{X}, \mathbf{Y}, \mathbf{U})} d\mathbf{Z} d\mathbf{H}$ includes an intractable marginal likelihood $p(\mathbf{X}, \mathbf{Y}, \mathbf{U})$, and convolution of $p(y, z|x, \theta)$ and $p(\theta|\mathbf{X}, \mathbf{Y}, \mathbf{U})$ could be intractable too. Nevertheless, approximation inference based on variational method or MCMC [1] is possible, and we will demonstrate it via Gibbs sampling in next section.

4   An Example of ILGLM - ILMNL

As stated in last section, ILGLM can be applied to both regression and classification with suitable GLM. Besides, the mapping from latent variables to covariates could be but not limit to any unsupervised latent variable models we have mentioned. The example discussed here focuses on classification in semi-supervised setting, and we call it ILMNL short for Infinite Latent MultiNomial Logit.

4.1 ILMNL Formulation

For simplicity, within each component, latent variable z_n is assumed to be generated by a multivariate Gaussian with parameter $\theta_{c_n,z} = (\mu_{c_n}, \lambda_{c_n})$:

$$p_z(z_n|c_n, \theta_z) = N(z_n|\mu_{c_n}, \lambda_{c_n}^{-1}) \tag{7}$$

The distribution of x_n conditioned on z_n and c_n parameterized by $\theta_x = (A, \Lambda)$:

$$p_x(x_n|z_n, c_n, \theta_x) = N(x_n|Az_n, \Lambda^{-1}) \tag{8}$$

Note that θ_x is shared by all components with A as a $D \times d$ loading matrix, and Λ as diagonal noise precision matrix. It may seem a bit different from ILGLM, but we could actually consider it a special case with $\theta_{k,x}$ being the same for all k.

Finally, the distribution of y_n conditioned on z_n and c_n with $\theta_y = (W_k, b_k)$ is modeled by multinomial logit:

$$p_y(y_n|z_n, c_n, \theta_y) = \frac{\exp(b_{c_n,y_n} + w_{c_n,y_n}^T z_n)}{\sum_{m=1}^M \exp(b_{c_n,m} + w_{c_n,m}^T z_n)} \tag{9}$$

where $b_{c_n} = [b_{c_n,1}, ..., b_{c_n,M}] \in \mathbf{R}^{1 \times M}$ and $W_{c_n} = [w_{c_n,1}, ..., w_{c_n,M}] \in \mathbf{R}^{d \times M}$.

The simplified(omitting some hyper-parameters) graphical model in semi-supervised setting is shown in Figure 2. We summarize the generative process as follows:

1. Generate d random samples $\{\tau_{A_1}, ..., \tau_{A_d}\}$ from $Gamma(\alpha_1, \beta_1)$ independently;

2. Generate loading matrix A column-wise with each column $A_{:,col} \sim N(0, \tau_{A_{col}}^{-1} I_D)$;

3. Generate diagonal noise precision matrix Λ with each diagonal element $\Lambda_{i,i}$ from $Gamma(\alpha_2, \beta_2)$;

4. Generate K mixing proportion $\{\pi_1, .., \pi_K\}$ with $\pi_k \sim Dir(\frac{\alpha}{K}, ..., \frac{\alpha}{K})$;

5. Generate d random samples $\{\tau_{W_1}, ..., \tau_{W_d}\}$, with $\tau_{W_i} \sim Gamma(\alpha_3, \beta_3)$;

6. Generate d random samples $\{\tau_{b_1}, ..., \tau_{b_d}\}$ with $\tau_{b_i} \sim Gamma(\alpha_4, \beta_4)$;

7. Generate parameters of multinomial logit $\{(W_1, b_1), ..., (W_K, b_K)\}$ for all components row-wise, with each row $W_{:,row} \sim N(0, \tau_{W_{row}}^{-1} I_M)$ and $b_{:,row} \sim N(0, \tau_{b_{row}}^{-1} I_M)$;

8. Generate parameters of multivariate Gaussian $\{(\mu_1, \lambda_1), ..., (\mu_K, \lambda_K)\}$ for all components from their conjugate prior Gaussian-Wishart distribution as $(\mu_k, \lambda_k) \sim N\left(\mu_k|\mu_0, (\kappa_0 \lambda_k)^{-1}\right) \mathcal{W}(\lambda_k|T_0, \nu_0)$;

9. Generate indicator vector **C** for labeled data with $c_n \sim Discrete(\pi_1, ..., \pi_K)$, and indicator vector **V** for unlabeled data in the same way;

10. Generate latent covariates matrix **Z** for labeled data with $z_n \sim N(\mu_{c_n}, \lambda_{c_n}^{-1})$, and latent covariates matrix **H** for unlabeled data in the same way;

11. Generate covariates matrix **X** for labeled data with $x_n \sim N(Az_n, \Lambda^{-1})$, and covariates matrix **U** for unlabeled data in the same way;

12. Generate response variables vector $\mathbf{Y}$ for labeled data with $y_n \sim$ $arg\,max_m\left(b_{c_n,m} + w_{c_n,m}^T z_n\right)$.

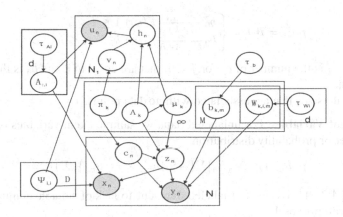

Fig. 2. Graphical model of ILMNL

4.2 Inference

Since the evaluations of both posterior distribution $p\left(\theta|\mathbf{X},\mathbf{Y},\mathbf{U}\right)$ and predictive distribution $p\left(y|x,X,Y,U\right)$ are analytically infeasible, we use Gibbs sampling to sample from $p\left(\theta|\mathbf{X},\mathbf{Y},\mathbf{U}\right)$, and approximate $p\left(y|x,X,Y,U\right)$ by a finite sum evaluated on these samples. For simplicity, We use notation $\cdot$ to denote all conditional variables. The algorithm of ILMNL is described in **Algorithm 1**.

Sample Indicators. Let $\mathbf{X}' = [\mathbf{X};\mathbf{U}]$, $\mathbf{C}' = [\mathbf{C};\mathbf{V}]$, and $N' = N + N_1$, and $\phi_k = \{\mu_k, \Lambda_k, W_k, b_k\}$. The conditional prior for $c_n' = k$ is:

$$p(c_n' = k|c_{-n}', \alpha) = \begin{cases} \frac{m_{n,k}}{N'-1+\alpha} & \exists j \neq n, c_j' = k \\ \frac{\alpha}{N'-1+\alpha} & \forall j \neq n, c_j' \neq k \end{cases} \tag{10}$$

The likelihood function $F(\phi_{c_n'})$ is different for labeled and unlabeled data. For labeled data,

$$F(\phi_{c_n'}) = N(x_n'|\mu_{c_n'}, \lambda_{c_n'}) \frac{\exp(b_{c_n',y_n} + w_{c_n',y_n}^T z_n)}{\sum_{i=1}^{M} \exp(b_{c_n',i} + w_{c_n',i}^T z_n)} \tag{11}$$

For unlabeled data,

$$F(\phi_{c_n'}) = N(x_n'|\mu_{c_n'}, \lambda_{c_n'}) \tag{12}$$

Using Gibbs sampling with auxiliary parameters(refer to Neal's algorithm 8), each iteration is as follows: • For $n = 1,...,N'$: Let k^- be the number of distinct c_j' for $j \neq n$, and let $h = k^- + m$. Label these c_j' with values in $\{1,...,k^-\}$. If there exists some $j \neq n$ such that $c_n' = c_j'$, for $k^- < i \leq h$, sample component parameters $\phi_i = \{\mu_i, \lambda_i, W_i, b_i\}$ from their corresponding prior distribution as described in section 4.1;

Otherwise, label c'_n with $k^- + 1$, and for $k^- + 1 < i \leq h$, sample $\phi_i = \{\mu_i, \lambda_i, W_i, b_i\}$ from their corresponding prior distribution as described in section 4.1. Draw a new value for c'_i from $\{1, ..., h\}$ using the following probabilities:

$$p(c'_n = i | \cdot) = \begin{cases} b_0 \frac{m_{-n,i}}{N'-1+\alpha} F(\phi_i) & 1 \leq i \leq k^- \\ b_0 \frac{\alpha}{N'-1+\alpha} F(\phi_i) & k^- < i \leq h \end{cases} \tag{13}$$

where $m_{-n,i}$ is the number of c'_j for $j \neq n$ that are equal to i, and b_0 is the normalizing constant.

- Remove those ϕ_i that are associated with no data points.

Sample Latent Variables. For unlabeled data, we sample latent variables one by one from its posterior probability distribution

$$p(h_n | \cdot) = N\left(h_n | \Sigma\{A^T \Lambda u_{v_n} + \lambda_{v_n} \mu_{v_n}\}, \Sigma\right) \tag{14}$$

where $\Sigma = (\Lambda + A^T \Lambda A)^{-1}$. For labeled data, due to lack of suitable conjugate prior, we use SIR with proposal

$$q(z_n) = N(z_n | \Sigma\{A^T \Lambda \mu_{c_n} + \lambda_{c_n} \mu_{c_n}\}, \Sigma\}) \tag{15}$$

and importance weight

$$\mathcal{IW}(z_n) = \frac{\exp(b_{c_n, y_n} + w^T_{c_n, y_n} z_n)}{\sum_{m=1}^{M} \exp(b_{c_n, m} + w^T_{c_n, m} z_n)} \tag{16}$$

to sample their latent variables.

Sample Component Parameters. We update parameters $\theta_{k,x} = \{\mu_k, \lambda_k\}$ and $\theta_{k,y} = \{W_k, b_k\}$ component-wise. $\theta_{k,x}$ is sampled from Gaussian-Wishart

$$p(\mu_k, \lambda_k | \cdot) = N(\mu_k | \mu_{k_0}, (\kappa_k \lambda_k)^{-1}) \mathcal{W}(\lambda_k | T_k, \nu_k) \tag{17}$$

where $\overline{z'_k}$ is the mean of data set $\mathbf{Z'_k}$, $\mu_{k_0} = \frac{\kappa_0 \mu_0 + N' \overline{z'_k}}{\kappa_0 + N'_k} \kappa_k = \kappa_0 + N'_k, \nu_k = \nu_0 + N'_k, T_k = T_0 + S_k + \frac{\kappa_0 N'_k}{\kappa_0 + N'_k}$, and $S_k = \sum_{n=1}^{N'_k} (z'_n - \overline{z'_k})(z'_n - \overline{z'_k})^T$.

The posterior of $\theta_{k,y}$ also has no suitable conjugate prior, and we sample it using Hybrid Monte Carlo(HMC) as [5]

Sample Loading Matrix. We sample loading matrix element-wise from its posterior:

$$p(A_{i,j} | \cdot) = N(A_{i,j} | \mu_{A_{i,j}}, \lambda_{A_j}^{-1}) \tag{18}$$

where $\lambda_{A_j} = \Lambda_{i,i} \mathbf{Z'_{j:}}^T \mathbf{Z}_{j:} + \tau_{A_j}$, $\mu_{A_{i,j}} = \frac{\Lambda_{i,i}}{\lambda_{A_j}} \mathbf{Z'_{j:}}^T \mathbf{E}_{i,:}$ and $\mathbf{E} = \mathbf{X'} - A\mathbf{Z'}$

Sample Diagonal Precision Matrix. We assume the diagonal precision matrix Λ is shared by all components, and each diagonal element $\Lambda_{i,i}$ is sampled from $p(\Lambda_{i,i} | \cdot) = Gamma(\alpha_2 + \frac{N'}{2}, \beta_2 + \sum_{n=1}^{N'} E_{i,n}^2)$

Sample Hyperparameters. Hyperparameters H_θ, such as α for $\mathcal{D}ir$, τ_{A_i} for loading matrix A are sampled with their corresponding conjugate priors.

Predication. We treat test data as unlabeled data and incorporate them to train the model. After a number of iterations or convergence conditions meet, we begin to estimate the predictive probabilities with samples drawn above. For test covariate x, its predictive probability given current samples ϑ^s is:

$$p(y = m|\vartheta_s) = \frac{\exp(b_{c_s,m} + w_{c_s,m}^T z_s)}{\sum_{m=1}^{M} \exp(b_{c_s,m} + w_{c_s,m}^T z_s)} \qquad (19)$$

And the final predictive probability approximated by a finite sum is as follows:

$$p(y = m|x) = b_0 \frac{1}{S} \sum_{s=1}^{S} p(y = m|\vartheta_s) \qquad (20)$$

where b_0 is the normalizing constant and S is the number of samples used to estimate $p(y = m|x)$

Algorithm 1. ILMNL

Input:
    max number of iterations $iterMax$
    number of accepted samples S;
    step size esp and frog Leap L for HMC;
    hyper-parameters H_θ.
Output:
    predicative distribution for each test data point
1: initialize model parameters θ, latent variables matrix $\mathbf{Z}$ and $\mathbf{H}$ randomly;
2: **while** $(iter \leq iterMax)$ **do**
3:     **for** $n = 1$ to N'
4:         sample indicator c_n';
5:     **end for**
6:     **for** $k = 1$ to K
7:         sample component parameters $\theta_{k,x}$ and $\theta_{k,y}$;
8:     **end for**
9:     sample loading matrix A;
10:    sample diagonal precision matrix Λ;
11:    sample hyperparameters H_θ;
12:    **if** $(iter > iterMax - S)$
13:       make predictions based on current samples;
14:    **end if**
15: **end while**
16: estimate final predictive distribution for each test data point.

---

5   Experiments

5.1   Dataset Description

The real world datasets we used come from various fields, and their dimensions range from tens to hundreds of thousands. Sonar, GasSensor and PEMS-SF are available from the UCI Machine Learning Repository[11]. ORLRAWS, CLL-SUB and GLA-BRA which are used as benchmark for feature selection are from Arizona State University[1]. The extracted object bank features of UIUC-Event is available from Computer Vision Lab, Stanford University[2]. And gene expression dataset Lung-Cancer for gene classification is from GEMS[3]. Each dataset is randomly divided into training set and test set according to the proportion of 1:2.

5.2   Experiment Setup

As discussed in section 3, the proposed ILGLM model can be used in both supervised and semi-supervised scenarios, we implement two versions of ILMNL, they are sILMNL (Supervised ILMNL) and ssILMNL (Semi-Supervised ILMNL) . As multinomial logit(MNL) is the linear model used in ILMNL, we will compare ILMNL with MNL and its Dirichlet process mixture extending - DPMNL. To make the comparison more comprehensive, we consider projecting data into a low-dimension latent space using either factor analysis(FA) or its supervised form(sFA) for both MNL and DPMNL. We use MNL-f-d and MNL-sf-d to denote MNL with data projected into d-dimension latent space using FA and sFA respectively. For example, MNL-f-5 stands for MNL tested on data projected into 5-dimension latent space using FA. And MNL-full stands for MNL tested on original input space, namely without dimension reduction. Notations for DPMNL are just alike. We also compare with SVM using both linear(L-SVM) and RBF kernel(R-SVM). When the number of classes is larger than 2, L-SVM and R-SVM use the one-vs-rest scheme. Source codes of MNL and DPMNL are available from author's homepage[4] [5] and for SVM we use LIBSVM[5]. FA can be found in Statistics toolbox, and we implement sFA according to [3].

Parameters affect the performance of ILMNL mainly include latent space dimension d, and parameters for HMC, i.e. esp and L. We fix $L = 200$ and tune esp to achieve an acceptance rate of between 60% and 95% for each dataset. For d, we try 5, 10, 20 and 30 for each dataset. The maximum iterations $iterMax$ is set to 1000, and the number of accepted samples S is set to 800. In other words, we discard the first 200, and only use the next 800 samples to estimate predictive distribution. The settings for L, $iterMax$ and S are also used for MNL and DPMNL, and we tune esp in the same way as ILMNL. Parameters for both FA and sFA are tuned according to [3]. Parameters for L-SVM and R-SVM are tuned under the guidance of LIBSVM. We repeat each experiment 30 times independently, and the results are illustrated in Table 1.

5.3   Accuracy Analysis

From Table 1, we can find out that, in most cases, ILMNL outperforms the competitive models with higher accuracy and smaller variance, and the advantage is more obvious

---

[1] http://featureselection.asu.edu/datasets.php
[2] http://vision.stanford.edu/projects/objectbank/
[3] http://www.gems-system.org/
[4] http://www.ics.uci.edu/ babaks/Site/Codes.html
[5] http://www.csie.ntu.edu.tw/ cjlin/libsvm/

Table 1. Classification accuracy of models (Standard deviations are provided in parentheses)

	Sonar	GasSensor	ORLRAWS	CLL-SUB	Lung-Cancer	UIUC-Event	GLA-BRA	PEMS-FS
L-SVM	0.717(0.000)	0.832(0.000)	**0.847(0.000)**	**0.588(0.001)**	**0.956(0.001)**	**0.698(0.001)**	**0.578(0.001)**	**0.734(0.001)**
R-SVM	0.710(0.000)	0.121(0.000)	0.132(0.000)	0.456(0.001)	0.691(0.001)	0.121(0.001)	0.289(0.002)	0.587(0.002)
MNL-Full	**0.739(0.059)**	**0.968(0.042)**	0.765(0.035)	0.474(0.027)	0.691(0.041)	0.552(0.043)	0.444(0.015)	0.653(0.041)
DPMNL-Full	0.716(0.026)	0.196(0.013)	0.082(0.022)	0.055(0.019)	0.730(0.052)	0.324(0.067)	0.133(0.023)	0.217(0.039)
MNL-FA-5	0.704(0.028)	0.877(0.026)	0.711(0.039)	0.534(0.032)	0.843(0.018)	0.507(0.016)	0.633(0.036)	0.474(0.026)
MNL-sFA-5	0.724(0.019)	0.805(0.003)	0.717(0.026)	0.470(0.030)	0.912(0.012)	0.525(0.019)	0.633(0.024)	0.497(0.031)
DPMNL-FA-5	0.702(0.014)	0.919(0.005)	0.682(0.038)	**0.582(0.023)**	0.916(0.020)	0.507(0.014)	0.644(0.030)	0.486(0.029)
DPMNL-sFA-5	0.717(0.026)	0.856(0.020)	0.178(0.050)	0.480(0.023)	0.888(0.014)	0.518(0.017)	0.433(0.015)	0.509(0.033)
sILMNL-5	0.733(0.011)	0.864(0.025)	0.805(0.026)	0.555(0.016)	0.901(0.019)	**0.540(0.027)**	0.664(0.029)	0.749(0.024)
ssILMNL-5	**0.748(0.012)**	**0.929(0.004)**	**0.863(0.022)**	0.570(0.016)	**0.920(0.014)**	**0.540(0.007)**	**0.672(0.011)**	**0.775(0.015)**
MNL-FA-10	0.740(0.009)	0.959(0.049)	0.764(0.049)	0.588(0.024)	0.922(0.020)	0.672(0.016)	0.644(0.034)	0.809(0.038)
MNL-sFA-10	0.755(0.033)	0.946(0.002)	0.757(0.060)	0.539(0.025)	0.956(0.012)	0.677(0.019)	0.622(0.026)	0.792(0.035)
DPMNL-FA-10	0.765(0.044)	0.958(0.015)	0.680(0.076)	0.593(0.036)	0.943(0.037)	0.673(0.026)	0.611(0.027)	0.809(0.043)
DPMNL-sFA-10	0.725(0.024)	0.955(0.005)	0.071(0.037)	0.495(0.055)	0.811(0.080)	0.662(0.019)	0.433(0.022)	0.769(0.057)
sILMNL-10	0.768(0.017)	**0.967(0.002)**	0.844(0.031)	0.636(0.041)	0.937(0.006)	0.685(0.026)	**0.667(0.028)**	0.808(0.022)
ssILMNL-10	**0.771(0.017)**	0.957(0.007)	**0.858(0.018)**	0.676(0.027)	**0.961(0.008)**	**0.688(0.009)**	**0.667(0.013)**	**0.815(0.019)**
MNL-FA-20	0.728(0.025)	0.955(0.012)	0.753(0.025)	0.630(0.032)	0.915(0.033)	0.741(0.025)	0.633(0.023)	0.827(0.036)
MNL-sFA-20	0.739(0.024)	0.957(0.033)	0.763(0.053)	0.631(0.022)	0.955(0.018)	0.750(0.029)	0.622(0.027)	0.815(0.037)
DPMNL-FA-20	0.746(0.033)	0.969(0.039)	0.678(0.061)	0.630(0.046)	0.923(0.017)	0.727(0.028)	0.633(0.018)	0.850(0.039)
DPMNL-sFA-20	0.760(0.028)	0.973(0.003)	0.136(0.036)	0.473(0.029)	0.711(0.019)	0.721(0.014)	0.433(0.017)	0.798(0.039)
sILMNL-20	0.717(0.026)	*0.982(0.002)	*0.864(0.022)	0.677(0.047)	0.938(0.007)	0.715(0.024)	0.699(0.028)	0.855(0.029)
ssILMNL-20	*0.779(0.015)	*0.982(0.001)	0.844(0.013)	0.666(0.031)	0.978(0.019)	0.748(0.007)	*0.711(0.014)	0.861(0.017)
MNL-FA-30	0.724(0.039)	0.953(0.029)	0.671(0.033)	0.600(0.032)	0.912(0.017)	0.777(0.014)	0.633(0.029)	0.786(0.037)
MNL-sFA-30	0.703(0.026)	0.975(0.013)	0.802(0.033)	0.643(0.013)	0.953(0.015)	0.762(0.020)	0.633(0.023)	0.849(0.035)
DPMNL-FA-30	0.685(0.029)	0.977(0.007)	0.484(0.115)	0.599(0.078)	0.871(0.014)	0.741(0.016)	0.611(0.027)	0.815(0.033)
DPMNL-sFA-30	0.705(0.022)	0.974(0.004)	0.131(0.055)	0.518(0.055)	0.954(0.009)	0.780(0.021)	0.650(0.024)	0.877(0.033)
sILMNL-30	0.753(0.023)	*0.982(0.002)	0.845(0.030)	0.659(0.037)	0.954(0.009)	0.780(0.021)	0.650(0.024)	0.877(0.033)
ssILMNL-30	0.758(0.019)	*0.982(0.001)	0.828(0.025)	*0.702(0.033)	*0.994(0.013)	*0.785(0.005)	0.657(0.008)	*0.882(0.020)

in case of high-dimensional datasets, such as ORLRAWS, GLA-BRA and PEMS-FS. For example, according to dataset PEMS-FS with 138672 features, ssILMNL performs the highest average accuracy of 0.882 when the latent dimension is set to 30, and the average accuracy of L-SVM, R-SVM, MNL-full and DPMNL-full are 0.734, 0.587, 0.653 and 0.217 respectively. As the most popular high-dimensional data classification algorithm, SVM performs well with high accuracy and small variance, but its performance is still lower than ILMNL most of time, because ILMNL perform classification in the low-dimensional latent space. When dealing with low-dimensional data such as Sonar, MNL-full and DPMNL-full perform well. However, when the dimension is very high, their performance get worse.

We also observe that, sILMNL and ssILMNL perform better than MNL/DPMNL-FA/sFA with the same latent dimension. As we know, MNL/DPMNL-FA/sFA performs dimension reduction and classification separately. The superior performance of ILMNL verify that it really makes sense to jointly learn latent variable model and infinite mixture of generalized linear model, which performs dimension reduction and classification simultaneously. Another interesting discovery is that, low-dimensional latent variables obtained through the projection of supervised factor analysis is a bit more discriminative than common factor analysis. However, DPMNL could not work well with sFA in most cases, especially when DPMNL learns more than one component. The reason is that the mapping function from latent variable to class label is global linear in sFA, while DPMNL expects a mixture of linear functions.

Most of time, the Semi-Supervised version of ILMNL (ssILMNL) performs better than Supervised ILMNL (sILMNL) with higher accuracy. For example, according to dataset Lung-Cancer, ssILMNL get the highest accuracy of 0.994, and sILMNL get the accuracy of 0.954 when the latent dimension number is 30. With other latent dimension number such as 5, 10 and 20, ssILMNL also performs obvious accuracy improvement compared to sILMNL according to dataset Lung-Cancer. This means unlabeled data information can be used effectively in ssILMNL to improve the accuracy of the ILMNL.

6   Conclusion

In this paper, we propose an Infinite Latent Generalized Linear Model (ILGLM), a flexible nonparametric Bayesian framework jointly learning latent variable and Dirichlet mixture of generalized linear model. This model can handle the high-dimension problem and multiple local linear classification problem jointly in latent space well. And because ILGLM is a general model framework, then we realize it based on Factor Analysis and MultiNomial Logit model, which results in the Infinite Latent MultiNomial Logit (ILMNL) model as an example of ILGLM. An approximate posterior inference algorithm for ILMNL based on Gibbs sampling is also developed. Experiment results on several real-world datasets show that the proposed model is promising to solve high-dimensional data classification problems in latent space.

Acknowledgements. This work is supported by the Fundamental Research Funds for the Central Universities under Project No. 12lgpy40 and Guangdong Natural Science Foundation under Project No. S2012010010390.

References

1. Andrieu, C., de Freitas, N., Doucet, A., Jordan, M.I.: An introduction to MCMC for machine learning. Machine Learning. Machine Learning 50(1-2), 5–43 (2003)
2. Antoniak, C.E.: Mixtures of Dirichlet processes with applications to Bayesian nonparametric problems. Annals of Statistics 2, 1152–1174 (1974)
3. Yu, S., Yu, K., Tresp, V., Kriegel, H.P., Wu, M.: Supervised probabilistic principal component analysis. In: KDD, Philadelphia, PA, USA, pp. 464–473. ACM (August 2006)
4. Rasmussen, C.E., Ghahramani, Z.: Infinite mixtures of Gaussian process experts. In: Vancouver, British Columbia, Canada, Vancouver, British Columbia, Canada, pp. 881–888. MIT Press (December 2001)
5. Shahbaba, B., Neal, R.M.: Nonlinear models using Dirichlet process mixtures. Journal of Machine Learning Research 10, 1829–1850 (2009)
6. Hannah, L., Blei, D.M., Powell, W.B.: Dirichlet process mixtures of generalized linear models. Journal of Machine Learning Research 12, 1923–1953 (2011)
7. Zhu, J., Chen, N., Xing, E.P.: Infinite svm: a Dirichlet process mixture of large-margin kernel machines. In: ICML, Bellevue, Washington, USA, June 2011, pp. 617–624. Omnipress (2011)
8. Bishop, C.M.: Pattern Recognition and Machine Learning. Springer (2006)
9. Rish, I., Grabarnik, G., Cecchi, G.A., Pereira, F., Gordon, G.J.: Closed-form supervised dimensionality reduction with generalized linear models. In: ICML, Helsinki, Finland, pp. 832–839. ACM (June 2008)
10. Neal, R.M.: Markov chain sampling methods for Dirichlet process mixture models. Journal of Computational and Graphical Statistics 9(2), 249–265 (2000)
11. Blake, C.L., Merz, C.J.: UCI repository of machine learning databases (1998)

ITCI: An Information Theory Based Classification Algorithm for Incomplete Data

Yicheng Chen, Jianzhong Li, and Jizhou Luo

Department of Computer Science and Technology
Harbin Institute of Technology
chenyicheng20@163.com, {lijzh,luojizhou}@hit.edu.cn

Abstract. In the field of data mining, classification is an important aspect which has been studied widely. However, most of the existing studies assumed the data for classification is complete, while in practice, a lot of data with missing values exists. When dealing with these data, deleting the incomplete instances will result in a reduction of available information and filling in missing values may introduce skew and errors. To avoid the above problems, it is of great importance to study how to classify directly with incomplete data. In the paper, an information theory based classification algorithm, ITCI, is proposed. ITCI calculates the initial uncertainty of each class and attributes' contribution to decrease class uncertainty in the training stage and then, in the testing stage, an instance is assigned to the class whose uncertainty is minimum after all of the attributes are taken into consideration. Extended experiments proved the effectiveness and feasibility of the proposed method.

Keywords: information theory, incomplete data, classification.

1 Introduction

Classification algorithms, as an important aspect of data mining, has been widely studied and applied to many fields. However, previous studies often assumed the available data is complete, thus did not take the missing values into account, but in real life, however, incomplete data is ubiquitous[1], for example, in an industrial test, part of the data may be lost because of mechanical or electronic failure; in medical field, doctors may not get all the required data due to lack of equipment or patients' physical condition; in a social survey, some respondents may refuse to provide part of information; for the lack of permission, database query can not get all the data needed etc. Thus, it is of great theoretical and practical importance to study how to classify directly with incomplete data.

The missing(incomplete) data mechanism can be divided into the following three groups[2]: missing completely at random(MCAR), missing at random(MAR), not missing at random(NMAR). MCAR occurs when the missing of a variable is independent of itself and any other external influences; the missingness of MAR is independent of the missing variables but traceable from other variables; NMAR happens when patterns of missingness is non-random and depends on the missing variables, which is the most common situation in real life.

F. Li et al. (Eds.): WAIM 2014, LNCS 8485, pp. 167–179, 2014.

Currently, the methods to deal with incomplete data in classification falls mainly into three aspects. (1) Deleting the incomplete instances[3], this method is simple, but it will lose the useful information contained in incomplete data. (2) Using statistical and machine learning methods to fill in values most likely to be[4][5][6], such as filling manually, filling with mean or median, regression filling, KNN filling, filling based on neural networks etc. In general, filling manually will brings small bias, but it is not feasible given a large dataset with many missing values; filling by mean or median does not fully reflect the data variability and ignores the association between attributes; regression filling assumes a regression relationship exists among complete items and missing items, which is often incorrect in practice; KNN filling needs to define a reasonable similarity measure and has a relatively high computational complexity; filling based on neural networks requires designing appropriate network architecture for specific missing modes and it is too complex and cumbersome to apply. (3) Training and classifying with incomplete data directly, such as those methods based on EM[7], decision tree[8], fuzzy C-means[9], support vector machines[10], Bayesian networks[11] and the nearest neighbor[12]. Those EM-based methods require that the probability density function and missing attributes must be given, besides, they are often complex to train and converge slowly; the ID3-based approach treats the missing values as a special one different from known ones, which does not fit the real world well and it is difficult to get optimum due to the lack of a global search; fuzzy C-means and support vector machine based methods need assumptions of missing data's distribution, which is often not available in practice, thus the application is limited; Bayesian networks based methods require domain knowledge and dependencies among variables must be known, otherwise, complex network structure will be produced, what's more, the network nodes will increase exponentially with the growth of variables, which will result in high maintenance cost; as for nearest neighbor based method, when data's dimension is high, the sample space will still appears to be sparse even the dataset is large and applying the method directly will result in poor performance.

Among the methods to classify directly with incomplete data, [13] found that Naive Bayes methods are most insensitive to missingness, but they rely on apriori probability density to make classification inferences, which results in a low accurate. [14] proposed a method named RBC, which estimates incomplete data by intervals. In this method, missingness mechanism is not required to meet MAR assumption because all possibilities of the incomplete values are considered. Though it has better classification accuracy, the calculation is relatively complicated. [15] proposed the NCC2 method, which has higher classification accuracy, but it requires the missingness mechanism is declared and assumes each attribute contributes to classification independently, however, when the assumption is not met, classification accuracy decreases sharply. Other studies for classification with incomplete data include rough set based methods[16][17], such methods don't require any assumptions of missingness mechanism, but they are inefficient and have poor scalability.

In this paper, an information theory based classification algorithm, ITCI, is proposed for incomplete nominal data. The basic idea of the algorithm is as follows. At first, an initial uncertainty is calculated for each class, then an instance's attributes are inspected one by one to reduce class uncertainty. When all attributes are used, the instance is assigned to the class whose uncertainty is minimum. During the training stage, ITCI estimates the initial uncertainty with the help of the incomplete records, meanwhile, it calculates attributes' attribution to decrease uncertainty and for missing attributes it gets expected contribution. In the classification stage, expected contribution is used to estimate the decrease of uncertainty for missing attributes. With these measures, ITCI need not to estimate missing values explicitly, at the same time, it makes full use of the information contained in incomplete instances. Extended experiments show that the accuracy and stability of the proposed method are significantly higher than RBC and NCC2, and the time complexity is comparable low.

The rest of the paper is organized as follows: Section 2 gives the related concepts of information theory, their properties and application in classification. Section 3 gives an information theory based classification algorithm, ITC, for complete data. Section 4 extends the methods presented in section 3 to get an algorithm, ITCI, for incomplete data. Section 5 presents the results and analysis of experiments; Section 6 is the conclusion.

2   Basic Concepts and Problem Definition

2.1   Basic Concepts of Information Theory

Definition 1. (*self-information*) *The self-information of a random event is defined as the negative logarithm of the event's probability, namely, if the probability of event x_i is $p(x_i)$, then its self-information is defined as:*

$$I(x_i) = -log p(x_i) \tag{1}$$

Definition 2. (*conditional self-information*) *For any events x_i and y_j in a join set XY, the conditional self-information of x_i given y_j is defined as:*

$$I(x_i|y_j) = -log p(x_i|y_j) \tag{2}$$

Definition 3. (*mutual information*) *For sets X and Y of discrete random events, the information x_i acquired given y_j is called mutual information, which is defined as:*

$$I(x_i; y_j) = log \frac{p(x_i|y_j)}{p(x_i)} \tag{3}$$

We can get from formula (3) that $I(x_i; y_j) = log\frac{1}{p(x_i)} - \frac{1}{p(x_i|y_j)}$, and then get

$$I(x_i; y_j) = I(x_i) - I(x_i|y_j) \tag{4}$$

Formula(4) implies that mutual information equals the result of subtracting conditional self-information from self-information, or in another way, mutual information is a measurement of decreased uncertainty, namely, mutual information equals the result of prior uncertainty $log\frac{1}{p(x_i)}$ subtracting remaining uncertainty $log\frac{1}{p(x_i|y_j)}$. Mutual information has the following properties:

(1) Reciprocity:

$$I\left(x_i; y_j\right) = I\left(y_j; x_i\right) \tag{5}$$

(2) When event x_i and event y_j are mutual independent, the mutual information is zero, namely, $I\left(x_i; y_j\right) = 0$.

(3) Mutual information may be positive or negative. When the value is positive, it means the appearance of event y_j will certainly contribute to the appearance of event x_i, on the contrary, it is disadvantageous.

(4) The mutual information between two events may not exceed the self-information of either one.

$$I\left(x_i; y_j\right) \leq I\left(x_i\right) \tag{6}$$

$$I\left(x_i; y_j\right) \leq I\left(y_j\right) \tag{7}$$

Definition 4. (*conditional mutual information*) *The conditional mutual information of x_i and y_j given z_k in join set XYZ is defined as:*

$$I\left(x_i; y_j|z_k\right) = log\frac{p\left(x_i|y_j z_k\right)}{p\left(x_i|z_k\right)} \tag{8}$$

The mutual information of x_i and $y_j z_k$ is defined as:

$$I\left(x_i; y_j z_k\right) = log\frac{p\left(x_i|y_j z_k\right)}{p\left(x_i\right)} \tag{9}$$

$$I\left(x_i; y_j z_k\right) = I\left(x_i; y_j\right) + I\left(x_i; z_k|y_j\right) \tag{10}$$

Formula(10) implies that given the appearance of a pair of events $y_j z_k$, the information x_i will get is $I\left(x_i; y_j z_k\right)$, which equals the information x_i get from the appearance of y_j, add the information x_i get from z_k when y_j is known.

The above four definitions are based on single event, similarly, they can be extended to event sets. We leave them out due to the limitation of the space.

2.2   Use Information Theory to Solve Classification Problems

In classification problems, an instance's feature can be represented by a n-dimensional vector $x = \{x_1, x_2, x_3, \ldots, x_n\}$. The classification task is to assign a label in label set $C = \{C_1, C_2, \ldots, C_K\}$ to each instance. Usually, the task includes two stages: classifier's training and testing. Considering the testing stage, for an instance x, we may assign any of the K labels to it and have some degree of uncertainty at the same time. The self-information of classes, $I(c_k)$, can be used to measure these initial uncertainty. Then attributes are taken into consideration one by one, meanwhile, the uncertainty of the classes will change with the adding of attributes. When all attributes are considered, we can get the final uncertainty of each class, namely $I(c_k|x_1x_2,\ldots,x_n)$, then the instance x is assigned to the class whose uncertainty is minimum. As for the training stage, we estimate self-information and conditional mutual information with the help of training instances and they will be used as arguments of the final classifier.

3 ITC: Information Theory Based Classification for Complete Data

Assume the input space $\mathcal{X} \subseteq R^n$ is a n-dimensional vector set and the output space is class label set $\mathcal{Y} = \{c_1, c_2, \ldots, c_K\}$. For each instance, classification algorithms take $x \in \mathcal{X}$ as input and get $y \in \mathcal{Y}$ as output. The training set is $T = \{(x^{(1)}, y^{(1)}), (x^{(2)}, y^{(2)}), \ldots, (x^{(N)}, y^{(N)})\}$, which has N instances. Let X be a random vector defined in input space $\mathcal{X}$, Y be a random variable defined in output space $\mathcal{Y}$. ITC builds a classifier by learning the self-information $I(c_k)$ of each class and the mutual information $I(c_k; x)$ $(k = 1, 2, \ldots, K)$ between class c_k and feature vector x.

Considering the estimation of $I(c_k)$, we need to get the probability $P(c_k)$ of class c_k , which can be estimated by the following formula:

$$P(Y = c_k) = \frac{1}{N} \sum_{i=1}^{N} I\left(y^{(i)} = c_k\right) \tag{11}$$

For $I(c_k; x)$, when x's dimension is 1, we can get the value following formula(3) and $P(Y = c_k | X = x) = \sum_{i=1}^{N} I\left(y^{(i)} = c_k, x^{(i)} = x\right) / \sum_{i=1}^{N} I\left(x^{(i)} = x\right)$ is the probability estimation. But when the dimension continues to grow, the number of parameters will increase exponentially, which means it is not feasible to estimate all of them efficiently. Let the number of different values for attribute x_i is p_i, $i = 1, 2, \ldots, n$, the number of possible values of Y is K, then the total count of arguments is $K \prod_{i=1}^{n} p_i$. Therefore, we take some approximation measures to simplify the estimation described above.

Denote the mutual information between c_k and feature vector x as $I(c_k; x)$:

$$
\begin{aligned}
I(c_k; x) &= log \frac{p(c_k|x_1, x_2, \ldots, x_n)}{p(c_k)} \\
&= log[\frac{p(c_k|x_1, x_2, \ldots, x_n)}{p(c_k|x_1, x_2, \ldots, x_{n-1})} \times \frac{p(c_k|x_1, x_2, \ldots, x_{n-1})}{p(c_k)}] \\
&= I(c_k; x_n|x_1, x_2, \ldots, x_{n-1}) + I(c_k; x_1, x_2, \ldots, x_{n-1})
\end{aligned}
\tag{12}
$$

Formula(12) implies that when the feature vector x is known, the decreased uncertainty $I(c_k; x)$ equals the sum of $I(c_k; x_1, x_2, \ldots, x_{n-1})$, which measures the decreased uncertainty get from the former $n-1$ dimensions, and the conditional mutual information $I(c_k; x_n|x_1, x_2, \ldots, x_{n-1})$, which measures the decreased uncertainty get from x_n when the former $n-1$ dimensions are given. Using formula(12) recursively, we can get:

$$I(c_k; x) = I(c_k; x_1) + \sum_{i=2}^{n} I(c_k; x_i | x_1, x_2, \ldots, x_{i-1}) \tag{13}$$

From the definition formula $I(c_k; x_i | x_1, x_2, \ldots, x_{i-1}) = log \frac{p(c_k | x_1, x_2, \ldots, x_i)}{p(c_k | x_1, x_2, \ldots, x_{i-1})}$, we can see the arguments also increased exponentially, here, we simplify it as follows:

$$I(c_k; x_i | x_1, x_2, \ldots, x_{i-1}) \approx I(c_k; x_i | x_{i-1}) \tag{14}$$

Formula (14) implies that when the former $i-1$ dimensions are given, the decreased uncertainty we get from x_i is approximated by the decreased uncertainty we get from x_i when x_{i-1} is given.

Theoretically, if the mutual information is estimated according to formula(13), the results will be sole no matter in which order the attributes are considered, however, they will differ from each other if estimated following formula(14). In fact, because of $I(c_k; x_i | x_1, x_2, \ldots, x_{i-1}) \leq I(c_k; x_i | x_{i-1})$, we should find an optimal order to make the expectation of $I(c_k; x)$ minimum, which will enable the bias as low as possible. Let $|x_1|, |x_2|, \ldots, |x_n|$ be the number of different values of $x_1, x_2, \ldots, x_n$, among which we denote the maximum one as x_{max}, so the complexity of enumeration and estimation is $O(Kn!x_{max}^2 n)$ and it will be $O(NKn!x_{max}^2 n)$ if we estimate the expectation of $I(c_k; x)$ additionally. When the feature vector's dimension is high, it is not hard to see that the calculation is costly, or even impossible, thus we proposed a heuristic attribute order.

Definition 5. (*expected mutual information*) *The expected mutual information between x_i (whose value can take any one of $x_{i1}, x_{i2}, \ldots, x_{ip}$) and class c_k is defined as:*

$$E(c_k; x_i) = \sum_{r=1}^{p} p(x_i = x_{ir} | c_k) I(c_k; x_i = x_{ir}) \tag{15}$$

Definition 6. (χ^2 *of attribute pair*) *Assume x_i and x_j can take any value from $x_{i1}, x_{i2}, \ldots, x_{ip}$, $x_{j1}, x_{j2}, \ldots, x_{jq}$ respectively and let n_{rs} denote the number of instances in class c_k that satisfies $x_i = x_{ir}, x_j = x_{js} (r=1,2,\ldots, p, s=1,2,\ldots, q)$, then we define the χ^2 value of attribute pair x_i and x_j as:*

$$\chi^2 = \sum_{r=1}^{p} \sum_{s=1}^{q} \frac{(n_{rs} - Np_{rs})^2}{Np_{rs}} \tag{16}$$

In the formula above, p_{rs} denotes the expected joint probability of x_{ir} and x_{js} when they independent of each other, which can be estimated by $p_{rs} = \frac{1}{N^2} \sum_{k=1}^{p} n_{ks} \sum_{k=1}^{q} n_{rk}$.

For a set of attributes, $A = \{x_1, x_2, \ldots, x_n\}$, we give a heuristic algorithm to find the optimal order S for class c_k as follows:

ATT_ORDER select the attribute with maximum mutual information as the first one for it can decrease the uncertainty largely. In the following process, it selects the attribute which has the largest χ^2 value with the last selected one, in that this can make $I(c_k; x_i | x_{i-1})$ closer to $I(c_k; x_i | x_1, x_2, \ldots, x_{i-1})$ than other choices, so the total approximation error is small.

Algorithm 1. ATT_ORDER

Intput: $A = \{x_1, x_2, \ldots, x_n\}$ is the attribute set to be ordered
Output: optimal attribute order S

1. Calculate the expected mutual information of c_k and $x_i (i = 1, 2, \ldots, n)$, which is denoted as $E(c_k; x_i)$, then, choose the attribute x_j with the maximum value and add it to S, set the last selected attribute x_{last} as x_j.
2. For $k = 2$ to $n-1$, calculate the χ^2 value between x_{last} and each of the left attribute in A, choose the attribute x_j with maximum value and add it to S, set x_{last} as x_j.

3. Add the only attribute left in A to S.

On the basic of ATT_ORDER, we can get an optimal attribute order from the training data. Here, an algorithm named ITC is given for complete data as follows. The algorithm is made up of two parts, ITC_LEARN, which is used for learning model arguments, and ITC_TEST, which is used for applying the learnt model to classify instances with unknown labels.

Algorithm 2. ITC_LEARN

Intput: training data set $T = (x^{(1)}, y^{(1)}), \ldots, (x^{(N)}, y^{(N)})$
Output: arguments $I(c_k), I(c_k; x_1), I(c_k; x_i | x_{i-1})$ $(k = 1, 2, \ldots, K, i = 2, 3, \ldots, n)$

1. Determine the optimal attribute order $S = x_1, x_2, \ldots, x_n$ by calling ATT_ORDER;

2. Calculate $I(c_k)$ using formula(1);
3. Calculate $I(c_k; x_1)$ using formula(3);
4. Calculate $I(c_k; x_i | x_{i-1})$ using formula(8);
5. Return all the calculated arguments.

---

4   ITCI: Information Theory Based Classification Algorithm for Incomplete Data

For incomplete data, we assume the missing mechanism to be MAR. The missing items can be any of the attributes of X or class label Y. In the same way, one or more attributes can be missing from feature vector X in the testing set.

We can get the algorithm, ITCI, based on ITC proposed in section 3. The main improvements include two parts: the estimation of statistic used to calculate model arguments; the estimation of decreased uncertainty of missing attribute. Once the estimations are acquired, ITCI can be trained and tested in the same way as ITC.

Algorithm 3. ITC_TEST

Intput: the feature vector x to be classified

Output: the predicted class label c of instance x

1. Calculate the initial uncertainty $I(c_1), I(c_2), \ldots, I(c_K)$ if x is classified to $c_1, c_2, \ldots, c_K$ without considering any attribute;
2. For each class, calculate $I(c_k; x_1), I(c_k; x_i | x_{i-1})(i = 2, 3, \ldots, n)$ according to the optimal attribute order $S = x_1, x_2, \ldots, x_n$;
3. Calculate the ultimate uncertainty by using formula: $I(c_k | x) = I(c_k) - I(c_k; x)$ for each class. $I(c_k; x)$ can be estimated by $I(c_k; x) = I(c_k; x_1) + \sum_{i=2}^{n} I(c_k; x_i | x_{i-1})$;
4. Return the class whose $I(c_k | x)$ is minimum.

4.1   Estimations of Statistic

It can be seen from formulas (1)(3)and(8) that the key point of estimating model arguments, which include $I(c_k)$, $I(c_k; x_i)$, $I(c_k; x_j | x_i)$, is the estimation of frequencies if we use frequency to approximate probability. The main estimation includes $f(c_k)$, $f(c_k x_{ir})$, $f(c_k x_{ir} x_{js})$, $f(x_{ir})$, $f(x_{ir} x_{js})$, we denote the estimated values as $g(c_k)$, $g(c_k x_{ir})$, $g(c_k x_{ir} x_{js})$, $g(x_{ir})$, $g(x_{ir} x_{js})$ respectively.

Let the non-empty values of x_i to be x_{i1}, $x_{i2}, \ldots, x_{ip}$ and the empty value to be x_{ip+1}, then denote the number of instances with value x_{i1}, $x_{i2}, \ldots, x_{ip}$, x_{ip+1} on x_i as $f(x_{i1})$, $f(x_{i2}), \ldots, f(x_{ip})$, $f(x_{ip+1})$. Due to the existence of missing values, we intend to replace the first p frequency with $g(x_{i1})$, $g(x_{i2}), \ldots, g(x_{ip})$, the estimation formula is as follows:

$$g(x_{ir}) = f(x_{ir}) + f(x_{ip+1}) \times f(x_{ir}) / \sum_{u=1}^{p} f(x_{iu}) \tag{17}$$

Let the non-empty values of x_j to be x_{j1}, $x_{j2}, \ldots, x_{jq}$ and the empty value to be x_{jq+1}. Denote the frequency of instances whose value is x_{ir} on attribute x_i and is x_{js} on x_j as $f(x_{ir} x_{js})(r = 1,2,\ldots,p, s=1,2,\ldots,q)$. Then the estimation formula to assign in proportion is:

$$g(x_{ir} x_{js}) = f(x_{ir} x_{js}) + f(x_{ir} x_{js}) \times f(x_{ir} x_{jq+1}) / \sum_{v=1}^{q} f(x_{ir} x_{jv}) +$$

$$f(x_{ir} x_{js}) \times f(x_{ip+1} x_{js}) / \sum_{u=1}^{p} f(x_{iu} x_{js}) + \tag{18}$$

$$f(x_{ir} x_{js}) \times f(x_{ip+1} x_{jq+1}) / \sum_{u=1}^{p} \sum_{v=1}^{q} f(x_{iu} x_{jv})$$

As for $f(c_k)$, $f(c_k x_{ir})$, $f(c_k x_{ir} x_{js})$, we estimate as follows. Assume G to be the set consist of T's instances whose class label is not missing and $T_{c1}, T_{c2}, \ldots, T_{cK}$ to be the sets get from partitioning G by class label. All the instances whose class label is missing are assigned to T_{cK+1}. Denote the number of instances in T_{ci} as $|T_{ci}|(i = 1,2,\ldots, K, K+1)$. Then the frequency estimation of c_k is:

$$g(c_k) = |T_{ck}| + |T_{cK+1}| \times |T_{ck}| / \sum_{i=1}^{K} |T_{ci}| \qquad (19)$$

The estimation of $f(c_k x_{ir})$ is related to T_{ck} and T_{cK+1}. We treat all of the instances' class label of T_{cK+1} as c_k and assign a weight as follows:

$$w_k = |T_{ck}| / \sum_{i=1}^{K} |T_{ci}| \qquad (20)$$

The frequency estimation $g_{ck}(c_k x_{ir})$ of T_{ck} can be estimated following formula (17), and the frequency of T_{cK+1} can be estimated as follows:

$$g_{cK+1}(c_{K+1} x_{ir}) = f_{cK+1}(x_{ir}) + f_{ck}(x_{ir}) \times f_{cK+1}(x_{ip+1}) / \sum_{u=1}^{p} f_{ck}(x_{iu}) \qquad (21)$$

The subscripts c_k, c_{K+1} indicate that the frequency is estimated based on the dataset T_{ck}, T_{cK+1}. Formula (21) implies that we get the frequency of instances whose class label is missing based on the proportion of complete instances. Combining(20)(21), we can get the final estimation of $g(c_k x_{ir})$ like:

$$g(c_k x_{ir}) = g_{ck}(c_k x_{ir}) + w_k \cdot g_{cK+1}(c_{K+1} x_{ir}) \qquad (22)$$

In the same way, the estimation of $f(c_k x_{ir} x_{jo})$ also contains two parts, we can get $g_{ck}(c_k x_{ir} x_{js})$ following formula(18) from dataset T_{ck} and get the weight w_k following formula(20), and then get the estimation $g_{cK+1}(c_{K+1} x_{ir} x_{js})$ from dataset T_{cK+1} in a similar way as formula(18), but the weight is acquired from complete instances. Finally, $g(c_{K+1} x_{ir} x_{js})$ can be estimated like:

$$g(c_k x_{ir} x_{js}) = g_{ck}(c_k x_{ir} x_{js}) + w_k \cdot g_{cK+1}(c_{K+1} x_{ir} x_{js}) \qquad (23)$$

4.2 The Arguments Estimation for Missing Attributes

Due to the existence of missing attribute, the arguments we need to estimate also include $I(c_k; x_{ip+1})$, $I(c_k; x_{jq+1}|x_{ir})$, $I(c_k; x_{js}|x_{ip+1})$, $I(c_k; x_{jq+1}|x_{ip+1})$. Here, we use the expected mutual information of all known attribute pairs as the estimations.

$$I(c_k; x_{ip+1}) = \sum_{u=1}^{p} p(x_{iu}|c_k) I(c_k; x_{iu}) \qquad (24)$$

$$I(c_k; x_{jq+1}|x_{ir}) = \sum_{v=1}^{q} p(x_{jv}|c_k x_{ir}) \qquad (25)$$

$$I(c_k; x_{js}|x_{ip+1}) = \sum_{u=1}^{p} p(x_{iu}|c_k x_{js}) I(c_k; x_{js}|x_{iu}) \qquad (26)$$

$$I(c_k; x_{jq+1}|x_{ip+1}) = \sum_{u=1}^{q} \sum_{v=1}^{p} p(x_{iu} x_{jv}|c_k) I(c_k; x_{jv}|x_{iu}) \qquad (27)$$

$I(c_k; x_{ir})$, $I(c_k; x_{js}|x_{ir})$ can be acquired based on the estimations given in section 4.1 and formula(1)(3)(8). The conditional probabilities $p(x_{ir}|c_k)$ can be estimated by the frequency of the instances whose value is x_{ir} on attribute x_i in class c_k. Other conditional probabilities can be estimated in a same way.

5   Experiment and Analysis

In order to evaluate the effectiveness of the algorithm, we did experiments on 12 datasets.The datasets are downloaded from the UC Irvine Machine Learning Repository and their basic information is given in table 1. Notice that all datasets are with missing values, among them, those whose names ended with 5 are acquired by randomly deleting 5% attributes from complete instances, others are incomplete initially. For continuous attributes, we discrete them in the preprocessing stage. All algorithms are implemented with java and run on a PC with 2.2GHZ cpu, 2GB memory and the operating system is Windows XP.

Table 1. The experimental datasets with missing values

dataset	instance number	class number	attribute number
breast_cancer	286	2	9
credit	690	2	15
cylinder	512	2	39
colic	368	2	22
mushroom_5	8124	2	22
wbdc_5	569	2	30
vote	435	2	16
crx_5	690	2	15
car_5	1728	4	6
nursery_5	12960	5	8
balance_5	625	3	4
vehicle_5	846	4	18

Table 2 shows the accuracy and standard deviation of the classifiers. In our experiments, the proposed algorithm ITCI and two classical classification algorithms dealing with incomplete data named RBC, NCC2 are compared. All the results are got from 10 times 10 fold cross-validation.

(1) We can see that the only dataset on which the accuracy of ITCI is lower than RBC is balance_5 and the difference is 2.90%. While on other 11 datasets, the accuracy is significantly higher, especially on vehicle_5 which exceeded more than 13.09%. Comparing ITCI and NCC2, we find that ITCI is lower on datasets credit and balance_5, while outstands significantly on the left 10. Especially on vehicle_5, it increased by 10.03%. For dataset balance_5, careful analysis found that the number of the three classes take proportions of 46.08%, 7.84%, 46.08% respectively. As the proportion of class 2 is too small, the initial uncertainty is relatively high, what's more, the number of attributes is too small to decrease

uncertainty during the classification process, so instances of class 2 are easily mis-classified and then the final accuracy is affected. In fact, attributes of balance_5 are numerical and the class label is determined by the difference of the product of the first two attributes and the product of the last two attributes.While in ITCI, we assumed the attribute is nominal, so it is this inconsistency led to a low classification accuracy.

(2) By comparing the standard deviation of ITCI, RBC and NCC2,we find that ITCI is much better except on datasets nursery_5 and balance_5. By Analyzing dataset nursery_5, we find it also has the problem of imbalance classes, which leads to a large deviation of the initial uncertainty, thus the standard deviation is relatively high. But on the left 10 datasets, the standard deviation is small than or equal to the latter two.

Table 2. The comparison of classification accuracy and standard deviation

dataset	RBC	NCC2	ITCI
breast_cancer	72.70 ± 7.39	73.72 ± 7.71	78.12 ± 6.24
credit	86.49 ± 3.74	87.09 ± 3.80	86.57 ± 2.56
cylinder	76.09 ± 5.68	75.37 ± 10.28	81.49 ± 7.57
colic	79.59 ± 5.87	80.32 ± 5.69	83.25 ± 3.07
mushroom_5	95.65 ± 0.71	99.18 ± 0.32	99.65 ± 0.13
wbdc_5	95.38 ± 2.71	96.09 ± 2.50	96.41 ± 1.29
vote	90.16 ± 4.23	90.33 ± 4.14	94.42 ± 1.68
crx_5	85.22 ± 4.25	86.09 ± 4.23	86.68 ± 2.01
car_5	83.87 ± 2.04	85.38 ± 2.05	90.36 ± 1.63
nursery_5	87.75 ± 0.86	87.85 ± 0.85	88.59 ± 2.98
balance_5	89.62 ± 1.87	92.82 ± 2.56	86.72 ± 3.91
vehicle_5	62.83 ± 4.45	65.89 ± 4.57	75.92 ± 4.49

(3) By comparing the total time consumption(Details are not presented due to space limitation), we find NCC2 has the highest efficiency, RBC has the middle and ITCI has the lowest. But we also notice the total running time of ITCI for 10 times 10 fold cross-validation is 19.547s on nursery_5 which has 12960 instances. On average, an experiment takes only 0.195s, which implies the efficiency is still relatively high. In fact, ITCI can get all the arguments needed by scanning datasets only once, that means the complexity is not high at all.

Combining the above three comparison, we can draw the conclusion that the proposed algorithm, ITCI, is more accurate and stable than RBC and NCC2. Although the efficiency of ITCI is lower than the latter two, the running time and complexity is still relatively low, so it is useful in practice.

6   Conclusion

In the paper, an information theory based classification algorithm for incomplete data, ITCI, was proposed. ITCI treats classification as a process of decreasing uncertainty, it calculates classes' initial uncertainty at first, then attributes are inspected one by one to decrease the uncertainty, and then an instance is assigned to the class whose uncertainty is minimum. In the training stage, ITCI weights frequencies by proportion, which makes full use of the information contained in incomplete instances. What's more, ITCI estimates the decreased uncertainty of missing attributes by expected mutual information. Experiments show that ITCI is more accurate and stable than existing ones and the time complexity is low, thus it is considered to be simple and practical. Our future work is to study classification with incomplete data for continuous attributes.

References

1. Gantayat, S.S., Misra, A., Panda, B.S.: A study of incomplete data – A review. In: Satapathy, S.C., Udgata, S.K., Biswal, B.N. (eds.) FICTA 2013. AISC, vol. 247, pp. 401–408. Springer, Heidelberg (2014)
2. Graham, J.W.: Missing Data Theory. Missing Data, pp. 3–46. Springer, New York (2012)
3. Little, R.J.A., Rubin, D.B.: Statistical analysis with missing data (2002)
4. Farhangfar, A., Kurgan, L.A., Pedrycz, W.: A novel framework for imputation of missing values in databases. IEEE Transactions on Systems, Man and Cybernetics, Part A: Systems and Humans 37(5), 692–709 (2007)
5. Zhang, S., Jin, Z., Zhu, X.: Missing data imputation by utilizing information within incomplete instances. Journal of Systems and Software 84(3), 452–459 (2011)
6. Garca-Laencina, P.J., Sancho-Gmez, J.L., Figueiras-Vidal, A.R.: Pattern classification with missing data: a review. Neural Computing and Applications 19(2), 263–282 (2010)
7. Zhang, X., Song, S., Wu, C.: Robust Bayesian Classification with Incomplete Data. Cognitive Computation, 1–18 (2013)
8. Quinlan, J.R.: C4. 5: programs for machine learning. Morgan Kaufmann (1993)
9. Ichihashi, H., Honda, K., Notsu, A., et al.: Fuzzy c-means classifier with deterministic initialization and missing value imputation. In: IEEE Symposium on Foundations of Computational Intelligence, FOCI 2007, pp. 214–221. IEEE (2007)
10. Chechik, G., Heitz, G., Elidan, G., et al.: Max-margin classification of incomplete data. In: Advances in Neural Information Processing Systems: Proceedings of the 2006 Conference, vol. 19, p. 233. The MIT Press (2007)
11. Wang, S.C., Yuan, S.M.: Research on Learning Bayesian Networks Structure with Missing Data. Journal of Software 7, 11 (2004)
12. Jonsson, P., Wohlin, C.: An evaluation of k-nearest neighbour imputation using likert data. In: Proceedings of the 10th International Symposium on Software Metrics, pp. 108–118. IEEE (2004)
13. Blomberg, L.C., Ruiz, D.D.A.: Evaluating the Influence of Missing Data on Classification Algorithms in Data Mining Applications. SBSI 2013: Simpiósio Brasileiro de Sistemas de Informacao (2013)
14. Ramoni, M., Sebastiani, P.: Robust bayes classifiers. Artificial Intelligence 125(1), 209–226 (2001)

15. Corani, G., Zaffalon, M.: Naive credal classifier 2: an extension of naive Bayes for delivering robust classifications. DMIN 8, 84–90 (2008)
16. Dai, J., Xu, Q., Wang, W.: A comparative study on strategies of rule induction for incomplete data based on rough set approach[J]. International Journal of Advancements in Computing Technology 3(3), 176–183 (2011)
17. Grzymala-Busse, J.W., Hippe, Z.S.: Mining Incomplete Data A Rough Set Approach. Emerging Paradigms in Machine Learning, pp. 49–74. Springer, Heidelberg (2013)

Discovering Informative Contents of Web Pages

Qifeng Fan, Chunwei Yan, Lifu Huang, and Lian'en Huang*

Shenzhen Key Lab for Cloud Computing Technology and Applications
Peking University Shenzhen Graduate School, Shenzhen, Guangdong, P.R. China
{fanqf1026,superjom,warrior.fu}@gmail.com, hle@net.pku.edu.cn

Abstract. The World Wide Web has become a huge information repository. However, besides informative contents, the Web pages also contain redundant contents, which are considered harmful for Web mining and searching systems. In this paper, we propose a new approach to discover informative contents from a set of Web pages within a single Web site. Our method works as follows: First, we propose a newly designed Site Style Tree, to capture the common presentation styles and the actual contents of the pages in the given Web site. The tree structure, which is different from the one formerly proposed, is built by aligning pages of the site. For each node of SST, informative contents are discovered based on entropy and threshold method. The proposed approach is evaluated with two mining tasks, Web page clustering and classification. The experimental performance shows a significant improvement when compared to previous template detection approaches.

Keywords: Template Detection, Information Extraction, Entropy.

1 Introduction

The World Wide Web has long become a huge container of information, which includes news and reports about politics, economics, culture, entertainment and others. Recently, developments of Web 2.0 have further greatly increased the magnitude of information. In the face of the large amount of information, many websites, especially commercial websites, provide web pages with much template content for many purposes, such as banner advertisements, navigation bars, copyright notices, decoration, etc. Gibson et al. [7] have found that 40-50% of the content on the Web is template content and the volume is still growing steadily. Although templates are helpful for users to browse the Web pages, they are harmful to many web mining and searching systems. They can decrease the precision of search, increase the size of index and impair the performance of applications that manipulate web pages. Thus it is very important to identify templates correctly and efficiently.

In this work, we research on discovering informative contents of Web pages based on the following observation: In a given Web site, templates usually share

---

* Corresponding author.

F. Li et al. (Eds.): WAIM 2014, LNCS 8485, pp. 180–191, 2014.

some common presentation styles. Moreover, the contents of templates tend to be similar or almost identical.

Many previous extraction methods we found in literature extract informative contents of Web pages based on single Web page analysis. These "page-level" methods have some drawbacks. Generally, "page-level" methods have lower accuracy than "site-level" methods. Some heuristics page-level methods rely on manual rules and extract template effectively only in specified sites. Our method is based on a site-oriented structure which is the alignment of DOM trees of pages within a site. Template detection methods proposed in [13] [16] are based on RTDM (Restricted Top-Down Mapping). While these methods are efficient, they are of limited use because of the following two reasons: First, in some Web sites, especially article-type sites, many informative contents have almost identical structures, and they tend to be detected as templates in these methods. Second, since these methods take no text contents of Web pages into consideration, they cannot utilize texts to distinguish informative contents from template ones. Yi et al. [19] and Li et al. [18] proposed a template detection method based on SST (Site Style Tree) by aligning the DOM (Document Object Model) trees to extract structural information of Web sites. The method leads to a high precision. However, it is unable to detect the informative contents in the multiple-template Web sites well, because the document collection of one SST node may be the combination of templates and informative contents. Moreover, although a node in the SST is meaningful as a unit, it may still contain some noisy items.

We observe that generally templates appear in the form of segments. For example, segments for navigation, copyright and advertisements appear as templates, and segments for main article appear as informative contents. In this paper, we use segment as the element node of the DOM tree. In addition, our approach set sentence as the granularity of informative content units, so it can effectively discover informative items within one content block.

The rest of this paper is organized as follows: In section 2, we give an overview of the problem of informative content extraction and review related work. Then we illustrate the representation of pages and Web sites in Section 3. In section 4, we describe the approach for informative contents extraction. Experiments are discussed in Section 5. Finally, we draw a conclusion of this study and outline directions in Section 6.

2   Related Work

The problem of detecting templates and extracting informative contents of Web pages has been addressed in many previous research papers. Most of the proposed methods are based on machine learning or heuristics.

The problem was first discussed by Bar-Yossef and Rajagopalan in [1], proposing two template detection algorithms based on DOM tree segmentation and segment selection. While their proposal was considered as a segmentation method, the goal was to detect noisy information. Their heuristics used the cues provided by hyperlinks. If an element contains more than K links, it is considered as a segment, otherwise it is counted as part of a segment containing it. Then template segments were selected by one of the above two template detection algorithms. A similar method was proposed by Ma et al. in [12], which segments pages by *table text chunk*. All table text chunks are identified as template table text chunks if their document frequency is over a determined threshold.

Wang et al. [17] proposed DSE (Data-rich Subtree Extraction) algorithm to recognize and extract the informative contents of Web pages by matching simplified DOM trees. Lin et al. [11] proposed a method to discover informative contents based on the heuristics that redundant blocks, opposite to informative blocks, appear more frequently. Reis et al. [13] presented a method based on the RTDM algorithm through generating patterns to identify templates. A similar solution was presented by Vieira et al. in [16], which detects templates through the operation of each step on the Tree Edit Distance of page structures.

Some other methods using the DOM tree of pages to extract informative contents were proposed in articles [19] [18] [7].

Yi et al. [19] and Li et al. [18] studied noise elimination by aligning the DOM trees to extract structural information of Web sites. They present a data structure called SST (Site Style Tree) to capture the actual contents and the common presentation styles of the Web pages in a Web site. Based on the constructed SST, the importance of each node is evaluated through information theory (or entropy). For each node in the SST tree, the more diversity of presentation styles and contents associated to it, the more likely the node is informative. On the contrary, less diversity indicates the node is likely to be noisy.

Gibson et al. [7] have conducted an extensive survey on the use of templates on the Web which revealed the rapid development of template usage. They also develop new randomized algorithms (DOM-based algorithm and Text-based algorithm) for template extraction. In DOM-based algorithm, for each node, a hash is computed by the content of the node and the start and end of offsets. The nodes are considered as templates if the occurrence counts of their hashes are within a specified threshold. In Text-based algorithm, the page is pre-processed to remove all HTML tags, comments, and text within <script> tags, and then a shingling procedure is performed on the result data through sliding a window of size W. Templates are identified by comparing the occurrence count of the fragment contained in the window to a specified threshold.

Some page-level algorithms of template detection have been proposed in [3] [9] [5] [8] [15].

Kao et al. [9] proposed a greedy algorithm operating on features derived from the page to segment a given Web page. Debnath et al. [5] also proposed a page-level algorithm ("L-Extractor") that uses various block-features and trains a Support Vector (SV) based classifier to identify an informative block versus a non-informative block. Kao et al. [8] utilized entropy-based Link Analysis on Mining Web Informative Structures. Song et al. [15] used VIPS (Vision-based Page Segmentation) algorithm to segment a Web page into blocks which are then judged on their salience and quality. Chakrabarti et al. [3] developed a framework for the page-level template detection problem based on two main ideas: the automatic generation of training data for a classifier that assigns templateness score to every DOM node of a page, and the global smoothing of per-node classifier scores. There are also other methods discussed in [4] [10].

Many authors have explored Web page segmentation. Although segmentation of pages and content extraction may appear different, there are many similarities between these two tasks, such as models and algorithms. We can also extract informative contents through segmentation. One of the best solutions found in literature, called Vision-based Page Segmentation algorithm (VIPS), was presented by Cai et al. in [2]. The method segments a page by simulating the visual perceptions of the users about that page, but it requires a considerable human effort. In [6] David addressed a method to segment Web pages of a site by aligning the DOM trees, which gives us inspiration.

We here propose a novel method based on SST for discovering informative contents. This method leads to a significant improvement, compared to DOM Based Algorithm in [7], Text Based Algorithm in [7], SST Based Algorithm in [19] [18] and RTDM Based Algorithm in [16]. Experiments indicate that our method outperforms these four baselines at F score.

3   Page and Site Representation

3.1   Representation of Web Pages

Unlike plain text documents, Web pages consist of text contents and tags. Each page can be represented as a DOM (Document Object Model) tree, in which tags are internal nodes and the detailed texts, images or hyperlink are leaf nodes. Based on the DOM tree, a page is partitioned into several segments in line with the following rules:

- Traverse the DOM tree through post-order.
- If a node and all its sibling nodes are leaf nodes and they also have the same tag, merge them into one.
- If a node has only one child, merge its child into it.

Thus, each node of the DOM tree can be considered as a segment.

3.2   Representation of Web Sites

A DOM tree is sufficient for representing the content and style of a single HTML page, but it is insufficient to study the overall presentation style and contents of a set of HTML pages. To address this problem, we create a hierarchical structure named SST tree that summarizes the DOM trees of the pages in a given Web site. Now we define a site style tree.

Definition: A node S in *site style tree* is the combination of DOM nodes having the identical *tag* and *attr*; it has seven components, denoted by (*tag*, *attr*, *collection*, *parent*, *children*, *counter*, *threshold*), where

- *tag* is the tagName of a DOM node.
- *attr* is the set of display attributes of TAG.

- *collection* is the document collection of segments containing it.
- *parent* is the pointer to its parent.
- *children* is the set of pointers to its children.
- *counter* is the number of pages containing it.
- *threshold* is the entropy-threshold which is used to detect templates.

4 Discovering Informative Contents

Figure 1 shows modules of our approach. Each module will be described in the following sub-sections.

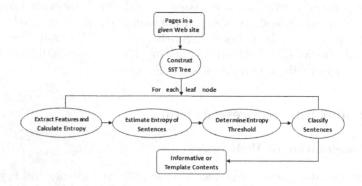

Fig. 1. The Modules of Discoverer

4.1 Constructing SST Tree

The first step of our algorithm is constructing the SST tree. The *Insert* and *Merge* algorithm is described in Algorithm 1 and Algorithm 2. In Algorithm 3, we describe the *Construct* algorithm. With the above 3 algorithms, the SST tree of the site can be easily gained and here we omit the detailed description.

Algorithm 1. Insert

Input:
- The parent of the SST node which is prepared to be created: S_{parent}
- The DOM node which is prepared to be inserted: N

1: **procedure** :
2:     Create a new SST node S.
3:     $S.tag \leftarrow N.tag$; $S.attr \leftarrow N.attr$; $S.counter \leftarrow 1$;
4:     Add $N.content$ to $S.collection$.
5:     **if** S_{parent} is not null **then**
6:         Add S to $S_{parent}.children$.
7:     **end if**
8:     **for** each child n in N **do**
9:         Insert(S, n);
10:     **end for**
11: **end procedure**

Algorithm 2. Merge

Input:
- The SST node which N is prepared to be merged into: S
- The DOM node which is prepared to be merged: N

1: **procedure** :
2:    $S.counter \leftarrow S.counter + 1$
3:    **for** each child n in N **do**
4:        **if** $\exists s \epsilon S.children$ and $n.tag=s.tag$ and $n.attr=s.attr$ **then**
5:            Add $n.content$ to $s.collection$.
6:            Merge(s, n);
7:        **else**
8:            Insert(S, n);
9:        **end if**
10:   **end for**
11: **end procedure**

Algorithm 3. Construct

Input:
- Web pages in a given Web site: C
Output:
- The root of SST tree: S

1: **procedure** :
2:    Create an empty SST tree S_t, and set S is the root of S_t.
3:    **for** each page β in C **do**
4:        $N \leftarrow$ the root node of DOM tree of β
5:        Merge(S, N);
6:    **end for**
7:    Return S;
8: **end procedure**

4.2   Extracting Features and Calculating Entropy of Features

Using SST tree, segments with the common presentation style can be aggregated together. Each node S of SST is a document collection, which contains many segments from different pages.

While constructing SST tree, features of all contents in SST are simultaneously extracted. In this paper, features correspond to meaningful keywords. Stop words are not included. The motivation of our approach is that: Features distributed in more pages in a Web site usually carry less information. In contrast, those appearing in fewer pages carry more information. Hence, we use entropy, which is determined from the probability distribution of features among the whole document sets, to represent the information strength of features. Shannon's information entropy [14] is applied on the feature-document matrix (F-D Matrix) which is generated from the feature extraction module to calculate the entropy. Thus, $S.collection$ form the F-D matrix.

The following is Shannon's famous general formula for uncertainly:

$$E = -\sum_{i=1}^{n} P_i * log_2 P_i \tag{1}$$

where p_i is the probability of $event_i$.

By normalizing the weight of the feature to be $[0, 1]$, the entropy of feature F_i is:

$$E(F_i) = -\sum_{j=1}^{n} w_{ij} * log_2 w_{ij} \tag{2}$$

in which w_{ij} is the weight of F_i in document D_i. w_{ij} is an entity in the feature-document matrix to represent the weight of a feature.

$$w_{ij} = \frac{tf_{ij}}{\sum_{k=1}^{n} tf_{ik}} \tag{3}$$

where tf_{ij} is the feature frequency of $feature_i$ in $document_j$. To normalize the entropy value of a feature to be range $[0, 1]$, the base of the logarithm is chosen to be the number of documents in the collection N. Equation (2) thus becomes:

$$0 \le E(F_i) = -\sum_{j=1}^{n} w_{ij} * log_n w_{ij} \le 1 \tag{4}$$

where $n=|C|$, C is the document collection.

4.3   Estimating Entropy of Sentences

We have observed that: A lot of times, one content block contains not only informative contents but also template contents. Sentence is a better fit for template detection. Hence, we use sentence as the granularity of informative content units. Features and their entropies are gained in the previous subsection. By instinct, the entropy of a sentence is the summation of its features' entropies. We then define the entropy of sentence S_i as the average entropy of all features in S_i below:

$$E(S_i) = \frac{\sum_{j=1}^{k} E(F_j)}{k} \tag{5}$$

where F_j is the jth feature among all k features of S_i.

4.4   Determining Entropy Threshold

Based on entropies, sentences can be divided into two categories: redundant and informative by a threshold T. If E(S) is higher than T, the sentence S is much more likely to be redundant, or else informative. It is difficult to determine the threshold since it would vary for different clusters or sites. The higher the threshold is, the higher recall is and the lower precision is. In order to address this problem, we have labeled sentences of 100 sampled pages as *redundant* or *informative* in each site. Based on these labeled data, F score can be gained easily. For each node S of SST, we start the threshold T from 0 to 1.0 with an interval of 0.1. With some a threshold T, if F score is the largest, T is determined as *S.threshold*.

4.5   Classifying Sentences

According to subsection 4.3, all sentences' entropies of Web pages can be calculated by features' entropies acquired before. We can identify informative sentences by checking if the entropy is larger than the specified threshold.

Algorithm 4. Discoverer

1: Given web pages of the site.
2: $SST \leftarrow Construct$
3: **for** each node S in SST **do**
4:      Extract features and calculate entropy.
5:      Estimate entropy of sentences in S.
6:      Determine the threshold T by labeled data.
7:      **for** each sentence s in $S.collection$ **do**
8:          **if** the entropy of s is less than T **then**
9:              Classify s into informative contents.
10:          **end if**
11:      **end for**
12: **end for**

Algorithm 4 summarizes all the steps of our discovering approach. Given a Web site, the system first constructs the SST. For each node of the SST, entropies of features and sentences are calculated. Then the threshold is determined by a labeled data set of the site. For all the pages in the site, informative contents can be discovered by comparing the entropy with threshold.

5   Experiments

This section evaluates our proposed informative contents discovering algorithm. Since the main purpose of Web content extraction is to improve Web data mining, we performed two data mining tasks, i.e., clustering and classification, to test our system. To validate our method, we compared our approach with these methods, i.e., Origin (results before cleaning), DOM Based Algorithm [7], Text Based Algorithm [7], SST Based Algorithm [18] [19] and RTDM Based Algorithm [16]. These methods have been introduced in the Related Work section.

In the following subsections, we first describe the data sets and evaluation measures used in our experiments. After that, we present the results of the experiments, and also give some discussions.

5.1   Data Sets and Evaluation Measures

In this paper, we crawled six distinct commercial Web sites: PCConnection[1], Amazon[2], CNet[3], J&R[4], PCMag[5] nd ZDnet[6]. The six Web sites contain Web

---

[1] http://www.pcconnection.com/
[2] http://www.amazon.com/
[3] http://www.cnet.com/
[4] http://www.jandr.com/
[5] http://www.pcmag.com/
[6] http://www.zdnet.com/

Table 1. Number of Web pages and their classes

Web sites	PCConnection	Amazon	CNet	J&R	PCMag	ZDnet
Notebook	560	410	431	60	145	198
Camera	156	230	206	150	138	139
Mobile	20	36	42	32	47	108
Printer	423	610	123	127	110	89
TV	267	589	146	171	56	72

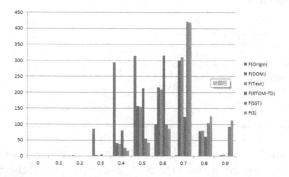

Fig. 2. Distribution of F scores

pages of many categories or classes of products. We choose the Web pages which focus on the following categories of products: Notebook, Camera, Mobile, Printer and TV. Table 1 lists the number of documents downloaded from each Web site, and their corresponding classes.

Since we test our method using clustering and classification, we use the popular F score measure to evaluate clustering and classification performance of our method and the baselines. F score is defined as follows:

$$F = \frac{2P*R}{P+R}$$

where P is the precision and R is the recall. $Fscore$ is the weighted harmonic mean of P and R, which reflects the average effect of both precision and recall.

5.2   Experimental Results

We now present the experimental results of Web page clustering and classification based on our method and the baselines.

For the experiments, we implemented the following algorithms for comparison, the parameters of which are set to the best value in line with the corresponding experiments:

- DOM Based Algorithm with parameters (F=100);
- Text Based Algorithm with parameters (W=32, F=50, D=0, P=1000);
- RTDM Based Algorithm;
- SST Based Algorithm with parameters (r=0.9, t=0.4);

Table 2. The average F scores

Method	F(Origin)	F(DOM)	F(Text)	F(RTDM)	F(SST)	F(S)
Avg	0.455	0.6275	0.631	0.583	0.701	**0721**

Table 3. Configuration of classification experiments

Configuration	Training Set	Test Set
1	Pages from categories p and q from site i	Pages from categories p and q from all sites except i
2	Pages from category p from site i and pages from category q from site $j \neq i$	Pages from categories p and q from all sites except i and j
3	Pages from category p from site i and pages from category q from site $j \neq i$	Pages from categories p and q from all sites except pages in the Training Set

In this paper, we construct the SST of each Web site by all the crawled pages for each site. After labeling sentences of sampled 100 Web pages, the entropy-threshold of each SST node is determined automatically.

Clustering. We stress the experimental procedure we use is the same as in [11], but the sets of pages used are not the same. We put all the 5 categories of Web pages into a big set, and use the popular k-means clustering algorithm to cluster them into 5 clusters. Since the k-means algorithm selects the initial cluster seeds randomly, we performed a large number of experiments (800) to show the behaviors of k-means clustering on our method and the baselines. Figure 2 shows 10 bins of F score from 0 to 1 with an interval of 0.1 and gives the statistics of the number of experiments whose F scores fall into each bin. The average F scores are plotted in Table 2.

F(Origin) represents the F score of clustering based on original noisy Web pages; F(DOM), F(Text), F(RTDM) and F(SST) represent the F scores of clustering based on DOM-Based Algorithm [7], Text-Based Algorithm [7], RTDM Based Algorithm [16] and SST-Based Algorithm [11] respectively; While F(S) represents the F score of clustering based on our method.

We can clearly observe that clustering results based on all the cleaning methods are dramatically better than results using the original noisy pages. Our method also performs better than the other baselines for Web page clustering.

Classification. The evaluation of our method over classification also follows the experimental procedure proposed in [11]. For classification, we use the Naive Bayesian classifier (NB), which has been shown to perform very well in practice by many researches. We experiment with three different configurations for classification tasks using all possible pairs of product categories. The configurations are summarized in Table 3. For each pair, we train the NB classifier using a Training Set and then run the classifier over the corresponding Test Set.

In Table 4, we can observe that classification results based on all the cleaning methods are dramatically better than the results using the original noisy Web

Table 4. F-measure of classification results for each configuration

| Categories | | Conf.1 | | | | | | Conf.2 | | | | | | Conf.2 | | | | | |
p	q	Orig	DOM	Text	RT	SST	S	Orig	DOM	Tex	RT	SST	S	Orig	DOM	Text	RT	SST	S
Not	Cam	0.901	0.965	0.971	0.976	**0.981**	0.978	0.620	0.832	0.847	0.853	**0.877**	0.875	0.469	0.655	0.678	0.685	**0.764**	0.762
Not	Mob	0.852	0.921	0.905	0.934	0.917	**0.944**	0.504	0.773	0.789	0.784	0.799	**0.821**	0.410	0.509	0.534	0.513	0.558	**0.588**
Not	Pri	0.901	0.987	0.989	0.988	0.979	**0.996**	0.618	0.852	**0.864**	0.848	0.856	0.863	0.467	0.651	0.702	0.664	0.783	**0.802**
Not	TV	0.900	0.958	0.961	0.985	**0.996**	0.989	0.592	0.802	0.813	0.816	0.862	**0.865**	0.441	0.679	0.753	0.651	**0.795**	0.794
Cam	Mob	0.854	0.899	0.902	0.901	**0.913**	0.907	0.531	0.868	0.897	0.882	**0.959**	0.936	0.428	0.657	0.732	0.637	**0.776**	0.768
Cam	Pri	0.876	0.904	0.943	0.963	0.971	**0.979**	0.667	0.821	0.834	0.865	0.873	**0.876**	0.504	0.701	0.746	0.699	0.805	**0.807**
Cam	TV	0.813	0.898	0.912	0.974	0.982	**0.991**	0.621	0.836	0.857	0.869	0.904	**0.923**	0.478	0.689	0.733	0.682	0.801	**0.813**
Mob	Pri	0.874	0.901	0.907	0.902	0.914	**0.917**	0.634	0.819	0.834	0.822	**0.880**	0.878	0.489	0.663	0.764	0.634	**0.792**	0.786
Mob	TV	0.783	0.896	0.947	0.870	0.968	**0.969**	0.569	0.855	0.928	0.802	**0.947**	0.944	0.442	0.631	0.768	0.624	0.784	**0.785**
Pri	TV	0.897	0.978	0.983	0.988	**0.996**	0.992	0.583	0.863	0.871	0.855	0.906	**0.912**	0.432	0.775	0.792	0.721	0.819	**0.823**
Avg		0.865	0.931	0.942	0.948	0.962	**0.966**	0.594	0.832	0.853	0.840	0.886	**0.889**	0.456	0.661	0.720	0.651	0.768	**0.773**

pages. Our method performs better than SST and much better than the other three baselines for Web page classification.

6   Conclusions

In this paper, we propose a novel method for information extraction based on SST. Given a Web site, SST tree is constructed from a collection of pages in the site. Then a method based on information theory is used to detect informative sentences in each node of SST.

Experiments conducted on 6 Web sites show that our approach produces quite strong result compared with other five typical algorithms (including original texts).

In future work, we will investigate the application of our method to the processing of short-text data, such as forum and micro blog dataset. Further, we plan to explore the improvement on how to discovery informative contents of a large scale of Web pages automatically.

Acknowledgments. We thank the anonymous reviewers for their valuable and constructive comments. This work is partially supported by NSFC Grant No.61272340.

References

1. Bar-Yossef, Z., Rajagopalan, S.: Template detection via data mining and its applications. In: Proceedings of the 11th International Conference on World Wide Web, pp. 580–591. ACM (2002)
2. Cai, D., Yu, S., Wen, J.R., Ma, W.Y.: Vips: a visionbased page segmentation algorithm. Tech. rep., Microsoft technical report, MSR-TR-2003-79 (2003)
3. Chakrabarti, D., Kumar, R., Punera, K.: Page-level template detection via isotonic smoothing. In: Proceedings of the 16th International Conference on World Wide Web, pp. 61–70. ACM (2007)

4. Davison, B.D.: Recognizing nepotistic links on the web. In: Artificial Intelligence for Web Search, pp. 23–28 (2000)
5. Debnath, S., Mitra, P., Pal, N., Giles, C.L.: Automatic identification of informative sections of web pages. IEEE Transactions on Knowledge and Data Engineerin 17(9), 1233–1246 (2005)
6. Fernandes, D., de Moura, E.S., da Silva, A.S., Ribeiro-Neto, B., Braga, E.: A site oriented method for segmenting web pages. In: Proceedings of the 34th International ACM SIGIR Conference on Research and Development in Information Retrieval, pp. 215–224. ACM (2011)
7. Gibson, D., Punera, K., Tomkins, A.: The volume and evolution of web page templates. In: Special Interest Tracks and Posters of the 14th International Conference on World Wide Web, pp. 830–839. ACM (2005)
8. Kao, H.Y., Chen, M.S., Lin, S.H., Ho, J.M.: Entropy-based link analysis for mining web informative structures. In: Proceedings of the Eleventh International Conference on Information and Knowledge Management, pp. 574–581. ACM (2002)
9. Kao, H.Y., Ho, J.M., Chen, M.S.: Wisdom: Web intrapage informative structure mining based on document object model. IEEE Transactions on Knowledge and Data Engineering 17(5), 614–627 (2005)
10. Kushmerick, N.: Learning to remove internet advertisements. In: Proceedings of the Third Annual Conference on Autonomous Agents, pp. 175–181. ACM (1999)
11. Lin, S.H., Ho, J.M.: Discovering informative content blocks from web documents. In: Proceedings of the eighth ACM SIGKDD International Conference on Knowledge Discovery and Data Mining, pp. 588–593. ACM (2002)
12. Ma, L., Goharian, N., Chowdhury, A., Chung, M.: Extracting unstructured data from template generated web documents. In: Proceedings of the Twelfth International Conference on Information and Knowledge Management, pp. 512–515. ACM (2003)
13. Reis, D.D.C., Golgher, P.B., Silva, A., Laender, A.: Automatic web news extraction using tree edit distance. In: Proceedings of the 13th International Conference on World Wide Web, pp. 502–511. ACM (2004)
14. Shannon, C.E.: A mathematical theory of communication. ACM SIGMOBILE Mobile Computing and Communications Review 5(1), 3–55 (2001)
15. Song, R., Liu, H., Wen, J.R., Ma, W.Y.: Learning block importance models for web pages. In: Proceedings of the 13th International Conference on World Wide Web, pp. 203–211. ACM (2004)
16. Vieira, K., da Silva, A.S., Pinto, N., de Moura, E.S., Cavalcanti, J., Freire, J.: A fast and robust method for web page template detection and removal. In: Proceedings of the 15th ACM International Conference on Information and Knowledge Management, pp. 258–267. ACM (2006)
17. Wang, J., Lochovsky, F.H.: Data-rich section extraction from html pages. In: Proceedings of the Third International Conference on Web Information Systems Engineering, WISE 2002, pp. 313–322. IEEE (2002)
18. Yi, L., Liu, B.: Web page cleaning for web mining through feature weighting. In: International Joint Conference on Artificial Intelligence, vol. 18, pp. 43–50. Lawrence Erlbaum Associates Ltd. (2003)
19. Yi, L., Liu, B., Li, X.: Eliminating noisy information in web pages for data mining. In: Proceedings of the Ninth ACM SIGKDD International Conference on Knowledge Discovery and Data Mining, pp. 296–305. ACM (2003)

Chinese Evaluation Phrase Extraction
Based on Cascaded Model

Yashen Wang, Chong Feng, Quanchao Liu, and Heyan Huang

School of Computer Science and Technology, Beijing Institute of Technology, Beijing, China
{yswang,fengchong,liuquanchao,hhy63}@bit.edu.cn

Abstract. With the development of social media, massive reviews are generated by users every day. The extraction of evaluative information, including opinion holder, comment target and evaluation phrase, is an important pre-task of opinion analysis and also in great need, especially for Chinese text. This paper proposes an efficient method for extracting Chinese evaluation phrase based on cascaded model and mainly makes three contributions: (i) to implement and evaluate the method, we construct an original annotated corpus for Chinese evaluation phrase of automobile; (ii) based on Conditional Random Fields, we identify the evaluation phrase which is in simple structure; (iii) three kinds of rule-based methods, such as parenthesis/preposition/adverb phrase rule, are designed to extract evaluation phrase in complex structure. According to the experiment results, the proposed method performs well. Meanwhile it contributes greatly to our subsequent tasks, such as sentiment analysis of social media.

Keywords: Information Extraction, Data Mining, Evaluation Phrase.

1 Introduction

With the development and maturity of the Internet, users begin to change their roles from "webpage readers" to "webpage writers". Therefore the Web contains a huge amount of opinions about products, politicians, and more, which are expressed in review sites, news posts, and elsewhere. As a result, researches on such opinion-rich information are conducted from different perspectives and granularity [1, 2]. Opinion mining for reviews is the field of study that analyzes people's evaluations and attitudes from natural language. As an important pre-task, the extraction of evaluative information, such as *opinion holder*, *comment target* and *appraisal word(s)*, is in great need, especially for Chinese text. This is because the number of Chinese users and comments grow rapidly. What's more, opinion mining of Chinese lags behind that of thoroughly studied language such as English.

This paper defines an *evaluation phrase* (EP) to be an elementary linguistic unit that refers to successive words, conveying an attitude towards some target. It could be a single word, successive words, or even a sentence, For example, "非常方便"(*English translation: very convenient*) and "极其无聊"(*English translation: extremely boring*). EP contains abundant emotional information, from which we then could derive useful

F. Li et al. (Eds.): WAIM 2014, LNCS 8485, pp. 192–203, 2014.

features for machine learning. Besides, EP not only contributes to detecting sentiment, but also could help people understand the corresponding event comprehensively, as the result of information extraction.

In the study of [3], the concept of *appraisal groups* was first proposed, which denotes coherent groups of words that express together a particular attitude. *Evaluative expression* was defined as a span of text describing evaluation [4]. EP is similar to the definition of *appraisal groups* and *evaluative expression*, but closer to the latter one.

Considering the efficiency of both statistical and rule-based extraction, as well as the difficulty of manually labeling corpus, we define and extract *simple evaluation phrase* (SEP) and *complex evaluation phrase* (CEP), respectively. Final extracted EPs are composed of both of them.

SEP mainly refers to the collocation of degree adverbs and evaluative words. And it usually occurs in simple structure and in fixed position (attribute, adverbial and complement etc.). The underlined parts below are SEP examples. In our cascaded model, a SEP can be used as an EP alone, and also be absorbed in a CEP after rule-based extension described in Section 5.

Example 1. 车型是那副优雅高贵的身材，内饰则有了很大的变化。

(*English translation: The model is elegant and noble, while the interior has changed a lot.*)

Example 2. 将内饰打造得更加典雅奢华。

(*English translation: making the Interior more elegant and luxury.*)

Many related works only focused on the simple collocation of adverb and appraisal words [3], [5], [6], [7], just like SEP defined here, or even a single appraisal word, but complicated phrase gets little attention. However there really exist many phrases which are in complicated structure in the text, and they also reflect abundant opinion information. For example, preposition phrase could release relation of comparison, and the recognition of comparative sentences has become an important task in recent *Chinese Opinion Analysis Evaluation* (COAE). This paper mainly focuses on three kinds of CEP: parenthesis phrase, preposition phrase and adverb phrase. And corresponding rules are designed respectively. Section 5 will discuss the definitions and instances for each kind of CEPs in detail. In many cases, A CEP may contain some SEPs as follows.

Example 3: CEP "在新的大灯造型的配合下显得时尚且动感" (*English translation: Looks stylish and dynamic with the new headlight shape*) contains two SEPs: "新的大灯造型的"(*English translation: the new headlight*) and "时尚且动感"(*English translation: stylish and dynamic*).

What's more, *sentiment dictionary* is believed to be a fundamental resource for opinion analysis [7, 8]. However, the combination of evaluation phrase and comment target truly determinates the sentiment polarity, rather than only considering the sentiment of evaluation phrase. So this paper does not discuss and use any dictionary in the extraction. But sentiment dictionary or domain dictionary will be referred in the subsequent tasks, such as comment target extraction and sentiment analysis.

This paper focuses on product reviews, though our method could apply to a broader range of opinions. As a cascaded model, the proposed method extracts Chinese evaluation phrase as shown in Fig. 1: (1) splits input sentence into *segments* (defined in Section 3) by the punctuations and removes these punctuations; (2) extracts SEPs based on CRFs; (3) employs rules to extract CEPs based on the extracted SEPs. Finally, results from (2) and (3) are the evaluation phrases that we want to extract.

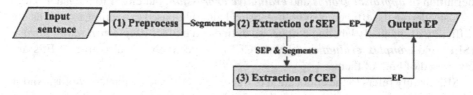

Fig. 1. The flow of the proposed method for extracting evaluation phrase (EP)

The paper is organized as follows. In Section 2, we provide a brief review of related work. Section 3 gives an overview of data description and related concepts. We design the method for extracting the SEP based on CRFs in Section 4, and the rule-based method for extracting the CEP in Section 5. In Section 6, experiments and results of the method on Chinese automobile review corpus are presented. Section 7, finally, concludes the paper.

2   Related Work

Extraction and analysis of human opinions from the Internet attract increasing interest, and various studies have been conducted for extraction and classification of evaluative information.

In the research domain of sentiment analysis, the study of [3] considered that, the basic unit of sentiment analysis should be the *appraisal group* instead of individual word, and presented a method for sentiment classification based on extracting and analyzing appraisal groups. Wilson et al. [9] presented a method to phrase-level sentiment analysis based on subjective expression. Extraction patterns were also used in [16,17]. Furthermore, other studies [10,11] presented methods for sentiment classification based on compositional semantics analysis. They state that the meaning of a compound expression is a function of the meaning of its parts and the syntactic rules by which they are combined.

Focusing on attributes and value expressions, a semi-automatic method that uses specific patterns of evaluated subjects was proposed in [12]. Nakagawa et al. [4] studied extraction and classification of both subjective and objective evaluative expressions on Japanese web documents using machine learning and evaluative word dictionaries. Inspired by this work, Wang et al. [13] constructed statistical method to extract evaluative expression and presented three new features (dependency feature, semantic class and distance feature) to improve the performance. Besides, there recently have been several tasks about Chinese opinion information extraction, such as COAE (2008-2013) and Chinese micro-blog evaluation task of CCF (2012-2013), attracting increasing attention.

3   Data Description

This paper focuses on product reviews, though our method applies to a broader range of opinions. We employ the proposed method to extract Chinese evaluation phrase in the domain of automobile. Unfortunately, there is no annotated Chinese review corpus of automobile publicly available. To train and test the proposed method, we designed a topic-focused web crawler, and extracted expert's evaluation articles and user's reviews from Tencent Automobile[1], Netease Automobile[2] and Phoenix Automobile[3] (2012-2013). Then the evaluation phrase corpus, consisting of Chinese automobile reviews, was constructed and published online[4]. The details of the corpus are shown in Table 1, and symbol #X represents total number of X.

Table 1. Information of the Chinese evaluation phrase corpus

#sentence	#segment	#SEP	#CEP	#EP	Average length of CEPs
2,872	14,606	6,450	4,887	8,787	10.4

To simplify the description, this paper defines following concepts:

Segment: the entire sentence is separated into many segments by the punctuations. For example, "经过努力，我们顺利完成了这项任务。"(*English translation: Through the efforts , we completed the task successfully* .) contains two segments (underline parts). We process only one segment and extract EPs from it every time.

Word collocation of a certain POS tag: one word or consecutive words (or connected by conjunctions) with the same POS tag. For example, the underlined part in the following segment is *a word collocation of adjective*.

将内饰打造得更加<u>典雅奢华</u>

(*English translation: making the Interior more <u>elegant and luxury</u>*)

Current word's neighborhood with radius R: the consecutive word collocation consisting of three aspects: the current word, R words before it and R words after it.

This paper uses the standard of POS tags of NLPIR. The following table explains the symbols in rules which will be proposed in Section 5.

Table 2. The explanation of symbols in rules

Symbol	Description	Symbol	Description
EP	SEP or CEP or EP	*CT*	comment target
p	preposition	*c*	conjunction
n	word collocation of noun	*uls*	auxiliary word such as "来说" "而言"
a	word collocation of adjective	*due3*	auxiliary word "得"
d	word collocation of adverb	*	any word
v	word collocation of verb	*f*	noun of locality

---

[1] http://auto.qq.com/
[2] http://auto.163.com/
[3] http://auto.ifeng.com/
[4] http://hlipca.org/corpus/ChineseEvluationPhrase.html

4   The Extraction of SEP Based on CRFs

4.1   The Introduction of CRFs

In [14], *Conditional Random Felds* (CRFs) are discriminative models and could thus capture many correlated features of the inputs. CRFs have a single exponential model for the joint probability of the entire paths given by the input sentence, and minimize the influences of the label and length bias. The structure of CRFs used for labeling sequence data is a simple chain as Fig. 2.

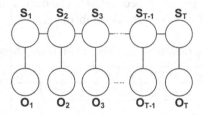

Fig. 2. The graphical structure of CRFs chain

$O = \{o_1, o_2, ..., o_T\}$ represents the observed sequence of input data and $S = \{s_1, s_2, ..., s_T\}$ represents the predicted sequence of states. When input is given and $\Lambda = \{\lambda_1, \lambda_2, ..., \lambda_K\}$ represents the parameters of CRFs, the conditional probability of the sequence of states is shown as the following formula:

$$P(S|O) = \frac{1}{Z_0} exp(\sum_{t=1}^{T} \sum_{k=1}^{K} \lambda_k f_k(s_{t-1}, s_t, o, t)) \; . \tag{1}$$

In the formula (1), f_k is an arbitrary characteristic function, and λ_k is a parameter learned from the training data and indicates the weight of the corresponding f_k. When CRFs model is defined by formula (1) and the sequence of input data O is given, we use formula (2) to obtain the most likely label sequence with a method similar to Viterbi algorithm in Hidden Markov Models (HMM).

$$S^* = arg\ max_S\ P(S|O) \tag{2}$$

4.2   The Extraction of SEP Based on CRFs

As mentioned above, CRFs are able to detect sequence boundary efficiently. Considering that SEP's structure is simple, and its position is also fixed, this paper regards the extraction as a *sequence labeling problem*, and employs CRFs model to recognize SEP. Features are shown in Table 3, wherein w_i and p_i represent the current word (i-th word) and its POS tag, respectively.

This paper chooses to use B/I/O tagging method [15] to label corpus. Wherein the word labeled *B* refers to the word at the beginning of a SEP, the word labeled *I* refers to the word in the middle of a SEP, and the word labeled *O* indicates that this word belongs to no SEP.

For example, after Chinese word segmentation and POS tagging, the segment, "2012/m 款/q 侧裙/nz 的/ude1 变化/vn 则/d 是/vshi 非常/d 细微/a 的/ude1" (*English translation: 2012 side skirt's change is very subtle*), is labeled as follows:

2012/m/B 款/q/I 侧裙/nz/I 的/ude1/I 变化/vn/O 则/d/O 是/vshi/O 非常/d/B 细微/a/I 的/ude1/I.

This segment contains two SEPs: "2012款侧裙的" (*English translation: 2012 side skirt's*) and "非常细微的"(*English translation: very subtle*). Meanwhile, for easier extraction of CEP in the next section, the words appeared in SEP are concatenated into a single-word, and its POS tag is set as **EP**, as follows:

2012款侧裙的/EP 变化/vn 则/d 是/vshi 非常细微的/EP.

Table 3. Feature template for SEP extraction in CRFs model

Feature	Description
w_i, p_i 　wherein i=-3,-2,-1,0,1,2,3 $w_{i-1}w_i$, $p_{i-1}p_i$ 　wherein i=-2,-1,0,1,2,3 $w_{i-1}w_iw_{i+1}$, $p_{i-1}p_ip_{i+1}$ 　wherein i=-2,-1,0,1,2	The current word's neighborhood with radius 3; their POS tags; their binary and ternary combined features.

5   The Extraction of CEP Based on Rules

The methods described in this section are based on the extraction results from Section 4, and final EPs are generated after using such rule-based methods.

On account of the characteristics of Chinese grammar and syntax, this paper summarizes three complex structure rules: parenthesis phrase rule, preposition phrase rule and adverb phrase rule. And each of them could be extended.

5.1   Parenthesis Phrase Rule

Since the content in parentheses includes annotation, instruction or evaluation information, such content could be extracted as an EP. And it modifies the comment target on its left-hand side. Namely, all the contents in parentheses are extracted as EP, and the corresponding comment target is the word collocation of noun on the left.

5.2   Preposition Phrase Rule

Simple preposition phrase often fails to express an emotional tendency. But combined with complement behind, it could do it. Although the internal structure of such phrase is complex, it is also easy to design rules due to its significant boundary features and context features.

For each segment, we search for preposition from right to left. After finding out a preposition, we analyze contents on its right side: if such contents match one of the following rules, we extract them (including the preposition itself) as a EP. Continue to search for preposition and repeat above process, until reaching the leftmost end.

This paper proposes eight basic preposition phrase rules as follows, and the symbols appearing in rules are described in Section 3:

Rule 1: *p + n + EP*

Description: If there is word collocation of noun and the word marked as EP sequentially on the right side of this preposition, we concatenate this preposition and these words into a single-word, and reset the POS tag as **EP**.

Example. 外观/n 上/f 将/d 会/v <u>比/p 传祺/nz 轿车/n 硬朗/EP</u>

(*English translation: The appearance is tougher than that of Trumpchi sedan*)

The extracted CEP is "比传祺轿车硬朗" (*English translation: is tougher than that of Trumpchi sedan*).

Rule 2: *p + n + d + v*

Description: If there is word collocation of noun, word collocation of adverb and word collocation of verb sequentially on the right side, we concatenate this preposition and these words into a single-word, and reset the POS tag as **EP**.

Example: 颇具层次感的/EP 镀铬/nz 栅格/nz <u>与/p 雪铁龙/nz LOGO/x 巧妙</u> <u>/ad 融合/vn</u>

(*English translation: Layered chrome grille fuses Citroën LOGO cleverly*)

The extracted CEP is "与雪铁龙LOGO巧妙融合" (*English translation: fuses Citroën LOGO cleverly*).

Other rules are shown in Table 4:

Table 4. Other basic preposition phrase rules

ID	Rule	ID	Rule
3	*p+n+v*	6	*p+n+f+EP*
4	*p+n+f+v*	7	*p+v+n*
5	*p+n+f+d+v*	8	*p+v+n+EP*

There are some methods to extend above basic rules:

1) The word collocation of noun in the rules can be replaced by word collocation of pronoun, and the rules are still valid.

2) The rules above could be extended to construct new, even more complicated, rules in some ways. For example, add the word marked EP before the word collocation of noun: *p+n+v→p+EP+n+v*.

3) Some Chinese words, such as "相比" and "对比", are able to indicate comparative relationship. The preposition in the rules can be replaced by such words, and the rules are still valid.

5.3   Adverb Phrase Rule

In practice, adverb can modify verb, adjective and even the entire sentence in many aspects. Full of sentiment information, modal adverbs ("究竟", "简直" etc.) and degree adverbs ("非常", "很" etc.) could indicate the tone and attitude of speaker.

This paper mainly focuses on the condition that the adverb phrase is used as predicate or complement. We search for adverbs form right to left in each segment, and extract words complying with above description as EP.

This paper proposes six basic adverb phrase rules as follows.

Rule 1: d + v + EP

Description: If there is word collocation of verb and the word marked EP sequentially on the right side of this adverb, we concatenate both this adverb and these words into a single-word, and reset the POS tag as **EP**.

Example. 内饰/nz 还/d 算/v 朴素大方/EP

(*English translation: interior is fairly simple and generous*)

The extracted CEP is "还算朴素大方" (*English translation: is fairly simple and generous*).

Rule 2: d + EP

Description: If there is the word marked EP on the right side of this adverb, we concatenate this adverb and the EP into a single-word, and reset the POS tag as **EP**.

Example. 厂家/n 仅仅/d 针对车身的包围进行了一定的调整/EP

(*English translation: Manufacturer only does certain adjustments for body surrounded.*)

The extracted CEP is "仅仅针对车身的包围进行了一定的调整" (*English translation: only does certain adjustments for body surrounded*).

Other rules are shown in the following table, and the basic rules could also be extended by using the methods mentioned above.

Table 5. Other basic adverb phrase rules

ID	Rule	ID	Rule
3	*d+n*	5	*d+v+n*
4	*d+v*	6	*d+z+v+n*

6   Experiments and Results

This section conducted experiments to confirm the effectiveness of proposed method. We use the Chinese evaluation phrase corpus described in Section 3 as the training and testing sets. NLPIR[5] is used to complete Chinese word segmentation and POS tagging. Meanwhile, we employ CRF++ (version 0.53)[6] to complete the extraction of SEP based on CRFs, and use both of unigram and bigram templates to describe

---

[5] http://ictclas.nlpir.org/
[6] http://crfpp.googlecode.com/svn/trunk/doc/index.html

features. The related system for ***extracting evaluation elements*** (evaluation phrase, comment target and opinion holder), including our work, is available online[7].

6.1   The Measurement of Extraction

The extraction of evaluation phrase is evaluated by **Precision (P)**, **Recall (R)** and **F-measure (F)**.

$$Precision = \left(\frac{N_3}{N_2}\right) * 100\% \tag{3}$$

$$Recall = \left(\frac{N_3}{N_1}\right) * 100\% \tag{4}$$

$$F - measure = \frac{(\beta^2+1)*p*R}{R+\beta^2*P}, where\ in\ \beta = 1 \tag{5}$$

Where N_1 is the total number of annotated EPs/SEPs/CEPs in the standard testing corpus, N_2 is the number of what we extract, and N_3 is the number of correct results among the extraction results. Meanwhile, we implement three criteria as follows [13], and the last two ones are also called ***lenient match***:

Extract match: extracted result is regarded as correct, if it exactly matches the standard one. It is also called ***strict match***.

Partial match: extracted result is regarded as correct, if it overlaps the first standard one.

Span partial match: if span of extracted phrase and span of standard one overlaps, ***span coverage*** is added to the number of correctly extracted results.

6.2   Results

According to the measurements in Section 6.1, the methods described in Sections 4 and 5 were adopted successively on the corpus. We performed 10-fold cross validation experiments and record the mean. The detailed performance information is shown in Table 6. The figures of SEP extraction are for the best combination of the features in CRFs, which will be discussed later. For the independent experiment of CEP extraction (Row 4), the input SEPs are the standard ones in corpus, rather than the results of SEP extraction.

As shown in Table 6, For SEP, both the precision and recall of exact match are relatively high. Because SEP's composition is regular and its grammatical elements are fixed. Besides, the boundary information also contributes to the accuracy. In CRFs model, the ***feature space*** is often very large, and the choice of features is a key task. Even for sample data, there are thousands of methods to select features, but not all of the features are useful. Based on the factors mentioned above, this paper constructs six feature templates shown in Table 7 for comparison, and w_i and p_i represent the current word and its POS tag respectively.

---

[7] http://hlipca.org:8080/extract

Table 6. The performance of extraction of Chinese evaluation phrases

		SEP	CEP	EP
Extract match	P	0.8597	0.4562	0.3921
	R	0.8634	0.3657	0.2148
	F	0.8615	0.4060	0.2776
Partial match	P	0.8953	0.7823	0.7603
	R	0.9570	0.7146	0.6908
	F	0.9252	0.7469	0.7238
Span partial match	P	0.8621	0.7436	0.6874
	R	0.8720	0.5943	0.4269
	F	0.8726	0.6606	0.5267

According to Table 7, performance of SEP extraction doesn't improve with features increasing, and the experiment achieves the best result when the radius of neighborhood is 3 (5th template in boldface). This is mainly because SEP's characteristics can't guarantee the validity of the features, and the performance might also be limited to the scale of corpus.

Table 7. CRFs feature templates and coresponding results

ID	Feature	P	R	F
1	w_i, p_i wherein i=-2,-1,0,1,2	0.8527	0.8845	0.8683
2	w_i, p_i wherein i=-2,-1,0,1,2 $w_{i-1}w_i, p_{i-1}p_i$ wherein i=-1,0,1,2	0.8489	0.9182	0.8822
3	w_i, p_i wherein i=-2,-1,0,1,2 $w_{i-1}w_i, p_{i-1}p_i$ wherein i=-1,0,1,2 $w_{i-1}w_iw_{i+1}, p_{i-1}p_ip_{i+1}$ wherein i=-1,0,1	0.8843	0.9396	0.9111
4	w_i, p_i wherein i=-3,-2,-1,0,1,2,3 $w_{i-1}w_i, p_{i-1}p_i$ wherein i=-2,-1,0,1,2,3	0.8602	0.9247	0.8913
5	**w_i, p_i wherein i=-3,-2,-1,0,1,2,3** **$w_{i-1}w_i, p_{i-1}p_i$ wherein i=-2,-1,0,1,2,3** **$w_{i-1}w_iw_{i+1}, p_{i-1}p_ip_{i+1}$ wherein i=-2,-1,0,1,2**	**0.8953**	**0.9570**	**0.9252**
6	w_i, p_i wherein i=-4,-3,-2,-1,0,1,2,3,4 $w_{i-1}w_i, p_{i-1}p_i$ wherein i=-3,-2,-1,0,1,2,3,4 $w_{i-1}w_iw_{i+1}, p_{i-1}p_ip_{i+1}$ wherein i=-3,-2,-1,0,1,2,3	0.8900	0.9510	0.9194

Let's turn to Table 6 again. For CEP and EP, the performances of exact match are not ideal, mainly because it is really difficult to identify the exact span of a Chinese evaluation phrase. So this paper uses partial match as the main measurement, and compares the proposed work with previous works, [4] and [13]. These works might not be compared directly, because they deal with different languages on different corpus. Unfortunately, such works cannot be re-implemented and reapplied on our corpus because some technical details were not revealed in the papers. But this paper still refers them to show that the proposed method achieves a competitive performance, because the task and concept definitions among ours are close.

Table 8. Comparison among the proposed work and the previous works

work	Evaluation criteria	P	R	F
The proposed work	Extract match	0.3921	0.2148	0.2776
	Partial match	0.7603	**0.6908**	**0.7238**
	Span partial match	0.6874	0.4269	0.5267
Ref. [13]	Extract match	0.3036	0.1746	0.2208
	Partial match	**0.7728**	0.4560	0.5734
	Span partial match	0.6289	0.3934	0.4832
Ref. [4]	Extract match	0.22	0.12	0.15

As shown in Table 8, the figures are taken from their papers. Especially, we choose the highest score among all choices of features in [13]. Through comparison, we could conclude that our recall is higher than previous work and the precision, which is an important metric in practice, is also competitive.

7   Conclusion and Future Work

Extraction of evaluation phrase plays an increasingly important role as the basis of Chinese opinion mining and sentiment analysis. In this paper, Chinese evaluation phrase was specified and the corresponding corpus was constructed. According to characteristic of Chinese, we conducted statistical and rule-based methods to extract Chinese evaluation phrase, and analyzed the experiment results. Through a series of experiments, the proposed method was demonstrated that it could achieve reasonably good performance, especially compared with previous works.

Dependency features and semantic class are known to be useful in evaluative information extraction [13], and we will utilize such information in the future. Besides, the proposed work mainly focuses on formal writing style, and we will do more experiments and developments on diverse writing styles.

Acknowledgements. We would like to thank Zhaoyu Wang from UIUC for useful conversations. The work was supported by the National Basic Research Program of China (973 Program, Grant No. 2013CB329605, 2013CB329303) and National Natural Science Foundation of China (Grant No. 61201351).

References

1. Pang, B., Lee, L.: Opinion Mining and Sentiment Analysis. Foundations and Trends in Information Retrieval 2(1-2), 1–135 (2008)
2. Liu, B.: Sentiment Analysis and Opinion Mining. Synthesis Lectures on Human Language Technologies 5(1), 1–167 (2012)

3. Whitelaw, C., Garg, N., Argamon, S.: Using Appraisal Groups for Sentiment Analysis. In: 14th ACM International Conference on Information and Knowledge Management, pp. 625–631. ACM Press, New York (2005)
4. Nakagawa, T., Kawada, T., Inui, K., Kurohashi, S.: Extracting Subjective and Objective Evaluative Expressions from the Web. In: 2th International Symposium on Universal Communication, pp. 251–258. IEEE Press (2008)
5. Bloom, K., Garg, N., Argamon, S.: Extracting appraisal expressions. In: Human Language Technologies: The Annual Conference of the North American Chapter of the Association for Computational Linguistics 2007, pp. 308–315. ACL Press, Rochester (2007)
6. Popescu, A.M., Etzioni, O.: Extracting Product Features and Opinions from Reviews. In: 2005 Human Language Technology Conference and Conference on Empirical Methods in Natural Language Processing, pp. 339–346. ACL Press, Vancouver (2005)
7. Yao, T.F., Nie, Q.Y., Li, J.C., Li, L.L., Lou, D.C., Chen, K., Fu, Y.: An Opinion Mining System for Chinese Automobile Reviews. In: Frontiers of Chinese Information Processing, pp. 260–281 (2006)
8. Nakagawa, T., Inui, K., Kurohashi, S.: Dependency Tree-Based Sentiment Classification Using CRFs with Hidden Variables. In: Human Language Technologies: The 11th Annual Conference of the North American Chapter of the Association for Computational Linguistics, pp. 786–794. ACL Press, Los Angeles (2010)
9. Wilson, T., Wiebe, J., Hoffmann, P.: Recognizing Contextual Polarity in Phrase-Level Sentiment Analysis. In: 2005 Human Language Technology Conference and Conference on Human Language Technology and Empirical Methods in Natural Language Processing, pp. 347–354. ACL Press, Vancouver (2005)
10. Moilanen, K., Pulman, S.: Sentiment Composition. In: Recent Advances in Natural Language Processing 2007, pp. 378–382. John Benjamins Publishing Company, Borovets (2007)
11. Choi, Y., Cardie, C.: Learning with Compositional Semantics as Structural Inference for Subsentential Sentiment Analysis. In: 2008 Conference on Empirical Methods in Natural Language Processing, pp. 793–801. ACL Press, Honolulu (2008)
12. Kobayashi, N., Inui, K., Matsumoto, Y., Tateishi, K., Fukushima, T.: Collecting Evaluative Expressions for Opinion Extraction. In: Su, K.-Y., Tsujii, J., Lee, J.-H., Kwong, O.Y. (eds.) IJCNLP 2004. LNCS (LNAI), vol. 3248, pp. 596–605. Springer, Heidelberg (2005)
13. Wang, Y., Kazama, J., Kawada, T., Torisawa, K.: Chinese Evaluative Information Analysis. In: 24th International Conference on Computational Linguistics, pp. 2773–2788. ACM Press, Mumbai (2012)
14. Lafferty, J., McCallum, A., Pereira, F.: Conditional Random Fields: Probabilistic Models for Segmenting and Labeling Sequence Data. In: 18th International Conference on Machine Learning 2001, pp. 282–289. ACM Press, Williamstown (2001)
15. Breck, E., Choi, Y., Cardie, C.: Identifying Expressions of Opinion in Context. In: 20th International Joint Conference on Artificial Intelligence, Hyderabad, India, pp. 2683–2688 (2007)
16. Riloff, E., Wiebe, J.: Learning Extraction Patterns for Subjective Expressions. In: 2003 Conference on Empirical Methods in Natural Language Processing, pp. 105–112. ACL Press, Sapporo (2003)
17. Choi, Y., Cardie, C., Riloff, E., Patwardhan, S.: Identifying Sources of Opinions with Conditional Random Fields and Extraction Patterns. In: 2005 Human Language Technology Conference and Conference on Human Language Technology and Empirical Methods in Natural Language Processing, pp. 355–362. ACL Press, Morristown (2005)

Tracking Topics on Revision Graphs
of Wikipedia Edit History

Bonan Li, Jianmin Wu, and Mizuho Iwaihara

Graduate School of Information, Production and Systems
Waseda University, Kitakyushu, Japan
`libonan@ruli.waseda.jp, jianmin.wu@moegi.waseda.jp,`
`iwaihara@waseda.jp`

Abstract. Wikipedia is known as the largest online encyclopedia, in which articles are constantly contributed and edited by users. Past revisions of articles after edits are also accessible from the public for confirming the edit process. However, the degree of similarity between revisions is very high, making it difficult to generate summaries for these small changes from revision graphs of Wikipedia edit history. In this paper, we propose an approach to give a concise summary to a given scope of revisions, by utilizing supergrams, which are consecutive unchanged term sequences.

Keywords: Wikipedia, topic summarization, edit history, supergram.

1   Introduction

Wikipedia is known as the largest online encyclopedia, to which users contribute articles, and edit articles by others. The edit history of each Wikipedia article is accessible from the public, providing users a precious resource for validating each change, finding errors and vandalism, and improving quality of the article. However, since the number of revisions in an edit history can grow rapidly, it is necessary to present a large and complex edit history with appropriate summarization.

A *revision graph* is a DAG (directed acyclic graph) where each node represents a revision and each directed edge represents a derivation relationship from the origin node to the destination node. Merges of revisions correspond to confluences of edges. But in real edit histories, such merges are seldom observed. Therefore, we focus on edit history that is modeled as a revision tree, as shown in Figure 1(a). In this paper, we call by a *branch* a subtree that is generated by removing the *mainstream*, which is the unique directed path from the initial revision to the current revision.

To easily understand how the content of an article has evolved, we may adopt existing topic detection methods[2,3]. Also, Latent Dirichlet Allocation (LDA) [1,4] is widely used for topic detection. However, the characteristics of Wikipedia revisions, such as heavy overlaps and minor changes, are quite different from scientific documents or news articles. In this paper, we discuss constructing easily understandable summaries by utilizing supergrams[5], which are consecutive unchanged term sequences. Through experiments, we show that our proposing method is superior to simple TF/IDF scoring on unigrams.

F. Li et al. (Eds.): WAIM 2014, LNCS 8485, pp. 204–207, 2014.

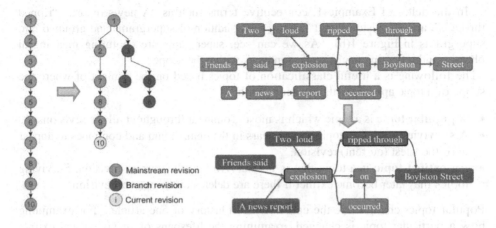

(a) Reconstructing a revision graph (tree)   (b) Path contraction to obtain supergrams

Fig. 1. Revision graph and supergrams

2   Summarizing Topics on Revision Deltas

Let us consider a simple idea of adorning edges of the revision graphs by portions of deltas (revision diffs), to recognize the trends of edits. However, this approach creates a flood of texts, or hard-to-read text fragments. We need to find appropriate summarization of deltas, such as phrases that capture topics of deltas should be extracted. Deltas are diverse in size; larger deltas can contain a complete sentence or a paragraph, but smaller deltas are insufficient to find phrases. In this case, we need to extract text surrounding the small delta.

Now we discuss extracting phrases from deltas, where deltas are taken from a branch or a scope in the mainstream, which is the path reaching to the current revision. In Figure 1(a), the nodes in the mainstream are colored in blue. Let us consider a real example of deltas from article "Boston Marathon bombings." Four deltas from one branch are shown below.

Example 1
D1: Explosion on Boylston Street.
D2: Two loud explosions on Boylston Street.
D3: Friends said explosion occurred on Boylston Street.
D4: A news report explosion ripped through Boylston Street.

Figure 1(b) shows a graph such that each node is labeled with a term, and each edge represents at least one consecutive occurrence of two terms (a bigram) within the scope. We find that"explosion" and "Boylston Street" appear in all of $D_1,..., D_4$. *Path contraction* is to merge two adjacent nodes such that one node is the sole destination or origin of another node.

Some term sequences keep appearing throughout all the deltas within a scope. We can group such unchanged consecutive term sequences into *supergrams*[5].

In the deltas of Example 1, consecutive terms such as "A news report," "ripped through", and "Boylston Street" satisfy the condition of supergrams and grouped into supergrams in Figure 1(b). As we can see, supergrams are capturing meaningful phrases and reflecting unchanged sequences within the scope.

The following is a useful classification of topics based on the context of where the scope of a topic appears in the revision graph:

- A **popular topic** is a topic which is most prominent throughout all the revisions.
- A **surviving topic** is a topic that appears in the mainstream and continues to appear until the latest (current) revision.
- An **extinct topic** is a topic that is not surviving to the current revision. Surviving topics may later become extinct if there are deletes in the current revision.

Popular topics characterize the entire revision history of one article. For examining how a particular topic is changed, examining the lifespans of surviving and extinct topics is effective. Below, we consider three algorithms for summarizing edit deltas in a given scope, by k terms, where k is a given size parameter for summaries.

1. **TF-IDF on deltas (Baseline)**, which simply merges deltas within the scope and selects top k terms based on TF-IDF.
2. **LDA on merged revisions**, which applies LDA[1,6] to the merged text of the revisions within the scope, and select top k terms having the highest term probabilities from the most popular topic. Since LDA can find topic distribution from a corpus larger than the other methods, LDA is expected to perform better than the cases applied on popular topics.
3. **TF-IDF on supergrams (proposed method)**, which applies TF-IDF scoring on supergrams generated from the scope and selects top k supergrams.

There are combinations of applying LDA to deltas or supergrams. But these are not performing well, because the sizes of the deltas or supergrams are sometimes insufficient, which causes that the results of LDA become close to TF.

3   Experimental Evaluation

To evaluate the quality of topics generated by the algorithms described in Section 2, we conducted human judgment evaluation. We used Wikipedia article "Nazi Germany" as benchmark, which has 1,093 revisions. We randomly assigned 10 scopes to the revision history and generated summaries. Due to space limitation, we only show a part of the summaries, as shown in Table 1. To evaluate the qualities of the summaries, we asked ten volunteers to rank the results. For each scope, each volunteer compared the outputs of the three algorithms and assigned symbols 'A', 'B', 'C' representing quality ranking, meaning that 'A' is the best, and 'C' is the worst.

As shown in Table 1, the algorithm by TF-IDF on supergrams has the best score, while LDA on revisions is the worst. As we can observe at the summaries of Scopes 1, 2, and 3, the summaries by TF/IDF on supergrams contain meaningful phrases and indicate topics specific to the scopes. On the other hand, the other two methods produce summaries which are fragmented into terms, and often contain topic terms common to most of scopes, such as "war" and "germany", making difficult to capture specific topics of the scopes. We also observed that if deltas in the scope are small, the results by Baseline and TF-IDF on supergrams tend to become identical, so that the rankings become split.

Table 1. Part of generated summaries and rankings by 10 human judges

Scope	Baseline (TF-IDF on deltas)	LDA on revisions	TF-IDF on super-grams
1	government, general, started, creation, genodic, policy	nazi, german, war, force, italian, invasion	genocide started creation general policies, government with nazis
2	operation, barbarossa, north, campaign, theatre, eastern	nazi, german, von, han, karl, war	this known african, south north theatre campaign
3	reich, nazi, third, war, tr, germany	war, nazi, karl, wilhelm, force, franz	reich, europe, adolf, empire, glorious, regime, power, republic failure
Rankings			
A(Best)	17	2	81
B	79	5	16
C(Worst)	4	93	3

4   Conclusion and Future Work

In this paper, we proposed a method for generating summaries for deltas of Wikipedia revision graphs, to indicate changes in the revision history of Wikipedia articles. Our approach is based on supergrams to find unchanged term sequences within a given scope of the revision history. We found that TF/IDF scoring on supergrams has the best performance in finding useful phrases. Our evaluation is preliminarily and we plan to conduct more extensive evaluations.

References

1. Blei, D.M., Ng, A.Y., Jordan, M.I.: Latent Dirichlet Allocation. J. Machine Learning Research 3, 993–1022 (2003)
2. Chen, Y., et al.: Emerging Topic Detection for Organizations from Microblogs. In: Proc. ACM SIGIR Conf. Research and Development in Information Retrieval. ACM (2013)
3. Georgescu, M., Pham, D.D., et al.: Temporal summarization of event-related updates in Wikipedia. In: Proc. 22nd Int. Conf. World Wide Web companion (WWW 2013 Companion), pp. 281–284 (2013)
4. Song, Y., Pan, S., Liu, S., Zhou, M.X., Qian, W.: Topic and Keyword Re-Ranking for LDA-Based Topic Modeling. In: Proc. 18th ACM CIKM, pp. 1757–1760 (2009)
5. Wu, J., Iwaihara, M.: Revision Graph Extraction in Wikipedia Based on Supergram Decomposition. In: Proc. Int. Symp. Open Collaboration (WikiSym + OpenSym 2013). ACM Press (August 2013)
6. Gibbs LDA(C++), http://gibbslda.sourceforge.net/

Holistic Subgraph Search over Large Graphs

Peng Peng, Lei Zou, Dong Wang, and Dongyan Zhao

Peking University, Beijing, China
{pku09pp,zoulei,WangD,zhaodongyan}@pku.edu.cn

Abstract. Due to its wide applications, subgraph matching problem has been studied extensively in the past decade. In this paper, we consider the subgraph match query in a more general scenario. We build a structural index that does not depend on any vertex content. Our method can be applied in graphs with any kind of vertex/edge content. Based on the index, we design a holistic subgraph matching algorithm. This paper is the first effort of the holistic pattern matching approach in the subgraph query problem. It can dramatically reduce the search space. Extensive experiments on real graphs show that our method outperforms state-of-the-art algorithms.

1 Introduction

In this work, we concentrate on subgraph query processing in a large general graph. In this model, each vertex/edge can be associated with any content. We integrate off-the-shelf systems with graph databases as the underlying storage. Meanwhile, the graph structure is stored in a native graph system. Furthermore, for speeding up the query processing, we propose an simple but efficient index structure, which does not rely on vertex contents.

Besides the index, we design a **holistic subgraph matching algorithm**. Generally speaking, we design a distance-based vertex coding strategy. In query processing, we maintain a priority queue (according to vertex codes) for each query vertex. When a vertex v needs to be dequeued, we find subgraph matches containing v directly. In order to speed up this step, we propose some pruning rules to reduce the search space.

Last but not the least, we evaluate our methods on graphs which have more than 100 millions edges. To the best of our knowledge, this is the largest data graph in the literature about subgraph query problem in a *single* machine.

2 Problem Definition

In this section, we review the terminology that we use throughout this paper.

Definition 1. *Data Graph. A data graph is denoted as $G = \{V(G), E(G), \sum, \Gamma\}$, where (1) $V(G)$ is a set of vertices; (2) $E(G) \subseteq V(G) \times V(G)$ is a set of undirected edges; (3) $\sum$ is a set of content; and (4) $\Gamma : V(G) \to \sum$ denotes the content assignment function, where $\forall v \in V(G)$, $\Gamma(v)$ is v's corresponding content.*

F. Li et al. (Eds.): WAIM 2014, LNCS 8485, pp. 208–212, 2014.
© Springer International Publishing Switzerland 2014

Definition 2. *Query Graph. A query graph is denoted as* $Q = \{V(Q), E(Q), F\}$, *where (1)* $V(Q)$ *is a set of vertices; (2)* $E(Q) \subseteq V(Q) \times V(Q)$ *is a set of undirected edges; (3)* $F = \{f_i\}$ *is a set of query criteria associated with vertex* $u_i \in V(Q)$, $i = 1, ..., |V(Q)|$.

Definition 3. *Subgraph Match. Given a data graph* $G = \{V(G), E(G), \sum, \Gamma\}$ *and a query graph* $Q = \{V(Q), E(Q), F\}$ *with n vertices* $\{u_1, ..., u_n\}$, *a subgraph M (of G) with n vertices* $\{v_1, ..., v_n\}$ *is called a* subgraph match *of Q over G if and only if the following conditions hold:*

1. $\Gamma(v_i)$ *satisfies the criterion* f_i *in query vertex* u_i, $i = 1, ..., n$.
2. $\forall u_i, u_j \in V(Q), \overline{u_i u_j} \in E(Q) \Rightarrow \overline{v_i v_j} \in E(G)$, $1 \le i, j \le n$.

A state $[v_1, ..., v_n]$ *is a serialization of a* subgraph match *M. Furthermore,* $M[u_i] = v_i$, $1 \le i \le n$.

Definition 4. (Problem Statement) *Given a data graph G and a query graph Q, where* $|V(Q)| \ll |V(G)|$, *the subgraph query problem is to find all* subgraph matches *(Definition 3) of Q in G.*

%vspace-0.3in

3 Distance-Based Index

In this section, we introduce a novel index only based on distance information. In particular, we first choose a vertex as a *pivot* and encode other vertices based on its distance to the pivot. We discuss how to encode the vertices in Section 3.1. How to select a good pivot is discussed in Section 3.2.

3.1 Pivot-Based Encoding Strategy

Here, we firstly assume that a pivot v^* is given. The vertex codes are defined in Definition 5.

Definition 5. *Vertex Code. Given a graph G and a pivot* $v^* \in V(G)$, $\forall v \in V(G)$, *the vertex code of v is* $L(v) = dist(v, v^*)$, *where* $L(v)$ *denotes the vertex code and* $dist(v, v^*)$ *denotes the shortest path distance between the two vertices.*

Given a query Q with n vertices $u_1, ..., u_n$, according to the query criteria in u_i, we can find candidate list $TL(u_i)$, where all vertices in $TL(u_i)$ satisfy the query criteria in u_i, $i = 1, ..., n$. The following theorem tells us which vertices should be considered when we find subgraph matches containing v.

Theorem 1. *Given a data graph G and a query graph Q, for a vertex v in* $TL(u_i)$, *when we find a match M containing v, for each* $TL(u_j)$, *the search space is* $\{v' | v' \in TL(u_j) \wedge |L(v) - L(v')| \le dist(u_i, u_j)\}$, *where* u_i *and* u_j *are two vertices in query Q* $(i \ne j)$.

Proof. If there exists a match M containing both v in $TL(u_i)$ and v' in $TL(u_j)$, $dist(v, v') \le dist(u_i, u_j)$. In addition, because of the triangle inequality, $|L(v) - L(v')| = |dist(v, v^*) - dist(v', v^*)| \le dist(v, v')$. Hence, $|L(v) - L(v')| \le dist(u_i, u_j)$. If $|L(v) - L(v')| > dist(u_i, u_j)$, v and v' cannot form a match.

3.2 Pivot Selection

In this section, we discuss how to select a *good* pivot. There are two requirements for a "good" pivot. First, the number of layers in its shortest path tree should be as large as possible. Second, the number of vertices in each layer should be as small as possible. According to Theorem 1, the search space is small in the above case. Considering the above two conditions, we introduce the *vertex entropy* (Definition 6) to evaluate the goodness. The vertex with the largest entropy is selected as the pivot v^*.

Definition 6. *Vertex Entropy.* *Given an vertex v, according to the vertex codes, there is a list $\{(code_1, fre_1), ..., (code_n, fre_n)\}$, where $code_i$ is a vertex code and fre_i denotes the number of vertices whose vertex codes are $code_i$, $i = 1, ..., n$. The* vertex entropy *is defined as follows:*

$$E(v) = -\sum_{i=1}^{i=n} \frac{fre_i}{|V(G)|} \log(\frac{fre_i}{|V(G)|}) \tag{1}$$

4 Holistic Subgraph Matching Algorithm

Based on vertex codes, we propose a *holistic subgraph matching* (HSM) algorithm. We discuss the data structures used in HSM algorithm in Section 4.1, followed by the algorithm in Section 4.2.

4.1 Data Structures

Each query vertex u_i is associated with three data structures in HSM: *candidate list* $TL(u_i)$, *cursor* $C(u_i)$ and *queue* $TQ(u_i)$.

Candidate list $TL(u_i)$ is a list of all candidate vertices satisfying the criteria in vertex u_i. Note that all candidate vertices are sorted according to the non-descending order of vertex codes (see Definition 5) in $TL(u_i)$.

Cursor $C(u_i)$ points to the vertex (in the candidate list $TL(u_i)$) that is currently accessed. For the simplicity of notations, we also use "$C(u_i)$" to denote the vertex that $C(u_i)$ points to, when the context is clear. Initially, each cursor $C(u_i)$ $(i = 1, ..., |V(Q)|)$ points to the first vertex in $TL(u_i)$. In each step, the cursor that points to the *minimal* one among all vertices $C(u_i)$ moves one step forward to the next vertex in the list.

Queue $TQ(u_i)$ stores the vertices that have been accessed. For a vertex v in $TQ(u_i)$, the search space for finding subgraph matches containing v is bounded by Theorem 1. If all vertices in the search space have been accessed, we will dequeue v from $TQ(u)$. The dequeuing rule is discussed in Section 4.2 in details. When a vertex v is dequeued from $TQ(u)$, we perform a graph exploration algorithm to find all subgraph matches containing v.

4.2 Algorithm

The key idea of our holistic subgraph matching algorithm is to repeatedly construct subgraph matches with some vertices that are close to each other. During query execution, all queues store some vertices close to each other. If a vertex in a queue is too far from any vertex in any queue (as discussed in Theorem 1), it dequeue from its queue.

When a vertex v dequeue from a queue, we employ graph exploration to search all queues and find subgraph matches containing v. Essentially, the graph exploration function is the same with VF2 subgraph isomorphism algorithm. The only difference is that our algorithm is beginning the search process from vertex a given vertex v_i.

5   Experimental Evaluation

In this section, we evaluate our method over two real datasets. In our experiments, We test two real-life datasets: US Patents and Yago2. US Patents has $3,774,768$ vertices and $16,518,948$ edges; and Yago2 has $10,557,345$ vertices and $130,447,832$ edges. We conducted all experiments on a computer with 2.0 GHz Intel Core 2 Duo processor and 32 GB memory running Linux. In experiments, we use MySQL (version 5.5.15.0) as the vertex content management system, i.e., using MySQL for finding candidate vertices satisfying the criteria in each query vertex.

Since most subgraph match algorithms can only work on vertex-labeled graphs, we only compare our solution with SQLs and the graph exploration method [4]. For SQLs, we store the graph structure by two tables, that are vertex table and edge table. A subgraph query can be modeled as a SQL query. In order to speed up query processing, we build the B^+-tree indices over all columns of tables. For graph exploration method, we cache the whole graph in memory.

Over US Patent dataset, Fig. 1(a) shows our method is faster than SQLs by an order of magnitude. As well, our method is twice as fast as graph exploration. Note that SQL cannot finish query processing within a reasonable time when $|E(Q)| > 8$. So does the graph exploration method when $|E(Q)| > 12$. We have the similar results over Yago2 as shown in Fig. 1(b). As well, SQL over Yago2 is too slow to finish the query when $|E(Q)| \geq 4$. This is because that SQL involves too manu expensive join steps.

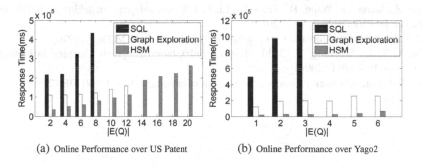

(a) Online Performance over US Patent   (b) Online Performance over Yago2

Fig. 1. Online Performance Comparison in Graphs With Rich-content Nodes

6   Related Work

Ullmann [5] and VF2 [1] are the two early efforts for subgraph isomorphism problem. In order to speed up query response time, rencent subgraph search methods [6,7] pre-compute some structural indexes to reduce the search space. They assume that the data graph is a vertex-labeled graph, i.e., each vertex has a single label. Obviously, these methods cannot be adopted in a data graph with various kinds of vertex-specific content. Jinsoo Lee et al. [3] re-implements some of the above methods and provide a fair comparison of them. Then, they present a solution, called $Turbo_{ISO}$ [2]. $Turbo_{ISO}$ divides all query vertices into some classes. When $Turbo_{ISO}$ find subgraph matches, only combinations for each class are generated.

7 Conclusions

In this work, we study subgraph search problem over a large general graph. We propose a distance-based vertex code and an associated holistic subgraph matching algorithm to handle subgraph search problem.

Acknowledgments. Peng Peng and Lei Zous work were supported by NSFC under grant No.61370055. Lei Zou was also supported by Beijing Higher Education Young Elite Teacher Project under grant No. YETP0016. Dongyan Zhao was supported by NSFC under grant No. 61272344 and 863 project under grant 2012AA011101.

References

1. Cordella, L.P., Foggia, P., Sansone, C., Vento, M.: A (sub)graph isomorphism algorithm for matching large graphs. IEEE Trans. Pattern Anal. Mach. Intell. 26(10) (2004)
2. Han, W.-S., Lee, J., Lee, J.-H.: $Turbo_{iso}$: towards ultrafast and robust subgraph isomorphism search in large graph databases. In: SIGMOD Conference, pp. 337–348 (2013)
3. Lee, J., Han, W.-S., Kasperovics, R., Lee, J.-H.: An in-depth comparison of subgraph isomorphism algorithms in graph databases. In: PVLDB, vol. 6(2), pp. 133–144 (2012)
4. Sun, Z., Wang, H., Wang, H., Shao, B., Li, J.: Efficient subgraph matching on billion node graphs. In: PVLDB, vol. 5(9) (2012)
5. Ullmann, J.R.: An algorithm for subgraph isomorphism. J. ACM 23(1) (1976)
6. Zhang, S., Li, S., Yang, J.: Gaddi: distance index based subgraph matching in biological networks. In: EDBT (2009)
7. Zhao, P., Han, J.: On graph query optimization in large networks. In: VLDB (2010)

Clustering Query Results to Support Keyword Search on Tree Data

Cem Aksoy[1], Ananya Dass[1], Dimitri Theodoratos[1], and Xiaoying Wu[2]

[1] New Jersey Institute of Technology, Newark, NJ, USA
[2] Wuhan University, Wuhan, China

Abstract. Keyword search conveniently allows users to search for information on tree data. Several semantics for keyword queries on tree data have been proposed in recent years. Some of these approaches filter the set of candidate results while others rank the candidate result set. In both cases, users might spend a significant amount of time searching for their intended result in a plethora of candidates. To address this problem, we introduce an original approach for clustering keyword search results on tree data at different levels. The clustered output allows the user to focus on a subset of the results while looking for the relevant results. We also provide a ranking of the clusters at different levels to facilitate the selection of the relevant clusters by the user. We present an algorithm that efficiently implements our approach. Our experimental results show that our proposed clusters can be computed efficiently and the clustering methodology is effective in retrieving the relevant results.

1 Introduction

Keyword search is a popular technique for retrieving information from tree data on the web. Keyword search has attracted a lot of attention as a querying technique in recent years because it frees the user from the need of learning the syntax of a complex query language and the schema of the data sources.

There are two major drawbacks of XML keyword search systems [8, 10]. The first one is that there are usually numerous candidate results which match the keyword queries. This is due to the fact that keyword queries are inherently imprecise and cannot express structural constraints. The second drawback is that the answer of keyword queries may contain different types of meaningful results even though the user is interested in only some of them. For example, consider the query $Q = \{Advanced, \ Database, \ Systems\}$ on the XML document of Figure 1, which represents a university database recording courses and seminars (Q is considered on the XML tree T of Figure 1 as the running example). There are courses with title "Advanced Database Systems" and a seminar whose topic also contains all the query keywords. Usually, the search systems do not group results and interleave in the answer results representing different concepts.

Result filtering [6, 13, 14] and ranking [4, 11, 12] have been proposed to address the problem of the large number of candidate results. The answers provided by these semantics are not satisfactory since still a large number of results may

F. Li et al. (Eds.): WAIM 2014, LNCS 8485, pp. 213–224, 2014.

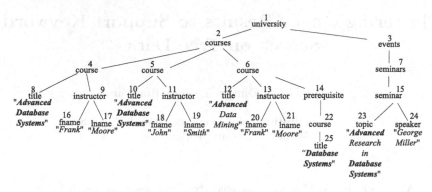

Fig. 1. An XML tree T

qualify to be in the answer and the quality of ranking is low [12]. Further, without disambiguation of the results, the users may still need to examine a large number of results which are not of interest to them. We propose applying clustering techniques to the results to overcome the previous problems. The main idea is that if the results can be clustered effectively with respect to the semantics they represent, the users can only examine the clusters which are of interest to them. Clustering has been used as an alternative method of retrieving results in information retrieval [5, 15]. Unfortunately, the techniques applied for clustering web search results are not directly applicable to XML keyword search results because XML documents are structured and the granularity of clustering is different.

In this paper, we introduce a three-level methodology for clustering the results of keyword queries on XML data. The clusters at every level are nested within the clusters of the higher level. The clusters of the first level consist of structural patterns of the keyword query results which provide alternative interpretations of the query results. Further, we define two relations, $\approx$ and $\prec_{dph}$ on patterns to determine the clusters at the second level (named classes) and the third level (named collections) of clustering, respectively.

In order to facilitate user navigation through the clusters, we also provide ranking for the clusters. Patterns within the same class are ranked with respect to their sizes. Classes which are nested within the same collection and the collections are organized into graphs using the $\prec_{dph}$ and $\prec_{pph}$ relations, respectively. The edges in the graphs represent a partial order on the clusters and this partial order is exploited to produce a ranking for the clusters of the graph. We designed an efficient algorithm to implement our clustering methodology. We run experiments on two datasets and showed that the clusters defined by our methodology can be computed efficiently and the clustering of the keyword search results helps the users to effectively retrieve the relevant ones.

2   Data Model and Query Semantics

We model XML documents as ordered node labeled trees. Nodes represent elements and attributes. Edges represent element to element and element to attribute relationships. The nodes may also have a content which is text. We allow keywords to match both the content and the label of a node. We say that a node n *contains* keyword k if the content of n contains k or is labeled as k. If n contains keyword k then, node n is an *instance* of k.

Keyword queries are embedded to XML trees. An *instance* of a query Q on an XML tree T is an embedding of Q to T (i.e., a function from Q to the nodes of T that maps every keyword k in Q to an instance of k in T).

Definition 1. *Let Q be a query, T be an XML tree, and I be an instance of Q on T. The* instance tree (IT) *of I is the minimum subtree S of T such that: (a) S is rooted at the root of T and comprises all the nodes of I, and (b) every node n in S is annotated by the keywords which are mapped by I to n. The minimum connected tree (MCT) of I is the minimum subtree of S that comprises the nodes of I.*

Consider our running example, Figures 2(a) and (b) show the IT and the MCT, respectively, of the instance $\{(Advanced, 12), (Database, 25), (Systems, 25)\}$ of Q on T. In the figures, the annotation of the nodes is shown between square brackets by the nodes. The root of the MCT is the *Lowest Common Ancestor* (LCA) of the nodes of the instance in T.

An IT of an instance of a query Q on an XML tree T is also called IT of Q on T. Several previous approaches return sets of LCAs as answers to the user. In our approach, the *result set* of a query Q on an XML tree T is defined to be the set of ITs of Q on T. An IT is much of a richer construct than an LCA in terms of the information it provides as it shows both: (a) how the keyword instances are combined under their LCA to form an MCT, and (b) how the LCA is linked to the root of the XML tree.

An alternative view to keyword search involves clustering the results using structural and semantic information [8, 10]. In this paper, we elaborate on an approach which clusters the results at different levels of granularity and then, exploits user input to navigate among the clusters in order to retrieve the relevant results.

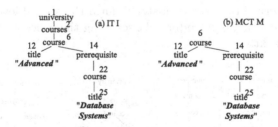

Fig. 2. (a) An IT and (b) its MCT

3 Query Result Clustering

Our approach clusters results in three different levels. The clusters at the lowest level partition the results. Clusters in higher levels contain clusters from lower levels. Every cluster has a representative. The users can navigate through the system by selecting clusters initially at the top level and by drilling down to their nested clusters and finally to the results. The selection of clusters at every level is facilitated by the ranking of the relevant clusters which is provided to the user.

Patterns. Different ITs are not usually of particular interest to the user as query results as long as they match the keywords in the same way and they share the same structural and semantic properties. Therefore, we use patterns (defined next) for clustering the results at the first level of our clustering scheme.

Definition 2. *A pattern P of a query Q on an XML tree T is a tree which is isomorphic (including the annotations) to an IT of Q on T. The MCT of a pattern P refers to P without the path that links the LCA of the annotated nodes to the root of P.*

A pattern has all the information of an IT except the physical location of that one in the XML tree. At the first clustering level, a cluster has a pattern as the representative and comprises all the ITs which comply with this pattern. Patterns are used as representatives for clusters at all levels. For our running example, Figure 3 shows four patterns (out of 32 in total) for our running example. Pattern P_2 has two ITs which comply with it: the IT of the query instance $\{(Advanced, 8), (Database, 8), (Systems, 8)\}$ and the IT of the query instance $\{(Advanced, 10), (Database, 10), (Systems, 10)\}$.

Classes. Different patterns can be similar in the sense that they match the keywords in the same way; they have the same root-to-annotated-node paths and they have the same LCA. These patterns are semantically very close since they only differ in the way they combine keyword instances to form partial LCAs. We put such patterns in the same cluster to form the second level of clustering, named classes. We formally introduce classes using the concept of the ≈ relation which is defined next.

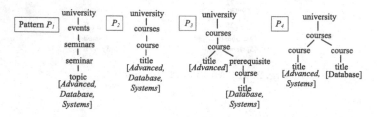

Fig. 3. Some patterns for our running example

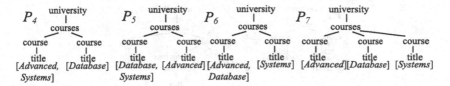

Fig. 4. A class of patterns consisting of four patterns

Definition 3. *Let P and P' be two patterns of a query on an XML tree. We say that $P \approx P'$ if all of the following holds: (a) the root-to-LCA-path of P and P' are the same, and (b) for every keyword in the query, the LCA-to-keyword instance path in P and P' is the same.*

Definition 4. *Given the set of patterns of a query Q on an XML tree T, a class of Q on T is an equivalence class of patterns with respect to the $\approx$ relation.*

Figure 4 shows four patterns. One can see that these patterns form a class in our running example.

We call *size* of a pattern P the number of edges in P. A pattern with the smallest size is chosen as a representative of a class (in case of a tie, a representative is chosen randomly among the patterns with the smallest size). We denote the representative of a cluster C (which is always a pattern) as $repr(C)$.

Collections. The $\approx$ relation identifies similarities between patterns by detecting identical path subpatterns between these patterns. However, similarities between patterns can also be identified by detecting common path subpatterns in a more relaxed way in the sense that the paths of one pattern can be embedded to the paths of another pattern. Embedding a path p_1 into a path p_2 means that the edges of p_1 are mapped to sequences of edges in p_2. That is, an edge in p_1 is viewed not as a child but as a descendant relationship between its nodes. We capture this type of similarity between patterns using the concepts of *descendant path homomorphism* and $\prec_{dph}$ relation. We then use the $\prec_{dph}$ relation to define the third level of clustering, and to cluster classes into collections.

Definition 5. *Let p_1 and p_2 be two pattern paths whose last nodes have the same label and are annotated by the same keyword. There is a descendant path homomorphism from p_1 to p_2 iff there is a function dph from the nodes of p_1 to the nodes of p_2 such that:*
(a) for every node n in p_1, n and dph(n) have the same labels.
(b) if n' is a child of n in p_1, dph(n') is a descendant of dph(n) in p_2.

For instance, in Figure 5, the path `courses/course/title`[*Database*] of pattern P_4 has a descendant path homomorphism to `courses/course/prerequisite/course/title`[*Database*] of pattern P_8. Similarly, the path `courses/course/title`[*Advanced*] of P_4 has a descendant path homomorphism to `courses/course/title`[*Advanced*] of P_8 since they are identical.

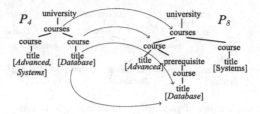

Fig. 5. Descendant path homomorphism from a path in P_4 to a path in P_8

We use the concept of descendant path homomorphism to define the $\prec_{dph}$ relation.

Definition 6. *Let P and P' be two patterns of a query Q on an XML tree. $P \prec_{dph} P'$ iff*
(a) P and P' share the same root-to-LCA path.
(b) for every keyword k in Q, the path from the LCA to the node annotated by k in P has a descendant path homomorphism to a path in the MCT of P'.

Consider the patterns P_4 and P_8 of Figure 5. As one can see, $P_4 \prec_{dph} P_8$ since for each one of the keywords *advanced, database, systems*, the path from **courses** to the node annotated by this keyword has a descendant path homomorphism to a path in the MCT of P_8.

We now use $\prec_{dph}$ to introduce the notion of collection.

Definition 7. *Consider the set of classes of a query on an XML tree. Let $\mathcal{R}$ be the set of class representatives. A collection is a set L of classes which contains exactly: (a) a class C whose representative is a minimal element in $\mathcal{R}$ w.r.t. $\prec_{dph}$, and (b) all the classes C' such that $repr(C) \prec_{dph} repr(C')$.*

That is, for L and C, we have that $\forall C' \in L, C \neq C', repr(C) \prec_{dph} repr(C')$ and for every class $C' \notin L, repr(C) \nprec_{dph} repr(C')$ and $repr(C') \nprec_{dph} repr(C)$.

Clearly, $repr(C)$ is the least element in the set of representatives of classes in L with respect to $\prec_{dph}$. We define the representative of collection L to be the representative of class C. That is, $repr(L) = repr(C)$.

There are as many collections for Q on T as there are minimal elements in the set of class representatives $\mathcal{R}$ w.r.t. $\prec_{dph}$. Note that the collections can overlap. However, they cannot overlap on a representative class (a class whose representative is also the representative of a collection). Therefore, no collection can be included into another collection.

Navigation among Clusters. We now explain how the user proceeds in order to find the relevant results of a query. In order to facilitate this process, we provide techniques for ranking the clusters: all collections are ranked at the top level, classes are ranked within collections, and patterns are ranked within classes.

Within a class, the patterns are ranked based on their size in ascending order. Patterns of the same size have the same rank. If the size of a pattern is smaller than the size of another pattern in the same class, it is assumed to be more relevant to the query as it more closely relates the query keyword instances.

The $\prec_{dph}$ relation is used to rank the classes within a collection. As Definition 7 determines, the representative of a collection is the least element in the set of representatives of classes in the collection with respect to $\prec_{dph}$. Therefore, every collection is a rooted DAG where the nodes correspond to classes and the direct edges correspond to $\prec_{dph}$ relation between the representatives of the classes. There is an edge from C_1 to class C_2 if $repr(C_1) \prec_{dph} repr(C_2)$. The class of the representative of the collection is the unique source of the DAG. We rank the classes within a collection so that the total order of the rank complies with the partial order defined by the DAG.

Collections are also organized into graphs in order to provide an ordering of them. For this purpose, we define another type of relation called *partial_path_homomorphism* relation ($\prec_{pph}$).

Definition 8. *Let P and P' be two patterns of a query on an XML tree, and p and p' be root-to-annotated-node paths of P and P' whose last nodes are annotated by the same keyword. We say $P \prec_{pph} P'$ iff there is a function pph from the nodes of p to the nodes of p' such that:*
(a) the root of P in p is mapped by pph to the root of P' in p',
(b) the root of the MCT of P in p is mapped by pph to a node which is a descendant (not self) of the root of the MCT of P' in p',
(c) for every node n in p, n and $pph(n)$ have the same labels, and
(d) if n_2 is a child of n_1 in p, $pph(n_2)$ is a child of $pph(n_1)$ in p'.

For our running example, Figure 6 shows two patterns, P_3 and P_4 where $P_3 \prec_{pph} P_4$. As one can see, P_3 represents a more closely connected relationship of the keywords because its MCT root is deeper in the XML tree.

The $\prec_{pph}$ relation is used to construct a graph of collections in which the nodes represent the collections. There is an edge from collection L to collection L' if $repr(L) \prec_{pph} repr(L')$. The collections are ordered in accordance with the partial order obtained from the graph.

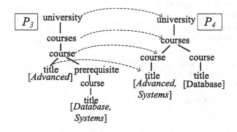

Fig. 6. The $\prec_{pph}$ relation between two patterns P_3 and P_4

4   Algorithm

Our implementation of the proposed clustering approach consists of two components. The first one is the pattern generation component which generates the query patterns (that is, the first level of clustering) and their associated ITs. We use the *PatternStack* algorithm [1] for this purpose. *PatternStack* takes as input a keyword query and the inverted lists of the keyword instances for the query keywords. The second component of our implementation constructs the classes and the collections by performing the necessary relation checks. It also compares collections, and constructs the final graph of collections for user navigation.

We exploit the transivity property of the $\approx$ relation ($\approx$ is an equivalance relation) to make the computation faster. Because of this property, it suffices to check for the existence of $\approx$ between a pattern and the representative of a class to decide if a pattern is a member of the class.

The outline of our algorithm is given in Algorithm 1. It takes as input the patterns produced by the *PatternStack* algorithm and updates the classes and collections, or constructs new ones. In lines 1-8, a pattern P_{cur} is compared with every collection representative that has the same root-to-MCT-path. There is no need to compare with the rest of the collections as both the $\prec_{dph}$ and $\approx$ relations have this structural requirement. After the checks are completed, all the modified collections are examined by **DiscoverMergeCollections** to discover if there are any new collections. If there is a new collection, it is compared with the rest of the collections in terms of $\prec_{pph}$ to update $G_{collection}$ (graph of collections with edges representing the $\prec_{pph}$ relation). In addition, if there are

---

Algorithm 1. Algorithm for constructing the clusters

```
 1  ClusterGraph(PList: list of patterns)
 2      foreach Pattern P_cur in PList do
 3          if G_collection is empty then
 4              collections.add(new Collection(new Class(P_cur)))
 5          else
 6              foreach Collection L in collections.getWithPath(P_cur.rootToMCTPath)
                do
 7                  Check( P_cur, L)
 8              DiscoverMergeCollections(P_cur)

 9  Check(Pattern P, Collection L)
10      foreach Class C in L.classes do
11          P' = C.representative
12          if P ≈ P' then
13              C.addPattern(P)
14              break
15          else if P ≺_dph P' then
16              Class c = new Class(P)
17              G_L.addEdge(c, C)
18          else if P' ≺_dph P then
19              Class c = new Class(P)
20              G_L.addEdge(C, c)
```

---

multiple collections which have P_{cur} as their representative, these collections are merged.

The function **Check** (lines 10-21) checks if a pattern under consideration belongs to a given collection. The pattern is compared with every class of a collection. Edges are added to the G_L graph appropriately between the class of the new pattern and the previously constructed classes to reflect the $\prec_{dph}$ relation. Because of space constraints, the details of the data structures that are used and of the implementation of **DiscoverMergeCollections** are omitted.

5 Experimental Evaluation

We performed experiments to assess the efficiency of our clustering algorithm and also, the retrieval effectiveness of our clustering methodology.

We use the Mondial and Sigmod datasets for the experiments [1]. The experiments were conducted on a 2.9 GHz Intel Core i7 machine with 3 GB memory running Ubuntu.

5.1 Metrics

We adapted the *reach time* metric [5, 8] to assess the retrieval effectiveness. This metric is used to quantify the time spent by a user to locate a relevant result (a pattern in our case). The reach time for a specific pattern is proportional to the number of elements that the user needs to examine while locating a relevant pattern through the clustering levels. The formal definition is given by Equation 1 where n_k is the number of patterns the user examines at the k-th step and l is the number of steps ($l = 3$ in our approach, since there are only three levels of clustering) that the user needs to take. Since the collections can overlap, there might be several paths leading to a relevant pattern. We consider the shortest path to this pattern to compute its reach time. We report the average reach time for a query over all of the relevant patterns for that query.

$$t_{reach} = (\sum_{k=1}^{l} n_k) \tag{1}$$

5.2 Effectiveness of Retrieval

We run the queries of Table 1 over Mondial and Sigmod databases. Figure 7 shows the reach time values measured for each query. According to these values, our clustering methodology almost always allows the user to find the relevant result in the first collection that they examine. This is because of the fact that $\prec_{pph}$ pushes down collections which contains less meaningful patterns to the lower ranks. In addition, within the collections, the ordering of the classes obtained from their graph organization also ranks lower the classes with patterns which have a looser connection between their keyword instances.

---

[1] http://www.cs.washington.edu/research/xmldatasets

Table 1. Queries used in the experiments

Dataset	Query ID	Keywords
Mondial	M1	*international, monetary, fund, established*
	M2	*religions, christian, muslim*
	M3	*european, union*
	M4	*japan, tokyo, population*
	M5	*north, atlantic, treaty, organization*
	M6	*massachusetts, population*
SIGMOD	S1	*divesh, srivastava, database*
	S2	*data, mining*
	S3	*query, optimization*
	S4	*database, volume, number*
	S5	*initpage, 7, article, endpage*
	S6	*asuman, pinar, article*

We present the number of patterns and the resulting number of collections for each query in Figure 8. Note that, the y-axis is in logarithmic scale. As one can see, even in the maximum case, the number of generated collections is less then 20 (query $M5$, which has 107 matching patterns). We can claim that the number of generated collections is reasonable and since it is significantly smaller than the number of patterns, examining the collections and drilling down to a specific collection would be more convenient for users than a flat interface.

5.3 Efficiency

We present the computation times of our algorithm for each query of Table 1 in Figure 9. Most of the execution times are smaller than 1 second even though further optimizations can be done for the computation. The response time of the system is comparable to those of real life applications.

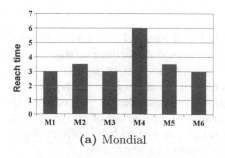

(a) Mondial

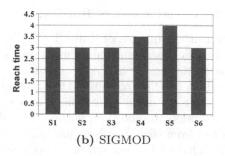

(b) SIGMOD

Fig. 7. Reach times for the queries of Table 1

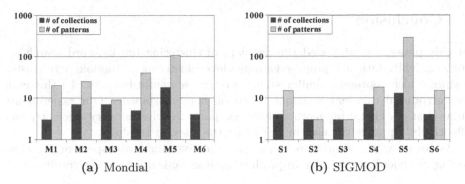

Fig. 8. Number of patterns and collections generated for the queries of Table 1

6   Related Work

Keyword search on XML is a complicated task since XML does not usually follow a strict schema. Several papers proposed filtering and ranking semantics for finding results which are relevant to the query and ranking them with respect to their relevancy, respectively. Filtering semantics, such as Smallest LCA [13], Exclusive LCA [4], Meaningful LCA [7] and VLCA [6] filters out results that do not follow some rules which are based on structural and/or semantic properties of the results. Ranking semantics exploits different statistical measures [2–4, 12] and semantic features [1].

Recently, different studies addressed the problems of keyword ambiguity and the multitude of results. XReal [2] proposed approaches to find the user intented result type for keyword queries. XSeek [9] addressed the problem of deciding upon the nodes to be included in the results. XMean [8], and Liu and Chen [10] address the problem of clustering XML keyword search results. XMean [8] introduced the conceptually related entity nodes as their semantics. Patterns are used to define clusters for the results. A hierarchy of clusters is also introduced in [8] by applying some relaxations to the patterns. Liu and Chen [10] clusters the results by using the types of keyword instance nodes (i.e., entity or attribute).

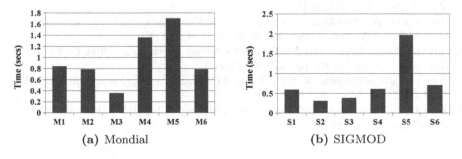

Fig. 9. Execution times (in secs) for the queries of Table 1

7   Conclusion

In this paper, we addressed the problem of clustering the keyword search results on XML data. We proposed a multi-level clustering methodology to cluster results which represent similar structural and semantic features. To this end, we introduced different relations based on homomorphisms. Our experiments show that our clustering methodology computes the clusters very efficiently and effectively supports the user in retrieving the relevant results.

As future work, we aim to investigate the diversification aspect of our clustering methodology in order to produce diverse ranked lists of the results.

References

1. Aksoy, C., Dimitriou, A., Theodoratos, D., Wu, X.: XReason: A semantic approach that reasons with patterns to answer XML keyword queries. In: Meng, W., Feng, L., Bressan, S., Winiwarter, W., Song, W. (eds.) DASFAA 2013, Part I. LNCS, vol. 7825, pp. 299–314. Springer, Heidelberg (2013)
2. Bao, Z., Ling, T.W., Chen, B., Lu, J.: Effective XML keyword search with relevance oriented ranking. In: ICDE, pp. 517–528 (2009)
3. Chen, L.J., Papakonstantinou, Y.: Supporting top-K keyword search in XML databases. In: ICDE, pp. 689–700 (2010)
4. Guo, L., Shao, F., Botev, C., Shanmugasundaram, J.: XRANK: Ranked keyword search over XML documents. In: SIGMOD, pp. 16–27 (2003)
5. Kummamuru, K., Lotlikar, R., Roy, S., Singal, K., Krishnapuram, R.: A hierarchical monothetic document clustering algorithm for summarization and browsing search results. In: WWW, pp. 658–665 (2004)
6. Li, G., Feng, J., Wang, J., Zhou, L.: Effective keyword search for valuable LCAs over XML documents. In: CIKM, pp. 31–40 (2007)
7. Li, Y., Yu, C., Jagadish, H.V.: Schema-free XQuery. In: VLDB, pp. 72–83 (2004)
8. Liu, X., Wan, C., Chen, L.: Returning clustered results for keyword search on XML documents. IEEE Trans. Knowl. Data Eng. 23(12), 1811–1825 (2011)
9. Liu, Z., Chen, Y.: Identifying meaningful return information for XML keyword search. In: SIGMOD, pp. 329–340 (2007)
10. Liu, Z., Chen, Y.: Return specification inference and result clustering for keyword search on XML. ACM Trans. Database Syst. 35(2), 10:1–10:47 (2010)
11. Nguyen, K., Cao, J.: Top-k answers for XML keyword queries. World Wide Web 15(5-6), 485–515 (2012)
12. Termehchy, A., Winslett, M.: Using structural information in XML keyword search effectively. ACM Trans. Database Syst. 36(1), 4 (2011)
13. Xu, Y., Papakonstantinou, Y.: Efficient keyword search for smallest LCAs in XML databases. In: SIGMOD, pp. 537–538 (2005)
14. Xu, Y., Papakonstantinou, Y.: Efficient LCA based keyword search in XML data. In: EDBT, pp. 535–546 (2008)
15. Zamir, O., Etzioni, O.: Web document clustering: A feasibility demonstration. In: SIGIR, pp. 46–54 (1998)

An Efficient Conditioning Method
for Probabilistic Relational Databases

Hong Zhu[1], Caicai Zhang[1], Zhongsheng Cao[1],
Ruiming Tang[2], and Mengyuan Yang[1]

[1] Huazhong University of Science and Technology
{whzhuhong,caicaizhng,sheepmengyuan}@gmail.com, caozhongsheng@163.com
[2] National University of Singapore
tangruiming@nus.edu.sg

Abstract. A probabilistic relational database is a probability distribution over a set of deterministic relational databases (namely, *possible worlds*). Efficient processing of updating information in probabilistic
databases is required in several applications, such as sensor networking,
data cleaning. As an important class of updating probabilistic databases,
conditioning refines probability distribution of possible worlds, and possibly removing some of the possible worlds based on general knowledge,
such as primary key constraints, functional dependencies and others.
The existing methods for conditioning are exponential over the number
of variables in the probabilistic database for an arbitrary constraint. In
this paper, a constraint-based conditioning algorithm is proposed by only
considering the variables in the given constraint without enumerating the
truth values of all the variables in the formulae of tuples. Then we prove
the correctness of the algorithm. The experimental study shows our proposed algorithm is more efficient comparing the work in the literatures.

Keywords: Database, Probabilistic data, Conditioning, Constraint.

1 Introduction

Many applications produce large amounts of *uncertain* data, such as, sensor data
management [10], information integration [2,5]. A probabilistic database is used
to manage uncertain data. Informally, a probabilistic database is a probability
distribution over a set of deterministic databases (namely, *possible worlds*).

Efficient processing of updating information in probabilistic databases is required in several applications. In the context of data cleaning [4], several rules
are specified to remove impossible value assignments of data. Continuous learning and survey introduce significant new rules. Updating probabilistic databases
based on new rules is important to a wide range of applications. For example, personal information systems may require updating databases based on an
additional rule of different persons having unique ids. In the context of sensor
networks [9], sensor readings are usually represented using correlated probabilistic models. Many applications benefit from updating correlated sensor data, for

F. Li et al. (Eds.): WAIM 2014, LNCS 8485, pp. 225–236, 2014.
© Springer International Publishing Switzerland 2014

example, obtaining the latest value of one sensor to find the values of the other correlated sensors. Similar applications exist in data integration [2].

Conditioning is to update the probability distribution of possible worlds of a probabilistic database. Specifically, conditioning probabilistic databases is an operation of removing possible worlds which do not satisfy a given condition from a probabilistic database and re-defining the new probabilities of the remaining possible worlds by their conditional probabilities. Several techniques have been proposed for *conditioning* probabilistic databases. Koch and Olteanu [8] and Tang et al. [11] have studied the conditioning problem for probabilistic relational databases. Koch and Olteanu develop *ws-trees* to capture constraint information. However, it is not easy to construct a tree which is efficient to compute the probability of the represented constraint and perform conditioning. Tang et al. devise efficient conditioning algorithms for some special classes of mutually exclusive constraints. However, one of the challenges in conditioning probabilistic databases is that the existing methods of enumerating possible worlds are exponential over the number of variables in the probabilistic database, for an arbitrary constraint. For a probabilistic database of n_R tuples and n_E variables involved in the formulae of tuples, the time complexity of conditioning is $O(n_R * 2^{n_E})$ for an arbitrary constraint.

We tackle this challenge under a general uncertainty models allowing for probabilistic correlations. Our techniques are based on two ideas: (1) considering the assignments of variables appeared in the constraint only, instead of enumerating possible assignments of all variables in the probabilistic database; and (2) applying variable-replaced mechanisms to update the formulae of tuples, while avoiding the replacement of every variable in the probabilistic database.

Our main contributions towards this goal are summarized as follows.

(1) We construct a constraint-based conditioning framework, which minimizes the set of tuples whose formulae have to be updated.

(2) We prove the correctness of the proposed constraint-based conditioning algorithm, by showing that it guarantees an equivalent representation of a conditioned probabilistic database.

(3) We conduct an extensive experimental study to evaluate our algorithm. The experiments evaluate our algorithm in different configurations, showing the efficiency and scalability of size of variables in the probabilistic database of our algorithm in different practical settings.

The remainder of this article is organized as follows. We review related works in Section 2. Then, we present in Section 3 necessary preliminaries on probabilistic databases and conditioning. Section 4 describes our proposed constraint-based conditioning algorithm for probabilistic relational databases. Experimental study and conclusion are shown in Section 5 and Section 6 respectively.

2   Related Work

While there has been a significant amount of work on uncertain databases in the past few decades, surprisingly not much work has focused on modifications.

Abiteboul and Grahne [1] consider various modification operations over incomplete databases, which are broadly defined as a transformation from one set of possible worlds to another. They study classes of modifications, such as inserting, deleting possible worlds, and constraining the set of possible worlds. The main result of the paper is to study expressiveness in representing results after modifications over different uncertain data models. Hegner [7] views the data modification problem as a programming language and maps programs to operations on possible worlds. While the language is quite expressive, no query answering techniques are presented for any uncertain data model.

Koch and Olteanu [8] are the first to study the conditioning problem for probabilistic relational databases. They adapt algorithms and heuristics for Boolean validity checking and simplification to solve the general NP-hard conditioning problem. They use the attribute-level model and develop ws-trees to capture data correlations. Confidence computation and conditioning benefit from small ws-tree decompositions. Unfortunately, if the number of constraints added into the probabilistic database is large, the represented ws-trees trend to be very deep, which causes the techniques used for conditioning to be inefficient.

Tang et al. [11] propose a framework for conditioning probabilistic relation instances. They have defined constraints as a part of the data model. They devise P-TIME algorithms for some special classes for mutually exclusive constraints, while we consider the arbitrary constraint case.

3 Preliminaries

3.1 Probabilistic Databases

Definition 1. *A probabilistic database in the extensional representation [9,3,6,12] is a discrete probability space $PDB = (W, P)$, where $W = \{w_1, \ldots, w_n\}$ is a set of traditional databases, called possible worlds, and $P : W \to [0, 1]$ is a function mapping from possible worlds to probability values, such that $\sum_{j=1}^{n} P(w_j) = 1$.*

The extensional representation of probabilistic databases can express any finite set of deterministic databases, however, it is not compact to enumerate all possible worlds and hence there is a need to have a concise representation formalism.

R_1		V_P	
RID	f	V	P
$t1$	e_1	e_1	0.6
$t2$	e_2	e_2	0.5
$t3$	$e_1 \wedge e_2$	e_3	0.8
$t4$	$e_1 \vee e_3$		
$t5$	$e_2 \wedge e_3$		
$t6$	e_3		

Fig. 1. A probabilistic database instance $\mathbb{D}$

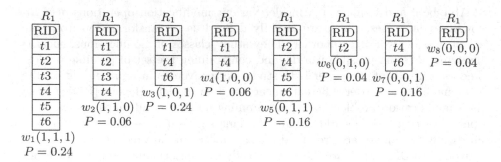

Fig. 2. Eight possible worlds of $\mathbb{D}$

Definition 2. *A probabilistic database in the intensional representation is a quadruple $\mathbb{D} = \langle \Re, E, P, f \rangle$, where $\Re$ is a traditional relational database, E is a set of Boolean independent variables $\{e_1, \ldots, e_m\}$, P specifies the probability value of each variable being true. f associates each tuple t with a variable or a Boolean expression composing of variables, namely $f(t)$, whose truth value determines the presence of t in the actual world and whose probability is defined by probabilities of composed variables [6].*

A possible world w_i is a traditional database instance such that the expression associated with the tuple in the possible world is true, the expression associated with the tuple not in the possible world is false. A joint value $V_j(E)(j \in [1, 2^m])$ of all variables in E determines a possible world w_i of the probabilistic database. $P(w_i) = \sum_{w_i \sim V_j(E)} P(V_j(E))$ and $\sum P(w_i) = 1$, where $w_i \sim V_j(E)$ represents the joint value $V_j(E)$ is a joint value leading to the possible world w_i.

Example 1. Assume a probabilistic database $\mathbb{D}$ in Fig.1 only includes one relation R_1. Table V_P in Fig.1 records a set of independent Boolean variables (in column V) and their probabilities being true. Column f in R_1 records Boolean formulae composting of variables in V_P associated to tuples. The probability of $f(t)$ being true is the existence probability of tuple t in the actual world.

Each assignment of variables in V_P represents one possible database instance of $\mathbb{D}$: R_1 containing all tuples whose associated formulae are satisfied with the given assignment. Let $w_i(x_1, x_2, x_3)$ be the ith possible world and $P(w_i)$ be the corresponding probability of w_i.

Definition 3. *Given an extensional probabilistic database $PDB = (W, P)$, and an intensional representation of a probabilistic database $\mathbb{D} = \langle \Re, E, P, f \rangle$, assume $W_{\mathbb{D}}$ is the set of possible worlds of $\mathbb{D}$ and $P_{\mathbb{D}}$ is the probability distribution of $W_{\mathbb{D}}$, we say $\mathbb{D}$ and PDB are equivalent representations, denoted by $\mathbb{D} \equiv_r PDB$, if and only if $W_{\mathbb{D}} = W$ and $\forall w_i \in W, P(w_i) = P_{\mathbb{D}}(w_i)$.*

In Example 1, $\mathbb{D}$ (in Fig.1) is an equivalent representation of a probabilistic database with eight possible worlds shown as in Fig.2.

3.2   Conditioning Probabilistic Databases

Conditioning allows users to start with a database of priori probabilities, to add some evidences, and enforce it to a posteriori probabilistic database. We call the new evidences as conditioning constraints, and call the generated posteriori probabilistic database as conditioned probabilistic database.

Definition 4. *Given a probabilistic database* $\mathbb{D} = \langle \Re, E, P, f \rangle$, *a conditioning constraint* C *is a well-formed logical expression composing of variables in* E, *with* $\vee, \wedge, \neg$.

Definition 5. *Given a probabilistic database* $\mathbb{D} = \langle \Re, E, P, f \rangle$, *a conditioning constraint* C, *assume* $\mathbb{D} \equiv_r PDB_{\mathbb{D}}(W_{\mathbb{D}}, P_{\mathbb{D}})$, *then the extensional representation of the conditioned probabilistic database* $PDB_{\{\mathbb{D},C\}}(W_{\{\mathbb{D},C\}}, P_{\{\mathbb{D},C\}})$ *is defined as:*

$$W_{\{\mathbb{D},C\}} = \{w_i \mid w_i \in W_{\mathbb{D}}, C \sim w_i\}, \ P_{\{\mathbb{D},C\}}(w_i) = P(w_i)/P(C)$$
where $P(C) = \sum_{w_i \in W_{\mathbb{D}}, C \sim w_i} P(w_i)$, *and* $C \sim w_i$ *donates* w_i *satisfies* C.

Example 2. In Example 1, if conditioning constraint $C = e_1 \vee e_2$, the conditioned probabilistic database $PDB_{\{\mathbb{D},C\}}$ is:

$$W_{\{\mathbb{D},C\}} = \{w_1, w_2, w_3, w_4, w_5, w_6\}, \ P(C) = 0.8$$
$$P_{\{\mathbb{D},C\}}(w_1) = 0.3, \ P_{\{\mathbb{D},C\}}(w_2) = 0.075, \ P_{\{\mathbb{D},C\}}(w_3) = P(w_3)/P(C) = 0.3$$
$$P_{\{\mathbb{D},C\}}(w_4) = 0.075, \ P_{\{\mathbb{D},C\}}(w_5) = 0.2, \ P_{\{\mathbb{D},C\}}(w_6) = 0.05$$

Definition 6. *Given a probabilistic database* $\mathbb{D} = \langle \Re, E, P, f \rangle$, *a conditioning constraint* C, *the problem of conditioning probabilistic databases is to find a probabilistic database* $\mathbb{D}' - \langle \Re, E', P', f' \rangle$ *such that* $\mathbb{D}' \equiv_r PDB_{\{\mathbb{D}, C\}}$.

Example 3. Fig.3 shows an intensional representation of the conditioned probabilistic database $PDB_{\{\mathbb{D},C\}}$ in Example 2 by the existing method [11](refer as *D_based*). In *D_based* approach, possible worlds in $W_{\{\mathbb{D},C\}}$ in Example 2 are enumerated. Since these possible worlds are mutually exclusive, their formulae are composing of a set of Boolean variables $U\{u_1, \ldots, u_5\}$ as in the left table. The probability values of variables in U (shown as in V_P) guarantees $P_{\{\mathbb{D},C\}}$ is the same as in Example 2. The formula of each tuple (shown as in R_1) is constructed by disjunction of formulae of possible worlds in which the tuple exists.

W	E
w_1	$E_1 : u_1$
w_2	$E_2 : \neg u_1 \wedge u_2$
w_3	$E_3 : \neg u_1 \wedge \neg u_2 \wedge u_3$
w_4	$E_4 : \neg u_1 \wedge \neg u_2 \wedge \neg u_3 \wedge u_4$
w_5	$E_5 : \neg u_1 \wedge \neg u_2 \wedge \neg u_3 \wedge \neg u_4 \wedge u_5$
w_6	$E_6 : \neg u_1 \wedge \neg u_2 \wedge \neg u_3 \wedge \neg u_4 \wedge \neg u_5$

	R_1
RID	f
$t1$	$E_1 \vee E_2 \vee E_3 \vee E_4$
$t2$	$E_1 \vee E_2 \vee E_5 \vee E_6$
$t3$	$E_1 \vee E_2$
$t4$	$E_1 \vee E_2 \vee E_3 \vee E_4 \vee E_5$
$t5$	$E_1 \vee E_5$
$t6$	$E_1 \vee E_3 \vee E_5$

V_P	
V	P
u_1	0.3
u_2	$3/28$
u_3	$12/25$
u_4	$3/13$
u_5	0.8

Fig. 3. The intensional representation of $PDB_{\{\mathbb{D},C\}}$ by *D_Base* approach

4   Constructing Conditioned Probabilistic Databases

In this section we present our proposed algorithm in detail.

Definition 7. *Given a set of Boolean variables or Boolean formulae $X = \{x_1, \ldots, x_n\}$, if $V(X) = (x_1 : v_1, \ldots, x_n : v_n)$ is a joint assignment for X, and all the possible $V(X)$ be denoted by $SV(X)$, then we say $I(X) = (v_1, \ldots, v_n)$ is an assignment vector of X, all the possible $I(X)$ be denoted by $SI(X)$.*

Example 4. Assume $E = \{e_1, e_2\}$ is a set of Boolean variables,
$SV(E) = \{(e_1 : 1, e_2 : 1), (e_1 : 1, e_2 : 0), (e_1 : 0, e_2 : 1), (e_1 : 0, e_2 : 0)\}$
$SI(E) = \{(1, 1), (1, 0), (0, 1), (0, 0)\}$

Theorem 1. *Given a probabilistic database $\mathbb{D} = \langle \Re, E, P, f \rangle$, and a conditioning constraint C, assume $S = \{e_1, \ldots, e_s\}$ is the set of variables existing in C. If $S' = \{e'_1, \ldots, e'_s\}$ is a mapping from S to Boolean formulae composing of a set of new independent Boolean variables U, and S' and U satisfy the following conditions: (a) $SI(S') = \bigcup_{C \sim I(S)} I(S)$; (b) $P(I(S')) = \sum_{V_i(U) \sim I(S')} P(V_i(U))$; (c) If $I(S') = I(S), P(I(S')) = P(I(S))/P(C)$, (where $C \sim I(S)$ represents C is true according to $I(S)$, and $V_i(U) \sim I(S')$ represents S' takes the assignment $I(S')$ when U takes $V_i(U)$), then the equivalent intensional representation $\mathbb{D}'$ of $PDB_{\{\mathbb{D}, C\}}$ can be generated by replacing S with U, and replacing each variable e_j in S with e'_j in each $f(t)$.*

Proof. By Definition 2, $\forall w_i \in W_{\mathbb{D}}$, $w_i = \{t \mid t \in \Re, f(t) \sim V_j(E)\}$
  $P(w_i) = \sum_{w_i \sim V_j(E)} P(V_j(E))$
  where $f(t) \sim V_j(E)$ represents that $f(t)$ is true according to $V_j(E)$.
  By Definition 5, $\forall w_i \in W_{\{\mathbb{D}, C\}}$, $w_i = \{t \mid t \in \Re, f(t) \sim V_j(E), C \sim V_j(E)\}$
  $P_{\{\mathbb{D}, C\}}(w_i) = \sum_{w_i \sim V_j(E), C \sim V_j(E)} P(V_j(E))/P(C)$
  where $C \sim V_j(E)$ represents the joint assignment $V_j(E)$ can make C be true.
  A joint assignment of variables in S determines the truth of C. Thus,
  $\forall w_i \in W_{\{\mathbb{D}, C\}}$, $w_i = \{t \mid t \in \Re, f(t) \sim V_j(E), V_j(E) \sim V(S), C \sim V(S)\}$
  where $V_j(E) \sim V(S)$ demonstrates $V(S)$ is the assignment value of variables in S when E takes the joint assignment $V_j(E)$ and $C \sim V(S)$ represents the assignment of S can make C be true.
  Let $E' = (E - S) \cup S'$, $f'(t)$ is generated by replacing each variable e_j in S with e'_j in $f(t)$. Since $SI(S') = \bigcup_{C \sim I(S)} I(S)$, for $\forall V(E) \in VS(E)$, and $C \sim V(S)$, $\exists V(E')$, $I(E) = I(E')$, $I(S) = I(S')$. Then $\forall t \in \Re$, $f'(t)$ is true when $f(t)$ is true. Therefore, if $w \sim V(E)$, then $w \sim V(E')$.
  E is a set of independent variables, and U and S are independent, thus,
  $P(V(E)) = \prod_{V(E) \sim V(e_i), e_i \in (E - S)} P(V(e_i)) * P(V(S))$
  $P(V(E')) = \prod_{V(E) \sim V(e_i), e_i \in (E - S)} P(V(e_i)) * P(V(S'))$
  Since $I(S) = I(S')$, $P(I(S')) = P(I(S))/P(C)$ and $P(V'(E)) = P(V(E))/P(C)$
  $\therefore \forall w \in W_{\mathbb{D}'}, P_{\mathbb{D}'}(w) = \sum_{w \sim V_j(E), C \sim V_j(E)} P(V_j(E))/P(C) = P_{\mathbb{D}}(w)/P(C)$
  $\therefore \mathbb{D}' \equiv_r PDB_{\mathbb{D}, C}$

4.1 New Variable Set

Property 1. Given a conditioning constraint C, assume $S = \{e_1, \ldots, e_s\}$ is the set of variables appeared in C, and $DNF_C = \{cf_1, \ldots, cf_k\}$ is the set of conjunctive clauses in the completed disjunctive normal form of C, any one joint assignment $V(S)$ of S makes at most one conjunctive clause in DNF_C be true, therefore, there exist k joint assignments of S that can make C be true.

These k assignment vectors of S are the set of assignment vectors of $S' = \{e'_1, \ldots, e'_s\}$ we need to construct.

Example 5. In Example 2, $C = e_1 \vee e_2$, $S = \{e_1, e_2\}$
$DNF_C = \{cf_1 : e_1 \wedge e_2, \ cf_2 : \neg e_1 \wedge e_2, \ cf_3 : e_1 \wedge \neg e_2\}$, $SI(S') = \{(1,1), (1,0), (0,1)\}$

We consider creating a mapping $DNF'_C = \{cf'_1, \ldots, cf'_k\}$ from $DNF_C = \{cf_1, \ldots, cf_k\}$ to Boolean formulae composing of a set of independent variables, and DNF'_C satisfies the following conditions: $(a)SI(DNF'_C) = \bigcup_{C \sim I(DNF_C)} I(DNF_C)$; (b)When $I(DNF'_C) = I(DNF_C)$, $P(I(DNF'_C)) = P(I(DNF_C))/P(C)$.

Each assignment vector of $DNF'_C = \{cf'_1, \ldots, cf'_k\}$ makes one and only one cf'_j be true. Thus, we construct the following mapping by a set of independent variables $U\{u_1, \ldots, u_{k-1}\}$ [11]:
$cf'_1 = u_1, \ cf'_2 = \neg u_1 \wedge u_2, \ \ldots, \ cf'_{k-1} = \neg u_1 \wedge \ldots \neg u_{k-2} \wedge u_{k-1}$
$cf'_k = \neg u_1 \wedge \ldots \wedge \neg u_{k-1}$
$P(u_1) = P(cf_1)/P(C), \ P(u_2) = P(cf_2)/(P(\neg u_1) * P(C)), \ \cdots$
$P(u_{k-1}) = P(cf_{k-1})/(P(\neg u_1) * \ldots * P(\neg u_{k-2}) * P(C))$

Example 6. In Example 5, the mapping of DNF_C is constructed as follows:
$DNF'_C = \{cf'_1, cf'_2, cf'_3\}$, $cf'_1 = u_1$, $cf'_2 = \neg u_1 \wedge u_2$, $cf'_3 = \neg u_1 \wedge \neg u_2$
$P(u_1) = P(cf_1)/P(C) = P(e_1 \wedge e_2)/P(e_1 \vee e_2)$
$P(u_2) = P(cf_2)/(P(\neg u_1) * P(C)) = P(\neg e_1 \wedge e_2)/(P(\neg u_1) * P(e_1 \vee e_2))$

4.2 Variable Mapping

Given a conditioning constraint C, if DNF'_C is constructed by mapping in Section 4.1, DNF'_C corresponds to all the assignments of S where C is true. If a variable $e_i \in S$ takes false in every assignment of S where C is true, e'_i corresponds to all the assignments of S where C is false. Else e'_i corresponds to all the assignments of S where e_i and C both are true. Therefore, the set of variable mappings $S = \{e'_1, \ldots, e'_s\}$ for $S = \{e_1, \ldots, e_s\}$ is constructed as follows:

$$e'_i = \begin{cases} \bigvee_{cf_j \wedge e_i \neq 0, \ cf_j \in DNF_C} cf'_j, & \exists cf_j \in DNF_C, \ cf_j \wedge e_i \neq 0 \\ \bigwedge_{cf_j \in DNF_C} \neg cf'_j, & \forall cf_j \in DNF_C, \ cf_j \wedge e_i = 0 \end{cases} \quad (1)$$

Example 7. In Example 6, $U = \{u_1, u_2\}$
$e'_1 = u_1 \vee (\neg u_1 \wedge u_2)$, $e'_2 = u_1 \vee (\neg u_1 \wedge \neg u_2)$

Theorem 2. *Given a conditioning constraint C, if $S = \{e_1, \ldots, e_s\}$ is the set of variables appeared in C, $S' = \{e'_1, \ldots, e'_s\}$ is constructed by Equation 1, then (a) $SI(S') = \bigcup_{C \sim I(S)} I(S)$; (b) When $I(S') = I(S)$, $P(I(S')) = P(I(S))/P(C)$.*

Proof. For any $V(S)$, $C \sim V(S)$, assume $V(S)$ makes cf_l be true. Since when $C \sim I(DNF_C)$, $I(DNF'_C) = I(DNF_C)$, cf'_l is true when cf_l is true. For any $e_i \in S$, if e_i is in its negative form in every conjunctive clause of DNF_C, then $v_i = 0$, $e'_i = \bigwedge_{cf_j \in DNF_C} \neg cf'_j$. Since there is one and only one cf'_j be true in DNF'_C, $e'_i = 0$. If e_i is in its positive form in at least one conjunctive clause of DNF_C, $e'_i = \bigvee_{cf_j \wedge e_i \neq 0, \ cf_j \in DNF_C} cf'_j$. When $e_i = 1$, then $cf_l \wedge e_i \neq 0, e'_i = cf'_l \vee \bigvee_{cf_j \wedge e_i \neq 0, \ cf_j \in DNF_C, \ i \neq j} cf'_j = 1$. When $e_i = 0$, then $cf_l \wedge e_i = 0, e'_i = 0$.
$\therefore I(S) = I(S')$.
Since when $I(DNF'_C) = I(DNF_C)$, $P(I(DNF'_C)) = P(I(DNF_C))/P(C)$
$P(cf'_i) = P(cf_i)/P(C)$
$\therefore SI(S') = \bigcup_{C \sim I(S)} I(S)$ and $P(I(S')) = P(I(S))/P(C)$.

4.3 Algorithm

Theorem 1 guarantees we obtain an equivalent representation of the conditioned probabilistic database if the new set of variables and the mapping satisfy some conditions, and Theorem 2 guarantees the mapping obtained by Equation 1 satisfies the conditions in Theorem 1. *Conditioning_Based*(for short, we call it *C_Based*) only enumerates the conjunctive clauses in DNF form of C. Each conjunctive clause corresponds to an assignment of variables appeared in C. Since these conjunctive clauses are mutually exclusive, we construct mapping for each conjunctive clause by creating a set of Boolean variables U. Then the mapping for each variable appeared in C is constructed as in Equation 1. *C_based* is shown as in Algorithm 1.

Given a probabilistic database $\mathbb{D}1 = \langle \Re, E1, P1, f1 \rangle$, and a conditioning constraint C, let n_R be the number of tuples in the database, n_C be the number of variables appeared in conditioning constraint C, and n_E be the number of variables in the formulae of tuples in the database. *D_Based* enumerates all the possible joint assignments of variables in E, and takes $O(n_R * 2^{n_E})$ time complexity. While *C_Based* only enumerates all the possible joint assignments of variables in constraint C, and takes $O(n_R + 2^{n_C})$ time complexity.

Example 8. In Example 2, $C = e_1 \vee e_2$, by the algorithm *C_Based*, the conditioned probabilistic database $\mathbb{D}2 = \langle \Re, E2, P2, f2 \rangle$ obtained by *C_Based* is shown as in Fig.4.

5   Experiments

In this section we conduct the experimental study of *C_based approach* and *D_based approach* for conditioning probabilistic databases.

Algorithm 1. Conditioning_Based

Input:
  $\mathbb{D}1\{\Re, E1, P1, f1\}$, the conditioning constraint C;
Output:
  $\mathbb{D}2\{\Re, E2, P2, f2\}$
1: $S \leftarrow$ the set of variables appeared in C
2: $DNF_C\{cf_1, \ldots, cf_k\} \leftarrow$ the set of conjunctive clauses in DNF of C
3: $U \leftarrow$ Create a new set of Boolean variables $\{u_1, \ldots, u_{k-1}\}$
4: $P2(u_1) = P1(cf_1)/P1(C)$
5: **for each** $i = 2, \ldots, (k-1)$ **do**
6:    $P2(u_i) = P1(cf_1)/(P2(\neg u_1) * \ldots * P2(\neg u_{i-1}) * P1(C))$
7: **end for**
8: $E2 = (E1 - S) \cup U$ /* Generating $E2$ */
9: $P2_{e \in (E1-S)}(e) = P1(e)$
10: $cf'_1 = u_1$
11: **for each** $i = 2, \ldots, (k-1)$ **do**
12:    $cf'_i = \neg u_1 \wedge \ldots \wedge \neg u_{i-1} \wedge u_i$
13: **end for**
14: $cf'_k = \neg u_1 \wedge \ldots \wedge \neg u_{k-1}$
15: **for each** variable $e \in VS(C)$ **do**
16:    **if** $\exists cf_j \in DNF_C, \; cf_j \wedge e_i \neq 0$ **then**
17:      $e'_i = \bigvee_{cf_j \wedge e_i \neq 0, \; cf_j \in DNF_C} cf'_j$
18:    **else**
19:      $e'_i = \bigwedge_{d_j \in D} \neg cf'_j$
20:    **end if**
21: **end for**
22: **for each** tuple $t \in R$ **do**
23:    $f2(t) \leftarrow$ Replace each e_i in S with e'_i in $f1(t)$ /* Generating $f2(t)$ */
24: **end for**
25: **return** $\mathbb{D}2\{\Re, E2, P2, f2\}$

We study the effect of n_R (the number of tuples in the database), n_C (the number of variables appeared in constraint C), and n_E(the number of variables in the formulae of tuples in the relation) on the efficiency of C_based and D_based approaches. We compare the efficiency of C_based and D_based by varying n_C, n_E, n_R, one at a time. Both of C_based and D_based approaches are implemented in Java, and all our experiments are conducted on a Pentium 2.5 GHz PC with 3G memory, running XP. We generate our synthetic data sets by varying the three parameters: fixing n_C and n_E, n_R varies from $1K$ to $10K$; fixing n_C and n_R, n_E varies from 5 to 14; fixing n_E and n_R, n_C varies from 5 to n_E.

Effect of Varying n_R: We evaluate the effect of varying the number of tuples on the efficiency of C_based approach and D_based approach. Fig.5 shows the execution time of C_based approach and D_based approach, with the number of tuples varying from $1K$ to $10K$ and $n_E = n_C = 10$. Increasing the number of tuples linearly degrades the performance of C_based approach and D_based approach.

	R_1
RID	f
$t1$	$u_1 \vee \neg u_2$
$t2$	$u_1 \vee u_2$
$t3$	u_1
$t4$	$(u_1 \vee \neg u_2) \vee e_3$
$t5$	$(u_1 \vee u_2) \wedge e_3$
$t6$	e_3

V_P	
V	P
u_1	$3/8$
u_2	0.4
e_3	0.8

Fig. 4. The intensional representation of $PDB_{\{\mathbb{D},C\}}$ by C_Base approach

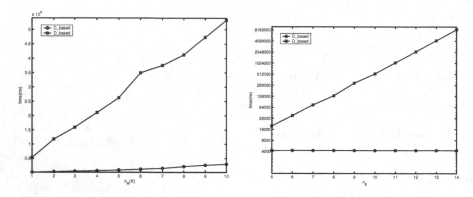

Fig. 5. Varying n_R. $n_E = n_C = 10$ **Fig. 6.** Varying n_E. $n_R = 1K$, $n_C = 5$

This is because these two algorithms perform conditioning on the database tuple by tuple, and the running time of updating the formula of each tuple is almost the same. Fig.5 also shows that C_based *approach* is much more efficient than D_based *approach*, because the number of possible worlds of the constraint C (as considered in C_based *approach*) is much smaller than the number of possible worlds of the database (as considered in D_based *approach*).

Effect of Varying n_E: We evaluate the effect of varying n_E on the efficiency of C_based *approach* and D_based *approach*, as shown in Fig. 6 (while $n_R = 1k$, $n_C = 5$). The execution time of D_based *approach* grows exponentially by increasing n_E. This is due to the fact that disjunctive normal form over the variables in database for each tuple is computed in D_based *approach*. Increasing n_E does not affect the performance of C_based *approach*. C_based *approach* computes the mapping for each variable appeared in C with $O(2^{n_C})$ time complexity, and replaces $f(t)$ for each tuple t with $O(n_R)$ time complexity. Therefore, n_E does not affect the efficiency of C_based *approach*. The performance of C_based *approach* remains stable when scaling n_E up to $1K$ and fixing n_R and n_C. We do not present the running time in figure since the running time of D_based *approach* is too large when n_E is greater than 20.

Effect of Varying n_C: We evaluate the effect of varying n_C on the efficiency of C_based *approach* and D_based *approach*. Fig.7 shows the execution time of

C_based approach and *D_based approach* when increasing n_C from 5 to 14 and fixing n_R to be $1K$ and n_E to be 14. The execution time of *C_based approach* increases slowly when n_C increases. n_C does not affect the efficiency of *D_based approach* since n_C is not a time-complexity factor of *D_based approach*. The maximum execution time when $n_C = 14$ is 8436 seconds for *D_based approach* and is 12.5 seconds for *C_based approach*. *C_based approach* is on average 100 times faster than *D_based approach*.

Real Datasets (MOV): We also perform experiments on a real-world probabilistic dataset [13], which stores uncertain data of movie-viewer ratings. We normalize the dataset by the data model introduced in Definition 2. This dataset contains 10037 tuples and 6693 variables. Figure 8 shows the execution time of *C_based approach* over this real dataset when n_C varies from 11 to 20. The execution time grows slowly when n_C increases from 11 to 17, and grows nearly exponentially when n_C increases from 17 to 20. The reason is that when n_C is greater than 17, the time for computing variables mapping (which is $O(2^{n_C})$) is much larger than the time for updating formulae of tuples (which is $O(n_R)$). The execution time is dominated by that of variables mapping when $n_C > 17$ over this dataset. We do not present the execution time of *D_based approach* in the figure since it is too large over this dataset.

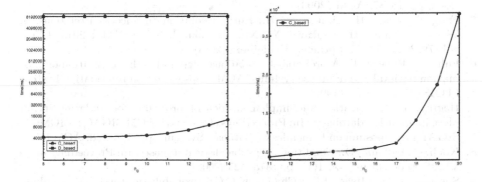

Fig. 7. Varying n_C. $n_R = 1K$, $n_E = 14$ **Fig. 8.** Varying n_C over real dataset MOV

C_Based approach avoids enumerating the truth vales of all the variables in the formulae of tuples, but only considers the variables in the given constraint, and applies variable-replaced mechanisms to update the formulae of tuples, while avoiding the replacement of every variable in the probabilistic database. Therefore, *C_Based* approach performs better than *D_Based* approach in every configuration.

6 Conclusion

In this paper, we proposed a framework for conditioning probabilistic relational databases efficiently. The proposed conditioning algorithm only focuses on variables in the conditioning constraint while avoiding involving the other variables.

We proved the correctness of the proposed constraint-based conditioning algorithm, by showing that it guarantees an equivalent representation of a conditioned probabilistic database. By conducting experimental studies, we found that (1) increasing the number of variables in the constraint does not contribute to large increases in the execution time of our algorithm; (2) scaling the number of variables in formulae of the tuples in the probabilistic relational database has no influence on conditioning; (3) conditioning time increases linearly when scaling the number of tuples in the probabilistic relational database.

References

1. Abiteboul, S., Grahne, G.: Update semantics for incomplete databases. In: Proceedings of the 11th International Conference on Very Large Data Bases, vol. 11, pp. 1–12. VLDB Endowment (1985)
2. Aggarwal, C.C., Yu, P.S.: A survey of uncertain data algorithms and applications. IEEE Transactions on Knowledge and Data Engineering 21(5), 609–623 (2009)
3. Cheng, R., Chen, J., Xie, X.: Cleaning uncertain data with quality guarantees. In: Proceedings of the VLDB Endowment, vol. 1(1), pp. 722–735 (2008)
4. Elnahrawy, E., Nath, B.: Cleaning and querying noisy sensors. In: Proceedings of the 2nd ACM International Conference on Wireless Sensor Networks and Applications, pp. 78–87. ACM (2003)
5. Feng, H., Wang, H., Li, J., Gao, H.: Entity resolution on uncertain relations. In: Wang, J., Xiong, H., Ishikawa, Y., Xu, J., Zhou, J. (eds.) WAIM 2013. LNCS, vol. 7923, pp. 77–86. Springer, Heidelberg (2013)
6. Fuhr, N., Rölleke, T.: A probabilistic relational algebra for the integration of information retrieval and database systems. ACM Transactions on Information Systems (TOIS) 15(1), 32–66 (1997)
7. Hegner, S.: Specification and implementation of programs for updating incomplete information databases. In: Proceedings of the Sixth ACM SIGACT-SIGMOD-SIGART Symposium on Principles of Database Systems, pp. 146–158. ACM (1987)
8. Koch, C., Olteanu, D.: Conditioning probabilistic databases. In: Proceedings of the VLDB Endowment, vol. 1(1), pp. 313–325 (2008)
9. Soliman, M.A., Ilyas, I.F., Chang, K.C.C.: Probabilistic top-k and ranking-aggregate queries. ACM Transactions on Database Systems (TODS) 33(3), 13 (2008)
10. Song, W., Yu, J.X., Cheng, H., Liu, H., He, J., Du, X.: Bayesian network structure learning from attribute uncertain data. In: Gao, H., Lim, L., Wang, W., Li, C., Chen, L. (eds.) WAIM 2012. LNCS, vol. 7418, pp. 314–321. Springer, Heidelberg (2012)
11. Tang, R., Cheng, R., Wu, H., Bressan, S.: A framework for conditioning uncertain relational data. In: Liddle, S.W., Schewe, K.-D., Tjoa, A.M., Zhou, X. (eds.) DEXA 2012, Part II. LNCS, vol. 7447, pp. 71–87. Springer, Heidelberg (2012)
12. Widom, J.: Trio: A system for integrated management of data, accuracy, and lineage. Technical Report (2004)
13. Moving rating system, http://infolab.stanford.edu/trio/code/index.html#examples

A Proactive Complex Event Processing Method Based on Parallel Markov Decision Processes[*]

Yongheng Wang and Kening Cao

College of Information Science and Engineering, Hunan University, Changsha, China 410082
yh.wang.cn@gmail.com, caokening@sina.com

Abstract. Large scale Internet of Things (IoT) produces enormous events. The key issue in IoT application is how to process the events. In this paper a proactive complex event processing method using parallel Markov Decision Processes is proposed for large-scale IoT. Based on a multi-layered adaptive dynamic Bayesian model, an accurate predictive analytics method is proposed. A parallel Markov decision processes model is designed to support proactive event processing. A state partition method and a reward decomposition method are used to support large-scale application. The experimental evaluations show that this method has good accuracy and scalability when used to process complex event proactively in large-scale internet of things.

1 Introduction

In Internet of Things (IoT) applications, signals can be processed as events. Complex Event Processing (CEP) [1,3] is used to process huge primitive events and get valuable high-level information. A proactive event processing system has the ability to mitigate or eliminate undesired future events, or to identify and take advantage of future opportunities, by applying prediction and automated decision making technologies [2].

The Predictive Analytics (PA) methods [4,5] based on complex events can predict the future values of some attributes of the monitored system based on the historical data. Markov Decision Processes (MDP) has been used for proactive event processing [2,6] but there are few papers about how to integrate MDP into proactive event processing and how to process the huge state space of MDP in this area. In this paper, we propose a proactive complex event processing architecture and method (Pro-CEP) for large scale proactive event driven systems. A novel parallel MDP method based on state partition and concurrent actions is proposed to support proactive event processing. A multi-layered Adaptive Dynamic Bayesian Network (mADBN) model is designed for predictive analytics.

[*] This work is supported by the National Natural Science Foundation of China (No.61371116) and the Hunan Provincial Natural Science Foundation (No.13JJ3046).

F. Li et al. (Eds.): WAIM 2014, LNCS 8485, pp. 237–241, 2014.

2   Proactive CEP

2.1   System Architecture and Predictive Analytics Method

The system architecture is shown in fig. 1. The probabilistic raw events are processed by Probabilistic Event Processing Network (PEPN) to get probabilistic complex events. The PA can predict the states of the system and proactive agents (PRA) execute the actions according to the decision based on MDP.

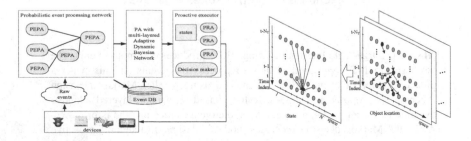

Fig. 1. The mADBN model   **Fig. 2.** The mADBN model

The mADBN model is shown in fig. 2. The model contains a state plane and a set of location planes. Each plane is represented by an adaptive dynamic Bayesian network with two dimensions: time and space. In the state plane, nodes denote the states of the system observed at different time instants and spatial locations, edges their probabilistic relations. Each location plane represents the location change of an objects (moving path). We use a search-and-score algorithm to learn the graph structure of the state plane based on analysis of the vehicle location planes.

Let $f_{i,t}$ represent the flow state of (i,t) and pa(i,t) represent the parent nodes of (i,t). N_P denotes the number of nodes in pa(i,t). The set of flow states for pa(i,t) is $F_{pa(i,t)}=\{f_{j,s}: (j,s) \in pa(i,t)\}$. According to the BN theory, the joint distribution of all nodes in the flow states network can be expressed as:

$$p(F) = \prod_{i,t} p(f_{i,t} \mid F_{pa(i,t)})$$
(1)

The conditional probability $p(f_{i,t}\mid F_{pa(i,t)})$ can be calculates as:

$$p(f_{i,t} \mid F_{pa(i,t)}) = p(f_{i,t}, F_{pa(i,t)}) / F_{pa(i,t)}$$
(2)

We model the joint distribution $p(f_{i,t}, F_{pa(i,t)})$ with Gaussian mixture model as:

$$p(f_{i,t}, F_{pa(i,t)}) = \sum_{m=1}^{M} \alpha_m g_m(f_{i,t}, F_{pa(i,t)} \mid \mu_m, C_m)$$
(3)

where M is the number of nodes and $g_m(\cdot \mid \mu_m, C_m)$ is the m-th Gaussian distribution with $(N_P + 1) \times 1$ vector of mean values μ_m and $(N_P + 1) \times (N_P + 1)$ covariance matrix C_m. Parameters $\{\alpha_m, \mu_m, C_m\}_{m=1}^{M}$ can be inferred from the historical data using EM algorithm.

2.2   Decision Making with Parallel MDP

In traditional MDP, actions are executed in sequence. Our MDP method is based on the following observation: the objects in large-scale IoT can be partitioned and the state of a node depends on the states of its neighbors. We call the nodes whose states we want to control proactively important nodes.

For a graph G=(V,E), neighbor state nodes are defined as $N(i)=\{j \in V \mid d_{i,j} \leq k\}$, where $d_{i,j}$ is the distance of node i and j, and k is a threshold value. A policy for node i at time t is defined as $\pi_{it}=\{a_{it1},a_{it2},...,a_{itm}\}$ where a_{itk} is an action aim to reduce the congestion level of node i at time t by guiding some vehicles to change their paths. We get:

$$P(s' \mid s, a) = \prod_{i=1}^{m} P(s_i' \mid s_i, a_i)$$

(4)

where s is the current state and s' is the next state. Assume there are n sub-state nodes and n corresponding actions, the total reward can be written as:

$$r(s, a) = \sum_{i=1}^{n} r(s_i, a_i) = \sum_{i=1}^{n} \sum_{k=1}^{|N(i)|} r_i \cdot (s_k, a_k)$$

(5)

In the implementation of transportation IoT, when an action a_i is executed, the traffic flow change of node i can be calculated by equation (6):

$$\Delta_i = S_i - \sum_{j_k \in Pa\_out(j_i)} S_k \cdot \alpha_{ki} \cdot \beta_{ki} \cdot p(j_k) - \sum_{j_m \in Pa\_In(j_i)} S_m \cdot \alpha_{im} \cdot \beta_{im} \cdot p(j_m)$$

(6)

where $Pa\_out(j_i)$ is the node set that try to reduce the flow of node i and $Pa\_in(j_i)$ is the node set that try to increase the flow of node i (can be predicted). α_{ki} is the proportion of S_k that flow to node i. β_{ki} is the proportion of S_k that affected by action a_i (notified by a_i to change path). $P(j_k)$ denotes the probability that vehicles in node k change their path according to action a_i. $P(j_k)$ can be learned from historical data. The total state transaction probability can be calculated based on $P(j)$ and Δ_i. we define the reward function of sub-action with Euclid distance and the total reward function is defined by equation (7):

$$\mathcal{R}(S,a,S')= \sum_{i=1}^{|g|} \left[(\bar{g}_i - \bar{g}'_i)^2 + (D(g_i) - D(g'_i))^2 \right]^{1/2}$$

(7)

Finally, the optimal policy can be calculated by equation (8):

$$\pi^*(S) = \arg\max_a \sum_S R(s, a, s') V(s')$$

(8)

3   Experimental Evaluations

We developed a transportation IoT simulation system based on the road traffic simulation package SUMO . In the experiment we selected 84 junctions from the map and set 80 thousand vehicles. We used 4 servers with Xeon E3 processor and 16GB memory as data processing servers and the operating system is Ubuntu 12. We first run the simulation for many times to get the historical data of the vehicle paths.

The result of the experiment is shown in figure 3-6 and table 1. From all the experiments we can see our method can support proactive complex event processing well in large-scale IoT applications. Based on the mADBN model, the PA method can get higher accuracy than traditional methods. The parallel MDP model has obvious effect for congestion control in large transportation IoT. The performance of the parallel MDP method decreases when the vehicle number and important node number becomes large but we can get better performance with more servers.

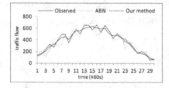

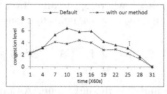

Table 1. Deviation of two methods

	deviation	
	ABN	Our method
Max	130	58
Min	6	13
Average	64.4	36.2
Average percent	16.67%	9.36%

Fig. 3. PA accuracy for a typical node

Fig. 4. Mean congestion level over time

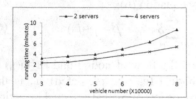

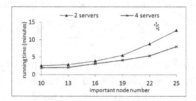

Fig. 5. PA accuracy for a typical node

Fig. 6. mean congestion level over time

4   Discussions and Conclusion

In this paper we proposed a proactive complex event processing method using parallel MDP for large-scale IoT. The performance and scalability of Pro-CEP still need to be improved. In the PA method, the EM algorithm needs to be parallelized to improve scalability. In the parallel MDP, the sub-states and sub-action space can also be very large. We need to develop new parallel or approximation algorithms for sub-state.

References

[1] Luckham, D.C.: The power of events: an introduction to complex event processing in distributed enterprise systems. Addison Wesley, Boston (2002)
[2] Engel, Y., Etzion, O.: Towards proactive event-driven computing. In: Proceedings of Fifth ACM International Conference on Distributed Event-Based Systems, DEBS 2011, New York, pp. 125–136 (2011)

[3] Etzion, O., Niblett, P.: Event Processing in Action. Manning Publications (2010)

[4] Pascale, A., Nicoli, M.: Adaptive Bayesian network for traffic flow prediction. In: Proceedings of the Statistical Signal Processing Workshop (SSP), pp. 177–180. IEEE (2011)

[5] Hofleitner, A., Herring, R., Abbeel, P.: Learning the Dynamics of Arterial Traffic from Probe Data Using a Dynamic Bayesian Network. ITS 13(4), 1679–1693 (2012)

[6] Engel, Y., Etzion, O., Feldman, Z.: A Basic Model for Proactive Event-Driven Computing. In: Proceedings of the 6th ACM International Conference on Distributed Event-Based Systems (DEBS 2012), pp. 107–118 (2012)

Social-Aware KNN Search
in Location-Based Social Networks

Huiqi Hu[1], Jianhua Feng[1], Sitong Liu[1], and Xuan Zhu[2]

[1] Department of Computer Science, Tsinghua University, Beijing, China
[2] Samsung R&D Institute, Beijing, China
`hhq11@mails.tsinghua.edu.cn`, `fengjh@tsinghua.edu.cn`, `liu-st10@gmail.com`,
`xuan.zhu@samsung.com`

Abstract. Location-based social network services have become widely available on mobile devices. It not only helps users to strengthen their social connections, but also provides useful information. An appealing application of using these information is helping users to find proper objects(points of interests) nearby with friends' visiting experiences. In this paper, we define friend based K nearest neighbor(F-KNN) query, which aims at finding objects near the query location as well as receiving high evaluations from user's friends. To answer F-KNN query efficiently, we propose a hybrid index called F-Quadtree index, which effectively combines the geographic coordinates of objects and user's evaluation. We develop an efficient searching algorithm on the index. To further accelerate the querying process, we refine the algorithm with user based partition and memory materialization. Experimental studies on real data sets show that our methods achieve high performance.

1 Introduction

In the past few years, location-based social network services such as Foursquare[1] and Jiepang[2], have seen increasing popularity, attracting millions of users. Supported by the capabilities of smart mobile devices and location-aware technology, GPS and Wi-Fi, people can easily share their locations, activities, and other information to their friends. The appearing LBSN services not only help users to strengthen their social connections, but also provide useful informations and generate appealing applications.

In this paper, we consider a new type of query, searching nearest objects(points of interests) with friends' visiting experiences provided by the LBSN system. We propose **F**riend based **K**-**N**earest **N**eighbor(F-KNN) query, which aims to find objects being spatially close to query location as well as receiving high evaluations from the friends of the user. The application of F-KNN query is obvious. For example, if we want to search a suitable restaurant for dinner, however, we have not been familiar with the surrounding restaurants, then all the restaurants

---

[1] `https://foursquare.com`
[2] `https://play.google.com/store/apps/details?id=com.jiepang.android`

F. Li et al. (Eds.): WAIM 2014, LNCS 8485, pp. 242–254, 2014.

are candidates and it is better for us to pick a restaurant which is near our place and received high evaluations from our friends. Utilizing friends' experiences in spatial objects search can benefit us from many aspects. First, it can guarantee the quality of searching results. Traditional query for k-nearest neighbors only concerns the spatial distance between location and objects. The quality of the provided objects can't be guaranteed. On the contrary, we can retrieve high quality objects if our friends have good evaluations on those objects. Second, it provides believable results to users. The query can provide good explanations to users and improve their user experience. Users are more tend to believe and choose objects once showing friends' experience or evaluations on those objects.

In this paper, we employ three essential information provided by the LBSN system. (1) The social links of a user, i.e. the friends of a user. (2) The geographic locations of the objects. (3) The user record which correlates a user with an object. We use a numerical score to represent the correlation between a user and and an object. Numerical scores are fundamental representations of interactions between users and objects, as users often leave five-star scores on objects and most recommendation system eventually recommend their result with a generated score. The F-KNN query returns close objects with higher scores given by the friends of the user. One big challenge to answer F-KNN query is to achieve high efficiency as it requires to support millions of spatial objects and more user records.

As far as we known, F-KNN query is first proposed in our work. In previous studies, people are mainly focused on the work of predicting a grade when given a pair of a user and an object [1,2,3]. For example, in [1], some complex metrics are proposed by combing the information of users and objects to calculate a numeric value for recommendation. These scores can be integrated into F-KNN query and our work. The main contributions of our work are, (1) we formulate the F-KNN query. (2) We propose effective index and algorithms to support F-KNN query. Moreover, we propose two efficient refine methods to accelerate the querying process. (3) Experimental results show F-KNN query is effective and our methods achieve high performance.

The rest of this paper is structured as follows: section 2 defines the preliminary concepts and introduces the baseline method. We introduce our index and algorithm for F-KNN query in section 3, followed by two refine methods in section 4. Experimental studies are shown in section 5. Section 6 talks about related works. Conclusions are given in section 7.

2   Preliminaries

2.1   Problem Formulation

We first define the three employed essential information(mentioned in section 1), then formulate the F-KNN problem.

Social Links. Let u denote a user in a LBSN system, all users are marked by their unique ids. The social links of u are all of his friends, which can be denoted as a set $S(u) = \{< u_i, w_i > \mid \sum_{i=1}^{|S(u)|} w_i = 1\}$, where u_i is a friend id and w_i

Social Links

u_1	$<u_2,0.8>,<u_3,0.2>$
u_2	$<u_1,0.67>,<u_3,0.33>$
u_3	$<u_1,0.67>,<u_3,0.33>$
u_4	$<u_5,1>$
u_5	$<u_4,1>$

User Record

u_1	u_2	u_3	u_4	u_5
o_7:0.4	o_7:0.2	o_6:0.2	o_2:0.4	o_2:0.2
	o_3:0.4	o_3:0.4	o_8:0.6	o_8:0.6
o_6:0.4	o_7:0.4	o_9:0.6	o_9:0.4	
o_1:0.6	o_5:0.6			
o_4:0.8	o_2:0.6			
o_5:0.8	o_8:0.6			
o_9:0.8	o_1:0.8			

Query $q\,(u_1,q.l)$

Spatial Score

objects	o_1	o_2	o_3	o_4	o_5	o_6	o_7	o_8	o_9
F_s	0.05	0.1	0.1	0.4	0.4	0.3	0.3	0.3	0.1

Fig. 1. An Example of F-KNN Query

represent the closeness u towards u_i or the confidence level of u on u_i. The total weight of closeness is constrained to be one. The weight can be evaluated from many metrics, such as the number of common friends, number of communicating messages and so on. The size of $S(u)$ is denoted as $|S(u)|$.

Objects. We use o to represent objects, o_i is an id of an object with its geographic coordinate.

User Record. we use $f_u(o_j)$ to represent a normalized score(with value between zero and one) of o_j given by user u.

Definition 1 (Spatial Score).

$$\mathcal{F}_s(l_i, l_j) = \frac{\mathcal{D}(l_i, l_j)}{\mathcal{D}_{max}}$$

$\mathcal{D}(l_i, l_j)$ is the euclidean distance between location l_i and l_j. The spatial score is normalized through the maximum spatial distance between objects($\mathcal{D}_{max}$). The closer two locations are, the lower spatial score they get.

Definition 2 (Friends Recommendation Score). *We compute the friends recommendation score of user u on objects o_j by reviewing all the advices of friends of u. i.e.*

$$\mathcal{F}_f(u, o_j) = \sum_{i=1}^{|S(u)|} w_i \cdot f_{u_i}(o_j)$$

Every friend's influence on u is weighted according to the closeness weight(w_i). Obviously $\mathcal{F}_f(u, o_j)$ is also a normalized score. For consistency with the spatial distance, we assume the lower the value of the user record $f_u(o_j)$, the higher evaluation o_j gets from u_i. For those objects that do not receive any evaluation from u, $f_u(o_j)$ is set to be the maximum value one. However, our methods can be easily modified to support the case that the user record is the higher the better.

We propose a F-KNN score to combine both the spatial score and friends recommendation score to qualify the objects.

Definition 3 (F-KNN Score).

$$\mathcal{F}(q, o_j) = \alpha \cdot \mathcal{F}_s(q.l, o_j) + (1 - \alpha) \cdot \mathcal{F}_f(q.u, o_j)$$

where $q.u$ is the query user and $q.l$ is his current geographic coordinate. α is tunable value to weight the importance of spatial score and friends recommendation score. Given a F-KNN query, it retrieves K objects with smallest F-KNN scores.

Example 1. Figure 1 illustrates an example of F-KNN query. Social links, user records and spatial objects are described in the figure. Suppose the objects in the figure are all restaurants. The query combines both the user's(u_1) location and his friends'(u_2 and u_3) visiting experience to return Top-K restaurants. The spatial scores between $q.l$ and objects are also showed in the figure. Let K=2, $\alpha = 0.5$, according to the F-KNN score mentioned before, the F-KNN score of o_6 can be calculated as $\mathcal{F}(q, o_6) = \alpha \cdot \mathcal{F}_s(q.l, o_6) + (1 - \alpha) \cdot (w_{u_2} \cdot f_{u_2}(o_6) + w_{u_3} \cdot f_{u_3}(o_6)) = 0.5 \times 0.3 + 0.5 \times (0.8 \times 0.4 + 0.2 \times 0.2) = 0.33$. o_7 and o_6 are the best restaurants with $\mathcal{F}(q, o_7) = 0.27$ and $\mathcal{F}(q, o_6) = 0.33$ for the query. o_1, though locates nearest the query place, is not selected since u_2 and u_3 give it low evaluations.

2.2   Baseline Method

The F-KNN problem is a special top K problem. We utilize the well-known threshold algorithm [4] as the baseline method and it's much more efficient than linearly checking all the objects. The algorithm is based on two types of accesses to data: sorted access and random access. In our baseline solution, a quadtree [5] is used to get the sorted access for spatial distance(by sequently find the nearest objects). And sorted inverted lists for user records are used to get sorted access for friends recommendation score. More details about the algorithm is described in [4].

3   F-Quadtree Based Algorithm

To solve the F-KNN query efficiently, we propose a Friend based-**Quadtree**(F-Quadtree) index by incorporating the user record and spatial objects into a quadtree. Then we propose a region-based pruning method based on the index. The index and the querying process algorithm are introduced in section 3.1 and 3.2 respectively.

3.1   F-QuadTree Index Structure

The F-Quadtree index is consisted of two parts, a quadtree index in memory and inverted lists in disk.

Quadtree. We use a quadtree [5] to index geographic information of objects. A quadtree can recursively partition the objects into different regions.

Inverted lists. We use inverted lists to index user records. We propose two types of inverted lists. (1) region based inverted index $L_g(u)$, $L_g(u)$ stores all the *regional user scores*. A regional user score is the minimum score among all the scores that the user has given to the objects in the region. Formally, we define $f_u(R)$ as the regional user record for u on region R, where

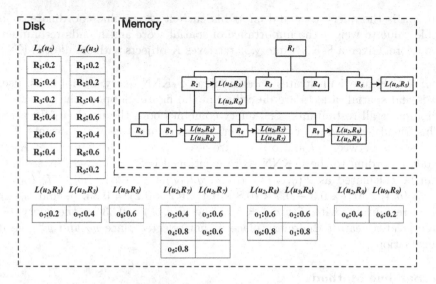

Fig. 2. F-Quadtree index structure

$$f_u(R) = \min_{o_i \in R} f_u(o_i)$$

all the regional user scores of u are inserted into $L_g(u)$. If user u hasn't been given scores for any object in region R_i, $f_u(R_i)$ is set as the maximum value and will not be stored in index. (2) user-object based inverted index $L(u, R)$, which stores the user record of objects in R given by u. The entry of $L(u, R)$ is a user u and a leaf region R. The items in the list are objects in R with receiving scores from u. The items are sorted in ascending order according to their scores.

Example 2. The F-Quadtree index for example 1 is in Figure 2. For example, the regional user score $f_{u_2}(R_7) = 0.4$. It is exactly same as $f_{u_2}(o_3)$, which is the smallest score given to the objects(o_3, o_4, o_5) in R_7 by u_2. All the regional user score for u_2 and u_3 are stored in $L_g(u_2)$ and $L_g(u_3)$ respectively. $L_g(u_1), L_g(u_4)$ and $L_g(u_5)$ are omitted due to the space limitation. For every leaf region of quadtree, such as $R_3, R_5, R_6, R_7, R_8, R_9$ in figure 2, we store user record in respect inverted lists. For instance, o_3, o_4, o_5 are in R_7, their related user records are stored in $L(u_2, R_7)$ and $L(u_3, R_7)$.

3.2 Querying Process

Region-Based Pruning. We can compute the lower bound of both spatial score and friends recommendation score. On one hand, given an object o_i in region R, we use the minimum spatial distance from query to the region as the spatial lower bound(denoted by $\mathcal{F}_s(q.l, R)$). If $o_i \in R$, $\mathcal{F}_s(q.l, R) = 0$; otherwise it is the minimum value among the distances from the query to the four corners and four boundaries of the region. On the other hand, the regional user score is a lower bound of friends recommendation score. Therefore, a lower bound of F-KNN score can be computed as

$$\mathcal{F}(q, R) = \alpha \cdot \mathcal{F}_s(q.l, R) + (1 - \alpha) \cdot \sum_{i=1}^{|S(q.u)|} w_{q.u_i} \cdot \mathcal{F}_{q.u_i}(R)$$

Algorithm 1: Query Process(q, K)

Input: q: $q.u$ is the user id and $q.l$ is the spatial location
       K: a given integer
       T_r:the F-Quadtree index

Output: Heap T

```
 1 begin
 2    for u_i ∈ S(q.u) do
 3       Load L_g(u_i) into memory;
 4    priority queue Q = ∅; Heap T = ∅; node N=T_r.root;
 5    Q.push(N);
 6    while |T| ≤ K or F(q, N) ≤ T_K do
 7       N = Q.top();
 8       if N is not a leaf node then
 9          for N' ∈ N.child do
10             Calculate F(q, N');
11             if F(q, N') ≤ T_K then
12                Q.push(N');

13       if N is a leaf node then
14          for u_i ∈ S(q.u) do
15             Load L(u_i, N) into memory;
16          SEARCHING LEAF REGION(q, T, N);

17    return T;
18 end
```

Fig. 3. F-Quadtree Based Algorithm

It's a weighted sum of lower spatial bound and regional user scores. For example, the lower bound for R_8 in example 1 is $\mathcal{F}(q, R_8) = \alpha \cdot \mathcal{F}_s(q.l, R_8) + (1 - \alpha) \cdot (w_{u_2} \cdot f_{u_2}(R_8) + w_{u_3} \cdot f_{u_3}(R_8)) = 0.3$, as we have listed all the spatial distance from query to all regions in figure 4 with $\mathcal{F}_s(q.l, R_8) = 0$. For any objects in R, the F-KNN score is larger than the lower bound $\mathcal{F}(q, R)$. We use the lower bound to prune unrelated regions, i.e. if the lower bound of a region is larger than the K-th smallest F-KNN value of objects, then the whole region can be pruned.

Querying Process. We employ a best-first traversal algorithm to process F-KNN query. Algorithm 1 shows the pseudo-code of the algorithm. A priority queue(line 4) is used to keep track of the nodes of quadtree and a min heap T is used to keep the temporary results. T_K is the k-th smallest value in the heap T. The algorithm begins by searching the root of F-Quadtree(line 5). When deciding which node to visit next, the algorithm picks the node N with the smallest regional F-KNN score value in the priority queue(line 7). If the picked node is not a leaf node, then it computes the lower bound $\mathcal{F}(q, R)$ for four of its children, and pushes the survived children into the queue(line 9 to 12). Otherwise, the algorithm starts to search the objects in the leaf region(line 16). The whole algorithm terminates when T_k is smaller than the smallest value of the lower bound of regions in the priority queue(line 6). If the picked node is a

Spatial Distance				Running Example

R	$F_s(q,R)$	R	$F_s(q,R)$
R_1	0	R_6	0.18
R_2	0	R_7	0.16
R_3	0.1	R_8	0
R_4	0.18	R_9	0.16
R_5	0.16		

Step1	Dequeue R_1,Enqueue{R_2,R_3,R_5} Q:{R_3:0.17,R_2:0.18,R_5:0.56} T:{}	Step4	Dequeue R_9,Searching R_9 Q:{R_7:0.28,R_8:0.3,R_5:0.56} T:{o_7:0.27,o_6:0.33}
Step2	Dequeue R_3,Searching R_3 Q:{R_2:0.18,R_5:0.56} T:{o_7:0.27}	Step5	Dequeue R_7,Searching R_7 Q:{R_8:0.3,R_5:0.56} T:{o_7:0.27,o_6:0.33}
Step3	Dequeue R_2,Enqueue{R_7,R_8,R_9} Q:{R_9:0.26,R_7:0.28,R_8:0.3,R_5:0.56} T:{o_7:0.27}	Step6	Dequeue R_8,Searching R_8 Q:{R_5:0.56} T:{o_7:0.27,o_6:0.33}

Fig. 4. An Example of Querying Process

leaf node, we need to search the leaf node(line 16). The basic method to search a leaf node is simply verifying the objects in the region one by one. However, we have stored the user-object based inverted lists of the region and user(i.e. $L(u,R)$) in the F-Quadtree index. Therefore, when searching a leaf region, we can optimize the function by using the threshold algorithm [4]. The details are omitted due to the space limitation.

Example 3. Again we consider the query in example 1. The algorithm starts by first pushing region R_1 into the priority queue with lower bound $\mathcal{F}(q, R_1) = 0.1$. It then executes the steps in figure 4. It will terminate and return T after step 6.

4   Refinements

4.1   User Based Partition

The F-Quadtree index groups objects with similar locations into spatial regions. Then the algorithm can directly prune unrelated regions during querying process. In this section, we further divide the objects in each leaf region into several partitions by considering social links and user records. The partition can accelerate the process of SEARCHING LEAF REGION in algorithm 1.

We divide the objects into k partitions with a clustering based method. We cluster the objects according to the user relationship and user record. The clustering approach takes three steps:

1. Instead of clustering objects, we first cluster the users into k partition (denoted as $P_1, P_2 \ldots P_k$). We cluster the users through their social links with a community detection(clustering) algorithm [6], provided by an open source tool Jung [7].

2. For every pair of object o_i and partition P_j, we calculate a score, where $g(o_i, P_j) = \sum_{u \in P_j} f_u(o_i)$.

3. For every o_i, we select the partition which owns the smallest value of $g(o_i, P_j)$. Thus all objects can be clustered into K partitions.

The clustering based approach owns a good explanatory. First, the users are clustered into k social circles and we will partition the objects into those circles. $g(o_i, P_j)$ calculates the sum of scores o_i received from the circle P_j. It can be viewed as the preference of circle P_j to object o_i. Finally the object will select the circle which gives the highest evaluation to the object. Thus all the objects are divided into k clusters according to their preferences of circles.

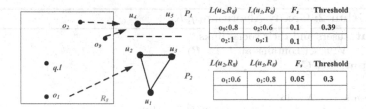

$L(u_2,R_8)$	$L(u_3,R_8)$	F_s	Threshold
o_9:0.8	o_2:0.6	0.1	0.39
o_2:1	o_9:1	0.1	

$L(u_2,R_8)$	$L(u_3,R_8)$	F_s	Threshold
o_1:0.6	o_1:0.8	0.05	0.3

Fig. 5. User Based Partition in R_8

During the querying process, when searching every leaf region(function SEARCHING LEAF REGION in algorithm 1), we separately run the threshold algorithm in every partition. The query will benefit since when given the query user, most of his friends will be in the corresponding circle and the objects preferred by the circle will be partitioned together. The objects in other clusters have very low probability to become results since they get user records mainly from other circles. The initial threshold generated by the threshold algorithm in each partition can help to prune unrelated partitions. When the initial threshold of a partition is larger than T_K, the whole partition is pruned.

Example 4. See the query in example 1. If we cluster the users into two clusters, $P_1 = \{u_4, u_5\}$ and $P_2 = \{u_1, u_2, u_3\}$ through their social links. Then objects in R_8 are partitioned according to the user circles. o_1 is partitioned into P_2 since it only receives evaluation from u_2 and u_3. o_2 and o_9 are partitioned into P_1 since they get evaluation mainly from u_4 and u_9. We create inverted lists separately for P_1 and P_2. During the query process, in step 6 of figure 4, when searching R_8, $T_K = 0.33$, o_2 and o_9 in P_1 will be skipped since the initial threshold for $P_1(\alpha \cdot 0.1 + (1 - \alpha) \cdot (0.8 \times 0.8 + 0.2 \times 0.6) = 0.39)$ is larger than T_K.

4.2   Improvement With Memory Materialization

All the user records are stored in the disk and loading the inverted lists into memory is the major cost of query processing. In this section, we improve the efficiency of F-Quadtree based algorithm by materializing some user records into memory in advance, which can significantly reduce the cost of loading unnecessary inverted lists.

We first discuss the "special" objects which will be loaded into memory. Recall the definition of regional user score $f_u(R)$(section 3), which is the smallest value of user record $f_u(o_i)$ in R. However, if we pick the object o_i with smallest value into memory and make $f_u(R)$ the second smallest value, and check o_i in memory before the F-Quadtree based algorithm, the querying process is still completely correct. Similarly, we can also pick objects with the smallest, second smallest values together (also the third one and so on) into memory, then replace the value of $f_u(R)$ with the remaining smallest value. This strategy can benefit us from two aspects: first, it increases the filtering ability of the quadtree nodes(regions) since it increases the lower bound of regions. Second, by checking the objects in memory first, it will provide an early smaller heap threshold, which will also help to reduce computing unrelated regions and objects. However, maintaining the

---

Algorithm 2: Greedy Algorithm(q, K)

Input: users and their social links
V: a set contains all $L^k(u, R)$

Output: Set C

```
1  begin
2  |    C = ∅;
3  |    while N ≤ Nm do
4  |    |    Select the Li(u, R) with largest Φ(Li(u, R));
5  |    |    C = C ∪ Li(u, R)
6  |    |    N = |C|
7  |    return C;
8  end
```

Fig. 6. Greedy Algorithm of Picking objects

objects in memory also leads to additional memory space cost and computations. The best choice is that we only pick a few effective objects into memory.

Example 5. Consider o_3 in example 1 and Figure 2. In R_7, o_3 is the smallest for u_2 and u_3. If we pick o_3 into memory, $f_{u_2}(R_7)$ becomes 0.8 and $f_{u_3}(R_7)$ becomes 0.6. During the querying process, the F-KNN score of o_3 in memory will be checked first. Then for the rest objects, we use algorithm 1, the regional F-KNN score of R_7 becomes 0.46 in step 3 of figure 4, then the whole region can be skipped, which means the step 5 in figure 4 is not necessary anymore.

Next we formally describe the approach to pick the "special" objects. Let $L_k(u, R)$ denote the k-th item in the inverted list $L(u, R)$ and $L_k(u, R).o$ is its object, $L_k(u, R).s$ is its score. We also define the set $L^k(u, R) = \{L_i(u, R).o | i \leq k\}$, as the set contains the front k objects, from the first to the k-th objects in $L(u, R)$. We formulate the pruning benefit of the set $L^k(u, R)$ as

$$\Phi(L^k(u, R)) = W(u) \cdot (L_{k+1}(u, R).s - L_k(u, R).s)$$

$W(u)$ represents the influence of a user. Apparently, the probability of a user appearing in a query becomes higher if he owns more friends. Therefore, we set $W(u)$ the number of user's friends, i.e. $W(u) = |S(u)|$. $L_{k+1}(u, R).s - L_k(u, R).s$ represents the increasing value from the k-th record score to the $(k + 1)$-th one. For example, in figure 2, $L^1(u_2, R_7) = \{L_1(u_2, R_7).o = o_3\}$ and $\Phi(L^1(u_2, R_7)) = |S(u_2)| \times (L_2(u_2, R_7).s - L_1(u_2, R_7).s) = 0.8$

We only load a limited number(denoted as N_m) of objects into memory. A greedy algorithm is utilized to choose the objects with largest pruning benefits. In every step, we greedily pick the set with largest pruning benefit, all the objects in the set will be chosen. The pseudo-code of the greedy algorithm is described in algorithm 2. We can get set $L^k(u, R)$ and compute $\Phi(L^k(u, R))$ easily by scanning the inverted lists $L(u, R)$.

During the query process, the objects in memory will be check first. The records in memory are also organized as inverted lists. The entry of the inverted lists is user and the list items are objects with corresponding scores. The threshold algorithm is first run on the inverted lists in memory and generates some results into heap, then we use the F-Quadtree based algorithm on rest objects.

5   Experimental Study

In this section,we present experimental studies of our methods to F-KNN queries.

Experiment Setup. We implemented all the methods with C++, on a PC with a 3.2Ghz CPU and a 2GB RAM and an ubuntu 12.0 system. We used two real data sets, a data set of Gowalla[3] and a data set crawled from Twitter(tweets with location information). We simulated the user record by the number of their check-ins($N_\mathcal{C}$) with the equation $\frac{1}{N_c+\mathcal{C}}$(we set constant $\mathcal{C} = 0.25$). The more number of a user checked at an object, the smaller the score was. The numerical closeness between friends were evaluated and normalized as their common friends number. The parameters of the two data sets are in table 1. For each data set, we randomly selected 2000 user check-ins as queries.

Table 1. Parameters of Data sets

	# of users	# of objects	# of records
Gowalla	0.19M	1.26M	6.44M
Twitter	2.68M	4.3M	13.5M

F-KNN Query Effectiveness. We first evaluated the effectiveness of F-KNN query. We used the real check-ins of querying user as ground truth, and compared the precision and recall of F-KNN query with the traditional Spatil Only(SO) query(i.e. return Top-K nearest objects). Figure 7 reflects the precision and recall on the two data sets with different K. The F-KNN query was far more effective than only considering spatial distance. For example, on Gowalla data, almost one of top-10 F-KNN(10%) results was in line with the real check-ins, and the top-40 F-KNN query could return 12% real check-ins.

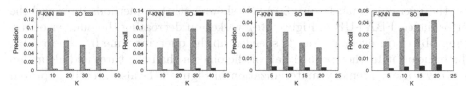

(a) Precision on Gowalla(b) Recall on Gowalla(c) Precision on Twitter (d) Recall on Twitter

Fig. 7. F-KNN Query Effectiveness

Efficiency Comparison. We compared three methods in our experiments. (a) The baseline method, **T**hreshold **A**lgorithm(TA). (b) The **F**-Quadtree **B**ased **A**lgorithm(FBA) without refinements. (c) **R**efined **F**-Quadtree **B**ased Algorithm (RFBA). We compared the three methods through different parameters of F-KNN query, K and α. Figure 8(a),8(c) evaluates the influence of K on the three algorithms when α is 0.5. Figure 8(b) and 8(d) evaluates the influence of α when

---

[3] http://snap.stanford.edu/data/loc-gowalla.html

K is 10. On both different parameters and data sets, FBA performs much better than TA. It took less than 100 microseconds to process a F-KNN query for FBA, however, RFBA can still speeding up over FBA by almost 60%.

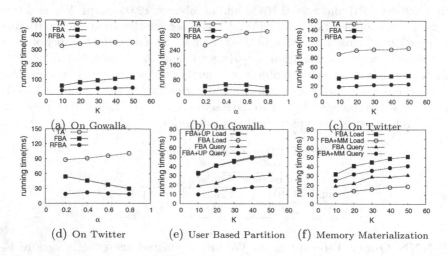

(d) On Twitter   (e) User Based Partition   (f) Memory Materialization

Fig. 8. Evaluation on Efficiency

Evaluation on refinements. We also evaluated the performance of refinements in section 4 separately. We only give the results on the Gowalla data set due to the space. We set $\alpha = 0.5$. We separately incorporated the two refine methods, user partition(denoted as FBA+UP, partition number was set to be 20) and memory materialization(FBA+MM). To see the result clearly, we divided the running time into loading time and querying time. The loading time included the process of loading $L_g(u)$ and $L(u, R)$ into memory and the rest was the querying time. Figure 8(e),8(f) shows both two methods improved the performance. FBA+UP in Figure 8(e) worked on decreasing the number of verifying objects in every region. It speeded up the querying time by almost 50%, which meant it further pruned half of the objects than the threshold algorithm in every region. Memory materialization in Figure 8(f) also greatly improved both the loading time and querying time.

Table 2. Index Size

	TA	RFBA	Memory
Gowalla	142M	274M	41M
Twitter	443M	1634M	124M

Index Size. We demonstrated the index sizes in table 2. The index size of inverted lists of F-Quadtree was 3-4 times larger than simple inverted lists of TA. We also given the memory cost for RFBA. The cost was small and apparently the quadtree index could easily support millions of spatial objects in memory.

6   Related Work

Spatial Objects Search. Traditional k-nearest neighbor queries(KNN) in spatial databases focuses on searching objects near the query location with euclidean distance. Usually R-tree index [8] and quadtree [5] index are used to assist the KNN queries. Recently, spatial search combining with some other information has attracted a lot of attention from research community. Most of the studies consider spatial search with text or keyword content [9,10]. Few related work consider both score and spatial information. Cong et. al. [11] and Senjuti et.al. [12] give single scores on objects. Spatial skyline [13,14] also considers numerical attributes(scores) of spatial objects. The scores are independent on users, i.e toward different users, objects have static scores. Obviously, F-KNN query differs from those problems. In an F-KNN query, an object can receives many scores from the friends of users. And those scores are different according to different querying users. To our knowledge, there are no similar work existing in spatial objects search field.

Friend Based Location Recommendation. Recommendation objects with friends' influence has aroused interest from data mining researches [2,1,3]. Their work focuses on predicting proper scores for different objects and users. These scores can be integrated into F-KNN query and our method.

7   Conclusion

We conclude our work in this paper. First, we address the new problem F-KNN query. Then we propose the F-Quadtree index and develope efficient algorithm to answer F-KNN query. Then we propose the user based partition and use memory materialization to further accelerate the query process. Experimental studies on two real data sets show our methods achieve good performance.

Acknowledgement. This work was partly supported by NSF of China (61272090), Tsinghua-Samsung Joint Laboratory, "NExT Research Center" funded by MDA, Singapore (WBS:R-252-300-001-490), and FDCT/106/2012/A3.

References

1. Ye, M., Yin, P., Lee, W.C., Lee, D.L.: Exploiting geographical influence for collaborative point-of-interest recommendation. In: SIGIR, pp. 325–334 (2011)
2. Ye, M., Yin, P., Lee, W.C.: Location recommendation for location-based social networks. In: GIS, pp. 458–461 (2010)
3. Zhou, D., Wang, B., Rahimi, S.M., Wang, X.: A study of recommending locations on location-based social network by collaborative filtering. In: Kosseim, L., Inkpen, D. (eds.) Canadian AI 2012. LNCS, vol. 7310, pp. 255–266. Springer, Heidelberg (2012)
4. Fagin, R., Lotem, A., Naor, M.: Optimal aggregation algorithms for middleware. J. Comput. Syst. Sci. 66(4), 614–656 (2003)

5. http://en.wikipedia.org/wiki/Quadtree
6. Wu, F., Huberman, B.A.: Finding communities in linear time: A physics approach. CoRR cond-mat/0310600 (2003)
7. http://jung.sourceforge.net/
8. Roussopoulos, N., Kelley, S., Vincent, F.: Nearest neighbor queries. In: SIGMOD Conference, pp. 71–79 (1995)
9. Zhang, D., Chee, Y.M., Mondal, A., Tung, A.K.H., Kitsuregawa, M.: Keyword search in spatial databases: Towards searching by document. In: ICDE (2009)
10. Cao, X., Cong, G., Jensen, C.S., Ooi, B.C.: Collective spatial keyword querying. In: SIGMOD Conference, pp. 373–384 (2011)
11. Cao, X., Cong, G., Jensen, C.S.: Retrieving top-k prestige-based relevant spatial web objects. PVLDB 3(1), 373–384 (2010)
12. Roy, S.B., Chakrabarti, K.: Location-aware type ahead search on spatial databases: semantics and efficiency. In: SIGMOD Conference, pp. 361–372 (2011)
13. Sharifzadeh, M., Shahabi, C.: The spatial skyline queries. In: VLDB, pp. 751–762 (2006)
14. Won You, G., Lee, M.W., Im, H., Won Hwang, S.: The farthest spatial skyline queries. Inf. Syst. 38(3), 286–301 (2013)

A Spatial-temporal Topic Segmentation Model for Human Mobile Behavior

Xingxing Xing[1], Man Li[2], Weisong Hu[2], Wenhao Huang[1],
Guojie Song[1,*], and Kunqing Xie[1]

[1] Key Laboratory of Machine Perception, Ministry of Education,
Peking University, Beijing, 100871, China
[2] NEC Labs, China
gjsong@pku.edu.cn

Abstract. Research on human mobile behavior is becoming more available and important. One of the key challenges is how to divide long and continuous trajectory sequences into meaningful segments which builds a foundation for user similarity measure, trajectory data management and routine mining. While in traditional research trajectory sequence is segmented on basis of fixed time window or spatiotemporal criteria. In this paper, we propose a probabilistic topic model considering the spatial property and temporal Markov property of human mobility to address the problem of topic segmentation in human mobile behavior: automatically segmenting trajectory sequence into meaningful segments. The trajectory segments reflect high-level semantics for understanding human mobile behavior and can be used for higher-level applications. We consider one synthetic dataset and one real-life human dataset collected by mobile phones to evaluate our model. Results show that our model has good results in segmentation and outperforms traditional methods for practical purposes especially in learning long duration routines.

Keywords: Topic segmentation, Human mobility, Spatial-temporal sequence.

1 Introduction

As mobile devices equipped with localization function become increasingly popular among people, the age of Mobile Big Data has come. The mass data has the potential of enabling the design of mobile intelligent applications, such as mobile advertisement, location-based social network and Life-log applications [1]. Thus, providing an efficient strategy to manage and mining such mass spatial-temporal trajectory data is becoming a challenging and promising task.

In this paper, we mainly focus on the trajectory segmentation problem: dividing a long and continues trajectory sequence into shorter and topically coherent segments. Each trajectory segment represents the user's regular mobility in a certain region and a certain time interval with particular spatial-temporal topic distribution reflecting high-level semantics of human mobile behavior. The motivation of trajectory segmentation is that it is a fundamental requirement to model the structure of human mobile behavior, especially in trajectory indexing, user similarity measure and routine mining.

---

* Corresponding author.

F. Li et al. (Eds.): WAIM 2014, LNCS 8485, pp. 255–267, 2014.

For instance, the management of massive human mobile behavior data requires dividing long trajectory into shorter segments to support effective and efficient trajectory indexing [16]. As for mining users' long-term activity routines from human mobile sequences, the trajectory segment can help better understanding human mobile behavior [3].

In traditional research the trajectory sequence is segmented on the basis of fixed time window or spatiotemporal criteria. Katayoun Farrahi [5] proposed a distant n-gram topic model (DNTM) to model trajectory sequence over pre-defined fixed time window for routines. As human routines have multiple timescales, the existing methods may destroy the structure of human mobile behavior. Buchin [17] addressed the problem of segmenting human GPS trajectory based on spatial-temporal criteria while ignoring the human mobile regularity. Another most similar series of work are topic segmentation model [9], which can divide the text into topically coherent segments with different topic distribution. But they only focused on lexical cohesion using bag of words without considering the spatial property and Markov property of human mobile behavior. Thus, we are confronted with following difficulties in human trajectory topic modeling compared with topic segmentation of other domains: the human mobility is affected by spatial distance and it has an obvious temporal Markov property beyond bag of words at the same time.

In this paper, we propose a probabilistic topic model considering the spatial property and the Markov property of human mobility to address the problem of topic segmentation in human mobile behavior. As in our model trajectory segment boundary is associated with significant change in the spatial-temporal topic distribution, our method could model the temporal persistence of human mobility. We assume that each word transition is attached with a particular topic shift tendency which is related to the spatial factor as supervising information according to Tobler's First Law of Geography [12]. As for the temporal Markov property of human mobility, bigram model [8] is jointly incorporated beyond bag of words. We derive the inference process using Markov Chain Monte Carlo (MCMC) sampling [11]. A comparative experimental analysis is performed compared with traditional method, providing better segmentation performance on a synthetic dataset. Lastly, we apply our model to a real-life dataset successfully finding typical long duration routines while previous methods failed.

This paper is organized as follows. The next section discusses related work. Problem definition is proposed in Section 3. We then discuss methodology in Section 4. We present and discuss the experimental results in Section 5. Section 6 draws the paper to conclusion.

2 Related Work

Most related works can be summarized as trajectory segmentation and topic segmentation model described as follows.

Trajectory Segmentation: Some researchers address the problem of trajectory segmentation based on spatiotemporal criteria. Buchin [17] requires that each segment is homogeneous in the sense that a set of spatiotemporal criteria are fulfilled. They define different such criteria, including location, heading, speed, velocity and so on. Researchers [15] propose a new partition-and-group framework for clustering trajectories of moving objects, which partitions a trajectory into a set of line segments, and then, groups similar line segments together into a cluster to discover common sub-trajectories from a trajectory database. For the first phase, they present a formal trajectory partitioning algorithm using the minimum description length (MDL) principle. This kind of research ignores the human mobile regularity focusing on the spatiotemporal criteria.

Trajectory segmentation has been used in extracting routines (e.g., work, home, entertainment). In these research trajectory sequence is segmented on the basis of fixed time window. Eagle and Pentland [2] use Principal Component Analysis (PCA) to identify the main components structuring daily human behavior by daily segmentation. PCA captures features over the entire day, whereas the method using topic models has the advantage of capturing characteristic trends occurring over part of the day. The state-of-the-art method is proposed by Katayoun Farrahi [5,6]. It proposes a distant n-gram topic model (DNTM) for location routines by modeling trajectory sequence over pre-defined fixed time interval of several hours. As human routines have multiple timescales, this kind of research may destroy the structure of human mobile behavior.

Topic Segmentation Model: Actually, topic segmentation has first been researched in natural language processing well and many models have been proposed. It is used to produce information which can be used to summarize, browse or retrieve information contained in text. And the recent models characterise the lexical cohesion using topic models. Lexical cohesion in this line of research is modeled by a probabilistic generative process. PLDA is presented by Purver et al [9]. PLDA is an unsupervised topic modeling approach for segmentation. It chains a set of LDAs [7] by assuming a Markov structure on topic distributions. A binary topic shift variable is attached to each text passage (i.e., an utterance in [9]. It is sampled to indicate whether the j_{th} text passage shares the topic distribution with the $(j - 1)_{th}$ passage. Using a similar Markov structure, SITS [10] chains a set of HDP-LDAs. Unlike PLDA, SITS assumes each text passage is associated with a speaker identity that is attached to the topic shift variable as supervising information.

In summary, traditional trajectory segmentation methods mostly focus on spatiotemporal criteria or fixed time window. While existing topic segmentation models only focus on lexical cohesion using bag of words ignoring the spatial property and Markov property of human mobility behavior. Our work can be viewed as the combination of two tasks together: segmenting trajectory sequence and then assigning topic distribution to trajectory segments.

3   Problem Definition

We use the GPS trajectory data collections throughout the paper, referring to entities such as "spatial-temporal word" and "trajectory segment". This is useful in that it helps to guide intuition, particularly when we introduce latent variables which aim to capture abstract notions such as "spatial-temporal topics".

Definition 1 (GPS trajectory). *A GPS point is a pair $p = (lng, lat)$, representing the longitude and latitude of the location. A GPS trajectory is a sequence of pairs*

$Traj = \langle (p_0, t_0), ..., (p_n, t_n) \rangle$, in which p_k is a GPS point and $t_k(k = 0...n)$ is a timestamp ($\forall 0 \leq k < n, t_k < t_{k+1}$).

Definition 2 (Visit point). *A visit point is a triple $VP = (p, t_{in}, t_{out})$, where p is a GPS point, t_{in} and t_{out} are timestamps, and the visit point stands for a location p which the user stays for longer than a time threshold (i.e. $t_{out} - t_{in} > \delta_{time}$).*

A reference place is a collection of visit points $P = \{VP_1, ..., VP_n\}$, in which they are close to each other. Reference places can be viewed as the significant places (e.g. home, work, etc.) where the user frequently visits. Because people's daily routes can be characterized by a significant probability to return to a few highly frequented locations, reference places can better model the structure of user mobility than raw GPS trajectories. Reference places are detected by [14] from all the visit points and indexed according to their frequency of occurrence.

Definition 3 (Spatial-temporal word). *A spatial-temporal word $w = (l, t)$ is composed of a location $l \in L$ where the user stopped over a time threshold and a time span of the day $t \in T$. $L = \{1, 2, ..., M\}$, where M is the number of reference places and l is the index. $T = \{1, 2, ..., D\}$, where D is is the number of discrete time spans of the day.*

For better data representation we transform raw GPS trajectory to a structured sequence of spatial-temporal words similar to a corpus.

Definition 4 (Spatial-temporal topic). *Each spatial-temporal word has a corresponding latent variable spatial-temporal topic. It can be modeled as a multinomial distribution over spatial-temporal words. A spatial-temporal topic can be interpreted as a human mobility interest considering spatial property and Markov property.*

A trajectory segment is a sequence of N words denoted by $Seg = \{w_1, ..., w_N\}$. It represents the user's regular mobility behavior in a certain region and a certain time interval with particular spatial-temporal topic distribution. The durations of trajectory segments are long and time varying because of characteristics of human mobility. Trajectory topic segmentation can be cast as an unsupervised machine learning problem: placing topic boundaries in unannotated trajectory sequence. The basic idea behind segmentation is that human trajectory sequence is made up of a set of continues segments. We associate a binary switching variable with each spatial-temporal word, indicating whether the topic distribution of mobility trajectory has changed. If it is true, a new segment starts from here. The variable is sampled by our spatial-temporal topic segmentation model.

4   Segmentation with Topic Models

In this section, we describe our method to tackle the problem. We first introduce our spatial-temporal topic segmentation model in Section 4.1. Like many existing topic models, our model also introduces new latent variables into the graphical structure. In Section 4.2, in order to derive the latent variables that best explain observed data, we use Gibbs sampling, a widely used Markov chain Monte Carlo inference technique.

Table 1. Notation description

Notation	Description
N	The length of the whole human trajectory sequence
K	The number of latent topics
V	The spatial-temporal vocabulary size
M	The number of distinct transitions
dis	The distance of the transition
Φ	The distribution of word given topic and previous word
Θ	The distribution of topics given segments
π	The tendency of topic shift of the transition
α, β, γ	The hyperparameter of Φ, Θ, π
Ω	The normalization constant of the function sampling topic shift
λ	The parameter describing the width of each trajectory segment activity area
ρ	A increasing function indicating importance of the distance of the transition to topic shift

4.1 Spatial-temporal Topic Segmentation Model

Modeling Topic Boundary: We endow each spatial-temporal word w_i with a binary switching latent variable c_i, called the topic shift indicator. The topic shift indicator shows whether the topic of the word with will change the spatial-temporal topic distribution of its predecessor segment. As a result the c_i indicates the probability that a new segment should start at i. As a result our method could model the temporal persistence of trajectory avoiding too frequent changes.

In fact, not every spatial-temporal word in the sequence could be the trajectory segment boundary. The basic principle is that there must be user's movement from one location to another between trajectory segments. So if the location of the spatial-temporal word is different from its predecessor, the word may be qualified to be the boundary. We call the two consecutive words a transition t_m "from word w_{i-1} to word w_i". Not every two adjacent spatial-temporal words correspond to a transition and the transitions are recurring.

Topic models typically do not model the spatial-temporal dynamics of human trajectory. Instead, we assume each transition is attached with a particular topic shift tendency. Tobler's First Law of Geography [12] shows that everything is related to everything else, but near things are more related than distant things. We make use of the spatial factor as supervising information in our model that the distance between two adjacent spatial-temporal words smaller, the smaller the shift tendency π_m of the m_{th} transition will be. And the frequency of the transition means a lot, too. The more frequent the transition is sampled as boundary, the probability of topic shift is bigger. In general, this variable π_m is intended to capture the propensity of a transition to affect a topic shift. It represents the probability that the transition t_m will change the topic distribution of a trajectory segment.

We assume the topic shift probability π_m of the word w_i (m is the index of the transition), a Beta distributed random variable, i.e., $\pi_m \sim Beta(\gamma_1, \gamma_2)$. Then, the boundary indicator variable c_i is sampled from $c_i \sim \text{Bernoulli}(\pi_m) * \rho(dis_m)$. dis_m is the distance of the m_{th} transition of t_m.

We have

$$\rho_0(dis) = \exp(-\lambda \tfrac{dis-\min}{\max-\min}) \quad \rho_1(dis) = \exp(-\lambda \tfrac{\max-dis}{\max-\min}) \tag{1}$$

In Equation 1, max and min are the maximum and minimum distance of the transitions. $\rho_0(dis_m)$ is a decreasing function and $\rho_1(dis_m)$ is a increasing function. The two functions imply the importance of the distance of the transition to topic shift. λ is a parameter describing the width of each trajectory segment activity area. As λ is decreased, the width increases. Probability $P(c_i|\pi_m, dis_m)$ can be writen as Equation 2.

$$P(c_i|\pi_m, dis_m) = \begin{cases} \frac{1}{\Omega_m}\pi_m * \rho_0(dis_m) & \text{if } c_i = 0 \\ \frac{1}{\Omega_m}\pi_m * \rho_1(dis_m) & \text{if } c_i = 1 \end{cases} \tag{2}$$

In Equation 2, $\Omega_m = \pi_m * \rho_0(dis_m) + \pi_m * \rho_1(dis_m)$ is the normalization constant.

To sample topic shift indicator c_i, we use a point-wise sampling algorithm. Consequently, a sequence of topic shift indicators defines a set of segments. For example, let a C vector as $(0, 0, 0, 0, 1, 0, 0, 1)$. It gives us two segments, which are $\{1, 2, 3, 4, 5\}$ and $\{6, 7, 8\}$.

Modeling Topic Structure: Since the segment boundary should be the associated with significant change in the topic distribution, we sample a new distribution over topics for the next segment if the c_i is sampled as 1.

From the research [4], people's mobile behavior is a kind of Markov structure which is beyond bag of words. What the user will do next is influenced by his current state(spatial and temporal). Here we introduce the bigram model [8] for modeling spatial-temporal words in the trajectory sequence. A spatial-temporal word in trajectory sequence is assumed to be also conditionally dependent on the previous word.

The model parameters are defined in Table 1.The graphical model for our model is illustrate in Figure 1. We use a probabilistic approach where observations are represented by random variables, highlighted in gray. The latent variable z corresponds to a spatial-temporal topic. The latent variable c corresponds to a topic shift. The generative process is defined as follows:

1. For each transition t_m in the sequence, draw corresponding topic shift tendency probability.
   $\pi_m \sim Beta(\gamma_1, \gamma_2)$.
2. For each spatial-temporal word w_i:
   (a) If it's a transition with word w_{i-1}, draw $c_i \sim \text{Bernoulli}(\pi_m) * \rho(dis_m)$.
   (b) If $c_i = 1$, draw a new topic distribution for the segment; otherwise, set $\theta_i = \theta_{i-1}$.
   (c) For each spatial-temporal word in the segment:
       i. Draw $z \sim \theta_i$
       ii. Draw $w \sim \Phi_{z,w_{i-1}}$

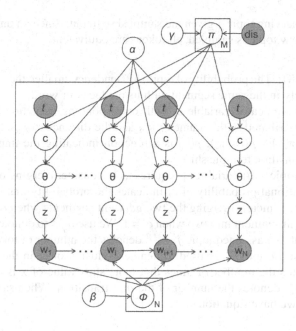

Fig. 1. The spatial-temporal topic segmentation model

4.2   Inference

To find the latent variables that best explain observed data, we use Gibbs sampling, a widely used Markov chain Monte Carlo inference technique [11]. The state space is latent variables for topic indices assigned to all tokens z and topic shifts assigned to c. We margin over all other latent variables.

Sample Topic: The conditional probability of z_i indicates the probability that w_i should be assigned to a particular topic, given other assignments, the current segmentation, and the words in the segment.

Use the Bayesian rule, we can get the conditional distribution for z_i is shown in Equation 3.

$$P(z_i = k | Z_{-i}, C, W) \propto \frac{n^j_{w_i, w_{i-1}} + \beta}{n^j_{*, w_{i-1}} + N\beta} \frac{n^{seg}_j + \alpha}{n^{seg}_* + K\alpha} \tag{3}$$

We use n^{seg}_j to denote the number of tokens in current segment *seg* assigned topic j; n^{seg}_* denotes the number of tokens in segment *seg*. $n^j_{w_i, w_{i-1}}$ denotes the number of times the topic j is assigned to the spatial-temporal word w_i given the previous word w_{i-1} in the vocabulary because of the sequence property. Marginal counts are represented with $n^j_{*, w_{i-1}}$.

In Equation 3, the first factor is proportional to the likelihood of observing word w_i given the sampled topic and the previous word; the second factor is proportional to the probability of observing topic j given the sampled segment. Since an uninformed prior is used, when a new topic is sampled, all tokens are equivalent.

Sample Topic Shift: The probability of topic shift tendency smaller, the spatial-temporal word is more likely in the same segment with its predecessor word.

The topic shift indicator variable c_i with corresponding transition t_m is sampled from Bernoulli distributed with parameter π_m and the distance dis_m of the transition, $c_i \sim$ Bernoulli$(\pi_m) * \rho(dis_m)$. $\rho(*)$ is a function indicating the importance of the distance of the transition to topic shift.

Sampling the topic shift variable c_i requires us to consider merging or splitting segments. The conditional probability of c_i indicates the probability that a new segment should start at w_i. It means merging the two adjacent segments when $c_i = 0$; it means splitting the current segment into two when $c_i = 1$. We use n_j^{seg} to denote the number of tokens in segment seg assigned topic j; n_*^{seg} denotes the number of tokens in segment seg. Again, the subscript C_{-i} is used to denote exclusion of c_i in the corresponding counts. n_m^x denotes the number of times that topic shift value of x is assigned to the m_{th} transition. n_m^* denotes the number of the m_{th} transition. When sampling c_i from this distribution, we have Equation 4.

$$
\begin{aligned}
&P(c_i|C_{-i}, Z, W) \\
&\propto
\begin{cases}
\dfrac{\prod\limits_{j=1}^{K} \Gamma(n_j^{seg}+\alpha)}{\Gamma(n_*^{seg}+K\alpha)} \quad \dfrac{n_m^0+\gamma}{n_m^*+2\gamma}\rho_0(dis_m)^{n_m^0}, & \text{if } c_i = 0 \\[2em]
\dfrac{\Gamma(K\alpha)}{\Gamma(\alpha)^K} \dfrac{\prod\limits_{j=1}^{K} \Gamma(n_j^{seg-1}+\alpha)}{\Gamma(n_*^{seg-1}+K\alpha)} \dfrac{\prod\limits_{j=1}^{K} \Gamma(n_j^{seg}+\alpha)}{\Gamma(n_*^{seg}+K\alpha)} \dfrac{n_m^1+\gamma}{n_m^*+2\gamma}\rho_1(dis_m)^{n_m^1}, & \text{if } c_i = 1
\end{cases}
\end{aligned} \tag{4}
$$

As for $c_i = 0$ in Equation 4, the first factor is proportional to the joint probability of all topics in segment seg and the second factor is proportional to the probability of assigning a topic shift of value 0 to transition m influenced the distance when $c_i = 0$.

As for $c_i = 1$ Equation 4, the first factor is proportional to the joint distribution of the topics in two adjacent segments seg_{-1} and seg. The second factor is proportional to the probability of assigning a topic shift of value 1 to transition m.

5   Evaluation

5.1   Data and Preprocessing

To inference the model, we sample the latent variables for many iterations, taking samples and then average the sampled c_i variables to derive an estimate for the posterior probability of a segmentation boundary at each transition. This probability is thresholded to derive a final segmentation which is compared to the manual annotations. Varying this threshold allows us to segment the trajectory in a more or less finer grained way.

We consider two different datasets for experiments. The representations for each are detailed below.

Synthetic Dataset. To solve the problem lacking of ground truth of human mobile behavior, we constructed a synthetic dataset: a sequence of spatial-temporal words chosen
from a vocabulary of 50. It is constructed according a typical user who has a regular weekly repeating mobility pattern. The trajectory sequence is 300 days long and
contains 14400 words. Each trajectory segment has a constant topic distribution over
spatial-temporal words. The user has a constant trajectory segment in Monday, a one
day long pattern in Tuesday and Wednesday, and two two-day long patterns in the next
four days in every week.

Nokia Dataset. We use real-life data collected by the Nokia N95 smart-phone from
2009.10.01 to 2010.09.01 corresponding to a 11 month period of the Lausanne Data
Collection Campaign [13]. We selected 10 from 110 volunteers. The phone has an application that collects location data on a quasi-continuous basis using a combination of
GPS and WiFi sensing, along with a method to reduce battery consumption. For data
representation, the day is divided into 48 discrete time spans with 30 minutes as an
interval. There are 105 reference places in total after place extraction.

5.2   Experiments in the Synthetic Dataset

To analyze the properties of our algorithm we first applied it to a synthetic dataset. We
compare our method with the most widely used and stable topic segmentation method
PLDA [9]. For the sampling, 10 randomly initialized Gibbs chains were used. Each
chain ran for 10,000 iterations with 5,000 for burn-in, with samples collected after every
200 iterations to minimize autocorrelation. Then 250 samples were drawn for 10 Gibbs
chains. The discount parameter $\lambda = 0.5$, and $\gamma_1 = \gamma_2 = 0.1, \alpha = 0.1$, the other on word
distributions $\beta = 0.01$.

We take one week long sequence segmentation results as an example. PLDA and our
algorithm were trained with 30 topics. Figure 2 shows an example of how the inferred
segment boundary probabilities at transitions compare with the gold-standard boundaries on the synthetic dataset. The gold-standard segmentation is {23, 71, 119, 232,
359, 407, 455, 568, 695, 743, 791, 904}, PLDA and our model infer {23, 69, 102, 230,
359, 435, 548, 701, 796,900} and {22, 72, 119, 226, 350, 412, 454, 572, 700, 741,
795, 908} respectively. PLDA miss the boundary after the 407_{th} and 743_{th} word. The
two transitions both have a quite long distance which in our method would be set the
segmentation boundary. Note the boundaries placed by our model are always within
20 words with respect to the gold standard. Thus, considering the properties of transitions and spatial property of human mobility might further improve the segmentation
performance.

5.3   Experiments with Nokia Dataset

To test the practical property of our algorithm we applied it with the Nokia smartphone
data. The results shown here are for the discount parameter $\lambda = 0.5$, and $\gamma_0 = \gamma_1 = 0.1, \alpha$
$= 0.1$, the other parameter on word distributions $\beta = 0.01$. We first train the model with
the user's trajectory to get the topic segmentation. Then we use k-means to cluster the

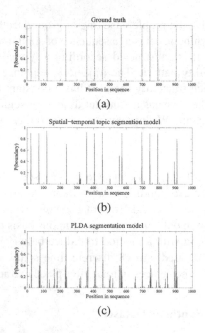

Fig. 2. Probability of topic shift: (a) the ground truth segmentation; (b) spatial-temporal topic segmentation results; (c) PLDA topic segmentation results

segments by spatial-temporal topic distribution with Euclidean distance into meaningful categories to find the user's typical routines and pattern. We will compare our results with DNTM [6].

We take one user as an example. The Figure 3 indicates his mobile history of 11 months in Lausanne, Switzerland,. And the point in the map is larger and redder, the place he visited more frequently.

After clustering, some group of trajectory segments indicate the user's typical routines shown in Figure 4. In Figure 4(a), it shows a kind of routine around working areas that the user usually works in the morning, eats in the midday and then goes back to work until late in the afternoon. The routine mainly occurs in the weekdays. It occurs in the same day. In Figure 4(b), it shows a kind of routine around home areas that the user stays at home for a long time from the night of the first day to the noon of the third day. This routine occurs across three days, usually in weekends because the users usually goes to home at Friday night and stays until Sunday noon.

Though only selected results are presented for the discussion here, our method is generally effective for all the regular users. The results show that our model outperforms DNTM [6] in finding long time duration routines. They modeling trajectory sequence over pre-defined fixed several hour time intervals. While our topic segmentation model could model the temporal persistence of trajectory to get long routines with multiple timescales as Figure 4(a) and Figure 4(b) show. As we can see from Figure 3 that the working area and the home area is far from each other. Our segmentation model can clearly segment the trajectory according the spatial factor of human mobility.

Fig. 3. Mobility history for a user

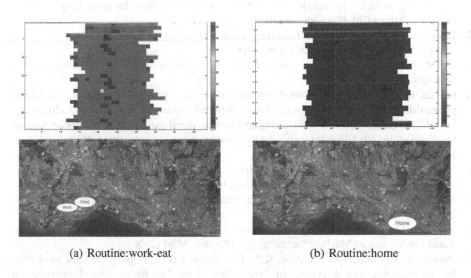

(a) Routine:work-eat   (b) Routine:home

Fig. 4. Topics and location details for one user. The x-axis is the time; the y-axis are trajectory sequence segment after clustering. Each unique color represents a unique spatial-temporal topic.

6 Conclusion

In this paper, we propose a probabilistic topic model to address the problem of topic segmentation in human mobile behavior. Our model takes the spatial property and the Markov property of human mobility into consideration to better model the structure of trajectory sequence. The trajectory segments here are meaningful and build a foundation for more intelligent applications. We performed a comparative experimental analysis with traditional method on a synthetic dataset with better segmentation performance. We also apply our model to a real-life dataset proving that through our model typical long duration routines can be successfully mined out while previous methods failed. As a result, our model performs well both in theory and practice.

There are several directions in the next. Firstly, we can make a comparison with the unsupervised generative model Hidden Markov Model in topic segmentation and long duration routine mining. Secondly, we can make use of the typical routines to measure user similarity among the people to group similar users together.

Acknowledgments. This work was supported by the National High Technology Research and Development Program of China ("863"Program) (No. 2014AA015103).

References

1. Gemmell, J., Bell, G., Lueder, R.: MyLifeBits: a personal database for everything. Communications of the ACM 49(1), 88–95 (2006)
2. Eagle, N., Pentland, A.S.: Eigenbehaviors: Identifying structure in routine. Behavioral Ecology and Sociobiology 63(7), 1057–1066 (2009)
3. Lv, M., Chen, L., Chen, G.: Mining user similarity based on routine activities. Information Sciences (2013)
4. Farrahi, K., Gatica-Perez, D.: Probabilistic mining of socio-geographic routines from mobile phone data. IEEE Journal of Selected Topics in Signal Processing 4(4), 746–755 (2010)
5. Farrahi, K., Gatica-Perez, D.: Extracting mobile behavioral patterns with the distant n-gram topic model. In: 2012 16th International Symposium Wearable Computers (ISWC), pp. 1–8. IEEE (2012)
6. Farrahi, K., Gatica-Perez, D.: A probabilistic approach to mining mobile phone data sequences. Personal and Ubiquitous Computing, 1–16 (2013)
7. Blei, D.M., Ng, A.Y., Jordan, M.I.: Latent dirichlet allocation. The Journal of Machine Learning Research 3, 993–1022 (2003)
8. Wallach, H.M.: Topic modeling: beyond bag-of-words. In: Proceedings of the 23rd International Conference on Machine Learning, pp. 977–984. ACM (2006)
9. Purver, M., Griffiths, T.L., Kording, K.P., Tenenbaum, J.B.: Unsupervised topic modelling for multi-party spoken discourse. In: Proceedings of the 21st International Conference on Computational Linguistics and the 44th Annual Meeting of the Association for Computational Linguistics, pp. 17–24. Association for Computational Linguistics (2006)
10. Nguyen, V.A., Boyd-Graber, J., Resnik, P.: SITS: A hierarchical nonparametric model using speaker identity for topic segmentation in multiparty conversations. In: Proceedings of the 50th Annual Meeting of the Association for Computational Linguistics: Long Papers, vol. 1, pp. 78–87. Association for Computational Linguistics (2012)

11. Gilks, W.R., Richardson, S., Spiegelhalter, D.J. (eds.): Markov chain Monte Carlo in practice, vol. 2. CRC Press (1996)
12. Miller, H.J.: Tobler's first law and spatial analysis. Annals of the Association of American Geographers 94(2), 284–289 (2004)
13. Kiukkonen, N., Blom, J., Dousse, O., Gatica-Perez, D., Laurila, J.: Towards rich mobile phone datasets: Lausanne data collection campaign. In: Proc. ICPS, Berlin (2010)
14. Montoliu, R., Gatica-Perez, D.: Discovering human places of interest from multimodal mobile phone data. In: Proceedings of the 9th International Conference on Mobile and Ubiquitous Multimedia, p. 12 (2010)
15. Lee, J.G., Han, J., Whang, K.Y.: Trajectory clustering: a partition-and-group framework. In: Proceedings of the 2007 ACM SIGMOD International Conference on Management of Data, pp. 593–604 (2007)
16. Yoon, H., Shahabi, C.: Robust time-referenced segmentation of moving object trajectories. In: Eighth IEEE International Conference on Data Mining, ICDM 2008, pp. 1121–1126 (2008)
17. Buchin, M., Driemel, A., van Kreveld, M.: An algorithmic framework for segmenting trajectories based on spatio-temporal criteria. In: Proceedings of the 18th SIGSPATIAL International Conference on Advances in Geographic Information Systems, pp. 202–211 (2010)

Real-Time Predicting
Bursting Hashtags on Twitter

Shoubin Kong[1], Qiaozhu Mei[2], Ling Feng[1], and Zhe Zhao[2]

[1] Tsinghua National Laboratory for Information Science and Technology
Dept. of CS&T, Tsinghua University
kongsb09@mails, fengling@tsinghua.edu.cn
[2] School of Information, University of Michigan
{qmei,zhezhao}@umich.edu

Abstract. A large number of hashtags are generated every day on Twitter. Only a few hashtags can become bursting topics. It is quite challenging to predict such bursting hashtags in real-time. In this paper, we provide the first definition of a bursting hashtag, and propose a solution to the real-time prediction problem of bursting hashtags. The experimental results show that the proposed method outperforms other related methods.

Keywords: hashtag, burst, real-time prediction.

1 Introduction

Twitter is a popular open platform of social network service, where users can follow anyone they are interested in and share anything they want. At the same time, real-time information diffusion makes Twitter an inherent social media where users can easily obtain and freely talk about various kinds of information, such as breaking news, real world events, interesting videos and so on.

Hashtags, starting with a # symbol, have been commonly used to identify topics on Twitter. In our dataset from Twitter, about 400,000 new hashtags are generated every day. However, only dozens of them can become bursting hashtags. There are two typical types of such hashtags. One type is breaking news or events happening in the real world, for example, #londonriots. The other type is memes, which motivates users to share their feelings following these topics, for example, #hashtagyourdreamschool. In this study, we concentrate on the real-time prediction of such bursting hashtags. Prediction of bursting hashtags are valuable to some practical applications, such as news or event discovery, information retrieval, microblog recommendation, public opinion analysis, etc. For example, it can be used to find rumors before they burst and spread widely. The contributions of this paper are as follows:

1. We provide the first definition of a bursting hashtag, and propose a solution to the real-time prediction problem of bursting hashtags.
2. Experiments are conducted on real datasets from Twitter, and the results show that the proposed solution outperforms other related methods.

F. Li et al. (Eds.): WAIM 2014, LNCS 8485, pp. 268–271, 2014.

2   Related Work

In [3], Gupta *et al.* used regression, classification and hybrid approaches to predict future popularity of current popular events. Analogously, Ma *et al.* [4,5] predicted popularity of hashtags in daily granularity. Tsur *et al.* [7] studied the effect of content on the spread of hashtags in weekly granularity. Most germane to this work are the studies from the same group [6,1], which focused on detecting trending topics by time series classification. Our work, conversely, provides the first definition of bursting hashtags and aims to predict the bursting hashtags in real-time before they burst.

3   Problem Definition

The lifecycle of a hashtag can be formed as a time series $< c_1, c_2, ..., c_t, ... >$. c_t denotes the count of tweets containing the hashtag at the t-th time interval. Considering the real-time characteristic of Twitter, the granularity of the time interval is set to 1 minute in this study. Definitions of bursting hashtags are as follows:

Definition 1. *Prediction-Trigger. A clear majority of hashtags will never get burst, and a substantial number of bursting hashtags have a long dormant period before they burst. The average time before a hashtag gets burst is about **8.72 days** since the hashtag first appears. Therefore we define a trigger to obtain a candidate set of hashtags to be predicted. For each hashtag, a five-minute sliding window is used to check the total count of tweets containing the hashtag within the consecutive five minutes, denoted as C_{slw}. If $C_{slw} > \phi$, the prediction is triggered.*

Definition 2. *Burst. We define the burst of a hashtag by referencing to the definition of spikes in [2]. Within 24 hours since the prediction was triggered, if c_t is greater than $max(c_1 + \delta, 1.5c_1)$, t is defined as the onset of burst. Such a hashtag is a bursting hashtag.*

There are two parameters in these definitions. δ, can be adjusted according to the statistics of real data and the requirement of practical applications. We have mentioned that in our dataset about 400,000 new hashtags can be identified every day. δ is set to 50 in this paper, which makes the proportion of bursting hashtags about 0.6%%, i.e., about 25 bursting hashtags can be found each day. If a larger proportion is required, the value of δ should be set smaller, and vice versa. ϕ is set equal to δ, which ensures that there are no bursting hashtags already burst before triggering prediction.

Based on the definitions above, the prediction task is defined as: among the hashtags triggering prediction, which ones will be bursting hashtags? This problem can be framed to a normal binary classification task.

Input: A set of candidate hashtags which triggered prediction $HT = \{ht_1, ht_2, ...\}$.

Algorithm 1. Training the optimal classification model.

Input:
     TNS: training set; TTS: training-test set;
     NPR: the ratio of negative samples to positive samples;
     $R_w = \{1, 2, ..., 2NPR\}$: the range of the weight for positive class;
Output:
     C_{opt}: the optimal classifier;
1. $F_1^{max} \leftarrow 0$;
2. **for all** $w \in R_w$ **do**
3.   Training a weighted SVM classifier C_w on TNS;
4.   Compute F_1-score F_1^w by applying C_w to TTS;
5.   **if** $F_1^w > F_1^{max}$ **then**
6. $C_{opt} \leftarrow C_w$;
7. $F_1^{max} \leftarrow F_1^w$;
8.   **end if**
9. **end for**
10. **return** C_{opt};

Table 1. Proportion of bursting hashtags in the dataset

5min	15min	30min	1h	3h	6h
12.39%	9.24%	6.27%	3.58%	1.27%	0.80%

Output: A class label for each hashtag $L(h_i), h_i \in HT$, indicating whether it will be a bursting one.

Solution: Since the dataset is unbalanced, we propose a weighted SVM-based method to solve this problem. An optimal weight for the positive class is needed to train the classification model. Algorithm 1 shows the process of optimizing the weight for the positive class. Since the dataset is unbalanced, F_1-score is used as the criteria for training the optimal model. At the same time, we also tried several related methods to evaluate the performance. The evaluation results are demonstrated in Section 4.

4   Experiments

Experiments are conducted on the datasets collected through the Twitter stream API with Gardenhose access, containing roughly 10% of all public statuses on Twitter. We have collected three datasets according to the requirements of the prediction problem, including a two-month historic set (2012.9-2012.10), a three-month training set (2012.11-2013.1) and a one-month test set (2013.3). We extracted seven types of features which may indicate the future trend of hashtags, including meme features, user features, content features, network features, hashtag features, time series features, and prototype features. Table 1 shows the distribution of hashtags triggering prediction. Since the dataset is largely unbalanced, F_1-score is used as the performance metric.

Table 2 shows the performance comparison for the prediction of bursting hashtags. Predictions were made at six representative time, which can be divided

Table 2. Performance comparison

Method	F_1-score					
	5min	15min	30min	1h	3h	6h
NaiveBayes	0.299	0.342	0.312	0.197	0.132	0.081
C4.5	0.211	0.238	0.235	0.118	0.043	0
LR	0.068	0.244	0.209	0.190	0.044	0.077
NN	0.170	0.202	0.210	0.250	0.185	0.194
SVM	0.207	0.308	0.329	0.277	0.187	0.125
Our Method	**0.326**	**0.372**	**0.368**	**0.360**	**0.290**	**0.250**

LR: Logistic Regression   NN: Neural Network

into three stages, early stage (5min, 15min), middle stage (30min, 1h) and late stage (3h, 6h). It can be seen that, our method significantly outperforms the other related methods in terms of F_1-score.

Acknowledgement. The work is supported by National Natural Science Foundation of China (61373022, 61073004, 60773156), and Chinese Major State Basic Research Development 973 Program (2011CB302203-2).

References

1. Chen, G.H., Nikolov, S., Shah, D.: A latent source model for nonparametric time series classification. In: Advances in Neural Information Processing Systems, pp. 1088–1096 (2013)
2. Gruhl, D., Guha, R., Kumar, R., Novak, J., Tomkins, A.: The predictive power of online chatter. In: Proceedings of the Eleventh ACM SIGKDD International Conference on Knowledge Discovery in Data Mining, pp. 78–87. ACM (2005)
3. Gupta, M., Gao, J., Zhai, C., Han, J.: Predicting future popularity trend of events in microblogging platforms. Proceedings of the American Society for Information Science and Technology 49(1), 1–10 (2012)
4. Ma, Z., Sun, A., Cong, G.: Will this# hashtag be popular tomorrow? In: Proceedings of the 35th ACM SIGIR International Conference on Research and Development in Information Retrieval, pp. 1173–1174. ACM (2012)
5. Ma, Z., Sun, A., Cong, G.: On predicting the popularity of newly emerging hashtags in twitter. Journal of the American Society for Information Science and Technology (2013)
6. Nikolov, S.: Trend or No Trend: A Novel Nonparametric Method for Classifying Time Series. PhD thesis, Massachusetts Institute of Technology (2012)
7. Tsur, O., Rappoport, A.: What's in a hashtag?: content based prediction of the spread of ideas in microblogging communities. In: Proceedings of the Fifth ACM International Conference on Web Search and Data Mining, pp. 643–652. ACM (2012)

Fast Approximation of Shortest Path
on Dynamic Information Networks*

Junting Jin, Xiaowei Shi, Cuiping Li, and Hong Chen

Renmin University of China, Beijing, China
jjt0901@126.com, shallwehh@163.com, {licuiping,chong}@ruc.edu.cn

Abstract. Computing shortest paths between a pair of nodes on a rapidly evolving network is a fundamental operation, and it is applied to wide range of networks. Classical exact methods for this problem can not extend to large-scale dynamic graphs. Meanwhile, existing approximate landmark-based methods can not be simply adapted for dynamic weighted graphs. In this paper, we consider four modifications on networks, including multi-edge insertions and deletions, weight increments and decrements. To address problems above, we present two improvements to existing methods: a novel method of indexing a weighted graph and high efficiency of approximating arbitrary pairs of nodes. Experimental results demonstrate that much faster and more precise estimation is achieved by our approaches.

Keywords: Dynamic Network, Shortest Path Tree, Landmark.

1 Introduction

A commonly used method of computing shortest distances is landmark embedding technique which is first proposed by [1] based on the triangle inequality. Furthermore, [2] presents a scalable sketch-based index structure by a sketch-based framework [3] deriving from approximate distance oracles [4], which not only supports estimation of shortest distances, but also computes corresponding SPs. However, this method can't be applied to weighted graphs.

Besides, when considering a weighted graph, one could repeatedly apply a static algorithm which takes almost hours to compute all SPs in an entire dynamic graph. Therefore, some researchers have studied dynamic algorithms to minimize re-computation time of SPs. [5] proposes algorithms to solve the single-source SP problem when only one edge changes. The noted contributions are [6] and [7], which processes multiple changes once. They shorten the computation

---

* This work was supported by National Basic Research Program of China (973 Program) (No. 2012CB316205), NSFC under the grant No.61272137, 61033010, 61202114, and NSSFC (No: 12 & ZD220). It was partially done when the authors worked in SA Center for Big Data Research in RUC. This Center is funded by a Chinese National "111" Project "Attracting International in Data Engineering Research".

F. Li et al. (Eds.): WAIM 2014, LNCS 8485, pp. 272–276, 2014.

time and improve accuracy in weighted graph, however, they answer queries of arbitrary node pairs with low efficiency.

We introduce two improvements based on landmark estimation methods. Firstly, we designed a new landmark embedding technology, that is, constructing weighted SPTs(Shortest Path Tree) and RSPTs(Reversed Shortest Path Tree) by rooting at each graph node and ending at all reachable landmarks, so SPT and RSPT can be updated in smaller range concurrently. The other improvement relates to extending existing dynamic SP algorithm MBallString and DynDijkstra for graph partitions, so that we can compute the SP of arbitrary pair of nodes by connecting common landmarks.

In this paper, we call the series of operations pre-computation, including landmarks selection, the construction of SPT and RSPT for each non-landmark node, as they can be calculated beforehand. Steps of updating SPT and RSPT and concatenating trees are referred as real-time computation. The rest of this paper is organized as follows. Section 2 presents our improved algorithms for handling 4 modifications. Meanwhile, SPs would be obtained by concatenating common landmarks. Section 3 displays experimental results. Finally, we draw a conclusion in Section 5.

2   Dynamic SP Algorithms

The structure of a graph changes in four ways, edge insertion and deletion, weight increase and decrease. Generally, edge insertion can be processed as weight decrease by reducing it from ∞ to the given value. Likewise, edge deletion is handled as weight increase by assigning it to ∞. As a result, the modifications on a graph can be simplified to weight update. In Section 2.1, we introduce some fundamental instructions and concepts used in $DASP$. In Section 2.2 and 2.3, algorithms $DASPInc$ and $DASPDec$ are presented. Finally, section 2.4 depicts the overall algorithm of approximating SP in a dynamic graph.

2.1   Basic Operations

Q is a priority queue. Q supports 3 operations. $ENQUEUE(Q, \langle ver, data, key \rangle)$ adds an item of ver to Q in ascending order of key. $EXTRACTMIN(Q)$ extracted an item with the minimum key and removes it from Q. $REMOVE(Q, ver)$ removes the item labeled ver from Q. We denote v in T_s as a $locally\text{-}affected$ node if $sp(v)$ changes. A locally-affected and unfinished node is said to be $boundary$ if it has either at least one locally-affected but finished parent or one not-locally-affected parent.

2.2   DASP Incremental Updates

Algorithm $DASPInc$ obtains T_s'(updated trees) by processing locally-affected nodes in T_s(original trees). **Stage 1:** Apply the set of edge weight changes to G, remove modified tree edges from T_s, and locate all locally-affected nodes. **Stage 2:** Enqueue boundary nodes with candidate distances. **Stage 3:** Consolidate and relax locally-affected nodes one by one.

2.3 DASP Decremental Updates

As the set of locally-affected nodes can't be predicted, algorithm $DASPDec$ starts from the first affected-node, then traverse all reachable nodes of them until no shorter distances are examined. **Stage 1:** Apply weight changes to G, and enqueue first affected-nodes. **Stage 2:** Dequeue and update current node's descendants. **Stage 3:** Consolidate and relax locally-affected nodes one by one.

2.4 Tree Concatenations

At last, we can estimate the shortest distance from s to t by connecting s's SPT and t's RSPT with common landmarks. The overall algorithm is presented in Algorithm 1.

Algorithm 1 Fast Dynamic Approximations

    Input: G is a directed and weighted graph, $s, t \in V$, T_s, $\overline{T}_t$.
    Output: $\widetilde{d}(s,t)$, $\widehat{sp}(s,t)$, T_s' and $\overline{T}_t'$
    01:    $DASPInc$ on T_s and $\overline{T}_t$, $DASPDec$ on $\widehat{T}_s$ and $\widehat{\overline{T}}_t$
    02:    $L \leftarrow$ the set of common landmarks of $\widehat{T}_s$ and $\widehat{\overline{T}}_t$
    03:    **for each** landmark $l \in L$ **do**
    04:        $\widehat{sp}(s,t) \leftarrow \widehat{sp}(s,l) \circ \widehat{sp}(l,t)$, $\widehat{d}(s,t) \leftarrow |\widehat{sp}(s,t)|$
    05:        Enqueue $\left\langle \widehat{d}(s,t), \widehat{sp}(s,t) \right\rangle$ to Q in an ascending order of $\widehat{d}(s,t)$
    06:    **end for**
    07:    **return** the top item of Q, T_s' and $\overline{T}_t'$

---

3   Experimental Evaluation

3.1   Experimental Setup

The experiments are conducted on a server with Intel(R) Xeon(R) CPU E5-2670 0 @ 2.60GHz processors, 264G of RAM. Algorithms are implemented in C.

3.2   Datasets

Pokec Social Network is the most popular on-line social network in Slovakia. A fraction named *Pokec SN* is used as a dataset, which has 6,998 nodes and 83,403 edges. Following Gaussian distribution, we add weights to graph edges to indicate users' intimacy.

California Road Network is road network of California. We randomly select a subset named as RN-Subset which has 10,000 nodes and 30,703 edges, and weigh each connected node pair in the subnet according to Poisson distribution. So the weights represent traffic conditions.

3.3 Results

In this section, we compare the efficiency of *Dijkstra*, *BallString*, *DynDijkstra*, *MFP* and *DASP*, between all pairs of nodes from the aspects of *pie* and *pde*, which are short for percentage of weight increased edges, percentage of weight decreased edges. We select 2% of graph nodes as landmarks to strike a better balance between high accuracy and high efficiency. All of the experimental results prove that *DASP* is the optimal algorithm to handle dynamic networks with directed and weighted edges, besides, *DASP* performs steady regardless of the various values of *pie* and *pde*.

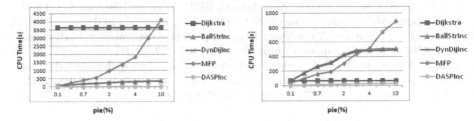

Fig. 1. Weight Increases in Pokec SN   **Fig. 2.** Weight Increases in RN-Subnet

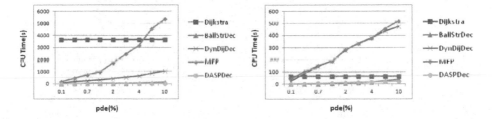

Fig. 3. Weight Decreases in Pokec SN   **Fig. 4.** Weight Decreases in RN-Subnet

4 Conclusion

In this paper, we present a dynamic algorithm with high efficiency to answer approximate *SP*s between arbitrary node pairs on a directed and weighted graph. Although the graph size is not large enough in our experiment, our algorithm can a prototype for further researches.

References

1. Tsuchiya, P.F.: The landmark hierarchy: A new hierarchy for routing in very large networks. In: SIGCOMM 1988: Proceedings on Communications Architectures and Protocols, pp. 35–42 (1988)

2. Gubichev, A., Bedathur, S.J., Seufert, S., Weikum, G.: Fast and accurate estimation of shortest paths in large graphs. In: CIKM 2010: Proceeding of the 19th ACM Conference on Information and Knowledge Management, pp. 499–508. ACM (2010)
3. Das Sarma, A., Gollapudi, S., Najork, M., Panigrahy, R.: A sketch-Based distance oracle for web-scale graphs. In: WSDM 2010: Proceedings of the 3rd ACM International Conference on Web Search and Data Mining, pp. 401–410 (2010)
4. Thorup, M., Zwick, U.: Approximate distance oracles. Journal of the ACM 52(1), 1–24 (2005)
5. Frigioni, D., Marchetti-Spaccamela, A., Nabbi, U.: Fully dynamic algorithms for maintaining shortest path trees. Algorithmica 34(2), 251–281 (2000)
6. Narvaez, P., Siu, K., Tzeng, H.: New dynamic SPT algorithm based on a ball and string model. ACM Transactions on Networking 9(6), 706–718 (2001)
7. Edward, P.F.: Chan and Yaya Yang: Shortest path tree computation in dynamic graphs. IEEE Transactions on Computers 58(4), 541–557 (2009)
8. Qiao, M., Cheng, H., Yu, J.X.: Querying shortest path distance with bounded errors in large graphs. In: Bayard Cushing, J., French, J., Bowers, S. (eds.) SSDBM 2011. LNCS, vol. 6809, pp. 255–273. Springer, Heidelberg (2011)

Detecting of PIU Behaviors Based on Discovered Generators and Emerging Patterns from Computer-Mediated Interaction Events

Yaxin Yu[1], Ke Yan[2], Xinhua Zhu[3], and Guoren Wang[1]

[1] College of Information Science and Engineering, Northeastern University, China
{Yuyx,Wanggr}@mail.neu.edu.cn
[2] College of Software, Northeastern University, China
Yanke1992.yk@gmail.com
[3] QCIS, University of Technology, Sydney, Australia
Xinhua.Zhu@uts.edu.au

Abstract. Internet provides many benefits to people, but yields a consequent disturbing phenomenon of obsession with Internet, which is called PIU(Pathological Internet Use) or IAD(Internet Addiction Disorder) in academia. PIU or IAD has negative effects on people's health of mind and body. Among tools of surfing Internet, computer is one of the most widely interactive medias. Therefore, it is significant to detect users PIU Behaviors(PIU-B) from human-computer interaction events. Emerging patterns(EPs) mining and application have becoming a major direction of contrast mining due to the ability of simplifying problems and classifying accurately. Further, generators are the shortest forms of EPs. In this light, two PIU-B detecting approaches, *i.e.*, Gen-based (**Gen**erator-based)and EP-based(**E**merging **P**attern-based) algorithms, are proposed respectively in this paper. Extensive experimental results show that both two methods are efficient and effective for detecting users PIU behaviors.

Keywords: emerging pattern, generator, PIU behavior, computer-mediated interaction, complex event.

1 Introduction

Since Internet has widely spread over the world, using computer to surfing Internet has been a basic part of our daily life. It is reported that the global consumer PC penetration per capita was 6 percent in 2006 and this number will reach 17 percent until 2015 [1]. Unfortunately, some heavy users suffer from extreme dependency on Internet, which affects their work, study and living severely. This phenomenon is named as Internet Addiction Disorder (IAD) by Goldberg early in 1996 [2] or Pathologica Internet Use (PIU) by Young [3]. With the popularity of Social Network Service in recent years, PIU or IAD phenomenon become more serious due to the exploration of Internet users in exponential scale. Here, only taking social game as an example, in August 2010, 21 percent of female users of social games on America's Facebook claimed to be addicted whereas on 21

F. Li et al. (Eds.): WAIM 2014, LNCS 8485, pp. 277–293, 2014.
© Springer International Publishing Switzerland 2014

percent of male users asserted this [1]. Therefore, it is significant to detect PC user's PIU Behaviors (PIU-B) from their daily computer-mediated interaction events to prevent them from addition.

In previous researches, a common approach to diagnosing PIU or IAD is based on diagnostic questionnaire made by medical or psychology specialists such as Young's 20 items questionnaire [4] and Beard's 5 criteria [5]. However, to our best knowledge, no existing work discusses how to diagnose PIU according to computer-mediated interaction information. In this light, by referring to the concepts of generator [6] and equivalence class [7] and the mining idea of discovering generators together with EPs in an efficient way [8], two PIU-B detecting algorithms are proposed in this paper. One is Gen-based (**Gen**erator-based) strategy and the other is EP-based(**E**merging **P**attern-based) strategy. The solutions of two strategies for detecting PIU-B are divided into two phases: building sample classifier(s) and contrasting a test behavior sequence with classifier(s) to judge whether it is a PIU-B or not. Gen-based method contrasts a test behavior sample with the mined PIU classifier consisting of PIU generators to detect PIU-B, while EP-based approach contrasts a test behavior sample with two discovered sample sets, PIU and NPIU (**N**on-**PIU**) classifiers corresponding to their respective EPs, to decide whether it is an instance of PIU-B or NPIU-B (**NPIU Behaviors**). The main contributions of this paper are summarized as follows.

1. Since behavior generator set is the lower bound of a behavior equivalence class, *i.e.*, generators are the simplest and shortest expression form of an equivalence class, we propose a Gen-based approach to building a PIU sample set and based on it to detect PC users PIU-B. The main advantage of this strategy is to achieve good precision satisfying us with less processing steps and memory space.

2. Since EPs can highlight the difference characteristics between two data sets, we propose an EP-based algorithm to build PIU and NPIU sample sets, and then base on them to detect which classifier that users behaviors belong to. Discovery of EPs is exploited by validating a generators to see if it's support value satisfies the threshold requirement of growth rate. The main advantage of this strategy is it has higher accuracy than Gen-based due to EPs instinct discrimination ability.

3. For dealing with excessive fragmentation of complex event's occurrence time and lasting time, Occurrence Time Mapping Strategy (OTMS) and Duration Rounding Strategy (DRS) are proposed in this paper. Based on these two strategies, multiple complex events with similar semantic meaning can be merged to a certain degree so as to reduce excessive numbers of frequent event item sets.

4. To the best knowledge of us, this is the first paper to address how to detect PC users PIU-B from computer's data mining viewpoint rather than from traditional medical and psychology perspective. This paper makes a try only in diagnosing addiction behaviors depending on computer-mediated interaction data, in fact, more and more mobile surfing tools are used in Internet.

Therefore, it is hoped that deep discussion about PIU-B detecting issues will start from here.

The rest of this paper is organized as follows. Section 2 introduces some preliminaries such as simple events, complex events, equivalence class, emerging patterns and generators. The details of two PIU-B detecting algorithms are discussed in Section 3. In Section 4, experimental results and evaluation are described. Section 5 gives some related work, while Section 6 concludes the paper.

2 Preliminaries

Before introducing two PIU-B detecting algorithms, some basic concepts such as simple events, complex events, emerging patterns, event equivalence class and event generators will be given in this section.

2.1 Simple Events

Events are real-world occurrences that unfold over space and time. In other words, an event is something notable that happens, owning a lasting time, occurs in a specific place, and typically will involve certain change of state. The formal definition about simple event and association rules to infer complex events are described in the following.

Definition 1. *Let $E(type_s, t_s, p_s, S_s)$ represent a simple event, where parameter $type_s$, t_s, and p_s represent type, time and place that a simple event occurs respectively. And S_s is an attribute set of different simple events, i.e., $S_s = (S_1, S_2, ..., S_n)$, where S_i ($1 \leq i \leq n$) is the ith attribute in S_s.*

Since all simple computer-mediated interactive events occur in a common place, parameter p_s can be omitted in this paper. In computer-mediated interactions, a simple event's S_s has an unique attribute depending on $type_s$. Therefore, both $type_s$ and different values of S_s's unique attribute can act as a monitoring measure together. Because of this, simple event $E(type_s, t_s, p_s, S_s)$ can be simplified into the form $E(Type_s, t_s)$, where parameter $Type_s$ not only reflects an event type but also gives its measurable value. Further, in order to facilitate discussing, we will use abbreviation E to replace $E(Type_s, t_s)$ in the following unless otherwise specified.

We focus on 8 aspects of simple events relating to computer-mediated interactions, which are 1) CPU utilization, 2) capacity of memory occupied, 3) the number of clicking mouse's left buttons, 4) the number of clicking mouse's right buttons, 5) the amount of moving pixel of mouse, 6) the number of pressing keyboards, 7) network flow and 8) a front running process of monitored applications. In fact, since there are lots of applications in real computer world, it is unrealistic to monitor all applications. Therefore, for simplifying, only 7 typical processes are selected, denoted by F_p ($1 \leq p \leq 7$), which are IE explorer, Google explorer, War3 (a real time strategy game), Trading Card Game Online

Table 1. Symbol and Meaning of $Type_s$

Symbol	Meaning of $Type_s$
A	CPU utilization
B	Capacity of memory occupied
C_1	Number of clicking mouse's left buttons
C_2	Number of clicking mouse's right button
C_3	Amount of moving pixel of mouse
D	Number of pressing keyboards
EF	Network flow
F_p $(1 \leq p \leq 7)$	p^{th} front running process of monitored applications

(a board game), Windows Media Player, QQ (an instant message software) and MSN. The different values of $Type_s$ are listed in Table 1. Based on the 8 aspects mentioned above, there are totally 8 types of simple events need to monitor, which are listed in the following.

1. Once A is over 50%, an instance of E, $e(A,t)$, is captured and created.
2. Once B is over 60%, an instance of E, $e(B,t)$, is captured and created.
3. Once C_1 is over 30, an instance of E, $e(C_1,t)$, is captured and created.
4. Once C_2 is over 10, an instance of E, $e(C_2,t)$, is captured and created.
5. Once C_3 is over 1600, an instance of E, $e(C_3,t)$, is captured and created.
6. Once D is over 100, an instance of E, $e(D,t)$, is captured and created.
7. Once EF is over 40 MB, an instance of E, $e(EF,t)$, is captured and created.
8. Once F_p is running, one of instances of E, $e(F_p,t)$, is captured and created.

2.2 Complex Events

Let $cE(type_c, t_c, p_c, S_c)$ represent a complex event, where each parameter's subscript c distinguishes complex events from simple events. Above all, parameter p_c can be omitted due to the same reason as that of simple events. Second, similar with simple event's characteristic of unique attribute, it is enough for complex events to let S_c only record their lasting time. Thus, based on two points just mentioned, $cE(type_c, t_c, p_c, S_c)$ can be simplified into the form $cE(type_c, t_c, dur)$, where the first two parameters represent the type and time that a complex event occurs and the last parameter dur represents event's duration time. In addition, substitute $Type_c$ for $type_c$ in order to keep coincident with the type expression of simple events. As a result, $cE(type_c, t_c, dur)$ is written into $cE(Type_c, t_c, dur)$. The different values and meanings of parameter $Type_c$ are listed in Table 2.

The association rules for identifying complex events are represented in a disjunctive normal form. Let R denote association rules set, then $R = (r_1 \lor r_2 \lor$

Table 2. $Type_c$ and Weight

$Type_c$	Meaning	Weight
$WVOn$	Watching Video Online	0.198
$WVOff$	Watching Video Offline	0.131
$PRTSG$	Playing Real Time Game	0.205
PBG	Playing Board Game	0.167
BWS	Browsing Web Site	0.155
COn	Chatting Online	0.143
DL	DownLoading	0.001

Table 3. T and Weight

Time Interval	T	Weight
$6a.m. - 11a.m.$	$morning$	0.14
$11a.m. - 2p.m.$	$noon$	0.16
$2p.m. - 6p.m.$	$afternoon$	0.14
$6p.m. - 11p.m.$	$evening$	0.23
$11p.m. - 6a.m.$	$before\ dawn$	0.33

... $\vee\ r_w$), ($1 \leq w \leq 7$), where $v_i's$ are the disjuncts. Each rule can be expressed in Formula (1).

$$r_v : (Condition_v) \rightarrow cE(Type_c^v, t, dur) \tag{1}$$

The left-hand side of the rule is called the rule antecedent or precondition. It contains a disjunctive normal of the conjunction of simple event tests, which is shown in Formula (2).

$$Condition_v = [e(Type_s^1, t) \wedge e(Type_s^2, t) \wedge ... \wedge e(Type_s^m, t)]$$
$$\vee [e(Type_s^1, t) \wedge e(Type_s^2, t) \wedge ... \wedge e(Type_s^h, t)]$$
$$\vee ...$$
$$\vee [e(Type_s^1, t) \wedge e(Type_s^2, t) \wedge ... \wedge e(Type_s^g, t)] \tag{2}$$

where $1 \leq (m, h, g) \leq 8$. Parameter m, g, h represent the number of simple events in a different conjunction normal. The right-hand side of the rule is called the rule consequent. If the precondition of r_v is satisfied, then r_v is said to be triggered, which results in the generation of $ce(Type_c^v, t, dur)$, i.e., an instance of a complex event. Association rules to deduce complex events are shown in Table 4. It is obvious that there are 7 association rules, which results in $1 \leq v \leq 7$. For example, if $e(A, t)$, $e(B, t)$ and $e(EF, t)$ are monitored simultaneously at time t, rule r_1 will be triggered. As a result, $ce(WVOn, t, dur)$ will be generated. In other words, we can induce that the computer user is watching video online at time t with dur=null because t is a time point. In order to obtain the duration time of $ce(Type_c^v, t, dur)$, an approach to obtaining complex event's lasting time will be exploited, which is introduced in next paragraph in detail.

During computing the lasting time of complex event, an interesting phenomenon is observed. Many complex events with a same event type are treated as different events just because their occurring time or lasting time is different. In fact, if time is limited to a reasonable range, these events have no obvious distinction. For example, given two complex events, "Surfing the Internet starting at 8:00 a.m. for 47 minutes" and "Surfing the Internet at 9:00 a.m. for 52 minutes", it is obvious that both of them have little semantic difference in real life,

Table 4. Association Rules of Generating Complex Events

r_1: $e(A, t) \wedge e(B, t) \wedge e(EF, t) \rightarrow ce(WVOn, t, dur)$
r_2: $e(A, t) \wedge e(B, t) \wedge e(EF_1, t) \rightarrow ce(WVOff, t, dur)$
r_3: $e(A, t) \wedge e(B, t) \wedge e(C_1, t) \wedge e(C_2, t) \wedge e(C_3, t) \wedge e(D, t) \wedge e(F_4, t)$
$\rightarrow ce(PRTSG, t, dur)$
r_4: $[e(A, t) \wedge e(B, t) \wedge e(C_1, t) \wedge e(C_3, t) \wedge e(F_3, t)]$
$\vee [e(A, t) \wedge e(C_1, t) \wedge e(C_3, t) \wedge e(F_3, t)] \rightarrow ce(PBG, t, dur)$
r_5: $[e(C_3, t) \wedge e(F_1, t)] \vee [e(C_3, t) \wedge e(F_2, t2] \vee [e(C_3, t) \wedge e(F_1, t) \wedge e(EF, t)]$
$\vee [e(C_3, t) \wedge e(F_2, t) \wedge e(EF, t)] \vee [e(C_1, t) \wedge e(C_3, t) \wedge e(F_1, t)]$
$\vee [e(C_1, t) \wedge e(C_3, t) \wedge e(F_2, t)] \vee [e(C_1, t) \wedge e(C_3, t) \wedge e(F_1, t) \wedge e(EF, t)]$
$\vee [e(C_1, t) \wedge e(C_3, t) \wedge e(F_2, t) \wedge e(EF, t)]$
$\rightarrow ce(BWS, t, dur)$
r_6: $[e(D, t) \wedge e(F_6, t)] \vee [e(D, t) \wedge e(F_7, t)] \vee [e(D, t) \wedge e(C_3, t) \wedge e(F_6, t)]$
$\vee [e(D, t) \wedge e(C_3, t) \wedge e(F_7, t)] \rightarrow ce(PBG, t, dur)$
r_7: $e(EF, t) \rightarrow ce(DL, t, dur)$

as they all happen in morning and the duration difference is not too much. However, two independent complex events are generated in event processing. This phenomenon results in the number of frequent complex events is too much, here, which is called Excessive Fragmentation of Time (EFT). For avoiding this issue, complex events can be merged together based on some coarse time granularity. Premise is this reduction has no negative effect on the precision of emerging pattern mining.

Considering the time semantic nature of real life, it is reasonable to partition a day with 24 hours into 5 time intervals, *i.e.*, (6:00 a.m. - 11:00 a.m.), (11:00 a.m. - 2:00 p.m.), (2:00 p.m. - 6:00 p.m.), (6:00 p.m. - 11:00 p.m.) and (11:00 p.m. - 6:00 a.m.) in this paper, and each of them represents "morning", "noon", "afternoon", "evening" and "before dawn" respectively. For avoiding EFT effects on both t_c and *dur*, two solution strategies, *i.e.*, Occurrence Time Mapping Strategy (OTMS) and Duration Rounding Strategy (DRS), are proposed in this paper. The basic idea of OTMS is to merge some complex events into a group, where one to one mapping relationship between groups and 5 time intervals need to satisfy. For emphasizing the semantic meaning of occurrence time, parameter t_c is replaced with T_c. The basic idea of DRS is to let duration is the multiple times of integer 10, here, time unit is minute, and if not, round it. Both OTMS and DRS all decrease the number of complex events dramatically. Finally, complex event's abstract form $cE(Type_c, t_c, dur)$ is represented as $CE(Type_c, T, Dur)$. Table 3 lists 5 values of T in detail. In addition, in order to facilitate discussing, we will use abbreviation CE and Ce, to replace $CE(Type_c, T, Dur)$ and its instance respectively, in the following unless otherwise specified.

2.3   Equivalence Class and Generators of Complex Events

A CE dataset is a set of CE transactions. A CE transaction is a non-empty set of CEs. The key idea of PIU-Miner is to mine a concise representation of equivalence classes of frequent CE set from a transactional CE database D. Formally, an equivalence class of CEs is defined as follows.

Definition 2. *Let EC_{CE} represent an equivalence class of CEs. EC_{CE} is a set of CEs that always occur together in some CE transactions of D. That is, $\forall X, Y \in EC_{CE}, \exists f_D(X) = f_D(Y)$, where $f_D(Z) = \{R \in D \mid Z \subseteq R\}$.*

Definition 3. *The support of an CE itemset P_{CE} in a dataset D, denoted by $sup(P_{CE}, D)$, is the percentage of CE transactions in D that contain P_{CE}.*

Definition 4. *Let X be a CE itemset of a CE dataset D. The equivalence class of X in D is denoted $[X]_D$. The maximal CE itemset and the minimal itemsets of $[X]_D$ are called the closed pattern and the generators of this CE equivalence class respectively. Generator of $[X]_D$ is represented as G. The closed patterns and generators of D are all the closed patterns and generators of their equivalence classes.*

Property 1. Let C_{CE} be the closed pattern of an equivalence class EC_{CE} and G_{CE} be a generator of EC_{CE}. Then all CE itemset X satisfying $G_{CE} \subseteq X \subseteq C_{CE}$ are also in this equivalence class.

Corollary 1. *An equivalence class EC_{CE} can be uniquely and concisely represented by a closed pattern C_{CE} and a set G_{CE} of generators, in the form of $EC_{CE} = [G_{CE}, C_{CE}]$, where $[G_{CE}, C_{CE}] - \{X \mid \exists g \in G_{CE}, g \subseteq X \subseteq C_{CE}\}$.*

Corollary 2. *The entire equivalence class can be concisely bounded as $EC_{CE} = [G_{CE}, C_{CE}]$, where G_{CE} is the set of generators of EC_{CE}, C_{CE} is the closed pattern, and $[G_{CE}, C_{CE}] = \{X \mid \exists g \in G_{CE}, g \subseteq X \subseteq C_{CE}\}$.*

In order to facilitating discuss, generator G_{CE} is abbreviated to G in the following unless otherwise specified.

2.4 Emerging Patterns

Assuming that we are given ordered pair of data sets D' and D, $sup_{D'}(X)$ and $sup_D(X)$ are their support respectively, some definitions are described as follows.

Definition 5. *The growth rate of an itemset X from D' to D is defined as*

$$GrowthRate_{D' \to D} = \begin{cases} 0 & , if\ sup_{D'}(X) = 0\ and\ sup_D(X) = 0 \\ \infty & , if\ sup_{D'}(X) = 0\ and\ sup_D(X) \neq 0 \\ \frac{sup_D}{sup_{D'}} & , otherwise \end{cases} \quad (3)$$

Definition 6. *Given $\rho > 1$ as a GrowthRate threshold, an itemset X is called a ρ emerging pattern from D' to D if $GrowthRate_{D' \to D}(X) \geq \rho$.*

A ρ emerging pattern is sometimes called ρEP or simply EP when ρ is understood. An EP from D' to D is sometimes described as "An EP in(or of) D", represented by $sup_D(X)$ simply, when D' is understood.

Definition 7. *A jumping Emerging Pattern (JEP) from D' to D is defined as an emerging pattern from D' to D with the growth rate ∞.*

Similarly, an JEP from D' to D is sometimes called "An JEP in(or of) D" when D' is understood.

Definition 8. *If $Gr_D(X) \geq \rho$, $sup_D(X) \geq MinSup$ and $\{\forall Y \subset X \mid Gr(Y) < sup_D(X)$ or*

$sup_D(Y) < MinSup\}$, then X is called an essential emerging pattern (eEP).

3   Two PIU-B Detecting Algorithms

Since the mining of frequent behavior generators is a prerequisite to exploit two PIU-B detecting algorithms, we firstly describe the discovery of frequent behavior generators in Section 3.1. And then, Section 3.2 and 3.3 give the detailed illustration about two algorithms. Here, each algorithm includes two phases: building sample classifier(s) and contrasting a test behavior sequence with classifier(s) to judge whether it is a PIU-B or not.

3.1   Discovering Frequent Generators of PIU-B Based on Equivalence Class

Generators are the lower bound of equivalence class, in other words, a set of generators is the shortest expression to represent an *EC*. Given that D is a formalized database of complex events, F_l is a set of behavior items, D_a is a condition database corresponding to some behavior item a_i in F_l, $Minsup$ is a predefined support threshold of frequent behavior generators need to discover, l is a frequent generator variable, FG is a frequent behavior generators set and $TempFG$ is it's temporary set. The procedure that how to discover frequent behavior generators, *i.e.*, Generator Mining algorithm, is illustrated in Algorithm 1. The main processing steps are described in the following.

Algorithm 1. Generator Mining Algorithm

Input: D, $Minsup$
Output: FG
1: F_l = Sort(D); Create a behavior set F_l, denoted as F_l = { a_1, a_2, ..., a_m }
2: Create a FP-tree(F_l);
3: **for** $a_i \in F_l$ **do**
4:     D_a = { $l | l \subset D_a$ }; Scan F_l's FP-tree and create a_i's D_a;
5:     if sup($l \cup a_i$, D) = sup (l, D) then $D_a = D_a - \{a_i\}$
6:     else $\{a_i \cup l\} \leftarrow TempFG$;
7:     **for** $\{a_i \cup l\} \in TempFG$ **do**
8:         if $\exists\, l' \subset \{a_i \cup l\}$ and sup(l',D)= sup($a_i \cup l$, D) then $TempFG$ = TempFG - $\{a_i \cup l\}$
9:         else $\{a_i \cup l\} \leftarrow FG$
10:         $FG = FG \cup D_a$
11:     **end for**
12: **end for**

---

1. Scan each item of D to generate item set F and record the number of each item appearing in database, *i.e.*, support, then sort all items according to their support values in descending order.
2. Build a FP-tree [8] aiming at the item set.
3. Visit the item with minimum support value in D and create D's condition database.
4. Aiming at each item from the condition database and all power sets assembled by these items, validate them to see if they are generators. If so, further to justify whether there is a subset with the same support as that of some generator. If exists, delete this generator, otherwise, reserve this generator and store it into a frequent behavior generator set *FG* until all items in condition database are finished. At last, output the frequent behavior generator set *FG*.

3.2   Generator-Based Algorithm for Detecting PIU-B

A computer-mediated interactive behavior can be represented as a set consisting of many frequent behavior generators. But not all these generators have the characteristics of PIU-B. As a result, the basic idea of Gen-based algorithm is to select some special generators representing PIU from frequent behavior generators.

Firstly, rank all frequent generators based on their scores to select PIU-B generators. Experts in PIU domain are involved in detecting work, scoring each one in generator set and rank the useful ones. According to knowledge and experiences, a formula for scoring every generator is given in Formula 4. Just like mentioned in Section 2.2, in Formula (3), $Type_c$ represents the type of frequent CE, T represents a CE's time interval projected by OTMS and Dur is CE's lasting time. The weights of $Type_c$ and T are listed in Table 2 and 3 respectively. Parameter k points out the number of CEs included in a generator G. After scoring all the generators, the generator set whose score less than threshold r_s will be pruned. The remaining ones are the final generators to represent PIU optimally. In other words, a PIU-B classifier is created based on these PIU generators.

$$Score(G_i\langle \bigcup_{k=1}^{n} CE_k\rangle) = \sum_{k=1}^{n} weight(Type_c^k) \times weight(T_k) \times Dur_k \qquad (4)$$

Secondly, next is to validate a test sample to see if it belongs to a PIU-B classifier by the sum of two PIU parameters, one is it's generator overlap rate and the other is it's duration time overlap rate. The former is denoted as *f(G)* and the latter is represented by *f(dur)*. In other words, if the sum of two overlap rates great than and equal to a given threshold, a behavior set is classified into PIU class, otherwise, it belongs to NPIU class. The larger the value of duration time overlap rate is, the higher probability of the behavior set belonging to a PIU class is. As a result, aiming at the $CE_m(type_m, t_m, dur_m)$ and $CE_n(type_n,$

t_n, dur_n), if $t_m=t_n$, $t_m=t_n$, and $dur_m \geq dur_n$, then CE_m and CE_n are thought to be a same event.

$$f(G) = \frac{count(FG_1 \cap FG_2)}{count(FG_1)} \tag{5}$$

$$f(dur) = \frac{\sum_{G_i \subset (FG_1 \cap FG_2)} \sum_{ce_k \in G_i} dur_k}{\sum_{G_i \subset FG_1} \sum_{ce_k \in G_i} dur_k} \tag{6}$$

Here, assume FG_1 is a set of frequent generators coming from a testing data set, FG_2 is that from a PIU sample set. The calculation of $f(G)$ and $f(dur)$ are given in Formula(5) and (6). The numerator in Formula5 is the intersection set of FG_1 and FG_2, i.e., the number of common generators and the denominator is the number of generators in the testing set. The ratio of two values is the percentage of PIU generators occurred in all testing generators, i.e., generator overlap rate. Before calculating $f(G)$, we must justify whether a behavior item ce_m maps a contrast item in a set of PIU generators, G_j. For example, assume that G_i is a generator from FG_1 and G_j is a generator from FG_2, for justifying whether a ce_m belongs to a contrast item, what need to do is to see if the ce_m has the same *Type* and t as those of the contrast item, and whether the dur of ce_m is greater than and equals to that of the contrast item. If so, then based on Formula(5), calculate generator overlap rate. In Formula(6), the numerator is the sum of duration time of PIU-B generators, and the denominator is that of testing generators. The ratio of two values is the overlap rate of PIU duration time.

3.3   EP-Based Algorithm for Detecting PIU-B

In first phase, EP-based algorithm classifies EPs, i.e., generators satisfying the threshold requirement of growth rate, to build PIU and NPIU classifiers by selecting their corresponding JEPs and eEPs. JEP's computation is exploited by set-difference operation and eEP's computation is implemented by set-intersection operation. After getting JEPs and eEPs, set-union operation is done to obtain final EPs of PIU and NPIU. In second phase, EP-based algorithm contrasts a test behavior sample with two classifiers respectively by calculating two individual scores and then judges which classifier belonging to based on the group decided by maximum score. The processing steps of EP-based algorithm are listed as follows.

1. **Computation of JEP.** JEP is the EPs whose growth rate is ∞, which can be obtained by *set-difference* operation between two sets FG_1 and FG_2. The result of FG_1-FG_2, i.e., the frequent generators existing in set FG_1 but not in FG_2, is D_1's JEP, written by JEP_{D_1/D_2}, while that of FG_2-FG_1 is D_2's JEP, written by JEP_{D_2/D_1}.
2. **Computation of eEP.** eEP is the EPs whose growth rate is greater than or equal to a given threshold $MinGr$. First, we implement *set-intersection* operation between two sets FG_1 and FG_2. Second, we validate each item of

result $FG_1 \cap FG_2$, f_i, to see if $f_i \geq MinGr$. If so, insert f_i into candidate set. Third, we identify whether there exists the sets having contain relationship. If exists, we select it's minimum subset and delete it's superset, otherwise, do nothing. As such, processed candidate set is the final result we need, *i.e.*, eEP.

3. **Computation of EP.** We implement *set-union* operation to obtain EPs. In other words, $eEP \bigcup JEP_{D_1/D_2}$ is EPs of D_1 and $eEP \bigcup JEP_{D_2/D_1}$ is the EPs of D_2.

4. **Computation of score.** After obtaining JEP and eEP, we calculate scores of class PIU and NPIU based on Formula(7).

$$Score(C \mid D_{i|j}) = \sum_{X \in RS} \frac{Gr(X, D_{j|i}, D_{i|j})}{Gr(X, D_{j|i}, D_{i|j}) + MaxGr_{RS}} * Sup_c(X)$$
$$+ \sum_{Y \in WS} \frac{Gr(Y, D_{i|j}, D_{j|i})}{Gr(X, D_{j|i}, D_{i|j}) + MinGr_{WS}} * Sup_c(Y) \tag{7}$$

In Formula(7), D_i and D_j are sample database, where i and j are class labels. Here, there are only two sample database, one is PIU and the other is NPIU. The former is represented by D_1 and the latter is denoted as D_2. Parameters X and Y represent generators of EPs in D_1 and D_2 respectively. RS is an positive object set which imposes positive effect on samples and WS is negative object set which brings negative effect on samples. $Gr(X, D_{j|i}, D_{i|j})$ represents X's growth rate from a data set to the other. For example, $Gr(X, D_2, D_1)$ represents X's growth rate from D_2 to D_1, while $Gr(X, D_1, D_2)$ represents X's growth rate from D_1 to D_2. $MaxGr_{RS}$ is the maximum growth rate threshold of object data set. $MinGr_{WS}$ is the minimum growth rate threshold of non-object data set. C is complex event set need to classify. $Sup_c(X)$ is the support of X in C and $Sup_c(Y)$ is the support of Y in C.

If a test data set C is taken as a *RS*, *i.e.*, a set of target class, we can depend on the first part of Formula(7) to determine classifying score. Otherwise, if C is thought to be a *WS*, *i.e.*, a set of non-target class, the the second part of Formula(7) is used to obtain score. In Formula(7), the first part of it describes the contributions of *JEPs* and *eEPs* to classify C, in particularly, parameter *MaxGr* emphasizes the special role of *JEPs* in calculating score. At the same time, the *EPs* in a non-target class also produce some effect on C's belonging to the target class. In other words, *EPs* in D_2, *i.e.*, Y, have some effect on identifying whether C belongs to D_1. For example, if C contains Y, the probability of C belongs to D_1 is $1/(Gr(X, D_2, D_1) + 1)$. When the value of $Gr(Y, D_2, D_1)$ is big, the contribution of Y to score is small, thus, Y can be omitted. But, when the value of $Gr(Y, D_2, D_1)$ satisfies a predefined threshold and is small, it will play important role in validating C belongs to D_1. In addition, considering the difference between *eEP* and *JEP*, the second part of Formula(7) is given. In a word, Formula(7) is a synthetical representation after much consideration.

4   Experiment Results and Evaluation

For testing efficiency and effectiveness of two PIU-B detecting algorithms, we gather real interactional behavior data from 20 masters in our lab by running event collector with trigger 24 hours a day in each student's computer. Among 20 students, we select 5 persons who have high PIU probability such as watching video online, playing board game, and Browsing Web Site by questionnaires as PIU sample objects and collected their daily computer interactions about 10 weeks as PIU sample samples. One third of 15 students are treated as normal NPIU samples. Two third of 15 students are treated as the testing data set and are feeded into two algorithms to justify whether an object has latent PIU behavior. In addition, considering scalability of two algorithms, synthetic data about 100 weeks are generated in a hybrid way based on some typical PIU-B generators. All the experiments are done with Intel Core i3 processor (2.53 GHz CPU with 4GB RAM) and operating system is Windows 7 professional edition.

4.1   Efficiency

Aiming at real and synthetic data, efficiency test of two PIU-B detecting algorithms, *i.e.*, running time and occupied memory space are measured. Experimental results are shown from Fig.1(a) to Fig.1(d) respectively.

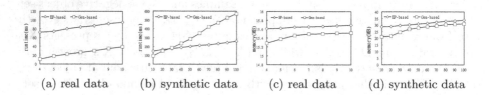

(a) real data   (b) synthetic data   (c) real data   (d) synthetic data

Fig. 1. Efficiency

Firstly, from the viewpoint of running time, Fig.1(a) and .1(b) show that two algorithms time increases as data scale(*i.e.*, the number of test days) increases, since the much data quantity is, the much the time consumes. Also, Fig.1(a) shows that Gen-based algorithm spends less time than EP-based algorithm since the latter includes more processing steps. But this advantage will be weakened as data scale enlarges. Especially, just like showing by crossing point where two lines intersected in Fig.1(b), when data volume exceeds approximate 30 days, EP-based algorithm is superior to Gen-based algorithm, *i.e.*, EP-based algorithm runs faster than Gen-based algorithm. This phenomenon is obvious in Fig.1(b) where the runtime's rising extent of EP-based algorithm is much slower than that of Gen-based algorithm. The reason lies in, with the persistent increment of data scale, the number of generators will reach to an unpredictable quantity, while the number of EPs will be kept in a stable range. In other words, ineffective generators will be filtered out due to threshold limitation of growth rate and characteristics of emerging pattern. As such, EP-based algorithm only need much less time than that of Gen-based algorithm.

Secondly, from the viewpoint of occupied memory space, both algorithms need more space with the increment of data scale. The reason is the much data quantity is, the much space consumes. In addition, EP-based algorithm need much more space than Gen-based algorithm due to dependence of generators for EPs discovery.

4.2 Effectiveness

Two algorithms effectiveness such as precision, false positive and false negative rates are tested in this paper. We discuss all possible affects that each of them imposes on two algorithms in the following. Besides these, how growth rate effects the EP-based algorithm's effectiveness is also addressed.

Firstly, from the viewpoint of precision rate, Fig.2(a) and Fig.2(b) show that both real data and synthetic data achieve high precision rate. In Fig.2(a), the average of EP-based algorithm's precision rate is 89.1% and that of Gen-based algorithm is 79.8%. In Fig.2(b), the average of EP-based algorithm's precision rate is 89.5% and that of Gen-based algorithm is 87.5%. These results show that data quantity of a test set is proportional to accurate rate of detecting results. In addition, EP-based algorithm's precision is superior to Gen-based algorithm no matter whether real data or synthetic data. The reason is that EP-based algorithm need to contrast with two sample sets and it's judgement principles is objective, fair and reasonable due to many factors to be considered. But Gen-based algorithm only depends on the contrast with one sample set and it's decision rules are more subjective since most of them come from specialists opinions.

Secondly, standing on the viewpoints of error rate, from Fig.2(c) to Fig.2(f), real data and synthetic data are in low level. Here, error rate includes false positive rate and false negative rate. For real data, as shown in Fig.2(c), average false positive of EP-based algorithm is 9.8%, while that of Gen-based algorithm is 16.4%. Also, as shown in Fig.2(d), the false negative rate of EP-based algorithm is 12.4% and that of Gen-based algorithm is 25.4%. For synthetic data, as shown in Fig.2(e), average false positive of EP-based algorithm is 9.5%, while that of Gen-based algorithm is 17.6%. At the same time, as shown in Fig.2(f), the false negative rate of EP-based algorithm is 7.8% and that of Gen-based algorithm is 8.6%. Based on these, one conclusion is that error rate of EP-based algorithm is lower than that of Gen-based algorithm. The reason lies in, compared with Gen-based algorithm, EP-based algorithm's classifier not only has concise expression form, but also includes the attribute characteristics with strong distinction ability. In addition, the other conclusion is error rates of two algorithms will descend as data scale increases since the much the data quantity is, the lower the error rate is.

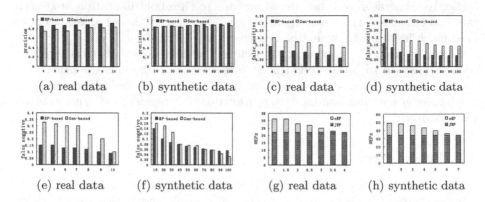

(a) real data	(b) synthetic data	(c) real data	(d) synthetic data
(e) real data	(f) synthetic data	(g) real data	(h) synthetic data

Fig. 2. Effectiveness

Finally, on the side of EP's growth rate, we discuss how it affect the EP-based algorithm's effectiveness. In Fig.2(g) and 2(h), axes X and Y represent growth rate and the number of EPs(written by #EPs) respectively. It is seen that the number of EPs will decrease as the growth rate increases until down to the number of JEPs. Just mentioned in Section 3.3, the number of EPs equals to an accumulation sum of the amount of JEPs and eEPs. While the number of JEPs is not influenced by growth rate due to its infinite growth rate. Therefore, in fact, the number of EPs is only associated with the amount of eEPs. When the number of EPs equals to the number of JEPs, the corresponding growth rate becomes a critical value, which means once growth rate exceeds this point, the number of EPs keeps invariable and equals to the number of JEPs all along. In this case, the classifier with only JEPs has the strongest distinguishing effectiveness.

5   Related Work

Main work associated with this paper has been introduced in Section 1. In the sequel, we give a brief review about research achievements related with our work.

Frequent Pattern Mining. Discovering frequent patterns from large database is meaningful and practical in association rule mining [9,10], correlation analysis [11], classification [12], emerging pattern [13] and other domains. Due to limited space, here, only some algorithms and ideas associated with emerging pattern are introduced. Li *et al.* [8] propose DPMiner to mine generators and equivalence classes. They construct a modified FP-tree which those frequent items with full support must be removed from the head table of FP-tree, thus both save running time and space in a great scale. Milton *et al.* [14] introduce fuzzy emerging patterns as an extension of emerging patterns to deal with numerical attributes using fuzzy discretization. Khan *et al.* [15] present the dual support Apriori for temporal data algorithm to discover EP and JEP from time series data using a sliding window technique. It dose not rely on itemsets borders with a constrained search space, thus requires less memory, minimum computational cost and very low data set accesses. Yu *et al.* [16] bridge causal relevance and EP discriminability to facilitate EP mining and propose a new framework of mining EPs from high-dimensional data, thus enables to extract a minimal set of strongly predictive EPs from an explosive number of candidate patterns.

Sequential pattern mining. Sequential pattern mining was first introduced by Agrawal and Strikant [17] used in data mining research field. SPADE [18], PrefixSpan [19] and SPAM [20] are quite popular ones. All these search strategies can be divided into two types [21] - breadth-first search, such as GSP and SPADE, generating many candidate patterns, and depth-first search, such as PrefixSpan and SPAM, iteratively partitioning the original data set. While most of previously developed closed pattern mining algorithms are inherently costly in both runtime and space usage when the support threshold is low or the patterns become long, Wang *et al.* [22] presented an efficient algorithm for mining frequent closed sequences, BIDE, without candidate maintenance. In recent years, the studying domain of sequential pattern mining has been extended. Since existing work of studying the problem of frequent sequence generator mining is rare, Gao *et al.* [23] present a novel algorithm, *i.e.*, FEAT, to perform this work.

Internet addiction disorder. Internet addiction was first introduced by Young [24], but the illustrate of term addiction is various between scholars and has greatly developed. In paper [2], Young defined Internet addiction as impulse-control disorder by using Pathological Gambling as a model. The behavior was first named as Pathological Compulsive Internet Usage (PCIU) and later changed to Pathological Internet Use (PIU), divided into 5 types - information overload, net compulsions, cyber-relationship addiction, cyber-sexual addiction and game addiction [25].

6   Conclusions

Be absorbed in Internet so addictively as to unable to control, this phenomenon is called PIU or IAD. PIU is a negative production of Internet popularizing and need to avoid. Aiming at this issue, two novel PIU-B detecting algorithms, Gen-based and EP-based approaches are proposed in this paper. The basic idea of the former is to mine generators only due to it's simplest representation of behavior equivalence classes. Based on mined frequent behavior generators, a PIU-B sample set, *i.e.*, a PIU classifier, is created, which can be used to detect PC users PIU-B. Taking growth rate between two data sets into account, the focus of EP-based algorithm is to discover EPs that highlight different characteristics between two data sets. Based on the EPs from generators satisfying the threshold requirement of growth rate, two sample sets, *i.e.*, PIU and NPIU classifiers, are built simultaneously so as to diagnose PIU-B and NPIU-B meanwhile. Extensive experimental results show that both two methods are efficient and effective.

References

1. http://www.statista.com
2. Goldberg: Internet addiction disorder, http://www.psycom.net
3. Young, K.: Internet addiction: Symptoms, evaluation and treatment. Journal of Innovations in Clinical Practice 17, 19–31 (1999)
4. Young, K.: Caught in the net. John Wiley and Sons, Inc., New York (1998)
5. Beard, W., Wolf, M.: Modification in the proposed diagnostic criteria for Internet addiction. Journal of Cyberpsychology Behavior 3, 377–383 (2001)
6. Pasquier, N., Bastide, Y., Taouil, R., Lakhal, L.: Discovring frequent closed itemsets for association rules. In: Proceedings of 7th International Conference on DataBase Theory, pp. 398–416 (1999)
7. Bastide, Y., Taouil, R., Pasquier, N., Stumme, G., Lakhal, L.: Mining frequent patterns with counting inference. Proceedings of SIGKDD Explorations 2(2), 66–75 (2000)
8. Li, J., Liu, G., Wong, L.: Mining statistically important equivalence classes and delta discriminative emerging patterns. In: Proceedings of the 13th ACM SIGKDD International Conference on Knowledge Discovery and Data Mining, pp. 430–439 (2007)
9. Agrawal, R., Srikant, R.: Fast algorithms for mining association rules. In: Proceedings of 20th International Conference on Very Large Data Bases, pp. 487–499 (1994)
10. Klemettinen, M., Mannila, H., Ronkainen, P., Toivonen, H., Verkamo, A.I.: Finding interesting rules from large sets of discovered association rules. In: Proceedings of the Third International Conference on Information and Knowledge Management, pp. 401–408 (1994)
11. Brin, S., Motwani, R., Silverstein, C.: Beyond market basket: Generalizing association rules to correlations. In: Proceedings ACM SIGMOD International Conference on Management of Data, pp. 265–276 (1997)
12. Liu, B., Hsu, W., Ma, Y.: Integrating classification and association rule mining. In: Proceedings of the Fourth International Conference on Knowledge Discovery and Data Mining, pp. 80–86 (1998)
13. Dong, G., Li, J.: Efficient mining of emerging patterns: Discovering trends and differences. In: Proceedings of the Fifth ACM SIGKDD International Conference on Knowledge Discovery and Data Mining, pp. 43–52 (1999)
14. Borroto, M., Trinidad, J., Ochoa, J.: Fuzzy Emerging Patterns for Classifying Hard Domains. Knowledge and Information Systems 28(2), 473–489 (2011)
15. Khan, M., Coenen, F., Reid, D.: A sliding windows based dual support framework for discovering emerging trends from temporal data. Knowledge-Based System 23(4), 316–322 (2010)
16. Yu, K., Ding, W., Wang, H., Wu, X.: Bridging causal relevance and pattern discriminability: Mining emerging patterns from high-dimensional data. IEEE Transactions on Knowledge and Data Engineering 25(12), 2721–2739 (2013)
17. Agrawal, R., Srikant, R.: Mining sequential patterns. In: Proceedings of the Eleventh International Conference on Data Engineering, pp. 3–14 (1995)
18. Zaki, M.: SPADE: An efficient algorithm gor mining frequent sequences. Journal of Machine Learning 42, 31–60 (2001)
19. Pei, J., Han, J., Mortazavi-Asl, B., Chen, Q., Dayal, U., Hsu, M.: PrefixSpan: Mining sequential patterns by prefix-projected pattern growth. In: Proceedings of the 17th International Conference on Data Engineering, pp. 215–224 (2001)

20. Ayres, J., Gehrke, J., Yiu, T., Flannick, J.: Sequential pattern mining using a bitmap representation. In: Proceedings of the Eighth ACM SIGKDD International Conference on Knowledge Discovery and Data Mining, pp. 429–435 (2002)
21. Feng, J., Xie, F., Hu, X., Li, P., Cao, J., Wu, X.: Keyword Extraction Based on Sequential Pattern Mining. In: The Third International Conference on Internet Multimedia Computing and Service, pp. 34–38 (2011)
22. Wang, J., Han, J.: BIDE: Efficient mining of frequent closed sequences. In: Proceedings of the 20th International Conference on Data Engineering, pp. 79–90 (2004)
23. Gao, C., Wang, J., He, Y., Zhou, L.: Efficient mining of frequent sequence generators. In: Proceedings of the 17th International Conference on World Wide Web, pp. 1051–1052 (2008)
24. Young, K.: Internet addiction: The emergence of a new clinical disorder. Journal of Cyberpsychology Behavior 1(3), 237–244 (1996)
25. Young, K.: Internet addiction: A new Clinical phenomenon and its consequences. Journal American Behavioral Scientist 48, 402–415 (2004)

Characterizing Tweeting Behaviors of Sina Weibo Users via Public Data Streaming

Kai Zhang[1], Qian Yu[1], Kai Lei[1,*], and Kuai Xu[2]

[1] Shenzhen Key Lab for Cloud Computing Technology & Applications (SPCCTA)
School of Electronics and Computer Engineering
Peking University
leik@pkusz.edu.cn
[2] School of Mathematical and Natural Sciences
Arizona State University

1 Introduction

Since the initial launch in August 2009, Sina Weibo, a Twitter-like microblogging service, has grown rapidly to become a major and influential site for millions of Internet users in China to disseminate news and urgent information, promote new productions, and express opinions and comments on controversial issues [4, 6]. However, unlike Twitter which attracts much attentions from the research community due to its popularity in United States and Europe, few studies have been done to characterize tweeting behaviors of Sina Weibo users.

In this paper, we present a first systematic study to infer the tweeting activities of Sina Weibo via a simple yet effective algorithm, which explores continuous public status streams and user status streams via Weibo open platform APIs. Through analyzing sampled tweets captured from public status streams and all complete tweets of two independent groups of Weibo users — a subset of users from the global Weibo user population and a subset of users from a local Weibo community, we first estimate the sampling rate of public status streams, and subsequently derive the overall Weibo tweeting activities using this sampling rate and the total tweets in sampled public status streams. In our experiments, we find that the sampling rates independently calculated from two groups are almost the same. In addition, the diurnal pattern of estimated tweeting activities well approximates the data in the public statistics released by Sina Weibo.

2 Data Collection of Public Data Streaming

Weibo provides real-time and sampled status streams of its all users via the public_timeline API, which returns a maximum of 200 latest randomly-selected tweets, which are also referred to as statuses or weibos. The real-time status streams, although a sample of all Weibo statuses, contain a rich set of valuable information on user tweeting behaviors and information cascading patterns over

---

* Corresponding author.

F. Li et al. (Eds.): WAIM 2014, LNCS 8485, pp. 294–297, 2014.
© Springer International Publishing Switzerland 2014

Weibo online social network. We collect public status streams from June 14, 2013 to July 2, 2013. During the 19-day data collection period, we have collected a total of over 91 million statuses.

In addition to providing sampled status streams of all users, Weibo also supports an API call that returns all statuses (or tweets) of a given user. However, due to the sheer size of Sina Weibo population, it is impractical to collect all statuses of all Weibo users. Sampling is a widely used technique to analyze and process vast amount of data in online social networks [1–3, 5], thus we adopt simple sampling approaches to collect complete tweeting activities of two independent groups of Weibo users — a subset of users from the global Weibo user population and a subset of users from a local Weibo community. For simplicity, we use *global cluster* and *local cluster* to refer to users in these two groups, respectively. For each user in the global cluster and local cluster, we launch Weibo user_timeline API calls to harvest its user status streams, i.e., complete tweeting activities of the user during the same time period as public status streams.

3   Inference of Sampling Rates for Public Status Streams

Before inferring the overall tweeting volumes of Sina Weibo, we first need to estimate the sampling rate used for public status streams. As Weibo randomly selects the latest tweets as a result for public status streams, the probability of any tweet being selected is the same, say p. Thus, our objective is to find an accurate inference of p based on data collected from Weibo. If the total number of sampled tweets during a given time period t is $n(t)$, then the estimated total number of tweeting activities, $N(t)$, is derived as $N(t) = \frac{n(t)}{p}$.

To calculate the sampling rate of public status streams, we develop a simple and intuitive approach by using the sampled tweets captured in the public status streams for global cluster and local cluster and their complete tweets obtained via separate user status_timeline API calls. Figures 1[a][b] illustrate strong linear correlations between the number of sampled tweets captured in public status streams and the number of complete tweets by users in the global cluster and local cluster during the first week of our data collection period, respectively. The clear linear relationships serve as a strong indication of the sampling rates for both clusters. Note that these two clusters are independently selected, thus consistent sampling rates are expected for any robust inference algorithm.

Let i denote the user i in a cluster with m user, and S_i and T_i represent the total number of tweets captured in the public status streams and the total number of complete tweets for the user i. To quantitatively calculate the overall sampling rates for each cluster, we use the following equation to infer the sampling rate, $s = \frac{\sum_{i=1}^{m} S_i}{\sum_{i=1}^{m} T_i}$, which essentially is the fraction of the total tweets in public status streams over the total number of actual tweets posted by these users. Using public status streams and the total complete tweets, the sampling rates are calculated as 0.2051 and 0.2018, for global cluster and local cluster, respectively. The consistent sampling rates across two independent groups indicate the robustness of our simple yet effective approach of estimating the sampling

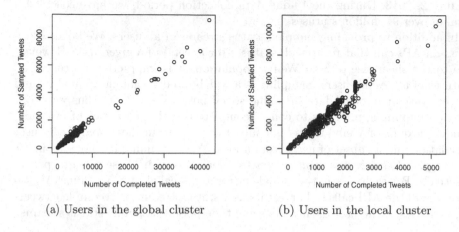

(a) Users in the global cluster   (b) Users in the local cluster

Fig. 1. Strong linear correlation between the number of tweets captured in public status streams and the number of complete tweets for sampled cluster and local cluster

rate of public status streams. The average sampling rate for the combined two clusters is 0.2063, which we use throughout the remainder of this paper.

4   Estimation of Sina Weibo Tweeting Activities

The availability of the estimated sampling rate allows to infer the total number of tweets posted by all Weibo users. For each time interval, we infer the overall Sina Weibo tweeting activities as the number of tweets captured in public status streams over the sampling rate 0.2063.

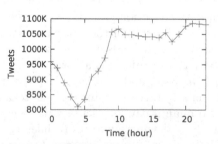

(a) Estimated Sina Weibo tweeting activity during a 24-hour time window

(b) Official Weibo release on average tweeting volumes posted from mobile devices during a 24-hour time window

Fig. 2. The inferred tweeting activities of Sina Weibo users and an official Weibo release on tweeting activities of Weibo users posted from mobile devices

Figure 2[a] illustrates the inferred number of actual tweets during a 24-hour time window. In this time window, we estimate that the total number of tweets posted by Weibo users is 23, 924, 607. Ideally, the validation of the estimation on inferred tweeting volumes of Weibo users is to use the public release data from the official Weibo announcements. As Weibo does not release such data, we find one public graph released in an official Weibo presentation, shown in Figure 2[b], which presents the average number of tweets posted by Weibo users from mobile devices during one-day cycle over March 2012. Although this public figure does not come with any real statistics, its shape and trend substantially match our estimation. This observation confirms that our proposed approach is a promising and reasonably accurate technique to estimate the tweeting activities of Sina Weibo or other microblogging services such as Twitter.

5   Conclusions and Future Work

This paper develops a simple yet effective inference algorithm to estimate the tweeting activities of Weibo users over time based on public status streams and complete user status streams of two independent groups. To the best of our knowledge, this paper presents the first effort to systematically infer Weibo tweeting volumes over time. We are currently in the process of analyzing the unstructured contents of tweeting messages posted by Weibo users for gaining an in-depth understanding of Weibo users' behaviors, interests and intents.

Acknowledgement. This research was financially supported by NFSC project (Grant No: 61103027), 973 project (No: 2011CB302305) and Shenzhen Gov Projects (JCYJ20130331144541058 and JCYJ20130331144416448), NSF grant CNS-1218212 and an ASU SRCA grant.

References

1. Ahmed, N.K., Berchmans, F., Neville, J., Kompella, R.: Time-based sampling of social network activity graphs. In: Proceedings of Mining and Learning with Graphs Workshop (2010)
2. Gjoka, M., Kurant, M., Butts, C.T., Markopoulou, A.: Walking in facebook: a case study of unbiased sampling of osns. In: Proceedings of IEEE INFOCOM (2010)
3. Kurant, M., Gjoka, M., Butts, C.T., Markopoulou, A.: Walking on a graph with a magnifying glass: stratified sampling via weighted random walks. In: Proceedings of the ACM SIGMETRICS (2011)
4. Lei, K., Zhang, K., Xu, K.: Understanding sina weibo online social network: A community approach. In: Proceedings of IEEE GLOBECOM (2013)
5. Papagelis, M., Das, G., Koudas, N.: Sampling online social networks. IEEE Transactions on Knowledge and Data Engineering 25, 662–676 (2013)
6. Wang, F., Wang, H., Xu, K., Wu, J., Jia, X.: Characterizing information diffusion in online social networks with linear diffusive model. In: Proceedings of IEEE International Conference on Distributed Computing Systems (ICDCS) (2013)

Time Series Classification Using Multi-Channels Deep Convolutional Neural Networks

Yi Zheng[1,2], Qi Liu[1], Enhong Chen[1], Yong Ge[3], and J. Leon Zhao[2]

[1] University of Science and Technology of China
xiaoe@mail.ustc.edu.cn, {qiliuql,cheneh}@ustc.edu.cn
[2] City University of Hong Kong, Hong Kong
jlzhao@cityu.edu.hk
[3] The University of North Carolina at Charlotte
yong.ge@uncc.edu

Abstract. Time series (particularly multivariate) classification has drawn a lot of attention in the literature because of its broad applications for different domains, such as health informatics and bioinformatics. Thus, many algorithms have been developed for this task. Among them, nearest neighbor classification (particularly *1*-NN) combined with Dynamic Time Warping (DTW) achieves the state of the art performance. However, when data set grows larger, the time consumption of *1*-NN with DTW grows linearly. Compared to *1*-NN with DTW, the traditional feature-based classification methods are usually more efficient but less effective since their performance is usually dependent on the quality of hand-crafted features. To that end, in this paper, we explore the feature learning techniques to improve the performance of traditional feature-based approaches. Specifically, we propose a novel deep learning framework for multivariate time series classification. We conduct two groups of experiments on real-world data sets from different application domains. The final results show that our model is not only more efficient than the state of the art but also competitive in accuracy. It also demonstrates that feature learning is worth to investigate for time series classification.

1 Introduction

As a large amount of time series data have been collected in many domains such as finance and bioinformatics, time series data mining has drawn a lot of attention in the literature. Particularly, multivariate time series classification is becoming very important in a broad range of real-world applications, such as health care and activity recognition [1–3].

In recent years, a plenty of classification algorithms for time series data have been developed. Among these classification methods, the distance-based method *k*-Nearest Neighbor (*k*-NN) classification has been empirically proven to be very difficult to beat [4, 5]. Also, more and more evidences have shown that the Dynamic Time Warping (DTW) is the best sequence distance measurement in most domains [4–7]. Thus, the simple combination of *k*-NN and DTW could

F. Li et al. (Eds.): WAIM 2014, LNCS 8485, pp. 298–310, 2014.

reach the best performance of classification in most domains [6]. Other than sequence distance based methods, feature-based classification methods [8] follow the traditional classification framework. As is known to all, the performance of traditional feature-based methods depends on the quality of hand-crafted features. However, unlike other applications, it is difficult to design good features to capture intrinsic properties embedded in various time series data. Therefore, the accuracy of feature-based methods is usually worse than that of sequence distance based ones, particularly 1-NN with DTW method. On the other hand, although many research works use 1-NN and DTW, both of them cause too much computation for many real-world applications [7].

Motivation. *Is it possible to improve the accuracy of feature-based methods? So that the feature-based methods are not only superior to 1-NN with DTW in efficiency but also competitive to it in accuracy.*

Inspired by the deep feature learning for image classification [9–11], in this paper, we explore a deep learning framework for multivariate time series classification. Deep learning does not need any hand-crafted features by people, instead it can learn a hierarchical feature representation from raw data automatically. Specifically, we propose an effective Multi-Channels Deep Convolution Neural Networks (MC-DCNN) model, each channel of which takes a single dimension of multivariate time series as input and learns features individually. Then the MC-DCNN model combines the learnt features of each channel and feeds them into a Multilayer Perceptron (MLP) to perform classification finally. To estimate the parameters, we utilize the gradient-based method to train our MC-DCNN model. We evaluate the performance of our MC-DCNN model on two real-world data sets. The experimental results on both data sets show that our MC-DCNN model outperforms the baseline methods with significant margins and has a good generalization, especially for weakly labeled data.

The rest of the paper is organized as follows. Section 2 depicts the definitions and notations used in the paper. In section 3, we present the architecture of MC-DCNN, and describe how to train the neural networks. In section 4, we conduct experiments on two real-world data sets and evaluate the performance of each model. We make a short review of related work in section 5. Finally, we conclude the paper and discuss future work in section 6.

2   Definitions and Notations

In this section, we introduce the definitions and notations used in the paper.

Definition 1 *Univariate time series is a sequence of data points, measured typically at successive points in time spaced at uniform time intervals. A univariate time series can be denoted as* $\mathbf{T} = \{t_1, t_2, ..., t_n\}$, *and n is the length of* $\mathbf{T}$.

Definition 2 *Multivariate time series is a set of time series with the same timestamps. For a multivariate time series* $\mathbf{M}$, *each element* $\mathbf{m}_i$ *is a univariate time series. At any timestamp t,* $\mathbf{m}_{\cdot t} = \{m_{1t}, m_{2t}, ..., m_{lt}\}$, *where l is the number of univariate time series in* $\mathbf{M}$.

As previous works shown [12], it's common to extract subsequences from long time series to do classification instead of classifying with the whole sequence.

Definition 3 *Subsequence is a sequence of consecutive points which are extracted from time series* $\mathbf{T}$ *and can be denoted as* $\mathbf{S} = \{t_i, t_{i+1}, ..., t_{i+k-1}\}$, *where* k *is the length of subsequence. Similarly, multivariate subsequence can be denoted as* $\mathbf{Y} = \{\mathbf{m}_{.i}, \mathbf{m}_{.i+1}, ..., \mathbf{m}_{.i+k-1}\}$, *where* $\mathbf{m}_{.i}$ *is defined in Definition 2.*

Since we perform classification on multivariate subsequences in our work, in remainder of the paper, we use subsequence standing for both univariate and multivariate subsequence for short according to the context. For a long-term time series, domain experts may manually label and align subsequences based on experience. We define this type of data as *well aligned and labeled data*.

Definition 4 *Well aligned and labeled data: Subsequences are labeled by domain experts, and different subsequences belonging to same pattern are well aligned.*

Fig.1 shows a snippet of time series extracted from BIDMC Congestive Heart Failure data set [13]. Each subsequence is extracted and labeled according to the red dotted line by medical staffs. However, to acquire the well aligned and labeled data, it always needs great manual cost.

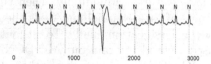

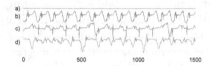

Fig. 1. A snippet of time series which contains two types of heartbeat: normal (N) and ventricular fibrillation (V)

Fig. 2. Four 1D samples of 3D weakly labeled physical activities: a) *'standing'*, b) *'walking'*, c) *'ascending stairs'*, d) *'descending stairs'*

In contrast to well aligned and labeled data, in practice, weakly labeled data can be obtained more easily [12, 1]. We define it as follows.

Definition 5 *Weakly labeled data: A long-term time series is associated with a single global label as shown in Fig.2.*

Due to the alignment-free property of weakly labeled data, it requires to extract subsequences by specific algorithm. The most widely used algorithm is sliding window [14]. By specifying *sliding* **step**, we can extract large amount of redundant subsequences from long-term time series.

In summary, in this paper, we will primarily concentrate on the time series of the same length and conduct experiments on both labeled data that is well aligned and weakly labeled data.

3   Multi-Channels Deep Convolutional Neural Networks

In this section, we will introduce a deep learning framework for multivariate time series classification: Multi-Channels Deep Convolutional Neural Networks (MC-DCNN). Traditional Convolutional Neural Networks (CNN) usually include two parts. One is a feature extractor, which learns features from raw data auto-matically. And the other is a trainable fully-connected MLP, which performs classification based on the learned features from the previous part. Generally, the feature extractor is composed of multiple similar stages, and each stage is made up of three cascading layers: filter layer, activation layer and pooling layer. The input and output of each layer are called *feature maps* [11]. In the previous work of CNN [11], the feature extractor usually contains one, two or three such 3-layers stages. Due to space constraint, we only introduce the components of CNN briefly. More details of CNN can be referred to [11, 15].

3.1 Architecture

In contrast to image classification, the input of multivariate time series classi-fication are multiple 1D subsequences but not 2D image pixels. We modify the traditional CNN and apply it to multivariate time series classification task in this way: we separate multivariate time series into univariate ones and perform feature learning on each univariate series individually. Then we concatenate a normal MLP at the end of feature learning to do classification. To be under-stood easily, we illustrate the architecture of MC-DCNN in Fig. 3. Specifically, this is an example of 2-stages MC-DCNN for activity classification. It includes 3-channels inputs and the length of each input is 256. For each channel, the input (i.e., the univariate time series) is fed into a 2-stages feature extractor, which learns hierarchical features through filter, activation and pooling layers. At the end of feature extractor, we flatten the feature maps of each channel and combine them as the input of subsequent MLP for classification. Note that in Fig. 3, the activation layer is embedded into filter layer in the form of non-linear operation on each feature map. Next, we describe how each layer works.

Filter Layer. The input of each filter is a univariate time series, which is denoted $\mathbf{x}_i^l \in \Re^{n_2^l}$, $1 \leq i \leq n_1^l$, where l denotes the layer which the time series comes from, n_1^l and n_2^l are number and length of input time series. To capture local temporal information, it requires to restrict each trainable filter $\mathbf{k}_{ij}$ with a small size, which is denoted m_2^l, and the number of filter at layer l is denoted m_1^l. Recalling the example described in Fig. 3, in first stage of channel 1, we have $n_1^l = 1$, $n_2^l = 256$, $m_2^l = 5$ and $m_1^l = 8$. We compute the output of each filter according to this: $\sum_i \mathbf{x}_i^{l-1} * \mathbf{k}_{ij}^l + b_j^l$, where the $*$ is convolution operator and b_j^l is the bias term.

Activation Layer. The activation function introduces the non-linearity into neural networks and allows it to learn more complex model. The most widely used activation functions are $sigmoid(t) = \frac{1}{1+e^{-t}}$ and $tanh(\cdot)$. In this paper, we adopt $sigmoid(\cdot)$ function in all activation layers due to its simplicity.

Pooling Layer. Pooling is also called subsampling because it usually subsam-ples the input feature maps by a specific factor. The purpose of pooling layer is to reduce the resolution of input time series, and make it robust to small variations for previous learned features. The simplest yet most popular method

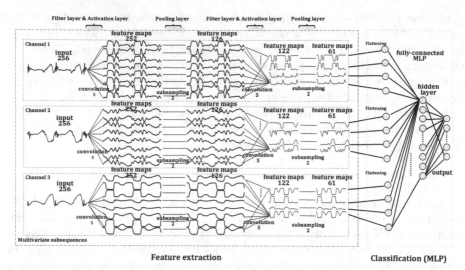

Fig. 3. A 2-stages MC-DCNN architecture for activity classification. This architecture consists of 3 channels input, 2 filter layers, 2 pooling layers and 2 fully-connected layers. This architecture is denoted as *8(5)-2-4(5)-2-732-4* based on the template *C1(Size)-S1-C2(Size)-S2-H-O*, where *C1* and *C2* are numbers of filters in first and second stage, *Size* denotes the kernel size, *S1* and *S2* are subsampling factors, *H* and *O* denote the numbers of units in hidden and output layers of MLP.

is to compute *average* value in each neighborhood at different positions with or without overlapping. The neighborhood is usually constructed by splitting input feature maps into equal length (larger than 1) subsequences. We utilize *average* pooling without overlapping for all stages in our work.

3.2   Gradient-Based Learning of MC-DCNN

The same as traditional MLP, for multi-class classification task, the loss function of our MC-DCNN model is defined as: $E = -\sum_t \sum_k y_k^*(t) \log(y_k(t))$, where $y_k^*(t)$ and $y_k(t)$ are the target and predicted values of t-th training example at k-th class, respectively. To estimate parameters of models, we utilize gradient-based optimization method to minimize the loss function. Specifically, we use simple backpropagation algorithm to train our MC-DCNN model, since it is efficient and most widely used in neural networks [16]. We adopt stochastic gradient descent (SGD) instead of full-batch version to update the parameters. Because SGD could converge faster than full-batch for large scale data sets [16].

A full cycle of parameter updating procedure includes three cascaded phases [17]: feedforward pass, backpropagation pass and the gradient applied.

Feedforward Pass. The objective of feedforward pass is to determine the predicted output of MC-DCNN on input vectors. Specifically, it computes feature maps from layer to layer and stage to stage until obtaining the output. As shown in the previous content, each stage contains three cascaded layers, and activation layer is embedded into filter layer in form of non-linear operation on each feature map. We compute output feature map of each layer as follows:

$$z_j^l = \sum_i x_i^{l-1} * k_{ij}^l + b_j^l, \quad x_j^l = sigmoid(z_j^l), \quad x_j^{l+1} = down(x_j^l)$$

where $down(\cdot)$ represents the subsampling function for *average* pooling, x_i^{l-1} and z_j^l denote the input and output of filter layer, z_j^l and x_j^l denote the input and output of activation layer, x_j^l and x_j^{l+1} denote the input and output of pooling layer.

Eventually, a 2-layer fully-connected MLP is concatenated to feature extractor. Since feedforward pass of MLP is standard and also the space is limited, more details of MLP can be referred to [16, 17].

Backpropagation Pass. Once acquiring predicted output y, the predicted error E can be calculated according to the loss function. By taking advantage of chain-rule of derivative, the predicted error propagates back on each parameter of each layer one by one, which can be used to work out the derivatives of them. We still don't present backpropagation pass of final MLP for the same reason of feedforward pass.

For pooling layer in the second stage of feature extractor, the derivative of x_j^{l-1} is computed by the upsampling function $up(\cdot)$, which is an inverse operation opposite to the subsampling function $down(\cdot)$ for the backward propagation of errors in this layer.

$$\frac{\partial E}{\partial x_j^{l-1}} = up(\frac{\partial E}{\partial x_j^l})$$

For filter layer in second stage of feature extractor, derivative of z_j^l is computed similar to that of MLP's hidden layer:

$$\delta_j^l = \frac{\partial E}{\partial z_j^l} = \frac{\partial E}{\partial x_j^l}\frac{\partial x_j^l}{\partial z_j^l} = sigmoid'(z_j^l) \circ up(\frac{\partial E}{\partial x_j^{l+1}})$$

where $\circ$ denotes element-wise product. Since the bias is a scalar, to compute its derivative, we should summate over all entries in δ_j^l as follows:

$$\frac{\partial E}{\partial b_j^l} = \sum_u (\delta_j^l)_u$$

The difference between kernel weight k_{ij}^l and MLP's weight w_{ij}^l is the weight sharing constraint, which means the weights between $(k_{ij}^l)_u$ and each entry of x_j^l must be the same. Due to this constraint, the number of parameters is reduced by comparing with the fully-connected MLP, Therefore, to compute the derivative of kernel weight k_{ij}^l, it needs to summate over all quantities related to this kernel. We perform this with convolution operation:

$$\frac{\partial E}{\partial k_{ij}^l} = \frac{\partial E}{\partial z_j^l}\frac{\partial z_j^l}{\partial k_{ij}^l} = \delta_j^l * reverse(x_i^{l-1})$$

where $reverse(\cdot)$ is the function of reversing corresponding feature map. Finally, we compute the derivative of $\mathbf{x}_i^{l-1}$ as follows:

$$\frac{\partial E}{\partial \mathbf{x}_i^{l-1}} = \sum_j \frac{\partial E}{\partial \mathbf{z}_j^l} \frac{\partial \mathbf{z}_j^l}{\partial \mathbf{x}_i^{l-1}} = \sum_j pad(\boldsymbol{\delta}_j^l) * reverse(\mathbf{k}_{ij}^l)$$

where $pad(\cdot)$ is a function which pads zeros into $\boldsymbol{\delta}_j^l$ from two ends, e.g., if the size of $\mathbf{k}_{ij}^l$ is n_2^l, then this function will pad each end of $\boldsymbol{\delta}_j^l$ with $n_2^l - 1$ zeros.

Gradients Applied. Once we obtain the derivatives of parameters, it's time to apply them to update parameters. To converge fast, we utilize *decay* and *momentum* strategies [16]. The weight w_{ij}^l in MLP is updated in this way:

$$w_{ij}^l = w_{ij}^l + \Delta w_{ij}^l$$

$$\Delta w_{ij}^l = momentum \cdot \Delta w_{ij}^l - decay \cdot \epsilon \cdot w_{ij}^l - \epsilon \cdot \frac{\partial E}{\partial w_{ij}^l}$$

where w_{ij}^l represents the weight between x_i^{l-1} and x_j^l, Δw_{ij}^l denotes the gradient of w_{ij}^l, and ϵ denotes the learning rate. The kernel weight $\mathbf{k}_{ij}^l$, the bias term b_j^l in filter layer and b^l in MLP are updated similar to the way of w_{ij}^l. The same as [18], we set $momentum = 0.9$, $decay = 0.0005$ and $\epsilon = 0.01$ for our experiments. It is noted that [19] claimed that both the initialization and the momentum are crucial for deep neural networks, hence, we consider how to select these values as a part of our future work.

4   Experiments

In this section, we will conduct two groups of experiments on real-world data sets from two different application domains. Particularly, we will show the performance of our methods via comparing with other baseline models in terms of both efficiency and accuracy.

To the best of our knowledge, indeed, there are many public time series data sets available, e.g., the UCR Suite [20]. However, we decide not using the UCR Suite for the following reasons. First, we focus on the classification of multivariate time series, whereas most data sets in UCR Suite only contain univariate time series. Second, data sets in UCR Suite are usually small and CNN may not work well on such small data sets [21]. Thus, we choose two data sets which are collected from real-world applications, and we will introduce the data sets in the next subsections.

We consider three approaches as baseline methods for evaluation: *1*-NN (ED), *1*-NN (DTW-5%) and MLP. Here, *1*-NN (ED) and *1*-NN (DTW-5%) are the methods that combine Euclidean Distance and Window Constraint DTW [7]) [1] with *1*-NN, respectively. Besides these two state-of-the-art methods, MLP is

---

[1] Following the discoveries in [7], we set the optimal window constraint r as 5%.

chosen to demonstrate that the feature learning process can improve the classification accuracy effectively. For the purpose of comparison, we record the performance of each method by tuning their parameters. Notice that some other classifiers are not considered here, since it is difficult to construct hand-crafted features for time series and many previous works have claimed that feature-based methods cannot achieve the accuracy as high as *1*-NN methods. Also, we do not choose the full DTW due to its expensive time consumption. Actually, at least more than a month will be cost if we use full DTW in our experiments.

4.1 Activity Classification (Weakly Labeled Data)

Data Set. We use the weakly labeled PAMAP2 data set for activity classification . It records 19 physical activities performed by 9 subjects. On a machine with Intel I5-2410 (2.3GHz) CPU and 8G Memory (our experimental platform), according to the estimation, it will cost nearly a month for *1*-NN (DTW-5%) on this data set if we use all the 19 physical activities. Hence, currently, we only consider 4 out of these 19 physical activities in our work, which are '*standing*', '*walking*', '*ascending stairs*' and '*descending stairs*'. And each physical activity corresponds to a **3D** time series. Moreover, 7 out of these 9 subjects are chosen. Because the other two either have different physical activities or have different dominant hand/foot.

 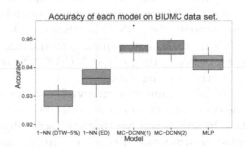

Fig. 4. Prediction time of each model on training sets with different size

Fig. 5. The box-and-whisker plot of classification accuracy on BIDMC data set

Experiment Setup. We normalize each dimension of **3D** time series as $\frac{x-\mu}{\sigma}$, where μ and σ are mean and standard deviation of time series. Then we apply the sliding window algorithm to extract subsequences from **3D** time series with different *sliding steps*. To evaluate the performance of different models, we adopt the *leave-one-out* cross validation (LOOCV) technique. Specifically, each time we use one subject's physical activities as test data, and the physical activities of remaining subjects as training data. Then we repeat this for every subject. To glance the impact of depths, we evaluate two models: MC-DCNN(1), MC-DCNN(2). They are 1-stage and 2-stages feature learning models, respectively.

Experimental Results To evaluate efficiency and scalability of each model, we get five data splits with different volumes by setting *sliding step* as 128, 64,

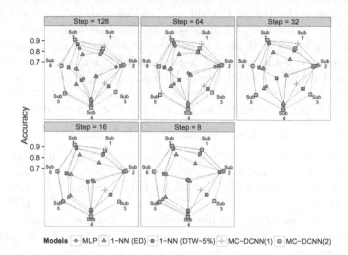

Fig. 6. Classification accuracy on each subject with different *sliding steps*

32, 16, 8, respectively. In addition, to ensure each subsequence to cover at least one pattern of time series, we set the sliding window length as 256.

As is well known, feature-based models have an advantage over lazy classification models (e.g., k-NN) in efficiency. As shown in Fig. 4, the prediction time of 1-NN model increases linearly as the size of training data set grows. In contrast, the prediction time of our MC-DCNN model is almost constant no matter how large the training data is.

We also evaluate accuracy of each model on these five data splits. Fig. 6 shows the detailed accuracy comparisons of each subject at different *step* settings. From this figure we can see that for each subject our MC-DCNN model is either the most accuracy one or very close to the most accuracy one. Especially, for subject 3, the 2-stages MC-DCNN leads to much better accuracy than other approaches. We suppose that 2-stages MC-DCNN may learn high-level and robust feature representations so that it has a good generalization. We also show the average and standard deviation of accuracy in Table 1. From the table we can see that our model leads to the highest average accuracy and the lowest standard deviation.

Table 1. Average and standard deviation of accuracy of each model at different *sliding step*. **Bold** numbers represent the best results.

Step	1-NN (DTW-5%)	MLP	1-NN (ED)	MC-DCNN(1)	MC-DCNN(2)
128	83.46 (0.063)	77.89 (0.076)	79.05 (0.076)	88.73 (0.057)	**90.34 (0.031)**
64	84.51 (0.070)	80.09 (0.098)	80.25 (0.089)	90.38 (0.050)	**91.00 (0.033)**
32	84.44 (0.080)	82.49 (0.096)	80.74 (0.094)	90.28 (0.063)	**91.14 (0.031)**
16	84.16 (0.094)	84.34 (0.104)	81.74 (0.096)	90.75 (0.062)	**93.15 (0.019)**
8	83.61 (0.104)	84.83 (0.115)	82.28 (0.103)	90.53 (0.065)	**93.36 (0.015)**

4.2   Congestive Heart Failure Detection (Well Aligned Data)

Data Set Well aligned BIDMC data set was downloaded from Congestive Heart Failure database [2] [13]. Long-term electrocardiograph (ECG) data was recorded from 15 subjects, each of them suffers severe Congestive Heart Failure. Different from PAMAP2 data, in BIDMC data set, each type of heart failure corresponds to a **2D** time series. In this experiment, we consider four types of heartbeats to evaluate all the models: 'N', 'V', 'S', 'r'.

Experiment Setup. We still normalize each univariate of **2D** time series as mentioned before. Different from weakly data, we extract subsequences centered at aligned marks (red dotted line in Fig. 1). And each subsequence still has a length of 256. Similar to the classification of individuals' heartbeats [12], we mix all data of 15 subjects and randomly split it into 10 folds to perform 10-folds cross validation. Because as [12] noted, it can be able to obtain huge amounts of labeled data in this way and a unhealthy individual may have many different *types* of heartbeats. To glance the impact of depths, we also evaluate two models: MC-DCNN(1), MC-DCNN(2). The former performs 1-stage feature learning, and the latter performs 2-stages. To determine the epochs, we separate one third of training data as validation set. As shown in Fig. 7, we set epoch to 40 and 80 for 1-stage and 2-stages MC-DCNN models respectively. Since the test error is stable when epochs are greater than them.

Experimental Results. We illustrate the accuracy of each model on BIDMC data set in Fig. 5. From this figure, we can see that accuracies of 1-stage MC-DCNN and 2-stages MC-DCNN models are 94.67% and 94.65%, which are also higher than the accuracies of *1*-NN(ED) (93.64%), *1*-NN(DTW-5%) (92.90%) and MLP (94.22%). Due to the space limit we do not report the prediction time of each model on BIDMC data set. However, the result is similar to Fig. 4 and it also supports that feature-based models have an advantage over lazy classification models (e.g., *k*-NN) in efficiency.

Fig. 7. Test error on validation set, left) 1-stage MC-DCNN model and right) 2-stages MC-DCNN model. The vertical purple line indicates the determined epoch.

5   Related Work

Many time series classification methods have been proposed based on different sequence distance measurements. Among these previous works, some researchers

---

[2] http://www.physionet.org/physiobank/database/chfdb/

claimed that *1*-NN combined DTW is the current state of the art [6, 7]. However, the biggest weakness of *1*-NN with DTW model is its expensive computation [7]. To overcome this drawback, a part of researchers explored to speed up the computation of distance measure (e.g., DTW) in certain methods (e.g., with boundary conditions) [7]. While another part of researchers tried to reduce the computation of *1*-NN by constructing data dictionary [12, 7, 14, 22]. When the data set grows large, all these approaches improve the performance significantly in contrast to simple *1*-NN with DTW. Some feature-based models have been explored for time series classification [2, 23], however, most of previous works extracted the hand-crafted statistical features based on domain knowledge, and achieved the performance not as well as sequence distance based models.

Feature learning (or representation learning) is becoming an important field in machine learning community in recent years [9]. The most successful feature learning framework is deep neural networks, which build hierarchical representations from raw data [10, 11, 24]. Particularly, as a supervised feature learning model, deep convolutional neural networks achieve remarkable successes in many tasks such as digit and object recognition [18], which motivates us to investigate the deep learning in time series field. In the literature, there are few works on time series classification using deep learning. Ref.[25] explored an unsupervised feature learning method with convolutional deep belief networks for audio classification, but in frequency domain rather than in time domain. Ref.[3] adopted a special time delay neural network (TDNN) model for electroencephalography (EEG) classification. However, their TDNN model only included a single hidden layer, which is not deep enough to learn good hierarchical features. To the best of our knowledge, none of existing works on time series classification has considered the supervised feature learning from raw data. In this paper, we explore a MC-DCNN model for multivariate time series classification and intend to investigate this problem in another way.

6   Conclusion and Future Work

Time series classification is becoming very important in a broad range of real-world applications, such as health care and activity recognition. However, most existing methods have high computational complexity or low prediction accuracy. To this end, we developed a novel deep learning framework (MC-DCNN) to classify multivariate time series in the paper. This model learns features from individual univariate time series in each channel automatically, and combines information from all channels as feature representation at final layer. A traditional MLP is concatenated to perform classification. We evaluated our MC-DCNN model on two real-world data sets. Experimental results show that our MC-DCNN model outperforms the competing baseline methods on both data sets, especially, the improvement of accuracy on weakly labeled data set is significant. Also, we showed that 2-stages MC-DCNN is superior to 1-stage MC-DCNN. It provides the evidence that the deeper architecture can learn more robust high-level features, which is helpful for improving performance of classification.

There are several research directions for future work. First, in this paper we simply use the 1-stage and 2-stages feature learning for better illustration, and in the future we plan to study and extend other deep learning models for multivariate time series classification on more data sets. Second, we also intend to perform unsupervised algorithms on unlabeled data to pre-train the networks.

Acknowledgement. This research was partially supported by grants from the National Science Foundation for Distinguished Young Scholars of China (Grant No. 61325010), the National High Technology Research and Development Program of China (Grant No.2014AA015203), the Anhui Provincial Natural Science Foundation (Grant No. 1408085QF110), the Science and Technology Development of Anhui Province, China (Grants No. 13Z02008-5 and 1301022064), and the International Science & Technology Cooperation Plan of Anhui Province (Grant No. 1303063008).

References

1. Reiss, A., Stricker, D.: Introducing a modular activity monitoring system. In: 2011 Annual International Conference of the IEEE Engineering in Medicine and Biology Society, EMBC, pp. 5621–5624 (2011)
2. Kampouraki, A., Manis, G., Nikou, C.: Heartbeat time series classification with support vector machines. IEEE Transactions on Information Technology in Biomedicine 13(4), 512–518 (2009)
3. Haselsteiner, E., Pfurtscheller, G.: Using time-dependent neural networks for EEG classification. IEEE Transactions on Rehabilitation Engineering 8(4), 457–463 (2000)
4. Batista, G.E.A.P.A., Wang, X., Keogh, E.J.: A complexity-invariant distance measure for time series. In: SIAM Conf. Data Mining (2011)
5. Ding, H., Trajcevski, G., Scheuermann, P., Wang, X., Keogh, E.: Querying and mining of time series data: experimental comparison of representations and distance measures. In: Proceedings of the VLDB Endowment, vol. 1(2), pp. 1542–1552 (2008)
6. Rakthanmanon, T., Campana, B., Mueen, A., Batista, G., Westover, B., Zhu, Q., Zakaria, J., Keogh, E.: Searching and mining trillions of time series subsequences under dynamic time warping. In: Proceedings of the 18th ACM SIGKDD International Conference on Knowledge Discovery and Data Mining, KDD 2012, p. 262 (2012)
7. Xi, X., Keogh, E.J., Shelton, C.R., Wei, L., Ratanamahatana, C.A.: Fast time series classification using numerosity reduction. In: International Conference on Machine Learning, pp. 1033–1040 (2006)
8. Xing, Z., Pei, J., Keogh, E.J.: A brief survey on sequence classification. Sigkdd Explorations 12(1), 40–48 (2010)
9. Bengio, Y., Courville, A., Vincent, P.: Representation learning: A review and new perspectives. arXiv preprint arXiv:1206.5538 (2012)
10. LeCun, Y., Bengio, Y.: Convolutional networks for images, speech, and time series. The Handbook of Brain Theory and Neural Networks 3361 (1995)

11. LeCun, Y., Kavukcuoglu, K., Farabet, C.: Convolutional networks and applications in vision. In: Proceedings of 2010 IEEE International Symposium on Circuits and Systems (ISCAS), pp. 253–256. IEEE (2010)
12. Hu, B., Chen, Y., Keogh, E.: Time Series Classification under More Realistic Assumptions. In: SIAM International Conference on Data Mining, p. 578 (2013)
13. Goldberger, A.L., Amaral, L.A.N., Glass, L., Hausdorff, J.M., Ivanov, P.C., Mark, R.G., Mietus, J.E., Moody, G.B., Peng, C.K., Stanley, H.E.: PhysioBank, PhysioToolkit, and PhysioNet: Components of a new research resource for complex physiologic signals. Circulation 101(23), e215–e220 (2000)
14. Ye, L., Keogh, E.: Time series shapelets: a new primitive for data mining. In: Proceedings of the 15th ACM SIGKDD International Conference on Knowledge Discovery and Data Mining, pp. 947–956. ACM (2009)
15. LeCun, Y., Bottou, L., Bengio, Y., Haffner, P.: Gradient-based learning applied to document recognition. Proceedings of the IEEE 86(11), 2278–2324 (1998)
16. LeCun, Y.A., Bottou, L., Orr, G.B., Müller, K.-R.: Efficient backProp. In: Orr, G.B., Müller, K.-R. (eds.) NIPS-WS 1996. LNCS, vol. 1524, pp. 9–50. Springer, Heidelberg (1998)
17. Bouvrie, J.: Notes on convolutional neural networks (2006)
18. Krizhevsky, A., Sutskever, I., Hinton, G.: Imagenet classification with deep convolutional neural networks. Advances in Neural Information Processing Systems 25, 1106–1114 (2012)
19. Sutskever, I., Martens, J., Dahl, G., Hinton, G.: On the importance of initialization and momentum in deep learning. In: Proceedings of the 30th International Conference on Machine Learning, ICML 2013, Atlanta, GA, USA, June 16-21, vol. 28 (2013)
20. Keogh, E., Zhu, Q., Hu, B., Hao, Y., Xi, X., Wei, L., Ratanamahatana, C.A.: The UCR Time Series Classification/Clustering Homepage (2011), http://www.cs.ucr.edu/~eamonn/time_series_data/
21. Pinto, N., Cox, D.D., DiCarlo, J.J.: Why is real-world visual object recognition hard? PLoS Computational Biology 4(1), e27 (2008)
22. Lines, J., Davis, L.M., Hills, J., Bagnall, A.: A shapelet transform for time series classification. In: Proceedings of the 18th ACM SIGKDD International Conference on Knowledge Discovery and Data Mining, pp. 289–297 (2012)
23. Nanopoulos, A., Alcock, R.O.B., Manolopoulos, Y.: Feature-based classification of time-series data. Information Processing and Technology 0056, 49–61 (2001)
24. Lee, H., Grosse, R., Ranganath, R., Ng, A.Y.: Convolutional deep belief networks for scalable unsupervised learning of hierarchical representations. In: Proceedings of the 26th Annual International Conference on Machine Learning, pp. 609–616. ACM (2009)
25. Lee, H., Largman, Y., Pham, P., Ng, A.Y.: Unsupervised Feature Learning for Audio Classification using Convolutional Deep Belief Networks. Advances in Neural Information Processing Systems 22, 1096–1104 (2009)

A Query Approach of Supporting Variable Physical Window in Large-Scale Smart Grid

Yan Wang[1,2,*], Qingxu Deng[1], Wei Liu[1], and Baoyan Song[2]

[1] School of Information Science and Engineering, Northeastern University,
Shenyang, P.R. China
dengqx@mail.neu.edu.cn
[2] School of Information Sci. and Tech., Liaoning University, Shenyang, P.R. China

Abstract. The region covered by monitoring system is vast in a large-scale smart grid. Traditional grid has the latency problem on the query of the historical data on server which cannot satisfy the needs of the smart grid real-time monitoring and real-time control. Furthermore, the global query in wireless sensor network (WSN) of the monitoring system cannot meet the requirement of users to query flexibly at any area, and needs high communication cost. Therefore, an efficient query for a large-scale smart grid becomes a key issue to be solved in the present smart grid research. On the basis of previous research, this paper proposes a query approach in support of variable physical window in large-scale smart grid. The approach uses physical window describing actual area of query which is variable according to user need. And through selecting query tree and node inside the query tree, physical window is modified to achieve high efficiency and flexible query. Experiments show that the approach can realize the queries for any physical regions, while reduce the cost of network communication and extend the lifetime for the monitoring system network.

Keywords: smart grid, physical window, monitoring system, data query.

1 Introduction

The so-called smart grid is a new type of power grid, which is formed by highly integrate of information technology, communication technology, computer technology, the original transport and power distribution infrastructure [1, 2]. The United States, the European Union and other countries put the smart grid rise to the height of national strategy, and expand the related technical research actively. According to the actual situation of power grid construction in China, State Grid Corporation of China first put forward to build the goal of "strong smart grid" in 2009.The realization of the smart grid depends on the on-line monitoring and real-time information control of important operating parameters in various parts of power

* This work was supported by National Key Technology R&D Program (No.2012BAF13B08, 2012BAK24B01) and Excellent Talent Support Program of Education Department of Liaoning Province (No. LR201017).

F. Li et al. (Eds.): WAIM 2014, LNCS 8485, pp. 311–322, 2014.

grid [1, 2, 3] . It is used for protecting the power grid transmission line from snow storms and other natural disasters, and providing supports for the whole life cycle management of devices.

In smart grid, when monitoring the high-voltage electrical installations and power transmission line, smart grid need widely deploy different function sensors, Intelligent Electronic Devices (IED) and etc. For example, with the ice disaster advance warning system, the centralized control center should send continuous query to sensor networks deployed in grid-power transmission line to find out the transmission line iced or will be iced in time. It can avoid wire icing or melt the ice within a short time through the way of increasing electric wire preheating. While query is executed through accessing historical data in Oracle or PI database in traditional power grid, it is efficient to analyze after the accident, but can't satisfy the needs of real-time monitoring and real-time control in large-scale smart grid [4, 5]. However, large-scale smart grid involves vast and complex geographical environment, the monitored objects increase by thousand magnitude and other factors. The traditional way of global query cannot satisfy the query in any area, and the communication cost is also large. Therefore, the research on energy efficient distributed query algorithm in large-scale smart grid monitoring system is in urgent need [6, 7].

Based on the requirement of data transmission, storage and query in large-scale smart grid, this paper proposes QPW, a query approach based on variable physical window. The method uses physical window describing actual area of query which is variable by the user need at any time. In query process, the strategy of selecting query tree and node inside the query tree is used for modifying physical window and querying flexibly at any area. The mapping array is used for selecting node inside the query tree instead of physical window, and reducing network communication cost when the query is issued. When collecting the query result, filter tuple approach is used for reducing the amount of data transmission in network. Experiments show that the approach can realize the queries for any physical regions, while reduce the cost of network communication and extend the lifetime for the monitoring system network.

2   Description of the Node in Large-Scale Smart Grid

As a rally point of transmission and distribution, the substation plays an important role in the smart grid. In the literature [8], it is the unit of large-scale smart grid which is divided into the equal size of two dimensional logical grids. Each grid covers the primary and secondary high-voltage devices in the substation, such as transformers, transformer, lightning arrester and switch equipment, and also covers all levels of transmission line region radiated by substations. A large number of sensors and IED nodes [9] are deployed in these devices.

2.1   Classification of the Node

The query in this paper is real-time, which is in view of the primary and secondary high-voltage device of transformer substation in large-scale smart grid, and its radiation sensors on each level of transmission line and IED node. There are detailed divisions and introductions of function on the sensor node and IED node as follows.

- Detecting node

Detecting node is a sensor which is used for sensing equipment working status and environment parameters, can deal with simple calculation.

- Gathering node

Gathering node is a sensor which is more powerful than detecting node. It can sense, participate in query issued and result gather.

- Logical root node

Logical root node cannot sense physical world, but possess high energy and computing power. It can communicate with the compute IED node, detecting node and gathering node.

- Compute IED node

There are a lot of Intelligent Electronic Devices (IED) nodes which are used for equipment life cycle management in smart grid. According to actual demand, each substation sets a compute IED node. The node integrates strong computation module and moderate storage module.

2.2   The Representation of Node Location

A large number of sensor devices are deployed in transmission lines and substations in smart grid. Their geographical locations are normal determined when installed. In this paper, the location of sensor node represents as <longitude, latitude > [10], that is, loc = <lon, lat>, and converts to the number which represents as the smallest unit in degrees. For example, the location of a sensor node is east longitude $23°27'30''$, south latitude$43°15'29''$. Then it represents as loc = <23.4, - 43.2> after conversion.

3   Initialization of Query

After compute IED node receiving the initialization order from the centralized control center, on the basis of self-organization network creation in WSN, the logic tree structure is formed according to initialization rules. All the nodes covered by a substation form a logic query tree which support variable physical window query. And all the logic query trees constitute a logic query forest.

3.1   The Attribute of Node

- **The level of node**

The query results collection phase determines the query collections path through the levels of node. The *level* can represent parent-child relationship of nodes. The values of *level* start from 0, and increase from small to large sequentially. In this paper, the *level* value of detecting node is 0, and that of gathering node is at least 1.   The value of compute IED node *level* is the highest and that of logical root node second, which is shown in Figure 1.

● **The maximum range of node**

Types of node are different in mart grid, and meanings of maximum range are also different. Range of the compute IED node is the possessed range of WSN [11]. Range of sensor node refers to the smallest rectangle range, which is received from the actual range of children node at a lower level after normalization. By traversing through the location of all children node in a sensor node, the maximum and minimum of longitude and latitude will be received, which is constituting the minimum rectangle range of the node, which expressed by *max-range*. *max-range* = *<long_max, long_min, lat_max, lat_min>*. The maximum range of leaf node in the tree structure is its geographical location. There is *long_max=long_min=long*, and *lat_max=lat_min=lat*.

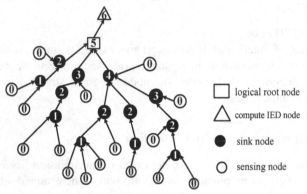

Fig. 1. Levels of different nodes

3.2   Data Structure of Node

In the initialization of query path and query issued, the compute IED node and sensor node need maintain different data structures. They are used for storing all kinds of information based on the variable physical window query what we need.

● Tuple of node

In smart grid the node maintains the normal information, such as tuple *node_info* = *<id, level, location, max-range>*. And the node need maintains route information in query process, such as tuple *node_rout* = *<id, level, status>*. The status in it represent whether the corresponding node relates to the query. If *status* equal to 1, the node need participate in query. If *status* equal to 0, the node doesn't do it. The initial value of *status* is zero.

● Table of node

With the establishing of query path, each sensor node will create information table *TB_Info* which is used for storing the *node_info* tuple sets of it and its children nodes. *TB_Info* is a static table that cannot change over the query, but only update when the network topology structure changes and query path is re-established. Besides, based on the query path, sensor nodes need build their own path table *TB_Route*. It is a dynamic table which constituted by node_route tuple. *TB_Route* of node in query path updates when query path changes.

3.3   Establishment of Logic Query Tree

● Creating information table of node

The compute IED node sends the order of establishing query path to the logical root node in a jump, which broadcasts the message in all the sensor nodes of the substation. After each sensor node receiving the establish query path message from root node, node_infor tuple as a feedback message will be formed, and sent to the next node which approaching root node in a jump range and owning a better energy transmission efficiency[12].

（1）Setting the level of node

If J is *node_info* tuple sets received by node S_i, we can know $J \geq 1$. The level of S_i is determined by the maximum level of sensor in J. The rule is shown as follows.

$S_i.level = MAX\{level \mid Sj.level, j{\neq}i \text{且} j {\in} J\} + 1$.

（2）Computing max maximal range of node

Node S_i calculates $S_i.max\text{-}range$ according to the *max-range* of *node_info* tuples collected. The calculation rules are as follows.

a) $S_i.max\text{-}range.lon\text{-}max = MAX\{lon \mid S_j.max\text{-}range.lon\text{-}max, j{\neq}i \text{ and } j{\in}J\}$
b) $S_i.max\text{-}range.lon\text{-}min = MIN\{lon \mid S_j.max\text{-}range.lon\text{-}min, j{\neq}i \text{ and } j{\in}J\}$
c) $S_i.max\text{-}range.lat\text{-}max = MAX\{lat \mid S_j.max\text{-}range.lat\text{-}max, j{\neq}i \text{ and } j{\in}J\}$
d) $S_i.max\text{-}range.lat\text{-}min= MIN\{lat \mid S_j.max\text{-}range.lat\text{-}max, j{\neq}i \text{ and } j{\in}J\}$

S_i creates table *TB_Info* according to the collected *node_info* tuples, which is shown in Figure 2. And S_i sends S_i .*node_info* toward the next node approaching root node in a jump range.

● Creating node path tables

S_i creates table *TB_Route* according to the *TB Info*, which is shown in Figure 3. *TB_Route* of node in query path will be updated when query path changing and play an important role in query processing. The logic query tree of each substation has been established, after creating *TB_Info* and *TB_Route*.

4   Query Based on the Variable Physical Window

4.1   Variable Physical Window

In the center, users can query sensors of the electric transmission line and other devices in any area of large-scale smart grid through query statement. The format of query statement is shown as follows.

SELECT Select_List
[FROM Physical_Window]
WHERE Predicate

Physical_Window （*PW*）is the actual physical range of specified query, that is PW = {*loc* = <*lon, lat*> | *loc* ∈ specified query range }. *PW* is determined according to the actual query area, and is a variable. Traverse geographical coordinates in the PW to determine the query range, that is *query_range*= <*lon_max, lon_min, lat_max, lat_min*>. The rules to determine query-range are similar to the max-range, here no longer say.

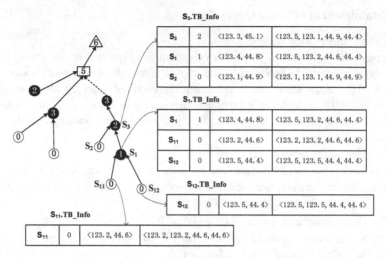

Fig. 2. Logic query tree

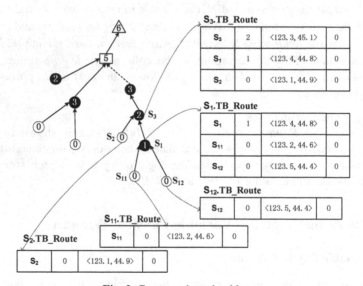

Fig. 3. Create node path tables

4.2   Selecting Query Tree

The center gets *query_range* through analyzing the query statement, compares it with the *max_range* uploaded by each compute IED node, and selects the right logic query tree for global or local query in entire query forest.

●   Selecting rules of query tree

(1) When *query_range* is disjoint from *IED$_i$.max-range*, it can infer that the *PW* mutually disjoints the substation's range, and query need not to be issued to the corresponding tree of the substation.

(2) When *query_range* intersects *IED$_i$.max-range*, it can't infer that the *PW* certainly intersects the substation's range. System need intercept the intersected minimum range among *query_range*, *IED$_i$.max-range* and *PW*, and then form *PW*'s subsets *PW'* and new *query_range'*. They will be delivered with query statement to the compute IED node of corresponding substation, and then execute selecting node inside the query tree.

(3) When *query_range* contains *IED$_i$.max-range*, it can't infer that the *PW* certainly contains this substation's range. And further judgments are need. If 4 vertexs of *IED$_i$.max-range* are all in the *PW*, it declares that this substation's range is certainly contained by *PW*. And the query need be sent to the substation for a global query. If 4 vertexs of *IED$_i$.max-range* are not all in the PW, it declares that this substation's range isn't contained by *PW*. System needs to intercept the intersected minimum range between *IED$_i$.max-range* and *PW*, and then form *query_range'* and *PW'*. They will be delivered with query statement to the compute IED node of corresponding substation, and then execute selecting node inside the query tree.

- Selecting query tree algorithm
  The time complexity of Algorithm 1 is $O(n)$.

Algorithm 1. SelectingQueryTree
Input: *query_range, IED$_i$.max-range*
Output: *PW', query_range'*
1: if (*query_range* disjoints *IED$_i$.max-range*)
2: {cut each node in this query tree (substation) }
3: else if(query_range intersects *IED$_i$.max-range*)
4: {return *PW', query_range'*}
5: else   // query_range contains *IED$_i$.max-range*
6: { if(4 vertices of *IED$_i$.max-range* are in the *PW*)
7: { global query in the logic query tree}
8: else //4 vertices are not in the PW
9: { return *PW', query_range'*} endif }
10: endif } endif

4.3   Selecting Node Inside the Query Tree

Wide query range of large-scale smart grid contains a large number of geographical coordinate points. If the query is sent with *PW* will greatly increase the network bandwidth, and also accelerate the energy cost of sensor. To solve this problem, QPW utilizes Bloom Filter (BF) [13] to encode the data set of *PW* to transmit. In the range of allowable error, it decreases communication cost, saves the energy of sensor.

● Setting the mapping array

First, define a bit array w, whose initial state is 0. Then define a set of hash function, $H\{ h(x) \mid h_i(x), i=n$, n is the number of hash functions$\}$, and each hash function is mutual independence. If there is a set of integers X, $|X|=m$, and each element of X are different. Select each element x_i ($i= 1, ..., m$) of X in turn. Then map $hi(x)$ into w separately and set the corresponding bit to 1, which is shown in Figure 4. To judge whether y is the element of X set or not, the k hash function is applied to y. If all the value of $h_i(y)$ is 1, then we think, y is the element of X, otherwise it's not. In the Figure 5, y1 doesn't belong to X, y2 may be the element of X, or just a false positive. Using the set H which is set up in advance, compute IED node sets a w for the received PW'. The w is used for selecting nodes inside the query tree instead of PW'.

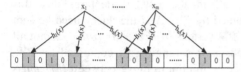 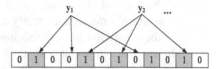

Fig. 4. Mapping array   **Fig. 5.** An example of mapping array

● Selecting the path branches

With the query is issued from root node to leaf node one by one, it traverses each TB_Info. Judge the intersection relation between $query\_range'$ and max_range of the S_i's children which $level>0$. When $query\_range'$ is disjoint from $S_j.max$-$range$, (S_j is a child of S_i) it can infer that the sub-tree which root is S_j is not in query range. So the path branch is cut in query process. Conversely, S_i sends w and $query\_range'$ to S_j. Meanwhile set $S_i.status$ and $S_j.status$ to 1, and $S_i.node\_route$ is sent to S_j. S_j stores it to its TB_Route (shown in Figure 6).

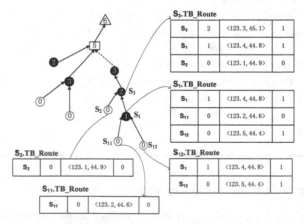

Fig. 6. Selecting node inside the tree

● Selecting the detecting node

S_i use the w of PW' to judge whether the S_j (S_j is a child of S_i) which $level=0$ is in the query range. If S_j' $location$ mapping in the w, set $S_i.status$ to 1 in $S_i.TB\_Route$ and send $S_i.node\_route$ tuple to S_j. S_j stores it to its TB_Route (shown in Figure 6).

● The algorithm of selecting node inside the query tree

The time complexity of Algorithm 2 is $O(n)$.

Algorithm 2. SelectingNodeInsideTree
Input: PW', $query\_range'$, TB_Info
Output: TB_Route (that is, mark the nodes which participate in query)
1: create a mapping array w for PW'
2: traverse children node in $S_i.TB\_info$
3: if $S_j.level>0$ // S_j is a children node in $S_i.TB\_info$
4: {if ($query\_range'$ disjoints $S_j.max\_range$)
5: {cut the path branch which root node is S_j }
6: else{transmit the w, $query\_range'$ and query statement, modify $S_i.TB\_route$ and $S_j.TB\_route$ }endif
7: else // $S_j.level=0$
8: if($S_j.location$ in the w) // S_j is in query range
9: {modify $S_i.TB\_Route$ and $S_j.TB\_Route$}
10: else{ cut S_j} endif} endif

4.4 Collecting the Query Result

When collecting the query result, a filter tuple f is generated in each sensor node of query path. Data dominated by f (that is, query needless data) will be discarded. Other data will be saved and transmitted in the data collections direction from the node. Sensor nodes get out a minimum value from the maximum of all its internal tuples, and regard it as the f of this node. Delete the data which is governed by f, and other data are transmitted to their parent node. The node received $S_i.node\_route$ finds its parent node through querying the TB_Route table. That is the node which $level$ is higher than S_i.

5   Performance Evaluation

In different network scale, we verify the performance of our work by comparing QPW against other methods in communication cost, negative error rate and transmission cost of different distributed data.

5.1 Physical Window Size and Transmission Cost

In the case of allowable error, the query approach proposed in this paper use mapping array to reduce the communication cost of network instead of actual physical window. When physical window size changes, the reduction rate of transmission cost is used to describe the reduction of transmission cost by using mapping array. Its calculation formula is shown as follows.

$$\text{reduction rate of transmission cost} = \frac{\text{physical window size - bits of mapping array}}{\text{physical window size}} \times 100\% \quad (1)$$

Here physical window size is bit number of physical window which equals n multiply the bits of sensor geographic coordinates (in this experiment, the assumption is 30), which n is the number of sensors in PW. Variable m represents the bits of mapping array. According to literature [14] we can obtain the optimal value of a mapping array when $k = 3$ in this experiment. According to different n, the reduction rate of transmission cost will be shown as follows.

Table 1. Transmission cost （m=2000 bit）

n	reduction rate
100	33.3%
200	66.7%
300	77.8%
400	83.3%
500	86.7%
600	88.9%
700	90.5%

Table 2. Transmission cost （m=4000 bit）

n	reduction rate
100	-33.3%
200	33.3%
300	55.6%
400	66.7%
500	73.3%
600	77.8%
700	81.0%
800	83.3%
900	85.2%

With the n constantly increasing, the reduction rate of transmission cost increases gradually when the number of mapping function is fixed. Through comparing table 1 with table 2, it shows that different values of m make different communication cost. System still has a reduction rate of 33.3% when m = 2000 and n = 100. However, the reduction rate decreases when m = 4000 and n = 100. It indicates that the query window size and number of mapping array bit have restrictive relations. In addition, when physical window increases to a certain extent, for example n = 800, the bits of a mapping array is at least 2654 according to $m \geq n \times 1.43 \times 2.32$. Then the mapping array which m=2000 cannot satisfy the PW in query. Therefore, using mapping array instead of actual physical window can save bandwidth in general, but it needs to set up the physical window size and bit number of mapping array reasonable.

5.2   Physical Window Size and Negative Error Rate

The above experiments show that mapping array can reduce the transmission cost of network. The limitations of mapping array cause that a few of sensors which are not in query range are considered in the range incorrectly, that is negative error rate. This experiment sets $k = 3$ which is the number of mapping function. And variable m is set 1000, 1500 and 2000 separately. Figure 7 shows that negative error rate is proportional to physical window size when m is fixed. Because the number of sensors in physical window increases, the number of 1 in mapping arrays increases, and negative error rate ascends. Furthermore, when the sizes of physical window are same, the negative error rate decreases as m increases. This is because that if the bits of mapping array and bits of 0 in the array increase, the probability of values, which are mapped to 1 by mapping function and don't belong to the physical window, is smaller . That is, the negative error rate is low. In the practical application, when use mapping arrays to reduce transmission cost and query time, the values of m and n should be set reasonable to reduce negative error rate.

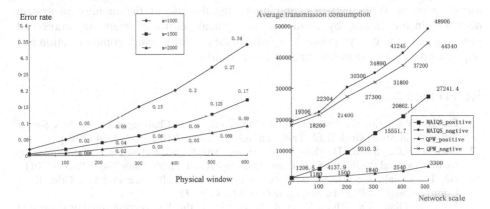

Fig. 7. Negative error rate   **Fig. 8.** Average transmission cost

5.3   Transmission Cost in different Correlation of Data

This experiment assumes that each sensor is unified deployed(1node/$10m^2$) and location are all fixed. This experiment sets 100m*100m physical range as a stepping, and increases the stepping from 100m*100m to 600m*600m. It assumes that the test data are distributed evenly in sensor nodes, each node produces a tuple data in every moment, and the attribute dimension of tuple is 3. Each attribute is 4 bytes, and this experiment regards the normalized data in [0, 1]. Data are transmitted in the form of packets which the biggest length is 48 bytes. The simulated data are all generated by standard dataset generator of skyline query [15]. This experiment compares the MAIQS algorithm [16] based on agency with QPW on the average transmission cost under the positive and negative correlation distribution (see Figure 8).

Figure 8 shows that the average transmission cost by using filter tuple in query approach is much lower than that normal query approach. The experimental comparison between positive correlation and negative correlation of sensor data in this paper shows that two attributes of data, which are distributed in positive correlation, are direct proportion. When data are distributed in positive correlation, transmission cost of two query approach is lower than that is in negative correlation. This is because that transmission cost is small when the distributed data skyline in positive correlation is small, while negative correlation is the opposite. With network size increasing, the average transmission cost also increases no matter that the data are distributed in positive correlation or negative correlation.

6   Conclusion

In large-scale smart grid the query range is vast and variety and the devices of referred sensor are numerous. This paper proposed a query approach in support of variable physical window in large-scale smart grid. In this approach, the selecting strategy of query tree and tree node is used for meeting users to query flexibly at any area. The mapping array is used for selecting strategy of tree node instead of physical window, and reducing network communication cost when query issued. When collecting the query results, filter tuple approach is used for reducing amount of data transmission in WSN network, and improving the life of the monitoring system network. In the future, by resolving the optimal setting of mapping arrays and improving skyline query approach, it reduces the cost of network communication and improves the life of the monitoring system network.

References

1. Gungor, V.C., Lu, B., Hancke, G.P.: Opportunities and Challenges of Wireless Sensor Networks in Smart Grid. IEEE Trans. on Ind. Electron. 57(10), 3557–3564 (2010)
2. Gungor, V.C., Sahin, D., Kocak, T., et al.: Smart grid technologies: communication technologies and standards. IEEE Transactions on Industrial Informatics 7(4), 529–539 (2011)
3. He, H., Kusiak, A.: Special Issue on Computational Intelligence in Smart Grid. IEEE Computational Intelligence Magazine 6(3), 12–64 (2011)
4. Yuan, X., Qian, Z.X., Zhou, Y., et al.: Discussion on the development trend of smart grid and its key technology. In: 2012 China International Conference on Electricity Distribution (CICED), pp. 1–8. IEEE (2012)
5. Erol-Kantarci, M., Mouftah, H.T.: Wireless sensor networks for smart grid applications. In: 2011 Saudi International Electronics, Communications and Photonics Conference (SIECPC), pp. 1–6. IEEE (2011)
6. Stokes, A.B., Fernandes, A.A.A., Paton, N.W.: Adapting to Node Failure in Sensor Network Query Processing. In: Gottlob, G., Grasso, G., Olteanu, D., Schallhart, C. (eds.) BNCOD 2013. LNCS, vol. 7968, pp. 33–47. Springer, Heidelberg (2013)
7. Wang, G., Xin, J., Chen, L., et al.: Energy-efficient reverse skyline query processing over wireless sensor networks. IEEE Transactions on Knowledge and Data Engineering 24(7), 1259–1275 (2012)
8. Wang, Y., Deng, Q., Liu, W., Song, B.: A Data-Centric Storage Approach for Efficient Query of Large-Scale Smart Grid. In: The 9th Web Information System and Application Conference, WISA 2012 (2012)
9. Liao, W.H., Shih, K.P., Wu, W.C.: A grid-based dynamic load balancing approach for data-centric storage in wireless sensor networks. Computers and Electrical Engineering 36(1), 19–30 (2010)
10. Fagin, R., Lotem, A., Naor, M.: Optimal aggregation algorithms for middleware. Journal of Computer and System Sciences 66(4), 614–656 (2003)
11. Lin, X., Yuan, Y., Wang, W.: Efficient skyline computation over sliding windows. In: Proceedings of the 21st International Conference on Data Engineering, pp. 502–513 (2005)
12. Song, B.: Data storage method supporting large-scale smart grid. Journal of Computer Applications 32(9), 2496–2499 (2012)
13. Mitzenmacher, M.: Compressed Bloom Filters. IEEE/ACM Transactions on Networking 10(5), 604–612 (2002)
14. Chen, H., Li, M., Jin, H., Liu, Y., Ni, L.M.: MDS: Efficient Multi-dimensional Search in Wireless Sensor Networks. In: IEEE RTSS 2008, Barcelona, Spain, November 30-December 3 (2008)
15. Borzsonyi, S., Stocker, K., Kossmann, D.: The Skyline operator. In: Proceedings of the 17th International Conference on Data Engineering, Heidelberg, Germany, pp. 421–430 (2001)
16. Wang, X.: MAIQS: Mobile Agent Based Intelligent Query Method 1for Wireless Sensor Network. Chinese Journal of Sensors and Actuators 21(9), 1613–1622 (2008)

P³RN:Personalized Privacy Protection Using Query Semantics over Road Networks

Xiao Pan, Lei Wu, Chunhui Piao, and Xiaoshuo Xu

Shijiazhuang Tiedao University, Shijiazhuang, 050043, China
{smallpx,wulei,pch,xuxs}@stdu.edu.cn

Abstract. Privacy protection has received considerable attention for location-based services. A lot of location cloaking approaches focus on the identity and location protection, but few algorithms pay attention to prevent the sensitive information disclosure using query semantics. In terms of personalized privacy requirements, all queries in a cloaking set, from some user's point of view, are sensitive. These users regard the privacy is breached. We call this attack as personalized homogeneity attacks. We show that none of the existing location cloaking approaches can effectively resolve this problem over road networks. We propose a (K, L, P)-anonymization model and a personalized privacy protection cloaking algorithm over road networks P³RN. The efficiency and effectiveness of P³RN are validated by a series experiments.

Keywords: Privacy protection, road networks, location based services, sensitive information.

1 Introduction

With advances in wireless communication and mobile positioning technologies, location-based services (LBSs) have seen wide-spread adoption. These applications provide users with a great convenience, and improve the quality of work and personal life significantly. However, the increasing collections of individual's information (*e.g.*, location) open the door for potential privacy disclosure.

In general, existing work on privacy preserving in LBSs protects users three kinds of information: *identity*, *location*, and *sensitive information*. In order to protect these information, different models and methods are proposed. To hide the user's identity, location k-anonymity model is the most acceptable model. For example, in Fig. 1(a), u_1, u_2 and u_3 constitute a cloaking set. The locations of u_1, u_2 and u_3 are represented by a segments set S. u_1 is indistinguishable from u_2 and u_3 in the cloaking set, thus the users identities are successfully protected.

To hide the exact locations, based on the location k-anonymity model, location cloaking is the popular methodology. Its main idea is to reduce the spatial and temporal resolution of the user's location. In Euclidean space, exact locations are usually extended to a rectangle or a circle [1,3,4]. While in a constrained space (*e.g.* road networks), exact locations are usually published as a segments

F. Li et al. (Eds.): WAIM 2014, LNCS 8485, pp. 323–335, 2014.

User	Published location	Query content
u_1	$S=\{<n_2, n_3>,$	Cancer Hospital
u_2	$<n_3, n_9>,$	Chest Hospital
u_3	$<n_2, n_9>\}$	Women's Hospital

(a)

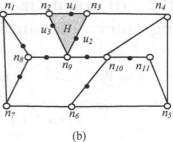

(b)

Fig. 1. Location 3-anonymity (a) query semantics (b) location semantics

set [2,13,10]. In the previous example, the adversary cannot sure where is the exactly location of u_1 on the segments in S.

Most existing work focuses on the identity and location protection, but few literatures pay attention to prevent the sensitive information disclosure. The sensitive information is disclosed using two kinds of published information: *location semantics* [3,7,16] and *query semantics* [14]. Fig. 1(b) shows an example of the first case. Considering the public geographical context, the segments in S are totally covered by a hospital. Thus, each user's visited place is disclosed. Fig. 1(a) illustrates the other case. When the adversary considers the query semantics, three users issue hospital-related queries. The attackers get the inquirer's health conditions. In this paper, we consider to protect the sensitive information using query semantics over road networks.

In order to protect the sensitive information, besides location k-anonymity model, a cloaking set also follows l-diversity model [9]. In our scenario, according to the l-diversity model, the query contents issued from a cloaking set are at least l different. However, location k-anonymity model and l-diversity model are not sufficient to protect the sensitive information.

User	Sensitivity setting		User	Published location	Query content
	secret	non-secret			
u_4	hospital, gay club, bank	tourist guides, shopping guides	u_4		hospital
u_5	hospital, gay club	bank, shopping guides, tourist guides	u_5	S	gay club
u_6	hospital	gay club, bank, tourist guides, shopping guides	u_6		bank

(a)

(b)

Fig. 2. Personalized homogeneity attack (a) privacy profiles (b) a cloaking set

As we known, privacy requirements are personalized. Whether a query is sensitive or not depends on the personalized privacy requirements. Fig. 2(a) shows an example of the personalized privacy profiles. Fig. 2(b) is a cloaking set following location 3-anonymity model and 3-diversity model. In u_4's side, three queries issued from the cloaking set are all sensitive. u_4 doesn't want any of these queries being associated with himself. The case, that Fig. 2(b) shows, is unacceptable for u_4. We call this kind of attacks as *personalized homogeneity attacks*, which consider personalized privacy profiles and query semantics.

This paper focuses on preventing personalized homogeneity attacks on road networks, which aiming at protection identity, location and sensitive information. We face one major challenge. What is the proper users partition strategy which incurs less re-computing? Different protection targets lead to re-compute cloaking sets iteratively. For example, some users in a candidate cloaking set satisfy location k-anonymity model originally, but other users in the same set fail to protect the sensitive information. However, when the sensitive information is protected by adjustment, some user fails to follow the location k-anonymity model.

To address the above issues, we propose a (K, L, P)-anonymity model to protect identity, location and sensitive information. Thanks to [10], a user location on the road network is mapped to a 1-D value. Our P³RN finds cloaking set based on the 1-D values considering the personalized privacy requirements. Specifically, users are first partitioned into groups as the maximum anonymity level. Then, conservative users shift from one group to another one considering the sensitivity requirements. Finally, the segments covered by each group are published to protect location information.

The contributions we make in this paper can be summarized as follows:

- We propose a (K, L, P)-anonymity model, which aims to protect user's identity, location and sensitive information.
- We propose a P³RN cloaking algorithm over road networks, which supports users personalized privacy requirements.
- Some preliminary experiments are conducted to evaluate the performance of our proposed algorithm.

The rest of the paper is organized as follows. We review the related work in Section 2. The problem under investigation is formally defined in Section 3. The cloaking algorithm P³RN is proposed in Section 4. Section 5 presents the preliminary performance evaluation results. Finally, the paper is concluded in Section 6.

2   Related Work

Hu *et al.* [4] first observed the problem of sensitive information disclosure, when the anonymized location data joins with a reference data set. [4] only follows location k-anonymity model. Location l-diversity model was introduced in [1] which requires that a cloaked region contains k mobile users and l places. However, [1] doesn't distinguish the place type. [15] refined the definition of l-diversity as the number of different places types. Then, [3] further classified the places into sensitive places and non-sensitive places. [16] presented an approach to the privacy preserving sharing of sensitive positions in urban settings. All of previous work protects individual sensitive information using location semantics.

[14] first pointed out the problem of the homogeneity attack using query semantics. A p-sensitive model is proposed, and a PE-tree is constructed for implementing the model. The drawback of [14] is as follows. First, the maintenance cost for the PE-tree is high. Second, every user has the same privacy requirement. Third, the system unifies to define whether a query is sensitive or not. [6] further categorized service attribute values (*e.g.*, burgers and pizzas are categorized as the fast food) and enhanced the cloaking algorithms by defining the query l-diversity concept in LBSs. [14] and [6] only apply to Euclidean space, whereas our work pays attention on privacy protection against personalized homogeneity attacks over road networks.

In terms of the anonymization methods on road networks, existing work is classified into four categories: tree-type index-based cloaking algorithms [8], graph traversal-based cloaking algorithms [13,10,2], mix-zone based cloaking algorithms [12], and PIR-based anoymization algorithms [11]. The cloaking sets, generated from the second method, reflect the feature of road networks, which is helpful for reducing query cost in LBSs server. Meanwhile, the cloaking set has k-sharing property, which is benefit for cases of high workloads, preventing query sampling attacks [10] and replay attacks [13]. Consequently, the graph traversal-based cloaking algorithms are the most widely used privacy preserving approach over road networks so far. Our proposed P^{3}RN algorithm is in line with the graph traversal-based methods.

3   Preliminary

3.1   Preliminary

Like most existing work, we employ a centralized system, which consists of mobile users, a trusted anonymizing proxy (TAP), and an un-trusted service provider. A category-sensitivity relation $CaSR$ is stored in the TAP, which is a many-to-one relation from a query categories relation to a sensitivity relation. Through $CaSR$, each query is labeled with a sensitivity[1]. $CaSR$ is defined as follows.

Definition 1. *(Category-Sensitivity Relation) Let $D(CaSet)$ and $D(SSet)$ be the domain of query categories and sensitivities respectively. Then,*

$$CaSR = \{(a,b)|a \in D(CaSet), b \in D(SSet),$$
$$if(a_1,b_1),(a_2,b_2) \in CaSR,\ a_1 = a_2, then\ b_1 = b_2\}$$

For example, $CaSR$={(sensitive location navigation, top secret), (emergency call, more secret), (location sensitive billing, secret), (infotainment services, less secret), (shopping guides, non-secret), (travel guides, non-secret)}, where $D(CaSet)$={sensitive location navigation, emergency call, location sensitive

---

[1] We assume that the category-sensitivity relation is pre-known. The method for constructing the relation is out of our scope.

billing, infotainment services, shopping guides, travel guides}, $D(SSet)=\{$top secret, more secret, secret, less secret, non-secret$\}^2$.

The relation $CaSR$ has two properties.

Property 1: For any two queries q_i and q_j, if q_i and q_j belong to a same category, $q_{s_i} = q_{s_j}$, where q_{s_i} is the sensitivity of the query q_i.

Property 2: For any two queries q_i and q_j, if $q_{s_i} \neq q_{s_j}$, the categories of q_i and q_j are different.

From Property 1 and Property 2, queries with the same contents are the special case of queries with the same sensitivity. Thus, in the following sections, we only focus on the query sensitivity instead of the specific query content.

In order to represent users locations on the road network, like the most existing work, we define a road network as an un-directional graph $G(V, E)$. V is a vertices set, including intersections $(d(v) \geq 3)$ and terminal points$(0 < d(v) \leq 2)$, where $d(v)$ is the vertex degree. E is an un-directional edges set, representing road segments between two vertices.

The privacy profile is formalized as follows.

Definition 2. *(Privacy Profile) In order to protect user's personalized privacy on road networks, each user specifies three kinds of parameters:*

- *Anonymity requirement k: It is the anonymity level in the location k-anonymity model, aiming at protection the user's identity;*
- *Location diversity requirement l: It is the minimum number of road segments[3] published by a cloaking set, which is to protect the user's location;*
- *Sensitivity requirements (t_s, p): t_s is the user's maximum tolerable query sensitivity. If the sensitivity of a query q is larger than t_s, from this user's side, q is a sensitive query; otherwise, q is a non-sensitive query. p is the user's maximum tolerable ratio of the sensitive queries in a cloaking set.*

Note that t_s is defined for one query, indicating whether a query is sensitive or not. p is based on a users set, implying the number of sensitive queries in the set. For example, a user u sets the privacy profile as $(3, 4,(0.5, 0.4))$. u requires that the cloaking set contains at least 3 users. At least 4 different road segments are published as the cloaked location. If a query sensitivity is larger than 0.5, u regards this query as a sensitive query. u requires that the portion of sensitive queries in the cloaking set is not larger than 0.4.

3.2 Attack Model and Privacy Model

Definition 3. *(Personalized Homogeneity Attack) Let CS be a users set. For $\forall u \in CS$, from u's point of view, the number of the sensitive query in CS is denoted as $Count\_SQ_u$. If*

$$\exists u, u \in CS, u.p < \frac{Count\_SQ_u}{|CS|}$$

---

[2] Besides the categorical values, a sensitivity could also be a normal number. Two kinds of the attribute values can be transformed into each other. For simplicity, we use a normal number in the following sections.

[3] A long road segment is divided into serveral segments.

where $|CS|$ is the number of users in CS. This attack is termed as personalized homogeneity attacks.

Fig. 3 shows a users set $CS = \{u_1, u_2, u_3\}$. Assume that $u_1.t_s$=0.25 and $u_1.p$=0.9. From u_1's view, all queries issued from CS are sensitive, which is un-acceptable. CS suffers from personalized homogeneity attacks.

Now let's define the (K, L, P)-anonymity model.

Definition 4. *((K, L, P)-anonymity model) Let CS be a cloaking set accompanied with a set of issued queries $QSet$ and the cloaked location RS. If*

- $K \leq |CS|$, where $K = \underset{\forall u \in CS}{MAX}\ u.k$;
- RS is a set of road segments, and $L \leq |RS|$, where $L = \underset{\forall u \in CS}{MAX}\ u.l$, and $|RS|$

  is the number of segments in RS;
- *For $\forall u \in CS, \frac{Count\_SQ_u}{|CS|} \leq u.p$.*

Users in CS satisfy (K, L, P)-anonymity model.

The first condition guarantees location K-anonymity model. The second condition ensures the diversity of the published location. Finally, the third one implies that the number of sensitive queries is acceptable for each user. Besides, users following the above three conditions has K-sharing proterty. Fig. 4(a) shows a cloaking set. Users privacy profiles are shown in Fig. 4(b). According to Definition 4, $\{u_1, u_2, u_3\}$ satisfies (K, L, P)-anonymity model.

User	Published location	q_s
u_1		1
u_2	S	0.5
u_3		0.5

User	Published location	q_s
u_1	$\{<n_2, n_3>,$	1
u_2	$<n_3, n_9>,$	0.5
u_3	$<n_2, n_9>\}$	0

User	k	l	t_s	p
u_1	2	3	0.25	0.75
u_2	3	2	0.75	0.5
u_3	3	3	1	0.5

(a)   (b)

Fig. 3. Personalized homogeneity attacks

Fig. 4. (K, L, P)-anonymity model (a) a cloaking set (b) privacy profiles

4   P³RN:Anonymization Algorithm for Protection Personalized Privacy on Road Networks

P³RN aims to protect the identity, location and sensitive information for each user. Therefore, we divide the problem into three sub-problems. First, to protect identities, we partition users into different groups according to the maximum anonymity level. Then, to protect sensitive information, several conservative users shift from one group to another according to each user's sensitivity requirement. Finally, to protect the exact locations, segments, which users in cloaking sets are on or near by, are published as the cloaked location. We elaborate each sub-problem as follows.

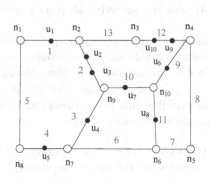

user	(k, l, t_s, p)	q_s
u_1	$(3, 3, 0.25, 0.5)$	0
u_2	$(3, 2, 1, 0.6)$	0.5
u_3	$(3, 4, 0.75, 0.4)$	0.5
u_4	$(2, 5, 0.5, 0.5)$	1
u_5	$(2, 5, 0.5, 0.5)$	0
u_6	$(2, 2, 1, 0.5)$	0.75
u_7	$(3, 4, 0.75, 0.4)$	0.5
u_8	$(3, 2, 1, 0.4)$	0.25
u_9	$(3, 2, 1, 0.4)$	1
u_{10}	$(2, 4, 1, 0.5)$	0.25

(a)   (b)

Fig. 5. 10 users on a road network (a) the road network (b) privacy profiles

4.1   User Identity Protection

As we known, the network distance computing is a costly work. Thus, we adopt users partition strategies on road networks in [10]. Let $moSet$ be the users set and $GK\_MAX = \underset{\forall u \in moSet}{MAX}\ u.k$. The baisc idea is to give each user an order number according to the road segments tranveral order. Thus, the user's location on the road network is mapped to a 1-D value. Based on the user orders, users are partitioned into $\lfloor \frac{|moSet|}{GK\_MAX} \rfloor$ groups. Each group has GK_MAX users except the last one with less than $2*GK\_MAX$ users. Obviously, the anonymity requirements for users in each group are satisfied.

Let's take the users in Fig. 5(a) as a running example. Suppose that the segment $< n_1, n_2 >$ is selected. From $< n_1, n_2 >$, do a depth-first traversal on the road network. The number on the edge is the order of edge visiting. Then, users on the edge are labeled with a number according to the edge visiting order. The subscript of each user is the user order. Fig. 5(b) shows the 10 users privacy profiles. $GK\_MAX=3$. 10 users are partitioned into three groups:$group_1=\{u_1, u_2, u_3\}$ and $group_2=\{u_4, u_5, u_6\}$, and $group_3 = \{u_7, u_8, u_9, u_{10}\}$.

4.2   Sensitive Information Protection

Intuitively, a user with the constrained sensitivity requirement (t_s, p) is hard to find the cloaking set. If these users are not anonymized successfully at the beginning of the algorithm, they are hardly to be anonymized later. Therefore, we build a min-heap for users with $t_s * p$ as the sorting key. Then, pop the top user $topu$ from the heap, and find the group $group_i$ where $topu$ is. Next, for each user u in $group_i$, compare $u.p$ with $\frac{Count\_SQ_u}{|group_i|}$. If there exists an un-safe user nu

in $group_i(\frac{Count\_SQ_{nu}}{|group_i|} > nu.p)$, users issuing non-sensitive queries, in nu's view, are found and inserted into $group_i$. Repeat the inserting step until $group_i$ has no unsafe users.

However, the inserting process would result in re-computing iteratively. Suppose that there exist an unsafe user nu and a safe user u in $group_i$. nu regards a query q as non-sensitivity, but u holds the opposite. If users issuing q are inserted into $group_i$, the safe user u may become unsafe with the number of sensitive queries increasing. In order to avoiding this case, conservative users are inserted into a group. We define conservative users as follows.

Definition 5. *(Conservative   Users)*   For   a   group $group_i$, $group_i.t_s = \underset{u \in group_i}{MIN} \ u.t_s$. *If a user u satisies the following conditions,*

- $u \notin group_i$;
- $u.q_s \le group_i.t_s$.

u *is called as a* conservative user *w.r.t.* $group_i$. In the previous example, $group_1 = \{u_1, u_2, u_3\}$. $group_1.t_s$=0.25. u_5, u_8, u_{10} are conservative users for $group_1$.

Theorem 1: For a group $group_i$, if $\underset{u \in uus}{MAX}(\frac{Count\_SQ_u}{u.p} - |group_i|)$ conservative users are inserted, the personalized homogeneity attack is prevented, where uus is the set of unsafe users whose number of sensitive queries is beyond expected.

Proof. With conservative users inserting, $Count\_SQ_u$ remains unchange for each user u, and $|group_i|$ increases. Therefore, for a safe user $u \in group_i - uus$, $\frac{Count\_SQ_u}{|group_i|} \le u.p$ still holds. For a unsafe user $u \in uus$, suppose x_u is the number of conservative users inserted. We hope that $\frac{Count\_SQ_u}{group_i + x_u} \le u.p$. After inequality derivation, $x_u \ge \frac{Count\_SQ_u}{u.p} - |group_i|$. For $group_i$, if $\underset{u \in uus}{MAX}(\frac{Count\_SQ_u}{u.p} - |group_i|)$ users are inserted, $group_i$ has no un-safe users. Proof done.

If conservative users are redundant, the nearest ones are inserted. Since the network distance computing is costly, we employ the order distance for proximate calculation. Specifically, we define the group order ord_{gp_i} as the average order of users in the group. The nearest conservative user is defined as the conservative user u with the minimum order distance $|ord_u - ord_{gp_i}|$. Still using the example in Fig. 5, u_1 is popped up from the min-heap, and $group_1=\{u_1, u_2, u_3\}$. u_1 is unsafe. The nearest conservative user u_5 is inserted into $group_1$. $\{u_1, u_2, u_3, u_5\}$ is a safe cloaking set. If there is not enough conservative users, in order to gurantee 100% success rates, dummies[4] with $q_s = 0$ and $t_s = 1$ are inserted.

Since some users (*e.g.* u_5) shift from one group (*e.g.* $group_2$) to another group (*e.g.* $group_1$), the user number of a group (*e.g.* $group_2$) decreases. Such that, the anonymity requirement may fail to be satisfied. For convenience, we call this kind of groups as *shrunk-groups*. We merge the shrunk-groups into its neighbor group. Specifically, suppose a shrunk-group is $group_i$. $group_i$ is merged with

---

[4] The locations of dummies are generated as the query history records.

$group_{i-1}$ or $group_{i+1}$ randomly. Repeat this merging process until each user's anonymity requirement is satisfied.

Continuing the previous example, the second top user popped from the min-heap is u_4. The group containing u_4 is $group_2$. Since u_5 shifted to $group_1$, $group_2$ is a shrunk-group. Therefore, merge users in $group_2$ with $group_3$. Finally, users in the merged group constitute a cloaking set.

The algorithm is shown in Algorithm 1. In order to improve the efficiency, we use the bucketization technique. Users who issuing queries with the same sensitivity are in a same bucket. For a top user u, the group number is computed (line 1). Then users in this group are found (line 2). If the group is a shrunk-group, merge it with the neighbor group (line 4 and line 5). Otherwise, check whether each user's sensitivity requirement is satisfied (line 8). If not, compute the number of consecutive users needed as Theorem 1 (line 9). If conservative users are enough, shift the max_nd nearest consecutive users to the $group_i$ (line 14); otherwise, dummies are generated and inserted into $group_i$ (line 16). Users in $group_i$ constitute a cloaking set.

4.3 Cloaked Locations Computing

The third target of P³RN is to generate the cloaked locations. Let CS be a cloaking set. We first extract the segments sg_{on}, where users in CS are. If $|sg_{on}| < ll_{max}$, where $ll_{max} = \underset{u \in CS}{MAX}(u.l)$, two kinds of segments are inserted successively. The first choice is the road segment whose nodes are both in sg_{on}. The second choice is the segment one of whose nodes is in sg_{on}. We continue running the example in Fig. 5. For the cloaking set $\{u_1, u_2, u_3, u_5\}$, segments $\{< n_1, n_2 >, < n_2, n_9 >, < n_7, n_8 >, < n_7, n_9 >, < n_1, n_8 >\}$ are published as the cloaked segments.

Algorithm 1. Algorithm for finding cloaking sets

---

1. $i = \lfloor \frac{order_u}{GK\_MAX} \rfloor$;
2. $group_i$=users with order from $(i-1) * GK\_MAX + 1$ to $i * GK\_MAX$;
3. $CK\_MAX = \underset{u \in group_i}{MAX} u.k$;
4. **if** $|group_i| < CK\_MAX$ **then**
5.   merge $group_i$ with a group from $group_{i-1}$, $group_{i+1}$ randomly;
6. **else**
7.   **for** each user uc in $group_i$ **do**
8.   **if** $\frac{Count\_SQ_{uc}}{|group_i|} > uc.p$ **then**
9. $max\_nd = max\_nd < \frac{Count\_SQ_u}{u.p} - |group_i|?\frac{Count\_SQ_u}{u.p} - |group_i| : max\_nd$;
10.   **if** $max\_nd > 0$ **then**
11. $group_i.t_s = \underset{u \in group_i}{MIN} u.t_s$;
12.   find conservative users con_users from buk;
13.   **if** $|con\_users| \geq |max\_nd|$ **then**
14.   insert max_nd nearest conservative users to $group_i$;
15.   **else**
16.   insert $|max\_nd| - |con\_users|$ dummies into $group_i$;

4.4   Strict Users Anonymization

Let $D(SSet) = \{S_1, S_2, \ldots, S_i, \ldots, S_n\}$ be the sensitvity domain. For any $S_i, S_j \in D(SSet)$ $(1 \le i < j \le n)$, $S_j > S_i$. Now let'e define strict users.

Definition 6. *(Strict Users) For a user u, if $u.p < \frac{|SSet|-i}{|SSet|}$, where i is the position of $u.t_s$ in $D(SSet)$, u is termed as a strict user.*

For example, $D(SSet) = \{0, 0.25, 0.5, 0.75, 1\}$, and a user u sets the sensitivity requirement (t_s, p) as $(0.25, 0.5)$. For any a query q, q has $0.6 (= 3/5)$ probability as a sensitive query to u. However, u only sets the sensivity profile $p = 0.5$. u is a strict user. From Theorem 1, the strict users need great conservative users to be anonymized togerther, as a result, the cloaking set is huge. Huge cloaking sets not only results in high query cost, but also draws the attacker's attention resulting in privacy disclosure.

In order to avoid huge cloaking sets, we bring strict users together for anonymization. Strict users $strU$ are sorted by their user orders ascendingly. Then, $strU$ is divided into $\lfloor \frac{strU}{GK\_MAX} \rfloor$ groups. For each group, if there exists unsafe users, we insert dummies as the conservative users. The algorithm is simple, so we omit the detailed algorithm for space limited.

In summary, P^3RN is shown in Algorithm 2. Line 1 employs [10] to sort mobile users as the segments traversal order. Then, insert users in $moSet$ into a min-heap with $t_s * p$ as the sorting key (line 2). Next, pop the top user u from the min-heap (line 3). If u is a strict user, insert u into the strict users set $strU$ (line 6); otherwise invoke Algorithm 1 to find the cloaking set (line 8). After, invoke the algorithm in Section 4.4 to find cloaking sets for users in $strU$ (line 9). Finally, applying the segment selecting strategy in Section 4.3 to generate cloaked locations (line 10 to line 13). Repeat line 4 to line 14 until the min-heap is empty.

Algorithm 2. P^3RN

---

1. mo_orders=users in $moSet$ are ordered as [10];
2. insert users into a min-heap with $t_s * p$ as the sorting key;
3. pop the top user u;
4. **while** the min-heap is not empty **do**
5.   **if** $u.t_s * u.p < strict\_threshold$ **then**
6.   insert u into $strU$;
7.   **else**
8. cs=invoke Algorithm 1 to find the cloaking set;
9.   invoke the algorithm in section 4.4 to find cloaking sets for $strU$;
10.   find the segments sg_{on} where users in cs are;
11.   **while** $|sg_{on}| < ll_{max}$ **do**
12. sg_{by}=segments one of whose nodes is in sg_{on};
13.   find segments in sg_{by} randomly and insert them into sg_{on};
14.   delete users in cs from the min-heap;

---

5 Experiments

We compare our algorithm P³RN with DF, which is the DFS-based cloaking algorithm [10] incuring the fewest query cost. DF cannot prevent personalized homogeneity attacks, but it is included for comparison to show the cost required for defending against personalized homogeneity attacks. Both cloaking algorithms are implemented in C++ and run on a desktop PC with a dual AMD 2.0GMHz processor and 2GB main memory.

Table 1. Default system settings

parameter	default setting
number of nodes	21,048
number of edges	21,693
number of users	32,399
$strict\_theshold$	0.15
$SSet$	{0,0.25,0.5,0.75,1}
k	Randomly chosen from [2, 10]
l	Randomly chosen from [2, 10]
p	Randomly chosen from [0.2, 1]
t_s	Randomly chosen from {0.25,0.5,0.75,1}

We adapt a real data set, California Road Network and Points of Interest (POI) [5], to validate the effectiveness of our cloaking algorithm. The road map contains 21,048 nodes and 21,693 edges; moreover, it is associated with a real dataset of 32,399 POIs. All the POIs are empoyed to simulate LBS queries. Table 1 lists the default system settings.

We investigate the impact of anonymity level k on the performance of cloaking algorithms by enlarging the anonymity level range. Enlarging the anonymity level range implies not only more constrained but also more diversified privacy requirements. In Fig. 6, we measure the information entropy [13] of anonymous locations, which indicates the protection strength. Larger information entropy implies that the attackers is more uncertain of the user's specific location. With k increasing, more users and segments are included in each cloaking set. Thus, the protection strength for two methods is both improved. However, P³RN provides better protection than DF at all settings.

Recall that we use dummy queries to achieve a 100% success rate. Thus, in Fig. 7, we measure the portion of dummys generated in the total mobile users . We observe that DF doesn't need dummies since the anonymity requiremented is only considered. In order to prevent personalized homogeneity attacks, about 7% dummies are generated by P³RN, which we think is acceptable.

We empoly the query cost model proposed in [2] to evaluate the average query cost, which is measured by the road segment length and the number of boundary nodes in the cloaked locations. From Fig. 8, the query costs for both algorithms increase with k increasing. Though the query cost of P³RN increases worse

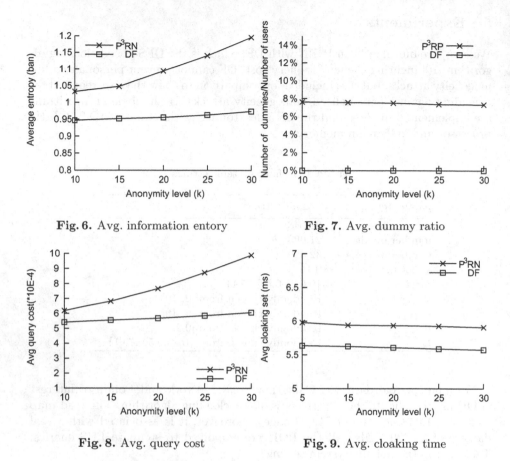

Fig. 6. Avg. information entory   Fig. 7. Avg. dummy ratio

Fig. 8. Avg. query cost   Fig. 9. Avg. cloaking time

than DF, 0.2‰ query cost is sacrificed for preventing personalized homogeneity attacks on average. Fig. 9 shows the effect of k on the average cloaking time. Since the time spending on traversing road segments and users, when users are sorted as the order of segments tranversal, dominates in the overall cloaking time, the time difference between DF and P³RN is slight (about $0.3ms$). The cloaking time of P³RN is only $6ms$ for each user on average.

6   Conclusion

In this paper, we investigated a cloaking algorithm that can protect privacy against personalized homogeneity attacks over road networks. To address this problem, we propose a (K, L, P)-anonymity model to protect the identity, location and sensitive information. Based on personalized privacy requirements, we have proposed a cloaking algorithm called P³RN to generate cloaking sets over road networks. A prelimiary of experiments has been conducted to evaluate the effectiveness and effciency of P³RN. Experimental results demonstrate that

P^{3}RN provides better protection strengh than the existing algorithm. The average cloaking time is only $6ms$. The price paid for defending against personalized homogeneity attacks is small.

Acknowledgement. This research was partially supported by the grant from the Natural Science Foundation of China (No. 61303017), and the Hebei Education Department (No. Q2012131), and Shandong Province Higher Educational Science and Technology Program (No. J12LN05).

References

1. Bamba, B., Liu, L., Pesti, P., Wang, T.: Supporting Anonymous Location Queries in Mobile Environments with PrivacyGrid. In: WWW, pp. 237–246 (2008)
2. Chow, C., Mokbel, M.F., Bao, J., Liu, X.: Query-aware Location Anonymization for Road Networks. Geoinformatical 15(3), 571–607 (2010)
3. Damiani, M.L., Bertino, E., Silvestri, C.: The PROBE Framework for the Personalized Cloaking of Private Locations. Transactions on Data Privacy 3, 123–148 (2010)
4. Hu, H., Xu, J., On, S.T.: Privacy-Aware Location Data Publishing. Transactions on Database Systems (TODS) 35(3) (2010)
5. Li, F.-F., Cheng, D., Hadjieleftheriou, M., Kollios, G., Teng, S.-H.: On Trip Planning Queries in Spatial Databases. In: Medciros, C.B., Egenhofer, M., Bertino, E. (eds.) SSTD 2005. LNCS, vol. 3633, pp. 273–290. Springer, Heidelberg (2005)
6. Liu, F., Hua, K.: Protecting User Privacy Better with Query l-diversity. International Journal of Information Security and Privacy 4(2), 1–18 (2010)
7. Lee, B., Oh, J., Yu, H., Kim, J.: Protecting Location Privacy Using Location Semantics. In: SIGKDD (2011)
8. Li, P., Peng, W.C., Wang, T., Ku, W.: A Cloaking Algorithm based on Spatial Networks for Location Privacy. In: SUTC (2008)
9. Machanavajjhala, A., Gehrke, J., Kifer., D.: l-Diversity: Privacy Beyond k-Anonymity. In: ICDE (2006)
10. Mouratidis, K., Yiu, M.L.: Anonymous Query Processing in Road Networks. TKDE 22(1), 2–15 (2009)
11. Mouratidis, K., Yiu, M.L.: Shortest Path Computation with No Information Leakage. In: 38th International Conference on Very Large Data Bases (2012)
12. Palanisamy, B., Liu, L., et al.: Location Privacy with Road network Mix-zones. In: MSN (2012)
13. Wang, T., Liu, L.: Privacy Aware Mobile Services over Road Networks. In: VLDB (2009)
14. Xiao, Z., Xu, J., Meng, X.: P-sensitivity: a semantic privacy-protection model for location-based services. In: PALMS, pp. 47–54 (2008)
15. Xue, M., Kalnis, P., Pung, H.K.: Enhanced privacy protection in location based services. In: LoCA, pp. 70–87 (2009)
16. Yigitoglu, E., Damiani, M.L., Abul, O., Silvestri, C.: Privacy-preserving sharing of sensitive semantic locations under road-network constraints. In: MDM (2012)

A Novel Privacy-Preserving Group Matching Scheme in Social Networks

Jialin Chi[1,2], Zhiquan Lv[1,2], Min Zhang[1], Hao Li[1], Cheng Hong[1], and Dengguo Feng[1]

[1] Trusted Computing and Information Assurance Laboratory, Institute of Software, Chinese Academy of Sciences, Beijing, China
[2] University of Chinese Academy of Sciences, Beijing, China
{chijialin,lvzhiquan,mzhang,lihao,hongcheng,feng}@tca.iscas.ac.cn

Abstract. The group service allowing users with common attributes to make new connections and share information has been a crucial service in social networks. In order to determine which group is more suitable to join, a stranger outside of the groups needs to collect profile information of group members. When a stranger applies to join one group, each group member also wants to learn more about the stranger to decide whether to agree to the application. In addition, users' profiles may contain private information and they don't want to disclose them to strangers. In this paper, by utilizing private set intersection (PSI) and a semi-trusted third party, we propose a group matching scheme which helps users to make better decisions without revealing personal information. We provide security proof and performance evaluation on our scheme, and show that our system is efficient and practical to be used in mobile social networks.

Keywords: Social Networks, Group Matching, Private Set Intersection.

1 Introduction

Social networks are changing our lifestyle and becoming an inseparable part of our daily lives. For example, Twitter [1] which is a well-known micro-blogging site enables users to share real-time information. The group service has been frequently used in social networks and allowed strangers with similar profiles to construct new relationships and share information. Generally, groups are consisted of users with common attributes, such as educational backgrounds and illness symptoms. In many situations, a group is only described by its classification, several keywords and a short introduction. These features may not be enough for users to decide which group is the most appropriate to join, especially when a few groups have similar keywords and introductions. In order to choose a suitable group to join, a stranger outside of the groups needs to collect profile information about each group member. In addition, attributes of users sometimes contain sensitive and private information, thus they don't want to disclose their profiles or exact matching results to untrusted users or any third party. Such a problem is referred to as *group matching* by Wang *et al.* in [2]. However, there are two main problems in existing works for group matching problem.

F. Li et al. (Eds.): WAIM 2014, LNCS 8485, pp. 336–347, 2014.
© Springer International Publishing Switzerland 2014

The first one is that only the stranger obtains the matching results while each group member learns nothing. In most practical applications, when a stranger who is an outsider of an existing group applies to join it, he just needs to simply send the reasons for application to the manager of this group. Since the reasons submitted by stranger may be incomplete or fake, it is inconvenient for the group manager to determine whether to agree to the application. In addition, sometimes other group members don't fully trust the group manager and they want to make their own decisions. In order to enable all group members to participate in the decision process and make a better decision jointly, each group member needs to learn more information about stranger's profile.

Another problem is that existing systems rely mostly on exponential operations and have high computation cost, so they are not lightweight and practical enough to be used in mobile social networks. The proliferation of networked portable devices such as smart phones and PADs, enables people to use social networking services anytime and anywhere. However, networked portable devices have limited computational abilities, and we have to consider computation cost in mobile social networks.

Our Contribution. In this paper, we focus on the above problems and propose a novel scheme to realize group matching by utilizing private set intersection (PSI) [3] and a semi-trusted third party. Our contributions can be summarized as follows:

- Our scheme helps both stranger and each group member to make better decisions. We take advantage of two kinds of matching information learnt by the stranger and each group member respectively: the intersection set between their attribute sets, and the size of their intersection set. By collecting different kinds of matching information, the stranger can make a better decision when choosing a suitable group to join and each group member can decide whether to agree to the stranger's application.

- We limit the risk of privacy exposure and only necessary information of each user's profile is exchanged. Our system protects each participator's private attributes and exact matching information between two entities. We provide thorough security analysis that our proposed scheme is secure under honest-but-curious (HBC) model and against several certain active attacks.

- We utilize a semi-trusted third party to improve the computation efficiency and our proposed scheme relies mostly on modular multiplication. We provide performance evaluation on our scheme. By comparison with an existing work, we show that ours is much more lightweight and efficient in computation to be used in mobile social networks.

Organization. The remainder of this paper is organized as follows. In Section 2, we discuss the related works. In Section 3, we present the system model and design goals. Section 4 describes the details of our scheme. We give the thorough security proof in Section 5 and analyze the efficiency of our scheme by comparison with an existing work in Section 6. Finally, we briefly conclude this paper in Section 7.

2   Related Work

Existing works related to our proposed scheme are mainly in the area of private set intersection (PSI) first introduced by Freedman*et al.* in [3]. Freedman *et al.* base their protocol on oblivious polynomial evaluation and the protocol is single-output, i.e., during the process, only one party learns the set intersection while the other one doesn't obtain any results. There have been other single-output PSI protocols. Based on oblivious polynomial evaluation, Dachman-Soled *et al.* [4] present an efficient two-party protocol which is robust in the presence of malicious adversaries. In [5], Hazay and Lindell claim a different protocol based on oblivious pseudo random functions and the proposed protocol is improved in complexity by Jarecki and Liu [6]. Cristofaro and Tsudik [7] propose protocols for plain and authorized private set intersection (PSI and APSI) and they base their protocols on blind RSA signatures. In [8], Agrawal *et al.* adopt another approach based on commutative encryption to realize private set intersection, which is extended by Vaidya *et al.* [9] to multiparty setting.

Above single-output protocols only allow one user to obtain the results, while in most situations, both of the two parties are desirable to learn the intersection of their attribute sets. Several mutual PSI protocols have been proposed. Kissner and Song exploit the first mutual PSI protocol in [10]. The proposed protocol builds upon oblivious polynomial evaluation and enables several set operations such as union, intersection, and element reduction operations. Camenisch and Zaverucha [11] have applied certified sets to private set intersection problem and ensured that all inputs are valid and bound to each protocol participant by utilizing a trusted third party. In [12], Kim *et al.* claim a more efficient mutual PSI scheme which is the first system with linear computational complexity in semi-honest model. Recent work in [13], Dong *et al.* present the first fair mutual PSI protocol by utilizing an offline semi-trusted third party arbiter which can resolve disputes blindly without obtaining any sensitive information from users. However, these mutual protocols can't be utilized in group matching problem directly. Users from the same group are familiar, and a group member may exchange the intersection between him and stranger with other group members to learn more about the stranger's private attributes. In addition, above protocols reveal the exact matching information which is undesirable in our work.

Based on private set intersection (PSI), there have been several practical systems designed for special purposes in social networks. The E-SmallTalker scheme [14] exploited by Yang *et al.* adopts iterative bloom filter (IBF) to denote attribute sets and enables a user to match people in physical proximity. Lu *et al.* [15] present a secure symptoms matching protocol by utilizing a trusted authority. The FindU scheme [16] proposed by Li *et al.* allows a user to find the one who best matches with him in mobile social networks. The proposed protocol is based on the FNP scheme [3], but they utilize secret sharing to calculate polynomial evaluation without using additive homomorphic encryption. Recently in

[2], Wang *et al.* introduce Gmatch that allows a user to find the most appropriate group to join without disclosing each user's private information and exact matching results. In the Gmatch system, only the stranger outside of the groups obtains the matching results while each group member learns nothing and the proposed scheme relies mostly on exponentiation operations.

3 Problem Definition

3.1 System Model

Our system is a mobile social network consisting of a stranger S, a group $\mathcal{P}$ and a semi-trusted third party C, and each user processes a networked portable device such as smart phones and PADs (as illustrated in Fig. 1). The stranger S, who launches the matching procedure, is an outsider of group $\mathcal{P}$ and has n attributes in his profile which is denoted as $\mathcal{A}_s = \{a_{s,1}, \ldots, a_{s,n}\}$. The group $\mathcal{P}$ has d group members $P_1, \ldots, P_d$ and P_i has m attributes in his profile which is denoted as $\mathcal{A}_i = \{a_{i,1}, \ldots, a_{i,m}\}$. For simplicity, we assume each group member has the same size of attribute set, i.e., $|\mathcal{A}_i| = m, 1 \leq i \leq d$. All attributes of every user's profile need to be kept private, and they are stored in local portable devices by each user. The semi-trusted third party C is a computation center with high computational ability to help users complete the matching process, but it doesn't access and collect each user's attributes.

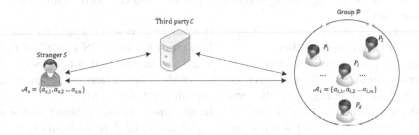

Fig. 1. In order to make better decisions, both stranger and each group member need to learn detail matching information between them

During the matching procedure, stranger S wants to collect the intersection set between him and each group member in order to decide whether group $\mathcal{P}$ is suitable and appropriate to join. When S applies to join $\mathcal{P}$, each group member in $\mathcal{P}$ wishes to learn the size of intersection set between him and stranger S to determine whether to agree to S's application. In this paper, we cite the definition of *matched attribute* and *matching degree* used in [2]. If an attribute in a group member's attribute set is also in stranger S's attribute set, it is called a matched attribute. Otherwise, it is an unmatched attribute. The total number of group members, who has the attribute equal to the attribute $a_{s,j}$ in stranger S's profile, is described as the matching degree D_j of $a_{s,j}$. The matching result learnt by stranger S can be denoted by the matching degree between S and $\mathcal{P}$, which is described as $\mathcal{D}(\mathcal{P}) = \{D_1, \ldots, D_n\}$.

3.2   Adversary Model

In this paper, we only consider attacks from insiders who are participators of the matching process. We assume all participators including stranger S, group members $P_1, \ldots, P_d$ and the third party C are honest-but-curious (HBC). That means all parties will honestly follow the scheme, but may try to obtain more information than allowed. We will prove our protocol's security under HBC model. We also consider several certain active attacks and analyze how the proposed scheme is secure against them. In addition, we assume that users from the same group are familiar and they may exchange information to learn more about stranger's private profile, while stranger S or any group member can't collude with the semi-trusted third party C.

3.3   Design Goals

Security Goals

Definition 1 (Security Goal 1 (SG-1)): When the scheme ends, stranger S only learns the matching degree $\mathcal{D}(\mathcal{P}) = \{D_1, \ldots, D_n\}$ from group $\mathcal{P}$ without knowing any unmatched attribute of group members and the exact matching information, i.e., each result's corresponding group member and whether two results are from the same user.

Definition 2 (Security Goal 2 (SG-2)): If stranger S doesn't apply to join the group $\mathcal{P}$, each group member in $\mathcal{P}$ will learn nothing about S's attributes, including the intersection set between them and the size of it. If S applies to join, each group member will only learn the size of the intersection set between him and S without knowing what the exact matching attributes are.

Definition 3 (Security Goal 3 (SG-3)): In any phase of our scheme, the semi-trusted third party C can't learn more than what can be derived from the values sent to him, his outputs and their corresponding group members.

Usability and Efficiency. For group matching in mobile social networks, it is better to require as few human interactions as possible. In this paper, stranger S only needs to determine which group is the most suitable and whether to join it, while group members in $\mathcal{P}$ need to decide whether to accede to S's application. In addition, networked portable devices have limited computational abilities, and our scheme should be lightweight and efficient enough in computation to be used in mobile social networks.

4   A Novel Privacy-Preserving Group Matching Scheme

In this section, we propose a novel scheme designed for group matching in social networks. The proposed scheme is based on the FNP protocol [3] and we take advantage of a semi-trusted third party C to help compute the polynomial evaluations without using additive homomorphic encryption. Our scheme consists of four phases: *Setup, Computation, Matching* and *Application*. The *Application* phase is only executed when stranger S applies to join the group $\mathcal{P}$. We assume that each party has a public/private key pair for secure communication and the encryption algorithms are denoted as $Enc_c, Enc_s, Enc_1, \ldots, Enc_d$. At first, all attributes in each user's profile are encoded in $\mathbb{Z}_p$. Details of each phase are listed as follows.

Setup. Stranger S first constructs a n-degree polynomial $f(x)$, whose n roots are all in his set of attributes and all his attributes are $f(x)$'s roots:

$$f(x) = (x - a_{s,1})(x - a_{s,2})\ldots(x - a_{s,n}) = \sum_{k=0}^{n} \alpha_k x^k. \tag{1}$$

After generating the polynomial $f(x)$, stranger S generates $\{r_{i,j}\}_{1 \leq j \leq m}$ and $\{\tau_{i,k}\}_{1 \leq k \leq n}$ randomly from $\mathbb{Z}_p$ for each group member $P_i \in \mathcal{P}$. Then he sends the encrypted values $\{Enc_c(\tau_{i,1}r_{i,j}\alpha_1), \ldots, Enc_c(\tau_{i,n}r_{i,j}\alpha_n)\}_{1 \leq j \leq m}$ to the semi-trusted third party C and $\{Enc_i(r_{i,j}\alpha_0)\}_{1 \leq j \leq m}$, $\{Enc_i(\tau_{i,k})\}_{1 \leq k \leq n}$ to group member P_i.

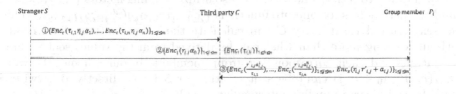

Fig. 2. Stranger S and group member P_i send the parameters used to calculate the matching results to third party C

Upon receiving the values $\{Enc_i(r_{i,j}\alpha_0)\}_{1 \leq j \leq m}$, $\{Enc_i(\tau_{i,k})\}_{1 < k \leq n}$ and decrypting them with the private key, group member P_i generates $\{r'_{i,j}\}_{1 \leq j \leq m}$ randomly from $\mathbb{Z}_p$. Then P_i sends $\{Enc_c(\frac{r'_{i,j}a_{i,j}^1}{\tau_{i,1}}), \ldots, Enc_c(\frac{r'_{i,j}a_{i,j}^n}{\tau_{i,n}})\}_{1 \leq j \leq m}$ and $Enc_s(r_{i,j}r'_{i,j}\alpha_0 + a_{i,j})_{1 \leq j \leq m}$ to the semi-trusted third party C (as illustrated in Fig. 2).

Computation. After decrypting the received values with his own private key, the third party C now learns $\{\tau_{i,1}r_{i,j}\alpha_1, \ldots, \tau_{i,n}r_{i,j}\alpha_n\}$ and $\{\frac{r'_{i,j}a_{i,j}^1}{\tau_{i,1}}, \ldots, \frac{r'_{i,j}a_{i,j}^n}{\tau_{i,n}}\}$ for each attribute $a_{i,j}$ in P_i's profile. C first calculates the intermediate result

$$z_{i,j} = \sum_{k=1}^{n} (\tau_{i,k}r_{i,j}\alpha_k)(\frac{r'_{i,j}a_{i,j}^k}{\tau_{i,k}})$$
$$= r_{i,j}r'_{i,j}f(a_{i,j}) - r_{i,j}r'_{i,j}\alpha_0, \tag{2}$$

and encrypts it with stranger S's public key. Then C packages the encrypted intermediate result $Enc_s(z_{i,j})$ and its corresponding $Enc_s(r_{i,j}r'_{i,j}\alpha_0 + a_{i,j})$, and sends all the packages $\{Enc_s(z_{i,j}), Enc_s(r_{i,j}r'_{i,j}\alpha_0 + a_{i,j})\}_{1 \le j \le m}$ to stranger S in random order.

Matching. Upon receiving the packages from the semi-trusted third party C, stranger S decrypts each $Enc_s(z_{i,j})$ and $Enc_s(r_{ij}r'_{ij}\alpha_0 + a_{ij})$ with his private key and computes

$$F_{i,j} = z_{i,j} + r_{i,j}r'_{i,j}\alpha_0 + a_{i,j}. \tag{3}$$

Because the value $f(a_{i,j})$ is randomized by random numbers $r_{i,j}$ and $r'_{i,j}$ generated by S and P_i respectively in *Setup* phase, stranger S will get an attribute in his profile or a random number from the result $F_{i,j}$. If value $F_{i,j}$ is equal to one attribute $a_{s,k}$ in S's profile, $a_{i,j}$ represents a matched attribute which equals $a_{s,k}$. Otherwise, $a_{i,j}$ is an unmatched attribute. Obviously, if $a_{i,j}$ is a matched attribute, it is a root of polynomial $f(x)$, i.e., $f(a_{i,j}) = 0$. Then $F_{i,j} = r_{i,j}r'_{i,j}f(a_{i,j}) + a_{i,j} = a_{i,j}$.

Since stranger S and group member P_i jointly randomize the value α_0 by generating $r_{i,j}$ and $r'_{i,j}$ respectively, and the results are sent by third party C in random order, S won't learn $F_{i,j}$'s corresponding group member and whether two results are from the same user. We also utilize blinding factors $\{\tau_{i,k}\}_{1 \le k \le n}$ to blind the parameters to compute functions $\{\sum_{k=1}^{n}(r_{i,j}\alpha_k)(r'_{i,j}a_{i,j}^k)\}_{1 \le j \le m}$. Thus the semi-trusted third party C can calculate the correct intermediate results without learning more than what can be derived from the values sent to him, his outputs and their corresponding group members. In our scheme, the value $Enc_s(r_{i,j}r'_{i,j}\alpha_0 + a_{i,j})$ can't be sent to stranger S by P_i directly, otherwise S won't know its corresponding intermediate result.

After computing all results $\{F_{i,j}\}_{1 \le i \le d, 1 \le j \le m}$ and comparing them with his own attributes, S learns each attribute $a_{s,k}$'s matching degree D_k and decides whether to join group $\mathcal{P}$. If stranger S determines to join it, the *Application* phase will be executed. Otherwise, the matching procedure is done.

Application. Stranger S first generates $\{\omega_{i,j}\}_{1 \le j \le m}$ randomly from $\mathbb{Z}_p$ for each group member P_i. Then he calculates $\{r_{i,j}\alpha_0 - \omega_{i,j}\}_{1 \le j \le m}$ and sends $\{Enc_c(r_{i,j}\alpha_0 - \omega_{i,j}), Enc_i(\omega_{i,j})\}_{1 \le j \le m}$ to the semi-trusted third party C. Group member P_i sends $\{Enc_c(r'_{i,j})\}_{1 \le j \le m}$ to C.

Upon receiving these values and decrypting them, third party C computes

$$
\begin{aligned}
z'_{i,j} &= \frac{z_{i,j}}{r'_{i,j}} + r_{i,j}\alpha_0 - \omega_{i,j} \\
&= r_{i,j}f(a_{i,j}) - \omega_{i,j},
\end{aligned}
\tag{4}
$$

and encrypts it with group member P_i's public key. Then C packages the intermediate result $Enc_i(z'_{i,j})$ and its corresponding $Enc_i(\omega_{i,j})$, and sends the packages $\{Enc_i(z'_{i,j}), Enc_i(\omega_{i,j})\}_{1 \le j \le m}$ to each $P_i \in \mathcal{P}$ in random order (as illustrated in Fig. 3).

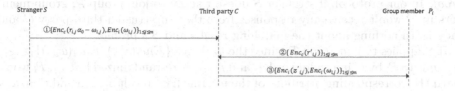

Stranger S Third party C Group member P_i

① $\{Enc_c(r_{i,j}\alpha_0 - \omega_{i,j}), Enc_i(\omega_{i,j})\}_{1 \leq j \leq m}$

② $\{Enc_c(r'_{i,j})\}_{1 \leq j \leq m}$

③ $\{Enc_i(z'_{i,j}), Enc_i(\omega_{i,j})\}_{1 \leq j \leq m}$

Fig. 3. When stranger S applies to join group $\mathcal{P}$, the *Application* phase will be executed

Group member $P(i)$ decrypts the received values with his private key and verifies

$$F'_{i,j} = z'_{i,j} + \omega_{i,j} \overset{?}{=} 0. \qquad (5)$$

Because the values $f(a_{i,j})$ is randomized by random number $r_{i,j}$ generated by S in *Setup* phase, $P(i)$ will get zero or a random number from the result $F'_{i,j}$. If the equation is valid, then $a_{i,j}$ is a matched attribute. Otherwise it's an unmatched attribute. This is because if $a_{i,j}$ is a matched attribute, it is a root of polynomial $f(x)$, i.e., $f(a_{i,j}) = 0$. Since stranger S has sent $r_{i,j}\alpha_0$ to group member $P(i)$ in the first phase, S should utilize $\omega_{i,j}$ to re-randomize it in *Application* phase. Otherwise, $P(i)$ will learn which attribute the intermediate result $z'_{i,j}$ correspondences to.

After calculating all the results $F'_{i,j}$ and comparing them with zero, $P(i)$ will learn the size of the intersection set between him and stranger S. Then $P(i)$ can decide whether to agree to S's application.

5 Security Analysis

5.1 Security Under the HBC Model

Theorem 1. Assuming the semi-trusted third party C sends all the packages $\{Enc_s(z_{i,j}), Enc_s(r_{i,j}r'_{i,j}\alpha_0 + a_{i,j})\}_{1 \leq j \leq m}$ to stranger S in random order and parameters $\{r'_{i,j}\}_{1 \leq j \leq m}$ are random, we can achieve SG-1.
proof: In our scheme, sending all the packages in random order by C will blind from S the correspondence between P_i and the intermediate results $z_{i,j}$. In addition, $f(a_{i,j})$ and α_0 are randomized by $r_{i,j}r'_{i,j}$, so $F_{i,j}$ is an attribute in S's profile or a random number, and S can learn nothing from $r_{i,j}r'_{i,j}\alpha_0$. Thus stranger S just learns whether $F_{i,j}$ represents a matched attribute and what the matching attribute is, but can't learn its corresponding group member, any unmatched attributes or whether two computing results are from the same group member. Note that, to realize SG-1, $\{r_{i,j}\}_{1 \leq j \leq m}$ don't have to be random.

Theorem 2. Assuming the semi-trusted third party C sends all the packages $\{Enc_i(z'_{i,j}), Enc_i(\omega_{i,j})\}_{1 \leq j \leq m}$ to group member P_i in random order and parameters $\{r_{i,j}\}_{1 \leq j \leq m}$ are random, we can achieve SG-2.

proof: In our protocol, if stranger S doesn't apply to join group $\mathcal{P}$, group members in $\mathcal{P}$ won't receive any responses from the semi-trusted third party C and they learn nothing about the matching results and S's profile.

If S applies to join group $\mathcal{P}$, since the packages $\{Enc_i(z'_{i,j}), Enc_i(\omega_{i,j})\}_{1\leq j\leq m}$ are sent to P_i by C in random order, and $r_{i,j}\alpha_0$ is re-randomized by $\omega_{i,j}$, P_i won't learn the corresponding attribute of the intermediate result $z'_{i,j}$. In addition, $\omega_{i,j}$ can't be equal to $r_{i,j}\alpha_0$ directly for P_i knowing the relationship between $r_{i,j}\alpha_0$ and its corresponding attribute. The value $f(a_{i,j})$ is randomized by $r_{i,j}$, so $F'_{i,j}$ is zero or a random number and group member P_i just learns weather $F'_{i,j}$ represents a matched attribute or not. Since $r_{i,j}$ is generated by stranger S for each attribute in group members' profiles, group members can learn nothing more than the matching results by exchanging information with each other. However, if the size of the intersection set between S and P_i equals to the size of P_i's own attribute set, i.e., $|\mathcal{A}_s \bigcap \mathcal{A}_i| = |\mathcal{A}_i|$, P_i will learn that all his attributes are in stranger S's profile. Note that, to realize SG-2, $\{r'_{i,j}\}_{1\leq j\leq m}$ don't have to be random.

Theorem 3. Assuming parameters $r\{_{i,j}\}_{1\leq j\leq m}$, $\{\tau_{i,k}\}_{1\leq k\leq n}$, $\{\omega_{i,j}\}_{1\leq j\leq m}$ are generated randomly, we can achieve SG-3.

proof: In any phase of our protocol, since the inputs received by the semi-trusted third party C are randomized, and some parameters used to calculate matching results are encrypted, C can learn nothing more than what can be derived from the values sent to him, his outputs and their corresponding group members. In the *Application* phase, even though C knows the values $r'_{i,j}$, it doesn't effect the security of our scheme.

5.2   Security Against Active Attacks

If Stranger S sets all coefficients $\{r_{i,j}\alpha_k\}_{0\leq k\leq n}$ of polynomial $r_{i,j}f(x)$ zero, the random numbers $r'_{i,j}$ in function $F_{i,j} = r_{i,j}r'_{i,j}f(a_{i,j}) + a_{i,j}$ won't work and then $F_{i,j}$ is equal to the attribute $a_{i,j}$. This kind of active attacks is referred to as zero polynomial attacks [3]. In our scheme, group member P_i sends $Enc_s(r_{i,j}r'_{i,j}\alpha_0 + a_{i,j})$ to stranger S, so merely setting $r_{i,j}\alpha_0$ zero, S can also realize zero polynomial attacks. To prevent this type of attacks, upon receiving the values $r_{i,j}\alpha_0$ from stranger S, P_i should first test $r_{i,j}\alpha_0 \stackrel{?}{=} 0$.

In order to increase the possibility of joining group $\mathcal{P}$, a malicious stranger S can use a large attribute set or launches the procedure many times to find out as many matched attributes in group members' profiles as possible. The former attack can be prevented by limiting the size of all users' attribute sets, the same approach as in [16]. The second attack can be prevented by auditing the times that stranger S runs the matching scheme to compute the intersection set with the same group in a short time by the semi-trusted third party C.

6   Performance Evaluation

In this section, we evaluate the performance of our scheme and compare it against the Gmatch scheme without batch verification [2]. We test the two schemes on the same hardware and OS, and our experimental environment is a 3.4GHz system with the OpenSSL library. We use RSA protocol to encrypt data to be transmitted and the length of the private/public key is 1024bits. In addition, we assume $|p| = 160$bits.

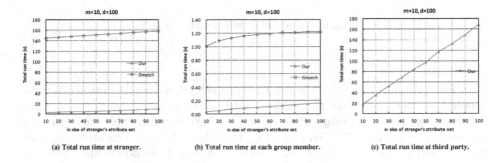

(a) Total run time at stranger.   (b) Total run time at each group member.   (c) Total run time at third party.

Fig. 4. Impact of n on total run time

(a) Total run time at stranger.   (b) Total run time at each group member.   (c) Total run time at third party.

Fig. 5. Impact of m on total run time

In experiments, we change the size of stranger S's profile n, the size of group member P_i's profile m and the number of group members d respectively and measure the total run time at stranger S, group member P_i and the semi-trusted third party C. Since the Gmatch scheme doesn't include a third party, Fig. 4(c), Fig. 5(c) and Fig. 6(c) only show our system's run time on C. As shown in Fig. 4(a), Fig. 5(a) and Fig. 6(a), we can see that, our scheme is more efficient than the Gmatch scheme on S's client. Especially in Fig. 6(a), when only changing the number of group members d, the total run time of S increases linearly with d in our scheme, while the Gmatch scheme increases exponentially. From Fig.

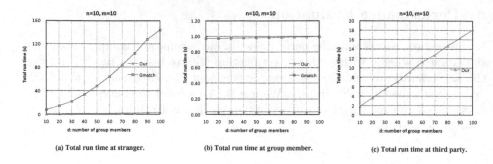

(a) Total run time at stranger.   (b) Total run time at group member.   (c) Total run time at third party.

Fig. 6. Impact of d on total run time

4(b), Fig. 5(b) and Fig. 6(b), the run time at each group member is only linearly affected by n and m, while the Gmatch scheme is affected by n, m and d. This is because we don't use ring signature and additive homomorphic encryption in our scheme. Fig. 4(c), Fig. 5(c) and Fig. 6(c) show that the semi-trusted third party C's run time increase linearly with n, m and d. Although the total run time at C is much larger than that at stranger S and group member P_i, it is acceptable for users.

The performance evaluation of the two schemes show that in all settings our scheme is much more efficient and faster than the Gmatch scheme, and it is practical and lightweight enough in computation to be used on networked portable devices. That is because our scheme take advantage of the semi-trusted third party C to help calculate the polynomial and send the results to stranger S instead of using additive homomorphic encryption and ring signature. Our system relies mostly on modular multiplication while the Gmatch scheme included many exponentiation operations and bilinear operations. Considering the semi-trusted third party utilized in our system is easy to access and can be provided by service providers, the assumption of the existing of a third party is realizable in social networks.

7   Conclusion

In this paper, we propose a novel protocol to realize group matching by utilizing private set intersection (PSI) and a semi-trusted third party. During our scheme, by collecting different kinds of matching information, the stranger outside of the groups can make a better decision when choosing the most suitable group to join, and each group member can decide whether to agree to the stranger's application. We provide the thorough security analysis on our scheme and prove its security under honest-but-curious (HBC) model and against several certain active attacks. By comparison with an existing work, we show our system is practical and efficient in computation to be used in social networks.

Acknowledgement. This work was supported by National Natural Science Foundation of China under Grant 61232005, 61100237, 91118006.

References

1. Twitter, http://twitter.com
2. Wang, B., Li, B., Li, H.: Gmatch: Secure and privacy-preserving group matching in social networks. In: GLOBECOM, pp. 726–731 (2012)
3. Freedman, M.J., Nissim, K., Pinkas, B.: Efficient private matching and set intersection. In: Cachin, C., Camenisch, J.L. (eds.) EUROCRYPT 2004. LNCS, vol. 3027, pp. 1–19. Springer, Heidelberg (2004)
4. Dachman-Soled, D., Malkin, T., Raykova, M., Yung, M.: Efficient robust private set intersection. In: Abdalla, M., Pointcheval, D., Fouque, P.-A., Vergnaud, D. (eds.) ACNS 2009. LNCS, vol. 5536, pp. 125–142. Springer, Heidelberg (2009)
5. Hazay, C., Lindell, Y.: Efficient protocols for set intersection and pattern matching with security against malicious and covert adversaries. In: Canetti, R. (ed.) TCC 2008. LNCS, vol. 4948, pp. 155–175. Springer, Heidelberg (2008)
6. Jarecki, S., Liu, X.: Efficient oblivious pseudorandom function with applications to adaptive ot and secure computation of set intersection. In: Reingold, O. (ed.) TCC 2009. LNCS, vol. 5444, pp. 577–594. Springer, Heidelberg (2009)
7. De Cristofaro, E., Tsudik, G.: Practical private set intersection protocols with linear complexity. In: Sion, R. (ed.) FC 2010. LNCS, vol. 6052, pp. 143–159. Springer, Heidelberg (2010)
8. Agrawal, R., Evfimievski, A., Srikant, R.: Information sharing across private databases. In: SIGMOD, pp. 86–97 (2003)
9. Vaidya, J., Clifton, C.: Secure set intersection cardinality with application to association rule mining. Journal of Computer Security 13(4), 593–622 (2005)
10. Kissner, L., Song, D.: Privacy-preserving set operations. In: Shoup, V. (ed.) CRYPTO 2005. LNCS, vol. 3621, pp. 241–257. Springer, Heidelberg (2005)
11. Camenisch, J., Zaverucha, G.M.: Private intersection of certified sets. In: Dingledine, R., Golle, P. (eds.) FC 2009. LNCS, vol. 5628, pp. 108–127. Springer, Heidelberg (2009)
12. Kim, M., Lee, H.T., Cheon, J.H.: Mutual private set intersection with linear complexity. In: Jung, S., Yung, M. (eds.) WISA 2011. LNCS, vol. 7115, pp. 219–231. Springer, Heidelberg (2012)
13. Dong, C., Chen, L., Camenisch, J., Russello, G.: Fair Private Set Intersection with a Semi-trusted Arbiter. IACR Cryptology ePrint Archive, p. 252 (2012)
14. Yang, Z., Zhang, B., Dai, J., Champion, A.C., Xuan, D., Li, D.: E-smalltalker: A distributed mobile system for social networking in physical proximity. In: IEEE ICDCS, pp. 468–477 (2010)
15. Lu, R., Lin, X., Liang, X., Shen, X.: A secure handshake scheme with symptoms-matching for mhealthcare social network. Mobile Networks and Applications 16(6), 683–694 (2011)
16. Li, M., Cao, N., Yu, S., Lou, W.: Findu: Privacy-preserving personal profile matching in mobile social networks. In: IEEE INFOCOM, pp. 2435–2443 (2011)

Highly Efficient Indexing for Privacy-Preserving Multi-keyword Query over Encrypted Cloud Data

Fangquan Cheng, Qian Wang, Qianwen Zhang, and Zhiyong Peng*

Computer School, Wuhan University, Wuhan 430072, China
{cheng,qianwang,qwzhang,peng}@whu.edu.cn

Abstract. This work presents a SSE scheme that provides a practical trade-off between performance and privacy. Our scheme supports highly efficient one-round multi-keyword query over symmetrically encrypted large-scale data. Meanwhile, our scheme provides semantic security against chosen-keyword attacks (IND2-CKA) and mitigates the leakage of search pattern. We present the security analysis and the query accuracy analysis of our scheme. Experimental evaluations conducted on large representative real-word data set show that compared to the state-of-the-art, the proposed solution indeed introduces high query performance and accuracy, and low overhead of index updates.

1 Introduction

Increasingly, outsourcing (potentially a huge amount of) data storage to remote servers (e.g., cloud) becomes a major industry trend and a promising solution for Database-as-a-Service (DaaS) [1]. Due to outsourcing, data owner can get rid of expensive local data storage and management costs and focus on their core business. As a consequence of data outsourcing, outsourced sensitive information such as personal profile, health records and financial data etc., will be put at risk [2]. To protect data privacy, a straightforward and effective solution is to encrypt the sensitive data locally before outsourcing to the cloud [1,2]. However, data encryption becomes a hindrance to data utilization, *i.e.*, it is difficult to retrieve data files based on their content as done in the plaintext search. The above problem motivates researchers to design advanced searchable encryption schemes that enable searching over encrypted data while keeping privacy guarantee of data contents and queries.

To allow data users to effectively retrieve encrypted data of interest from remote server (i.e., database), two distinct models have been proposed and investigated for privacy-preserving search in recent years: symmetric-key searchable encryption (SSE) [3–12] and public-key searchable encryption (PEKS) [13,14]. From the perspective of efficiency, while PEKS has its unique application scenarios to which SSE cannot apply, they are usually more computational expensive.

---

* Corresponding author.

F. Li et al. (Eds.): WAIM 2014, LNCS 8485, pp. 348–359, 2014.

In this paper, we focus on SSE based solutions where the best possible search efficiency can be obtained. The first notable work along this direction was done by Song et al. [3], where the search time is linear in the length of the file collection. By associating an index to each file, constructions proposed by Goh [4] and Cao et al. [6] can achieve search time that is proportional to the number of files. Cao et al. [6] proposed a multi-keyword ranked search scheme using the secure kNN computation technique in [8]. However, their approach is limited by the low search efficiency. For the first time, Liesdonk et al. [9] proposed an SSE construction supporting incremental data file update and practical search ability. However, efficient deletion update of data files can not be supported and the size of the encrypted index is relatively large. Recently, Kamara et al. [12] extended the inverted index approach [5] and designed a two-level dynamic encrypted index which allows for sublinear query performance and both addition and removal of data files. However, they only support single-keyword search and reveal a non-trivial amount of information in the single data file adding operation. Cash et al. [7] proposed a construction for boolean query over encrypted data, but their construction requires multiple rounds of interactions for each query and puts too much computational burden on clients. Moreover, one common limitation of the state-of-the-art SSE solutions is that the search pattern during query execution is still at risk without any protection.

In summary, none of existing SSE solutions can satisfy all desirable properties in terms of one-round search, efficient search time, multi-keyword query, practical privacy protection and efficient data dynamic operations. From a practical point of view, we propose to design a more practical scheme that meets all the above-mentioned properties. More specifically, we make the following contributions.

- We explore a probabilistic encoding/decoding (PED) method based on multi-PRF (psuedo-random function) mapping, and design a compact and highly efficient indexing scheme for the exact conjunction and ranked disjunction queries over encrypted cloud data. Compared to the inverted index structure for plaintext search, the scheme provides a practical trade-off between performance and privacy without introducing additional costs in terms of query, storage and communication.
- We propose a non-trivial index-faking scheme to achieve higher level of security (semantic security against adaptive chosen-keyword attack, IND2-CKA) without sacrificing the query efficiency and with negligible query accuracy loss. The risk of search pattern can be mitigated by leveraging the indistinguishability property of the PRF. An appealing property of our indexing scheme is that it can be naturally extended to support efficient data dynamic operations, including addition and removal of data files.
- We present a formal security analysis and a thorough query accuracy analysis. Experimental evaluations conducted on large representative real-word data set show that compared to the state-of-the-art the proposed scheme indeed introduces high query performance and accuracy, low communication overhead, practically strong privacy guarantee and efficient index updates.

2   Preliminaries and Definitions

Problem statement and notations. For ease of exposition, in the following statement when we mention a *client*, we refer it as the data owner or an authorized user. In a secure cloud-based data service system, the data owner outsources a large-scale collection of data files $\mathbf{d} = (d_1, \ldots, d_{\#\mathbf{d}})$ to the cloud service provider in the encrypted form $\mathbf{c} = (c_1, \ldots, c_{\#\mathbf{c}})$. To enable the query service over $\mathbf{c}$ for the support of effective data utilization, the data owner needs to build an encrypted index γ and outsources γ and $\mathbf{c}$ to cloud. Later, the client can generate the trapdoor τ, generated from a multi-keyword query request $\mathbf{q} = \{w_1, \ldots, w_{\#\mathbf{q}}\}$, to the remote cloud server. After receiving τ, the cloud executes the query on γ, $\mathbf{c}$ and returns the all needed encrypted data files $\mathbf{c_q}$ containing all query keywords for the AND logic query, or returns the (possible top-k) ranked encrypted data files $\mathbf{c_q}$ for the OR logic query. Finally, the client can locally decrypt the encrypted results and obtain the data content in plaintext form. Formally, such a scheme can be defined as below.

Definition 1. *An indexing scheme for encrypted data search consists of four polynomial-time algorithms (KeyGen, BuildIndex, Trapdoor, Query) such that*

$sk \leftarrow$ KeyGen(1^k): *is a probabilistic algorithm run by the data owner. It takes as input a security parameter k and outputs the secret key sk.*

$\gamma \leftarrow$ BuildIndex($\mathbf{d}, sk$): *is a (possibly probabilistic) algorithm run by the data owner. It takes as input the secret key sk and a data file collection $\mathbf{d}$, and outputs an encrypted index γ.*

$\tau \leftarrow$ Trapdoor($\mathbf{q}, sk$): *is a (possibly probabilistic) algorithm run by the client. It takes as input the secret key sk and a query request $\mathbf{q}$, and outputs the query trapdoor τ.*

$\mathbf{c_q} \leftarrow$ Query($\tau, \gamma, \mathbf{c}$): *is a deterministic algorithm run by the cloud. It takes as input a trapdoor τ, the encrypted index γ and the encrypted data file set $\mathbf{c}$, and outputs a sequence of encrypted data files $\mathbf{c_q}$.*

Definition 2. In addition, Let $\mathrm{id}(d_j)$ denote the identifier of the data file d_j. The keyword universe which contains all distinct keywords extracted from $\mathbf{d}$ is denoted as $\mathbf{w} = (w_1, \ldots, w_{\#\mathbf{w}})$. $\mathrm{div} \in \{0, 1\}^{\#\mathbf{d}}$ is denoted as a data identifier vector (DIV), and its jth entry is 1 if d_j is included; 0 otherwise. Given a vector div, we refer its jth element as $\mathrm{div}[j]$. Given w_i, we use $\mathrm{div}_i \in \{0, 1\}^{\#\mathbf{d}}$ or $\mathrm{div}_{w_i} \in \{0, 1\}^{\#\mathbf{d}}$ to denote the data identifier vector that includes all data files containing w_i. For a vector div, if there exists a keyword $w_i \in \mathbf{w}$ and its data identifier vector div_i such that $\mathrm{div} = \mathrm{div}_i$, we call div a *real data identifier vector*. Additionally, we use $\phi(\mathrm{div})$ to denote the number of "1"s in div. Standard bitwise Boolean operations are defined on binary vectors: bitwise OR (union) "$\bigvee$" and bitwise AND (intersaction) "$\bigwedge$", which take the same notations for single bit OR and AND operations respectively.

Security model. We follow the widely-accepted security model of SSE in [4,5,9, 12]. We will focus on the practical value of SSE schemes in real-world cloud-based data service systems, while providing semantic security against IND2-CKA and further mitigating the search pattern leakage in multi-keyword queries.

Definition 3. *(Search Pattern Privacy π). Given s query requests $Q = \{q_1, \ldots, q_s\}$, the search pattern privacy is defined as a symmetric binary $s \times s$ matrix π, where the element $\pi[i,j] = 1$ if $q_i = q_j$ and $\pi[i,j] = 0$ otherwise.*

Definition 4. *(Semantic security against IND2-CKA [4]) Suppose the challenger $\mathcal{C}$ gives the adversary $\mathcal{A}$ two files d_0 and d_1 (potentially with unequal size) together with an index γ_b ($b \xleftarrow{\$} \{0,1\}$). If $\mathcal{A}$ cannot determine which file is encoded in the index (i.e., output b' as its guess for b) with probability non-negligibly different from $\frac{1}{2}$, we say the index is semantically secure, or say the index is indistinguishable.*

3 Highly Efficient PED-Based Indexing Scheme

From a practical perspective, we first present an indexing scheme for highly efficient query over encrypted cloud data. We design a probabilistic encoding/decoding (PED) method over a compact indexing structure, which encodes the inverted index structure of all data files. Compared to the inverted index structure of plaintext, an appealing property of the indexing scheme is that a certain level of index privacy is guaranteed without introducing additional costs in terms of query, storage and communication. We state our indexing scheme under the framework in Definition 1.

KeyGen(1^k): Given a security parameter k, generate $sk \leftarrow$ SKE.KeyGen(1^k), where SKE is a PCPA-secure symmetric encryption scheme.

BuildIndex($sk, \mathbf{d}$): Extract the global distinct keyword collection $\mathbf{w} = \{w_1, \ldots, w_{\#\mathbf{w}}\}$ and generate the corresponding data identifier vectors $\{\mathsf{div}_1, \ldots, \mathsf{div}_{\#\mathbf{w}}\}$. We construct the index γ, a data structure consisting of m buckets. Each bucket of γ is initially set to be empty. There is a collection of K independent pseudo-random functions (PRFs) $F_{i=[1,K]} = \{F_i | i \in [1,K]\}$, where $F_i : \{0,1\}^k \times \{0,1\}^* \rightarrow \{0,1\}^k$. The PRFs are polynomial-time computable functions that cannot be distinguished from random functions by any polynomial-time adversary.

For each unique keyword w_i for $i \in [1, \#\mathbf{w}]$, the client computes the K bucket positions $\{y_1 = F_1(w_i, sk), \ldots, y_K = F_K(w_i, sk)\}$. For each bucket at position y_j for $j \in [1, K]$, insert div_i into it according to the following rule: if the bucket at position y_j is empty, store div_i in the bucket, which is denoted as $\gamma[y_j] = \mathsf{div}_i$; otherwise update the bucket by storing the "union" of div_i and the "old" DIV previously stored in the bucket, which is denoted as $\gamma[y_j] = \gamma[y_j] \bigvee \mathsf{div}_i$. Finally, the client outsources γ and $\mathbf{c}$ (encryptions of $\mathbf{d}$) to the remote cloud server.

To search for data files according to the given query keywords $\mathbf{q} = \{w_1, \ldots, w_{\#\mathbf{q}}\}$, it suffices the client to generate an encrypted search trapdoor. The server can then use the trapdoor to locate the buckets in the index γ and decode the identifiers of all needed data files according to the user-specified query logic.

Trapdoor($sk, \mathbf{q}$): For each keyword w_i for $i \in [1, \#\mathbf{q}]$, the client computes the K bucket positions $\{y_1 = F_1(w_i, sk), \ldots, y_K = F_K(w_i, sk)\}$ and adds all positions into the trapdoor τ. Then the client submits τ which has no redundant position items and the query logic information (AND or OR) to cloud server.

Query($\gamma, \mathbf{c}, \tau$): After receiving the trapdoor τ, the server executes the query over γ according to the user-specified query logic:

1)Conjunction logic query. The server computes the data identifier vector div_τ from the buckets at the all positions in τ, which is denoted as follows: $\mathsf{div}_\tau = \bigwedge_{y \in \tau} \gamma[y]$. Then the server recovers the data identifiers from div_τ and returns the corresponding encrypted data files to the client. To state the correctness, we let τ_w be the sub-trapdoor for a single keyword w in $\mathbf{q}$. We have $\tau_w \subset \tau$ and

$$\mathsf{div}_\tau = \bigwedge_{w \in \mathbf{q}} \bigwedge_{y \in \tau_w} \gamma[y] = \bigwedge_{w \in \mathbf{q}, y \in \tau_w} \gamma[y] = \bigwedge_{y \in \tau} \gamma[y] \qquad (1)$$

That is, the result of the multi-keyword conjunction query is equivalent to that of the post-processing of single keyword query results.

2)Disjunction logic query. The server computes the similarity score, denoted as $score(d_j, \mathbf{q})$, between the data file d_j and the query $\mathbf{q}$.

$$score(d_j, \mathbf{q}) = \sum_{y \in \tau} \gamma[y][j]. \qquad (2)$$

In our scheme, the keyword which is contained by both the data d_j and the query $\mathbf{q}$ will be mapped to the same buckets in index construction and query. Therefore, $score(d_j, \mathbf{q})$ is simply the number of common buckets between the buckets mapped by the keywords of d_j and $\mathbf{q}$. If a data file d_j contains at least one query keyword, there must be $score(d_j, \mathbf{q}) \geq K$. Due to this property, then the server can send back all ranked encrypted data items with $score(d_j, \mathbf{q}) \geq K$, or the ranked encrypted data items with the top-k highest scores. The time complexity of the trapdoor generation at the client side is $O(K\#\mathbf{q})$ and that of the query at the cloud side is no more than $O(K\#\mathbf{q})$ (as bucket positions of some query keywords may overlap).

Query accuracy analysis of the index scheme. We next show that the probability of false decoding can be reduced to almost a negligible level by properly choosing system parameters m, K. We define exp as the expectation of the number of the falsely returned data files for a given query $\mathbf{q}$. Thus, the false positive rate of a query $\mathbf{q}$ can be denoted as

$$\neg p_\mathbf{q} = \begin{cases} \frac{exp}{\phi(\bigwedge_{w_i \in \mathbf{q}} \mathsf{div}_i) + exp}, & \text{for the conjunction logic} \\ \frac{exp}{\phi(\bigvee_{w_i \in \mathbf{q}} \mathsf{div}_i) + exp}, & \text{for the disjunction logic} \end{cases} \qquad (3)$$

We next state the computation of exp and first discuss the tricky case where $\mathbf{q} \subset \mathbf{w}$. *For each data $d_j \in \mathbf{d}$, we use p_j to denote the probability that d_j is falsely returned.* For ease of presentation, we start from a single keyword w_i ($w_i \in \mathbf{q}$ but $w_i \notin d_j$). Let $\hat{p}_j$ denote the probability of that w_i is falsely considered as a membership of d_j. Let n_j be the number of distinct keywords in d_j, we have

$$\hat{p}_j = (1 - (1 - \frac{1}{m})^{Kn_j})^K \approx (1 - e^{-Kn_j/m})^K. \tag{4}$$

Now we discuss p_j as follows. 1) The conjunction logic query. For $\forall d_j \in \mathbf{d}$, d_j will not be falsely returned if and only if there is always a keyword in $\mathbf{q}$ which is not contained and hit by d_j. Thus, we have $p_j = 1 - (1 - \hat{p}_j) = \hat{p}_j$. 2) The conjunction logic query. Similarly, for $\forall d_j \in \mathbf{d}$, d_j will not be falsely returned if and only if none of the keywords in $\mathbf{q}$ is hit by d_j. Thus, we have $p_j = 1 - (1 - \hat{p}_j)^{\#\mathbf{q}}$. Then we can compute the expectation $exp = \sum_{d_j \in \hat{\mathbf{d}}} p_j$ where $\hat{\mathbf{d}}$ denotes the collection of data files which might be falsely returned. When considering the case that $\mathbf{q} \not\subseteq \mathbf{w}$, we compute exp over $\mathbf{d}$, i.e., $\hat{\mathbf{d}} = \mathbf{d}$.

Security analysis of the index scheme. From the perspective of the adversary (or say the cloud server), he may try to launch guessing attacks to compromise a small number of real data identifier vectors by observing the index. Consider the following decoding approach in Query: Given the index γ and the single keyword trapdoor τ, the server can find the position y such that $\phi(\gamma[y])$ is a minimum. Then the server can regard $\mathtt{div} = \gamma[y]$ as a real data identifier vector. Fortunately, such attacks would only lead to the limited privacy leakage.

Claim. Given the index γ, the computational complexity that an adversary can distinguish ϵ real data identifier vectors from γ with non-negligible probability is approximately $\mathcal{O}(2^\epsilon)$.

The proof of the claim is straightforward. Given all vectors $\{v_1, \ldots, v_\#\}$ which are extracted from the buckets of γ and ranked in ascending order of the number of '1's. The adversary try to distinguish if each vector is real data identifier vector in ascending order. When guessing the ϵth vector v_ϵ, the adversary has to consider if there exists a union of two or more vectors of $\{v_1, \ldots, v_\epsilon\}$ which is equal to v_ϵ. Such problem could be deduced to the well-known subset sum problem. Its possible computational space is $\sum_{i=2}^{\epsilon} \binom{i}{\epsilon} = 2^\epsilon$. That is, the computational complexity is $\mathcal{O}(2^\epsilon)$, thus proving the claim. In fact, although the adversary performs $\mathcal{O}(2^\epsilon)$ guessing computation against v_ϵ, the probability of successful attack might be negligible if the false positive decoding is considered. Therefore, *we define ϵ as the maximum leakage threshold of real data identifier vector.*

4   Index-Faking Scheme with IND2-CKA Security

In order to make the indexing scheme semantically secure against adaptive chosen keyword attack (IND2-CKA), we next state a index-faking protocol of the PED-based indexing scheme. Before stating the details , we define the sub-index γ_{d_j} associated the data file d_j as follows: for $\forall y \in [1, m]$, if $\gamma[y] \neq null \wedge \gamma[y][j] = 1$, then $\gamma_{d_j}[y] = 1$. Otherwise, $\gamma_{d_j}[y] = 0$. Our index-faking protocol needs to achieve the following two goals: 1) The vectors in the buckets are indistinguishable, i.e., it is impossible to determine whether the vector stored in a bucket is a real data identifier vector or not, and 2) The sub-indices of data files are indistinguishable, *i.e.*, it is impossible to determine whether the sub-index of an encrypted data file encodes more distinct keywords than the others or not.

The key idea is to fake γ in BuildIndex before outsourcing it to the cloud server. To achieve the first goal, the client will find all non-empty buckets of γ at positions $pos = \{y_1, \ldots, y_{\hat{m}}\}$, which are sorted such that $\forall i < j$, $\phi(\gamma[y_i]) \leq \phi(\gamma[y_j])$. Then, the client fakes the buckets at these positions to ensure none of $\{\gamma[y_1], \ldots, \gamma[y_\epsilon]\}$ is real data identifier vector, where ϵ is the maximum leakage threshold defined above section. This is to ensure it is computationally difficult for an adversary to launch a successful guessing attack.

To achieve the second goal, the client will make different sub-indices of different data files encoded by the same number of keywords. Specifically, for all data files, the client can compute a globe bound ω for the maximum number of distinct keywords in a data file. Then, for each data d_j, the client selects $(\omega - \#d_j)K$ buckets uniformly at random to fake the identifier of d_j in these buckets (possibly with replacement). To ensure more strict index update security, the globe bound ω needs to be encrypted using a symmetric encryption algorithm under the secret key sk, and the encrypted globe bound $\mathsf{SKE.En}(\omega, sk)$ will be stored in the cloud together with the encrypted index γ.

While faking the index, we should also guarantee the query accuracy. That is, we should ensure the false rate of queries after faking is equal or less than an acceptable level (pre-defined by the clients), denoted as λ ($\lambda \in [0, 1)$). To do so, for any bucket $\gamma[y]$, the faking operation should meet the following rule.

Definition 5. *(The faking rule). Let $\boldsymbol{div}$ be the real inverted data vector in the original bucket $\gamma[y]$ at the position y and $\widetilde{\boldsymbol{div}}$ be the one in the faked bucket $\gamma[y]$. Denote $\hat{\boldsymbol{w}}$ as the set of keywords encoded into $\gamma[y]$. For $\forall w \in \hat{\boldsymbol{w}}$, we have*

$$\frac{\phi(v_2) - \phi(v_1)}{\phi(v_2)} \leq \lambda, \ where \ \begin{cases} v_1 = \bigwedge_{i \in [1,K], F_i(w,sk) \neq y} \gamma[F_i(w, sk)] \bigwedge \boldsymbol{div} \\ v_2 = \bigwedge_{i \in [1,K], F_i(w,sk) \neq y} \gamma[F_i(w, sk)] \bigwedge \widetilde{\boldsymbol{div}}. \end{cases} \quad (5)$$

Here, λ can be considered as a system parameter, which is adjusted to meet users' requirements. Algorithm 1 shows the index-faking protocol over γ in detail.

Theorem 1. *The index-faking based indexing scheme is an $(t, \epsilon, g/2)$-IND2-CKA secure, i.e, the advantage of an adversary $\mathcal{A}$ wins the index guessing game is $Adv_{\mathcal{A}} = |\Pr[b = b'] - 1/2| < \varepsilon$ after $\mathcal{A}$ takes at most t time and makes $g/2$ trapdoor queries to the challenger $\mathcal{C}$.*

Proof. Due to the space limitation, we only provide a sketch of the proof here. We follow the existing proof strategy in [4]. Before discussing the security, we introduce an auxiliary concept. Given K positions $pos = \{y_1, \ldots, y_K\} \subset [1, m]$, we call pos a valid position set, which is denoted as $\psi(pos) = true$, if there exists a keyword w ($w \in \mathbf{w}$) such that $pos = \{F_1(w, sk), \ldots, F_K(w, sk)\}$.

We extract the sub-index γ_{d_j} for each data d_j. We show that if the problem of distinguishing between the index for files d_0 and d_1 is hard, then deducing at least one keyword that d_0 and d_1 do not have in common must also be hard.

Algorithm 1. The Index-Faking Protocol

Input: The original index γ
Output: The faked index γ

1 Let $\nu = \{\nu_1, ..., \nu_{\#d}\} = \{\omega - \#d_1, ..., \omega - \#d_{\#d}\}$;
2 **while** *there exists* $\nu_j \in \nu$ *such that* $\nu_j > 0$ **do**
3   Sort all non-empty buckets of γ as $\{\gamma[y_1], ..., \gamma[y_{\hat{m}}]\}$, where for $\forall i \le j$, $\phi(\gamma[y_i]) \le \phi(\gamma[y_j])$;
4   Initialize a temporary bucket array $\gamma_{tp} = NULL$ and a counter $cnt = 0$;
5   **for** *each* $y_i \in [1, \hat{m}]$ **do**
6   **for** *each* $\gamma[y_k] \in \gamma_{tp}$ *where* $\gamma_{tp} \ne NULL$ **do**
7   **if** $\phi(\gamma[y_k]) \le \phi(\gamma[y_{i+1}])$ **then**
8                   $cnt = cnt + 1$;
9   Delete $\gamma[y_k]$ from γ_{tp} ;
10   **if** $cnt = \epsilon$ **then**
11   Break;
12   **if** $\gamma[y_i]$ *is not real data identifier vector* **then**
13              $cnt = cnt + 1$;
14   **else**
15   Generate a random security value $\kappa \in [1, \sharp d - \phi(\gamma[y_i])]$;
16   Select randomly at most κ data in which $\nu_j > 0$ for each selected data d_j;
17   **for** *each selected data* d_j **do**
18   Fake d_j into $\gamma[y_i]$ ($\gamma[y_i][j] = 1$) while meeting the *Faking rule*;
19   **if** *the data* d_j *is faked* **then**
20                      $\nu_j = \nu_j - 1$;
21   **if** $\phi(\gamma[y_i]) \le \phi(\gamma[y_{i+1}])$ **then**
22                  $cnt = cnt + 1$;
23   **else**
24   Add $\gamma[y_i]$ to γ_{tp};
25   **for** *each data* d_j *(in which* $\nu_j > 0$*)* **do**
26   Select $\frac{\nu_j}{2}$ buckets to be used to fake the data d_j;
27   **for** *each selected bucket* $\gamma[y]$ **do**
28   Fake the data d_j into $\gamma[y]$ ($\gamma[y][j] = 1$) while meeting the *Faking rule* ;
29   **if** *the data* d_j *is faked* **then**
30                  $\nu_j = \nu_j - 1$;

To achieve a high level of security, in our construction, we randomize the bucket locations using PRFs and fake randomly-selected empty buckets with randomly-selected vectors. Here, the index-faking operation is to ensure the sub-indices of γ_0 and γ_1 are encoded by the same number of keywords, whose mapping positions are randomized. Hence, from the γ_b ($b \in \{0,1\}$), it is impossible to distinguish d_0 and d_1 by observing the number of similar keywords in data files themselves. We prove the theorem by contradiction. The proof sketch is as follows: Suppose $\mathcal{A}$ can break the index, we build a simulator $\mathcal{B}$ that uses $\mathcal{A}$ to distinguish a PRF F and a random function. By following this assumption and the index guessing game defined in Definition 4, $\mathcal{B}$ simulates perfectly the challenger $\mathcal{C}$ and then $|\Pr[B^{F_{k_1}(\cdot)} = 0|k_1 \leftarrow \{0,1\}^k] - 1/2| \geq \epsilon$ (I). Assume $\#\mathbf{w} = g/2$, after making at most $g/2$ queries, $\mathcal{A}$ decides d_0 and d_1 such that $|d_0 \triangle d_1| = (d_0 - d_1) \cup (d_1 - d_0) \neq 0$. Based on the properties of random functions, we can show that it is impossible for $\mathcal{A}$ to correlate trapdoors across the index. Under the assumption that $\mathcal{A}$ is not allowed to query $\mathcal{C}$ for the trapdoors of keyword in $d_0 \triangle d_1$ after issuing the challenge, we can conclude that $\mathcal{A}$ learns nothing about $d_0 \triangle d_1$ from other (encrypted) data files and their indexes. Thus, by only considering the challenge set, we can further show that if F' is a random function, $|\Pr[B^{F'} = 0|F' \leftarrow \{F : \{0,1\}^* \rightarrow \{k\}\}| = 1/2$ (II). Based on (I) and (II), we have $|\Pr[B^{F_{k_1}(\cdot)} = 0|k_1 \leftarrow \{0,1\}^k] - \Pr[B^{F'} = 0|F' \leftarrow \{F : \{0,1\}^* \rightarrow \{k\}\}| \geq \epsilon$ (i.e, PRFs are distinguishable from random functions), which contradicts the definition of PRF. Similarly, by virtue of the properties of PRFs, given K positions $pos = \{y_1, \ldots, y_K\}$ of the non-empty buckets, it is hard to distinguish whether or not $\psi(pos) = true$. Thus proving the theorem.

Discussion of search pattern privacy mitigation. Let τ_1, τ_2 be the trapdoors of two queries $\mathbf{q}_1, \mathbf{q}_2$. As defined in Definition 3, mitigating the search pattern leakage means that we should make $\mathbf{q}_1$ indistinguishable from $\mathbf{q}_2$ even though $\tau_1 = \tau_2$. We next discuss the indistinguishability issue by analyzing the probability, denoted as p_{sp}, of the case $\tau_1 = \tau_2$ but $\mathbf{q}_1 \neq \mathbf{q}_2$. We use $\hat{\mathbf{q}}$ to denote the supper set of all queries. For ease of analysis, we consider the following equivalent case: Given $\mathbf{q}_1 \subset \hat{\mathbf{q}}$ and τ_1, the probability of that there exists $\mathbf{q}_2 \subset \hat{\mathbf{q}}$ ($\mathbf{q}_2 \neq \mathbf{q}_1$) such that $\tau_1 = \tau_2$.

For $\forall w \in \hat{\mathbf{q}} - \mathbf{q}_1$ and its trapdoor τ_w, there must exist one position $y \in \tau_1$ with two possible cases: 1)$\tau_w \subset \tau_1/\{y\}$; 2)$y \in \tau_w$ and there is at least one position of τ_w that locates in the position set $\{1, \ldots, m\}/\tau_1$. Therefore, we can compute $p_{sp} = 1 - [(\frac{\#\tau_1 - 1}{m})^K + \frac{K}{m}(1 - (\frac{\#\tau_1}{m})^{K-1})]^{\#\mathbf{q} - \#\mathbf{q}_1}$, where the notation $\#$ presents the number of membership items of the object. Undoubtedly, $K, \#\tau_1, \#\mathbf{q}_1 \ll m, \#\hat{\mathbf{q}}$ and $\hat{\mathbf{q}} \supseteq \mathbf{w}$, then $p_{sp} \geq \frac{1}{2}$ since n is relatively large. Therefore, for all probabilistic polynomial-time adversaries, there exists no probabilistic polynomial time simulator that can distinguish τ_1 from τ_2.

Discussion of data dynamics support. Our indexing scheme can easily support efficient data dynamics without exposing any privacy information and needing to either re-index the entire data collection or make use of expensive dynamization techniques. To add new data files $\mathbf{d}_u$, the client locally encrypts the data files and constructs the adding information τ_a (which is essentially the index of $\mathbf{d}_u$) used to update the index γ or say merge into γ stored on the cloud. To delete existing data files $\mathbf{d}_u$, upon receiving the deleting information τ_d (which is essentially the identifiers of $\mathbf{d}_u$), the server checks all non-empty buckets of γ by virtue of some auxiliary indices, and resets the bits of each bucket corresponding to the file identifiers to 0. Due to the space limitation, we will not provide the detailed presentation of the index update here.

5   Performance Evaluation

To demonstrate the efficiency of our solution, this section reports the status of the proposed scheme implementation and performance evaluation. We also implement the state-of-the-art [12] (called SSE) to present a comparative analysis.

Prototype and data set. We implement the schemes in Java programming language and execute them in 64-bit OpenJDK 1.7. To evaluate the efficiency, we separate the cost of the index related operations from the system cost which will vary with different underlying systems. The index building and the queries are evaluated in memory. The experiments are performed on a 64-bit Windows 7 operation system with Intel Core i3-2330M processor 2.20GHz and 4GB RAM. To show the practical viability of our solution, we choose a real-word dataset for the experiments: Enron Email Dataset(EED) [15], where an email is regarded as a data file to be outsourced.

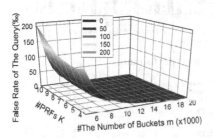
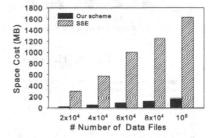

Fig. 1. False rate of the query   **Fig. 2.** The space cost of the index

Fig. 1 shows the experimental evaluation of the false rate of the query. In the experiment, we set $\#d = 2000$ and $\#w = 18000$. Then, we build different indices for different values of m and K. In the experiments, each data point is the average of 10^3 queries. From Fig. 1, we can see that the false rates of the proposed scheme are approximately identical, and extremely small or even negligible by choosing suitable parameters m and K.

Fig. 2 shows the space cost of the encrypted index versus different sub-datasets. In the SSE scheme [12], the index needs to store the encryptions of all the keyword-data pairs and distinct keywords. While in our scheme, each keyword-data pairs is recorded in the index by using a bit ("0" or "1"). Therefore, our scheme has an obvious advantage over [12].

Fig. 3 shows the time cost of query performance versus the number of data files from $\#d = 2 \times 10^4, \ldots, 10^5$, and versus the number of query keywords from $\#q = 1, \ldots, 25$, where each data point is the average of the 5000 queries and the query keywords are selected randomly from 5000 most frequent keywords. The query performance consist of two components: trapdoor generation on the client and the query execution on the cloud. In SSE scheme [12], the time cost needed for the query depends on the prevalence of the query words in data files, and is considered as a simple joint of the time cost needed for the single keyword queries because the multi-keyword query is the post-processing of all single keyword queries. In our scheme, the time cost is only related to the number of hit buckets and the intersection operations (bitwise boolean operation) of the vectors corresponding to the hit buckets. As can be seen from Fig. 3, our scheme achieves highly efficient query performance and thus it is much more efficient than the SSE scheme [12] which needs to decrypt the encrypted items corresponding to all hit data files.

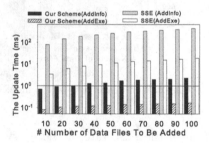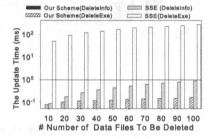

Fig. 3. The time cost of query performance

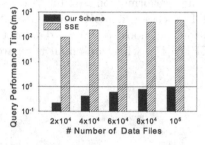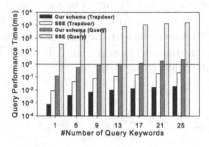

Fig. 4. The time cost of index update

Fig. 4 shows the execution time for adding and deleting the files versus the number of the data files to be updated from $10, \ldots, 100$, respectively. The experiment is performed over a sub-dataset with size $\#d = 10^4$. The cost of update operation are two-fold: the update information generation at the client side and the encrypted index update execution at the cloud side. As shown in Fig. 4, our scheme is much more efficient than [12] because 1)for the adding update, only the merge operations are involved in our scheme 2)for the deletion update, our scheme only refer to the bit value reset after finding the right positions.

6 Conclusion

In this paper, we presented the highly efficient indexing scheme for supporting practical multi-keyword query over encrypted cloud data, while proving semantic security against chosen-keyword attacks (IND2-CKA) and mitigating the leakage of search pattern. Thorough analysis and experimental evaluations showed that compared to the state-of-the-art, the proposed scheme indeed introduces IND2-CKA security, high query performance and accuracy, and efficient index updates.

Acknowledgements. This work was partially supported by the Key Program of National Natural Science Foundation of China (No. 61232002), National Natural Science Foundation of China (No. 61373167,61202034), Natural Science Foundation of Hubei Province of China (No. 2013CFB297), and Wuhan High-Tech Industrial Science and Technology Innovation Team Training Program of China (No. 2014070504020237).

References

1. Hacigms, H., Lyer, B., Li, C., Mhrotra, S.: Executing sql over encrypted data in the database-server-provider model. In: SIGMOD, pp. 216–227. ACM (2002)
2. Kamara, S., Lauter, K.: Cryptographic cloud storage. In: Sion, R., Curtmola, R., Dietrich, S., Kiayias, A., Miret, J.M., Sako, K., Sebé, F. (eds.) RLCPS, WECSR, and WLC 2010. LNCS, vol. 6054, pp. 136–149. Springer, Heidelberg (2010)
3. Song, D.X., Wagner, D., Perrig, A.: Practical techniques for searches on encrypted data. In: IEEE Symposium on Security and Privacy, pp. 44–55. IEEE (2000)
4. Goh, E.J.: Secure indexes. In: IACR (2003)
5. Curtmola, R., Garay, J.A., Kamara, S., Ostrovsky, R.: Searchable symmetric encryption: Improved definitions and efficient constructions. In: ACM CCS, pp. 79–88. ACM (2006)
6. Cao, N., Wang, C., Li, M., Ren, K., Lou, W.: Privacy-preserving multi keyword ranked search over encrypted cloud data. In: INFOCOM, pp. 829–837. IEEE (2011)
7. Cash, D., Jarecki, S., Jutla, C., Krawczyk, H., Roşu, M.-C., Steiner, M.: Highly-scalable searchable symmetric encryption with support for boolean queries. In: Canetti, R., Garay, J.A. (eds.) CRYPTO 2013, Part I. LNCS, vol. 8042, pp. 353–373. Springer, Heidelberg (2013)
8. Wong, W.K., Cheung, D.W.-L., Kao, B., Mamoulis, N.: Secure knn computation on encrypted databases. In: SIGMOD, pp. 139–152. ACM (2009)
9. van Liesdonk, P., Sedghi, S., Doumen, J., Hartel, P., Jonker, W.: Computationally efficient searchable symmetric encryption. In: Jonker, W., Petković, M. (eds.) SDM 2010. LNCS, vol. 6358, pp. 87–100. Springer, Heidelberg (2010)
10. Kuzu, M., Islam, M.S., Kantarcioglu, M.: Efficient similarity search over encrypted data. In: ICDE, pp. 1156–1167. IEEE (2012)
11. Hu, H., Xu, J., Ren, C., Choi, B.: Processing private queries over untrusted sata cloud through privacy homomorphism. In: ICDE, pp. 601–612. IEEE (2011)
12. Kamara, S., Papamanthou, C., Roeder, T.: Dynamic searchable symmetric encryption. In: ACM CCS, pp. 965–976. ACM (2012)
13. Boneh, D., Di Crescenzo, G., Ostrovsky, R., Persiano, G.: Public encryption with keyword search. In: Cachin, C., Camenisch, J.L. (eds.) EUROCRYPT 2004. LNCS, vol. 3027, pp. 506–522. Springer, Heidelberg (2004)
14. Boneh, D., Kushilevitz, E., Ostrovsky, R., Skeith III, W.E.: Public key encryption that allows pir queries. In: Menezes, A. (ed.) CRYPTO 2007. LNCS, vol. 4622, pp. 50–67. Springer, Heidelberg (2007)
15. Cohen, W.W.: Enron email dataset, http://www.cs.cmu.edu/~enron/

Parallel Mining of OWL 2 EL Ontology
from Linked Data

Huiying Li and Qiang Sima

School of Computer Science and Engineering, Southeast University,
Nanjing 210096, P.R. China
{huiyingli,220131534}@seu.edu.cn

Abstract. The Linked Data is a rich common resource with billions of triples available in thousands of datasets. One of the challenges to integrate, query and reuse the Linked Data is to know about the ontology to which the datasets conform. Although there are many ontologies built manually, there are also many RDF datasets published without any prescribed schema to adhere. In this paper, we propose a parallel approach to generate ontologies from large RDF datasets based on statistical data analysis. We divide the large RDF dataset into blocks and allocate them to parallel hardware, then we obtain the statistical data by SPARQL queries, finally the OWL 2 EL axioms are generated based on statistical data analysis. The evaluations tested on two kinds of DBpedia datasets (Mapping-based Dataset with ontology and Raw Infobox Dataset without ontology) show the effectivity and efficiency of our approach.

1 Introduction

The abundance of Linked Data brings many challenges to the integration, querying and maintenance of RDF datasets [1]. One of the challenges is to know about the ontology to which the RDF data conforms, especially when the data schema is absent. Although there are many ontologies built manually, there are also many RDF datasets published without any prescribed schema to adhere. The generation of ontologies from formal and semi-formal data is still a problem despite it has been studied for several years within the Semantic Web community.

The more RDF datasets that are available without existing schema, the more important that the ontology mining problem becomes. For ontology mining, it means all those activities that allow to discover hidden knowledge from ontological knowledge bases [2]. There are two purposes to the discovered knowledge. For the dataset that does not come with any existing schema, the discovered knowledge can be considered as the customized ontology for the dataset. For the dataset that does conform to existing schema, the discovered knowledge can be used to enrich the existing schema and to detect the degree to which the dataset conforms to the existing schema.

In this paper, We propose a parallel approach to generate ontologies from large RDF datasets based on statistical data analysis. Firstly, we propose an non-parallel approach which implements SPARQL queries to obtain ontologies

F. Li et al. (Eds.): WAIM 2014, LNCS 8485, pp. 360–371, 2014.

and describe them by EL profile of OWL 2 Web Ontology Language. Secondly, to deal with large RDF datasets such as DBpedia, we study the problem of parallel ontology mining. We divide and allocate the large data to parallel hardware, generate the OWL 2 EL ontology based on statistical measures on data. Finally, we evaluate our approach on two kinds of DBpedia datasets (Mapping-based Dataset with ontology and Raw Infobox Dataset without ontology). The evaluations tested on DBpedia with ontology show that the mined knowledge is a supplement to the existing ontology. Meanwhile, the overlap between the mined knowledge and the existing ontology shows the degree to which the dataset conforms to the existing ontology. The evaluations tested on DBpedia without ontology show that the mined knowledge can be considered as a customized ontology for the dataset. Furthermore, the evaluations on scalability confirm that our parallel approach can deal with large Linked Data datasets.

The rest of the paper is organized as follows. Section 2 introduces the related work. Section 3 introduces the SPARQL-based statistical measure approach for ontology mining. Section 4 presents the steps to parallel ontology mining from large RDF datasets. Section 5 details the experimental results of our approach. Section 6 concludes the study.

2   Related Work

There are some research work in the field of ontology generation from formal and semi-formal data by using Machine Learning techniques. Several methods have been proposed for constructing ontology classes by means of Machine Learning techniques from positive and negative examples. Research work [3] and [4] are tailored for small and medium size knowledge bases, while they cannot be directly applied to large knowledge bases due to their dependency on reasoning methods. [5] presents an approach for leveraging Machine Learning algorithms for learning of ontology class descriptions in large knowledge bases.

Other research work focus on learning ontologies from text documents by the use of association rules. Research work [6] uses association rules to discover causal relations in RDF-based medical data. [7] considers containment relationships between sets of class instantiations for producing alignments among several Linked Data repositories, including DBpedia. While their approach could as well be used to suggest refinements for a single ontology, they currently only acquire mappings which express subsumption or equivalence between so-called restriction classes.

Statistical Schema Induction (SSI) [8] is the most similar work to our approach. SSI can generate ontologies from RDF datasets based on association rule mining. SSI acquires the terminology by posing SPARQL queries to the repositorys endpoint. The result of this step is a set of relational database tables containing the URIs of all those RDF resources corresponding to classes and properties. Then, the transaction tables are constructed to mine the dataset for the various kinds of OWL axioms. Although the experiments on DBpedia datasets show that the time needed to finally compute the association rules is

less than 5 seconds for the largest transaction table. The process to construct the
association tables is very time-consuming and space-consuming. [9] is a following
work to mine RDF data for various types of property axioms.

Compared with SSI and [9], our approach is distinct in three aspects. Firstly,
we propose a non-parallel approach which avoids the constructing of transaction
tables. Since the process to construct transaction tables is time-consuming and
space-consuming, our approach is more efficient than SSI. Secondly, we propose
an approach which supports the concurrent ontology mining on parallel hardware. This parallel approach improves the efficiency further. Thirdly, we test our
non-parallel and parallel approach on both the RDF data with ontology and the
RDF data without ontology. The evaluations show that our approach is effective
and efficient.

3   Ontology Mining Based on Statistical Data Analysis

3.1   EL Profile of OWL 2

OWL 2 EL, which is based on the description logic $\mathcal{EL}++$, is particularly
useful in applications employing ontologies that contain very large numbers of
properties and/or classes. This profile captures the expressive power used by
many such ontologies and is a subset of OWL 2 for which the basic reasoning
problems can be performed in time that is polynomial with respect to the size
of the ontology.

Table 1. Syntax and semantics of $\mathcal{EL}++$ without nominal, concrete domain and
assertions

Name	Syntax	Semantics
top	$\top$	Δ^I
bottom	$\bot$	$\emptyset$
conjunction	$C \sqcap D$	$C^I \cap D^I$
existential restriction	$\exists r.C$	$\{x \in \Delta^I \mid \exists y \in \Delta^I : (x,y) \in r^I \wedge y \in C^I\}$
general concept inclusion	$C \sqsubseteq D$	$C^I \subseteq D^I$
role inclusion	$r_1 \circ \ldots \circ r_k \sqsubseteq r$	$r_1^I \circ \ldots \circ r_k^I \subseteq r^I$
domain restriction	$\mathsf{dom}(r) \sqsubseteq C$	$r^I \subseteq C^I \times \Delta^I$
range restriction	$\mathsf{ran}(r) \sqsubseteq C$	$r^I \subseteq \Delta^I \times C^I$

In $\mathcal{EL}++$, concepts are inductively defined from a set N_C of concept (or
class) names, a set N_R of role (or property) names, and a set N_I of individual
names. We use C and D to refer to concepts, r to refer to a role name, a and
b to refer to individual names. The semantics of $\mathcal{EL}++$ concepts is defined in
terms of an *interpretation* $I = (\Delta^I, \cdot^I)$. The *domain* Δ^I is a non-empty set of
individuals and the *interpretation function* $\cdot^I$ maps each concept name $A \in N_C$

to a subset A^I of Δ^I, each role name $r \in N_R$ to a binary relation r^I on Δ^I, and each individual name $a \in N_I$ to an individual $a^I \in \Delta^I$. The syntax and semantics of $\mathcal{EL}++$ are listed in Table 1.

With the axioms listed in Table 1, $\mathcal{EL}++$ generalizes several means of expressivity such as role equivalences can be expressed as $r \sqsubseteq s$, $s \sqsubseteq r$; transitive roles can be expressed as $r \circ r \sqsubseteq r$; the bottom concept in combination with general concept inclusions (GCIs) can be used to express disjointness of complex concept descriptions.

3.2 Ontology Mining

In this subsection, we show the approach to obtain the OWL 2 EL axioms based on statistical measures on data. Firstly, we use a set of indicators, their related interpretations in this paper are described in Table 2 [1].

Table 2. Indicators used for statistical measures on data

Indicator	Description
$Res(\{C_i\})$	$\{s_i \mid s_i \; rdf : type \; C_i.\}$
$SubjectRes(r)$	$\{s_i \mid s_i \; r \; o_j.\}$
$SubjectRes(C_i, r)$	$\{s_i \mid s_i \; rdf : type \; C_i. \; s_i \; r \; o_j.\}$
$SubjectRes(r, C_j)$	$\{s_i \mid s_i \; r \; o_j. \; o_j \; rdf : type \; C_j.\}$
$SubjectRes(C_i, r, C_j)$	$\{s_i \mid s_i \; rdf : type \; C_i. \; s_i \; r \; o_j. \; o_j \; rdf : type \; C_j.\}$
$ObjectRes(r)$	$\{o_j \mid s_i \; r \; o_j.\}$
$ObjectRes(r, C_j)$	$\{o_j \mid s_i \; r \; o_j. \; o_j \; rdf : type \; C_j.\}$
$ResPair(\{r_m\})$	$\{< s_i, o_j > \mid s_i \; r_m \; o_j.\}$
$ResPair(r_m \circ r_n)$	$\{< s_i, o_j > \mid s_i \; r_m \; i_k. \; i_k \; r_n \; o_j.\}$

Notice that, there can be more than one classes in indicator $Res(\{C_i\})$. For example, $Res(\{C_i, C_j\})$ denotes the set of resources typed both as class C_i and C_j. Also, there can be more than one properties in indicator $ResPair(\{r_m\})$.

Table 3 shows how we get the axioms with the statistical measures on data. For example, $|SubjectRes(C, r)|/|SubjectRes(r)| = 1$ means that every instance that has property r is also typed as class C, then we can induce that the domain of property r is a subclass of class C. According to the descriptions of indicators, the numerical values in Table 3 can be obtained by SPARQL querying. For instance, a SPARQL query like "$SELECT \; count(distinct \; ?s) \; WHERE \; \{?s < rdf : type > < C_i > . \; ?s < r > ?o. \; ?o < rdf : type > < C_j > .\}$"is issued to get the cardinality of set $SubjectRes(C_i, r, C_j)$. After querying the classes and properties in a dataset, many SPARQL queries are issued to get the numerical values, then the OWL 2 EL axioms listed in Table 3 are generated based on statistical measures.

---

[1] The namespace rdf denotes $http : //www.w3.org/1999/02/22 - rdf - syntax - ns\#$.

Table 3. Axioms computed by statistical measures on data

Axiom Type	Statistical Measures on Data
$C_i \sqsubseteq C_j$	$\|Res(\{C_i, C_j\})\|/\|Res(\{C_i\})\| = 1$
$C_i \cap C_j \sqsubseteq C_k$	$\|Res(\{C_i, C_j, C_k\})\|/\|Res(\{C_i, C_j\})\| = 1$
$C_i \sqsubseteq \exists r.C_j$	$\|SubjectRes(C_i, r, C_j)\|/\|Res(\{C_i\})\| = 1$
$\exists r.C_j \sqsubseteq C_i$	$\|SubjectRes(C_i, r, C_j)\|/\|SubjectRes(r, C_j)\| = 1$
$\mathsf{dom}(r) \sqsubseteq C$	$\|SubjectRes(C, r)\|/\|SubjectRes(r)\| = 1$
$\mathsf{ran}(r) \sqsubseteq C$	$\|ObjectRes(r, C)\|/\|ObjectRes(r)\| = 1$
$r_m \sqsubseteq r_n$	$\|ResPair(\{r_m, r_n\})\|/\|ResPair(\{r_m\})\| = 1$
$r_m \circ r_n \sqsubseteq r_l$	$\|ResPair(\{r_m \circ r_n, r_l\})\|/\|ResPair(r_m \circ r_n)\| = 1$

4   Parallel Ontology Mining

We find the approach proposed in Section 3 is time-consuming when applied to large scale dataset. It is natural to consider the idea of concurrent ontology mining on parallel hardware. In this section, we introduce how to divide the dataset into blocks and perform parallel ontology mining.

When observing the axiom types in Table 3, we find that the indicator descriptions to compute the first six axioms include at most one property. It is reasonable to consider dividing the data by property. If we divide the triples with same property into one block and add instance type triples to every block, the statistical data computed in blocks will be same to the results computed in the whole data. However, the indicator descriptions to compute the last two axiom types involve more than one properties. It may lead to wrong statistical data if we divide the dataset simply by property. Suppose that $r_m \sqsubseteq r_n$, if we divide triples with property r_m and r_n into different blocks, the computation of $\|ResPair(\{r_m, r_n\})\|$ in one block may be wrong, this axiom can not be mined consequently. So, we must make sure that those properties might have subproperty relationship are divided into one block. Similarly, considering axiom $r_m \circ r_n \sqsubseteq r_l$, we must divide those properties might have subproperty relationship or might compose property chains into one block. In OWL 2 EL, the property inclusion axiom with property chains is limited to only object property. Hence, we deal with data property and object property separately.

For data property, those properties might have subproperty relationship must be ensured in one block. We go through the RDF data and find out subproperty candidates. The subproperty candidate denotes a property pair which has common instance pair. Such as property pair r_i and r_j where s r_i o and s r_j o. Considering the property as node, the subproperty candidate as an undirected edge, the definition of G_{dp} is as follows.

Definition 1. $G_{dp} = (V, E, \rho)$ *is an undirected graph. Each node in V represents a data property,* $(v_i, v_j) \in E$ *iff property pair (v_i, v_j) is subproperty*

candidate. ρ is a function from V to positive integer, $\rho(v_i) = |ResPair(\{v_i\})|$ denotes the instantiation number of property v_i .

The connected subgraphs of G_{dp} help us to divide data property into blocks. The nodes in a connected subgraph denotes the properties should be divided into one block, the sum of these nodes' weight denotes the corresponding triple number. While the experiment shows that there may has one connected subgraph contains most of the triples. For instance, one connected subgraph of DBpedia Raw Infobox Dataset 3.9 contains more than 75% triples. It leads to the problem of unbalance when allocating data to parallel processing unit.

To break the connectivity of graph G_{dp} and get more subgraphs, we try to cut edges in graph G_{dp}. If property node v_i and v_j in G_{dp} are proved that they can not have subproperty relationship, edge (v_i, v_j) can be removed. We use SPARQL queries to verify whether two properties have subproperty relationship or not. Since verifying all subproperty candidates is infeasible, we focus on verifying the subproperty candidates which could help the increase of the connected subgraphs number or the decrease of triple number in one block. To increase the number of connected subgraphs, we verify all the cut edges in graph G_{dp} firstly. Then, we verify the adjacent edges of the node (or top-k nodes) with largest degree. To decrease the triple number in one block, we verify the adjacent edges of the node (or top-k nodes) with the largest weight. Finally, the verified edges are removed from graph G_{dp}, the BFS algorithm runs to get all connected subgraphs. Based on these property blocks, we divide the RDF data into blocks and allocate them to parallel processing unit.

For object property, those properties (or property chains) might have subproperty relationship must be ensured in one block. The method to deal with object property is similar to the one for data property.

5 Experimental Study

We test our approach on two kinds of datasets (the datasets with ontology and the datasets without ontology). We evaluate the effectivity and efficiency of our approach. Section 5.1 describes the two kinds of datasets in the experiments. Section 5.2 gives the evaluations on those datasets. We use Jena toolkit (jena.sourceforge.net) to manage RDF data for SSI and our approach. The experiments were developed within the Eclipse environment and on two 64bit quad Core ThinkStations with 3.10 MHz and 16 GB of RAM (of which 14 GB was assigned to the JVM).

5.1 Datasets

We choose the real world RDF data DBpedia from Linked Data to evaluate our approach. DBpedia contains extracted data from Wikipedia, we concentrate on the infobox subset. The DBpedia project has different datasets including Mapping-based Datasets and Raw Infobox Dataset.

The Mapping-based Datasets are extracted based on hand-generated mappings of Wikipedia infoboxes/templates to a DBpedia ontology. The instance data within the Mapping-based Dataset is much cleaner and better structured. For comparison, we consider three newest versions of DBpedia Mapping-based Datasets, named DBpedia Dataset 3.7, DBpedia Dataset 3.8 and DBpedia Dataset 3.9.

The Raw Infobox Dataset extracts all properties from all infoboxes and templates within all Wikipedia articles. Property names in Raw Infobox Dataset are not cleaned or merged. There is no consistent ontology for Raw Infobox Dataset. We consider the Raw Infobox Dataset of DBpedia 3.9 as test datasets without ontology.

Table 4 shows the statistics about the different versions of DBpedia: its name, the number of classes, data properties and object properties, the number of triples it contains. Notice that, the statistics are collected from the DBpedia data not from the DBpedia ontology. For comparison, we list the corresponding numbers obtained from DBpedia ontology in brackets except the Raw Infobox Dataset 3.9. Because Raw Infobox Dataset do not have consistent ontology.

Table 4. The statistics about different versions of DBpedia

Dataset	# of classes	# of data properties	# of object properties	# of triples
DBpedia Dataset 3.7	326 (319)	724 (893)	655 (750)	26,988,054
DBpedia Dataset 3.8	348 (359)	627 (975)	624 (800)	33,742,024
DBpedia Dataset 3.9	434 (529)	688 (1,406)	685 (927)	41,804,710
Raw Infobox Dataset 3.9	434	45,257	5,032	89,093,202

For the Mapping-based Datasets, most of the classes and properties collected from datasets are declared in the ontology except those with other namespaces such as property $foaf : familyName$ [2]. For the Raw Infobox Dataset, the numbers of properties are much larger than the numbers in Mapping-based Datasets. The Raw Infobox Dataset is relatively noisy, for example there are many misspellings like the property $http : //dbpedia.org/property/deathplce$ which supposed to be $http : //dbpedia.org/property/deathplace$.

5.2 Effectivity Evaluation Results

We apply our approach to different versions of DBpedia datasets. There are some interesting mining results, for instance, we generate the axioms that property $dbpedia : lowestMountain$ is a subproperty of $dbpedia : lowestPlace$ and the class which exits a property $dbpedia : editor$ to class $dbpedia : Writer$ is

---

[2] The namespace $foaf$ denotes $http : //xmlns.com/foaf/0.1/$.

a subclass of *dbpedia* : *Work* from DBpedia dataset 3.7. We get the axiom that property chain *dbpedia* : *wineProduced* to *dbpedia* : *country* is a sub-property of *dbpedia* : *location* from DBpedia dataset 3.8. We also obtain the axiom that the domain of property *dbpedia* : *designCompany* is a subclass of *schema* : *CreativeWork* from DBpedia dataset 3.9 [3]. However, these axioms are not declared in the corresponding DBpedia ontology.

Table 5. Axioms numbers obtained from different versions of DBpedia

Axiom Type	# of DBpedia Dataset 3.7	# of DBpedia Dataset 3.8	# of DBpedia Dataset 3.9	# of Raw Infobox Dataset 3.9
$C_i \sqsubseteq C_j$	1,520	1,521	1,926	1,926
$C_i \sqcap C_j \sqsubseteq C_k$	6,246	6,689	8,490	8,006
$C_i \sqsubseteq \exists r.C_j$	0	0	0	0
$\exists r.C_j \sqsubseteq C_i$	7,189	8,982	11,184	56,780
$\mathrm{dom}(r) \sqsubseteq C$	499	183	237	23,798
$\mathrm{ran}(r) \sqsubseteq C$	493	414	448	1,663
$r_m \sqsubseteq r_n$	29	33	29	32,087
$r_m \circ r_n \sqsubseteq r_l$	969	1,152	1,267	129

Table 5 lists the axioms numbers obtained from different versions of DBpedia datasets. We find that the three versions of Mapping-based Datasets have similar mining results. While there are much more axioms about $\exists r.C_j \sqsubseteq C_i$, $\mathrm{dom}(r) \sqsubseteq C$, $\mathrm{ran}(r) \sqsubseteq C$ and $r_m \sqsubseteq r_n$ for Raw Infobox Dataset 3.9. The reason is that the property numbers in Raw Infobox Dataset are larger than the numbers in Mapping-based Dataset. We also find that the numbers of axiom $C_i \sqsubseteq \exists r.C_j$ are both zero for all the datasets. It means that we do not find that all instances of class C_i have the property r and the value instance is typed as C_j.

To show the effectivity of our approach, we compare our mining results from Mapping-based Datasets with the axioms declared in DBpedia ontology in Figure 1. The numbers from 1 to 8 in x-coordinate denotes the axiom type from $C_i \sqsubseteq C_j$ to $r_m \circ r_n \sqsubseteq r_l$. The numbers in y-coordinate denotes the percentages distribution. For every axiom type, suppose the axiom number generated by our approach is n_{mined}, the axiom number declared in DBpedia ontology is n_{onto}, the common axiom number both generated by our approach and declared in DBpedia ontology is n_{com}. The yellow bar denotes the percentage $\frac{n_{onto}}{n_{mined}+n_{onto}-n_{com}}$, the blue bar denotes the percentage $\frac{n_{com}}{n_{mined}+n_{onto}-n_{com}}$, the cyan bar denotes the percentage $\frac{n_{mined}}{n_{mined}+n_{onto}-n_{com}}$.

Figure 1 shows that the results are similar when applying our approach to three versions of DBpedia datasets. For the axioms $C_i \sqcap C_j \sqsubseteq C_k$, $\exists r.C_j \sqsubseteq C_i$

---

[3] The namespace *dbpedia* denotes $http://dbpedia.org/ontology/$, namespace *schema* denotes $http://schema.org/$.

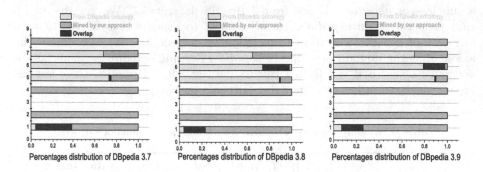

Fig. 1. Percentages distribution of Mapping-based Datasets

and $r_m \circ r_n \sqsubseteq r_l$, the percentage of the axioms mined by our approach is 100% because there are no such statements in DBpedia ontology.

For the axiom $C_i \sqsubseteq C_j$, the axioms mined by our approach not only cover most of the axioms declared in the ontology but also contain many subclass axioms which are not declared in the ontology.

For the axiom $\mathsf{dom}(r) \sqsubseteq C$, the percentage of the common axioms and the percentage of the axioms mined by our approach are relatively small. There are many domain statements declared in DBpedia ontology are not generated by our approach. One of the reasons is that there are many properties with domain statement in ontology are not used in the dataset. For instance, there are $1,406$ data properties declared in DBpedia ontology 3.9 while only 688 data properties are used in the dataset. The other reason is that one property can be applied to different classes in the Mapping-based Dataset. For instance, the property $dbpedia : volume$ can be applied to both the class $dbpedia : lake$ and the class $dbpedia : planet$. This affects the mining results of the domain axiom type.

For the axiom $\mathsf{ran}(r) \sqsubseteq C$, most of the axioms mined by our approach are declared in ontology, there are still many range axioms declared in ontology but not mined by our approach. One of the reason is also because there are many properties with range statement in ontology are not used in the dataset. The other reason is that some instances as the value of a property do not have a class type.

In summary, we apply our approach to both datasets with ontology and dataset without ontology to test the effectivity. The evaluations on the datasets with ontology show that our approach can be applied to obtain axioms effectively. The comparison between the mined axioms and the axioms from ontology shows that our approach can generate not only many axioms declared in the ontology but also many axioms which are not declared in the ontology. These axioms can be considered as a supplement to the ontology. Moreover, the overlap between mined axioms and ontology axioms can be used to show the degree to which the dataset conforms to the ontology. The more overlap means the better that the dataset conforms to the ontology. The evaluations on dataset without ontology show that our approach can be applied to obtain axioms, these axioms can be considered as a customized ontology for the dataset.

5.3   Efficiency Evaluation Results

To show the efficiency of our approach, we conduct a performance evaluation. We compare SSI with our non-parallel approach and our parallel approach. We test these approaches on four different datasets, the time taken by different approaches are measured.

The premise for SSI is to construct the transaction tables which are used to mine for various kinds of OWL axioms. So, the time taken by SSI is the sum of the transaction tables constructing time and the association rules mining time. The running time of our non-parallel approach is the time to apply our approach to the whole dataset. The running time of our parallel approach is the time to apply our approach to the separated datasets on 8 parallel computing units.

Figure 2(a) shows the comparison results on DBpedia 3.7, 3.8, 3.9 and Raw Infobox 3.9. Notice that, the y-coordinate is logarithmic coordinates. The yellow column denotes the running time of SSI, the blue bar denotes the running time of our non-parallel approach, the cyan bar denotes the running time of our parallel approach.

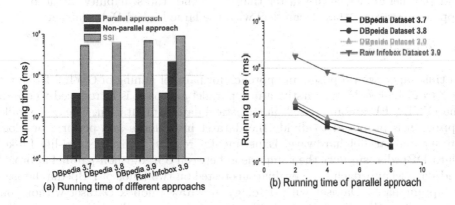

Fig. 2. Efficiency evaluation results

We find that the running time of SSI is longer than our approach even the non-parallel approach. That is because SSI needs to construct 6 transaction databases in order to obtain the axioms. Although the time needed to obtain axioms is less than few seconds, the premise is that it takes a lot of time to construct the transaction tables. And there also needs a large space to store the transaction databases. Take the DBpedia dataset 3.7 for example, the space needed is about 56GB. Since our approach only query the dataset for statistical data, there is no need to construct the transaction databases. Even the non-parallel approach is faster than SSI. The parallel approach separates the dataset on 8 parallel computing units, the running time is much faster than SSI and the non-parallel approach. It confirms the scalability of our approach to other large Linked Data repositories.

Notice that, the running time for Raw Infobox dataset 3.9 is much longer than other datasets. The reason is that Raw Infobox dataset 3.9 has much more properties than other datasets. We query the dataset to obtain the statistical data by SPARQL queries. The number of these SPARQL queries is decided by the number of the classes and properties. Since there are much more properties in Raw Infobox dataset 3.9 than other datasets, we need much more SPARQL queries to obtain the statistical data. It leads to the longer running time on Raw Infobox dataset 3.9 than other datasets.

To test our parallel approach, the running time on different numbers of computing units are depicted in Figure 2(b). The black line with square shows the running time tested on DBpedia Dataset 3.7. The blue line with circle shows the running time tested on DBpedia Dataset 3.8. The cyan line with upper triangle shows the running time tested on DBpedia Dataset 3.9. The yellow line with lower triangle shows the running time tested on Raw Infobox Dataset 3.9. Notice that the y-coordinate is logarithmic coordinates. The tests show that the running time of our parallel approach is approximately halved when the number of computing units is doubled for up to 8 compute units.

In summary, the efficiency test shows that both our non-parallel approach and parallel approach are faster than SSI. And, the scalability of the parallel approach confirms that it can deal with the large Linked Data datasets.

6   Conclusion

In this paper, we proposed an approach for parallel mining of OWL 2 EL ontology from Linked Data. Firstly, a non-parallel approach is introduced to obtain the OWL 2 EL axioms based on statistical data computation. Then, a parallel approach is proposed to divide the dataset into blocks and perform ontology mining on parallel hardware. Experimental results, using the real-life Linked Data DBpedia, support the comparison between SSI, our non-parallel approach and parallel approach. The evaluations tested on two kinds of DBpedia datasets (Mapping-based Dataset with ontology and Raw Infobox Dataset without ontology) show the effectivity and efficiency of our approach.

The effectivity evaluation results show that our approach is effective regardless applying to the dataset with ontology or the dataset without ontology. There are two advantages when applying our approach to the dataset with ontology. One is that our approach can generate not only many axioms declared in the ontology but also many axioms which are not declared in the ontology. These axioms can be considered as a supplement to the ontology. The other is that the overlap between mined axioms and ontology axioms can be used to show the degree to which the dataset conforms to the ontology. The more overlap means the better that dataset conforms to the ontology. When applying our approach to the dataset without ontology, the most important advantage is to obtain axioms. These axioms can be considered as a customized ontology for the dataset.

The efficiency evaluation results show that both our non-parallel approach and parallel approach are faster than SSI. Moreover, the tests show that the running time of the parallel approach is approximately halved when the number of computing cores is doubled. The scalability of the parallel approach confirms that it can deal with the large Linked Data datasets.

Acknowledgments. The work is supported by the National Natural Science Foundation of China under grant No. 61170165 and No. 61003055.

References

1. Bizer, C., Heath, T., Berners-Lee, T.: Linked Data - The Story So Far. IJSWIS 5(3), 1–22 (2009)
2. d'Amato, C., Fanizzi, N., Esposito, F.: Inductive learning for the semantic web: What does it buy? Semantic Web 1(1-2), 53–59 (2010)
3. Lehmann, J., Hitzler, P.: A refinement operator based learning algorithm for the [Equation image] description logic. In: Blockeel, H., Ramon, J., Shavlik, J., Tadepalli, P. (eds.) ILP 2007. LNCS (LNAI), vol. 4894, pp. 147–160. Springer, Heidelberg (2008)
4. Lehmann, J., Hitzler, P.: Foundations of refinement operators for description logics. In: Blockeel, H., Ramon, J., Shavlik, J., Tadepalli, P. (eds.) ILP 2007. LNCS (LNAI), vol. 4894, pp. 161–174. Springer, Heidelberg (2008)
5. Hellmann, S., Lehmann, J., Auer, S.: Learning of OWL class descriptions on very large knowledge bases. International Journal on Semantic Web and Information Systems 5(2), 25–48 (2009)
6. Nebot, V., Berlanga, R.: Mining association rules from semantic web data. In: García Pedrajas, N., Herrera, F., Fyfe, C., Benítez, J.M., Ali, M. (eds.) IEA/AIE 2010, Part II. LNCS, vol. 6097, pp. 504–513. Springer, Heidelberg (2010)
7. Parundekar, R., Knoblock, C.A., Ambite, J.L.: Linking and building ontologies of linked data. In: Patel-Schneider, P.F., Pan, Y., Hitzler, P., Mika, P., Zhang, L., Pan, J.Z., Horrocks, I., Glimm, B. (eds.) ISWC 2010, Part I. LNCS, vol. 6496, pp. 598–614. Springer, Heidelberg (2010)
8. Völker, J., Niepert, M.: Statistical Schema Induction. In: Antoniou, G., Grobelnik, M., Simperl, E., Parsia, B., Plexousakis, D., De Leenheer, P., Pan, J. (eds.) ESWC 2011, Part I. LNCS, vol. 6643, pp. 124–138. Springer, Heidelberg (2011)
9. Fleischhacker, D., Völker, J., Stuckenschmidt, H.: Mining RDF Data for Property Axioms. In: Meersman, R., Panetto, H., Dillon, T., Rinderle-Ma, S., Dadam, P., Zhou, X., Pearson, S., Ferscha, A., Bergamaschi, S., Cruz, I.F. (eds.) OTM 2012, Part II. LNCS, vol. 7566, pp. 718–735. Springer, Heidelberg (2012)
10. Baader, F., Brandt, S., Lutz, C.: Pushing the $\mathcal{EL}$ envelope. In: Proceedings of the Nineteenth International Joint Conference on Artificial Intelligence IJCAI, Edinburgh, UK. Morgan-Kaufmann Publishers, San Francisco (2005)

TraPath: Fast Regular Path Query Evaluation on Large-Scale RDF Graphs

Xin Wang[1,2], Guozheng Rao[1,2,*], Longxiang Jiang[1,2], Xuedong Lyu[1,2],
Yajun Yang[1,2], and Zhiyong Feng[1,2]

[1] School of Computer Science and Technology, Tianjin University, Tianjin, China
[2] Tianjin Key Laboratory of Cognitive Computing and Application, Tianjin, China
{wangx,rgz,yjyang,zyfeng}@tju.edu.cn, {lxjiang2012,lxd.1990}@gmail.com

Abstract. Regular path queries, or RPQs, are basic querying mechanisms on graphs that play an increasingly important role over the past
decade. In recent years, large amounts of RDF data are published on the
Web since the development of Linked Data. Such a large-scale of data
has posed serious challenges to the efficiency of RPQs. In this paper,
we devise a double-layer bi-directional index structure that has a linear
space complexity, and propose a novel traversal-based algorithm TraPath
that achieves the fast evaluation of RPQs by using the index structure.
We conduct extensive experiments to evaluate and compare the performance of our prototype system and the Sesame RDF repository with a
real-world RDF dataset from DBpedia. The experimental results show
that TraPath significantly outperforms the state-of-the-art methods.

Keywords: regular path query, RDF graph, large-scale, index structure.

1 Introduction

The Semantic Web is considered as the next-generation of the current Web,
on which information is machine-understandable. The standard data model of
the Semantic Web is the Resource Description Framework (RDF) [1], which
describes resources with triples of the form (s, p, o) where s is the subject, p the
predicate, and o the object. Since each triple states a relation from its subject to
object, an RDF dataset, consisting of a set of triples, is represented as a directed
labeled graph. Queries on RDF graphs belong to subgraph matching queries or
path queries, rather than relational queries. Therefore, traditional RDBMS is
not applicable for the management of large-scale RDF data. As a basic querying
mechanism in RDF databases, RPQs are recognized as an essential operation
to explore more complex relationships between recourses in RDF graphs. In
particular, researchers in some areas that have been equipped with relatively rich
RDF datasets, such as bioinformatics [2] and social networking [3], have tried
to use different forms of RPQs to gain new knowledge from large RDF graphs.
In addition, as the current standard query language for RDF, SPARQL 1.1 [4]

---

* Corresponding author.

F. Li et al. (Eds.): WAIM 2014, LNCS 8485, pp. 372–383, 2014.
© Springer International Publishing Switzerland 2014

has introduced a new feature called *property paths* to realize the functionality of RPQs. Therefore, the efficiency of evaluating RPQs on large RDF graphs is of great importance.

Mendelzon and Wood [5] have proven that the evaluation of RPQs by simple paths on graph is NP-complete, which indicates regular path queries have a high complexity. In addition, the RDF graphs have proliferated significantly with the development of Linked Data [6], which posed serious challenges to graph data management. As a consequence, traditional methods based on triple indexes and sort-merge joins, which need to load large amounts of triples into memory, exhibit low performance and cannot be adapted to the big data scenario.

In this paper, we propose an efficient method for answering RPQs on large-scale RDF data. The contributions of our paper are summarized as follows:

- We devise a *double-layer bi-directional* index structure that covers all regular path queries and has a linear space complexity.
- Based on this index structure, we propose a novel *traversal-based* algorithm, named TraPath, for searching paths on large RDF graphs, which achieves the efficient evaluation of RPQs.
- We perform extensive experiments to evaluate and compare the performance of our method and Sesame. The experimental results show that TraPath significantly outperforms the state-of-the-art methods.

The rest of the paper is organized as follows. After a review of related work in Section 2, we introduce the necessary definitions and formalize the RPQ problem in Section 3. In Section 4, we present the double-layer bi-directional index structure that covers all regular path queries and has a linear space complexity. In Section 5, we describe our algorithms which traverse RDF graphs bi-directionally in parallel. In Section 6 we evaluate our work by a series of experiments. Finally, we conclude the paper in Section 7.

2   Related Work

We focus on regular path queries on large-scale RDF graphs and review related work from two aspects separately, i.e., RDF indexes and approaches to RPQs.

Indexes for RDF data can be divided into two categories, B^+-tree-based indexes and Bigtable-based indexes. RDF-3X [7] introduces the concept of sextuple indexing based on B^+-tree, and processes triple pattern queries efficiently. However, RDF-3X also employees the multi-way join operations to implement more complex SPARQL queries, which may incur the high time overhead due to the large number of intermediate results. CumulusRDF [8] and Jingwei+ [9] both implement triple indexes based upon Bigtable, but their performance is limited since the mechanism of the super-column may reduce the efficiency of the operations on triple indexes.

SPARQL is the W3C recommended query language for RDF. However, the functionality of RPQs has not been proposed until the *property paths* are introduced in SPARQL 1.1. In the past few years, researchers had concentrated on studying SPARQL including implementations, speeding up queries, and extensions to support RPQs. Sesame and Jena are two state-of-the-art single-machine implementations of SPARQL, while they provide weak support for RPQs. PSPARQL [10] is an RPQ extension to SPARQL, but it does not provide the implementation method for the language. SPARQLeR [11] also extends SPARQL with regular paths, and Koschmieder and Leser [12] propose to use rare labels for answering RPQs. However, both of these approaches are implemented as a bi-directional breadth-first search, which is obviously different from our approach. Besides, both approaches employ the *counting paths* semantics that has a PSPACE complexity.

Our approach differs from the above work significantly. On one hand, we use the flat structure to construct indexes on Bigtable, which achieves both high performance and scalability. The *double-layer bi-directional* index structure is built specifically for RPQs, which has superior performance for joining triples. On the other hand, our approach to answering RPQs is implemented as a *depth-first* search that has the tremendous performance advantages for finding paths on large-scale RDF graphs. In addition, we simplify the problem by restricting the semantics of queries, which can be tackled in polynomial time.

3   Definitions

Before introducing our work in detail, we give several definitions of RDF graphs, in/out-degree nodes of a predicate, and the syntax of our RPQ language. In this paper, we define an RDF graph as a set of triples that can be mapped to a general graph of the form $G = (V, E)$.

Definition 1. *An RDF graph is defined as $T = \{(s, p, o) \mid s \in S, p \in P, o \in O\}$, in which we define the set of subjects as $S = \{s \mid s = lab(v), v \in V, \forall v_i \in V, \langle v, v_i \rangle \in E\}$, the set of predicates as $P = \{p \mid p = lab(\langle v_i, v_j \rangle), \forall v_i, v_j \in V, \langle v_i, v_j \rangle \in E\}$ and the set of objects as $O = \{o \mid o = lab(v), v \in V, \forall v_i \in V, \langle v_i, v \rangle \in E\}$. $lab()$ is the function that returns the label of a vertex or an edge.*

Definition 1 describes the logical model of an RDF graph that differs from a general graph. If we define an RDF graph in the form of both T and G, the following properties hold: (1) $lab(V) = S \cup O$, and $lab(E) = P$; (2) $|V| = |S \cup O|$, but $|E| \geqslant |P|$, since a subject s or an object o of a triple in T corresponds to a unique label of a vertex in G, while a predicate p may correspond to more than one edges in G.

In an RDF graph, a path is composed of consecutive edges, each edge is represented as a triple (s, p, o). If we consider the predicate p as a vertex, then it has both in-degree (subjects) nodes and out-degree (objects) nodes.

Definition 2. *We define in-degree subjects $PS_{p'} = \{s \mid \forall s \in S, \forall o \in O, (s, p', o) \in T\}$, in-degree subjects with specified object $POS_{p'o'} = \{s \mid \forall s \in S, (s, p', o') \in T\}$, out-degree objects $PO_{p'} = \{o \mid \forall s \in S, \forall o \in O, (s, p', o) \in T\}$, and out-degree objects with specified subject $PSO_{p's'} = \{o \mid \forall o \in O, (s', p', o) \in T\}$.*

The formulas in Definition 2 describe the in/out-degree nodes of a specified predicate p, from which we obtain: (1) For a predicate $p' \in P$, $PSO_{p's}$, $POS_{p'o}$, $PS_{p'}$ and $PO_{p'}$ are all not empty, iff $\exists s \in S, \exists o \in O, (s, p', o) \in T$; (2) $POS_{p'o} \subseteq PS_{p'}$, $PSO_{p's} \subseteq PO_{p'}$; and (3) $POS_{p'o} = PS_{p'}$, iff $PO_{p'} = \{o\}$, and similarly, $PSO_{p's} = PO_{p'}$, iff $PS_{p'} = \{s\}$.

Definition 3. *The regular expression of path queries is defined as:* $r = p \mid -r \mid (r/r) \mid (r|r) \mid r^*$, $p \in P$. *We define the syntax of our RPQ queries as* $Q = (?x|s, r, ?y|o)$, $s \in PS_{p_1}, o \in PO_{p_n}$, p_1 *and* p_n *is the first and last edge of* r *respectively.*

Definition 3 gives a recursive definition to the regular paths, which covers all the possible forms of RPQs. However, the complexity of RPQs with counting paths semantics is PSPACE-complete [5], which is considered infeasible for large-scale RDF graphs. To simplify the problem, in our RPQ semantics, we just find one satisfiable path for an RPQ (i.e., not counting paths), and we also allow non-simple paths as the answers. Definition 3 also introduces the syntax of our RPQ language. For example, if Tom wants to find that whether there exists a person who is a friend of his friends and that person also has a pet, then the query is expressed as: $(Tom, isFriend/isFriend/hasPet, ?y)$.

4   Index Structures

This section presents the double-layer bi-directional index structure that is abbreviated as DB-Index. First we give a detailed introduction to DB-Index. Then we describe the procedure for the index construction.

4.1   DB-Index

Definition 2 describes our new perspective of the primitive path edge. Under normal circumstances, there is no need to specify the in/out-degree nodes, while in the context of big data, the scale of triples under the same predicate p might be extremely large. Therefore, we specify these nodes and separate them into smaller units, which is also called the subject/object refinement. However, for a triple in RPQs, we do not know s or o in most cases, as a consequence, POS_{po} and PSO_{ps} do not seem to work. For example, we want to access all objects in PSO_{ps}, but we get nothing if only p is specified.

To compensate for this defect, we define PS_p and PO_p to co-work with POS_{po} and PSO_{ps}. As a matter of fact, DB-Index is constructed on the basis of the formulas in Definition 3, in which we regard POS_{po} and PSO_{ps} as primary indexes, PS_p and PO_p as secondary indexes. The structure of DB-Index is shown in Fig. 1, of which the space complexity is $O(|T|)$, and $|T|$ is the size of the RDF graph T. We separate RPQs into atomic units, as mentioned before, which is actually the subset of triple pattern query of the form $(?s|s, ?p|p, ?o|o)$. However, the predicates of RPQs are never variables, as a consequence, there are four possibilities for a primitive edge of RPQs $(?s|s, p, ?o|o)$, all of which DB-Index covers. For example, $(s, p, ?o)$ can be obtained by PSO_{ps}, $(?s, p, o)$ by POS_{po}, and $(?s, p, ?o)$ by PSO_{ps} with PS_p. There is no need to query (s, p, o), since we have already obtained all terms of this triple. However, RPQs are much more complex than triple pattern queries, and later in Section 5, we will present the algorithms.

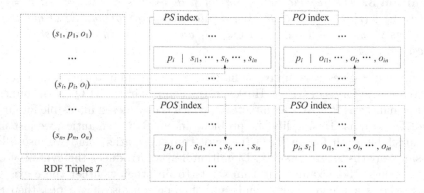

Fig. 1. The structure and construction of DB-Index

4.2   Index Construction

In order to construct DB-Index, for each triple $(s, p, o) \in T$, s is inserted into POS_{po} and PS_p, o into PSO_{ps} and PO_p, as demonstrated in Fig. 1 and Algorithm 1. If s already exists in POS_{po} and PS_p, we skip this step, the same as o. The complexity of Algorithm 1 is $O(|T|)$, in which $|T|$ is the size of the RDF graph T. As shown in Fig. 1, we ensure the indexes are ordered by inserting elements into the appropriate positions, which are to be used in the sort-merge join operations.

Algorithm 1. Constructing DB-Index

Input: An RDF graph T
Output: DB-Index
1. **constructIndex**(T)
2. **for** each triple $(s, p, o) \in T$ **do**
3.   **if** s does not exist in PS_p and POS_{po} **then**
4.   Insert s into PS_p and POS_{po};
5.   **end if**
6.   **if** o does not exist in PO_p and PSO_{ps} **then**
7.   Insert o into PO_p and PSO_{ps};
8.   **end if**
9. **end for**
10. **return**

---

5   TraPath Algorithms

In this section, we introduce a series of maintenance algorithms, called TraPath, which are based upon DB-Index, which is composed of three parts, the traversal-based search algorithm, the parallel evaluation algorithm and the scheduling algorithm. First, we present the traversal-based search algorithm, which is the core algorithm of this paper. Then, we introduce the parallel evaluation algorithm that accelerates our query. Finally, we describe the scheduling algorithm. The following theorems lay the theoretical foundation for the TraPath algorithms.

Theorem 1. *For two arbitrary edges* $p_1, p_2 \in P$. *There exists a path on them, if* $PO_{p_1} \cap PS_{p_2} \neq \emptyset$.

Proof. There exists a path on two arbitrary edges, iff there exist a common vertex between them, i.e., the union of out-degree objects of p_1 and in-degree subjects of p_2 is not an empty set.

Theorem 2. *For two arbitrary edges* $p_1, p_2 \in P$, *the in-degree of* p_1 *is specified with* s. *There exists a path on them, if* $\exists e \in PSO_{p_1 s'}$, *s.t* $PSO_{p_2 e} \neq \emptyset$.

Proof. If $PSO_{p_2 e} \neq \emptyset$, then $\exists (e, p_2, o') \in T$, $e \in PS_{p_2}$, we know $PSO_{p_1 s'} \subseteq PO_{p_1}$, and $e \in PSO_{p_1 s'}$, then $e \in PO_{p_1}$ obviously. $PO_{p_1} \cap PS_{p_2} \neq \emptyset$, according to Theorem 1, we have a path on p_1 and p_2.

The traversal-based search is devised on the basis of Theorem 2, and Theorem 1 plays an important role in the parallel evaluation algorithm.

5.1   Traversal Based Search

The traversal based search algorithm (abbreviated as TBS) takes advantage of the depth-first traversal on the basis of Theorem 2, in which nested-loops join is used to bridge various edges. For each step of TBS of a forward search, two terms of the triple are known of which the form is $(s, p, ?o)$. s is obtained in different ways, from the previous step, PS_p, or specified by users. We access objects from PSO_{ps}, and then push an o into the next step, in which o is treated as s', and $PSO_{p's'}$ is invoked again with the next predicate p'. We save the intermediate state, and push a next o if nothing obtained from $PSO_{p's'}$. The algorithm prints a result when a path is found, and exits when all elements of PSO_{ps} have been traversed, in which p is the label of the first edge of RPQs.

The pseudo code of the traversal based search is shown in Algorithm 2, in which we traverse recursively on an RDF graph. The stacks are chosen to store regular path expressions due to their belonging to sequential sets. Line 4-8 makes it possible for us to search bi-directionally. For each step, the algorithm traverses once at least, and $|list|$ times in the worse case. If we traverse $|r|$ steps, then the complexity of Algorithm 1 is $O(n^{|r|})$, where n is the max size of PSO_{ps} or POS_{ps}.

Sort-merge joins can be also taken into account in the TBS. For example, there are two sequential predicates p_1 and p_2, as described in Theorem 1, PO_{p_1}

Algorithm 2. Traversal based search

Input: $StackC$ with sequential predicates, $StackP \leftarrow \emptyset$, and start node s
Output: Path matching the given regular expression
1. **traversalBasedSearch**($StackC, StackP, s$)
2. $p \leftarrow$ pop($StackC$);
3.   push($StackP, p$)
4.   **if** $flag$ is $forward$ **then**
5. $list \leftarrow PSO_{ps}$;
6.   **else**
7. $list \leftarrow POS_{ps}$;
8.   **end if**;
9.   **if** $list$ is not empty **then**
10.   **for** each element $e \in list$ **do**
11.   **if** C is $\emptyset$ **then**
12.   Put the result into the queue $Queue$;
13.   Push($StackC$, pop($StackP$));
14.   **return**
15.   **else**
16.   **traversalBasedSearch**($StackC, StackP, e$);
17.   **end if**
18.   **end for**
19.   **end if**
20.   push($StackC$, pop($StackP$));
21.   **return**

---

is joined with PS_{p_2}. This the complexity of TBS is $O(m+n)$, which is better than the nested-loop join's $O(m \cdot n)$. However, for a predicate p in a large-scale RDF graph T, the size of PS_p appears to be extremely large. All of the terms in both PO_{p_1} and PS_{p_2} need to be loaded into the memory, which exhibits a inferior performance. Our investigation and analysis of the real-world RDF graph, which will be introduced in detail in Section 6, indicates $|PSO_{ps}| \ll |PS_p|$ for the same predicate p and arbitrary s. As a consequence, for both methods of constituting a path as represented in Theorem 1 and 2, nested-loops join achieves superior performance than sort-merge join on large-scale RDF graphs, which prompts us to choose the nested-loops join as our basic method.

5.2 Parallel Evaluation

The parallel evaluation algorithm parallelizes TBS to expedite the execution by separating paths. In the parallel evaluation process, there are several threads to process sub-paths invoking the TBS algorithm, cooperating with the master thread to collect partial results and bridge them together. For each TBS, counting paths are printed into the result queues, nevertheless, we search for an existing path in the master. Algorithm 3 and 4 gives the pseudo code of our parallelization strategies. Each $subPathProcessor$ processes a sub-path on the basis of TBS, the $pathGenerator$ gathers partial results and generates the path matching the regular expression r. The complexity of $pathGenerator$ is $O(m \cdot n)$, in which m is the max size of $Queue[i]$. The $pathGenerator$ achieves superior performance, on one hand, the max $|Queue[i]|$ could be quite small if the path is divided appropriately, on the other hand, $pathGenerator$ starts with $subPathProcessors$ simultaneously, which means they work in parallel.

Algorithm 3. Parallel producer

Input: $StackC$ with sequential predicates, $StackP \leftarrow \emptyset$
Output: Partial results $Queue$, $n \geq 2$
 1. **subPathProcessor**($StackC$, $StackP$)
 2. $p \leftarrow$ get($StackC$);
 3. **if** $flag$ is $forward$ **then**
 4. $list \leftarrow PS_p$;
 5. **else**
 6. $list \leftarrow PO_p$;
 7. **end if**;
 8. **if** $list$ is not empty **then**
 9.   **for** each element $e \in list$ **do**
10.   **traversalBasedSearch**($StackC$, $StackP$, e);
11.   **end for**
12. **end if**
13. **return**

Algorithm 4. Parallel comsumer

Input: partial results $Queue$, $n \geq 2$
Output: Path matching the given regular expression
 1. **pathGenerator**($StackQ$)
 2. **for** $Queue[i] \in Queue$ **do**
 3. $Path[i] \leftarrow$ pop($Queue[i]$);
 4. **end for**
 5. **while** $true$ **do**
 6.   **if** a path $Path$ is found **then**
 7.   **return** $Path$;
 8.   **end if**
 9.   find the smallest $Path[i] \in Path$;
10. $Path[i] \leftarrow$ pop($Queue[i]$);
11. **end while**
12. **return**

Here we have a *trade-off* between parallel algorithms with higher complexity and serial algorithms with lower complexity. We choose the former since parallel algorithms cannot only withstand the pressure of large-scale data, but also have a good scalability, of which we could take advantage to expand our index and algorithm schemes.

5.3   Scheduling

The parallel evaluation algorithm only works when no starting or ending node is initialized, of which the query expression is $(?x, r, ?y)$. It is worth mentioning that, if s or o is specified by users, it is more efficient to invoke the TBS algorithm directly. Therefore, the algorithms need to be scheduled in accordance to the form of the query syntax. The pseudo code of the scheduling algorithm is shown in Algorithm 5, which schedules different algorithms under various conditions.

Algorithm 5. Scheduling

Input: Query $Q = (?x|s, r, ?y|o)$
Output: Path matching given regular expression
1. **queryScheduling**(Q)
2. **if** $Q \in (?x, r, ?y)$ **then**
3.   Separate r into sub-paths $SubPaths$
4.   **for** each sub-path $sp \in SubPaths$ **do**
5.   Push each edge of sp into $StackC$; $StackP \leftarrow \emptyset$;
6.   **subPathProcessor**$(StackC, StackP)$;
7.   **end for**
8. **else if** $Q \in (?x, r, o)$ **then**
9. $flag \leftarrow inverse$;
10.   Push each edge of r into $StackC$; $StackP \leftarrow \emptyset$;
11.   **traversalBasedSearch**$(StackC, StackP, o)$;
12. **else**
13. $flag \leftarrow forward$;
14.   Push each edge of r into $StackC$; $StackP \leftarrow \emptyset$;
15.   **traversalBasedSearch**$(StackC, StackP, s)$;
16. **end if**;

5.4   Alternation and Kleene Star

The alternation and Kleene star operators with the existing path semantics can be both transformed to the basic paths of which the regular expressions are of the forms $r = p \mid (r/r)$. We search an alternation path by permutation and combination. For example, for a query $Q = (?s, (p_1|p_2)/p_3, ?o)$, the path p_1/p_3 and p_2/p_3 are both taken into account. For a path contains a Kleene star, the search begins with an empty path of the Kleene star, extends the path by repeating predicates, and ends with either a path obtained or no path found with a circle formed.

6   Experiments

In this section, we carry out a series of experiments to evaluate our index structures and algorithms. All experiments are conducted on a Dell OptiPlex 990 PC with a 3.10 GHz Intel i5-2400 quad-core CPU. We use Cassandra 1.1.6 as our underlying repository, and implement the algorithms with Java.

6.1 Storage Performance

We estimate the storage performance from two aspects including execution time of data loading and size of index structures, and use Lehigh University Benchmark (LUBM) as the datasets of our storage performance experiments.

The Sesame repository can be deployed only in a stand-alone environment, for the sake of fairness, we load data in a single node to compare with Sesame. As shown in Fig. 2, our approach and Sesame are equally matched in terms of loading performance. Fig. 3 displays the size of our index structure and Sesame. The space of primary indexes is larger relatively, which carries most of the information of paths. As speculated, the storage performance experiments indicates that DB-Index has a linear space complexity.

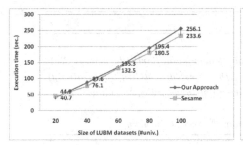
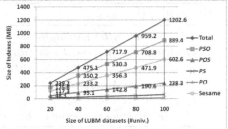

Fig. 2. Load time   **Fig. 3.** Index size

6.2 Query Performance

We use DBpedia 3.6 as the dataset for query performance experiments since it is a real-world RDF data extracted from Wikipedia. For the in-depth understanding of the internal structure of DBpedia, we have analysed the experimental dataset, and the in/out-degree size distribution of predicates is demonstrated in Fig. 4. As is shown in the figure, the sizes of primary indexes mostly distribute in the range of e^0 to e^2. It means almost all of POS_{po} and PSO_{ps} have a small size, which verifies our assumption. If we describe an instance with RDF triples as the form $(s, p, ?o)$, there will not be too many objects. For example, if we have a query $(Aristotle, name, ?o)$, and we can only obtain one object, since Aristotle has only one name. It is similar that we change the predicate as long as the subject is specified. As opposed to the primary indexes, the sizes of secondary indexes distribute more evenly, from e^0 to e^{12}, which signifies that a lot of PS_p and PO_p have a large scale. For the above example, if we replace $Aristotle$ with a variable, then the query becomes $(?s, name, ?o)$, and PS_{name} is extremely large, since almost all of the instances in DBpedia have the predicate $name$.

We have devised several RPQ test cases, among which Q_2 is generated by random walking, Q_3 by predicates random selecting, and the rest by artificial design. As shown in Table 1, Q_2, Q_4 and Q_6 have the specified in/out-degree

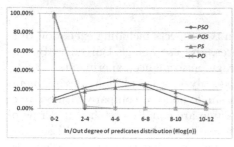 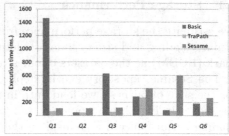

Fig. 4. Distribution of DB-Index size   **Fig. 5.** Query performance experiments

Table 1. RPQ test cases

#Q RPQ test cases
Q_1 $(?x, country/largestCity/name, ?y)$
Q_2 $(Aristotle, influenced/beatifiedPlace/leaderName/birthPlace/name, ?y)$
Q_3 $(?x, purpose/(birthPlace|deathPlace)/honours/largestCity/birthDate, ?y)$
Q_4 $(Alabama\_Army\_National\_Guard, (country/largestCity)^{100}/name, ?y)$
Q_5 $(?x, (lowestPlace/depth)^{100}, ?y)$
Q_6 $(?x, (govenment/govenmentElevation)^{100}/area, Kentucky)$

node, Q_4-Q_6 belong to long paths. As mentioned before, the Sesame repository provides weak support for RPQs, for the fairness of the evaluations, we apply long paths instead of Kleene star.

We have implemented TraPath based upon DB-Index, in which there are two *subPathProcessor*s. For the purpose of comparison, we have removed the parallel algorithms so that there is only a serial algorithm with TBS. Besides, we have deployed Sesame 2.6.0 with its own native repository in our experimental environment. The experimental results are shown in Fig. 5. As shown in the figure, our parallel approach exhibits better performance than Sesame, especially for longer RPQ paths.

7   Conclusion

In this paper, we devise a double-layer bi-directional index structure, called DB-Index, which covers all possible forms of RPQs, and propose a novel traversal-based algorithm, named TraPath, which achieves the efficient execution of RPQs on large-scale RDF graphs by using the DB-Index. The experiment results show that DB-Index has a linear complexity and TraPath outperforms the state-of-the-art methods.

Acknowledgments. This work is supported by the National Natural Science Foundation of China (Grant No. 61100049, 61373165) and the National Hightech R&D Program of China (863 Program) (Grant No. 2013AA013204).

References

1. Klyne, G., Carroll, J.J., McBride, B.: RDF 1.1 Concepts and Abstract Syntax. W3C Recommendation (2014)
2. Jupp, S., Malone, J., Bolleman, J., et al.: The EBI RDF platform: Linked Open Data for the Life Sciences. Bioinformatics (2014)
3. Breslin, J., Decker, S.: The Future of Social Networks on the Internet: the Need for Semantics. IEEE Internet Computing 11(6), 86–90 (2007)
4. Harris, S., Seaborne, A.: SPARQL 1.1 Query Language. W3C Recommendation (2013)
5. Mendelzon, A.O., Wood, P.T.: Finding Regular Simple Paths in Graph Databases. SIAM Journal of Computing 24(6), 1235–1258 (1995)
6. Bizer, C., Heath, T., Berners-Lee, T.: Linked Data - The Story So Far. International Journal on Semantic Web and Information Systems 5(3), 1–22 (2009)
7. Neumann, T., Weikum, G.: RDF-3X: A RISC-Style Engine for RDF. In: Proceedings of the VLDB Endowment, vol. 1(1), pp. 647–659 (2008)
8. Ladwig, G., Harth, A.: CumulusRDF: Linked Data Management on Nested Key-Value Stores. In: 7th International Workshop on Scalable Semantic Web Knowledge Base Systems, pp. 30–42 (2011)
9. Wang, X., Jiang, L., Shi, H., Feng, Z., Du, P.: Jingwei+: A distributed large-scale RDF data server. In: Sheng, Q.Z., Wang, G., Jensen, C.S., Xu, G. (eds.) APWeb 2012. LNCS, vol. 7235, pp. 779–783. Springer, Heidelberg (2012)
10. Alkhateeb, F., Baget, J.F., Euzenat, J.: Extending SPARQL with Regular Expression Patterns (for Querying RDF). Journal of Web Semantics 7(2), 57–73 (2009)
11. Kochut, K.J., Janik, M.: SPARQLeR: Extended SPARQL for Semantic Association Discovery. In: Franconi, E., Kifer, M., May, W. (eds.) ESWC 2007. LNCS, vol. 4519, pp. 145–159. Springer, Heidelberg (2007)
12. Koschmieder, A., Leser, U.: Regular Path Queries on Large Graphs. In: Ailamaki, A., Bowers, S. (eds.) SSDBM 2012. LNCS, vol. 7338, pp. 177–194. Springer, Heidelberg (2012)

BF-Matrix: A Secondary Index for the Cloud Storage*

Xu Cheng[1,3], Hongyan Li[2,3], Yue Wang[4], Tengjiao Wang[1,3],
and Dongqing Yang[1,3]

[1] Key Laboratory of High Confidence Software Technologies(Peking University),
Ministry of Education, China
[2] Key Laboratory of Machine Perception(Peking University),
Ministry of Education, China
[3] School of Electronics Engineering and Computer Science, Peking University,
Beijing, China
[4] Department of Computer Science, School of Information,
Central University of Finance and Economics, Beijing, China
{chengxu,eecswangyue,tjwang,dqyang}@pku.edu.cn, lihy@cis.pku.edu.cn

Abstract. Although people have proposed many kinds of NoSQL databases, also referred as Key-Value stores, there is still lack of an efficient solution for the problem of non-key attribute queries. In this paper, we propose BF-Matrix, a hierarchical index composed of bloom filter and B+ tree. Faced with the massive data and the large scale cluster, the layered solution could shorten the search path and make the best of scattered resources. Moreover, it is able to scale up and scale back according to the changes of data size and cluster scale, and isolate the job of update and retrieval in a limited scope. To eliminate the risk of false negative and to ensure our index "look like consistent", two rules are given to specify the behavior of index update and data retrieval . Experimental results demonstrate that our solution not only outperforms the state of the art, but also is flexible enough to adapt to the cloud environment.

Keywords: index, cloud storage, key-value store, NoSQL.

1 Introduction

The wave of big data is leading a profound change to information infrastructure and is reshaping the landscape of storage systems. As an important complement of traditional database, cloud data store provides a possibility to cope with the ever-growing data volume and adapt to the real-time variable storage demand. For those people who are willing to maximize value from big data, it is appealing to deploy their applications on the cloud. However, while some of the Key-Value

---

* This work was supported by Natural Science Foundation of China (No.60973002 and No.61170003), the National High Technology Research and Development Program of China (Grant No. 2012AA011002), and MOE-CMCC Research Fund.

F. Li et al. (Eds.): WAIM 2014, LNCS 8485, pp. 384–396, 2014.

stores such as BigTable[1], Cassandra[2] and Dynamo[3] have already served as the back-end for many large scale applications, the lack of built-in support for secondary index restrains more users from migrating to the cloud storage.

Even though MapReduce has acted as some systems' execution engine, like Hive[4], it is not able to provide low latency response for on-line interactive applications.

Faced with such a short slab, the traditional index techniques could neither be directly applied to the massive data nor be simply migrated to the cloud platform[11]. Although some distributed indexes have been proposed for the cloud storage, the following challenges have not been well addressed:

(1)In the face of the massive data, the current solutions did not explicitly discuss the scalability and the elasticity of their indexes, which are the major reasons for the successful and widespread adoption of cloud data store.

With the increase of data volume, the cloud storage may provide continuously growing storage space and computing power by means of scale-out. As a critical component in the cloud software stack, the index itself is also needed to be scalable. It is not only required that the index structure is capable of being scaled, but also need the index could make the most of distributed resources and keep the overhead of retrieval under control.

Furthermore, elasticity is critical to minimize operating cost while ensuring good performance during high loads. In the presence of fluctuant workloads, users may adjust available resources to match the current demands. This may result in the change of cluster scale, and also lead to frequent index expansion and contraction. Accordingly, the index should be flexible enough to adapt to the dynamic environment of cloud storage.

(2)As NoSQL databases used to achieve high availability and low latency at the expense of strong consistency, none of the existing works explored the index consistency in the absence of transaction guarantee.

Be different from the relational database, cloud data store usually does not provide cross-row, cross-table ACID-compliant transactions. In consideration of the availability and access delay, most of them opt for eventually consistency. Atomic access is supported only at the granularity of single keys. Even if some systems make effort to provide 'local' transaction guarantee, the transaction semantics are still confined to a limited range of data set. As for the job of index update, we need to write the basic table and the index table at once. This goes beyond the semantics of single key access. So the difficulty is how to ensure the index consistency.

With above observations, our motivation of this work is to give a scalable and elastic secondary index. Without the support of distributed transactions, it should still be consistent in the eyes of applications.

In this paper, to shorten the length of access path and make the best of scattered resources, we resort to the hierarchical structure to organize our index. It is a loose coupling solution. For each data node, a B+ tree is created as the local index. At the master node, a Bloom Filter Matrix(BF-Matrix) is used as the global index. In order to make our solution scalable and elastic, a vector of counting bloom filter is used to replace the native one. Moreover, we also describe the algorithms for data retrieval and index update. By presented two rules, they can avoid the problem of false negative and ensure the index table "look like consistent". While the B+ tree index local to a datanode obey ACID semantics, the global index have looser consistency.

The rest of this paper is structured as follows. Section 2 discusses the related work. Section 3 describes the problem and presents the topology of our distributed index. In Section 4, we proposed the BF-Matrix solution. In Section 5, we give the algorithms for retrieval and update. Experimental evaluation of our method is presented in Section 6. We conclude the paper in Section 7.

2   Related Work

Since the topology of distributed index usually depend on the underlying network architecture, to discuss the existing solutions, we incline to classify them according to the organization of data nodes. (1)In a peer to peer configuration, the key-value stores of this kind, such as Dynamo[3], Cassandra[2] and Riak[1], represent the cluster as a circle space or ring, and rely on the consistent hashing to locate the correct node. while the indexes of P2P structure benefit from high scalability, they are not compatible with the MapReduce paradigm and are difficult to support richer query semantics. (2)For the Master/Slave mode, a minority of nodes play the role of coordinators. As Fig. 1 illustrated, theoretically, there are three methods to construct an index for a huge mass of data. ①The first way is to create a centralized index on the master. However, it introduces a performance bottleneck and suffers from the single point of failure. In reality, people choose to distribute the whole index on multiple master nodes. BigTable[1], HBase[2], Scalable distributed B-Tree[5], ITHBase[3] are of this type. ②The second approach turn to the decentralized autonomous solution. By this way, each data node or each data split maintains an separate index by itself, and the master node is only responsible for forwarding messages to them. This kind of index can be used for scan operation and selective MapReduce jobs but fail to support random probe. HAIL[6], hindex[4], Trojan Index[7], IHBase[5] belong to this type. ③The last scheme gives consideration to both centralized and decentralized interests. Like the data chunks, the index segmentations are disseminated in the cluster. Each data node builds some local index for its data, and one or more master nodes collectively maintain a global index. So, it is actually a two-layered distributed index. While the global index indicate which datanode kept the user required data, the local index concretely point to the position in the file. Examples include CG-index[8], RT-CAN[9], EMINC[10].

---

[1] http://docs.basho.com/riak/latest/
[2] http://hbase.apache.org/
[3] https://github.com/hbase-trx/hbase-transactional-tableindexed
[4] https://github.com/Huawei-Hadoop/hindex
[5] https://github.com/ykulbak/ihbase

Fig. 1. Distributed Index of Master-Slave Architecture

3 Problem Description

To design a secondary index, as Fig. 2 illustrated, we resort to a two-layered architecture to organize our distributed index. It is a hierarchical solution, but the global index and the local index are independent and loosely coupled. On the upper layer, we use a Bloom Filter Matrix to save the knowledge of data distribution. Each row of the matrix represents a bloom filter. It corresponds to a datanode and indicates whether a data object is stored on that node. At the same time, a B+ tree is built on each datanode, which only takes charge of the local storage. There are as many rows in the matrix as there are datanodes in the cluster. All the bloom filters are homogeneous and share the same group of hash functions.

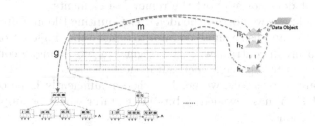

Fig. 2. The Topology of BF-Matrix

With above solution, the process of non-key attribute queries could be divided into two steps: (1)locate the relevant datanodes; (2)look up the local index.

Since the second subproblem is not different from the traditional index techniques and has been well studied by the database community, the real difficulties lie in how to construct a scalable and elastic global index, which supports the membership query, and propose effective algorithms to defend the consistency of it.

4 Index Structure

For our solution, to create a secondary index, a Standard Bloom Filter [12][13][14](SBF) is built for each datanode.

As illustrated in Fig. 3, supposing the universal set contains d possible data objects, and a total of h objects are stored on one of the nodes. According to the

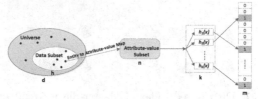

Fig. 3. Bloom filter

value of the indexed column, the data subset is mapped to an attribute-value subset of n elements. A bloom filter for representing the data subset is described by a vector of m bits, initially all set to 0. It uses k independent hash functions $h_1, h_2, \ldots, h_k$ to map the attribute value of a data object to a random number uniform over the range $\{1, \ldots, m\}$.

If the false positive rate is denoted as f, it can be deduced that $f = (1 - e^{\frac{-kn}{m}})^k$. Generally, given n and f, the optimal k and m can be calculated as $k_{opt} = \frac{m}{n} ln2$, and $m_{opt} = -\frac{n \cdot lnf}{(ln2)^2}$, which minimizes the false positive rate[12].

4.1   Deletable Index

Although SBF offers a compact probabilistic way to represent the data subset of a datanode, it does not support the removal of elements.

To fix the weakness, we borrow the idea from Counting Bloom Filter[15](CBF), as shown in Fig. 4, and use a small counter to replace the single bit of each entry in the bloom filter. In contrast with the original m bits string, a counter vector of length m is used to represent each datanode.

To avoid counter overflow, we need to choose sufficiently large counters. As the work of CBF[15] has revealed, 4 bits per counter would be amply sufficient for most applications.

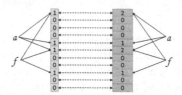

Fig. 4. Counting Bloom Filter

4.2   Scalable and Elastic Index

For a shared-nothing distributed system, if a datanode join or depart from the cluster, what we need to do is just to add a row into or delete a row from our BF-Matrix. In this case, the cost is small and the impact is limited. But if it is the data volume continues to grow, neither the SBF nor the CBF could expand with the increasing data set.

Fig. 5. Dynamic Counting Bloom Filter

Without knowledge of the upper bound on the size of data set, a target false positive probability threshold cannot be guaranteed unless the bloom filter is rebuilt from scratch each time the set cardinality changes[16][17]. In our solution, we employ a variable number of counting bloom filters to represent a dynamic data set, named Dynamic Counting Bloom Filter(DCBF). As illustrated in Fig. 5, a DCBF is made up of one or more CBFs. There are two states of a CBF: active and full. A CBF is called active only if its volume does not reach a designed upper bound, and the DCBF only inserts items of a set into the active CBFs. At the outset, an empty CBF is allocated for each datanode and the initial CBF is active; as more and more data objects are stored on the datanode, when the existing CBFs get full due to the limit on the fill ratio, a new one is added. On the contrary, if an item is removed from a datanode, a full filter may also become active. Querying is made by testing for the presence in each filter, if one of the CBFs return true, the item is assumed to be stored on the corresponding datanode. While a DCBF could scale up along with the growth of data volume, it can also constrict as the set cardinality decreases through merge operations, which is defined by the union of two bloom filters. Given two sets S_1 and S_2, suppose we have bloom filters B_1 and B_2 to represent them respectively. Assuming their length and the related hash functions are all of the same, and the sum of the volume of the two filters is not larger than the capacity of one filter. The two active CBFs could be replaced by the union of them. Then a filter B that represents the union $S = S_1 \cup S_2$ is created by taking the counter-wise *addition* of the original CBFs $B = B_1 \oplus B_2$. So, the merged filter B will report any element belonging to S_1 or S_2 as belonging to set S.

Apart from the volume growth, a DCBF could also bring the change of false positive rate. As discussed above, a DCBF with $\lceil n/c \rceil$ CBFs can represent a dynamic set S with n items. When we use the DCBF instead of S to answer a membership query, the false match probability of the DCBF can be calculated in a straightforward way.

$$f_d = 1 - (1 - f)^{\lceil n/c \rceil} = 1 - \left[1 - (1 - e^{-kc/m})^k \right]^{\lceil n/c \rceil}$$

So, the false positive rate will grows linearly with the number of CBFs contained in a DCBF. For example, suppose the false ratio of a CBF $f = 0.01$, after the DCBF expand to 10 times its original size $\lceil \frac{n}{c} \rceil = 10$, then $f_d \approx 0.1$. This is also acceptable and meaningful for our index circumstance.

4.3   False Positive and False Negative

To test whether an element is a member of a set, SBF may introduce false positive but not false negative. For the case of our distributed index, when a filter report that an element stored on a datanode by mistake, the cost is to forward a redundant query message to the datanode, and trigger a local retrieval via B+Tree index on that node. This will not destroy the correctness of search behavior or place a real burden on the system.

Although SBF never yield a false negative by itself, its variant DCBF may bring the mistake. A false negative in the index scenario indicates that even a data object is stored on a datanode, it is still possible that the filter return false. This is not allowed for cloud storage.

In order to avoid the risk of false negative, we present the first rule that restrict the behavior of index operations: To delete an index entry, first read back the data object so as to ensure the existence of data item; To insert an index entry, save the filter label with the data object so as to identify the correct filter that record the index entry.

5   Data Retrieval and Index Maintenance

Given the structure of BF-Matrix, in this section, we present the algorithms for data retrieval and index maintenance, and discuss how to ensure the consistency of our index.

5.1   Data Retrieval

It is clear that a BF-Matrix consisted of multiple DCBFs, and each DCBF is composed of several CBFs. All of the CBFs have the same length and share the same number of hash functions. To answer a membership query based on the DCBF instead of the set S, the first step is to calculate k hash values. The query operation iterates the set of DCBFs, and for each DCBF traverse over the set of CBFs. If all the *hash* counters of a CBF are set to a nonzero value, it returns true. An item is a member of a DCBF, if one of its CBFs contain the element. An item is not a member of a DCBF, if it is not found in all CBFs. After obtained the subset of DCBFs that may contained the element, we forward the query message to the corresponding datanodes, and collect result from them.

Algorithm 1. Single Attribute Query

Input: an attribute value d on column c of table T, a BF-Matrix $M_{[i][j]}$ for column c of table T, $i \in \{0, 1, \ldots, I - 1\}$ and $j \in \{0, 1, \ldots\}$, where I is the number of datanodes, the filter size is m, and the number of hash functions is k, two seed hash functions $h_1(x)$ and $h_2(x)$

Output: a set of tuples satisfied the query condition

1: $h_1 \leftarrow h_1(d)$, $h_2 \leftarrow h_2(d)$
2: **for** $t = 0$ **to** $k - 1$ **do**
3: $\quad\quad g_t(d) = (h_1 + t \times h_2 + t^2) mod\ m$
4: **for** $i = 0$ **to** $I - 1$ **do**
5: $\quad\quad$ **foreach** CBF in vector $M_{[i][]}$
6: $\quad\quad\quad$ **if** $\exists j$ for $\forall t \in \{0, 1, \ldots, k - 1\}$, $M_{[i][j]}(g_t(d)) > 0$, **then**
7: $\quad\quad\quad\quad$ forward the query message to datanode i, add label i into set W
8: collect results from the datanodes in set W

5.2   Index Maintenance

(1)Index Consistency

When it comes to the consistency of our hierarchical distributed index, the B+ tree index is treated as separate indexes for each datanode and is co-located with the primary table, it is not difficult to ensure the strong consistency of the local index. But the same problem for the global index is nontrivial.

However, although we could not keep the global index consistent with the primary table at any time, we may make the global index "look like consistent" even through sometimes it actually deviate from the user table. In other words, if an index only improve the query efficiency but not affect the results that a user read from the primary table, then we say the index table "look like consistent". It is a weaker level than the strong consistency and may lead to inconsistency between the two tables, but will not change the correct behavior of the storage system.

If we wrap the index update operations in a transaction, to implement "look like consistent", the following three conditions must be satisfied:

①after a data object has been inserted into a primary table, the users must be able to find it along the index;②an index update transaction is idempotent;③partial index transactions can be detected at read time, so the atomicity can be guaranteed.

Since most of the cloud storages employ MVCC(Multi-Version Concurrency Control) to increase concurrency, we also assume that there are multiple versions of a data object, and each version is marked by a timestamp. Thus the idempotent property of index transaction is satisfied. Fig. 6 shows the execution order of index operations. For example, Upon select, by choosing the latest version of a data object, we can filter the redundant index entries at read time, and do not need to clean them. They must be from a partial index update - either from a concurrent update that is still in progress, or from a partially failed update. This is how we tolerate partial index update and assure atomicity at read time. In conclusion, the intuition behind above operations is that, it is allowable to have an extra index entry, but not to have an entry in the main table and not in the index table. In this way, we meet the three conditions and could achieve a "look like consistent" global index. At last, we introduce the second rule that restrict the behavior of index operations: Before delete the index entry, the data item should first be removed; Before insert a data item, the index entry should first be inserted.

(2)Insert Operation

Given above two rules, Algorithm 2 contains the details regarding the process of insert operation. Due to the space limitation, we omit the description of the whole process.

(3)Delete Operation

Similarly, the pseudo-code of delete operation is given in Algorithm 3 without further explanation.

(4)Update Operation

With regard to the update operation, as illustrated in Fig. 6, it involves an index insertion and an index deletion successively. So, the job can be done by call the

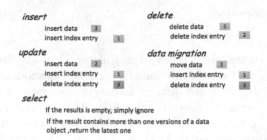

insert
  insert data 2
  insert index entry 1

delete
  delete data 1
  delete index entry 2

update
  insert data 2
  insert index entry 1
  delete index entry 3

data migration
  move data 2
  insert index entry 1
  delete index entry 3

select
  If the results is empty, simply ignore
  If the result contains more than one versions of a data
  object ,return the latest one

Fig. 6. The Execution order of Index Operations

Algorithm 2 and Algorithm 3, and we no longer present a separate one for it.
(5)Data Migration
For the purpose of load balance and fault tolerance, the data objects in cloud
storage need to be moved from one node to another node from time to time[18],
which will also trigger the index update. Similar to the update operation, it can
be solved with above algorithms as well.

6 Evaluation

In this section, we evaluate BF-Matrix, and compare it with ITHBase, hindex
and parallel full table scan(MR-FTS) from two aspects: retrieval performance

---

Algorithm 2. Insert

Input: a data object x_a with column value d, d is the attribute value on column c, y is the capacity of CBF, l is the datanode label which the inserted tuple to be stored on, a BF-Matrix $M_{[i][j]}$ for column c of table T, $i \in \{0, 1, \ldots, I-1\}$ and $j \in \{0, 1, \ldots\}$, where I is the number of datanodes, the filter size is m, and the number of hash functions is k, two seed hash functions $h_1(x)$ and $h_2(x)$
Output: true or false
1: $h_1 \leftarrow h_1(d)$, $h_2 \leftarrow h_2(d)$
2: **for** $t = 0$ **to** $k - 1$ **do**
3:     $g_t(d) = (h_1 + t \times h_2 + t^2) mod\ m$
4: $s \leftarrow M_{[l][\ ]}.length()$, $ActiveBF \leftarrow Null$
5: **for** $u \leftarrow 0$ **to** $s - 1$ **do**
6:     **if** $M_{[l][u]}.count() < y$ **then**
7:         $ActiveBF \leftarrow M_{[l][u]}$
8: **if** $ActiveBF$ is $Null$ **then**
9:     $ActiveBF \leftarrow CreateCBF(m, k)$
10:     $M_{[l][s]} \leftarrow ActiveBF$
11:     $M_{[l][\ ]}.length() \leftarrow s + 1$
12: **for** $t = 0$ **to** $k - 1$ **do**
13:     $ActiveBF(g_t(d)) \leftarrow ActiveBF(g_t(d)) + 1$

and scalability. Furthermore, we also study the overhead of index maintenance. The experiments are implemented on Hadoop 0.20.2 and HBase 0.90.4.

6.1 Experimental Setup

We perform the tests on an in-house cluster of 20 nodes connected by a $1Gbit$ Ethernet switch. Each node has 4 cores, $16GB$ of main memory, and $1.4TB$ SATA disks. For the data set, a manmade table is generated which is composed of 11 columns. There are totally $500million$ records and per record size is $270bytes$, so the whole table size is $125.8GB$. As shown in the shading part of the table schema, we create index on five columns and declare a $32 - bits$ integer to save the 'file number', which acted as the RowKey.

(**file number**, name, gender, age, occupation, native place, identity card number , email address , microblog username , skype number , drive license number)

6.2 Results and Discussion

For the first experiment, we keep the size of a 7 nodes cluster unchanged and increase the data volume from 2 million to 500 million. As illustrated in Fig. 7, the auxiliary indexes indeed greatly improve the response time of non-key attribute queries. With the growth of data volume, the MapReduce based MR-FTS confront with increased delay. This is because the fixed cluster scale lead to unchanged computing power and network bandwidth. Since hindex is 100% server side implementation with Coprocessors and the index table regions are collocated with primary table regions, when the data size is less than 15 million, the decentralized hindex outperform the centralized ITHBase. But as the data size continue to increase, the datanodes will spend more time to scan the region wise index data, the distributed method also encounter synchronous enlarged delay. By contrast, our hierarchical solution all along keeps stable performance.

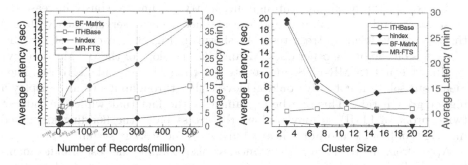

Fig. 7. Query Performance of Increased Data Set

Fig. 8. Query Performance of Enlarged Cluster

Algorithm 3. Delete

Input: a data object x_a with column value d, d is the attribute value on column c, y is the capacity of CBF, a BF-Matrix $M_{[i][j]}$ for column c of table T, $i \in \{0, 1, \ldots, I-1\}$ and $j \in \{0, 1, \ldots\}$, where I is the number of datanodes, the filter size is m, and the number of hash functions is k, two seed hash functions $h_1(x)$ and $h_2(x)$

Output: true or false

1: $h_1 \leftarrow h_1(d)$, $h_2 \leftarrow h_2(d)$

2: **for** $t = 0$ **to** $k - 1$ **do**

3: $\qquad g_t(d) = (h_1 + t \times h_2 + t^2) mod\ m$

4: **for** $i = 0$ **to** $I - 1$ **do**

5: $\qquad$ **foreach** CBF in vector $M_{[i][\ ]}$

6: $\qquad\qquad$ **if** $\exists j$ for $\forall t \in \{0, 1, \ldots, k-1\}$, $M_{[i][j]}(g_t(d)) > 0$, **then**

7: $\qquad\qquad\qquad$ forward the query message and data object x_a to datanode i

8: $\qquad\qquad\qquad$ add label i into set W

9: collect results from all the datanodes in set W

10: identify the datanode that really stored x_a, saved as $l \in W$

11: identify the CBF that actually record the index entry of x_a, saved as r

12: $ActiveBF \leftarrow M_{[l][r]}$

13: **for** $t = 0$ **to** $k - 1$ **do**

14: $\qquad ActiveBF(g_t(d)) \leftarrow ActiveBF(g_t(d)) - 1$

In the second experiment shown in Fig. 8, we choose to scale out the cluster under a constant data set. In the literature, more nodes could share the overhead of computing and storage, MR-FTS evidently benefit from the enhanced cluster capability. At the same time, as more nodes join the cluster, the index segmentation and the data partition of the same record are more likely to be scattered on different datanodes. This introduces more cross-node interactions, which amplify the access latency. Since hindex could make the most of cluster resources, the average latency at first decreases. However, for each query, the decentralized method has to interact with all the datanodes and wait for their responses. After the number of nodes is over 11, the wait time increase. Compared with the previous methods, BF-Matrix could always restrict the query processing in a limited scope, and has a good scalability.

To explore the maximum throughput, we try to simultaneously increase the cluster size and data volume, and use another 5 client server, each with 100 threads, to perform a stress test. As shown in Fig. 9, the curve of the BF-Matrix is on top of the other ones. Meanwhile, the slope of the BF-Matrix curve is almost equal to MR-FTS, and is steeper than hindex and ITHBase. They mean our solution not only could achieve higher throughput but also display a higher growth rate. This can be explained by the fact that while the indirect twice probe of ITHBase may prolong the access path, the flooding forwarded messages of hindex also put a heavy burden on the storage system.

Apart from the retrieval performance, we also examine the cost of index maintenance. To illustrate the average extra run time, we compare the latency of different operations on non-indexed table with the latency on indexed table, and show the differences in Fig. 10. Since ITHBase implement transactions support to

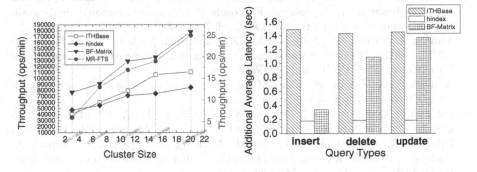

Fig. 9. The Maximum Thoughput   **Fig. 10.** The Cost of Index Maintenance

guarantee that all secondary index updates are consistent, it triggers the biggest latency in the three methods. On the contrary, hindex bring an almost constant delay across different operations. For the BF-Matrix, the delete operation actually contains a read to confirm the correct datanode. So its cost is larger than the insert. With regard to the update operation, the overhead roughly equate to an insert plus a delete.

7   Conclusion

In this paper, we try to address the problem of non-key attribute queries on the huge mass of data. The goal is to offer a secondary index in the cloud data store.

The main contributions of our work include:

✓To achieve a scalable and elastic solution, we utilize bloom filter and traditional B+ tree to construct a hierarchical index. It not only adapts to the continuous growth of data set, but also is able to cope with the dynamic change of cluster size.

✓The counting bloom filter and the dynamic bloom filter bring the problem of false negative, which is unacceptable for the storage system. By giving the first behavior rule, we eliminate the possibility of false negative.

✓In the absence of transaction guarantee, we introduce a weaker concept than the strong consistency and give the second behavior rule. It makes the global index "look like consistent" so as to assures the correct behavior of applications.

References

1. Chang, F., Dean, J., Ghemawat, S., Hsieh, W.C., Wallach, D.A., Burrows, M., Gruber, R.E.: Bigtable: A distributed structured data storage system. In: Proc. of 7th OSDI, pp. 305–314 (2006)
2. Lakshman, A., Malik, P.: Cassandra: a decentralized structured storage system. ACM SIGOPS Operating Systems Review 44(2), 35–40 (2010)

3. DeCandia, G., Hastorun, D., Jampani, M., Kakulapati, G., Lakshman, A., Pilchin, A., Sivasubramanian, S., Vosshall, P., Vogels, W.: Dynamo: amazon's highly available key-value store. In: Proc. of SOSP, vol. 7, pp. 205–220 (2007)

4. Thusoo, A., Sarma, J.S., Jain, N., Shao, Z., Chakka, P., Anthony, S., Liu, H., Wyckoff, P., Murthy, R.: Hive: a warehousing solution over a map-reduce framework. In: PVLDB, vol. 2(2), pp. 1626–1629 (2009)

5. Aguilera, M.K., Golab, W., Shah, M.A.: A practical scalable distributed b-tree. In: PVLDB, vol. 1(1), pp. 598–609 (2008)

6. Dittrich, J., Quian-Ruiz, J.A., Richter, S., Schuh, S., Jindal, A., Schad, J.: Only aggressive elephants are fast elephants. In: PVLDB, vol. 5(11), pp. 1591–1602 (2012)

7. Dittrich, J., Quian-Ruiz, J.A., Jindal, A., Kargin, Y., Setty, V., Schad, J.: Hadoop++: Making a yellow elephant run like a cheetah (without it even noticing). In: PVLDB, vol. 3(1-2), pp. 515–529 (2010)

8. Wu, S., Jiang, D., Ooi, B.C., Wu, K.L.: Efficient b-tree based indexing for cloud data processing. In: PVLDB, vol. 3(1-2), pp. 1207–1218 (2010)

9. Wang, J., Wu, S., Gao, H., Li, J., Ooi, B.C.: Indexing multi-dimensional data in a cloud system. In: Procs. of the 2010 ACM SIGMOD International Conference on Management of Data, pp. 591–602. ACM, NY (2010)

10. Zhang, X., Ai, J., Wang, Z., Lu, J., Meng, X.: An efficient multi-dimensional index for cloud data management. In: Procs. of the CloudDB 2009, pp. 17–24. ACM, NY (2009)

11. Lu, P., Wu, S., Shou, L., Tan, K.L.: An efficient and compact indexing scheme for large-scale data store. In: 2013 IEEE 29th International Conference on Data Engineering (ICDE), pp. 326–337 (2013)

12. Bloom, B.H.: Space/time trade-offs in hash coding with allowable errors. Communications of the ACM 13(7), 422–426 (1970)

13. Broder, A., Mitzenmacher, M.: Network applications of bloom filters: A survey. Internet Mathematics 1(4), 485–509 (2004)

14. Tarkoma, S., Rothenberg, C.E., Lagerspetz, E.: Theory and practice of bloom filters for distributed systems. IEEE Communications Surveys & Tutorials 14(1), 131–155 (2012)

15. Fan, L., Cao, P., Almeida, J., Broder, A.Z.: Summary cache: a scalable wide-area web cache sharing protocol. IEEE/ACM Transactions on Networking (TON) 8(3), 281–293 (2000)

16. Almeida, P.S., Baquero, C., Preguica, N., Hutchison, D.: Scalable bloom filters. Information Processing Letters 101(6), 255–261 (2007)

17. Guo, D., Wu, J., Chen, H., Yuan, Y., Luo, X.: The dynamic bloom filters. IEEE Transactions on Knowledge and Data Engineering 22(1), 120–133 (2010)

18. Wang, T.J., Lin, Z.Y., Yang, B.S., et al.: MBA: A market-based approach to data allocation and dynamic migration for cloud database. Science China Information Sciences 55(9), 1935–1948 (2012)

A Cluster Based Schema Design
for Multi-tenant Database

Jiacai Ni and Jianhua Feng

Department of Computer Science and Technology,
Tsinghua University, Beijing 100084, China
njc10@mails.thinghua.edu.cn, fengjh@tsinghua.edu.cn

Abstract. Existing multi-tenant database schema design studies either
emphasize high performance and scalability at the expense of limited
customization or provide enough customization at the cost of low performance and scalability. In this paper, we propose a cluster based schema
design for multi-tenant database which supports full customization without sacrificing performance and scalability. We devise the hierarchical
agglomerative clustering algorithm and multi-tenancy integration algorithm to integrate the customized schemas based on the schema and instance information. Our method can be easily applied to existing
databases with minor revisions. Experimental results show that our
method achieves excellent performance and high scalability with schema
customization property.

1 Introduction

The multi-tenant database system can amortize the cost of hardware, software
and professional services to a large number of tenants and has become one significant trend to provide the database service. Therefore, it has attracted more
and more attention from both industrial and academic communities.

The multi-tenant system usually predefines some tables which are called *predefined tables*. In predefined tables, the system provides some attributes which
are called *predefined attributes*. In order to design the schema which fully satisfies tenants' requirement, the system must allow tenants to configure their
customized tables or add some new customized attributes in the *predefined tables*. The tables newly configured are called *customized tables*, and the attributes
configured both in predefined and customized tables are called *customized attributes*. Providing high customization is one of the important features in the
multi-tenant database in order to be competitive in the service. Aulbach et al.
[2] emphasized that a high-level customization is very attractive to tenants.

To our best knowledge, state-of-the-art approaches on the multi-tenant schema
design can be broadly divided into three categories [4]. The first one is Independent Tables Shared Instances (ITSI). It maintains a physical schema for each
customized schema. This method has poor scalability [1] since it needs to maintain independent tables for each tenant. The second is Shared Tables Shared
Instances (STSI). In this method different customized tenants share only one

F. Li et al. (Eds.): WAIM 2014, LNCS 8485, pp. 397–401, 2014.

table. This method has poor performance and consumes more space since it contains large numbers of NULLs. The third simplifies the service system and does not allow tenants to precisely configure their private schema in order to achieve high performance, low space and better scalability. To address these limitations, we devise hierarchical agglomerative clustering algorithm and multi-tenancy integration algorithm to integrate schemas based on schema and instance information. Experimental results show that our method with full schema customization property achieves high performance and good scalability with low space.

2   Cluster Based Schema Design

2.1   Basic Idea

The schema design for the multi-tenant database can be divided into three aspects. (1) The predefined attributes in predefined tables are often the compulsory part. Thus we can just put them together, the new tables are very dense and performance can be guaranteed. In addition, the scalability is excellent. So the design for the predefined attributes in predefined tables is straightforward and easy. (2) For the customized attributes in predefined tables, if we put them into the predefined tables, the predefined tables can be extremely wide and sparse or even cannot hold all the customized attributes. The performance significantly degrades. If we maintain one extension table for the customized attributes, the number of extension tables can be really large. The scalability becomes poor. (3) For customized tables, similar to the customized attributes in predefined tables, if we maintain one table for each of them, the scalability becomes the bottleneck. Current studies have no effective solutions for customized tables. To sum up, the first category is easy to solve, the second and the third are both the customized schemas and pose great challenges. In this paper we focus on how to redesign high-quality schemas for customized schemas.

2.2   Cluster Based Schema Design

The number of customized tables is large, customized tables is changing, we need to optimize the schema periodically. We propose the cluster based methods to integrate customized schemas. Different from traditional schema integration, the integrated schemas in the multi-tenant database reflect the physical data storage. Instead of directly integrating attributes among all the customized tables we need to avoid the integrated attributes distribute too far. Thus, we first cluster similar customized tables and then integrate the schemas in each cluster.

Clustering Customized Tables: Salesforce is a well-known company which adopts the multi-tenancy architecture to provide the software service. In Salesforce, the data type are very expressive. Besides traditional data types *boolean, date, double, int*, it also defines *textarea, email, phone, url, percent* and some other concrete data types which can directly reflect the content of the attributes especially for the various string fields. Similarly, in the Microsoft Dynamics system accurate data types are also defined. In the SAP Business Suit the data types are even more detailed. Then we can use the data type information to build the feature vector for each customized table and then cluster.

Definition 1 (Table Feature Vector). *The dimension of the vector is the number of the data types supported in the system. The ith bit value in the vector is the number of the attributes with the ith data type in the customized table.*

After we define the table feature vector for each customized table, we next utilize Cosine function to compute the similarity value between two customized tables. Considering that we will integrate the customized schemas in each cluster, we want to avoid the "noisy" customized schema and we do not know the number of clusters in advance. We adopt the hierarchical agglomerative clustering algorithm. Initially, each table feature vector is one single cluster. Then if the similarity of the most similar pair of clusters is larger than the predefined threshold τ_{tab} we merge them into a new cluster. The clustering process stops when the similarity between the most similar clusters is smaller than τ_{tab}. Different threshold τ_{tab} mean different clustering quality and build different numbers of clusters. Among the classic methods, the Elbow Method is the usual way because it is effective and obvious. In customized, we adopt the same method to determine the threshold. We graph the percentage of variance of cluster number against τ_{tab}. At some τ_{tab}, the number of clusters increases sharply and give an angle, then the threshold τ_{tab} is set as " Elbow".

Multi-Tenancy Attribute Integration: Based on the clustering results, next we integrate the customized schemas into one integrated table for each cluster. Considering that the data type information in the multi-tenant applications is very expressive to represent the content of the attribute. Thus in the attribute integration process, the attributes can only be integrated between the ones with the same data type. In order to further ensure the accuracy, we take the constraint and instance information into account. Li et al. [3] utilized many specification information to integrate attributes. Besides data type, among the constraint information we refer to uniqueness, nullable and data length. Among the instance information, data cardinality information can reflect the content and is easily obtained by calling the aggregate function. Next we define the Attribute Feature Vector for each customized attribute to illustrate all the information above.

Definition 2 (Attribute Feature Vector). *In Attribute Feature Vector, the first bit A_0 represents uniqueness, unique is 1 otherwise 0. The second bit A_1 is represents nullable, 1 means this bit may have NULL value, or else 0. The third A_2 is the data length bit, and the fourth bit A_3 is the data cardinality.*

Then the similarity between attributes A_i and A_j is computed as follows:

$$S(A_i, A_j) = s(A_{i_0}, A_{j_0}) + s(A_{i_1}, A_{j_1}) + s(A_{i_2}, A_{j_2}) + s(A_{i_3}, A_{j_3})$$

In the equation above, if only one of the two attributes are unique, then $s(A_{i_0}, A_{j_0})$ is 0; otherwise it is 1. $s(A_{i_1}, A_{j_1})$ is computed similarly. For the latter

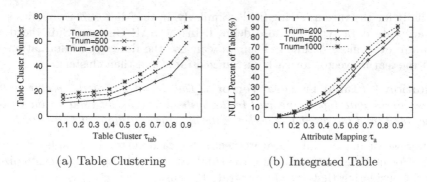

(a) Table Clustering   (b) Integrated Table

Fig. 1. Schema Integration

two parts, we both use the ratio of the minor value to the max. For each customized table t we maintain the attribute matrix M_t, each column corresponds
to one attribute feature vector. The attribute integration algorithm firstly selects
the customized table with most attributes as *base cluster*. Then it iteratively selects the table most similar to the base cluster. Given one customized table, for
each attribute we integrate it to the attribute with the same data type. Further
we must integrate it to the most similar integrated attribute above the threshold.
Instead, we add a new column in the integrated table.

3   Experimental Study

Based on the typical tables and data type information in Salesforce we design our
fully customized multi-tenant benchmark. When generating customized tables
we define following parameters. The number of total tenants is T_{num}. The average
number of customized tables for each tenant is Tab_{num}. The average number of
attributes for each customized table satisfies the normal distribution $N(\mu, \sigma^2)$,
where μ represents the average number of attributes and σ is variance and set
as 2. We use MySQL to conduct experiments on Windows machine. Figure 1(a)
shows the number of clusters under different values of τ_{tab}. In our experiments,
Tab_{num} is set 10 and μ is set 15. We see that when the number of tenants is 200
and 500, there is a sharp increase of the cluster number when τ_{tab} is altered from
0.8 to 0.9. Thus τ_{tab}=0.8 is the "Elbow" point. When we use the cluster based
method to integrate the schema, we set τ_{tab}=0.8. After determining the number
of clusters, we integrate the customized schemas in each cluster. Figure 1(b)
shows the average NULL percent of each integrated table. We find as τ_a increases
the integrated tables become more sparse. When τ_{tab} is above 0.6, more than
half of the tables are NULLs. Thus we should choose appropriate τ_a.

4   Conclusion

In this paper, we propose cluster based method and multi-tenancy attribute
integration algorithm to integrate the customized schemas with tenant configuration, schema and instance information so as to enhance the integration quality.

Experimental results show that our framework with fully customized property achieves high performance, good scalability and low space as well.

References

1. Aulbach, S., Grust, T., Jacobs, D., Kemper, A., Rittinger, J.: Multi-tenant databases for software as a service: schema-mapping techniques. In: SIGMOD Conference, pp. 1195–1206 (2008)
2. Aulbach, S., Seibold, M., Jacobs, D., Kemper, A.: Extensibility and data sharing in evolving multi-tenant databases. In: ICDE, pp. 99–110 (2011)
3. Li, W.-S., Clifton, C.: Semantic integration in heterogeneous databases using neural networks. In: VLDB, pp. 1–12 (1994)
4. Ni, J., Li, G., Zhang, J., Li, L., Feng, J.: Adapt: adaptive database schema design for multi-tenant applications. In: CIKM, pp. 2199–2203 (2012)

Tenant-Oriented Composite Authentication Tree for Data Integrity Protection in SaaS

Li Lin[1,2], Li Qingzhong[1,2,*], Kong Lanju[1,2], and Shi Yuliang[1,2]

[1] School of Computer Science and Technology, Shandong University, Jinan, China
[2] Shandong Provincial Key Laboratory of Software Engineering
lilinsdu@126.com, {lqz,syl,klj}@sdu.edu.cn

Abstract. SaaS is an emerging model that allows tenants to host computation and data to cloud service providers. However, untrustworthy service providers might violate tenants' data integrity by deleting, modifying and falsifying tenants' data due to some benefits. So it is important for tenants to verify their query results integrity. In this paper, we propose a tenant query result authentication structure, called MTAS(Multi-tenant Authentication Structure), that applies to multi-tenant shared pivot-universal storage model. MTAS separates indexes with authentication structures to support isolation and customization characteristics of multi-tenant application. And we present composite authentication tree-PUA tree for tenant data in pivot table and universal table in MTAS. Comparing with traditional authentication tree approaches, PUA tree only needs one tree travel to get verification object(VO) corresponding to query results in pivot table and universal table. And PUA tree saves about 30% hash computing at verification stage.

Keywords: SaaS, Multi-tenancy, Integrity, Authentication.

1   Introduction

In SaaS mode, by leasing the application and putting data onto the service providers, tenants can pay more attention to their business. However, the malicious insiders of service providers may violate tenant data integrity for financial or other benefits. So tenants should be able to validate their query results are both correct and complete. Correctness implies that the results indeed come from the tenant's legitimate original data and have not been changed in anyway. Completeness means that all qualified results are included in the result set.

Being a new software delivery mode, SaaS has many significantly differences compared with traditional software application in actual design and deployment. And those differences bring new challenges to multi-tenant data integrity protection.

For example, in order to make full use of resources, most SaaS service providers adopt the single-instance multi-tenancy strategy[1] in which multiple tenants' data are stored in one physical table such as universal table and data of different types may be

---

F. Li et al. (Eds.): WAIM 2014, LNCS 8485, pp. 402–414, 2014.

stored into a flex column based on tenants' customization [2].To the best of our knowledge, how to construct authenticated data structures for multi-tenants in a shared data table has not been considered in related works, as it is commonly assumed that each data table is hosted by a single user. Meanwhile, in order to speed up query process and ensure performance isolation in large multi-tenants database, adequate pivot tables [3] are established to store tenants' index data. So the integrity protection of those index data can not be ignored in SaaS. It is an important part of tenants' data integrity protection consideration and has not been discussed in related work.

Additionally, independent tenants may customize different integrity requirements based on their needs. Therefore, how to support customization and isolation characteristics among tenants is another challenge for SaaS.

In this paper, we present a tenant oriented data authentication structure (MTAS). MTAS provides data integrity assurance for multi-tenant data in pivot table and universal table. By separating indexes with authentication structures, MTAS preserves tenants' isolation and customization characteristics. And we propose a new authentication structure *PUA tree* (Pivot and Universal table Authentication tree) which composite separate authentication trees built for pivot table and universal table into a single tree based on the characteristic of pivot-universal storage model. So we can get the VO corresponding to queries data in pivot table and universal table in one *PUA tree* travel. *PUA tree* saves about 30% hash computing at VO verification. Also, *PUA tree* can handle dynamic structure adjustments for tenant data update operations, such as data insertion, deletion and modification.

The rest of this paper is organized as follows. The next section covers related works. In Section 3 present the system model, storage model and attack model. Section 4 introduces MTAS for tenant data authentication and experiment is presented in Section 5. Section 6 gives the conclusion of this paper.

2   Related Work

Now, multi-tenant data integrity protection research is on the beginning. But in the relevant areas such as outsourced database, many techniques have been proposed to address data integrity issues.

One relevant prior work researched integrity issues in outsourced databases and suggested solutions using authenticated data structures based on MHT [4]. MHT is a binary tree originally proposed for efficient authentication of equality queries in a database sorted on the query attribute. Reference [5] incorporated MHT into a B+-tree [8] called MB tree to facilitate verification. The MB-Tree was basically a B+-Tree that hierarchically organized digests, in the same way as the MHT. In addition to the actual result, the VO transmitted to the client contained two boundary records as well as a set of digests by which the client can reconstruct the digest of the root. Based on [5], Reference [6] promoted EMB^--tree and EMB^*-tree to reduce the VO size. Reference [7] focused on relational stream environments and provided CADS index scheme to minimize the processing and transmission overhead. Reference [9] presented index authentication scheme iBigTable that provides data integrity assurance for BigTable. Reference [10] aimed at aggregation queries and promoting index authentication structures on SUM queries. However, traditional index authentication structures lack the ability to discern the tenant and it is hard for them to guarantee performance isolation of different tenants.

Reference [12] proposed a partially materialized digest scheme in which split the authentication structure from the data index to provide users with the option to receive a non-authenticated answer, without any performance degradation due to verification structures. This structure is attractive for multi-tenant database. Because it is easier to guarantee performance isolation by constructing separate authentication structure for different tenants. But [12] did not consider how to deal with the situation that the data table is shared by multiple tenants meanwhile different data types are stored into a flex column.

In addition to authenticated data structures, signature aggregation is another solution for data integrity protection. Reference [13] investigated the notion of signature aggregation which enables bandwidth efficient integrity verification of query replies. Base on this scheme, Reference [14] used signature aggregation and chaining to achieve correctness and completeness of query replies. Reference [11] discussed a signature aggregation protocol that provided freshness guarantees. Reference [15] introduced a scheme for users to verify that their query results were integrity with access control policies. However, signature aggregation approach was inefficient compared with authenticated data structures because signature aggregation involves modular multiplication, signing and verification operations that were 100, 10,000 and 1,000 times slower than hashing[11].

Besides the above approaches, Reference [16, 17] inserted certain fake tuples into the real data and verified query integrity by checking the fake tuple in the result. Reference [18] presented the dual encryption approach, where certain data are encrypted with different keys and query integrity could be checking by "cross examination". But they only provided probabilistic assurance for the owner data and were not suitable for multi-tenant data authentication.

For multi-tenant data integrity protection, traditional index authentication structures could not guarantee the performance isolation of different tenants. While the aggregation approach was inefficient for its high resources consumption. In this paper, we introduce the general model of MTAS for multi-tenant data integrity. As in [12], MTAS separates authentication structure with index to ensure tenant's isolation and customization characteristic. And we propose composite data authentication structure *PUA tree* for tenant data in pivot table and universal table to promote the efficiency of VO construction and verification.

3   Preliminary

This section gives the over view of system model, storage model and attack model in SaaS.

System model. The system model includes three entities: tenant, trusted third party and service provider[19].

Tenant T: T is a client that customizes and consumes SaaS applications. In SaaS mode, tenants rely on the service provider for data maintenance and computation.

The trusted third party (TTP): The trusted third party is used to assist tenants for their secret key and integrity policy information management. TTP can prohibit unauthorized parties from getting tenant's privacy information.

Service provider(SP): Service provider is responsible for the operation of SaaS platform. SaaS platform is a mechanism that has significant computing resource and storage space to maintain the tenants' applications and data storage.

Based on the research on trust platform[20, 21], we assume that the SaaS platform is trusted, and we explore an integrity protection module(IPM) in platform. IPM can assist tenants to customize their data integrity policies and verify the returned results. We assume that all communications go through a secure channel between the SP, TTP and tenants.

Storage model. In this paper, we mainly discuss the scenario that multiple tenants share a single application with logical view $R(A_1, A_2, ..., A_n)$. Tenants' data stored in universal table in one data node and adequate pivot table are set up to serve as the index data. We mainly aim at the case that searching key has no duplicate. Here we introduce some basic definition of the Pivot-Universal storage model.

Definition 1. Universal table[3]. Universal table stores all the tenant data. *Universal Table*= U *<GUID, TenantID, { U value$_i$,i=1...n}>*. The *GUID* (globally unique identifier) column is the primary key of a universal table, *TenantID* is the identifier of a tenant, *value$_i$* corresponding to A_i in R.

Definition 2. Pivot table[3]. Pivot table stores tenant index data, *Pivot table* =*<TenantID&indexID:I>*, I=U *<value$_j$, GUID)*. *indexID* is the identifier of a tenant index, *value$_j$* is the index data that tenants customized.

Attack model. Based on the above assumptions, we concentrate on the analysis of attack behaviors form the *SP* malicious insiders. For example, insiders violate tenant data integrity by deleting the record of a tenant in universal table, changing the data records in pivot table or universal table or forging some non-existence records into tenants' data.

4   Tenant Oriented Data Integrity Protection

For data authentication in Pivot-Universal storage model, one way is to set up traditional authentication trees for pivot table and universal table according to all query attribute. In Fig.1., we use a variant of *DMH-Tree*[7] to set up the authentication structures for pivot table and universal table. For convenience, we call them *PA tree* and *UA tree* below. From Fig.1., we can see if a tenant customizes m query attributes, we have to pre-compute and store m *PA trees* and m *UA trees* separately to prove integrity of query replies. This results in a higher storage overhead in the SP and much more complex VO construction and verification. Also storing multiple trees for the same universal table increases the cost of updates.

To resolve those problems, in this section, we first present the general model of MTAS with composite data authentication structure *PUA tree* for tenant data in pivot table and universal table. Then we give the VO construction and verification algorithms in *PUA tree*. Next we discuss the dynamic data operation in MTAS and the security and cost of MTAS.

4.1   Multi-tenant Data Authentication Structure-MTAS

Here we first introduce the general model of MTAS. As shown in Fig.2., MTAS contains two levers: the upper is a index built on *TenantID&PUATreeID*, and the lower is a collect of composite data authentication structure *PUA tree*. Here, MTAS separates *PUA tree* with data index to ensure tenant's isolation and customization characteristic. In this way, query answering and VO generation can be executed paralleling. So the authenticated process of some tenants can not cause performance degradation to those tenants don't customize the integrity protection.

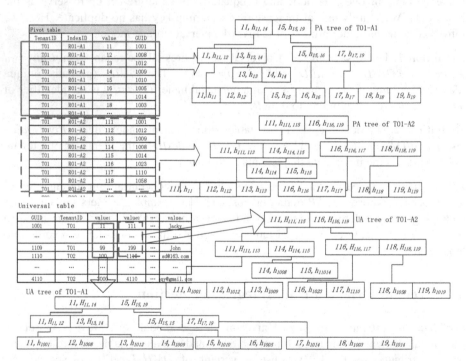

Fig. 1. An example of the basic solution for Pivot-Universal storage model

Definition 3. Multi-tenant Data Authentication Structure(MTAS). MTAS=< *TenantID&PUATreeID: {∪ PSA tree_i, i = 1…k}>*. PUA tree is a composite tree that merge two *DMH-Tree* into a single tree based on index data element *I. PUA tree=< {∪Lnode_i}, {∪Inode_j}, Root-PUA>, i=1..n,j=1..m,*

(i) *Lnode* presents leaf nodes, $Lnode=<k,guid,H_{guid}>$, k is the searching key of *PUA tree*, *guid* is the identifier corresponding to k in I. H_{guid} is record hash of Universal table.

(ii) *Inode* is internal nodes, $Inode=<k,h,H,p>$, where k is the search key of the first record in the subtree of *Inode*, p is a pointer to the corresponding child node. H is the digest of its children hash concatenation on H_{guid} . The value of h depends on the level. For the adjacent upper level *(AUL)* of *Lnode*, h is a hash value on the concatenation of all $<k,guid>$s in the node pointed by p; for the upper levels, h is computed on the concatenation of the hash values of the children entries pointed by p.

(iii) *Root-PUA* is a value H_{root} by hashing the concatenation of the hash values contained in the root of the tree $H_{root}=h(TenantID\&indexID\&h_{root})$. The symbol & denotes string concatenation. *Root-PUA* is signed with the tenant's private key.

From definition 2, we can see pivot table is a two level index, in which *TenantID&indexID* can be treated as shared information and element I contains correspondence relationship between pivot table and universal table. Based on those characteristics, *PUA tree* combines *PA tree* and *UA tree* into a single tree to reduce the number of authentication structures. We can get the VO corresponding to queries in one *PUA tree* travel instead of double travels of *PA tree* and *UA tree* which greatly reduces the complexity of the VO construction.

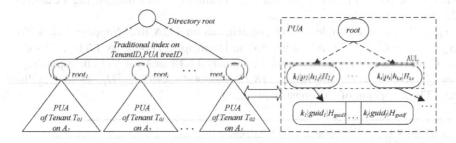

Fig. 2. General model of MTAS

The query process with integrity assurance of MTAS model is :first tenant sends query request and then the *SP* return the result set of pivot table and universal table called *set(QI)* and *Set(QS)* along with corresponding VO in one *PUA* tree round-trip. When IPM receives result set, he runs verification algorithm to verify the integrity of the received data. If verification is success, IPM returns the result to tenant through application.

Notice that the *PUA tree* is a composite tree with paratactic hashes on pivot table and universal table instead of constructing *PUA' tree* in a simple way that mix the pivot table and universal table together to set up a *DMH tree* with $Lnode=<k, h>$ and h is hash value of record concatenation of pivot table and universal table where *pivot.GUID=universal.GUID*. Because in some circumstance tenant' expected result maybe get in *set(QI)*, so the *SP* does not need to go in to the universal table, the *PUA* tree could generate the *VO* for *set(QI)* separately without the message on universal table.

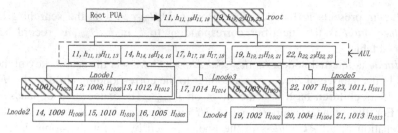

Fig. 3. An example of *PUA tree* of T_{01}

4.2   VO Construction and Authentication

The VO construction on *PUA tree* is similar to *DMH*. The main difference is we have to deal with two hash values in the *Inode*. Suppose tenant T_{01} request a query Q :[q_L, q_U], $q_L(q_U)$ is the lower(upper) bound of Q. The SP performs two top-down traversals on PUA tree to locate the *Lnodes* L_l and L_u that contains q_{L^-} and q_{U+} immediately before q_L and after q_U, respectively to get the sibling hashes on the path.. These boundary nodes are necessary to enforce completeness. The *VO* includes: (i) the signed *PUA tree* root , (ii) all left sibling hashes to the path of q_{L^-}, (iii) all right sibling hashes to the path of q_{U+}, (iv) auxiliary information (*AI*) that indicate the location of a node in the tree structure.

Fig.3. shows an example of VO construction on PUA tree. Suppose tenant T_{01} a query on A_1 as $Q:[12,17]$. According to the lower(upper) boundary 12(17), *SP* locates *Lnode1* and *Lnode3* for the expanded boundary 11,18 and get VO corresponding to query Q contains $<11,1001,H_{1001}>,<18,1003,H1003>,<19,h_{19,23},H_{19,23}>$ as gray lined nodes. The algorithm of VO construction is:

Algorithm 1. *VO-PUA*(Expanded Q; PUA tree T)

1: Append *Root* to the *VO*
2: Extends Q and get the q_{L^-} and q_{U+}
3: *Range PUA* (T.root, q_{L^-}) //get the left sibling hashes on the path of q_{L^-}
4: *Range PUA* (T.root, q_{U+})//get right sibling hashes to the path of q_{U+},
RangePUA (Node *N*, key *k*)
5: For each node *n* in *N*
6: If *n* is a Inode and n.k<k and n+1.k>k // n may contains results
7: RangePUA (n.p, k)
8: Else append *n* to the *VO*
9: Else If *n* is a Lnode and n.k<k or n.k=k
10: //return all the sibling node with left boundary node that have the same father node
11: Append n to *VO*
12: If n is a Lnode and n.k=k or n.k>k// n is the right boundary node
13: Append n to *VO*

The verification process at the IPM utilizes auxiliary information (*AI*) that indicate the tree-structure information to compute the hash value H_{root}. Upon receipt of the query result *set(QI)* and *Set(QS)*, IPM combines it with VO components (ii) and (iii), to reconstruct the(missing) part of the *PUA tree* between the paths of q_{L-} and q_{U+}. Then, she verifies with the owner's public key whether the component (i) of the VO matches the locally computed H_{root}. If they match, *Set(QI)*and *Set(QS)* are deemed both complete and authentic. The verification algorithm is:

Algorithm 2. VO verification (Verification Object VO, *Set(QS)*, *Set(QI)* , *AI*)

1: Retrieve q_{L-} and q_{U+} nodes in VO
2: For each element in *set(QS)* compute $H._{GUID}$
3: Sort *set(QI)* and $H._{GUID}$ on searching key get *Lnode* set L
4: Combine VO and L with *AI* to get reconstruct queue *VO-L*
5: *h'root= Reconstruct(VO-L)*
6: verify *h'root* with VO.*Root* or reject
Reconstruct (Object VO-L)
7:   *S=Null, P=Null*;
8:   While *VO-L* still has entries
9:     Remove next entry e from *VO-L*
10: If *e* is a *Inode* Append *e.h* to S and *e.H* to P
11: If *e* is *Lnode* Append *e.k* and *e.guid* to S
12: Append *e.H* to P
13: If *e* is left boundary Append *Reconstruct (e)*
14: If *e* is right boundary, return *hash(S)*, and *hash (P)*

4.3   Dynamic Data Operation

MTAS can support integrity protection adjustment for dynamic data operations such as insertion, modification and deletion. Due to the space limit, we focus on discussing how to insert a new record into the universal table. Before insertion, IPM determines *PUA trees* in MTAS that involved in this update and makes adjustment on those *PUA trees*. Here we mainly explain adjustment process of a single *PUA tree*.

When tenant T_{01} insert a new record in his logical view *R*, *SP* inserts corresponding physical record *r'* into universal table and updates pivot table with *I'*.

In a *PUA tree*, IPM locates the *Lnode N* that element *I'* belongs, computes *H'* =$h(r')$ and insert *<I',H' >* into node *N*. After insertion, if the number of keywords is no more than the *PUA tree* fanout *f*, IPM computes *N'* father node in *AUL* with *<I',H' >* directly and computes bottom-up recursively until root node *Root'*. Otherwise if *N* overflows after insertion, node *N* splits into two nodes *N', N''* each contains $\lceil (f+1)/2 \rceil$ keywords. Then *PUA tree* adjusts bottom-up recursively until root node *Root'*. Fig.4. shows a simple example of *PUA tree*(*f*=3) adjustment after insertion *r'* with searching *k*=25.

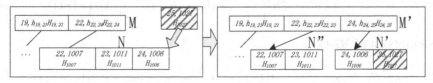

Fig. 4. Data Insertion

The data deletion operation is similar with data insertion. For data modification, there are two types of modification: one is occurring on the specified searching key, the other is not. The first category, we treat is as a deletion followed by an insertion as in[7]. For the second category, we only need to update the hash value of *PUA tree* on the corresponding node to the modified record.

4.4   Security Analysis

The correctness of *Set(QI)*and *Set(QS)*can be guaranteed by *PUA tree* due to the security of collision-resistance hash functions and the public key digital signature for the hash value of the root node. The completeness of *Set(QI)*can be assured by the sorted leaves and the boundary leaves that enclose the select range. Then we will prove that *PUA tree* can guarantee the completeness of *Set(QS)*.

Lemma 1. If the *Set(QI)* is correct and complete, any deletion on the universal table of *Set(QS)* can be checked by *PUA tree*.

Proof. As the query process can be treated as equal-join query between pivot table and universal table with join condition *Pivottable.GUID= Universaltable.GUID* in their respective attribute. And the *GUID* attribute is the globally unique identifier for record level locating, it is a one-to-one correspondence between Pivot table and Universal table. The verification process on *PUA tree* can be treated as comparing *Set(QI)* with *Set(QS)* on *GUID*, if *{GUID/ Set(QI)}={GUID/ Set(QS)}*,we can say that *Set(QS)* is complete. So give the precondition that *Set(QI)* is correct and complete, we can check the completeness of *Set(QS)*.

4.5   Cost Analysis

In this section, we compare our work with set up separate authentication trees approach (for simple we call it SA in the following) for pivot table and universal table. Here we assume that the fan-out of *PA tree, UA tree* and *PUA tree* is f. And tenant has N records in logical view R and N_Q is the result number of Q. $/k/$, $/p/$, $/h/$ and $/s/$ denote the size of the searching key, pointer, hash value and signature respectively. The signature and hash cost are C_v and C_h. Here we mainly compare the Storage cost, VO construction size and Verification cost.

Storage Cost. The storage cost of SA can be treated as the sum up of *PA tree* and *UA tree*.

$$C_s^{SA} = \sum_{i=1}^{n}\left(C_s^{PA} + C_s^{UA}\right) \approx 2n \cdot \left(\frac{Nf-1}{f-1} \cdot (|k|+|h|) + \frac{N-1}{f-1} \cdot |p|+|s| \right) \quad (1)$$

The storage cost of *MTAS* is:

$$C_s^{MTAS} \approx \sum_{i=1}^{n} C_s^{PUA} \approx n \cdot \left(\frac{Nf-1}{f-1} \cdot (|k|+|h|) + \frac{N-1}{f-1} \cdot (|p|+|h|) + N \cdot |k|+|s| \right) \quad (2)$$

The storage cost is the total cost of a tenant with n searching attributes. From the above formula we can get the conclusion that storage cost of MTAS is less than SA Next we will discuss the VO construction cost on one searching attribute.

VO Construction. In SA the VO involves digest path to $q_{l-}$and q_{u+} of *PA tree* and *UA tree*. VO also includes auxiliary information (*AI*) about their position in the *PA tree* and *UA tree* to facilitate proper verification at the *IPM* side. Similar to [6], we ignore this information in our analysis, as it is negligible compared to the size of the hashes in the VO. The max size of the SA is:

$$C_s^{SA} = 4 \cdot (\log_f N - 1) \cdot (f - 1) \cdot (|k| + |p| + |h|) + 2 \cdot |s| \tag{3}$$

The max size of the VO on MTAS is:

$$C_s^{MTAS} = 2 \cdot \left[(\log_f N - 2) \cdot (f - 1) \cdot (|k| + 2|h| + |p|) + 2(f - 1) \cdot (2|k| + |h|) \right] + |s| \tag{4}$$

Verification cost. Given the *set(QI)* and *Set(QS)* and the VO, the IPM has to compute the missing digests and combine them with the VO to retrieve the root of the *PA tree* , *UA tree* and *PUA tree*. Suppose d and d' is the height from root to the root of query results, $d = (\log_f N - \log_f N_Q)$. The VO verification cost of *PA tree* and *UA tree* is:

$$C_c^{PA} \approx C_c^{UA} \approx \left((N_Q \cdot f - 1)/(f - 1) + d \cdot (f - 1) \right) \cdot C_h + C_v \tag{5}$$

The VO verification cost of *PUA tree* is:

$$C_c^{PUA} = \left((N_Q \cdot (f + 1) - 2)/(f - 1) + 2 \cdot d \cdot (f - 1) \right) \cdot C_h + C_v \tag{6}$$

From above cost analysis, we can see SA makes the VO much complicated at construction and verification compared with MTAS. Because it needs to deal with two separate authentication trees, which obviously cause performance degradation at initial set up, VO construction and verification.

5   Experiment

We carry out a serious simulation experiments to demonstrate our analysis of the SA approach and MTAS. The development environment is Eclipse-SDK-4.3.1-win 32 Bit, operating system is Windows XP Professional Service Pack 3, CPU is Inter Core (TM)2 2,33GHz, and the memory is 2GB. We utilize RSA signatures that are typically 128 bytes in size and SHA1with20-byte outputs.

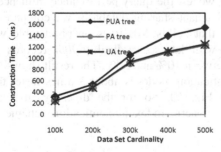

Fig. 5. Total time for construction

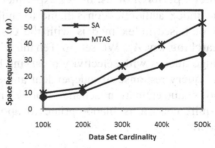

Fig. 6. Hash for VO verification

Construction Cost. First we give an analysis on the fanout f for *PUA tree*. We set the data cardinality to *500K*, if the block size is 4k the maximum of *PUA tree* is 84. The tree height and VO size are shown in Table 1. From the result we can see as the f increases the *PUA tree* height decreases but the VO size grows sharply, when $f=2$ the VO has the smallest VO size while remains the highest tree height, so in our following experiment we adopt the eclectic $f=8$ as *PUA tree* fanout so is *PA tree* and *UA tree*. We set up the initial construction of one tenants T_{01} with data set cardinality of *100K, 200K, 300K, 400K* and *500K* records on the T_{01}-A_1.

Fig.5 describes the initial construction time for *PA tree*, *UA tree* and *PUA tree*. From the result we can see *PA tree* and *UA tree* require roughly the same total time while *PUA tree* needs a little more. Because *PUA tree* needs to deal pivot table with universal table together. Fig.6 shows the total storage cost of SA and MTAS, and MTAS has a lower storage cost. Because SA has to store both the *PA tree* and the *UA tree*, so the overall storage cost is larger than MTAS.

Table 1. The height and VO size of PUA tree

Fanout	2	4	8	16	32	64
Height	19	10	7	5	4	4
VO size(Byte)	1736	2576	3992	5768	9776	18272

Verification Cost. To investigate the verification cost, we create workloads of random 100 range queries with selectivity σ varying between 10% and 50% on data cardinality *300K* on T_{01-A1}. Fig.7 and Fig.8 shows the total hash operations VO verification times of SA and MTAS. From Fig.7 and Fig.8, we can see MTAS reduces about 30% hash computing compared with SA. The reason for this is SA has to deal with not only the *PA tree* but also the *UA tree* to ensure the integrity, so the overall performance is lower than MTAS.

Update Cost. For the dynamic data operation, we experiment the time consumption on data insertions between SA and MTAS on an insertion parameter σ varying between 10% and 50% of data cardinality *300K* on T_{01-A1}. The insertions are processed simultaneously, Fig.9 shows the overall update time. From the result we can see MTAS has a faster total update speed for he only has to deal with a single *PUA tree*.

Query performances. In this experiment, we test the query performance influence of the index authentication scheme to multi tenant sharing database. We compare the pivot based index models with the case that builds MB trees on pivot table with searching key A_1. We set up Tenants T_{01} with data set cardinality *300k* records and range queries with selectivity σ varying between 10% and 50%. The result shows that the query response of independent authentication model is about 3 times faster than index authentication scheme, shown in Fig. 10. So for the diversity of tenant requirement, it is inappropriate to apply generic index authentication scheme for tenants.

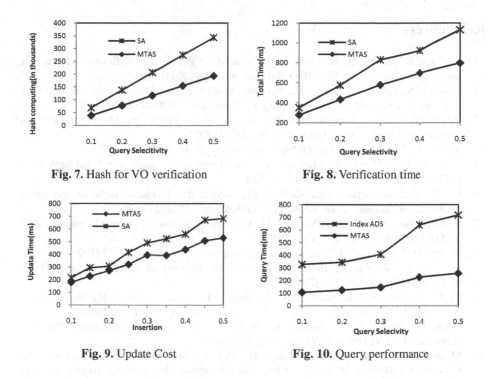

Fig. 7. Hash for VO verification   **Fig. 8.** Verification time

Fig. 9. Update Cost   **Fig. 10.** Query performance

6   Conclusion and Future work

In this paper, we propose a authentication structure MTAS that accommodate the multi-tenant properties perfectly by establishing composite authentication *PUA trees* for each tenant based on their integrity demands. Comparing with traditional approaches MTAS shows advantage at storage cost, VO construction and verification. In the future, we will go on our research on dealing with tenant's data verification structure on different sharing storage modes like chunk table.

Acknowledgement. National Natural Science Foundation of China under Grant No.61272241, No.61303085, No. 61303005; Natural Science Foundation of Shandong Province of China under Grant No.ZR2013 FQ014; Science and Technology Development Plan Project of Shandong Province No. 2012GGX10134; Independent Innovation Foundation of Shandong University under Grant No.2012TS075, No.2012TS074; Shandong Province Independent Innovation Major Special Project No.2013CXC30201; the Shandong Distinguished Middle-aged and Young Scientist Encouragement and Reward Foundation NO.BS2012DX015.

References

1. Aulbach, S., Jacobs, D., Kemper, A., Seibold, M.: A comparison of flexible schemas for software as a service. In: SIGMOD 2009, pp. 881–888 (2009)
2. Aulbach, S., Grust, T., Jacobs, D., Kemper, A., Rittinger, J.: Multi-Tenant Databases for Software as a Service: Schema-Mapping Techniques. In: SIGMOD (2008)
3. Weissman, C.D., Bobrowski, S.: The Design of the Force.com Multitenant Internet Application Development Platform. In: SIGMOD (2009)
4. Merkle, R.C.: A Certified Digital Signature. In: Crypto, pp. 218–238 (1989)
5. Devanbu, P.T., Gertz, M., Martel, C.U., Stubblebine, S.G.: Authentic Third-party Data Publication. In: DBSec, pp. 101–112 (2000)
6. Li, F., Hadjieleftheriou, M., Kollios, G., Reyzin, L.: Dynamic authenticated index structures for outsourced databases. In: SIGMOD, pp. 121–132 (2006)
7. Papadopoulos, S., Yang, Y., Papadias, D.: Continuous authentication on relational streams. VLDB J (VLDB) 19(2), 161–180 (2010)
8. Comer, D.: Ubiquitous B-Tree. ACM Computing Surveys 11(2), 121–137 (1979)
9. Wei, W., Yu, T., Xue, R.: iBigTable: practical data integrity for bigtable in public cloud. In: CODASPY, pp. 341–352 (2013)
10. Li, F., Hadjieleftheriou, M., Kollios, G., Reyzin, L.: Authenticated Index Structures for Aggregation Queries. ACM Trans. Inf. Syst. Secur (TISSEC) 13(4), 32 (2010)
11. Pang, H., Zhang, J., Mouratidis, K.: Scalable Verification for Outsourced Dynamic Databases. In: PVLDB, vol. 2(1), pp. 802–813 (2009)
12. Mouratidis, K., Sacharidis, D., Pang, H.: Partially Materialized Digest Scheme: An Efficient Verification Method for Outsourced Databases. International Journal on Very Large Data Bases 18(1), 363–381 (2009)
13. Mykletun, E., Narasimha, M., Tsudik, G.: Authentication and integrity in outsourced databases. TOS 2(2), 107–138 (2006)
14. Narasimha, M., Tsudik, G.: Authentication of Outsourced Databases Using Signature Aggregation and Chaining. In: Li Lee, M., Tan, K.-L., Wuwongse, V. (eds.) DASFAA 2006. LNCS, vol. 3882, pp. 420–436. Springer, Heidelberg (2006)
15. Pang, H., Jain, A., Ramamritham, K., Tan, K.-L.: Verifying Completeness of Relational Query Results in Data Publishing. In: ACM SIGMOD, pp. 407–418 (2005)
16. Xie, M., Wang, H., Yin, J., Meng, X.: Integrity Auditing of Outsourced Data. In: Proceedings of the 33rd International Conference on Very Large Data Bases (VLDB 2007), pp. 782–793 (2007)
17. Xie, M., Wang, H., Yin, J., Meng: Providing, X.: freshness guarantees for outsourced databases. In: EDBT 2008, pp. 323–332 (2008)
18. Wang, H., Yin, J., Perng, C., Yu, P.: Dual encryption for query integrity assurance. In: Proceedings of the 17th ACM Conference on Information and Knowledge Management (CIKM 2008), pp. 863–872 (2008)
19. Shi, Y., Zhang, K., Li, Q.: Meta-data Driven Data Chunk Based Secure Data Storage for SaaS. JDCTA 5(1), 173–185 (2011)
20. Brown, A., Chase, J.S.: Trusted platform-as-a-service: a foundation for trustworthy cloud-hosted applications. In: CCSW 2011, pp. 15–20 (2011)
21. Alsouri, S., Feller, T., Malipatlolla, S., Katzenbeisser, S.: Hardware-based Security for Virtual Trusted Platform Modules. CoRR abs/1308.1539 (2013)

Efficient Graph Similarity Join with Scalable Prefix-Filtering Using MapReduce

Jun Pang[1], Yu Gu[1], Jia Xu[2], Yubin Bao[1], and Ge Yu[1]

[1] College of Information Science and Engineering,
Northeastern University, Liaoning, 110819, China
pangjun@research.neu.edu.cn, {guyu,baoyubin,yuge}@ise.neu.edu.cn
[2] School of Information System and Management,
National University of Defense Technology, Changsha, 410073, China
xujia.neu@gmail.com

Abstract. The graph similarity join retrieves all pairs of similar graphs on graph datasets. In this paper, we propose an efficient MapReduce-friendly algorithm tackling with the graph similarity join problem on large-scale graph datasets. In particular, the efficiency of our algorithm is guaranteed by: 1) scalable prefix-filtering suitable for q-gram alphabet that is beyond the memory; 2) an effective candidate reduction strategy that greatly cuts down the data communication cost; 3) a two-round data access proposal that reduces the data access overhead. Extensive experiments on large-scale real and synthetic datasets demonstrate that our proposal outperforms the state-of-the-art method with higher system scalability and faster speed.

1 Introduction

With the quick growth of graph data generated and collected by many applications in social networks, bioinformatics and chemistry, there is a huge demand of developing effective analysis tools on the big graph datasets. The graph similarity join provides an indispensable functionality to such analysis tasks. However, most previous graph similarity join algorithms are in-memory algorithms, being incompetent to analyze the graph datasets with large sizes. Worse still, the graph similarity functions, e.g. graph edit distance, is commonly computationally expensive [1], making the performance of the graph similarity join faced with large-scale sets a serious concern.

To solve the problems above, a potential solution is to resort to the popular distributed computation paradigms, such as MapReduce [2][3]. However, to our best knowledge, it has not been reported that the works of large-scale graph similarity joins based on MapReduce. In this paper, we implement the *GSimJoin* algorithm [6] in parallel, that is the state-of-the-art centralized graph similarity join method with edit distance constrains. In particular, we optimize this parallel algorithm with scalable prefix-filtering and compression techniques.

We propose the progressive *MR-GSimJoin* algorithm in Section 2. Extensive experimental results are reported in Section 3. We discuss related work in Section 4 and Section 5 concludes this paper.

F. Li et al. (Eds.): WAIM 2014, LNCS 8485, pp. 415–418, 2014.

2 *MR-GSJ* Algorithm

This paper focuses on undirected label graphs and employs the popular graph edit distance as the graph similarity metric.

Definition 1. *Given two graph sets R and S, and a similarity threshold τ, the graph similarity join retrievals all pairs of similar graphs from each graph set, i.e., $\{ (r, s)|GED(r, s) \leq \tau, r \in R, s \in S \}$.*

For ease of description, this paper focuses on self-joins scenario. Our solution can also be extended to implement the general graph similarity join between any two different graph sets.

We first propose a naive MapReduce implementation named N-GSJ based on $GSimJoin$. First, one MapReduce job is utilized to compute the q-grams global frequency. Second, another MapReduce job is started to implement candidate filtering and verifying. Although N-GSJ algorithm seems simply, it will suffer serious performance issues due to the same problems as algorithms in [6][8].

In order to solve problems above, we propose the MR-GSJ algorithm, which sorts q-grams of every graph according to the global frequency without computing the q-grams global order to improve the efficiency and the scalability of the prefix filtering. Moreover, the MR-GSJ utilizes the *graph id* to produce the candidate id pairs, then substitutes the *graph id* with the graph data through two-round dataset access to relieve the data explosion of the reduce phase. The dataflow of the MR-$GSimJoin$ framework is illustrated in Fig. 1. The map and reduce functions are shown in Table 1.

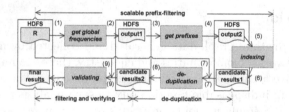

Fig. 1. The data flow of MR-$GSimJoin$ framework

Table 1. MR-$GSimJoin$ map functions and reduce functions

Job.	F	Input	Output	Job.	F	Input	Output
getf	M	$<id_r, r>$	$set< hc, (id_r, vertices)>$	de-du	M	$<id_r, r>, <*, list(id)>$	$<id_r, r>, <*, list(id)>$
	R	$<hc, set(id, vertices)>$	$<hc, [sum, set(id, vertices)]>$		R	$<id_{r_1}, data_{id_{r_1}}>,$ $list<id_{r_1}, set(id_{r_2})>$	$<id_{r_1}, [set(id_{r_2}), data_{id_{r_1}}]>$
getp	M	$<hc, [sum, set(id, vertices)]>$	$set< (id, sum), (hc, vertices)>$	validating	M	$<id_{r_1}, r>$	$<(id_r, 0), r>$
	R	$<(id, s), set(hc, vertices)>$	$<id, set(hc)>$			$<id_{r_1}, [set(id_{r_2}), data_{id_{r_1}}]>$	$set<(id_{r_2}, 1), (id_{r_1}, data_{id_{r_1}})>$
indexing	M	$<id, set(hc)>$	$set< hc, id>$		R	$<(id_{r_2}, 0), data_{r_2}>, set<$ $(id_{r_2}, 1), (id_{r_1}, data_{id_{r_1}})>$	$set<(id_{r_1}, id_{r_2}),$ $GED(id_{r_1}, id_{r_2})>$
	R	$<hc, set(id)>$	$<*, list(id)>$				

3 Experiments

The MR-$GSimJoin$ algorithm is compared with the N-GSJ proposal and the state-of-art proposal SSJ-2. The following parameters were compared: 1)running time, 2)scaleup and 3)speedup.

The **real** dataset *AIDS* contains 40,000 graphs, one of which represents an AIDS chemical compound ($http://dtp.nci.nih.gov/docs/aids/aids\_data.html$).

All **synthetic** graph datasets are generated with the GraphGen [4]. Statistics of the SYN_1 synthetic dataset are described as follows: $ngraphs$=10,000,000, $graph\ size$=20, $nnodel$=100, $nedgel$=6, and $density$=0.3.

We employ a cluster of 30 nodes, one of which has the same configuration: two 3.1GHz CPUs, 8GB RAM, 500GB hard disk, Redhat 4.4.4-13 and Hadoop-0.20.2.

The running time of all three proposals is compared over sample datasets from *AIDS* and SYN_1. The results described in Table 2 show that *MR-GSimJoin* outperforms both *N-GSJ* and *SSJ-2* over sample datasets of *AIDS*. Because *N-GSJ* need to filter and verify duplicate results, which introduces unnecessary expense. *SSJ-2* generates candidate pairs and then remotely accesses the DFS for achieving graph data in its third MapReduce job.

The speedup of *MR-GSimJoin* is analyzed on the SYN_1 dataset as shown in Fig. 2. Fig. 2(a) shows that the *MR-GSimJoin* running time reduces with the node increasing from 6 to 30, which displays a satisfactory speed up. Fig. 2(b) depicts that a higher parallelization can better accelerate the processing.

The scaleup of *MR-GSimJoin* is evaluated on the SYN_1 dataset as shown in Fig. 3. The running time of *job one-four* slightly increases with the growth of *scalefactor*, while *validating* displays a obvious upward trend. This is because the growth rate of candidate pairs is beyond the dataset growth rate, yielding more and more candidate results in the process of scalable prefix-filtering with the augment of dataset size.

Table 2. Running time of *N-GSJ*, *SSJ-2* and *MR-GSimJoin* on a 6-node cluster

dataset	R10K			R20K			R30K			S10K			S20K			S30K		
algorithm	N-GSJ	SSJ-2	MR-GSimJoin	N-GSJ	SSJ-2	MR-GSimJoin	N-GSJ	SSJ-2	MR-GSimJoin	N-GSJ	SSJ-2	MR-GSimJoin	N-GSJ	SSJ-2	MR-GSimJoin	N-GSJ	SSJ-2	MR-GSimJoin
jobOne time(s)	22	22		25	25	29	25	25	29	33	33	20	36	36	33	47	47	37
jobTwo time(s)	1,027	30	29	4,238	31	32	9,844	34	38	53	47	41	158	196	53			81
jobThree time(s)		>1,800	26		>5,400	26		>10,800	26		26	26		26	26			26
jobFour time(s)			26			30			32			26			26			26
jobFive time(s)		62			183			554				26			26			26
total time(s)	1,049	>1,852	172	4,262	>5,456	300	9,869	>10,859	679	86	106	148	194	258	164			196
alphabet size(MB)	0.05	0.05		0.10	0.10		0.11	0.11		53	53		105	105		159	159	

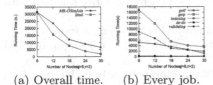

(a) Overall time.   (b) Every job.

Fig. 2. Running time for joining SYN_1 dataset on m-node cluster(where m=6, 12, 18, 24 and 30)

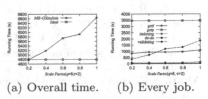

(a) Overall time.   (b) Every job.

Fig. 3. Running time for joining SYN_1*n datasets(where n=0.2, 0.4, 0.6, 0.8 and 1.0) on a 30*n-node cluster

4   Related Work

The graph similarity join problem has been widely studied in recent years[1][5][6]. However, none of these solutions can handle increasingly large-scale data sets due to the lack of scalability. Many algorithms based on MapReduce are proposed to perform similarity joins over large-scale datasets of various data types, such as documents, sets, etc[7][8][9][10][11][12]. Unfortunately, these existing algorithms are not directly applied to tackle similarity joins on large-scale graph datasets.

5   Conclusion

This paper proposes a parallel $MR\text{-}GSimJoin$ algorithm based on the MapReduce framework to perform similarity joins on large-scale graph datasets. A scalable prefix-filtering technique is proposed to adapt the $MR\text{-}GSimJoin$ algorithm to suit large q-gram alphabets. The candidate results are compressed to reduce the communication cost and enhance the performance. The data access costs are decreased through a two-round $full\ copy$ method. Extensive experiments on real and synthetic datasets demonstrate that our algorithm outperforms the state-of-the-art method in efficiency and scalability.

Acknowledgments. This work is supported by the National Natural Science Foundation of China(Nos. 61272179,61173028).

References

1. Zeng, Z., Tung, A.K.H., Wang, J., et al.: Comparing stars: On approximating graph edit distance. In: PVLDB, vol. 2(1), pp. 25–36 (2009)
2. MapReduce, http://hadoop.apache.org/
3. MapReduce, http://en.wikipedia.org/wiki/MapReduce
4. A graph synthetic generator, http://www.cse.ust.hk/graphgen/
5. Wang, G., Wang, B., Yang, X., et al.: Efficiently indexing large sparse graphs for similarity search. In: TKDE, pp. 440–451 (2010)
6. Zhao, X., Xiao, C., Lin, X.M., et al.: Efficient graph similarity joins with edit distance constraints. In: ICDE, pp. 834–845 (2012)
7. Vernica, R., Carey, M.J., Li, C.: Efficient parallel set-similarity joins using MapReduce. In: SIGMOD, pp. 495–506 (2010)
8. Baraglia, R., De Morales, G.F., Lucchese, C.: Document similarity self-Join with MapReduce. In: ICDM, pp. 731–736 (2010)
9. Metwally, A., Faloutsos, C.: V-SMART-Join a scalable MapReduce framework for all-pair similarity joins of multisets and vectors. In: PVLDB, vol. 5(8), pp. 704–715 (2012)
10. Afrati, N.F., Sarma, D., et al.: Fuzzy joins using MapReduce. In: ICDE, pp. 498–509 (2012)
11. Elsayed, T., Lin, J., Oard, D.W.: Pairwise document similarity in large collections with MapReduce. In: ACL, pp. 265–268 (2008)
12. Silva, N.Y., Jason, M., et al.: Exploiting MapReduce-based similarity joins. In: SIGMOD, pp. 693–696 (2012)

SSD-Aware Temporary Data Management Policy for Improving Query Performance

Zhiliang Guo, Jiangtao Wang, Chunling Wang, and Xiaofeng Meng

Renmin University of China, Beijing, China
{guozhiliang,jiangtaow,wangchunling,xfmeng}@ruc.edu.cn

Abstract. As data volume grows rapidly and queries become more complex, database engine has to deal with larger amount of temporary data when performing operators such as sort and join. Generally, these temporary data are organized as small temporary files stored in HDD. Due to the poor access performance on HDD, processing time of I/O operations on these files has a direct impact on the response time of queries. Compared to HDD, solid-state drive (SSD) offers more random IOPS and comparable sequential bandwidth. So, using SSD to replace HDD may be an ideal pattern when dealing with temporary data. In this paper, we find out that query processing performing improvement is unsatisfactory if we store temporary files in SSD directly. Thus, we propose a SSD aware temporary data management policy called STDM. STDM takes both the I/O behaviors of temporary data and SSD physical characteristics into account. Temporary data are cached in memory at first, and then written to SSD in an append-only fashion. In this way, we reduce random writes and take full advantage of fast random reads on SSD. We implement a prototype based on PostgreSQL and evaluate its performance on TPC-H benchmark. Experiment result shows that STDM improves query processing performance dramatically compared to the traditional method for organizing temporary data.

Keywords: SSD, Temporary Data, Query Processing, Hybrid Storage.

1   Introduction

Complex queries often need to be dealt with in database applications such as On-Line Analytical Processing (OLAP) and Geographical Information Systems (GIS). When performing these queries, large amount of temporary data are generated by query processing operators. For examples, hash join algorithm partitions each input table into hash buckets, and then process these buckets one by one; sort algorithm generally partitions an input data set into smaller runs, sorts the runs separately, and then merges them into a single sorted file. As data volume grows rapidly, temporary data can no longer be totally maintained in memory. Generally, they are organized as temporary files storing in disk and a lot of random I/Os are invoked during both their generation and consumption phases. However, because of the rotational latency, the cost of random I/Os on hard disk drives (HDDs) is sufficiently large, if temporary data are stored in HDDs, random access time will dominates the query processing time and limits the

F. Li et al. (Eds.): WAIM 2014, LNCS 8485, pp. 419–430, 2014.

overall performance. Thanks to the emergence of solid-state drives (SSDs) in recent years, we have a great alternative to HDD for storing temporary data. Compared to HDDs, SSDs can offer more random IOPS and comparable sequential bandwidth [1]; moreover, they have many other significant advantages like lighter weight, better shock resistance and lower power consumption. After years of development, SSD has been widely adopted in personal and server computers [2, 3].

Since the price per gigabyte of SSD is much more expensive than HDD, enterprise users usually adopt hybrid systems which consist of SSDs and HDDs. To design a hybrid storage system that uses SSD to store temporary data, several problems should be considered. First, SSD offers poor random write performance because of write amplification. Write amplification refers to additional writes caused by garbage collection and wear leveling. It is an undesirable phenomenon associated with flash memory and results in sub-optimal performance improvement if SSD is used as a simple drop-in replacement for disk [4]. Thus, it is necessary for us to redesign the temporary data management policy to take advantage of fast random reads and reduce random writes on SSD. Second, we must provide several efficient in-memory index structures for locating data in memory or SSD, and reclaiming SSD space quickly.

In this paper, we propose a novel temporary data management policy called STDM (SSD aware temporary data management policy). STDM takes both the I/O behaviors of temporary data and SSD physical characteristics into account. In order to identify the effectiveness of STDM, we implement a prototype system based on PostgreSQL 9.2, which is a world-famous, open source object-relational database system. The key contributions of this work are summarized as follows.

— We observe the process of temporary data generation and compare system performance improvements on TPC-H benchmark of two approaches: one is enlarging the size of shared memory buffer; the other is storing temporary data files on SSD directly. Our analysis further confirm the conclusion that processing time of I/O operations on temporary data has a direct impact on the response time of queries and limits the overall performance of database [12, 13]. Test results also show that traditional temporary data management policy doesn't suit for SSD.

— We propose STDM, a novel temporary data management policy which aims to improve the performance of processing complex queries by writing temporary data to SSD in an append-only fashion. And with other efficient measures, STDM can take advantage of fast random reads and reduce random writes on SSD.

— We implement a prototype system based on PostgreSQL 9.2 and evaluate its performance on TPC-H benchmark. Experiment result shows that STDM can manage temporary data efficiently and get obvious performance improvement over original PostgreSQL.

The rest of this paper is organized as follows. Section 2 reviews the previous work on improving query processing performance with SSD. In Section 3, we run tests to observe the generation of temporary data files and the influence of I/Os on these files; and then we discuss the importance of changing the way they are organized. In Section 4, we describe the basic framework and key components of STDM. Section 5 contains performance evaluations and result analyses. Section 6 is the conclusion of this paper.

2   Related Work

In recent years, SSD is expected to gradually replace HDD as the primary permanent storage media in both personal and server computers because of its outstanding advantages mentioned above. This tendency has also attracted researchers to improve database performance at the aspect of optimizing query processing with SSD.

Some works focus on optimizing query algorithms and page layout, RARE-join is designed for a column-based page layout on flash [5]. It uses a mixture of random reads and sequential I/O. When only a fraction of the rows or columns are needed, it leverages the fast random reads of SSD to retrieve and processes less data and thereby improves performance. D. Tsirogiannis introduces FlashJoin, a general pipelined join algorithm that minimize accesses to database and intermediate relational data [6]. Similarly, DigestJoin optimizes non-indexed join processing by reducing the intermediate result size in row-based database [8]. Some works utilizing the internal parallelism of SSDs, B+-tree variant PIO B-tree presents a new I/O request concept called psyncI/O that can exploit the internal parallelism of SSD in a single process [9]. Another work proposes a parallel table scan operator called ParaScan and a hash join operator called ParaHashJoin to take full advantage of internal parallelism of SSDs [10]. From the same viewpoint in our work, hStorage-DB optimizes temporary data management by utilizing semantic information. It classifies requests to different types and temporary data are assigned a corresponding QoS to ensure they are cached once generated and evicted out at the end of their lifetimes [11]. However, different from hStorage-DB, STDM uses SSD to store temporary data only, rather than mixing them with data of query results.

3   Problem Definition

In general, each query operator preserves a small size of memory to buffer temporary data blocks. When the buffer is full, it creates a file in HDD and writes the blocks into it. Note that for a complex query, several operators might be running in parallel; each operator will be allowed to use as much memory as this size before it starts to write data into temporary files. Also, several running sessions could be doing such operations concurrently. Therefore, the total memory used could be many times the size of this memory. Since the available size of memory for database system is limit, it is necessary to keep it small enough to support more concurrent connections. However, the negative impact is that a large amount of small files will be generated.

To better understand generations of temporary data files', we run all 22 queries in TPC-H at a scale factor of 30. TPC-H illustrates decision support systems that examine large volumes of data, execute queries with a high degree of complexity, and give answers to critical business questions.

Table 1 shows the test result. Note that some queries only generate a very small amount of temporary files, so we ignore them in the table. The first row represents the query number of the 22 queries; the second row represents the number of temporary files generated and the third row represents the total size of temporary data generated by the query. The file numbers and sizes are calculated by detecting changing of all generated files in temporary data directory. Specifically, we can see that the number of temporary files depends on the complexity of corresponding query so much that they varies from tens to thousands. The total file size of each query is not in proportion to its file number; this is because in sort algorithm, large sorted file by merging small sorted runs are calculated too.

Traditional temporary data management policy is obviously inefficient and not friendly to secondary storages. There are two main reasons: On the one hand, cost for managing such a large number of files cannot be ignored. What's more, most operating systems restrict the number of concurrently opened file descriptors because they need memory to manage each opened file. If there is no limit, a userland software will be able to create files endlessly until the server goes down. On the other hand, because random I/Os exist either in generation or consumption phase during lifetimes of these files, as the count of concurrent connections increases, performance of file system will fall badly. Popular industry practice is merging multiple small files into a large file to reduce the number of files. In this paper, we learn from this approach and optimize it for SSD.

Table 1. Count and size of Temporary files

Query	2	3	5	7	8	9	10	14	16	18	19	21
Number	15	317	196	324	551	32	2171	511	126	258	510	8191
Size(MB)	988	2530	6507	2521	2812	5528	1918	223	550	124	277	3300

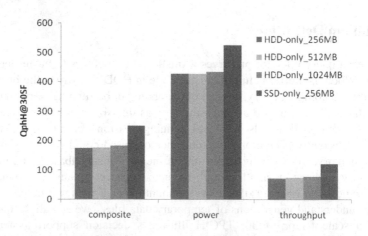

Fig. 1. Performance comparation of SSD-only and HDD-only with different shared buffer sizes

It is generally expected that database performance can be improved by increasing the size of shared buffer, which is used to cache query results. In this way, more query results can be cached and then future disk reads or writes can be avoided. It always works well for OLTP applications. We conducted experiments with TPC-H benchmark to see if shared buffer size also plays an important role in OLAP applications. In the experiments, shared buffer sizes in 3 HDD-only tests are 256MB, 512MB and 1024MB, in SSD-only test it is set to be 256MB. HDD-only represents the approach of storing temporary data on HDD, while SSD-only represents the approach of storing them on SSD. The only difference between them is the storage device, while the way temporary files organized didn't changed. The result is presented in Fig. 1. We can see that increasing shared buffer size doesn't make large contribution to system performance and SSD-only is superior to HDD-only. It is because when processing complex queries, I/O latencies are mainly on accessing temporary data, not query results. So, processing time of I/O operations on temporary data has a direct impact on the response time of queries. Moreover, the disparity between HDD-only and SSD-only is much smaller than the I/O performance disparity between HDD and SSD. This provides conclusive evidence that traditional temporary data management policy doesn't suit for SSD and then cannot make full use of SSD.

4   System Details

4.1   Basic Framework

Fig. 2 shows the basic framework of our hybrid storage system STDM. In STDM, SSD serves as the secondary storage medium, which is used to store temporary data. STDM doesn't change the logic of temporary data generation, that is to say, from the view of query operators, the existence of multiple small temporary files and the way temporary data blocks written to them never changed. However, these temporary files

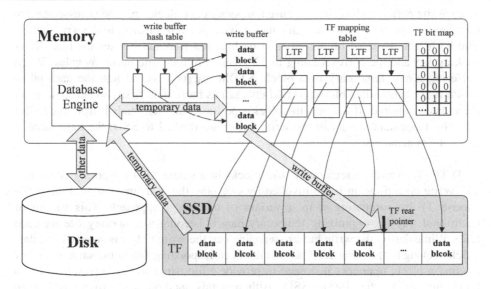

Fig. 2. Basic framework of STDM

are created logically rather than physically in STDM. In this paper, they are referred as logic temporary files (LTFs) and the single big temporary file in SSD is referred as temporary file (TF). We organize temporary data in TF and provide a mapping scheme for the transition from LTF to TF. The transition is transparent to query operators. Each LTF is assigned a unique id called fileID. Also, each data block is assigned a unique id called blockID, which consists of the fileID and its offset in the LTF.

4.2   System Components

This system consists of five main components: in-memory write buffer, write buffer hash table, TF, TF mapping table and TF bit map.

1) In-memory write buffer: This component is used to delaying materialization. When a temporary data block generated, instead of writing it to SSD immediately, STDM buffers it in memory at first; and then writes these buffered blocks to SSD once the write buffer is full. Data blocks are appended to the write buffer sequentially in spite that they come from different operators even different running sessions. In this way, STDM transforms several random writes to a sequential one. As abundant random writes may degrade I/O performance on SSD significantly, the write buffer is helpful for reducing random writes. What calls for special attention is that the write buffer is shared by all operators in all running sessions. To ensure data consistency, we maintain some shared and exclusive locks in concurrency situations. Another important issue to be considered is the write buffer size, i.e. how many data blocks should be cached in main memory. It will be discussed in more detail in Section 5.

2) Write buffer hash table: Because temporary data blocks may be accessed before they are written to SSD, we design a hash table to lookup for them in the write buffer. The hash table is based on a consistent hashing function and it takes the blockID as hashing key and the corresponding block position in the write buffer as value. When buffering a new temporary data block, STDM inserts a record into the hash table. When requesting a data block, STDM checks this hash table firstly to detect whether it has been buffered; if so, gets the block's position and then reads it from the buffer directly. Once data blocks in the write buffer are flushed to SSD, the hash table is cleared simultaneously.

3) TF: We store all temporary data blocks in a single SSD temporary file called TF. Write operations on it are invoked by swapping the in-memory write buffer. We choose TF by comparing it to a variant of traditional approach. This variant of traditional way for organizing temporary data is keeping a temporary file for each LTF. While during the write buffer swapping out, the write buffer is scanned and data blocks belong to the same LTF are written to corresponding file in the same time. For reading a block, operators just need to provide a file name and an offset; to delete a LTF, just delete the file in SSD. Although this method leads to no additional management cost because the file system is responsible for space allocation and reclamation, it has a fatal flaw that it doesn't take concurrency into account. With

count of concurrent query operators increasing, a lot of random writes on SSD may be caused. The reason is that the write buffer size is limit and probability of a block belonging to a file is uniform, so blocks in the write buffer may belong to a number of LTFs; and then incur a lot of SSD writes for writing them into different files. Since I/O performance degeneration is unacceptable in our system, we abandon this method. In our approach, to make use of the significant performance of sequential writes and random reads on SSD, we write all blocks to a single file called TF in an append-only fashion; thus, TF is just like a circular linked list that all incoming blocks are enqueued to the position indicated by a rear pointer. This particular write pattern is known to be perfect match with flash memory, and helps the system yield the best performance.

4) TF mapping table: Since data blocks coming from different LTFs are scattered over the TF, the first function of this data structure is locating them by their blockIDs. The other function is to link blocks belonging to the same LTF together. It is absolutely necessary because when upper-layer requests to delete a LTF, all its blocks can be found immediately. TF mapping table is an in-memory hash table which uses fileID as hashing key and a pointer to an array list as value. Each element in this array list is the physical address of corresponding block in TF. We pre-allocate this array with size 256 and if it is exhausted, we reallocate its size as double of its previous value. Before blocks in the write buffer swapping out, for each of them, if there is no record in the mapping table for the LTF it belongs to, a record is inserted and a new block list is initialized; then the position where this block will be placed in the TF is put into the new block list. Otherwise, the position is put into the corresponding block list directly. We also use a consistent hashing to ensure the stability of the looking-up time.

5) TF bit map: This data structure is an essential part for space reclamation. Its size is the same as the size of TF and each bit represents the state of corresponding block in TF. Value 0 means that the block is free and value 1 means the block has been occupied. When TF is full, several free blocks in its front-end need to be reclaimed, the procedure is detailed in the next section.

4.3   Data Management Policy on SSD

1) Temporary data generation phase: Temporary data blocks may be migrated from memory owned by their operators to the write buffer and then written to the TF. As described in Algorithm 1, when a query operator requests to create a temporary file, instead of creating a physical file in SSD, we create a LTF (lines 1 to 2). If the write buffer is full, for each block in it, we insert an item into the TF mapping table, and update corresponding bit in the TF bit map (lines 4 to 6).Then we migrate the write buffer to the TF (lines 7). At line 8, we relocate the TF rear pointer to the new end. The following data blocks will be put into the write buffer if it is not full (10 to 11).

Algorithm 1: Temp_Data_Create(logic_file f, block b)

Input: *file*: logic file, *block*: temporary data block

1: **if** f hasn't been created **then**
2:      create a LTF and assign it a unique fileID
3: **if** the write buffer is full **do**
4:      **for each** block B in write buffer
5:          insert an item into the TF mapping table for B
6:          update corresponding bit in the TF bit map as 1 /* i.e. occupied*/
7:      swap the write buffer out to the rear end of TF
8:      relocate the TF rear pointer
9: **end if**
10: put b into the rear end of the write buffer
11: insert an item into the write buffer hash table for b

---

2) Temporary data consumption phase: The algorithm of accessing a temporary data block is described in Algorithm 2. As temporary blocks may stay in the write buffer or the TF, when operator issues a read request to a block, we check the write buffer hash table at first. If it locates at the write buffer, we get its offset and read it directly (lines 2 to 4). Otherwise, we need to calculate the fileID and offset. It can be easily done because the block's ID is simply a concatenation of them (line 5). Firstly, we search for the block list of which LTF the block belongs to (line 6). Next, we get the nth value of the block list, where n represents offset and the value is the physical address of the block in TF (line 7). Then, we initiate a SSD read to load the block into memory (line 8).

Algorithm 2: Temp_Data_Read(block b)

Input: *block*: temporary data block

1: get position p of b by looking up the write buffer hash table
2: **if** p is not null **then**
3:      read block b from the write buffer
4: **else**
5:      calculate the fileID and the block offset
6:      look up for the block list in the TF mapping table
7:      get physical address of b
8:      read block b from TF
9: **end if**

---

3) Temporary data deletion logic: When an operator has completed all its operations, SSD spaces occupied by temporary data it generated have to be remarked as invalid. It is important because they will no longer be used and SSD space they occupied should be reclaimed. The detail of deletion process is described in

Algorithm 3. When an operator requests to delete a logic file, we look up its block list in TF mapping table (line 1).   Then we traverse the block list. For each block, if it exists in the TF, we update corresponding value in the TF bit map to 0 (line 3 to 4); otherwise, we find and delete it from the write buffer (lines 5 to 8). At last, we remove the LTF from the TF mapping table (line 10).

Algorithm 3: Temp_Data_Delete(logic_file *f*)

Input: *file*: logic file

  1: lookup for block list of *f* in the TF mapping table
  2: **for each** block *b* in block list
  3:    **if** *b* exists in the TF
  4:      update corresponding bit in the TF bit map to 0 /* i.e. free*/
  5:    **else**
  6:      get position *p* of *b* by looking up the write buffer hash table
  7:      delete *b* from write buffer hash table
  8:      delete *b* from the write buffer
  9:    **end if**
10: delete *f*   from TF mapping table

---

4) TF space reclamation logic: Because of our append-only approach and the limit size of TF, it is unavoidable to reclaim TF space frequently. In consideration of the lifetime of temporary data, when TF is full, TF bit map will be scanned from position pointed by the TF rear pointer. We limit the scan depth to 20% of the size of TF, so the reclamation cost is bounded. Blocks within the scan depth may be valid because LTFs they belong to haven't been deleted yet when we performing the reclamation. So, for invalid blocks, we just skip them without additional operations; while valid blocks will be loaded to memory and enqueued back to the write buffer. Then SSD space occupied by these blocks can be reused.

5   Experiment Evaluation

In this section, we will first describe the experiment setup and STDM prototype implementation; and then evaluate performance of STDM based on TPC-H benchmark; at last we investigate impacts of the write buffer and the TF sizes.

5.1   Experimental Setup

Our experiments all performed on a DELL PERC H700_H800 Server. The configurations are: 12 Intel Xeon(R) CPU E5-26200 @ 2.00GHz Processors, 16 GB of main memory, Linux 3.2.0, a Seagate 15K RPM 146 GB HDD, a Samsung 840 Pro Series 128GB SATA III SSD.

5.2   Prototype Implementation

We implement a prototype of STDM by modifying PostgreSQL 9.2; several in-memory data structures have been added such as a byte array for caching temporary data blocks, a bit map for representing states of blocks in TF, a hash table that maps blockID to data cached in the write buffer and another hash table that maps LTF's fileID to all data blocks belong to it in the TF. The most relevant functions in the temporary data management module are BufFileDumpBuffer and BufFileLoadBuffer. The original logic of BufFileDumpBuffer is writing a data block to its temporary file in HDD directly. We have modified it to write data to the write buffer firstly and then put an entry into the write buffer hash table. What's more, we add a new function to deal with swapping the write buffer out to SSD and the TF space reclamation. The original logic of BufFileDumpBuffer is reading a data block from its temporary file in HDD. It also has been modified to locate data by looking up the write buffer hash table or the TF mapping table, then read data from the write buffer or the TF.

5.3   TPC-H Evaluation

In our experiments, we use TPC-H as our OLAP benchmark.

1) Performance of STDM: In this section, we perform a set of experiments to demonstrate the effectiveness of our approach. We choose scale factors of 10 and 30, which corresponding total dataset sizes are 16GB and 50GB. The write buffer size is fixed at 256KB; the TF size is fixed at 2GB and the shared memory is fixed at 1GB. The experiment results are shown in Figure 3 and Figure 4. In power test, queries are submitted by a single stream, while in throughput test, they are submitted by multiple concurrent sessions.

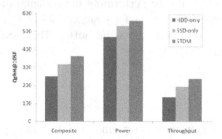   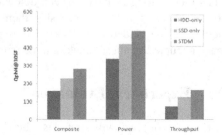

Fig. 3. Power and throughput test (SF=10)   **Fig. 4.** Power and throughput test (SF=30)

According to the results in both configurations, STDM improves query processing performance by approximately 45 percent and 78 percent compared to HDD-only, and by approximately 14 percent and 23 percent compared to SSD-only (HDD-only and SSD-only are defined in section 3). Specifically, the speed up of throughput tests is more obvious than power test. This is because concurrent query processing in throughput test issues a lot of random I/Os relative to power test. STDM can transform multiple random writes to a sequential write and takes full advantage of fast random reads of SSD. The result shows that STDM is excellent for managing temporary data.

2) Impact of various write buffer sizes: In order to evaluate the impact of write buffer sizes on system performance, we run a set of TPC-H tests with 30SF, 2GB TF size and the write buffer size varies from 64KB to 1MB. The test results are shown in Fig. 5. We can see that throughput achieves its best performance at 256KB and degenerates when we increase or decrease the write buffer size. When we adopt a small write buffer size, swapping all blocks in it may become too frequently, more SSD I/Os will be needed. On the other hand, when the write buffer is set to a large size, operators in the temporary data generation phase have to wait too long until the swapping is finished. Thus, to optimize STDM performance, the write buffer size should not be too small or too large.

3) Impact of various TF sizes: We can infer that the larger TF size is, the higher TPC-H performance can be achieved. This is reasonable due to that when we performing TF space reclamation, blocks within the scan depth are more likely to have already reached their lifetimes. So, the number of valid blocks is fewer and then I/O cost for loading them back to the write buffer is reduced. With fixed write buffer size and various TF sizes, we run a series of tests. The result is shown in Fig. 6. On one hand, the result matches our inferior as when we increase the TF size, performance gets better. However, on the other hand, it will not increase any more if the TF size is larger than 2GB. It is because for scale factor 30, 2GB is large enough to make most blocks within scan depth reach their lifetimes before next space reclamation invoked. Thus, system performance almost cannot be improved any more by increasing the TF size. The appropriate TF size may be various for different workloads. So, it is unnecessary for STDM to use too large a SSD space.

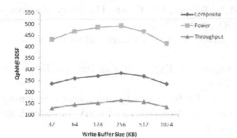 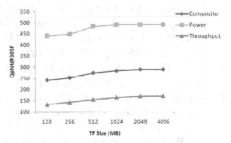

Fig. 5. Impact of various write buffer sizes   **Fig. 6.** Impact of various TF sizes

6   Conclusion

In this paper, we verified the conclusion that processing time of I/O operations on temporary data has a direct impact on performance of database and the inefficient of traditional temporary data management policy with SSD. Then we propose a novel temporary data management policy called STDM. STDM improves performance of query processing by buffering temporary data in memory and then writing them to SSD in an append-only fashion. Thus, it can takes advantage of fast random reads and reduces random writes on SSD. We implement a prototype system based on

PostgreSQL and evaluate its performance on TPC-H benchmark. Experiment result shows that STDM is more suits for SSD and get obvious performance improvement over traditional approach.

Acknowledgements. This research was partially supported by the grants from the Natural Science Foundation of China (No. 61379050, 91224008); the National 863 High-tech Program (No. 2013AA013204); Specialized Research Fund for the Doctoral Program of Higher Education(No. 20130004130001), and the Fundamental Research Funds for the Central Universities, and the Research Funds of Renmin University(No. 11XNL010).

References

1. Canim, M., Bhattacharjee, B., Mihaila, G.A., Lang, C.A., Ross, K.A.: An object placement advisor for db2 using solid state storage. In: PVLDB, pp. 1318–1329 (2009)
2. Agrawal, N., Prabhakaran, V., Wobber, T., Davis, J.D., Manasse, M.S., Panigrahy, R.: Design tradeoffs for ssd performance. In: USENIX Annual Technical Conference 2008, pp. 57–70 (2008)
3. Yang, Q., Ren, J.: I-cash: Intelligently coupled array of ssd and hdd. In: HPCA 2011, pp. 278–289 (2011)
4. Lee, S.-W., Moon, B.: Design of flash-based DBMS: An in-page logging approach. In: SIGMOD, pp. 55–66 (2007)
5. Shah, M., Harizopoulos, S., Wiener, J., Graefe, G.: Fast scans and joins using flash drives. In: Proc. of DaMoN Conf., pp. 17–24. ACM Press, New York (2008)
6. Tsirogiannis, D., Harizopoulos, S., Shah, M.A., Wiener, J.L., Graefe, G.: Query processing techniques for solid state drives. In: SIGMOD, pp. 59–72 (2009)
7. Shah, M.A., Harizopoulos, S., Wiener, J.L.: andG. Graefe. Fast scans and joins using flash drives. In: DaMoN, pp. 17–24 (2008)
8. Li, Y., On, S.T., Xu, J., Choi, B., Hu, H.: DigestJoin: Exploiting Fast Random Reads for Flash-based Joins. In: MDM, pp. 152–161 (2009)
9. Roh, H., Park, S., Kim, S., Shin, M., Lee, S.-W.: B+-tree index optimization by exploiting internal parallelism of flash-based solid state drives. Proceedings of the Very Large Data Base (VLDB) Endowment 5(4), 286–297 (2012)
10. Lai, W., Fan, Y., Meng, X.: Scan and Join Optimization by Exploiting Internal Parallelism of Flash-Based Solid State Drives. In: Wang, J., Xiong, H., Ishikawa, Y., Xu, J., Zhou, J. (eds.) WAIM 2013. LNCS, vol. 7923, pp. 381–392. Springer, Heidelberg (2013)
11. Luo, T., Lee, R., Mesnier, M.P., Chen, F., Zhang, X.: hStorage-DB: Heterogeneity aware data management to exploit the full capability of hybrid storage systems. In: PVLDB, pp. 1076–1087 (2012)
12. Lee, S.W., Moon, B., Park, C., Kim, J.M., Kim, S.W.: A case for ash memory ssd in enterprise database applications. In: SIGMOD Conference 2008, pp. 1075–1086 (2008)
13. Lee, S.-W., Moon, B., Park, C.: Advances in flash memory SSD technology for enterprise database applications. In: SIGMOD, pp. 863–870 (2009)

A Novel Index Structure for Multi-key Search

Dongyu Wei[1], Xin Pan[1], Chuan Shi[1,*], and Yueguo Chen[2]

[1] Beijing University of Posts and Telecommunications, Beijing, China 100876
[2] Renmin University, Beijing, China

Abstract. The linear storage model is widely used to support in-memory multi-key search running on small devices of limited computing capacity, simply because it avoids the maintenance of space-costly and energy-costly indexing structures. However, it only supports sequential multi-key scan which is slow and energy-consuming. We design an index structure called D-Tree to address the problem.

Keywords: Storage model, Multi-key search, Space-sensitive, Energy-sensitive.

1 Introduction

Multi-key search has been an important function in database systems running on small devices [3]. It is to search tuples within a table constrained by two or more keys. Modern database systems use various indexing techniques (e.g., B+-tree [2], kd-Tree [1] and Bitmap [4]) to support efficient query processing of multi-key search queries. However, for applications running on small devices which have critical physical constraints (e.g., space and energy), the advanced indexes are often not applicable due to their excessive cost in space consumption. Quite often, linear list is more favored by many database systems designed for small devices with physical limitations. Although linear storage model has no extra space overhead and very few maintenance cost, the sequential scan process is however slow and energy-consuming due to the vacancy of indexing supports. In this paper, we address the multi-key search problem in space-limited memory and propose a novel lightweight index named D-Tree for multi-feature datasets.

2 The D-Tree Index Structure

2.1 Construction Algorithm of D-Tree

Inspired by the dominance tree [5], we propose the D-Tree index structure to support such a multi-key search query with small extra space cost but efficient search speed. A dominance tree [5] is a binary tree, where the left-link field links to its left subtree whose root node is dominated by that node, and the right-link filed links to its right sub-tree whose root node is non-dominated by that node. However,different from dominance tree, the D-Tree has just the child field which point to the dominated nodes, whose data structure can be defined as follows.

---

* Corresponding author.

F. Li et al. (Eds.): WAIM 2014, LNCS 8485, pp. 431–434, 2014.

```
typedef struct DT{
    int id;  //coordinate in the sibling list
    struct DT *child; //point to the dominated nodes
}
```

Algorithm 1. addinTree(DT *pNode, DT *newNode)

Insert a new node (newNode) into pNode's left subtree when newNode is dominated by pNode.

```
1. if pNode has child then
2.     addinList(pNode, pNode.child, newNode);
3. else
4.     pNode.child = newNode;
5. end if
```

Algorithm 2. addinList(DT *pNode, DT *cNode, DT *newNode)	**Algorithm 3.** Equality Search

When newNode is inserted into pNode's left subtree, newNode is compared with cNode, a node in the pNode's left child list.
 	
while TRUE do
  if cNode is nondominated with newNode then
    if cNode is the last one in the list then
      append newNode to the list;
      return;
    else
      cNode = next node in the list;
    end if
  else if cNode dominates newNode then
    addinTree(cNode, newNode);
    return;
  else if cNode is dominated by newNode then
    remove cNode from the list;
    addinTree(newNode, cNode);
    if cNode is the last node in the list then
      append newNode to the list;
      return;
    else
      cNode = next node in the list;
    end if
  end if
end while
	Require: query;
	  SN = CreateSN(query);
	  Create an empty stack; CN = root;
	  while CN != NULL do
	    if Better(SN,CN) == 1 then
	      prune the left branch of CN; CN=CN's right neighbor;
	    else if Better(SN,CN) == -1 then
	      push CN's left child to stack; CN=CN's right neighbor;
	    else if Better(SN,CN) == 0 then
	      if SN match CN then
	        find a target tuple;
	        push CN's left child to stack; CN=CN's right neighbor;
	      else
	        prune the left branch of CN; CN=CN's right neighbor;
	      end if
	    end if
	    if CN == NULL then
	      if stack == NULL then
	        return;
	      else
	        CN = Pop(stack);
	      end if
	    end if
	  end while

To construct a D-Tree, nodes are inserted one by one. When a node (called *newNode*) is compared with an existing node (called *cNode*), there are three possible results. 1) *newNode* dominates the *cNode*. *cNode* is removed and inserted into the *newNode*'s left branch. *newNode* will be further compared with original *cNode*'s right neighbors. 2) *newNode* is dominated by *cNode*. The *newNode* will be inserted into the *cNode*'s left branch for further comparison. 3) *newNode* is non-dominated with *cNode*. *newNode* will be further compared with the *cNode*'s right neighbors. The node insertion algorithm of D-Tree is shown in Algorithm 1 and 2.

Right-link Sort. In many cases, a key (called hot key) is usually visit many times in queries.We propose the right-link sort method to boost the performance of queries containing a hot key. We will show that such a variation can improve search performance in experiments.

Equality Search Here we consider the simplest operation (i.e., "=") for the multi-key search algorithm.The algorithm is shown in Algorithm 3. The search process can be roughly separated into three steps. 1) Input an array list (a sibling chain). 2) Iteratively compare SN with CN through the list. 3) Determine whether go back to 1) or not.

3   Experiments

We test the equality search performances on synthetic datasets. Three structures, FS, D-Tree and srD-Tree, are included in the experiments. FS means the multi-key sequential scan on linear list. D-Tree means the multi-key search on D-Tree without right-link sort adjustment. srD-Tree means the multi-key search on D-Tree with right-link sort. One search key is always selected as hot key.

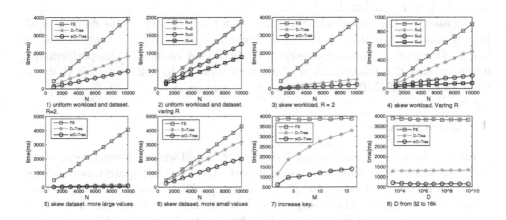

Fig. 1. Time efficiency experiments on equality search. N represents the number of tuples, M represents the number of keys indexed by D-Tree, R represents the the number of keys specified in Search Node, D represents the value domain of each key

In the results of Fig. 1(1), the data of a table with four keys are indexed. Each key has a domain of [1,50]. All queries contain 2 search key (i.e., $R = 2$). The results show that D-Tree consistently outperforms the solution of the sequential multi-key scan. srD-Tree is the fastest one. In experiments of Fig. 1(2), we vary the number of search keys from 1 to 4. As shown in the figure, the more search keys used, the faster D-Tree and srD-Tree are. This is because when more values are filled in the search node (SN), SN has larger possibility to be non-dominated with CN.

Different from Fig. 1(1) and (2) which use uniform workload, Fig. 1(3) and (4) generate skew workload using Zipfian distribution[1] with parameter 1.0. Smaller values have larger possibility to be searched. Similar to the experimental setting in Fig. 1(1) and (2), we use 2 search keys in Fig. 1(3) and vary the number of keys in Fig. 1(4). As we expected, D-Tree and srD-Tree further improve the performance. srD-Tree in Fig. 1(4) is over 2 orders of magnitude faster than the sequential scan. The reason is that, when the values in the query become smaller, the target tuples stay closer to the root.

Instead of generating skew workload, we generate skew dataset using Zipfian distribution with parameter 1.0 in Fig. 1(5) and Fig. 1(6). Besides, 3 search keys

---

[1] http://en.wikipedia.org/wiki/Zipf's_law

are used in queries. In Fig. 4(5), large values has large possibility to be generated, while in Fig. 1(6) small values have large possibility of being generated. srD-Tree is nearly two orders of magnitude faster than FS.

In experiments of Fig. 1(7), the impact of M is tested. As shown in Fig. 1(7), when more keys are indexed, the performance of D-Tree and srD-Tree degrade. The reason is that when more keys are indexed, nodes have large possibility to be non-dominated with each other. And the sibling chain, which cannot be pruned by D-Tree, gets longer. Even though D-Tree degrades, it still outperforms FS when 15 keys are indexed.

We also test the impact of value domain in Fig. 1(8). The value domain of all keys is enlarged together. As the value domain is enlarged, fewer duplicate values occur. However, we do not observe the significant change of the performance when varying the value domain of all keys from 32 to 16k. It means that the time efficiency of D-Tree is not sensitive to domain size.

4   Conclusions

Data management engines on small devices are faced with common special physical restrictions and limited energy support. They are not quite time-sensitive but have high requirements on space and energy. We propose the novel D-Tree, an in-memory lightweight storage model for multi-key search. The D-Tree can effectively store the dominance relationships of tuples with small extra index space. We design efficient search algorithms based on D-Tree for multi-key search. Extensive experiments show that D-Tree can achieve 2 orders of magnitude improvement than linear scan with very small extra space cost. It indicates that D-Tree can be an effective substitution of linear list on the extreme space limit scenarios, such as smart card and sensor database.

Acknowledgement. This work is supported by the National Basic Research Program of China (2013CB329603). It is also supported by the National Natural Science Foundation of China (No. 61375058, 60905025, 61074128) and Ministry of Education of China and China Mobile Research Fund (MCM20123021).

References

1. Bentley, J.L.: Multidimentional binary search trees used for associative searching. Communications of the ACM 18(9), 509–517 (1975)
2. Comer, D.: The ubiquitous b-tree. ACM Computing Surveys 11(2), 121–137 (1979)
3. Li, X., Kim, Y.J., Govindan, R., Hong, W.: Multi-dimensional range queries in sensor networks. In: SenSys, pp. 63–75 (2003)
4. O'Neil, P., Quass, D.: Improved query performance with variant indexes. In: SIGMOD, pp. 38–49 (1997)
5. Shi, C., Yan, Z., Lu, K., Shi, Z., Wang, B.: A dominance tree and its application in evolutionary multi-objective optimization. In: Information Sciences, pp. 3540–3560 (2009)

Load-Balanced Breadth-First Search on GPUs

Zhe Zhu, Jianjun Li, and Guohui Li

School of Computer Science & Technology,
Huazhong University of Science & Technology, China
luokezhu@gmail.com, Jianjunli,Guohuili@hust.edu.cn

Abstract. Breadth-first search (BFS) is widely used in web link and social network analysis as well as other fields. The Graphics Processing Unit (GPU) has been demonstrated to have great potential in accelerating graph algorithms through parallel processing. However, BFS is difficult to parallelize efficiently due to the irregular workload distribution, leading to load imbalance between threads. Previous work has proposed several strategies to alleviate the load imbalance but none of them solves this issue in general.

This paper presents a new GPU BFS algorithm that focuses on full load balance. Each BFS iteration is decoupled into two phases: work redistribution and neighbor gathering. Work redistribution phase reorganizes the irregular workloads in order for the neighbor gathering phase to visit the vertices in a load-balanced way. The evaluation results show that the proposed approach achieves speedups of up to 39x and 1.42x over CPU sequential implementation and state-of-the-art GPU implementation respectively.

Keywords: Breadth-first search, GPU, load balance, graph algorithms, parallel algorithms

1 Introduction

Graph algorithms are becoming increasingly important, with applications ranging from web link analysis to computer-aided design to machine learning. Breadth-first search (BFS) is an important low-level operation that serves as a fundamental building block for more complicated graph algorithms. Thus efficient parallelization of BFS has gained much attention.

Unfortunately, exploiting the nested parallelism in BFS is challenging. Assigning the workloads to each thread evenly is non-trivial because the work distribution patterns are determined by the structure of the input graph.

Modern GPUs have become popular general computing devices due to their high memory and computational throughput, low costs and power efficiency. However, accelerating BFS on GPUs requires much more attention. The wide SIMD architecture of GPUs is particularly sensitive to load imbalance [3]. Inadequate handling of this issue can lead to a significant performance hit.

Prior work has proposed several parallelization approaches [7,8,10,11]. They mainly rely on overlapped execution of massive amount of threads, local reorga-

F. Li et al. (Eds.): WAIM 2014, LNCS 8485, pp. 435–447, 2014.
© Springer International Publishing Switzerland 2014

nization of workloads and work stealing to limit load imbalance to some extent. However, none of them eliminates this issue in general.

In this paper, we present a load-balanced GPU BFS algorithm, which decouples each BFS iteration into two phases: work redistribution and neighbor gathering. Work redistribution phase serves as a preprocessing operation, employing a parallel expansion to reorganize the nested and irregular workloads of a BFS iteration. Neighbor gathering phase then subsequently assigns the workloads to threads uniformly and visits each neighbor in a load-balanced way.

Specifically, we make the following contributions:

- We propose a load-balanced GPU BFS algorithm. To the best of our knowledge, ours is the first BFS implementation on GPUs that achieves fully load-balanced neighbor gathering.
- We analyze the coupling possibilities between different phases of the algorithm for optimal performance. Coupling separate procedures into one kernel reduces I/O overhead but may amplify load imbalance. We show that a hybrid coupling strategy has the best performance.
- Our approach delivers great performance on a wide diversity of real-world graphs, achieving speedups of up to 39x and 1.42x over CPU sequential implementation and state-of-the-art GPU implementation, respectively.

2 Background and Motivation

In this section, we first introduce some unique properties of GPU architecture. Then we review existing BFS algorithms on GPUs and motivate our approach.

2.1 Modern GPU Architecture

In order to deliver high computational throughput, modern GPUs adopt a wide SIMD architecture[3], meaning threads within a *warp* execute the same instructions synchronously. Control flow divergence among these threads will result in serialization of different execution paths. Warps are grouped into *cooperative thread arrays* (or CTAs). Threads within a CTA can communicate through a local *shared memory*, and GPU hardware treats the CTA as the unit of scheduling. A program running on the GPU is called a *kernel*.

This hierarchical model introduces several types of workload imbalance. The SIMD execution within a warp will cause thread load imbalance and underutilization if control flow diverges. Within a CTA, the warp with the highest workload will cause other completed warps to sit idle and prevent the completion of the CTA, which in turn will prevent other CTAs in the wait queue from being scheduled. Likewise, few CTAs taking too much time to complete can extend the completion time of the kernel. Figure 1 illustrates these three types of workload imbalance.

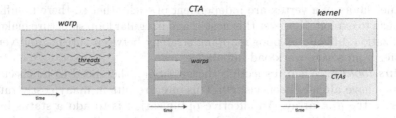

Fig. 1. (Left to right) thread imbalance, warp imbalance and CTA imbalance

---

Algorithm 1. Linear-work parallel BFS

Input: v_0 , input queue inQ and output queue $outQ$
Output: Array $dist[0...n-1]$ holding the distance from s to each vertex
1 initialize all elements in $dist[0...n-1]$ to ∞ and empty inQ
2 $dist[v_0] \leftarrow 0, iteration \leftarrow 0$
3 $inQ.Enqueue(v_0)$
4 **while** inQ *not empty* **do**
5   empty $outQ$
6   **foreach** $v \in inQ$ **in parallel do**
7   **foreach** *neighbor* of v **in parallel do**
8   **if** $StatusLookup(neighbor) = valid \cap dist[neighbor] = \infty$ **then**
9                  $dist[neighbor] \leftarrow iteration + 1$
10                 $outQ.Enqueue(neighbor)$
11     $iteration + +$
12   switch inQ and $outQ$

2.2 Existing BFS Algorithms on GPUs

Given a source vertex v_0, the BFS process traverses the vertices in breath-first order and label each vertex with its distance from v_0. Other variants of BFS may record other attributes such as the parent of each vertex.

Earlier GPU BFS research mainly focuses on work-inefficient parallelization [7,8] which has quadratic work complexity ($O(n^2 + m)$ or $O(mn)$, n and m represent the vertex and edge numbers, respectively). Luo et al. [10] present the first linear work BFS ($O(m+n)$) and achieve much better performance. In this paper, we will focus on work-efficient algorithms.

The skeleton of the linear-work BFS algorithm on the GPU is similar to the standard serial BFS on the CPU [9], which is listed as Algorithm 1. On each iteration, vertices are taken out of the input queue, and their neighbors are visited and inserted into the output queue for next iteration. However, there are two main differences between CPU and GPU BFS algorithms, which are also the main challenges of GPU BFS:

Parallel neighbor gathering. The neighbor gathering process read in all the neighbors of the input vertices. Both the vertices in the input queue and all the neighbors of a vertex are independent of each other so there is sufficient parallelism to exploit. However this nested and irregular loop structure makes the parallelization difficult. A poor mapping strategy between threads and vertices will suffer from severe workload imbalance.

Status lookup. When inspecting the neighbors, they need to be checked to see if they have already been visited. This often results in many costly random accesses to the *dist* array. An effective optimization is to add a status lookup process and use a bitmap array to check the status, leading to reduced global memory overhead and improved cache hit rate.

We will focus on the neighbor gathering process, as it is where load imbalance happens and can easily become the bottleneck of the whole BFS algorithm.

The simplest strategy is to map each thread to a vertex in the input queue, having each thread inspect the neighbors of the assigned vertex serially. Harish et al. [7] and Luo et al. [10] use this strategy. It only exploits the parallelism of the outer loop, and can lead to severe thread imbalance within a warp for graphs having non-uniform degree distributions. Moreover, the arbitrary memory accesses from each thread result in terrible coalescing too.

A better strategy is to map a whole warp or CTA to a vertex in the input queue, which is adopted by Hong et al. [8] and Merrill et al. [11]. In this way, the whole warp or CTA gather the adjacency list of the vertex in parallel. This approach provides good thread balance for vertices having large numbers of neighbors. However for vertices with the adjacency list sizes smaller than the warp/CTA width, some threads in the warp/CTA will go unused, imposing underutilization of the warp/CTA. Furthermore, there may exist warp imbalance or CTA imbalance if the adjacency list sizes vary significantly.

Another scan-based strategy introduced by Merrill et al. [11] maps a CTA to a certain number of vertices in the input queue. The CTA first constructs a shared array of neighbor locations corresponding to the concatenation of the assigned adjacency lists. Then the CTA reads in the locations from the shared array and gather the neighbors iteratively. Compared to the CTA mapping approach, this strategy solves the CTA underutilization problem at the cost of additional concatenating operations, which is efficient for vertices having small sizes of adjacency lists. Since each thread constructs its part of the shared array serially and its workload is proportional to the size of the assigned adjacency list, large adjacency lists can impose thread imbalance and inefficiency.

Each of the above mapping strategies is suitable for certain types of graphs. Merrill et al. [11] therefore adopt a hybrid approach. For vertices having more neighbors than the CTA width, CTA mapping is applied. For vertices having the number of neighbors smaller than the CTA width but larger than the warp width, warp mapping is applied. Finally, scan-based mapping is performed on the remaining vertices. This hybrid approach limits thread imbalance and warp imbalance, which is the current state of the art on GPU BFS.

Other works explore general graph algorithms on GPUs [12,16]. They focuses on flexibility and clarity but lacks specific optimization. Their BFS implementations are inefficient.

2.3   Motivation of This Work

All the existing parallelization strategies suffer from load imbalance issues. They cannot achieve consistent performance over various graphs. The hybrid CTA+warp+scan approach has been shown to perform efficiently. However, this solution is not good enough for the following reasons:

(1) It does not solve the load imbalance problem in general, but only limits thread imbalance and warp imbalance to some extent. CTA imbalance is not addressed. Instead, it relies on work stealing to alleviate CTA imbalance.
(2) The neighbor gathering and status lookup process must be put in separate kernels for optimal performance because fusing these two processes would amplify the CTA imbalance. This leads to additional global data movement.
(3) It is complicated and unintuitive. Work partitioning and neighbor gathering logic are mixed up, resulting in an algorithm difficult to understand.

To address these problems, we present a load-balanced BFS algorithm. It is decoupled into two phases: work redistribution and neighbor gathering. Moreover, in the absence of CTA imbalance we get to fuse neighbor gathering and status lookup into one kernel and further improve performance.

3   Parallel Expansion

The nested and highly irregular parallelism shown in BFS, together with the static thread creation mechanism of GPUs, make a balanced work partitioning very difficult. The latest NVIDIA GPU architecture GK110 supports *dynamic parallelism* [3] in order to ease this problem, which enables the GPU kernel to launch other kernels itself. However, this does not solve this issue in general because the number of newly allocated threads does not match the problem size very well. Vertices with few neighbors would be provisioned entire CTAs, leading to underutilization. To address this problem, we preprocess the input to reorganize the workloads, eliminating the nested parallelism. In this section, we introduce the *expand* operation which is the basis of the workload reorganization, and the parallelization of *expand*.

3.1   The *expand* Operation

To get rid of the nested workload structure, we pack the neighbor gathering work produced by each input vertex together into a single sequence, with each element of the sequence representing the gathering address. In this way, threads can be uniformly mapped to this sequence and do the neighbor gathering in a load-balanced fashion.

In order to generate this sequence, we first define a basic operation. As illustrated in Fig. 2, taking the degree of each vertex in the queue as input, this operation outputs an array whose length is equal to the total number of neighbors to be produced. Each element in the array represents the index of the vertex

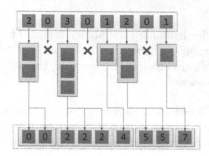

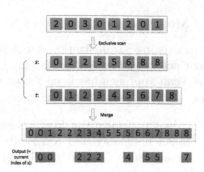

Fig. 2. The *expand* operation

Fig. 3. Converting *expand* into *merge* operation

in the queue that will produce it so in the subsequent gathering phase we can find that vertex and its neighbors.

We will call this operation *expand*, which is a useful pattern in data-parallel algorithms. Using *expand*, the nested loop structure is reorganized and flattened, which is the key to achieving load balance. Obviously, serial implementation of *expand* has $O(m)$ time complexity, thus efficient parallelization of the *expand* operation is the basis of high performance of the whole BFS algorithm.

3.2   Parallelization of *expand*

The *expand* operation can actually be converted to a merging of two sorted arrays. As demonstrated in Fig. 3, we first run an exclusive scan [13] on the inputs, obtaining the result array s and the sum *total*. We then construct an array t of length *total* filled with $[0...total - 1]$ and merge s and t. The difference compared to a normal *merge* is that we only output *total* elements, and the value of each output element equals to the current index of array s. In practice, the array t is not necessary because the indices and values are the same. Algorithm 2 shows the sequential implementation of the *expand* operation.

The parallelization of the *merge* operation has been studied for decades [15,6]. Basically, the input sequences are partitioned into non-overlapping segments, and the independent pairs of segments are merged in parallel. Odeh et al. [14] present a *merge path* algorithm that achieves a perfectly load-balanced partitioning. As depicted in Fig. 4, the two input sequences are listed perpendicularly. The merge process can be seen as the traversal of a path from the upper left corner to the bottom right corner, and each step represents a comparison operation. This path is partitioned by equispaced cross diagonals, and the intersection points are computed using binary searches (search along the diagonal for the dividing point between $s > t$ and $s \leq t$). In this way, each segment of the path contains exactly the same number of merge steps (except the last segment, which we can handle through padding), resulting in a load-balanced partitioning.

This partitioning scheme can be easily applied to GPUs. We first employ a coarse-grained CTA-wide partitioning, assigning each CTA with the same num-

---

Algorithm 2. Sequential *expand*

Input: Array *in* with each element representing the number of elements to be
        produced, the length *in_count* of array *in*
Output: Array *out* with each element representing the index of the input
         element that produced it
Function: *ExclusiveScan(input)* returns the scan result array and the sum of
the input elements

1   $(s, total) \leftarrow ExclusiveScan(in)$
2   $si \leftarrow 0, ti \leftarrow 0$
3   **while** $si < in\_count \cap ti < total$ **do**
4   **if** $ti < s[si]$ **then**
5         $out[ti] \leftarrow si - 1$
6         $ti + +$
7   **else** $si + +$;
8   **if** $si = in\_count$ **then**
9      $out[ti...total - 1] \leftarrow si - 1$

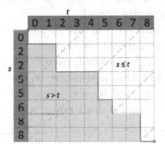

Fig. 4. Merge path partitioning

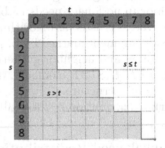

Fig. 5. Partitioning merge path vertically

ber *TILE_SIZE* of elements to process, which is a tunable constant. This is
done by placing each cross diagonal at a distance of *TILE_SIZE* steps. After
that, each CTA runs a similar fine-grained local partitioning to further assign
TILE_SIZE/CTA_SIZE input elements to each thread. When threads have
obtained their independent segments of the input elements, they can run the
sequential *expand* in parallel, leading to a high-performance parallel expansion.

4   Load-Balanced BFS

Having explained the parallel expansion in detail, we now use it as a work redistribution scheme to construct the full algorithms for a BFS iteration. We also
explore the coupling possibilities of work redistribution and neighbor gathering.

    To efficiently utilize the GPU memory model, we use the well-known com-
pressed sparse row (CSR) format to store the graph in GPU main memory,
which contains two arrays, namely column-indices C and row-offsets R.

4.1 Perfect Balance + Global Data Movement

The most straightforward approach is to separate the work redistribution phase and neighbor gathering phase into different kernels. In work reditribution phase, we read in the vertices in the input queue and construct the array $offsets$ holding the starting index of each vertexs adjacency list. Then we compute the adjacency list size $neighbor\_num$ of each vertex, and run the parallel expansion using $neighbor\_num$ as input. With the expansion output $vertex\_index$ we can further compute the location of each neighbor to be gathered as:

$$gather\_location = neighbor\_index - scan\_results[vertex\_index]$$
$$+ offsets[vertex\_index]. \tag{1}$$

These locations are then written to the output queue.

In neighbor gathering phase, the locations of all the neighbors are read back in. We then gather the neighbors at these locations and perform status lookup. Finally the valid vertices are output for distance update. As mentioned in Sect. 2.3, since the neighbor gathering is now fully load-balanced, the subsequent status lookup no longer needs to be put in a separate kernel.

This process is listed as algorithm 3, which requires at least five kernel launches (each code fragment marked by *in parallel* indicates a separate kernel). The work redistribution and neighbor gathering are both load-balanced: each CTA always processes $TILE\_SIZE$ elements, making the algorithm insensitive to the difference in graph structure. But the net slowdown caused by writing and reading the work redistribution results will limit the obtained overall performance.

4.2 Imbalanced Redistribution + Balanced Gathering

A natural optimization is to fuse work redistribution and neighbor gathering. Unfortunately this will compromise the load balance property. Assume that the input queue has $input\_total$ vertices and they generate $output\_total$ neighbors. The redistribution phase will take $input\_total + output\_total$ elements as input but only output $output\_total$ elements for the next phase, which makes the thread mapping policy inconsistent across the two phases if we fuse them together. In normal cases however, the performance gain through reduced I/O overhead can often make up for the impact of the load imbalance.

When coupling the redistribution and gathering, we may choose the thread mapping policy so that it benefits either process. We first focus on balanced gathering because it is the more time-consuming phase. Assuming each CTA processes $TILE\_SIZE$ elements, the algorithm will assign $output\_total/TILE\_SIZE$ CTAs to perform the gathering. To achieve the coupling, the same number of CTAs should be assigned to the redistribution, and each CTA should output $TILE\_SIZE$ elements at the end of the redistribution process which are then fed into the gathering process on the fly.

We use a different redistribution approach to fulfill these requirements. In the coarse-grained partitioning step, we partition the merge path vertically rather

---

Algorithm 3. Perfectly load-balanced BFS iteration

Input: inQ, $outQ$, inQ length in_total, column-indices array C and row-offsets array R

Output: Array $dist$

Function: $MergePathPartition(range, dist, k)$ partitions the merge path of the input $range$ diagonally at a distance of $dist$, and returns the kth independent range.

1 **foreach** $v \in inQ$ **in parallel do**
2    $(start, end) \leftarrow (R[v], R[v+1])$
3    $offsets[v\_index] \leftarrow start$
4    $neighbor\_num \leftarrow end - start$
5    $(inQ[v\_index], out\_total) \leftarrow ExclusiveScan(neighbor\_num)$

6 $cta\_num \leftarrow (in\_total + out\_total)/TILE\_SIZE$
7 **for** $thread\_id \in [0, cta\_num - 1]$ **in parallel do**
8    $coarse\_range[thread\_id] \leftarrow$
     $MergePathPartition(whole\_input, TILE\_SIZE, thread\_id)$

9 **for** $thread\_id \in [0, cta\_num * CTA\_SIZE - 1]$ **in parallel do**
10    $fine\_range \leftarrow MergePathPartition(coarse\_range[cta\_id],$
     $TILE\_SIZE/CTA\_SIZE, thread\_id\%CTA\_SIZE)$
11   shared array $indices \leftarrow Expand(fine\_range)$
12   **foreach** $v\_index \in indices$ **do**   /* strided */
13      $gather\_loc \leftarrow n\_index - inQ[v\_index] + offsets[v\_index]$
14      $outQ[n\_index] \leftarrow gather\_loc$

15 **foreach** $loc \in outQ$ **in parallel do**
16    $neighbor \leftarrow C[loc]$
17   **if** $StatusLookup(neighbor) = valid$ **then** scatter $neighbor$ to inQ

18 **foreach** $neighbor \in inQ$ **in parallel do**
19   update $dist$ array

---

than diagonally and at a distance of $TILE\_SIZE$ as illustrated in Fig. 5. In this way, it is guaranteed that each CTA produces $TILE\_SIZE$ outputs. We then do a fine-grained partitioning and expansion within each CTA.

This work redistribution process is relatively inefficient and imbalanced, because a CTA does not know the number of inputs it will process a priori. Fortunately, since the gathering process is the more time-consuming phase and is perfectly load-balanced, this approach can achieve better overall performance from the reduced I/O overhead.

4.3 Balanced Redistribution + Imbalanced Gathering

Another coupling strategy is to focus on balanced work redistribution. In this way, $(input\_total + output\_total)/TILE\_SIZE$ CTAs are assigned to run the redistribution process as in the first approach, the outputs are then fed into the gathering process immediately. Since each CTA will take $TILE\_SIZE$ elements as input but produce less than $TILE\_SIZE$ outputs, subsequent gathering can suffer from CTA imbalance: each CTA will gather 0 to $TILE\_SIZE$ neighbors.

When $output\_total$ is much larger than $input\_total$, the gathering load imbalance is negligible, and this approach can be more efficient than previous strategy because of the balanced redistribution process. However, if they are close, the imbalance problem can make the parallelism drop down by half, compromising the overall performance significantly.

4.4   Hybrid

The hybrid strategy combines the advantages of the *imbalanced redistribution + balanced gathering* and *balanced redistribution + imbalanced gathering* approaches. We define $expand\_factor$ as the ratio of $output\_total$ versus $input\_total$. If $expand\_factor$ is larger than a threshold f_0 for a given BFS iteration, we invoke the *balanced redistribution + imbalanced gathering* approach because the gathering imbalance will be small enough to be safely ignored, achieving the best performance. Otherwise we invoke the *imbalanced redistribution + balanced gathering* approach to guarantee the efficiency of the gathering process, which dominates the overall performance.

By selecting an appropriate f_0, this strategy can ensure that the neighbor gathering phase is always load-balanced while the efficiency of the work redistribution phase is maximized.

5   Experimental Results

In this section, we evaluate the performance of the proposed BFS algorithms. Our algorithms are implemented using CUDA 5.5 [3], and all experiments are run on a host machine with 4GB memory, an Intel 3.4 GHz Core i7 2600k CPU and an NVIDIA Geforce GTX 580 GPU. For each graph, the BFS performance is measured by the average traversal throughput (edges per second) across 100 randomly-sourced traversals.

5.1   Strategy Evaluation

We first compare different coupling strategies of work redistribution and neighbor gathering. We use Random and R-MAT graphs generated with GTgraph [5], and adjust the average degree d to see its impact on each strategy. Fig. 6 and Fig. 7 plot the traversal throughput for each graph and strategy. All the graphs have 2 million vertices while d ranges from 2 to 64, and we choose the threshold f_0 to be 10 for the hybrid strategy. As anticipated, the *balanced redistribution + imbalanced gathering* strategy excels at traversing graphs with large d because the $expand\_factor$ tends to be large for the BFS iterations. On the contrary, the *imbalanced redistribution + balanced gathering* strategy performs better on graphs with small d. The *hybrid* approach outperforms or is as good as the others in all the tests.

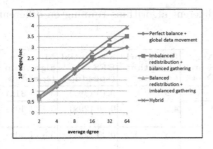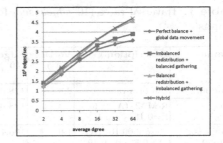

Fig. 6. Comparison of different strategies on random graphs **Fig. 7.** Comparison of different strategies on R-MAT graphs

5.2 Comparison with Other Algorithms

To compare with the CPU implementation, we implement our own efficient sequential BFS according to the standard algorithm in [9], which has optimal single-threaded performance. For the GPU implementation, we compare our approach to that from Merrill et al. [11] which is the current state of the art and achieves the highest published performance for GPU BFS. Note that we compile and run Merrills source code on the same platform under the same configuration so the results are comparable.

Our benchmark suite incorporates twelve graphs listed in Table 1. In addition to *random* and *rmat*, the rest are from the Graph500 Competition [2], the 10th DIMACS Implementation Challenge [1] and the University of Florida Sparse Matrix Collection [4].

The results are presented in Table 1. As we can see, our hybrid approach performs very well compared to the CPU sequential BFS. For the majority of the graphs, our approach provides traversal speedups of an order of magnitude, and at the extreme, we achieve a 39x speedup for the *random* graph. We do not have an available parallel CPU BFS implementation, but we can simply assume a perfect 8x speedup for our 4-core/8-thread CPU, and our GPU approach still outperforms this theoretical performance upper bound for almost all the tests.

The state-of-the-art GPU implementation [11] employs a hybrid CTA+warp+scan strategy to do neighbor gathering. Through full load balance, our approach outperforms theirs for most of the tests, and we obtain up to 1.42x speedup. The last few tests also reveal the limitation of our approach. The advantage of load balance comes at the price of additional preprocessing of the input vertices. When the search depth is small, the number of vertices examined each iteration is large enough to fully utilize the high throughput of GPU hardware, making the preprocessing overhead negligible. However, when the search depth gets high, the overhead of the work redistribution process and additional kernel launches starts to become significant. As a result, the performance gain through load-balanced gathering is outweighed by the preprocessing penalty and we observe a slowdown for *germany.osm* and *hugebubbles-00020* datasets. In practice, we can choose to apply direct gathering for graphs with large diameters.

Table 1. Traversal rates ($10^9 edges/sec$) for different algorithms running on different graphs

Name	Description	Vertices (10^6)	Edges (10^6)	Avg. Search Depth	Our hybrid algorithm ($f_0 = 10$)	Sequential BFS (speedup)	Merrill's algorithm (speedup)
random	Uniform random	2.0	128.0	6	3.9	0.10 (39x)	3.0 (1.31x)
rmat	R-MAT (A=0.45, B=0.15, C=0.15)	2.0	128.0	6	4.7	0.16 (29x)	3.4 (1.40x)
kron_g500-logn20	Graph500 R-MAT (A=0.57, B=0.19, C=0.19)	1.0	100.7	7	4.5	0.20 (22x)	3.3 (1.36x)
hollywood-2009	Hollywood movie actor network	1.1	113.9	10	4.1	0.18 (23x)	3.1 (1.32x)
flickr	2005 crawl of flickr.com	0.8	9.8	12	3.6	0.18 (20x)	3.2 (1.13x)
eu-2005	small web crawl of .eu domain	0.9	32.3	14	4.3	0.47 (9.1x)	3.7 (1.16x)
wikipedia-20070206	Links between Wikipedia pages	3.6	45.0	20	2.6	0.069 (38x)	1.8 (1.42x)
FullChip	Circuit simulation	3.0	26.6	38	3.1	0.26 (12x)	2.4 (1.29x)
audikw1	Automotive finite element analysis	0.9	76.7	62	3.3	0.65 (5.1x)	3.2 (1.03x)
wb-edu	Links between *.edu web pages	9.8	57.2	143	2.9	0.16 (18x)	2.7 (1.07x)
germany.osm	Germany road network	11.5	24.7	5345	0.32	0.034 (9.4x)	0.38 (0.84x)
hugebubbles-00020	Adaptive numerical simulation mesh	21.2	63.6	6200	0.37	0.031 (12x)	0.48 (0.77x)

6   Conclusion

Load balance is a key factor in designing efficient algorithms for GPUs. We have demonstrated a GPU BFS algorithm that leverages the unique wide SIMD GPU architecture and achieves fully load-balanced neighbor gathering. We have showed that our approach achieves very high performance on a broad range of graphs, and outperforms current state-of-the-art implementation.

In order to exploit the nested and irregular parallelism on a BFS iteration, we have introduced a work redistribution process to flatten the nested workloads. It utilizes a parallel expansion to compute the gathering location of each neighbor so the neighbor gathering process can visit the neighbors in a load-balanced way. We have also explored the coupling possibilities of different phases, and proposed a hybrid approach that yields the best overall performance.

Acknowledgements. We thank all the anonymous reviewers for their valuable comments. This work is substantially supported by National Natural Science Foundation of China under Grants No.61300045, and China Postdoctoral Science Foundation under Grant No.2013M531696.

References

1. 10th dimacs implementation challenge, http://www.cc.gatech.edu/dimacs10/index.shtml
2. The graph 500 list, http://www.graph500.org/
3. Nvidia cuda, http://www.nvidia.com/cuda/
4. University of florida sparse matrix collection, http://www.cise.ufl.edu/research/sparse/matrices/
5. Bader, D.A., Madduri, K.: Gtgraph: A synthetic graph generator suite, Atlanta, GA (February 2006)
6. Deo, N., Sarkar, D.: Parallel algorithms for merging and sorting. Information Sciences 56(1), 151–161 (1991)
7. Harish, P., Narayanan, P.J.: Accelerating large graph algorithms on the GPU using CUDA. In: Aluru, S., Parashar, M., Badrinath, R., Prasanna, V.K. (eds.) HiPC 2007. LNCS, vol. 4873, pp. 197–208. Springer, Heidelberg (2007)
8. Hong, S., Kim, S.K., Oguntebi, T., Olukotun, K.: Accelerating cuda graph algorithms at maximum warp. In: Proceedings of the 16th ACM Symposium on Principles and Practice of Parallel Programming, pp. 267–276. ACM (2011)
9. Leiserson, C.E., Rivest, R.L., Stein, C., Cormen, T.H.: Introduction to algorithms. The MIT Press (2009)
10. Luo, L., Wong, M., Hwu, W.M.: An effective gpu implementation of breadth-first search. In: Proceedings of the 47th Design Automation Conference, pp. 52–55. ACM (2010)
11. Merrill, D., Garland, M., Grimshaw, A.: Scalable gpu graph traversal. In: ACM SIGPLAN Notices, vol. 17, pp. 117–128. ACM (2012)
12. Nasre, R., Burtscher, M., Pingali, K.: Data-driven versus topology-driven irregular computations on gpus. In: 2013 IEEE 27th International Symposium onParallel & Distributed Processing (IPDPS), pp. 463–474. IEEE (2013)
13. Nguyen, H.: Gpu gems 3. Addison-Wesley Professional (2007)
14. Odeh, S., Green, O., Mwassi, Z., Shmueli, O., Birk, Y.: Merge path-parallel merging made simple. In: 2012 IEEE 26th International Parallel and Distributed Processing Symposium Workshops & PhD Forum (IPDPSW), pp. 1611–1618. IEEE (2012)
15. Shiloach, Y., Vishkin, U.: Finding the maximum, merging, and sorting in a parallel computation model. Journal of Algorithms 2(1), 88–102 (1981)
16. Zhong, J., He, B.: Medusa: Simplified graph processing on gpus. IEEE Transactions on Parallel and Distributed Systems 99, 1 (2013) (PrePrints)

Leveraging Attributes
and Crowdsourcing for Join

Jianhong Feng, Jianhua Feng, and Huiqi Hu

Department of Computer Science, Tsinghua University, Beijing 100084, China
{fengjh11,hhq11}@mails.thu.edu.cn, fengjh@tsinghua.edu.cn

Abstract. Join operation is usually hard to achieve high quality with
machine alone. We adopt crowdsourcing to improve the quality of join.
Depending on the number of generated pairs, the overall cost can be
expensive for hiring workers to do the verification. We propose a hybrid
approach to generate pairs by leveraging attributes, which combines category, sorting and clustering techniques, called CSCER. We also propose
an adaptive attribute-selection strategy to efficiently generate pairs based
on attributes. Experiments on a real crowdsourcing platform using real
datasets indicate that our approaches save the overall cost compared to
existing methods and achieve high quality of join results.

1 Introduction

Join operation is an essential task in data cleaning. The quality of join results is
usually hard to guarantee with machine alone since it is difficult for machine to
correctly find out all records refer to the same entity from different databases.
We leverage crowdsourcing to improve the quality of join. We generate a set of
likely-matching pairs by machine, and then assign them on the crowdsourcing
platforms, such as Crowdflower [1]. The financial cost of hiring workers to do
join operation depends primarily on the quantities of generated pairs. A naive
method to generate pairs is using pairwise comparison of the entire records,
which generates $O(n^2)$ pairs for n records.

We can filter obviously non-matching pairs in order to save the cost. First,
we apply category-based and sorting-based methods to generate pairs. Then we
propose CSCER, a hybrid pairs generation approach, to reduce the number of
generated pairs. CSCER combines category, sorting and clustering techniques
for not only remaining true matching records but also filtering non-matching
records as many as possible. We evaluate our approaches with two real data sets
on a crowdsourcing platform. The experiments indicate that CSCER minimizes
the number of generated pairs and join operation achieves high quality.

2 Problem Formulation

Crowdsourcing-based Join. Given a set of tables. These tables consist of a
set of Records: $R = \{R_1, R_2, \ldots R_n\}$. Crowdsourcing-based join is to find all
matching records in these tables depending on workers.

F. Li et al. (Eds.): WAIM 2014, LNCS 8485, pp. 448–452, 2014.

Attribute-based pairs generation. A record in the table is associated with a fixed number of attributes, denoted as A_i. Attribute-based pairs generation utilizes attributes of records to prune non-matching records.

3 The CSCER

3.1 Category-Sorting Approach

The records usually can be classified according to a specific attribute. Records in the same bucket do pairwise comparison. The number of pairs generated using category-based is denoted by C_i. However it still leads to cost problem with large records when records are classified in just a few category buckets.

Sorting-based approach is another method to filter the likely non-matching pairs. This approach have two steps: the first step is to sort records based on an appropriate attribute. The second step is to set a window with fixed size and generate pairs by sliding the window. The number of pairs generated by sorting is denoted by S_i. However this approach may miss true matching records.

To overcome the limitations of category-based and sorting-based methods, we propose a hybrid method which combines category with sorting. Category is used first to obtain category bucket. Then sorting is applied for every bucket.

3.2 Category-Sorting Improvement

We apply clustering to eliminate those pairs which are significantly dissimilar. We use distance measure to cluster records. First, attributes' value need to be normalized. Then we calculate the distance between adjacent records. Given a distance threshold θ, if the calculated distance is larger than θ, the "larger" of two records leads to a new cluster. Finally, we generate pairs with the the the sliding window technique for each cluster.

3.3 Adaptive Attribute Selection

In general, category-attribute chosen principle is to choose attributes whose value is both unchangeable and enumerable. Sorting records by numeric is a appropriate choice. However, we find that some attributes are appropriate for both approaches. Thus, we propose the strategy of adaptive attribute selection.

If there is an attribute A_s suitable for category and sorting both, which method we choose for A_s depends on the rest of attributes. If we can find an attribute A_t which is suitable for sorting among the rest of attributes, then A_s is used for category. Otherwise, we need to compare C_i and S_i. There are three possible cases for C_i and S_i: **Case 1**: $C_i \leq S_i$. A_i is chosen for category. **Case 2**: $C_i > S_i$ and $C_i - S_i <$ threshold ε. A_i is selected for category. **Case 3**: $C_i > S_i$ and $C_i - S_i \geq \varepsilon$. If A_i is the only one which satisfies this constraint, A_i is used for sorting. Otherwise, the process is developed for this situation as follows. Each round leaves one attribute as a reference for comparison, and the rest of attributes are used for category. We calculate S_i of the left attribute per round, then find the smallest S_i by comparing them. For the attribute with the smallest S_i, we compute $C_i - S_i$ and estimate which case this attribute is.

3.4   CSCER Workflow

According to the above approaches , we propose CSCER to generate pairs. Algorithm 1 illustrates the pseudo code of CSCER workflow. Let A^c denote a set of attributes which is only suitable for category, and A^s denote those only suitable for sorting and A^t denote those suitable for both category and sorting.

Algorithm 1. CSCER

   Input: R; w; θ; ε

   Output: a set of pairs

1 **begin**

2     $|A^c|=|A^s|=|A^t|=0$;

3   **for** *each A_i* **do**

4   **if** *A_i is only suitable for category* **then** add A_i to A^c;

5   **if** *A_i is only suitable for sorting* **then** add A_i to A^s;

6   **if** *A_i is suitable for both category and sorting* **then** add A_i to A^t;

7   **if** $|A^c| > 0$ **then**

8   Classify Records with all attributes in A^c;

9   **if** $|A^s| = 0$ *and* $|A^t|=0$ **then**

10   generate pairs based on Category-based approach;

11   **if** $|A^s| > 0$ **then**

12   classify Records with all attributes in A^t;

13   generate pairs based on sorting and clustering approach with w and θ;

14   **else**

15   generate pairs based on adaptive attribute-selection approach with A^t and ε;

---

4   Experiments and Results Analysis

Datasets. Product denotes two datasets of electronic products, which has a total of 338 records. 177 records of them are taken from *Suning*[1] and the other 161 records are taken from *Amazon*[2]. Product generates $\frac{338*337}{2} = 56,953$ pairs by using naive method. There are 63 pairs of them, which are true matching.

4.1   Pairs Generation Analysis

First, we generate pairs by using category-based and sorting-based methods separately. Sorting-based method uses different w to generate pairs. Table 1 shows the number of pairs generated by these two methods. `Recall` is the percentage of matching pairs found by these approaches on total matching pairs. We can observe that both methods greatly reduce the number of pairs. We also see that sorting-based filters some true matching pairs while category-based remains all true matching pairs. Then, we examine the number of pairs generated

---

[1] http://www.suning.com/

[2] http://www.amazon.cn/

by CSCER. As shown in Table 2, CSCER reduces the number of pairs involving less losses of `Recall`. It also can be observed that the larger the w is, the larger the `Recall` is. We can see that the larger θ can further reduce the number of pairs with less impact on `Recall`. The above results indicate that CSCER method performed better than category-based and sorting-based methods.

4.2 Accuracy of Join

We assign 163 pairs generated by CSCER with $\theta = 0.3$ and $w = 4$ on Crowdflower. We infer the final results using Weighted MV and MV respecively. The weights reflect the quality of workers. Table 3 shows the comparison of join results. Weighted MV performed well with only one pair which was not be verified correctly. We come to the conclusion from Weighted MV that the inference method combining workers' quality can improve the quality of final results.

Table 1. # pairs generated by Category-based and Sorting-based methods

Method	# Generated Pair	# Matching Pair	Recall
Category-based	1,219	63	100%
Sorting-based with $w = 3$	673	16	25.4%
Sorting-based with $w = 4$	1,008	18	28.6%

Table 2. # pairs generated by CSCER with different θ and w

θ	w	# Generated Pair	# Matching Pair	Recall
0.2	3	116	46	79%
0.2	4	127	49	77.8%
0.3	3	147	58	92.1%
0.3	4	163	62	98.4%

Table 3. Comparison of join results

Inference method	# True Mathing	# True non-matching
MV	59	98
Weighted MV	62	100

5   Related Work

Some researches have integrated human intelligence to achieve join. Marcus et al.[2] proposed several different task forms to reduce the financial cost. CrowdER[3] proposed cluster-based task generation to reduce the number of tasks. Wang et al.[4] utilized transitive relations to devise a order of verifying pairs.

6   Conclusion

We have studied the problem of leverage attributes and crowdsourcing for join. We proposed a hybrid approach with the aim of a significant reduction in number of generated pairs, which contained three steps: category step, sorting step and cluster step. We developed adaptive attribute-selection strategy to efficiently generate pairs based on attributes. Experimental results on Crowdflower using real-world datasets showed that our approach achieved good performance and high accuracy of join resluts at lower cost.

References

1. http://crowdflower.com/
2. Marcus, A., Wu, E., Karger, D.R., Madden, S., Miller, R.C.: Human-powered sorts and joins. PVLDB 5(1), 13–24 (2011)
3. Wang, J., Kraska, T., Franklin, M.J., Feng, J.: Crowder: Crowdsourcing entity resolution. Proceedings of the VLDB Endowment 5(11), 1483–1494 (2012)
4. Wang, J., Li, G., Kraska, T., Franklin, M.J., Feng, J.: Leveraging transitive relations for crowdsourced joins. In: SIGMOD Conference, pp. 229–240 (2013)

Truth Discovery Based on Crowdsourcing

Chen Ye, Hongzhi Wang, Hong Gao, Jianzhong Li, and Hui Xie

Harbin Institute of Technology, Harbin, China
sunnyleaves0228@qq.com, {wangzhi,honggao,lijzh}@hit.edu.cn,
xiehuixb@gmail.com

Abstract . Truth discovery is an important component of data cleaning and information integration. However, in the absence of knowledge, some truth could not be found from databases themselves. A possible solution is to involve crowds to find all the truth with the knowledge of crowds. In this paper, we propose a truth discovery framework based on active learning model with crowdsourcing. First, we give the basic voting algorithm BVote. Then we present the simple crowding-based truth discovery framework STDA based on BVote. Experimental results show that the STDA framework for truth discovery has improved significantly in accuracy with minimal efforts of workers.

Keywords: truth discovery, crowdsourcing, active learning.

1   Introduction

Information from various data sources may provide conflicting information, such as different specifications for the same product. To solve data conflicts among different data sources, truth discovery techniques are in demand.

Due to its importance, much relevant research work has been proposed for truth discovery for conflicts. Existing truth discovery methods only consider the value confidence [1], the data source fidelity [2]. Even though they can resolve conflicts for many applications, the major problem is that voting by the most reliable data sources is not very trustworthy. Copy case among different data sources is extremely common on the web [3].

For this challenge, we use crowdsourcing [4] for the cases required human knowledge as well as avoiding the expensive cost for experts. Costs are queried for crowdsourcing. To reduce such cost, we use active learning [5] to minimize the training set. As the integration of active learning and crowdsourcing, we propose a framework for more effective truth discovery in this paper. The contributions of this paper are summarized as follows.

(1)We present the framework STDA for truth discovery. STDA assumes that the contribution of each worker to each tuple is 1. It uses the basic voting algorithm *BVote* to generate true values. In order to ensure the correctness of the true value, we design an active learning model to confirm the candidate true values from workers.

(2)The STDA framework generates candidates with active learning model based on crowdsourcing. The workers feedback from crowdsourcing is used to re-training

F. Li et al. (Eds.): WAIM 2014, LNCS 8485, pp. 453–458, 2014.

learning component which improves the correctness of predicted machine learning models predicted greatly. The experimental results show that the worker feedback mechanism improves the accuracy of data recovery significantly. Our framework can keep the balance between workers involvement and repaired accuracy.

The rest of this paper is organized as follows. Section 2 introduces the background and required definitions. Section 3 presents our simple truth discovery framework STDA based on BVote. Section 4 discusses the experimental studies. Finally, we conclude the whole paper in Section 5.

2   Preliminary

2.1   Problem Definition

Given a data source collection T consisting of the data sources $S_1, ..., S_n$, each data source S_i is considered as a relation with schema $(A_1, ..., A_m)$. Given an tuple O_i and an attribute A_j, the task is to find the true value of attribute A_j for tuple O_i from the tuples in the n sources. Given an answer collection W consisting of workers $W_1, ..., W_n$, each worker W_k chose a value of attribute A_j for tuple O_i from the conflicting values.

Definition 1: The confidence degree $c(v)$ of a value v of a tuple O is the probability that v is the true value.

Definition 2: The quality $q(W)$ of a worker W is the average degree of all the values self-confidence provided by W.

There may be conflict for the same tuple in different values for the same attribute. In order to confirm the true value, we attempt to solve the problem in two steps:

For each tuple O_i and attribute A_j workers give its candidate true value $u = < O_i, A_j, v >$.

In order to determine the confidence of the true value, we design an active learning algorithm to produce the true value based on $BVote$.

2.2 $BVote$ Algorithm

In general, we believe that a true value is provided by the majority of workers, and an error value is only provided by a few specific workers, so that we can apply the voting algorithm, the value provided by most workers is selected as the true value. The pseudo code is shown in Algorithm 1.

Algorithm1. BVote(W,O,A)

Input: workers answers collection W, tuple O and attribute A

Output: the true value of attribute A_j for tuple O_i, represents as $< O, A, v >$

1: Find the candidate collection$\{v_1, ..., v_N\}$ of attribute A_j for tuple O_i from worker collection W

2: find the value v which appear the most frequency

3: return $< V, A, v >$

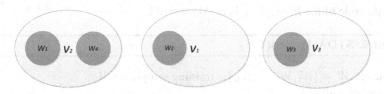

Fig. 1. Tuple O_1 Vote Schematic

Example: In Figure 1, we want to confirm the true value of an attribute of the tuple O_1. The value worker W_1 and W_4 provided is v_2, the value worker W_2 provided is v_1 and the value worker W_3 provided is v_3. Thus $vote(O_1) = \{v1:1,\ v2:2,\ v3:1\}$. where v_2 has the largest number of votes. As a result, the true value is v_2.

3 STDA Framework

We choose overlapping partition to create committees to reduce cost. First we partition a training dataset D into N disjoint sets $D_1, D_2, \cdots, D_n$. Then, the $i - th$ committee is trained with the dataset $D - D_i$. In this way, each member is trained on $(1 - \frac{1}{n})th$ fraction of the data. For each attribute $A_i \in attr(S)$, we train a series of committees $\{M_{A_1}, M_{A_2}, \cdots, M_{A_n}\}$ by active learning.

Then we use entropy on the fraction of committee members to calculate the uncertain score of a candidate. The score computation is shown as follows.

$$Usc(l)=prlog(pr) \qquad (1)$$

where $p_r = the\ frequency\ of\ r = \dfrac{\text{the number of committees whose feedback result is r}}{\text{the number of total committees}}$,
r is the result of committee prediction. With such score function, we can selection n examples with the highest uncertain score.

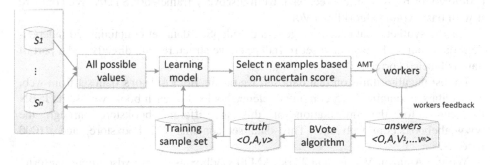

Fig. 2. STDA framework

STDA assumes that the contribution of each worker to each tuple is 1. STDA framework uses the basic voting algorithm $BVote$ to generate true values. In order to ensure the correctness of the true value, it creates an active learning model to confirm the candidate true values from workers. The STDA framework is shown in Figure 2. The pseudo code of the framework is in Algorithm 2.

Algorithm 2. STDA_Process(T,U)

Input: sources $W = \{W_1, W_2, \ldots, W_N\}$, training sample set U
Output: correct gather S
1: For each attribute $A_i \in attr(S)$, train a series of committees
$AL = \{M_{A_1}, M_{A_2}, \cdots, M_{A_n}\}$ through training sample set U
2:   collect all different values for tuple O's attribute A as $v_1, \ldots, v_n$
3:   while workers have prepared and there exists tuple which has uncertain value
do4:   confirm all the candidate true values by machine learning model
5: select n most uncertain candidate true values from candidate true values set
6:     workers give selections to the candidate true values
7:     For each attribute A of each tuple O, $BVote$ is invoked to generate candidate true value v
8:     add the candidate true value new labeled to the sample set to retrain the machine learning model
9:     add the candidate true value that machine learning model confirmed to the correct gather S
10:     update $TruthList$
11:   return S

---

4   Experiments and Analysis

4.1   Experimental Setting

We combined into a Synthetic data sets and a set of true value set to represent the efficiency of naïve workers feedback truth discovery framework STDA. We compare it with basic voting algorithm BVote.

For the synthetic data sets, we generate IndepSet data set containing 20 independent data source. In each data set in IndepSet, we structure data directly and a part of data is labeled correct.

To test the algorithms on real data set, we obtain BookAuthors data set from web. This data set contains 1115 computer science books. For each book, we use its ISBN to search for the information of this in different bookstore through the www.abebooks.com web site. This data set contains 712 bookstore, and 41000 records.

We use Amazon Mechanical Turk (AMT) sandbox as the crowdsourcing platform. All the algorithms are implemented in python, all the experiment runs on intel® Core™i32110 (3.30GHz) CPU and 4GB memory machine.

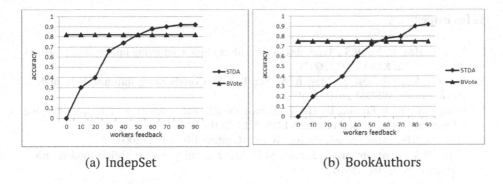

(a) IndepSet   (b) BookAuthors

Fig. 3. The accuracy of BookAuthors works on different frameworks

We use accuracy to test the performance of proposed algorithm. Accuracy is defined as the proportion of true results in the correct data gather.

4.2   Experimental Results

To test the effectiveness of our framework, we vary the number of workers feedback from 0 to 90 and compare our methods with BVote. The experimental results are shown in Figure 3. We observe that ITDA framework's effect is better than BVote algorithm when the number of feedback is more than 50%.

In Figure 3, we assessed the utility of STDA framework through the BookAuthors data set. Result shows that our framework have an obvious good effect on true value finding, i.e. generate tuple's correct true value. BVote algorithm is basic voting algorithm, here we use it as a benchmark to compare the performance of STDA framework Abscissa of workers feedback represents the proportion of the update number of accepting workers feedback in the candidate update number.

5   Conclusions and Future Work

In this paper, we present the STDA framework for truth discovery based on Crowdsourcing. STDA is simple truth discovery framework based on BVote algorithm which adds active learning component. To a certain extent it improved the accuracy of the truth discovery. Our study represents an initial attempt to use crowdsourcing platform to solve human problems. There are many further researches to explore.

(1) The degree of confidentiality of information that we should make more profound constraints. With the permission of owners, information can be provided to the workers, but the information belongs to privacy should be masked out.

(2) The cost of different quality workers is different. We intend to find a way to balance the cost and the accuracy.

Acknowledgement. This paper was partially supported by NGFR 973 grant 2012C B316200, NSFC grant 61003046, 61111130189 and NGFR 863 grant 2012AA011004.

References

1. Yin, X., Han, J., Yu, P.S.: Truth discovery with multiple conflicting information providers on the web. In: KDD 2007 (2007)
2. Galland, A., Abiteboul, S., Marian, A., Senellart, P.: Corroborating information from disagreeing views. WSDM 2010 (2010)
3. Dong, X.L., Berti-Equille, L., Hu, Y., Srivastava, D.: Global detection of complex copying relationships between sources. In: VLDB 2010 (2010)
4. Howe, J.: The rise of crowdsourcing. Wired Magazine 14(6), 1–4 (2006)
5. Cohn, D.A., Ghahramani, Z., Jordan, M.I.: Active learning with statistical models. arXiv preprint cs/9603104 (1996)

Sarcasm Detection in Social Media
Based on Imbalanced Classification

Peng Liu[1,2,3], Wei Chen[1,2,*], Gaoyan Ou[1,2], Tengjiao Wang[1,2],
Dongqing Yang[1,2], and Kai Lei[3]

[1] Key Laboratory of High Confidence Software Technologies,
Ministry of Education, Beijing 100871, China
[2] School of Electronics Engineering and Computer Science, Peking University,
Beijing 100871, China
pekingchenwei@pku.edu.cn
[3] The Shenzhen Key Lab for Cloud Computing Technology and Applications,
Peking University Shenzhen Graduate School, Shenzhen 518055, China

Abstract. Sarcasm is a pervasive linguistic phenomenon in online documents that express subjective and deeply-felt opinions. Detection of sarcasm is of great importance and beneficial to many NLP applications, such as sentiment analysis, opinion mining and advertising. Current studies consider automatic sarcasm detection as a simple text classification problem. They do not use explicit features to detect sarcasm and ignore the imbalance between sarcastic and non-sarcastic samples in real applications. In this paper, we first explore the characteristics of both English and Chinese sarcastic sentences and introduce a set of features specifically for detecting sarcasm in social media. Then, we propose a novel multi-strategy ensemble learning approach(MSELA) to handle the imbalance problem. We evaluate our proposed model on English and Chinese data sets. Experimental results show that our ensemble approach outperforms the state-of-the-art sarcasm detection approaches and popular imbalanced classification methods.

Keywords: sarcasm detection, sarcasm features, imbalanced classification, ensemble learning.

1 Introduction

Sarcasm is a sophisticated communicative act which allows a speaker to express sentiment-rich viewpoints in an implicit way. Sarcastic writing is common in many online texts. This linguistic phenomenon has much significance for tasks such as sentiment analysis, opinion mining and advertising [1].

Sarcasm is classically defined as the rhetorical process of intentionally using words or expressions for uttering a meaning different (usually the opposite) from the one they have when used literally [2]. Consider the following tweet on Twitter[1], which includes the words "yay" and "lucky" but actually expresses

---

[*] Corresponding author.
[1] http://www.twitter.com/

F. Li et al. (Eds.): WAIM 2014, LNCS 8485, pp. 459–471, 2014.

a negative sentiment: "Yay! It's a holiday weekend and i'm on call for work! Couldn't be more lucky! #sarcasm." In this case, the hashtag #sarcasm reveals the intended sarcasm, but we don't always have the benefit of an explicit sarcasm label.

Automatic sarcasm detection is considered as a simple text classification problem. They use lexical and syntactic features to detect sarcasm in text. For example, lexical cues interjections, emoticons and N-grams. However, they are commonly used features for text classification, not specifically for detecting sarcasm. Sarcasm detection is a complex and challenging task, which needs a set of explicit features cutting through every aspect of language, from pronunciation to lexical choice, syntactic structure, semantics and conceptualization.

In recent years, most researches [3–5] adopt supervised or semi-supervised learning approaches to perform sarcasm detection. However, they normally assume the balance between sarcastic and non-sarcastic samples, which may not keep in practice. Actually, many sarcasm classification applications involve imbalanced class distributions in that the sample number of sarcasm in the training data is much smaller than the non-sarcasm.

To have a better understanding of the imbalanced class distribution phenomenon in sarcasm detection, Table 1 gives the statistics on the number of sarcastic and non-sarcastic sentences occurring in experimental datasets. For more details about data sources, please refer to Section 4.

Table 1. Class distributions of samples across six experimental datasets

Data Sets	Sample Size	MI(sarcastic sentence)	MA(non-sarcastic sentence)	Imbalance Ratios(MA/MI)
Amazon data set	5491	471	5020	10
Twitter data set	40000	3200	36800	11.5
News article set	4233	233	4000	17
Sina data set	3859	238	3621	15
Tencent data set	5487	359	5128	14
Netease data set	10356	546	9810	17

Where MI represents sarcasm sample set, MA represents non-sarcasm sample set. As we can see from Table 1, all the class distributions are imbalanced, with the imbalanced ratios (MA/MI) ranging from 10 to 17, which are greater than the imbalance rate 2.333 of "German", a famous UCI imbalanced dataset (Merz & Murphy, 1995) on Germany credit scoring [6].

The imbalanced class distribution can cause many serious problems in the training process of sarcasm detection. Firstly, datasets suffer from class overlapping and are short of rare sarcastic samples, which make the classifier learning difficult. Furthermore, evaluation criterion, which guides learning procedure, tends to ignore minority samples (treating them as noise) and makes the induced classifier lose its classification ability in this scenario. These indicate the necessity of dealing with imbalanced problem in sarcasm detection.

In this paper, we propose a novel multi-strategy ensemble classification algorithm and a comprehensive sarcasm feature set to automatic detection of sarcastic sentences in both English and Chinese social media. Compared to existing methods, the main advantages of our model can be summarized as follows: (i) our classification algorithm combines weighted random sampling with ensemble of multiple classifiers based on information entropy to guarantee the high diversity of the involved member classifiers and the weighted voting being fairer. (ii) the sarcasm feature set is a multidimensional model, including lexical, syntactic, semantics and constructions. The sarcasm model can automatically differentiate an sarcastic text from a non-sarcastic one in both English and Chinese. As far as we know, it's first time to explore the sarcasm detection in chinese social media and the imbalanced class distribution problem of sarcasm detection.

The rest of the paper is organized as follows. Section 2 introduces the related work. In section 3, we present our model which is made up of sarcasm feature set and multi-strategy ensemble learning algorithm. Section 4 describes the data sets and sarcastic dictionary. Section 5 shows our experiment results. Section 6 concludes the research and gives directions for future studies.

2   Related Work

2.1   Sarcasm Detection

Sarcasm is a well-studied phenomena in linguistics, psychology and cognitive science [7]. But in the text mining literature, automatic detection of sarcasm is considered a difficult problem and has been addressed in only a few studies. Generally, sarcasm detection methods can be divided into two categories: supervised [3–5] and semi-supervised [8].

In the supervised mothods, Burfoot and Baldwin use SVM to determine whether newswire articles are true or sarcastic [3]. They introduce the notion of validity which models absurdity via a measure somewhat close to PMI. Validity is relatively lower when a sentence contains unusual combinations of named entities. González-Ibáñez et al. explored the usefulness of lexical and pragmatic features for sarcasm detection in tweets [4]. They found that positive and negative emotions in tweets have a strong correlation with sarcasm. Liebrecht et al. explored N-gram features from 1 to 3-grams to build a classifier to recognize sarcasm in Dutch tweets [5]. They made an interesting observation from their most effective N-gram features that people tend to be more sarcastic towards specific topics such as school, homework, weather, public transport, etc.

Tsur et al. presented a semi-supervised learning framework that exploits syntactic and pattern based features in sarcastic sentences of Amazon product reviews [8]. They observed correlated sentiment words such as "yay!" or "great!" often occurring in their most useful patterns.

To the best of our knowledge, no existing methods consider the class imbalance problem in sarcasm detection.

2.2   Imbalanced Classification

Many techniques are proposed to solve classification problems based on imbalanced data sets. There are two major categories of techniques developed to address the class imbalance issue. One is at data level and the other is at algorithmic level.

At the data level, different forms of re-sampling, such as over-sampling and under-sampling, are proposed. Specifically, over-sampling aims to balance the class populations through replicating the MI samples [9] while under-sampling aims to balance the class populations through eliminating the MA samples [10].

At the algorithmic level, the solutions mainly include cost-sensitive learning, one-class learning, and ensemble learning. Many cost-sensitive learning methods have been proposed [11, 12]. They intentionally increase the weights of samples with higher misclassification cost in the training process. Tax and Duin proposed a support vector algorithm for one-class classification, using kernels to obtain a tight boundary around normal samples in high dimensions [13]. Sun et al. investigate cost-sensitive boosting algorithms for advancing the classification of imbalanced data and propose three cost-sensitive boosting algorithms by introducing cost items into the learning framework of AdaBoost [14]. In addition, Maalouf et al. proposes a robust weighted kernel logistic regression. It can correct the bias of logistic regression in imbalanced classification [15].

However, their methods focus on either data processing or learning algorithm. Our approach combines them together to improve the classification accuracy.

3   The Proposed Model

A schematic representation of the proposed imbalanced sarcasm detection approach is shown in Fig.1. In general, our approach consists of two main components: (1) feature extraction and (2) multi-strategy ensemble learning. In this

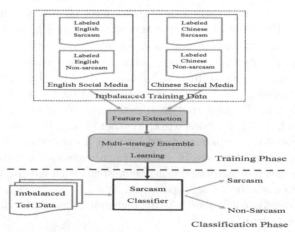

Fig. 1. A schematic representation of the imbalanced sarcasm detection approach

section, we first present a set of English sarcasm features which are important for sarcasm detection. Then we proposed a Chinese sarcasm feature set according to the differences between the English and Chinese sarcastic sentence. Finally we describe our multi-strategy ensemble learning algorithm handling imbalanced class distribution.

3.1 English Sarcasm Features

In spoken discourse, we are usually able to detect a variety of external clues that enable the perception of sarcasm. In written text, a set of explicit linguistic strategies is also used to express sarcasm. According to the research efforts on sarcasm detection, we summarize the following three categories of linguistic features that are related to the expression of sarcasm:

1) Punctuation symbols: they are focused on explicit marks which reflect a sharp distinction of sarcasm in text. These punctuation symbols include punctuation marks (, ;, ?, !, :), emoticons, quotes and capitalized words [3][4][8].

2) Lexical features: lexical features play an important role in the delivery and detection of sarcasm. According to the work in [4], it contains different aspects features such as counter-factuality(e.g. *yet*), temporal compression(e.g. *suddenly*), recurring sequences(n-grams, skip-grams, polarity s-grams) and emotional scenarios(activation, imagery, pleasantness).

3) Syntactic features: these features contain recurrent sequences of morphosyntactic patterns and degree of opposition in text with respect to the information profiled in the present and past tenses. According to the work in [16], they are composed of POS grams(e.g. *ADV / ADJ / N*) and temporal imbalance(e.g. *hate* and *didn't*).

Semantic factor has received little attention in the existing studies. However, semantic factor is a critical feature to express the context contradiction in sarcasm detection. Thus, we propose semantic imbalance rate as a new feature.

4) Semantic imbalance rate: this is intended to capture inconsistencies within a context. The intuition here is: the smaller the semantic inter-relatedness of a text, the greater its semantic imbalance (sarcastic text); the greater the semantic inter-relatedness of a text, the lesser its semantic imbalance (non-sarcastic text). In order to measure this feature, we define semantic imbalance rate as follows:

Definition 1 *Semantic imbalance rate (SIR) is the reciprocal of a text's semantic relatedness, which is summing the maximum semantic similarity scores (across different senses of words in text) and dividing by the length of the text.*

$$SIR = \frac{N}{\sum_{i=1}^{N} \max_{j=1...N \& j \neq i} Sim(w_i, w_j)} \tag{1}$$

where N is the length of text, $Sim(w_i, w_j)$ denotes the semantic similarity between word w_i and word w_j.

3.2   Chinese Sarcasm Features

In Chinese culture, sarcasm as a rhetorical device has a long history. It is more commonly used than English sarcasm. However, there is no work on automatic sarcasm detection in Chinese document. Since the Chinese grammar is different from English, we have to investigate Chinese sarcasm features independently.

By comparing the English and Chinese sarcastic sentence in our experimental data sets, we find the following differences between them:

Table 2. The differences between English and Chinese sarcasm

English	Chinese
People prefer to use hyperbole in English	People frequently use proverbs in Chinese
People like cause-and-effect reasoning	People are good at analogical reasoning
People use capital letters to express stress	People use quotation marks to stress
It does not exist in English	People use honorifics and "我朝", "咱朝", "天朝", "本朝" (my/our/great dynasty) some special appellations, etc
It does not exist in English	People use homophony to accomplish a sarcastic effect
People use very positive, absolute words to express sarcasm in English	People use extreme negative words except them in Chinese

According to Table 2, we propose a Chinese sarcasm feature set including punctuation symbols, recurring sequences, semantic imbalance rate, rhetorical feature, homophony and construction. Among them, the first three features are the same as the English features. The remaining features are explained as follows:

1) Rhetorical feature: rhetorical feature includes extreme positive or negative nouns, extreme adjectives, adverbs of degree, demonstratives, honorifics and proverbs. This feature carries on valuable information about authors' emotions.

2) Homophony feature: homophony is commonly used in Chinese. It can be sarcastically used for expressing an insult or depreciation towards the entity they represent. For example, "我把他们称为'审查猪'(审查组)。"(I call them "examining pigs"(examining groups)). Pigs and groups are homophonic in Chinese.

3) Construction feature: construction feature is inherently tied to a particular model of the 'semantics of understanding', which offers a way of structuring and representing meaning while taking into account the relationship between lexical meaning and grammatical patterning. For example, 比N还N (as more as), adverb + adjective + interjection, 拿A当B (take A as B), etc.

3.3   Imbalanced Classification

In order to fully utilize the training data and to guarantee the diversity of the multiple classifiers, we propose a novel multi-strategy ensemble learning approach (MSELA) for imbalanced sarcasm detection, which integrates sample-ensemble strategy, classifier-ensemble strategy and weighted voting strategy.

The basic motivation of MSELA is as follows. Firstly, sample-ensemble strategy uses weighted random sampling which can fully utilize features of the whole training data and cost less time. Secondly, classifier-ensemble strategy guarantees the diversity of multiple classifiers. Thirdly, weighted voting strategy considers both the accuracy of different classification algorithms and the importance of sample subsets to achieve fairer weighted voting and higher prediction accuracy.

A. Sample-Ensemble Strategy

Weighted random sampling is an effective method to deal with class imbalanced problem. In sample-ensemble strategy, the small samples are set various weights and merged with large samples into new datasets. With a weighted random sampling for each new dataset, N balanced datasets are obtained. These balanced datasets are used to train sub-classifiers which will vote for the last result.

B. Classifier-Ensemble Strategy

Besides the sample-ensemble strategy, we also propose a classifier-ensemble strategy to guarantee the diversity of the involved classifiers. The classifier-ensemble strategy involves more than one classification algorithm. That is to say, multiple classification algorithms are prepared and the classifier of each dataset is trained with a random selected classification algorithm. In our experiments, we employ three algorithms: Naive Bayes, SVM and Maximum Entropy.

C. Weighted Voting Strategy

Information entropy can measure the uncertainty of a random variable. The greater the information entropy of the data set is, the more uniform its class distribution will be. The smaller the information entropy, the more consistent the class distribution of data set will be. Assume that data set S contains c classes, then the information entropy $E(S)$ of data set S is:

$$E(S) = - \sum_{i=1}^{c} P_i log_2 P_i \tag{2}$$

where P_i is the proportion of class i in the whole data set.

Following the simple weight calculation method in [17], our weight formula of sub-classifier C_i is:

$$w_i = \frac{acc_i |Sub_i| E(i)}{\sum acc_i |Sub_i| E(i)} \tag{3}$$

where acc_i is the accuracy of sub-classifier C_i, $|Sub_i|$ is the size of data subset D_i', $E(i)$ is the information entropy of data subset D_i'. Our multi-strategy ensemble learning approach is illustrated in Algorithm 1.

Algorithm 1 Multi-strategy ensemble learning algorithm

Input: The MI training set S_{MI} and the MA training set S_{MA};
Output: A combination classifier C_n;
1. Initialize the sample set $A = S_{MA}$, $B = S_{MI}$;
2. $M = sizeof(A)$, $N = sizeof(B)$, $K = M/N$;
3. **for** $i = 0$; $i < \lfloor K \rfloor$; $i + +$ **do**
4.   Initialize sample set $S_i = \varnothing$;
5.   **for** $j = 0$; $i < N$; $j + +$ **do**
6.   Set the weight of sample $B(j)$ to $i + 1$, and put it into S_i;
7.   **end for**
8. $D_i = S_i \cup A$;
9.   Perform weighted random sampling for D_i to get sample set D_i';
10.   As for D_i', randomly choose a classification algorithm to train sub-classifier C_i;
11.   Calculate the information entropy of sample set D_i' according to Eq.(2),
         and then calculate the weight of sub-classifier C_i according to Eq.(3);
12. **end for**
13. Integrate all sub-classifiers $C_i(i = 1, 2, .., \lfloor K \rfloor)$ into classifier C_n by weighted voting;
14. **return** C_n;

MI represents sarcasm sample set, MA represents non-sarcasm sample set. K denotes the imbalanced ratio. The small samples are set various weights(step 6), and merged with large samples into new datasets D_i(step 8). With a weighted random sampling for each new dataset D_i, N balanced datasets $D_i'(i = 1, 2, ..., N)$ can be obtained(step 9). These balanced datasets are used to train different classifiers(step 10), which will vote for the last result(step 13). The information entropy is used to measure the differences among various N balanced datasets D_i', and then the weight for each classifier is computed using Eq.(3) (step 11). The above process (step 4 to step 11) is executed $\lfloor K \rfloor$ rounds.

4   Experimental Setup

We use two corpora to evaluate our proposed model. The first corpus is used to verify English Sarcasm Features, which contains a Amazon data set provided in [8], a Twitter data set from [16] and a News article set from [3]. The second corpus is used to verify Chinese Sarcasm Features, which contains three data sets obtained by crawling different topic comments from Sina Weibo[2], Tencent Weibo[3] and Netease BBS[4]. The class distributions of these data sets are described in Section 1.

Since the datasets are highly skewed, we use Area under the Curve (AUC) for the ROC curve as the measure of evaluation. AUC is a common evaluation metric that has been shown to be more resistant to skew dataset than F-score, due to using true positive rate (TPR) rather than precision.

---

[2] http://huati.weibo.com/
[3] http://t.qq.com/
[4] http://bbs.163.com/

To extract rhetorical feature, we create a sarcastic dictionary with manually annotated sarcastic keywords and phrases. The sarcastic dictionary is composed by 1230 adjectives, 1096 nouns, 189 adverbs of degree, 65 demonstratives, 120 honorifics and 2322 proverbs. Among them, the adjectives, nouns and adverbs of degree come from Dalian University of Technology Affective Lexicon Ontology[5].

5   Experiments

In this section, we evaluate the performances of our proposed model with three experiments. In the first experiment, we compare our multi-strategy ensemble learning algorithm with popular imbalanced classification methods. The second experiment evaluates feature correlation of our model. In the third experiment, we compare our method with the state-of-the-art sarcasm detection model.

5.1   Imbalanced Sarcasm Classification

In this subsection, we report the performances of ensemble methods for imbalanced sarcasm classification. The Maximum Entropy classifier and Naive Bayes classifier are implemented using the Mallet tool[6], while SVM is implemented using the SVM-light tool[7]. For thorough comparison, various kinds of ensemble methods are implemented including:

1. Full-training (FullT): directly throwing all the training data for training.
2. Bagging [9]: creating sub-classifiers on data subsets that are uniformly sampled from original data, then using a linear combination to aggregate them.
3. AdaBoost [9]: generating weak classifiers sequentially and then combining them into a strong one.
4. SMOTEBoost [10]: combining the over-sampling method SMOTE and boosting together. More specifically, a SMOTE over-sampling procedure is performed first in every round of boosting, and then the traditional AdaBoost method is used to update the weight distributions of all samples.
5. Random Forest [9]: containing many decision tree classifiers where each tree is trained on a randomly sampled feature subset from original data.
6. Undersampling + multi-classifiers ensemble (EUM): performing clustering-based stratified under-sampling as proposed by [17].
7. Logistic Regression: correcting bias of logistic regression as proposed by [15].

Among them, bagging, adaboost, smoteboost and random forest are implemented using the Weka tool[8]. Since most methods involve random selection of samples, we run 10 times for each method and report the average performance of the 10 runs.

---

[5] http://ir.dlut.edu.cn/EmotionOntologyDownload.aspx/
[6] http://mallet.cs.umass.edu/
[7] http://www.csie.ntu.edu.tw/ cjlin/libsvm/
[8] http://www.cs.waikato.ac.nz/ml/weka/

Table 3 compares the seven methods with our multi-strategy algorithm. It shows that almost all the specifically designed methods outperform full-training. The reason is that full-training fails to take the sample imbalance into account. Consistently, ensemble-learning always performs much better than the other approaches, especially in the Chinese corpus. Among the ensemble approaches, undersampling plus multi-classifiers ensemble learning significantly outperforms sample ensemble learning or classifier ensemble learning alone. Our approach (multi-strategy) achieves the best performance on average, which confirms its effectiveness to enlarge the diversities of the member classifiers and keep balance between positive and negative samples.

Table 3. The AUC scores of the eight imbalanced classification algorithms

Data sets	Amazon Data Set	Twitter Data Set	News Article Set	Sina Data Set	Tencent Data Set	Netease Data Set
FullT	0.683	0.638	0.624	0.708	0.713	0.687
Bagging	0.816	0.803	0.796	0.853	0.869	0.830
AdaBoost	0.809	0.810	0.781	0.846	0.857	0.819
SMOTEBoost	0.823	0.814	0.805	0.859	0.862	0.837
Random Forest	0.801	0.794	0.779	0.820	0.834	0.809
EUM	0.842	0.826	0.819	0.874	0.883	0.856
Logistic Regression	0.767	0.749	0.742	0.793	0.786	0.771
MSELA	**0.854**	**0.840**	**0.833**	**0.889**	**0.897**	**0.873**

We also analyze the cost of time among the seven ensemble methods for imbalanced sarcasm classification, which is shown in Fig.2.

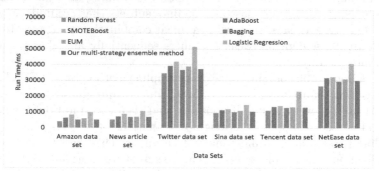

Fig. 2. Comparison of running time among different classification algorithms

From Fig.2, we can see that Random Forest takes the least time while Logistic Regression takes the most time. Our multi-strategy ensemble learning always takes less time than the undersampling plus multi-classifiers ensemble learning.

5.2 Feature Correlation Verification

Our model operates with a series of features, each type of feature can be analyzed in terms of information gain to determine its individual contribution to the discrimination power of the system. Fig.3 and Fig.4 present the results of an information gain filter (Y axis) on each type of our features (X axis).

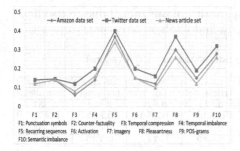

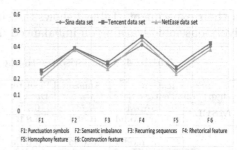

Fig. 3. Feature correlation verification in English corpus

Fig. 4. Feature correlation verification in Chinese corpus

As shown by the results in Fig.3 and Fig.4, this relevance is a function of the kinds of texts that are to be discriminated. For example, the features recurring sequences, pleasantness and semantic imbalance appear to be sufficiently indicative to represent sarcastic sentences from the three English datasets. While, the features semantic imbalance, rhetorical feature and construction feature appear to be indicative to represent sarcastic sentences from the three Chinese datasets.

We analyze the performance improvements when each time a new type of feature is added. We use tenfold cross validation and results are shown in Fig.5.

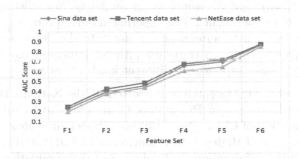

Fig. 5. Classification accuracy with the addition of features

Fig.5 indicates an acceptable performance on the automatic Chinese sarcasm detection. The model evidently improves its performance in almost all cases each time a new type of feature is added.

5.3   Comparison with the State-of-the-Art

For comparison, we implement a state-of-the-art supervised model (SASM) for sarcasm detection, as proposed by [18], which takes n-grams, POS n-grams, funny profiling, positive/negative profiling, affective profiling, and pleasantness profiling as their feature set. They assess classification accuracy employing three classifiers: Naive Bayes, Support Vector Machine and Decision Tree.

We compare our sarcasm feature set with theirs, successively using our multi-strategy ensemble learning(MSELA) and decision tree(DT), which performs best in their experiment. Table 4 shows the superiority of our model due to its effectively ensemble learning on imbalanced data and explicit sarcasm feature set.

Table 4. Comparison with the state-of-the-art (AUC score)

Classification Algorithm&Feature Sets		Amazon Data Set	Twitter Data Set	News Article Set	Sina Data Set	Tencent Data Set	Netease Data Set
MSELA	SASM	0.741	0.733	0.715	0.629	0.642	0.637
	Our model	**0.854**	**0.840**	**0.833**	**0.889**	**0.897**	**0.873**
DT	SASM	0.613	0.596	0.585	0.554	0.581	0.568
	Our model	**0.685**	**0.653**	**0.639**	**0.712**	**0.733**	**0.696**

6   Conclusion

Sarcasm is one of the most subjective phenomena related to linguistic analysis. Automatic sarcasm detection is a real challenge, not only from a computational perspective but from a linguistic one as well. In this paper, we present multi-strategy ensemble learning, a novel algorithm for detection of sarcastic sentences in both Chinese and English social media. Our method can guarantee the high diversity of the sub-classifiers and take full advantage of training data. We propose an explicit sarcasm feature set which attempts to describe salient characteristics of both Chinese and English sarcasm. They intend to symbolize low and high level properties of sarcasm. We compare our model against existing approaches using three different experiments. The experiments give promising results.

In the future, we plan to automatically annotate the sarcastic samples and study new features in order to come up with an improved model capable of detecting more sarcastic sentences in different kinds of texts.

Acknowledgments. This research is supported by the National High Technology Research and Development Program of China (No. 2012AA011002), Natural Science Foundation of China (No. 61300003), Specialized Research Fund for the Doctoral Program of Higher Education(No. 20130001120001) and Research Foundation of China Information Technology Security Evaluation Center (No. CNITSEC-KY-2013-018).

References

1. Liu, B., Zhang, L.: A survey of opinion mining and sentiment analysis. In: Mining Text Data, pp. 415–463. Springer (2012)
2. Carvalho, P., Sarmento, L., Silva, M.J.: Clues for detecting irony in user-generated contents: oh..!! it's so easy;-). In: Proceedings of the 1st International CIKM Workshop on Topic-Sentiment Analysis for Mass Opinion, pp. 53–56. ACM (2009)
3. Burfoot, C., Baldwin, T.: Automatic satire detection: Are you having a laugh? In: Proceedings of the ACL-IJCNLP 2009 Conference Short Papers, pp. 161–164. Association for Computational Linguistics (2009)
4. González-Ibáñez, R., Muresan, S., Wacholder, N.: Identifying sarcasm in twitter: A closer look. In: ACL (Short Papers), pp. 581–586. Citeseer (2011)
5. Liebrecht, C., Kunneman, F., van den Bosch, A.: The perfect solution for detecting sarcasm in tweets# not. In: WASSA 2013, p. 29 (2013)

6. Blake, C., Merz, C.J.: Uci repository of machine learning databases (1998)
7. Gibbs Jr, R.W., Colston, H.L.: Irony in language and thought: A cognitive science reader. Psychology Press (2007)
8. Tsur, O., Davidov, D., Rappoport, A.: Icwsm-a great catchy name: Semi-supervised recognition of sarcastic sentences in online product reviews. In: ICWSM (2010)
9. Chawla, N.V., Bowyer, K.W., Hall, L.O., Kegelmeyer, W.P.: Smote: synthetic minority over-sampling technique. arXiv preprint arXiv:1106.1813 (2011)
10. Yen, S.J., Lee, Y.S.: Cluster-based under-sampling approaches for imbalanced data distributions. Expert Systems with Applications 36(3), 5718–5727 (2009)
11. Elkan, C.: The foundations of cost-sensitive learning. In: International Joint Conference on Artificial Intelligence, vol. 17, pp. 973–978. Citeseer (2001)
12. Wang, B.X., Japkowicz, N.: Boosting support vector machines for imbalanced data sets. Knowledge and Information Systems 25(1), 1–20 (2010)
13. Tax, D.M., Duin, R.P.: Support vector data description. Machine Learning 54(1), 45–66 (2004)
14. Sun, Y., Kamel, M.S., Wong, A.K., Wang, Y.: Cost-sensitive boosting for classification of imbalanced data. Pattern Recognition 40(12), 3358–3378 (2007)
15. Maalouf, M.: Trafalis: Robust weighted kernel logistic regression in imbalanced and rare events data. Computational Statistics & Data Analysis 55(1), 168–183 (2011)
16. Reyes, A., Rosso, P., Buscaldi, D.: From humor recognition to irony detection: The figurative language of social media. Data & Knowledge Engineering 74, 1–12 (2012)
17. Chen, H., Du, Y., Jiang, K.: Classification of incomplete data using classifier ensembles. In: Systems and Informatics (ICSAI), pp. 2229–2232. IEEE (2012)
18. Reyes, A., Rosso, P.: Making objective decisions from subjective data: Detecting irony in customer reviews. Decision Support Systems 53(4), 754–760 (2012)

Centroid-Based Classification
of Categorical Data

Lifei Chen and Gongde Guo

School of Mathematics and Computer Science,Fujian Normal University, China
{clfei,ggd}@fjnu.edu.cn

Abstract. The traditional centroid-based classifiers cannot be directly applied to categorical data classification due to the undefined concept of centroid for a categorical class, and the lack of an effective distance measure for categorical objects. In this paper, two centroid-based classifiers are proposed for categorical data classification. We propose a new formulation for the centroid of categorical classes to address the first problem, while two weighted distance measures are defined for the second problem. The experimental results conducted on real-world data sets show the effectiveness of the proposed methods.

Keywords: Categorical data, classification, centroid, weighted distance.

1 Introduction

In recent years, centroid-based (alternative known as prototype-based) classification [1], as one of the popular instance-based methods, has gained increasing attention owing to their inherent simplicity and linear-time complexity. In such a classifier, a centroid vector is computed to represent the training samples of each class; then, test sample is assigned to the class that corresponds to its most similar centroid, according to the distances between test sample and the centroid vectors. Though simple, the experimental results reported in the literature have validated that the performance of a centroid-based classifier could outperform other algorithms for a wide range of applications [2].

However, the traditional methods cannot be applied to categorical data classification. The difficulties are two-folds. First, when the data are categorical the set *mean* is an undefined concept, since the attributes can only take discrete values. Consequently, the class centroid, generally defined on the mean of numeric attributes, also becomes undefined for categorical data. Second, the commonly used distance measures, such as the Euclidean distance designed for numeric data, cannot be used for categorical objects. A few alternative measures have been suggested, including the common simple matching coefficient distance [3]. However, they are ineffective in practice because, essentially, they assume that all the attributes are equally important. The assumption hardly holds in many real-world applications: for example, with high-dimensional data where there are typically a considerable number of noisy attributes that do not contribute to class prediction [4].

F. Li et al. (Eds.): WAIM 2014, LNCS 8485, pp. 472–475, 2014.
© Springer International Publishing Switzerland 2014

In this paper, two new approaches are proposed to make the centroid-based method effective for categorical data classification. We reformulate the definition of centroids for categorical classes, allowing the centroids to be learned from categorical attributes. We propose two attribute-weighting approaches for the computation of weighted distance between categorical object and the centroid. The attribute weights are learned in terms of their contributions to class prediction, in a supervised manner.

2   The New Centroid-Based Classifiers

Suppose that we are given a training data set tr consisting of N objects each with one of the K pre-defined classes. The kth class, where $k = 1, 2, \ldots, K$, is denoted by c_k containing $|c_k|$ objects. A data object is denoted by $\mathbf{x} = (x_1, x_2, \ldots, x_D)$ where D is the number of categorical attributes. The set of categories taken by attribute $d = 1, 2, \ldots, D$ is denoted by S_d, and an arbitrary category in the set is denoted by $s \in S_d$. The dth attribute of $\mathbf{x}$ can be represented in the *vector representation* involving only numeric values, given by $\mathbf{v}(x_d) = < I(x_d = s_{d1}), \ldots, I(x_d = s_{dl}), \ldots, I(x_d = s_{d|S_d|}) >$, where $I(\cdot)$ is an indicator function. This suggests a new formulation for the class centroid in categorical data.

Definition 1. The centroid of c_k on the dth attribute is the vector $\mathbf{m}_d^{(k)} = < m_{d1}^{(k)}, \ldots, m_{dl}^{(k)}, \ldots, m_{d|S_d|}^{(k)} >$ that minimizes $J(\mathbf{m}_d^{(k)}) - \sum_{\mathbf{x} \in c_k} ||\mathbf{v}(x_d) - \mathbf{m}_d^{(k)}||_2^2$ subject to $\sum_{l=1}^{|S_d|} m_{dl}^{(k)} = 1$, where $|| \cdot ||_2$ stands for the Euclidean norm.

Using the Lagrangian multiplier technique, the optimization problem defined in Definition 1 can be transformed into an unconstrained problem. Thus, the centroid can be solved by taking derivatives to the objective function with respect to the variables and the Lagrangian multiplier, yielding

$$m_{dl}^{(k)} = \frac{1}{|c_k|} \sum_{\mathbf{x} \in c_k} I(x_d = s_{dl}) \tag{1}$$

which is precisely the relative frequency of s_{dl} appearing in the dth attribute of c_k. Then, we measure the distance between $\mathbf{x}$ and the class c_k by

$$dist(\mathbf{x}, M_k, \mathbf{w}) = \sum_{d=1}^{D} w_d \times ||\mathbf{v}(x_d) - \mathbf{m}_d^{(k)}||_2^2 \tag{2}$$

where $M_k = \{\mathbf{m}_d^{(k)}\}_{d=1}^{D}$ is the class centroids of c_k and $\mathbf{w} = < w_1, \ldots, w_d, \ldots, w_D >$ the weighting vector. Here, each attribute-weight $w_d > 0$ is defined to measure the contribution of attribute d to class prediction. The greater the contribution, the larger the weight.

We estimate the weights in a supervised way and present two weighting approaches. The first approach is based on the *mutual information*, one of the widely used measures to define dependency of variables [5], computed by

$$w_d^{(\text{MI})} = \sum_{s \in S_d} \sum_{k=1}^{K} p(s, c_k) \log \frac{p(s, c_k)}{p(s) p(c_k)} \tag{3}$$

where $p(c_k) = \frac{|c_k|}{N}$, $p(s) = \frac{1}{N}\sum_{\mathbf{x} \in tr} I(x_d = s)$ and $p(s, c_k) = \frac{1}{N}\sum_{\mathbf{x} \in c_k} I(x_d = s)$. This will be called mutual-information-based (MI-based for short) weighting approach. Another measure that has been popularly utilized to rank categorical attributes is the *Gini index* [3]. For $s \in S_d$, the Gini index is computed by $GI(s) = 1 - \sum_{k=1}^{K} [p(c_k|s)]^2$ with $p(c_k|s) = \sum_{\mathbf{x} \in c_k} I(x_d = s)/\sum_{\mathbf{x} \in tr} I(x_d = s)$. Then, we define the Gini-index-based (GI-based for short) weight as

$$w_d^{(GI)} = 1 - \sum_{s \in S_d} p(s) \times GI(s) = \sum_{s \in S_d} p(s) \sum_{k=1}^{K} [p(c_k|s)]^2. \qquad (4)$$

The methods presented above are applied to derive new centroid-based classifiers, called FCC (Frequency-Centroid-based Classification). In the training phase of the new classifiers, the centroids are learned in terms of Eq. (1) and the attribute weights are learned according to Eq. (3) or Eq. (4), depending on the the MI-base or GI-based weighting approach used. The testing algorithm is similar to the traditional centroid-based classifier [1] but using Eq. (2) as the distance measure. That is, for each test sample $\mathbf{x}$, its class label is predicted as the most similar class in terms of the K distances, using the weights and the centroids learned by the training algorithm. The classifiers are named FCC/MI and FCC/GI, respectively, according to the weighting approach used.

3   Experimental Evaluation

We evaluate the performance of FCC/MI and FCC/GI on six widely used real-world data sets, all of which were obtained from the UCI Machine Learning Repository (available at ftp.ics.uci.edu: pub/machine-learning-databases), as Table 1 shows. Our new classifiers will be compared with the traditional centroid-based classification method, called CBC, which is an extension to the the traditional method [1] that makes use of Definition 1 to define the class centroids. Note that the distance measure used in CBC can be regarded as a special case of Eq. (2) where the weights are equally set to 1. We also carried out the classifications using C4.5 from the WEKA system [6] to provide a reference point.

Table 1. Comparison of classification accuracies (Macro-F1) on real-world data sets

DATA SET	CBC	FCC/MI	FCC/GI	C4.5
LUNGCANCER	0.140±0.057	**0.569±0.287**	0.477±0.285	0.330±0.263
PROMOTERS	0.333±0.018	**0.925±0.077**	**0.925±0.073**	0.788±0.129
DERMATOLOGY	0.048±0.003	**0.973±0.029**	0.967±0.032	0.934±0.042
BALANCE	0.049±0.003	0.590±0.052	**0.598±0.050**	0.445±0.029
SOYBEAN	0.003±0.000	**0.920±0.031**	**0.920±0.030**	0.913±0.040
BREASTCANCER	0.396±0.000	0.948±0.026	**0.955±0.023**	0.940±0.031

The classification performance of different classifiers were measured using Macro-F1 measure [2]. Each data set was classified by each classifier for 20 executions using ten-fold cross validation, and the average performances are reported in the format *average* $\pm$ 1 *standard deviation*. Tables 1 shows the results. In the tables, the highest accuracy is marked in bold typeface, for the algorithm on each data set comparing with others, using the paired t-test with significance level 0.05.

The tables show that the centroid-based method can be made much more effective for categorical data classification by our new approaches used in FCC/MI and FCC/GI. We observe the surprising improvements of the two new classifiers comparing with the traditional method CBC. In fact, CBC fails in classifying all eight data sets, indicating that simple extensions to the traditional method is ineffective. However, when combined with the weighted distance measures, as used in FCC/MI and FCC/GI, the new classifier becomes much more accurate. This can be explained by the observation of [4] that a linear transformation of input features can lead to significant improvement in an instance-based method. The tables also show that our new centroid-based classifiers do significantly better than C4.5 on the data sets. The results largely dues to the attribute-weighting methods of *FCC*, which are able to recognize the different importance of attributes in discriminating the classes.

4   Conclusion

In this paper, we proposed two new classifiers in order to make the traditional centroid-based method effective for categorical data classification while retaining their numerous strengths in terms of classification efficiency. We proposed a new formulation for the centroid of categorical classes, such that the traditional method can be extended for categorical data. We also proposed two effective distance measures for the dissimilarity computation of test samples and the training classes. We defined the new distance measures using mutual-information-based and Gini-index-based weighting approaches, which in effect perform soft feature-selection for the categorical attributes during the training process.

References

1. Han, E.-H(S.), Karypis, G.: Centroid-based document classification: Analysis and experimental results. In: Zighed, D.A., Komorowski, J., Żytkow, J.M. (eds.) PKDD 2000. LNCS (LNAI), vol. 1910, pp. 424–431. Springer, Heidelberg (2000)
2. Chen, L., Ye, Y., Jiang, Q.: New centroid-based classifier for text categorization. In: Proceedings of the AINAW, pp. 1217–1222 (2008)
3. Sen, P.: Gini diversity index, hamming distance and curse of dimensionality. Metron - International Journal of Statistics LXIII(3), 329–349 (2005)
4. Weinberger, K., Saul, L.: Distance Metric Learning for Large Margin Nearest Neighbor Classification. Journal of Machine Learning Research 10, 207–244 (2009)
5. Peng, H., Long, F., Ding, C.: Feature selection based on mutual information: Criteria of max-dependency, max-relevance, and min-redundancy. IEEE Transactions on Pattern Analysis and Machine Intelligence 27, 1226–1238 (2005)
6. Hall, M., Frank, E., et al.: The weka data mining software: An update. SIGKDD Explorations 11 (2009)

ChronoSAGE: Diversifying Topic Modeling Chronologically

Tomonari Masada[1] and Atsuhiro Takasu[2]

[1] Nagasaki University, 1-14 Bunkyo-machi, Nagasaki 8528521, Japan
masada@nagasaki-u.ac.jp
[2] National Institute of Informatics, 2-1-2 Hitotsubashi, Chiyoda-ku, Tokyo 1018430, Japan
takasu@nii.ac.jp

Abstract. This paper provides an application of sparse additive generative models (SAGE) for temporal topic analysis. In our model, called ChronoSAGE, topic modeling results are diversified chronologically by using document timestamps. That is, word tokens are generated not only in a topic-specific manner, but also in a time-specific manner. We firstly compare ChronoSAGE with latent Dirichlet allocation (LDA) in terms of pointwise mutual information to show its practical effectiveness. We secondly give an example of time-differentiated topics, obtained by ChronoSAGE as word lists, to show its usefulness in trend detection.

1 Introduction

In text mining, topic modeling approach prevails, because it can give a compact representation of topics found in a document set as word lists. Latent Dirichlet allocation (LDA) [1] represents each topic as a word probability distribution and extracts a predefined number, say K, of topics from a document set. Consequently, LDA provides a summarizing view of documents as K word lists, each expressing a particular subject in a human-readable way (cf. Figure 8 in [1]).

However, recent applications of text mining require using metadata to make topic modeling results more persuasive. Especially, timestamps are often considered due to their importance in SNS posts, newswire documents, academic articles, etc. We thus use timestamps so that per-topic word probability distributions are time-dependent. We make our approach based on sparse additive generative models (SAGE) [2] and call it *ChronoSAGE*. We denote the LDA-type SAGE in its simplest form, given in Section 4 of [2], as *vanilla SAGE*. ChronoSAGE has three types of parameters for defining word probabilities: the parameters for topics, those for timestamps, and those for pairs of a topic and a timestamp. The parameters of the first type only follows the original rationale of LDA and vanilla SAGE. Those of the second type are intentionally introduced to filter out the words showing non-informative time-dependency, e.g. "Sunday", "May", "2001", etc. Further, those of the third type give time-differentiated word probabilities for each topic and make our approach attractive. ChronoSAGE outputs as many human-readable word lists as timestamps for each topic. That is, when the number of timestamps is T, ChronoSAGE outputs TK word lists.

F. Li et al. (Eds.): WAIM 2014, LNCS 8485, pp. 476–479, 2014.

Table 1. Top seven words sorted by η_{tw}s for each of 18 timestamps in case of TDT4

edt paralymp lebanon 32nd wild-card u.s china	kippur 10-13 lebanon china palestinian text join
10-14 10-16 10-18 10-15 10-19 10-17 10-20	10-24 10-23 10-22 10-25 10-21 10-26 10-27
10-29 10-28 10-31 10-30 11-3 leipzig lebanon	11-10 11-8 11-9 11-6 11-7 11-5 convuls
11-17 11-16 11-11 11-14 11-15 11-12 11-13	11-18 11-19 11-24 11-22 11-23 11-20 11-21
11-25 11-27 11-28 11-26 11-30 11-29 seclus	12-8 12-6 12-5 12-7 12-3 537-vote 12-4
12-12 12-15 12-14 12-10 12-13 12-11 12-9	12-17 12-18 12-21 12-20 12-19 12-22 12-16
12-24 12-28 12-29 12-23 12-27 12-26 12-25	309 tabasco 2001 1-5 vy 12-0 free-agent
presid-elect's 1-12 1-8 1-11 1-9 1-10 1-7	1-14 1-13 1-19 1-18 1-17 1-16 1-15
1-21 1-26 1-25 1-22 1-20 1-23 1-24	1-28 1-31 1-30 1-27 1-29 dawosi bhuj

In the evaluation, we compare ChronoSAGE and vanilla SAGE with LDA in terms of pointwise mutual information [4] to show the basic competence of the SAGE-type approaches. Further, we present an example of time-dependent word lists extracted by ChronoSAGE and discuss them from a qualitative viewpoint.

2 ChronoSAGE

ChronoSAGE is an application of the multi-faceted SAGE [2] for using document timestamps. We give its generative description, but omit the inference details.

- With respect to each word w, draw a variance parameter τ_{kw} for each topic k, τ_{tw} for each timestamp t, and τ_{tkw} for each pair of a timestamp t and a topic k from the improper Jeffrey's prior distribution $p(\tau) \propto 1/\tau$. Further, draw parameters η_{kw}, η_{tw}, and η_{tkw} from the zero-mean Gaussian distributions $\mathcal{N}(0, \tau_{kw})$, $\mathcal{N}(0, \tau_{tw})$, and $\mathcal{N}(0, \tau_{tkw})$, respectively.
- Obtain the probability ϕ_{tkw} that the word w is used to express the topic k in the documents having the timestamp t as $\phi_{tkw} \propto \exp(m_w + \eta_{kw} + \eta_{tw} + \eta_{tkw})$, where m_k corresponds to the background probability of the word w.
- For each document d, draw a multinomial parameter $\boldsymbol{\theta}_d = (\theta_{d1}, \dots, \theta_{dK})$ from the symmetric Dirichlet prior Dirichlet(α). Further, for the ith word token of d, draw a topic z_{di} from the multinomial Multi($\boldsymbol{\theta}_d$) and draw a word x_{di} from the multinomial Multi($\boldsymbol{\phi}_{y_d z_{di}}$), where y_d is the timestamp of d.

ChronoSAGE has three types of parameters defining word probabilities. η_{kw}s make word probabilities dependent on topics. Vanilla SAGE only has this type of parameters. η_{tw}s are introduced to find the words showing non-informative time-dependency. Table 1 presents top seven words sorted by η_{tw}s for each timestamp t in TDT4, a document set used in our evaluation (cf. Section 3). In TDT4, we gave the same timestamp to the documents belonging to the same range of seven days, e.g. from December 14 to 20, 2000. In Table 1, most words are just the dates falling in the corresponding range. Since η_{tw}s remove non-informative time-dependency in this manner, η_{tkw}s can make per-topic word probabilities time-differentiated and can give informative word lists as we show later in Fig. 1.

3 Evaluation

We firstly compare ChronoSAGE and vanilla SAGE with LDA in terms of PMI to show the competence of SAGE-type approaches and secondly evaluate the

Table 2. Specifications of the three document sets used in the experiment

	# documents	# words	# timestamps (T)	average document length
DBLP	2,093,913	10,694	22	5.2
NSF	128,181	19,066	13	95.9
TDT4	96,246	15,153	18	156.6

timestamped word lists extracted by ChronoSAGE from a qualitative viewpoint. We used the three document sets given in Table 2. DBLP is a set of paper titles in DBLP CS bibliography[1]. NSF is available at the UCI ML repository[2]. We regarded publication years as timestamps in DBLP and NSF. TDT4 is a corpus for the topic detection and tracking evaluation by LDC[3]. The number K of topics was set to 100 or 300. For each of the compared approaches, i.e., LDA, vanilla SAGE, and ChronoSAGE, we initialized topic assignments randomly, ran Gibbs sampling [3] as pretraining, and ran a variational inference ten times. We thus obtained ten topic modeling results for each approach and for each setting of K.

We adopted an external evaluation measure, called pointwise mutual information (PMI) [4], to make our evaluation realistic. The reference corpus for PMI was the entire English Wikipedia downloaded on June 6, 2013, containing 7,298,899 entries. We selected top 10 words ($w_1, \ldots, w_{10}$) sorted by η_{kw}s for each k and calculated PMI for every word pair as $\mathrm{PMI}(w_i, w_j) = \ln \frac{p(w_i, w_j)}{p(w_i)p(w_j)}$, for $i, j \in \{1, \ldots, 10\}$. $p(w_i)$ is defined as R_i/R, where R_i is the number of documents containing w_i in the reference corpus, and R is the size of the reference corpus. $p(w_i, w_j)$ is defined as R_{ij}/R, where R_{ij} is the number of documents containing both w_i and w_j in the reference corpus. We compared the three approaches by the median of the PMIs calculated for all word pairs. A larger median is better.

The left panel of Fig. 1 summarizes the comparison in terms of PMI. The ten medians obtained from the ten different runs of the inference are plotted for each approach and for each setting of K. The horizontal axis represents PMI. As this panel shows, ChronoSAGE gave almost the same medians as vanilla SAGE. Further, both methods worked better than LDA for NSF and TDT4 and at least gave a result comparable with LDA for DBLP. Therefore, it can be concluded that SAGE-type topic modeling is a better choice than LDA in terms of PMI.

Next, we give an example of the time-differentiated word lists obtained by ChronoSAGE on the right panel of Fig. 1. This example was obtained for DBLP when $K = 300$. These word lists seemingly related to mobile communications. On the top of the panel, top 15 words are given in the order of their η_{kw}s. These words represent the *time-independent* content of the topic. The size of an ellipse behind each word indicates the magnitude of η_{kw}. Below these top 15 words, we present top 10 words sorted by η_{tkw}s for each timestamp t. The size of a circle behind each timestamp indicates the largest η_{tkw} for each t. In this example, we can find clear trends. The word "GSM", mainly related to 2G networks, appears

---

[1] Downloaded from `http://dblp.uni-trier.de/xml/` on June 11, 2013.
[2] `http://archive.ics.uci.edu/ml/`
[3] `http://www.itl.nist.gov/iad/mig/tests/tdt/2004/`

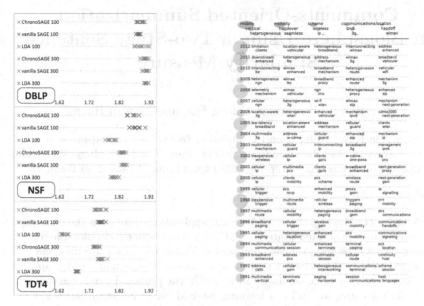

Fig. 1. Comparison of ChronoSAGE with vanilla SAGE and LDA in terms of PMI (left) and timestamped word lists ChronoSAGE extracts from DBLP (right)

in the word lists of earlier years. The word "GPRS" comes after it and appears in the lists of 2001 and 2002. The word "LTE" only appears in the word lists of the most recent years. In this manner, ChronoSAGE can extract clear trends, because it diversifies topic modeling results chronologically by using η_{tkw}s.

4   Conclusions

In this paper, we proposed ChronoSAGE, an application of the multi-faceted SAGE standing on its own merit. The evaluation led to two conclusions. Firstly, ChronoSAGE and vanilla SAGE were superior to LDA in terms of PMI. Secondly, ChronoSAGE extracted informative timestamped word lists, though vanilla SAGE cannot do this extraction due to its model construction. Our important future work is to explicitly model the inherent dependency among timestamps.

References

1. Blei, D.M., Ng, A.Y., Jordan, M.I.: Latent Dirichlet allocation. JMLR 3, 993–1022 (2003)
2. Eisenstein, J., Ahmed, A., Xing, E.P.: Sparse additive generative models of text. In: ICML, pp. 1041–1048 (2011)
3. Griffiths, T.L., Steyvers, M.: Finding scientific topics. PNAS 101(suppl. 1), 5228–5235 (2004)
4. Newman, D., Karimi, S., Cavedon, L.: External evaluation of topic models. In: ADCS, pp. 11–18 (2009)

Comments-Oriented Summarization in Blogsphere Using a Two-Stage Sentence Similarity Measure

Hongjie Li, Lifu Huang, Qifeng Fan, and Lian'en Huang*

Shenzhen Key Lab for Cloud Computing Technology and Applications
Peking University Shenzhen Graduate School, Shenzhen, Guangdong, P.R. China
{lihongjie99,warrior.fu,fanqf1026}@gmail.com, hle@net.pku.edu.cn

Abstract. The popularity of Web 2.0 applications promotes the emergence of user generated content (UGC), e.g., the comments in blogsphere, and the UGC reflects the viewpoints of web users towards a specific event or product. In this paper, we propose a summarization model which applies a novel sentence similarity measure. In the proposed two-stage similarity measure, we utilize a learning method based on an optimization perspective to combine different types of similarity for a refined similarity measure. Both standard cosine similarity and topic based similarity measure are explored to compute the preliminary similarity. In order to evaluate the novel similarity measure, we conduct experiments on a real-world blog data set and the result proves the effectiveness of our proposed method.

Keywords: Comments-oriented Summarization, Blogsphere, Sentence Similarity.

1 Introduction and Related Work

Most studies of summarization focused on plain text and a variety of methods have been proposed to summarize a single document or multiple documents. The increasing availability of online information, especially the mushroom of user generated content (UGC), puts forward a new challenge in the area of automatic text summarization.

Comments-oriented blog summarization is a specific task in social summarization which considers not only the content of a blog entry but also the comments from the readers. Hu et al. [1] defined comments-oriented blog summarization and proposed an algorithm to select summary sentences containing representative words from comments. In [2], authors first exploited three relations contained in comments to score comments. Yang [8] proposed a model to capture both the informativeness of sentences and interests of social users. Authors in [7] proposed a summarization framework to summarize the differences in microblog. Hu et al. [3] proposed to improve online news summarization. Our work is also related to

---

* Corresponding author.

F. Li et al. (Eds.): WAIM 2014, LNCS 8485, pp. 480–483, 2014.

sentence similarity measure. [4] proposed an algorithm that takes into account both semantic information and word order information of sentences. Quan et al. [6] proposed a method to compute short text similarity based on the common terms and the relationship of the distinguishing terms of two short text snippets. The relationship of distinguishing terms are captured by their probabilities on topics which are extracted by Gibbs sampling.

In this study, we make full use of the UGC to generate the comments-oriented summary. Our method refines sentence similarity between sentences in a blog entry by combining preliminary sentence similarity with the similarity relation between a sentence and a comment. We define the similarity refinement process from an optimization perspective and utilize an iteration method proposed in [5] to solve the optimization problem. We conduct the experiments on a real-world data set, and the results prove the effectiveness of our proposed similarity measure according to the comparison between our two-stage similarity measure and traditional methods.

The rest of the paper is organized as follows. The similarity computing is discussed in Section 2. The experiments and results are represented in Section 3 and finally we conclude this paper in Section 4.

2   Two-stage Similarity Computing

2.1   Stage 1: Preliminary Similarity

Standard cosine similarity and topic based similarity are explored as the preliminary measure in this stage. But we focus on topic based similarity because standard cosine similarity has been well known. In this study we directly apply the similarity measure proposed in [6] to capture the topic-level relation between sentences. We derive a set of probabilistic topics $T = \{t_1, t_2, ..., t_z\}$ on a blog entry. Aiming to exploit the influence of non-co-occurring terms on sentence similarity, we only take into consideration the distinguishing words of two sentences and the distinguishing term set is defined as:

$$Dist(s^{(i)}) = \{w|w \in s^{(i)}, w \notin s^{(j)}\}, Dist(s^{(j)}) = \{w|w \in s^{(j)}, w \notin s^{(i)}\}. \quad (1)$$

After obtaining the novel topic based vectors, we use the cosine method to compute the similarity between sentences.

2.2   Stage 2: Refined Similarity

In order to refine the similarity between sentences using the preliminary similarity value, we first define three types of similarity in the blogsphere : similarity between sentences in blog entries (**SS**), similarity between sentences in comments (**CC**) and similarity between sentences in blog entries and sentences in comments (**SC**).

After defining these similarity relations, we refine the similarity from an optimization perspective and the objective formula is defined below:

$$\min_{w} \ \alpha \cdot \sum_{i,j \in SS} (ss_{i,j} - ss_{i,j}^*)^2 + (1-\alpha) \cdot \sum_{a,b \in SS} \sum_{c,d \in CC} sc_{a,c} sc_{b,d} (cc_{a,b} - cc_{c,d})^2, \quad (2)$$

where $ss_{i,j}^*$ denotes the preliminary similarity weight between sentence i and j in a blog entry, $sc_{i,j}$ is the SC relation between sentence i and comment j, and $cc_{i,j}$ represents the CC relation between comment i and j. In addition, the first part of Formula (2) aims to ensure the refined similarity between sentences remain close to the preliminary similarity while the second part tries to minimize the dissimilarity between sentences in the blog entry and sentences in the comments. We use the balance parameter $0 < \alpha < 1$ to control the tradeoff between the two parts. It is worthy noting that if $\alpha = 1$, then the solution is the preliminary similarity itself.

However, solving the optimization problem shown in Formula (2) is time-consuming. In [5], an iteration method has been proposed and we apply it in this study directly. Specifically, we use W_{SS} to denote the similarity between sentences in a blog entry and W_{CC} to denote the similarity between sentences in comments. W_{SC} is used to denote the cross-graph similarity, which is between sentences in a blog entry and its comments. Then the refined similarity can be computed as the following iterative method:

$$\begin{aligned}
W_{SS}^{(t)} &= \alpha \cdot W_{SS}^* + (1-\alpha) \cdot (W_{SC} \cdot W_{CC}^{(t-1)} \cdot W_{SC}^T) \\
W_{CC}^{(t)} &= \alpha \cdot W_{CC}^* + (1-\alpha) \cdot (W_{SC}^T \cdot W_{SS}^{(t-1)} \cdot W_{SC}),
\end{aligned} \quad (3)$$

where $W_{SS}^{(t)}$ denotes the similarity between sentences in t^{th} iteration and W_{SC}^T is the transposition of W_{SC}. W_{SS}^* and W_{CC}^* are the SS and CC preliminary similarity, respectively.

3   Experiments

In the experimental study, we apply the data used in [2]. We have totally implemented four different systems in our experiments. All the four summarization systems are compared on ROUGE measures and a brief description is introduced as follows.

Standard + LexPR (System 1): This system applies standard cosine similarity in LexPageRank to generate the summary.

Topic + LexPR (System 2): In this system, topic based similarity is used in LexPageRank model.

Two-stage Standard + LexPR (System 3): This system utilizes our two-stage similarity measure by using standard cosine similarity as the preliminary similarity.

Two-stage Topic + LexPR (System 4): Our two-stage similarity measure is adopted in this system and the preliminary similarity is topic based similarity.

Table 1. Comparison results in ROUGE

Systems	ROUGE-1	ROUGE-2	ROUGE-SU4
System 1: Standard + LexPR	0.57143	0.43858	0.44063
System 2: Topic + LexPR	0.60884	0.47268	0.46879
System 3: Two-stage Standard + LexPR	0.61862	0.48197	0.48322
System 4: Two-stage Topic + LexPR	**0.62669**	**0.49147**	**0.49335**

Table 1 shows the comparison results on ROUGE measures. It can be seen that System 4 outperforms all the other three systems on different evaluation measures and the results indicate the effectiveness of the proposed two-stage similarity measure. This comparison also indicates that topic based similarity can effectively capture the relation of sentences in similarity measure.

4   Conclusion

In this paper, we propose a summarization model which applies a novel sentence similarity measure. A two-stage algorithm is constructed to refine the sentence similarity between sentences with UGC information. The experiment results show that our method achieves better performance compared with traditional methods.

References

1. Hu, M., Sun, A., Lim, E.: Comments-oriented blog summarization by sentence extraction. In: Proceedings of the Sixteenth ACM Conference on Conference on Information and Knowledge Management, pp. 901–904. ACM (2007)
2. Hu, M., Sun, A., Lim, E.: Comments-oriented document summarization: understanding documents with readers feedback. In: Proceedings of the 31st Annual International ACM SIGIR Conference on Research and Development in Information Retrieval, pp. 291–298. Citeseer (2008)
3. Hu, P., Ji, D., Sun, C., Teng, C., Zhang, Y.: Improving document summarization by incorporating social contextual information. In: Salem, M.V.M., Shaalan, K., Oroumchian, F., Shakery, A., Khelalfa, H. (eds.) AIRS 2011. LNCS, vol. 7097, pp. 499–508. Springer, Heidelberg (2011)
4. Islam, A., Inkpen, D.: Semantic text similarity using corpus-based word similarity and string similarity. ACM Transactions on Knowledge Discovery from Data (TKDD) 2(2), 10 (2008)
5. Muthukrishnan, P., Radev, D., Mei, Q.: Edge weight regularization over multiple graphs for similarity learning. In: 2010 IEEE 10th International Conference on Data Mining (ICDM), pp. 374–383. IEEE (2010)
6. Quan, X., Liu, G., Lu, Z., Ni, X., Wenyin, L.: Short text similarity based on probabilistic topics. Knowledge and Information Systems 25(3), 473–491 (2010)
7. Wang, D., Ogihara, M., Li, T.: Summarizing the differences from microblogs. In: Proceedings of the 35th International ACM SIGIR Conference on Research and Development in Information Retrieval, pp. 1147–1148. ACM (2012)
8. Yang, Z., Cai, K., Tang, J., Zhang, L., Su, Z., Li, J.: Social context summarization. In: Proceedings of the 34th ACM SIGIR Conference, pp. 255–264. ACM (2011)

Popularity Prediction of Burst Event in Microblogging

Xiaoming Zhang[1], Zhoujun Li[1], Wenhan Chao[1], and Jiali Xia[2]

[1]School of Computer Science and Engineering, Beihang University, Beijing, China, 100191
[2] School of Software, Jiangxi University of Finance & Economics, Nanchang, China
{yolixs,lizj}@buaa.edu.cn

Abstract. Every day, thousands of burst events are generated in microblogging first, and then affect the public opinion to a large degree. Thus, it is quite necessary to find out "how hot the burst event will be in the future". In this paper, we propose a prediction model which combines the analysis of event content and users' interest to predict the volume of the burst event in the implicit network. Particularly, it is assumed that different user has different influence power and different interest in the burst event. The popularity of an event depends on the volumes produced by the users infected in the past and its historical popularity. Experimental results show the superior performance of our approach.

Keywords: Burst event, Information spread, Event popularity, Event detection.

1 Introduction

Every day, many news events are even first reported in microblogging sites, such as the event "Hurricane Sandy hitting New York in Oct 2012" was extensively reported by Twitter users. It is very interest to predict how popular the event will be. Analyzing the spread of burst event is very important in many fields. There have been many works on modeling information spread of social media stream. These works can be categorized into two groups. One group of works focus on the topology of social graph, investigating what topologies and what activation patterns facilitate efficient propagation of information [2], [3], [5]. However, in many scenarios, the underlying network is in fact implicit or even unknown. The other category of researches mainly focuses on the effect of the content of the idea on the information propagation [1], [4].These approaches have no consideration on different users' affection, which make them depend heavily on the analyzing of content features in a macro-level. We model the spread of the event based on the analysis of event content and users' social relation information. Our approach doesn't require the complete topology of social network to be known. To accomplish this, a linear spread prediction function, based on user's influence power, users' interest in the event, and event's historical popularity information, is proposed to predict the popularity of the event in the next time point.

2 Linear Spread Model for Burst Event

Given a continuous stream of incoming micro-blogs, we split it use a time unit with a given length l. We define the volume, $V_e(t)$, as the number of micro-blogs that discuss

F. Li et al. (Eds.): WAIM 2014, LNCS 8485, pp. 484–487, 2014.

the event e at the tth time unit after it is detected as a burst event. We model the volume $V_e(t+1)$ in the next time unit as a linear function:

$$V_e(t+1)= \sum_{u \in In(t)} Sim(u,e)*F_u(t-t_u)+\chi_{t+1}H_e(t+1) \tag{1}$$

where $F_u(d)$ denotes user u's influence power d time units after he is infected and it is a signature that, once this user adopts an event, how many other users will be infected by this user to adopt the event and then how many new micro-blogs will be produced, $Sim(u,e)$ denotes user u's interest in event e, $H_e(t+1)$ denotes the volume introduced by its historical popularity, $In(t)$ denotes the set of already infected users and t_u is the time when user u is infected. The key problem is how to estimate the parameters, i.e., $Sim(u,e)$, $F_u(d)$, and $H_e(t+1)$.

Usually, a user's interest is reflected by the micro-blogs that are uploaded by this user in the past. Thus, we aggregate all the micro-blogs generated by the same individual user into a profile that corresponds to a document in the LDA model. Then, each user's interest is modeled by a vector of probability distribution over latent topics as: $\overrightarrow{\theta_u} =< p_{u,1}, p_{u,2},...,p_{u,K} >$, where $p_{u,k}$ represent the probability of topic k occurring in user u's profile, and it is estimated using the Gibbs sampling rule. In the other side, we aggregate some hot micro-blogs selected from the event into an event profile. Then, the trained topic model is used to infer the probability distribution over topics of the event profile. Finally, the cosine similarity between the probability distribution vectors of event profile and user profile is used to measure user's interest in the event.

The introduced volume $H_e(t+1)$ is produced by the historical popularity information. The Newton interpolation is applied to estimate the introduced volume as follows:

$$H_e(t+1)=\sum_{j=a}^{t}(\prod_{i=a,i \neq j}^{t} \frac{I_{t+1}-I_i}{I_j-I_i}) \cdot V(j) \tag{2}$$

where $0 \leq a \leq t$ is a index number which indicates the start time point from which the volume values is used in the Newton interpolation equation, and I_i is the index of the time unit.

We use the approach that is similar to the one used in the previous model [5] to infer user influence power $F_u(d)$ and the other parameter χ. Assume there are U users and data about M different events spread in the network over time. Each event can infect any number of users. We also use a additional indicator function $R_{u,m}(t)$ to represent the infection of event e_m on user u in the model. $R_{u,m}(t)>0$ if user u got infected by event e_m at time unit t, and $R_{u,m}(t)=0$ otherwise. Then, the volume $V_m(t)$ of event e_m at time unit t is represented as follows:

$$V_{e_m}(t+1)=\sum_{i=1}^{U}\sum_{d=0}^{D-1}R_{u_i,e_m}(t-d) \cdot Sim(u_i,e_m) \cdot F_{u_i}(d+1)+\chi_{t+1}H_{e_m}(t+1) \tag{3}$$

where D denotes the amount of time units, meaning that the influence power of a user can be neglected D time units after the user is infected. In the training process, we record T volume values of T sequential time units for each event. As there are M

different events, the number of these volume equations denoted by Eq. (3) is $M \cdot T$. To infer these parameters, we can represent these equations in a matrix form:

$$\vec{V} = \mathbf{R}\vec{F} + \vec{H} \tag{4}$$

where $\vec{V}$ is a volume vector of length $M \cdot T$, $\mathbf{R}$ is a matrix of size $M \cdot T \times U \cdot D$, $\vec{F}$ is a influence power vector of length $U \cdot D$, $\vec{H}$ is the vector of introduced volumes weighted by the parameter vector χ. Now, the parameters can be inferred by solving a matrix equation as Eq. (4). An approximate value of $\vec{F}$ can be inferred by minimizing the prediction error measured by the Euclidean distance between the true and the predicted volumes of the events:

$$\min \| \vec{V} - (\mathbf{R}\vec{F} + \vec{H}) \|_2^2$$
$$\text{subject to } \vec{F} \geq 0 \tag{5}$$

3   Experiments

To validate the performance of our approach, we conduct a set of experiments on a datasets collected from *Sina Weibo*. Sina Weibo recommends 10 hottest topics in its website every day. We collected about 4,137,892 micro-blogs and 159,782 users between January 2012 and July 2012. The total number of collected event is 678. The metric of relative error is used to evaluate the performance of prediction:

$$Err = \frac{\sqrt{\sum_{m,t} (V_m(t+1) - TV_m(t+1))^2}}{\sqrt{\sum_{m,t} TV_m(t+1)^2}} \tag{10}$$

where $TV_m(t+1)$ is the true volume of event m at time unit $t+1$. We compare our approach (named *EventPRE*) with *HTall* [1] and *BassPre* [4] respectively.

In the first experiment, we test the effect of parameter D on the performance of prediction. Fig.1. shows the relative error of our approach with D varied from 1 to 7, which shows that our approach achieve the best performance when the length of D is 4. When D is too short, each user's influence power is estimated based on a short period, which may decrease his influence power and hence affect the prediction. When D is too long, the out of date information may be noisy to the estimation of user's influence power.

Then, we analyze the effect of time unit length l on the prediction performance and compare the performance of different approaches. Fig.2. shows the relative errors of different approaches. It shows that our approach outperform other two approaches consistently. This is because that our approach uses users' interest information to distinguish different users' contributions in the spread of event, while other approaches assume that all the users have the same contribution. Moreover, event's historical popularity information in previous time units is also used in our approach. It also shows that all the approaches achieve better prediction as the length of time unit grows. This can be attributed to the deceased standard deviation of event volumes.

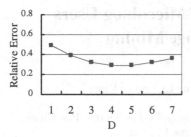 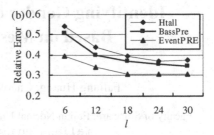

Fig. 1. Effect of parameter of D **Fig. 2.** Comparision of prediction

4   Conclusion

In this paper, we present a novel approach predict the spread of burst event in online communities. Compared to other approaches, our approach needn't to know the complete topology. Moreover, we proposed a revised author-topic model to mine user's interest which is neglected by other approaches. Experimental results demonstrate the superiority of our approach over other approaches. This task is very interesting from both commercial and psychological perspectives.

Acknowledgements. This work was supported by the National Natural Science Foundation of China (No 61370126. No. 61170189, and No. 61202239), the Fundamental Research Funds for the Central Universities (YWF-13-T-RSC-072), and the Fund of the State Key Laboratory of Software Development Environment (No. SKLSDE-2013ZX-19).

References

1. Jin, X., Gallagher, A., Cao, L., Luo, J., Han, J.: The Wisdom of Social Multimedia: Using Flickr For Prediction and Forecast. In: Proceedings of the 18th ACM Multimedia (2010)
2. Leskovec, J., Adamic, L.A., Huberman, B.A.: The dynamics of viral marketing. ACM Transactions on the Web 1(1), 5 (2007)
3. Saito, K., Kimura, M., Ohara, K., Motoda, H.: Selecting information diffusion models over social networks for behavioral analysis. In: Balcázar, J.L., Bonchi, F., Gionis, A., Sebag, M. (eds.) ECML PKDD 2010, Part III. LNCS, vol. 6323, pp. 180–195. Springer, Heidelberg (2010)
4. Tsur, O., Rappoport, A.: What's in a Hashtag? Content based Prediction of the Spread of Ideas in Microblogging Communities. In: Proceedings of the 5th ACM International Conference on Web Search and Data Mining (2012)
5. Yang, J., Leskovec, J.: Modeling information diffusion in implicit networks. In: Proceedings of the 10th IEEE International Conference on Data Mining (2010)

Identifying Gender of Microblog Users
Based on Message Mining

Faliang Huang, Chaoxiong Li, and Li Lin

Faculty of Software, Fujian Normal University, Fuzhou 350007, P.R. China
faliang.huang@gmail.com

Abstract. Microblog messages display gender tendency to some extent, so automatic identification of gender of microblog users with message content mining techniques is studied. A novel approach is proposed to identify microblog user gender. The proposed approach extracts three types of features, i.e., characteristic item features, stylometry features and medium diversity features, from microblog messages with high gender-relatedness, and utilizes a series of pattern recognition techniques, such as feature normalization, feature selection and SVM, to detect microblogger gender. Massive experiments demonstrate that the effectiveness of the proposed approach.

Keywords: Microblog mining, Gender identification,cyberspace forensics.

1   Introduction

Over the past few years, microblogging services such as Twitter, Sina-Weibo and etc have become increasingly popular platforms for web users to communicate with each other. Popularity of microblogging has greatly enriched the way people communicate, while providing convenience for malicious users to engage in criminal activities such as identity theft and advance-fee fraud with its access openness and identity virtualness. Many of these online crimes are related with gender camouflage [1]. The famous "Myspace mom" case is the clear example. Therefore, it becomes imperative to design efficient methods for gender identity tracing in cyberspace forensics.

Unlike traditional documents, messages have their own characteristics such as 1) length of a message is short; 2) presence of slang, abbreviations, special characters and grammatical errors; 3) use of hyperlinks to other users and to external resources. It is these characteristics that prevent some standard text mining tools to be employed with their full potentials. To the best of our knowledge, very few studies have been devoted to microblog user gender identification with content analysis techniques.

Although author gender identification(AGI)[1,2,3,4] and short text classification (STC) [5,6,7,8] are closely related with microblogger gender detection and have attracted much attention, almost existing AGI techniques ignore the properties such as colloquial and personalization of microblog messages and all the STC approaches aim at topic classification and yet not gender recognition of authors of short texts.

F. Li et al. (Eds.): WAIM 2014, LNCS 8485, pp. 488–493, 2014.

In this paper, we will show that true gender of the microblog users can be effectively detected based on their own messages.

2 Problem Formulation

Microblogger gender identification problem can be considered as a binary classification problem of messages, i.e., assigns a microblogger with many related messages to one of two given classes {male, female}. Mathematically, the problem can be formalized as learning an effective classifier y=f(x) from a training set of microbloggers with true gender label $D = \{(v_1, g_1), (v_2, g_2), \cdots, (v_N, g_N)\}$, two-tuple (v_i, g_i) denotes microblogger i, $v_i = (v_{i1}, v_{i2}, \cdots, v_{id})^T$ is a d-dimensional gender-specific characteristic vector resulted from a given feature extraction and selection approach, $g_i \in \{-1, +1\}$ is gender label encoding class female (-1) or class male (+1) of the microblogger i, and N is the number of instances in the dataset D.

3 Gender-Specific Features in Message

3.1 Term Features

Here we propose a method to extract topic-sensitive terms from the messages posted by a microblogger to form characteristic item features, as can be detailed as below:

Suppose training message set be $M = \{m_1, m_2, \cdots, m_N\}$, where $m_i = \{m_{i1}, m_{i2}, \cdots, m_{iN_i}\}$ denotes the messages posted by microblogger i. After preprocessing such as stop words filtering and stemming, a characteristic item set $T = \bigcup_{i=1}^{N} T_i$ can be generated, where $T_i = \{t_{ij} | t_{ij}$ is a term from $m_i \land j = 1, 2, \cdots, r\}$ denotes the terms from the messages of microblogger i. Each term of T_i can be computed according to formula(1), and m_i can be further vectorized as $vec(m_i) = (w_1, w_2, \cdots, w_{|T|})$.

$$w_i = \begin{cases} freq(t_i) & if\ t_i\ occurs\ in\ m_i \\ 0 & otherwise \end{cases} \tag{1}$$

where $freq(t_i)$ denotes the occurrence frequency of t_i in message m_i.

3.2 Linguistic Features

Based on human psychology research and extensive experimentation, we design three sub-types of gender-related linguistic features: (1) function words; (2) punctuation; (3) others. Table 1 lists all the 406=(383+10+13) features we designed. Note that some symbols and formulae are described as following: W denotes total number of words, V denotes the number of different words, V_i denotes the number of different words that occur i times, Hapax Dislegomena denotes the words that occur only twice, Hapax Legomena denotes the words that occur only once. Function $F1(x)$ denotes average number of x per message of a microblogger.

Table 1. Designed features and remarks

Type	Features:(count)
Function words (383)	F1(article):(3); F1 (pro-sentence): (4); F1 (pro-sentence): (4); F1(pronoun): (74); F1(auxiliary-verb): (47); F1(conjunction): (22); F1(interjection):(109); F1(adposition): (124)
Punctuation (10)	F1(single quotes) : (1); F1(commas) : (1); F1(periods) : (1); F1(colons) : (1); F1(semi-colons) : (1); F1(question) : (1); F1(multiple question): (1); F1(exclamation) : (1); F1(multiple exclamation) : (1); F1(ellipsis) : (1)
others(13)	F1(words) : (1); F1(characters) : (1); F1(Upper character) : (1); F1(digital) : (1); Vocabulary richness : (1); F1(long words) : (1); F1(short words) : (1); F1(abbreviation) : (1); Yule' s K measure:(1); Simpson's D measure:(1); Sichel's S measure:(1); Entropy measure:(1); Honores' R measure

3.3   Medium Diversity Features

Further observation of messages, we can find that messages posted by microbloggers are different from the traditional texts. They have their own particularities, for example, microbloggers can freely add some images and videos in the messages to give others the immersive feeling. Different gender may show some preferences of use such multimedia. To measure the medium diversity of messages, we invented three features, namely, *#music /#total, #video /#total, # image /#total. #image, #video* and *#music* respectively denote the number of micro-blogs containing *image*, *video* and *music*.

4   Experiments

4.1   Dataset Preparation

To experimentally evaluate our approach, we download messages posted by 1000 active users consisted of 486 males and 514 females during the 100 days from September to December in 2012; these registered Twitter users all use true gender information. During this timeframe we gathered a total of 3,643,697 public tweets.

Since the computed values of the invented features could range from 0 to more than 500. To ensure all features are treated equally in the classification process, we use max-min normalization method to make all feature values fall in interval (0,1), and mutual-information-based feature selection method for dimension reduction. We use MATLAB SVM Toolbox (http://www.isis.ecs.soton.ac.uk/resources/svminfo/) for gender classifier training.

4.2   Experimental Results

A.   Impact of Microblogger Feature Type on Classification Accuracy

To make clear how the three types of features (Term feature(T), Linguistic feature(L), and Medium diversity feature(M)) respectively influence the classification quality, we construct 7 training datasets, and each of those contains one type of 8 combination features, i.e., T, L, M, T+L, T+M, L+M and T+L+M, of randomly selected 300 male and 300 female microbloggers. Moreover, we produce 7 testing datasets, each of which consists of randomly selected 150 male and 150 female microbloggers.

Table 2 shows the comparisons of classification accuracy in the 7 datasets of different combination features. A comparative analysis is given below: First, comparing column Total Accuracy(%) of row T, row L and row M, classifier trained with L (linguistic feature) can produce higher classification accuracy than the other two classifiers resulted from dataset of T (term feature) or M (medium diversity feature). Second, in general, the more types of features are used in training SVM classifiers, the higher detection accuracy of microblogger gender is obtained. However, the two is by no means correlated linearly. For example, recognition accuracy of the classier coming from the dataset with combination feature (T+M) is less than the dataset with feature L. Thirdly, further comparison between classification accuracy of male and female in all datasets, i.e., column Male Accuracy(%) and column female Accuracy(%), can discover that, identical type feature or combination feature possess different gender discriminating ability, in other words, the gender difference between male and female microbloggers exists in discriminating ability of the three types of features.

Table 2. The Impact of microblogger feature type on gender detection accuracy

Message Type	Male Accuracy (%)	Female Accuracy(%)	Total Accuracy(%)
T	72.43	72.78	72.61
L	76.72	75.39	76.06
M	64.67	63.45	64.06
T+L	80.74	79.12	79.93
T+M	75.81	74.98	75.39
L+M	81.25	81.26	81.255
T+L+M	82.27	84.05	83.16

B.   Impact of Feature Set Size on Classification Accuracy

Section 4 indicates that in contrast with linguistic feature and medium diversity feature, the number of term features is much larger while representing a microblogger. So here we just test how the number of term features selected for training a SVM classifier influences the classification accuracy. Experimental results are shown in figure 2. From the figure, we can see that gender detection accuracy does not increase with the number of term feature set. And its evolution can be described as below: the detection accuracy begins with 0.532 when just choosing the top 10000 term features, and rises gradually with the number of term features and reaches its maximum at the point of selecting top 2500000 term features, then drop to 0.688 step by step. The experimental result can be explained like following: on one hand, too more term features can add noise data while representing a microblogger, on the other hand, too less term features may miss some valuable characteristics of a microblogger, both cases can lead to worse performance.

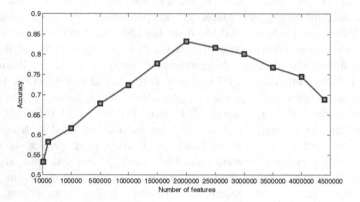

Fig. 1. The impact of the number of term feature set on gender detection accuracy

5   Conclusion

In this paper, we have studied and formalized the problem of microblogger gender identification and proposed a new approach, which extracts different types of features from microblog messages and utilizes a series of pattern recognition techniques, such as feature normalization, feature selection and SVM, to detect microblogger gender. Experimental results show that our approach is effective and promising.

Acknowledgement. This work is supported in part by Humanity and Social Science Youth foundation of Ministry of Education of China under grant 12YJCZH074; Foundation of Fujian Educational Committee under grant JA13077.

References

1. Köse, C., Özyurt, Ö., Amanmyradov, G.: Mining Chat Conversations for sex Identification. In: Washio, T., Zhou, Z.-H., Huang, J.Z., Hu, X., Li, J., Xie, C., He, J., Zou, D., Li, K.-C., Freire, M.M. (eds.) PAKDD 2007. LNCS (LNAI), vol. 4819, pp. 45–55. Springer, Heidelberg (2007)

2. Cheng, N., Chandramouli, R., Subbalakshmi, K.P.: Author gender identification from text. Digital Investigation 8, 78–88 (2011)
3. Schler, J., Koppel, M., Argamon, S., et al.: Effects of age and gender on blogging. In: AAAI Spring Symposium: Computational Approaches to Analyzing Weblogs, pp. 199–205 (2006)
4. Miller, Z., Dickinson, B., Hu, W.: Gender Prediction on Twitter Using Stream Algorithms with N-Gram Character Features. International Journal of Intelligence Science 2(24), 143–148 (2012)
5. Sriram, B., Fuhry, D., Demir, E., et al.: Short text classification in twitter to improve information filtering. In: SIGIR, pp. 841–842 (2010)
6. Phan, X.H., Nguyen, L.M., Horiguchi, S.: Learning to classify short and sparse text & web with hidden topics from large-scale data collections. In: WWW 2008, pp. 91–100 (2008)
7. Chen, M., Jin, X., Shen, D.: Short text classification improved by learning multi-granularity topics. In: IJCAI 2011, pp. 1776–1781 (2011)
8. Sun, A.: Short text classification using very few words. In: SIGIR 2012, pp. 1145–1146 (2012)

Continuous Temporal Top-k Query over Versioned Documents

Chao Lan[1,2], Yong Zhang[2], Chunxiao Xing[2], and Chao Li[2]

[1] Department of Computer Science and Technology
[2] Research Institute of Information Technology
Tsinghua University, Beijing 100084, China
lanc11@mails.tsinghua.edu.cn,
{zhangyong05,xingcx,li-chao}@tsinghua.edu.cn

Abstract. The management of versioned documents has attracted researchers' attentions in recent years. Based on the observation that decision-makers are often interested in finding the set of objects that have continuous behavior over time, we study the problem of continuous temporal top-k query. With a given a query, continuous temporal top-k search finds the documents that frequently rank in the top-k during a time period and take the weights of different time intervals into account. Existing works regarding querying versioned documents have focused on adding the constraint of time, however lacked to consider the continuous ranking of objects and weights of time intervals. We propose a new interval window-based method to address this problem. Our method can get the continuous temporal top-k results while using interval windows to support time and weight constraints simultaneously. We use data from Wikipedia to evaluate our method.

1 Introduction

Versioned documents are documents that may retain multiple versions as time processes. Versioned documents on the Web includes large-scale datasets of versioned text such as archives of websites, wikis, blogs, micro-blogs, and business records, which have become abundant as the Internet evolves. Broadly speaking, objects continuously evolving during time can be treated as versioned documents as well, such as opinions from the same speakers, articles from the same authors, etc. Versioned documents can be used to support numerous applications such as analysis of financial market, public opinions, historical data, sensor data and other time-sensitive data. Thus querying versioned documents has been widely studied such [3,1,2,4]. We study the problem of *continuous temporal top-k* query over versioned documents. Because the problem can be widely adopted to many applications thus many related works were proposed about top-k query over versioned document other temporal data sets [6,7,5,8]. Continuous temporal top-k retrieves the top-k objects that not only rank in top-k position in some time point but hold the position for at least a specific time period. In that case, our problem focuses on the sustainable ranking of the documents. Thus, both the

F. Li et al. (Eds.): WAIM 2014, LNCS 8485, pp. 494–497, 2014.

ranking position and the time interval of each version of the documents need to be considered.

Let q be the query submitted by a user to conduct a continuous temporal top-k search on a document set $\mathbb{D}$. Query q is defined as $q : (\mathcal{W}, \lambda, k, r, \Omega)$. $\mathcal{W}$ denotes a set of keywords. λ denotes to the query interval. $q.\lambda = [s, e)$, which indicates that the query needs to find the results in $\mathbb{D}$ that are valid in $q.\lambda$. k denotes the number of results a top-k query needs to find and $0 < k \le |\mathbb{D}|$. r is a *relaxing factor* that is specified by the user to indicate how frequently the documents needs to rank in top-k to become continuous temporal top-k results, and $0 < r \le 1$. Ω denotes to the set of weights of sub-intervals of λ.

Let $\mathbb{D}$ be the target document sets. Given a query q, the continuous temporal top-k search finds the following document set:

$$\left\{ d | d \in \mathbb{D} \wedge \left(\sum_{i=1}^{n} m_i \times \omega_i \ge r \right) \right\},$$

where ω_i is the weight of each atomic interval and $m_i = 1$ if d is a top-k result in that atomic interval, otherwise $m_i = 0$.

We have the following contributions: (1) we define the problem for continuous temporal top-k query over versioned documents. (2) we propose a new algorithm called Interval Window based Algorithm (IWA). (3) we evaluate and demonstrate our method by a conducting a set of experiments.

2   Interval Window Based Algorithm

The basic idea of IWA is to modify NRA to support continuous temporal top-k search. In order to accomplish that, we modify the procedure of NRA by maintaining the lattice-based windows to gradually segment the continuous query space into sub-query spaces in the time dimension. Versions are treated as objects of NRA. Postings are sorted accessed from the lists in parallel. A new posting (v, p) is access and the lower bound of the version v is updated. Then v is added to the windows set and the window set is maintained. After that, for each modified window: i) if less than k versions have been seen in the current window, the next window is checked; ii) if more than k versions have been seen, the lowest lower bound l of versions is extracted. The highest upper bound u of versions is extracted. iii) If the $l \ge u$, the window is marked as *finalized*. *Finalized* means that versions with top-k aggregated score (i.e., lower bound) of the window have been found or current window no longer need to be maintained. This means top-k versions in this window have been determined and top-k results of each atomic intervals in this window can be derived. Then we check whether all the windows are marked as *finalized*. If they are, top-k versions of each atomic interval have been determined, then continuous temporal top-k result of the query are generated. Otherwise the algorithm procedure goes back to the beginning. If all the postings have been visited and there are still non-finalized windows, we only check the finalized windows to see if continuous temporal top-k results can be generated with these windows.

3   Experiments

We evaluate the performance of our algorithms on the Wikipedia revision history data set. We randomly selected 7 sub-sets with different sizes ranging from 50 GBytes to 8 TBytes. We adopted the same form of inverted index proposed in [3]. The average length of posting list is 1 million in 1 TBytes data set and 10 million in the 8 TBytes data set. The baseline method we implemented is DAA [8]. We extend DAA to support our problem and refer to it as EDAA (extended DAA). We evaluated the performance and scalability of EDAA and IWA. 5 queries were chosen for each test instance. The result of each test instance is the average result of the 5 queries.

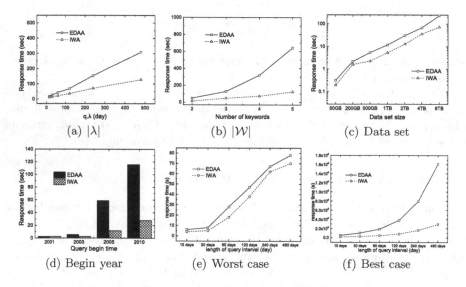

(a) $|\lambda|$ (b) $|\mathcal{W}|$ (c) Data set

(d) Begin year   (e) Worst case   (f) Best case

Fig. 1. Latency of queries with different query parameters

We tested the memory usage of EDAA and IWA. Figure 1(a) shows latency of a method as the query interval $q.\lambda$ varies. When $|\lambda|$ becomes larger, both methods become more expensive since a larger problem space needs to be handled. IWA responds much faster than EDAA, and the growth of response time gets slower as length of $|\lambda|$ increases. Figure 1(b) shows the response time as k varies. Response time of both algorithms increase as k grows. EDAA is more time consuming and more sensitive to the growth of k, especially when k gets larger. Figure 1(c) plots the response time as a function of the size of the data set. EDAA is more sensitive to the size of posting lists. Figure 1(d) shows the response time of both methods as the query start time t_b varies. It reveals the fact that IWA performs better than EDAA as the number of version grows. Figure 1(e) and 1(f) show the latency as the $q.\lambda$ varies of extreme cases of the inputs. Comparing the results of 1(e) and 1(f) with previous results, we see that the advantage of IWA becomes more obvious as the complexity of query grows and the size of the data set increases.

In conclusion, IWA consistently outperforms EDAA in terms of both memory the time consumption. IWA has lower latency and better scalability than EDAA, which makes it more suitable for real applications.

4   Conclusion

We discussed the problem of continuous temporal top-k query. The problem is abstracted from the application scenarios that require to perform top-k queries on versioned documents considering the sustaining ranking position of objects and the weight of different intervals. We proposed the interval window based algorithm to solve this issue. We implemented the methods and tested our solutions in contrast to the baseline method on large-scale document sets, which were selected from Wikipedia with complete revision history from 2001 to 2013. Our experiment results show that IWA outperforms the baseline method EDAA in all test cases.

Acknowledgment. This work was supported by National Basic Research Program of China (973 Program) under Grant No.2011CB302302, Key S&T Projects of Press and Publication under Grant No. GXTC-CZ-1015004/02, the Support Program of the National '12th Five-Year-Plan' of China under Grant No. 2012AA09A408 and Tsinghua University Initiative Scientific Research Program.

References

1. Anand, A., Bedathur, S., Berberich, K., Schenkel, R.: Efficient temporal keyword search over versioned text. In: CIKM, pp. 699–708. ACM, New York (2010)
2. Anand, A., Bedathur, S., Berberich, K., Schenkel, R.: Index maintenance for time-travel text search. In: SIGIR, pp. 235–244. ACM, New York (2012)
3. Berberich, K., Bedathur, S., Neumann, T., Weikum, G.: A time machine for text search. In: SIGIR, pp. 519–526. ACM, New York (2007)
4. Huo, W., Tsotras, V.J.: A comparison of top-k temporal keyword querying over versioned text collections. In: Liddle, S.W., Schewe, K.-D., Tjoa, A.M., Zhou, X. (eds.) DEXA 2012, Part II. LNCS, vol. 7447, pp. 360–374. Springer, Heidelberg (2012)
5. Lee, M.L., Hsu, W., Li, L., Tok, W.H.: Consistent top-k queries over time. In: Zhou, X., Yokota, H., Deng, K., Liu, Q. (eds.) DASFAA 2009. LNCS, vol. 5463, pp. 51–65. Springer, Heidelberg (2009)
6. Li, F., Yi, K., Le, W.: Top-k queries on temporal data. VLDB J. 19(5), 715–733 (2010)
7. Mouratidis, K., Bakiras, S., Papadias, D.: Continuous monitoring of top-k queries over sliding windows. In: SIGMOD, pp. 635–646. ACM, New York (2006)
8. Leong Hou, U., Mamoulis, N., Berberich, K., Bedathur, S.: Durable top-k search in document archives. In: SIGMOD, pp. 555–566. ACM, New York (2010)

Theme-Aware Social Strength Inference from Spatiotemporal Data

Ningnan Zhou, Xiao Zhang*, and Shan Wang

Key Laboratory of Data Engineering and Knowledge Engineering,
Ministry of Education, Renmin University of hina, Beijing 100872, China
School of Information, Renmin University of China, Beijing 100872, China
{zhouningnan123,zhangxiao,swang}@ruc.edu.cn

Abstract. The popularity of location based services has resulted in rich spatiotemporal data that indicates whether persons have social connections. This valuable indication can be used in a wide range of applications such as friend recommendation in social networks or target advertisement for Internet companies. The state-of-the-art approach only considers people who visit non-popular locations together are more socially related. Further, none of existing methods notices that themes of co-occurrence behavior, e.g. dating, of every pair of persons can be used to infer their social strength. In this paper, we novelly introduce the theme to measure the social strength of two persons. A theme, mathematically, is in the form of a probabilistic distribution over Spatiotemporal Windows(SWs), the unit for co-occurrence. In this paper, we propose a Theme-Aware social strength Inference(TAI) approach that mines themes from co-occurrence behaviors consisting of SWs and trains each theme with its contribution to social strength. We employ tf-idf concept for SW and design a novel dynamic programming algorithm to find proper SWs. Extensive experiments are conducted on real dataset and the results show that our method can significantly improve the effectiveness, i.e. more than 5% to 15% in precision under the same recall over the state-of-the-art approach.

Keywords: social strength inference, spatiotemporal data, theme aware.

1   Introduction

A variety of geo-social networks such as Foursquare[8] have been emerged with ubiquity of smart mobile devices and popularity of location based services such as check-in service. As rich spatiotemporal data, which consists of people's record with location tag and timestamp are being produced, behaviors of human beings associated with them make it possible to investigate social strength, which is the likelihood to be friends, between a pair of persons. The knowledge of social strength among people can be used to a wide range of applications such as

---

* Corresponding author.

F. Li et al. (Eds.): WAIM 2014, LNCS 8485, pp. 498–509, 2014.

friend recommendation in social networks or target advertisement for Internet companies[10].

There has been much efforts devoting to inferring social strength from spatiotemporal data. The general idea is that visiting locations together contributes to strong social strength and especially in the state-of-the-art method, visiting public locations together contributes less to social strength and visiting non-public locations implies strong social strength. In Figure 1(a), two persons visited two public locations together and thus are regarded as owning weak social strength. In fact, they are good friends and first played together in amusement park in the morning and spent evening hours in cafe as Figure 1(b) illustrated. From this example, we can see existing methods make wrong estimation because (1)they ignore not only non-public locations implies strong social strength. Public locations also indicate strong social strength in this example. (2)they ignore the visiting time implies social strength. For example, co-occurrence at cafe at morning may mean that persons are co-workers and own weak social strength. On the other hand, at night, visiting cafe together implies strong social strength. These two drawbacks yield the wrong estimation in the example above.

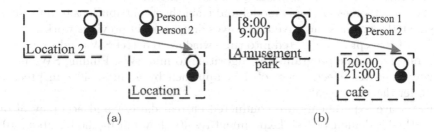

(a)   (b)

Fig. 1. Good friends visited amusement park at morning and cafe at night

In this paper, we proposed a Theme Aware social strength Inference approach(TAI) to overcome drawbacks in the previous approaches. In TAI, we first define co-occurrence as visiting in the same SW. Judging co-occurrence by SW rather than exact location and timestamp can avoid judging mistakes at persons who actually co-occurred but reported their location at different time. For each pair of persons, their co-occurrence behavior indicates the probability for them to co-occur at SWs. The core of TAI approach is identifying that if a few themes take different proportion, they can generate various co-occurrence behaviors. Here, each theme has different tendencies to co-occur at certain SWs. We use topic model[18] to mine themes by regarding themes as topics, co-occurrence behaviors as documents and SWs as words. Further, we exploit social connection graph to train each theme with its contribution to social strength. In this way, we can infer social strength by first decompose co-occurrence behavior into themes with different proportion and sum the weighted contribution of themes. For example, there are two themes like "playing" and "dating", both strongly contribute to social strength. The former tends to visit amusement park and the latter tends to visit cafe at night. Co-occurrence behavior in the example above can be considered to be generated by these two themes uniformly and the inferred social strength is the average of contribution of these two themes to social strength and thus leads to correct strong social strength. Therefore, TAI approach determines location's contribution in context of co-occurrence behavior, and visiting time is considered in SW and thus drawbacks are overcome.

Naturally, a critical issue in mining themes is to detect spatiotemporal window(SW). For each location, we define SW by assigning them with several disjoint time windows. The time window in each SW can neither be too long or too short. To find proper SWs, we adopt the concept of tf-idf of time window[15] and propose a novel dynamic programming algorithm. Extensive experiments are conducted on real dataset including both location data and social connection graphs. Our method to find optimal spatiotemporal points is shown to improve effectiveness of social strength inference; that is about 10% more precise than the state-of-the-art approach.

Our contributions are as follows:

− For the first time, we introduce themes to generate co-occurrence behaviors and exploit their contributions to social strength.
− We mine themes via topic model and train their contributions to infer social strength, which is intuitive and solves drawbacks in previous works.
− Tf-idf concept is integrated into time window to detect SWs and we propose a novel dynamic programming algorithm to find SWs. Finding SWs in this way improves effectiveness of TAI approach by at most 40% in precision under the same recall.
− Extensive experiments are conducted on real dataset and accuracy of our method is demonstrated. Experiment results show our method is about 10% more precise than the state-of-the-art approach.

The rest of this paper is organized as follows. Section 2 presents problem definition and related works. Section 3 describes our TAI approach. Section 4 shows how to detect SWs. Section 5 conducts extensive experiments. At last, we conclude this paper in Section 6.

2 Preliminary

In this section, we first present formal definition of social strength inference problem and then provide a brief summary of existing methods.

2.1 Problem Statement

Location data reported by check-in service is a triplet $\langle User, Location, Timestamp \rangle$, which contains person identifier, location identifier[1] and a timestamp. After obtaining SWs that are described in Section 4, we obtain co-occurrence

---

[1] Location identifier can be either location id or longitude and latitude or both.

behavior between a pair of persons by counting co-occurrence at every SW and dividing them by the total count. For example, for persons in Figure 1, their co-occurrence behavior is $(\frac{1}{2}, \frac{1}{2}) = (0.5, 0.5)$.

Assume there are N SWs and we denote probability to co-occur at SW i as p_i, the formal definition of social strength inference problem from spatiotemporal data is as follows:

Definition 1. *Given co-occurrence behavior of a pair of persons $(p_1, ..., p_n)$, the problem is to infer a value to tell the likelihood for this pair of persons to be friends.*

2.2 Related Works

Previous studies can be categorized into trajectory based methods and location based methods. The common idea among them is to exploit co-occurrence.

trajectory Based Methods. Trajectory based methods[1][2][3] relax the concept of co-occurrence and use similarity in trajectory to measure likelihood of friendship of a pair of persons.

[1] predicts friendship between a pair of persons by measuring similarity of their location sequence pattern. It summarizes each person's location sequence pattern into several stay points, each of which stands for a set of record referring to the same person within a given size of region and a time interval. Then, each person's stay points are clustered into several hierarchies and similarity is measured based on this hierarchical cluster structure. [2] associates each stay point with a pre-defined tags and similarity are measured based on these tags. [3] assigns each stay point to different tags with different weights.

These methods reveal that similarity in trajectory indicates people's friendship. However, similarity in trajectory does not necessarily mean co-occurrence and such methods suffer from mistakes in detecting stay points.

Location Based Methods. Location based methods[6][4][5] utilize exact co-occurrence behavior.

[6] confirms co-occurrence contributes positively to social ties by conducting experiments over 38 million geo-tagged photos from a geo-social network Flickr[7]. [4] therefore seeks for features of co-occurrence to infer friendship. It finds that the number of distinct locations and duration length of co-occurrence are useful features. Furthermore, [4] finds that frequent co-occurrence at public location may come from coincidence and features above suffers from data sparseness. To tackle these issues, location entropy is introduced. That is, each location is assigned a weight, which is less if more people visit this location. [5] further infers social strength rather than just classifying a pair of people as friends or not. [9] considers different location weights by introducing manually categorized tags for each location.

Location based methods have been shown to improve accuracy of social relationship estimation than trajectory based methods because of exploitation features of co-occurrence and characteristics of locations. However, effectiveness

of existing methods suffers from determining location's contribution simply by
its popularity and ignoring effect of time in co-occurrence. In contrast, our approach determines location's contribution in themes and visiting time is taken
into consideration and therefore overcomes the drawbacks in previous works.

3   Social Strength Inference

In this section, we first introduce an overview for social strength inference. Next,
details about mining themes are presented. Finally, we discuss how to infer social
strength between a given pair of persons.

3.1   Overview

As Figure 2 illustrates, the inference of social strength consists of four phases:
(1)detect SWs; (2)extract themes; (3)train each theme with its contribution to
social strength and (4)given co-occurrence behavior of a pair of persons, calculate
their social strength.

  We discuss SW detection in section 4 and the following paragraphs assume we
have obtained proper SWs and can determine co-occurrence behaviors among
persons. Next, we extract themes that can generate the set of all co-occurrence
behaviors. This routine will be discussed in detail in Section 3.2 and briefly, each
theme has its own flavor to co-occur at certain SWs. Then, for training contribution for each theme, co-occurrence behaviors of persons in training data are
decomposed into themes with different proportion. Themes'contribution to social strength are obtained from social strength calculated from social connection
graph of persons in training data as Section 5.3 discusses. Finally, as presented
in Section 3.3, social strength is inferred given a pair of persons' co-occurrence
behavior.

  In the example above, suppose there are only 2 SWs such as $sw_1 = \langle$amusement
park,$[8:00, 9:00]\rangle$ and $sw_2 = \langle$cafe,$[20:00, 21:00]\rangle$ and only two themes are ex-
tracted such as $\{\langle sw_1, 0.9\rangle, \langle sw_2, 0.1\rangle\}$ and $\{\langle sw_1, 0.1\rangle, \langle sw_2, 0.9\rangle\}$. Co-occurrence
behavior between person 1 and 2 can be represented by $\{0.5, 0.5\}$ and can be
viewed as generated by two themes with 50% proportion each respectively. If the
former theme contributes 0.7 to social strength and the latter contributes 0.9,
we can estimate the social strength by $50\% \times 0.7 + 50\% \times 0.9 = 0.8$, which is a
relative strong indication for both person to be friends.

  So far, we have introduced the TAI approach and in the next sections, we will
describe the four routines outlined in the framework in detail.

3.2   Mining Themes

In this section, we describe the generation of co-occurrence behaviors by themes
and show the topic model used to mine themes.

  We formally define theme, co-occurrence behavior and the generation phase
by notations listed in Table 1. Themes are different probabilistic distributions

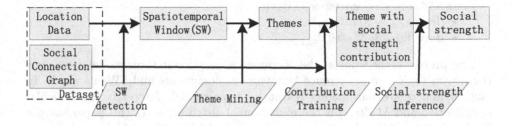

Fig. 2. Framework for social strength inference

over all SWs. Assume there are K themes and N SWs, we represent themes in a $K \times N$ matrix β, where β_k represents the kth theme and $\beta_{k,n}$ indicates the probability for theme k to generate a co-occurrence at SW n. If there are D co-occurrence behaviors, we represent them in a $D \times N$ matrix w, where w_d is dth pair of persons' co-occurrence behavior and $w_{d,n}$ stands for the probability for them to co-occur at SW n. We mention co-occurrence behavior is generated by themes with different proportion, and it can be represented by a $D \times K$ matrix θ, where θ_d is the theme proportion that generates the w_d and $\theta_{d,k}$ is the proportion the theme k takes in the dth co-occurrence behavior. In addition, SWs in co-occurrence behaviors own a $D \times N$ matrix z, where z_d is theme assignment for co-occurrence behavior d's SWs, where $z_{d,n}$ is the theme assigned to the SW n in co-occurrence behavior d.

Table 1. Notations

Symbol	Interpretation
N	The total number of SWs
sw_n	The nth SW
K	The total number of themes
β	A $K \times N$ matrix, where $\beta_{k,n}$ indicates the probability co-occur at SW n in theme k
D	The total number of co-occurrence behaviors
w	A $D \times N$ matrix, where $w_{d,n}$ indicates the probability for the dth pair of persons to co-occur at SW n
θ	A $D \times K$ matrix, where $\theta_{d,k}$ indicates the proportion for theme k takes in generating the dth co-occurrence behavior
z	A $D \times N$ matrix, where $z_{d,n}$ indicates the theme assigned for the dth pair of persons's co-occurrence on SW n

Given notations above, the following generative model describes how themes are used to generate co-occurrence behaviors.

- For each theme k, draw $\beta_k \sim Dir(\eta)$.
- Given co-occurrence behavior d, draw $\theta_d \sim Dir(\alpha)$.

- For SW n in co-occurrence behavior d,
  - draw $z_{d,n} \sim Multi(\theta_d)$;
  - draw $w_{d,n} \sim Multi(\beta_{z_{d,n}})$.

The process is the same as the document generation in topic model, if we view themes as topics, co-occurrence behaviors as documents and SWs as words. In fact, η and α are parameters learnt by topic model, $Dir(\cdot)$ is the Dirichlet distribution and $Multi(\cdot)$ is the multinomial distribution. In this paper, we use Gibbs sampling[14] to implement the learning of LDA[13], a widely used variant of topic model.

After obtaining themes and their proportion on each pair of persons, we will describe how to train the themes with their contribution to social strength and use them to infer social strength in the next section.

3.3 Social Strength

In this section, we first introduce how to assign a contribution to each theme and infer social strength then.

Geo-social network provides both check-in records and social connection graph. In section 5.3, we will show several methods to obtain social strength based on purely the graph. Here, we just assume we have obtained a D-dimension vector ss, where ss_d is a calculated social strength for co-occurrence behavior d. Then, we simply use linear regression to calculate the K-dimension contribution vector ct, where ct_k is the contribution score of the theme k. That is, we use incremental gradient descent algorithm[17] to obtain such contribution vector $\langle ct_1, ..., ct_K \rangle$ that minimizes $\sum_{d=1}^{D}(ss_d - \sum_{k=1}^{K} ct_k\theta_{d,k})^2$.

Given a pair of persons and their co-occurrence behavior $w = \langle w_1, ..., w_N \rangle$, we first decompose it into combination of themes. Namely, we obtain theme's proportion vector $\langle tp_1, ..., tp_K \rangle$ via linear regression[17] such that $|w - \sum_{k=1}^{K} tp_i\beta_i|$ is minimized. Then, the corresponding social strength is inferred as $\sum_{i=1}^{K} tp_i ct_i$.

4   Spatiotemporal Window Detection

In this section, we introduce tf-idf concept to determine proper SWs and prove there exists a optimal substructure for finding proper SWs and thus propose a dynamic programming algorithm to implement the detection.

4.1   Tf-idf for Time Window

For each location, we use several disjoint SW to cover all timestamps on it. Naturally, the time window in each SW should not be too large so to contain all timestamps and too small to contain no timestamp. A tf-idf weight for time window is used to measure how many timestamps a time windows contained in unit length[15]. We partition the time line into several disjoint time windows for each location and thus define SW correspondingly. The goal is to maximize the sum of tf-idf weighting. For example, if we determine $\{[t_1, t_2], ..., [t_k, t_{k+1}]\}$ is the final time windows for a location, the sum of tf-idf weighting $\sum_{i=1}^{k} [t_i, t_{i+1}].tf\_idf$ should be larger than any other possible time window partitions. Therefore, the problem of detecting SWs reduces to the problem of Time Window Determination.

4.2 Time Window Determination

In this section, we first show that the Time Window Determine problem exists an optimal substructure. Thus, we propose a dynamic programming algorithm to find the time windows that maximize the sum of tf-idf weighting.

Theorem 1. *For a set of timestamps* $\{t_0, ..., t_m\}$, *if time windows* $\{[t_0, t_k], [t_k, t_m]\}$ *maximize the sum of tf-idf weighting over all possible time windows,* $[t_0, t_k]$ *and* $[t_k, t_m]$ *maximize tf-idf weighting over all possible time windows on* $\{t_0, ..., t_k\}$ *and* $\{t_k, ..., t_m\}$ *respectively.*

Proof. This can be proofed by contradiction. If another time windows $\{[t_{k1}, t_{k2}], ..., [t_{km}, t_{km+1}]\}$ maximize the sum of tf-idf weighting for $\{t_0, ..., t_k\}$, which means $\sum_{i=k1}^{km} [t_i, t_{i+1}].tf\_idf > [t_0, t_k].tf\_idf$. Therefore, time windows $\{[t_{k1}, t_{k2}], ..., [t_{km}, t_{km+1}]\} \cup [t_k, t_m]$ is better because $\sum_{i=k1}^{km} [t_i, t_{i+1}].tf\_idf + [t_k, t_m].tf\_idf > [t_0, t_k].tf\_idf + [t_k, t_m].tf\_idf$, which conflicts with assumption.

Theorem 1 shows the problem of finding optimal time windows has an optimal substructure and therefore allows us to propose a dynamic programming algorithm to find the optimal partition as Algorithm 1 shows.

Algorithm 1. Time Windows Finding Algorithm

FindTimeWindow($\{t_0, ..., t_m\}$)

1. **for** i := 0 to m **do**
2.   **for** j := i + 1 to m **do**
3.   **for** k := i to j **do**
4.   **if** (t[i, k].tf_idf + t[k, j].tf_idf) > t[i, j].$tf_i df$ **then**
5.   t[i, j] := t[i, k] $\cup$ t[k, j]
6.   **end if**
7.   **end for**
8.   **end for**
9. **end for**

Though the algorithm above, we obtain a binary tree rooted by t[0, m]. The mid-order traversal started from t[0, m] can obtain the final time windows at leaf node.

5   Experiment

In this section, we first describe the dataset and measurement of social strength inference approaches. Next, details of calculating social strength from purely social connection graph and the phase for model training are discussed. Finally, we compare TAI approach with several existing promising methods to demonstrate its effectiveness.

5.1   Settings

We conduct our experiments on a real dataset collected from Foursqaures[16], a location based social network, where users report their location and timestamp by check-in service. This dataset contains more than 200 million check-in records of over 180 thousands people in the form of ⟨UserId, Location Id, latitude, longitude, timestamp⟩. Besides, this dataset contains social connection graph among these users.

5.2   Methodology

Given social connecting graph, we pick up a person and his friends to test whether approaches can estimate the friendship among him and his real friends. We use three metrics to measure effectiveness of our approach. They are precision, recall and friendship percentage. Since our method only produce a social strength score, we set a threshold S such that when the estimated social strength score is larger than S, the corresponding pair of persons is estimated to be friends. The threshold S is determined by varying from 0 to 1 with a step of 0.001. The value that results in optimal accuracy in training dataset is picked up as S. The third metric, friendship percentage is used to measure effectiveness of social strength. Since larger social strength score means larger likelihood to be friends, we check the friendship percentage over each social strength score and evaluate effectiveness of social strength score.

5.3   Social Strength in Training Data

In this part, we present how to calculate social strength to assign contribution scores to themes in Section 3.3. Jaccard's index, Adamic/Adar similarity and Katz score are used to calculate social strength by the graph. Jaccard's Index measures social strength by the ratio of common friends over total friends. Adamic/Adar similarity further considers the popularity of common friends. Katz score calculate social strength between persons without direct common friends.

5.4   Model Training

Model training consists of two phases:(1)Gibbs sampling is used to obtain the themes and their proportion over co-occurrence behaviors;(2)Pick one of the three social strength scores calculated in previous section to assign contribution to themes.

Gibbs Sampling. Gibbs sampling[14] is a robust and widely used technique to train topic model. We just need to specify a proper value as the number of themes. In this paper, we vary the number of themes, from 5 with a step of 5. For each value, we train a model for 1000 iterations.

Figure 3 illustrates that 25 is an appropriate value for the number of themes in our dataset. It can be seen that all 6 curves peak with 25 to be the number of themes. In fact, this indicates people's co-occurrence behaviors consists of 25 different themes. When we use less than 25 themes to train the model, different themes that contributes differently to social strength will be merged and error will happen. When the number is larger than 25, precision and recall suffers from training model without sufficient data.

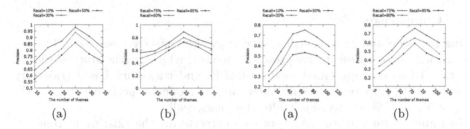

(a)   (b)   (a)   (b)

Fig. 3. Effect of the number of themes   **Fig. 4.** Effect of SW

Effect of SW. In this section, we show regarding people visiting locations in the same SW as co-occurrence improves effectiveness than judging co-occurrence by exact timestamp in check-in records. Figure 4 shows that without SW, both precision and recall are much lower than experiment results in Figure 3 and the number of themes where the precision gets peak also increases. Figure 4 shows, without SW, more themes are mined because the number of themes where precision gets peak increases. However, the consequence is that data become insufficient to train such model and thus both precision and recall deteriorate.

Social Strength Score. In this part, we measure effectiveness of social strength score. The criteria is that higher social strength score should yield more percentage of actual friends. The experiment results show social strength score obtained by Katz score outperforms other two methods.

Figure 5 illustrates the experiment results. First, we can see that social strength is consistent with actual friendship. That is, in all three methods, higher social strength score yields higher friendship percentage. On the other hand, because Kazt score is better than the other two methods in terms of social strength estimation, curves in Figure 5(c) are more smooth and estimation by the model trained by it is more accurate. Because Jaccard's Index is the worst approach, its curves are most fluctuated.

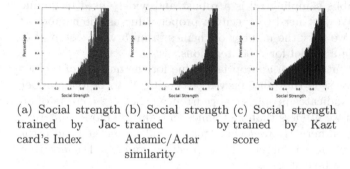

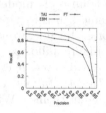

(a) Social strength (b) Social strength (c) Social strength
trained   by   Jac- trained   by trained   by   Kazt
card's Index   Adamic/Adar   score
                    similarity

Fig. 6. Precision vs Recall

Fig. 5. Friendship percentage over social strength

5.5   Comparison

Using precision and recall metrics, we compare the effectiveness of our TAI approach with two representative promising models, which are the-state-of-the-art location based entropy based model(EBM)[5] and trajectory based trajectory model(TR)[1]. Experiment results show our TAI model outperforms other methods by 5% to 15% in precision under the same recall.

In figure 6, we can see TAI approach outperforms the existing methods in terms of both precision and recall. Under the same recall, TAI is 5% to 15% more precise than others. This is because it considers the contribution of location in context of themes and visiting time is considered in inference.

6   Conclusion

In this paper, we propose a Theme-aware social strength Inference (TAI)approach to infer social strength. TAI approach mines themes that generate co-occurrence behaviors by different proportion and train each theme with its contribution to social strength, which fits intuition and overcomes drawbacks in previous works. TAI approach employs tf-idf concept for time window and proposes a novel dynamic programming algorithm to obtain optimal SWs for a location. Extensive experiments are conducted to show TAI approach outperforms other representative methods by 5% to 15% in precision under the same recall.

In the future, we will investigate further in detecting SWs with overlapping time window rather than disjoint time window in this paper. In addition, we will try other topic models such as probabilistic latent semantic indexing.

Acknowledgments. We thank Jialong Han for his helpful discussions. This work was partially supported by the Fundamental Research Funds for the Central Universities, and the Research Funds of Renmin University of China under grant No. 14XNH115, the National Key Basic Research Program (973 Program) of China under grant No. 2014CB340403 and the Fundamental Research Funds for the Central Universities, and the Research Funds of Renmin University of China (10XNI018).

References

1. Li, Q., Zheng, Y., Xie, X., Chen, Y., Liu, W., Ma, W.-Y.: Mining user similarity based on location history. In: Proceedings of the 16th ACM SIGSPATIAL, pp. 1–34. ACM, New York (2008)
2. Ying, J.J.-C., Lu, E.H.-C., Lee, W.-C., Weng, T.-C., Tseng, V.S.: Mining user similarity from semantic trajectories. In: Proceedings of the 2nd ACM SIGSPATIAL International Workshop on Location Based Social Networks, pp. 19–26. ACM, New York (2010)
3. Xiao, X., Zheng, Y., Luo, Q., Xie, X.: Inding similar users using category-based location history. In: Proceedings of the 18th SIGSPATIAL International Conference, pp. 442–445. ACM, New York (2010)
4. Cranshaw, J., Toch, E., Hong, J., Kittur, A., Sadeh, N.: Bridging the gap between physical location and online social networks. In: Proceedings of the 12th ACM International Conference on Ubiquitous Computing, pp. 119–128. ACM, New York (2010)
5. Pham, H., Shahabi, C., Liu, Y.: EBM: an entropy-based model to infer social strength from spatiotemporal data. In: Proceedings of the 2013 ACM SIGMOD International Conference, pp. 265–276. ACM, New York (2010)
6. Crandall, D.J., Backstrom, L., Cosley, D., Suri, S., Huttenlocher, D., Kleinberg, J.: Inferring social ties from geographic coincidences. J. PNAS. 107, 22436–22441 (2010)
7. Flickr, http://www.flickr.com/
8. Foursquare, https://foursquare.com/
9. Lee, M.-J., Chung, C.-W.: A User Similarity Calculation Based on the Location for Social Network Services. In: Yu, J.X., Kim, M.H., Unland, R. (eds.) DASFAA 2011, Part I. LNCS, vol. 6587, pp. 38–52. Springer, Heidelberg (2011)
10. Machanavajjhala, A., Korolova, A., Sarma, A.D.: Personalized social recommendations: accurate or private. J. Proc. VLDB Endow. 4, 440–450 (2011)
11. Blei, D.M., Lafferty, J.D.: Dynamic topic models. In: Proceedings of the 23rd International Conference on Machine Learning, pp. 113–120. ACM, New York (2006)
12. Hofmann, T.: Probabilistic latent semantic analysis. In: Proceedings of the 22nd Annual International ACM SIGIR Conference, pp. 50–57. ACM, New York (1999)
13. Blei, D.M., Ng, A.Y.: Latent dirichlet allocation. J. Mach. Learn. Res. 3, 993–1022 (2003)
14. Porteous, I., Newman, D., Ihler, A., Asuncion, A., Smyth, P., Welling, M.: Fast collapsed gibbs sampling for latent dirichlet allocation. In: Proceedings of the 14th ACM SIGKDD International Conference, pp. 569–577. ACM, New York (2008)
15. Khodaei, A., Shahabi, C., Khodaei, A.: Temporal-Textual Retrieval: Time and Keyword Search in Web Documents. J.IJNGC. 3, 288–312 (2012)
16. Foursquare Dataset, http://www.public.asu.edu/~hgao16/dataset.html
17. Bishop, M.C., Nasrabadi, M.N.: Pattern recognition and machine learning. Springer, Berlin (2006)
18. Blei, D.M.: Probabilistic topic models. J.Commun. ACM. 55, 77–84 (2012)

Organizing Sightseeing Tweets
Based on Content Relatedness and Sharability

Qiang Ma and Keisuke Hasegawa

Kyoto University, Kyoto, Japan 606-8501
qiang@i.kyoto-u.ac.jp, k.hasegawa@db.soc.i.kyoto-u.ac.jp

Abstract. A large amount of tweets about user experiences such as sightseeing appear on Twitter. These tweets are fragmented information and not easy to share with other people as a total experience. In this paper, we propose a novel method for finding and organizing such fragmented information by considering well the content relatedness to the target sightseeing experience and sharing worth with friends. The experimental results demonstrate the efficiency of our method.

Keywords: Twitter, Sightseeing, Fragmented Information, Share, User Experience.

1 Introduction

Consumer generated media (CGM), such as microblogs and SNS, through which users publish information are rapidly becoming popular on the Internet. Although a large amount of information about user experience is accumulated in CGM, developing technology to organize and utilize the stored information has not been progressing quickly enough.

Twitter is a typical microblogging service that has two particular features: The first one is that users can freely post information on their experiences in real time. The second one is the length limitation, i.e., a tweet must be less than 140 characters. These features make that many users tweet their experience fragmentally and in real time. For example, many users tweet their impression of visiting sightseeing spots with photographs shot around there; each tweet is just a piece of information about the whole sightseeing experience.

Developing technology for organizing, storing, searching, and sharing experiences posted on Twitter becomes one of the key challenges. Many users have come to post information on their experiences mainly to Twitter, and they often do not repost what they have tweeted to blog and SNS diary service again. This circumstance make it difficult to share the whole user experience because it is published in a fragment manner. In addition, although a certain experience is often described in many tweets, these tweets do not contain keywords that directly express it. As a result, even if one wants to find tweets about a sightseeing experience, conventional keyword-based methods are not enough to find all target tweets without omissions; i.e., the recall is poor.

F. Li et al. (Eds.): WAIM 2014, LNCS 8485, pp. 510–521, 2014.

In this paper, we propose a novel method of organizing tweets that describe sightseeing experiences to help users organize, share and store their experiences. To this, we estimate the *relatedness* relationship of a tweet with the given sightseeing experience, and its *sharability* value to store and share with others.

We label a sightseeing experience as a sequence of spot names. For example, when a user visits Kyoto, s/he may visit famous spots such as Yasaka-jinja, Kodai-ji, and Kiyomizu-dera. In this case, we label this sightseeing experience as a sequence of spot names, "Yasaka-jinja, Kodai-ji, Kiyomizu-dera". Given the label of a user's sightseeing experience (a series of spot names) and the tweet archive of that user[1], to find the tweets describing this experience, we estimate the relatedness of a tweet with that sightseeing experience from the following two aspects.

– Content relatedness

   We utilize the co-occurrence relationship of spot names and keywords in a tweet to compute content relatedness. It is based on the idea that if a tweet containing many words related to a spot, the relatedness of that tweet and that spot is high.

– Context relatedness

   We also consider well the effect of surrounding tweets. It is based on the idea that the tweets before and after a tweet related to a sightseeing experience may have high probabilities of describing the same experience, although they may not contain any word related to any spot the user visited.

On the other hand, we notice that many tweets posted during a sightseeing event have little worth to share with other people although they are related to that experience. For example, a user may just tweet information about his/her current location such as "I'm at Kyoto station". Although this tweet is related to his/her sightseeing in Kyoto, it has little worth to share with other people in most cases. Therefore, we propose a novel notion called *sharability* to estimate whether a sightseeing tweet is worth sharing. Currently, we compute the sharability of a sightseeing tweet by considering its content quality with linguistic and media features.

The input of our method includes 1) the label of sightseeing experience, which is a sequence of spots the user visited, 2) the sightseeing period, and 3) the tweet id of the user who want to organize his/her sightseeing tweets. At first, we use the user id and time period to collect the candidate tweets, and then rank and filter them by using the relatedness and sharability criteria. The output is a sequence of tweets describing the experience with the high value of storing and sharing with other people.

The major contributions of this paper can be summarized as follows.

– We propose a novel method for organizing fragmented information (tweets) on a certain sightseeing experience.

---

[1] In fact, we do not need to compute the entire archive. We can use the time-based search function to obtain the tweets posted during the time period of the target sightseeing.

Because a tweet is short (only a few words) and may not contain keywords directly related to the spots a user visited, it is difficult to determine the relatedness of a tweet to the given sightseeing experience by using conventional keyword-appearance-based methods (e.g., tfidf). To address this issue, we propose a novel method that takes into account the spatio-temporal continuity of user behaviors. That is, we calculate the relatedness of a tweet to a sightseeing experience by considering 1) the co-occurrence of a term in a tweet and spot names (Section 3.1), and 2) the relation among surrounding tweets (Section 3.2).

- We propose a novel notion called *sharability* to discover meaningful and shareable sightseeing tweets (Section 4).
  Sightseeing tweets may contain both valuable descriptions and general/usual information. How to discover tweets with high worth to store and share with other people is one of the key challenges in organizing sightseeing content. To the best of our knowledge, this study is the first attack on this issue.
- We carried out experiments to evaluate the efficiency of our method for organizing tweets on sightseeing experiences. The experimental results on relatedness revealed the efficiency of organizing tweets (Section 5.2). The results from the experiment on sharability showed that sharability is useful for excluding tweets containing little value for sharing (Section 5.3).

2   Related Work

There are much research on Twitter in recent years. Fujisaka et al. propose a method of estimating the regions of influence of social events found from geo-tagged tweets[1]. Wu et al. investigate how information is propagated from celebrities and bloggers to ordinary users[2]. Castillo et al. propose automatic methods of assessing credibility of tweets based on features of user behaviors[3]. However, to the best of our knowledge, there are few studies on mining and organizing user experiences from tweets. Arimitsu et al. and Hasegawa et al. propose methods of searching user experiences[8][12]. They define user experiences as several actions carried out in a particular sequence. In contrast, we discover more valuable information to share with other people by considering the sharing worth of tweets.

Many researchers focus on mining user experiences from blog entries and lifelog [4][5]. Kurashima et al. propose a method of extracting rules between five attributes of a person's experience[4], i.e., time, location, activity, opinion, and emotion, from a large set of blog entries. Their method focuses on how to mine frequent patterns from user experiences. In contrast, our method mines and organizes the sightseeing experiences of each user. We also consider the worth for sharing with other people to organize fragmental information.

Several methods of discovering geographical topics have been proposed. Yin et al. propose a method of discovering geographical topics from geo-tagged photos in Flickr by using Latent Geographical Topic Analysis (LGTA), which combines location-driven and text-driven models[6]. Hong et al. propose a more accurate method of discovering geographical topics from geo-tagged tweets[7]. GPS-associated information is necessary in these methods; however, only 0.77% of tweets are geo-tagged in Twitter[2].

Data quality is an important issue in big data era[9]. Agichtein et al. [10] propose a method of automatically identifying high-quality content from Yahoo! Answers by considering intrinsic content quality, user relationships, and usage statistics. For intrinsic content quality, they use semantic features, such as punctuation density, capitalization errors, number of words, and distribution of word n-grams and part-of-speech sequences. Iwaki et al. propose a method of extracting "persuasive" images in a collection of regional photos on the Web[11]. They formulate persuasiveness with approximate subject goodness estimated from Web 2.0 content and with a previously proposed method of photo quality assessment. Our notion of sharability is related to content quality. However, to the best of our knowledge, our work is the first attempt at computing the sharing worth of tweets.

3   Relatedness

3.1   Content Relatedness

We utilize the co-occurrence relationship to estimate the content relatedness of a candidate tweet and a certain sightseeing experience. It is based on the idea that if a tweet containing more words related to a spot, it has high probability of describing that spot. Usually, two terms having a strong co-occurrence relationship means they are strongly related to each other because they appear together often. Currently, we apply the method proposed by Hasegawa et al. [12] to construct a co-occurrence dictionary to support this task.

The content relatedness of tweet t_i to sightseeing experience $E = (p_1, \cdots, p_n)$ is defined as follows. Here, E is a series of location names where the user visited.

$$Rc(t_i) = \sigma + \sum_{j=1}^{n} \sum_{k=1}^{m} co(p_j, w_k) \tag{1}$$

where $w_1, \cdots, w_m$ are the keywords appearing in tweet t_i, and σ is a parameter to prevent the content relatedness score from becoming 0 when we cannot find proper co-occurrence relationships by using the dictionary which may not cover all related words. $co(p_j, w_k)$ denotes the co-occurrence relationship of words p_j and w_k, and is specified in the co-occurrence dictionary.

3.2   Context Relatedness

In Twitter, information on an experience may be described in many tweets and a certain topic may be divided into a series of tweets due to the 140 characters

---

[2] http://techcrunch.com/2012/07/30/analyst-twitter-passed-500m-users-in-june-2012-140m-of-them-in-us-jakarta-biggest-tweeting-city/

limitation. As a result, a tweet may not contain words directly related to the sightseeing experience, although it is a part of the description. In such case, one of the considerable solutions is to consider the effect of surrounding tweets. Intuitively, because a sightseeing experience exhibits the continuity of user behaviors and tweets are posted in real time, surrounding tweets may describe the same experiences.

By considering the effect of preceding and following X tweets, the context relatedness score of tweet t_i, $Rx(t_i)$, is calculated as follows.

$$Rx(t_i) = \sum_{x=-X}^{X} \tau_{i+x} \cdot Rc(t_{i+x}) \tag{2}$$

$$\tau_{i+x} = e^{-\mu_t |time(t_i) - time(t_{i+x})|} \tag{3}$$

where $time(t_i)$ is the publishing time of the i-th candidate tweet t_i in time order. τ_{i+x} is the weighting representing the temporal continuity calculated by applying the exponential degradation model [13][14]. μ_t is a parameter for determining the rate in degradation of the similarity of content with increasing time intervals.

4   Content Sharability

After we select the sightseeing tweets with high relatedness scores, the next step is to determine their worth for sharing and storing. As we mentioned above, there are many sightseeing tweets that only provide information of the user's current location (e.g., "I'm at Kyoto now"), daily behavior (e.g., "lunch time!"), and so on. To discover tweets containing valuable information, we propose a novel notion of content sharability.

Generally, the sharability of a tweet can be estimated from three aspects such as 1) content quality, 2) social support from other users (e.g., frequency of retweets), and 3) authority of the Twitter user (e.g., the tweets written by a famous person may have high sharability). However, because we currently focus on organizing a user's own sightseeing experiences, authority is not a valid measure (the author is the user himself/herself!). Also, because social support depends on the online social network of that user and is the measure based on other users' viewpoint, it is more useful when a user want to search for another one's experience rather than organizing experiences of herself/himself. Hence, in this paper, we focus on the way of estimating sharability of a tweet from the aspect of content quality.

We assume that more informative and impressive tweets have high worth to share. In other words, we assume that a tweet containing many descriptive words and multimedia contents is informative; a tweet with emotional words is impressive. Currently, we compute the content sharability by considering linguistic and media features of tweets. Intuitively, from the linguistic aspect, the more descriptive and emotional words a tweet contains, the more sharing worth it has. On the other hand, from the media aspect, a tweet containing multimedia content, such as pictures and videos, will be assigned a higher content sharability score.

The content sharability $S(t_i)$ of tweet t_i is defined as follows.

$$S(t_i) = \log(words(t_i) + \phi \cdot pic(t_i)) \qquad (4)$$

$$words(t_i) = \alpha \cdot modifier(t_i) + other(t_i) \qquad (5)$$

where $modifier(t_i)$ is the total number of adjectives, adjective verbs, adverbs, and adnominal adjectives in tweet t_i, and $other(t_i)$ is the number of the other words in t_i. As mentioned before, we assume that a tweet consisting of more adjectives, adjective verbs, adverbs, and nominal adjectives are more impressive, and long tweets are more informative. $\alpha(>1)$ is a weighting parameter to increase the importance of these types of words. Currently, we use the Japanese linguistic tool MeCab[3] to identify these types of words. $pic(t_i)$ is the number of URLs of pictures and videos attached to tweet t_i. We empirically count a picture (video) as ϕ words to simplify the computation. This is based on the assumption that a tweet with a picture (video) is more impressive and informative.

5   Experiments

We carried out three experiments to evaluate the efficiency our method: 1) experiment on relatedness, 2) experiment on content sharability, and 3) experiment on organizing tweets.

5.1   Experimental Method

We focused on the sightseeing experience in Kyoto as our experimental target because Kyoto is the most famous sightseeing city in Japan. In our experiments, we used the names of sightseeing spots in Kyoto as keywords to collect tweets to build the co-occurrence dictionary. We used MeCab as the Japanese morphological tool to extract keywords and spots names from the tweets.

We used the sightseeing spots' names (Yasaka-jinja, Kiyomizu-dera, etc.) to find the Twitter users who have been to Kyoto. Finally, we randomly selected three users U_a, U_b and U_c, and their one-day sightseeing experiences E_a, E_b, and E_c as our test targets. We respectively collected these three user's tweets posted in a week (the day s/he visited Kyoto and three days before/after that day) as the test data.

Based on the preliminary experimental results, the parameters are specified as follows: $\sigma = 0.01$ (the parameter for calculating the content relatedness score in Eq. (1)), and $\phi = 50, \rho = 0.1, \alpha = 5$ (the parameter for calculating the content sharability score in Eq. (4), (5)).

We used the results of morphological analysis from MeCab to count the number of words in tweets and identify adjectives, adjective verbs, adverbs, and nominal adjectives. To calculate $pic(t_i)$, the number of pictures of tweet t_i, we assumed that if a tweet, except official retweets, has one URL, one picture or video is attached in that tweet.

---

[3] http://mecab.googlecode.com/svn/trunk/mecab/doc/index.html

Table 1. Precision and recall ratios with maximum F-measure (Relatedness)

Experience	Scorering Method	Precision	Recall	F-measure
	$Rc(t_i)$	0.5287	0.4792	0.5027
	$Rx(t_i\|X=1, \mu_t=1)$	0.6484	0.6146	0.6310
E_a	$Rx(t_i\|X=1, \mu_t=10)$	0.6860	0.6146	0.6484
	$Rx(t_i\|X=2, \mu_t=10)$	**0.7582**	**0.7188**	**0.7380**
	Baseline	0.7273	0.0833	0.1495
	$Rc(t_i)$	0.5952	0.4167	0.4902
	$Rx(t_i\|X=1, \mu_t=1)$	0.4607	**0.6833**	0.5616
E_b	$Rx(t_i\|X=1, \mu_t=10)$	0.4659	**0.6833**	0.5541
	$Rx(t_i\|X=2, \mu_t=10)$	0.4940	**0.6833**	**0.5734**
	Baseline	**1.0000**	0.1000	0.1818
	$Rc(t_i)$	0.3874	**1.0000**	0.5584
	$Rx(t_i\|X=1, \mu_t=1)$	0.4111	0.8605	0.5564
E_c	$Rx(t_i\|X=1, \mu_t=10)$	0.4805	0.8605	0.6167
	$Rx(t_i\|X=2, \mu_t=10)$	0.5263	0.9302	**0.6723**
	Baseline	**1.0000**	0.0930	0.1702

We used recall, precision, and F-measure as the evaluaion measures. We computed the scores by using our method for the tweets on each user experience (E_a, E_b, and E_c). If a tweet's score is higher than the threshold, we regard it as one of the results returned by our method. We also manually evaluated the relatedness and sharing worth of each candidate tweet with the five-point scale method to construct the relevant result sets.

Each candidate tweet was manually ranked into one of five categories based on two aspects respectively; whether each tweet is related to the target sightseeing experience and each tweet is worth sharing, where 5 means excellent relatedness (sharing worth) and 1 denotes poor relatedness (sharing worth).

5.2   Experiment on Relatedness

We selected the tweets that manually scored greater than 4 as relevant results. We used the results from the keyword-based OR search "Yasaka-jinja or Kiyomizu-dera" as the baseline method. "Yasaka-jinja, Kiyomizu-dera" is the sequence of spot names indicates our target sightseeing experience.

Table 1, Figure 1(a), 1(b), and 1(c) summarize the experimental results.

As shown in Table 1, compared to the baseline method, the proposed method by considering content relatedness improved the F-measure by an average of three times, and that by context relatedness improved it by an average of four times. Also, in our experiment, when $X=2, \mu_t=10$, our context relatedness based method achieved the best performance.

Fig. 1(a), 1(b), and 1(c) are the precision-recall curves of the four methods for each sightseeing experience.

The F-measure of context relatedness $Rx(t_i)$ was on average 20 percent higher than that of content relatedness $Rc(t_i)$. The precision ratio of $Rx(t_i)$ was higher

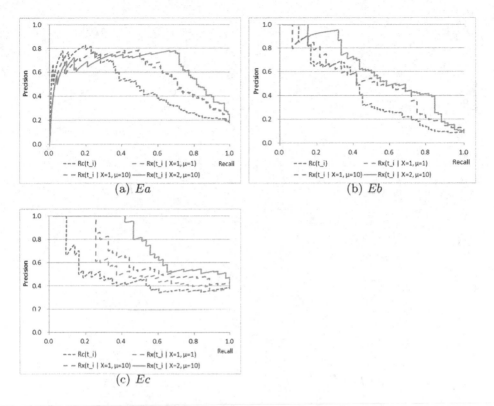

Fig. 1. Precision-recall curve of relatedness

than that of $Rc(t_i)$ when the recall ratio was higher than about 0.4 for all the sightseeing experiences.

These results indicated that using content and context relatedness can improve efficiency of discovering tweets related to a certain sightseeing experience. These results also revealed the importance of considering surrounding tweets (context) in twitter search systems.

5.3   Experiment on Content Sharability

We evaluated the efficiency of our method to calculate content sharability. In this experiment, we selected tweets that manually scored greater than 3 (sharability) as relevant results for calculating precision, recall, and the F-measure. Based on preliminary experimental results, we let $\phi = 50$ and $\alpha = 5$. We used the results of morphological analysis by using MeCab to count the number of words in tweets and identify adjectives, adjective verbs, adverbs, and nominal adjectives.

Figure 2(a), 2(b), and 2(c) are the precision-recall curves of S_{wp}, S_w, and S_p for sightseeing experiences E_a, E_b, and E_c , respectively.

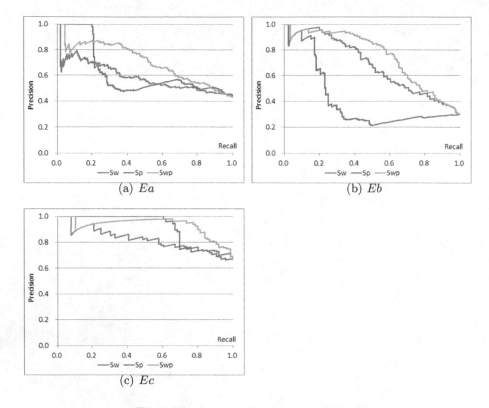

(a) Ea

(b) Eb

(c) Ec

Fig. 2. Precision-recall curve of sharability

Table 2 summarizes the experimental results. Compared to the baselines (S_p and S_w), in three sightseeing experiences, the proposed method by considering both linguistic and media features improved the precision and F-measure by an average of about 1.5 times and 1.1 times, respectively. These results indicated that the effective of considering both the feature of words in a tweet and the number of pictures to estimate sharability.

5.4   Efficiency of Organization by Combination of Relatedness and Sharability

We compared the efficiency of our method of organizing sightseeing tweets by combining the context relatedness and sharability scores with that of a baseline. We used the results from the keyword-based OR search described in the experiment in Section 5.2 as the baseline.

In this experiment, we calculated an integrated score $RS(t_i)$ of relatedness and sharability as following.

$$RS(t_i) = Rx(t_i) \times S(t_i) \tag{6}$$

Table 2. Precision and recall ratios with maximum F-measure (Sharability)

Experience	Scoring Method	Precision	Recall	F-measure
	Our Method : S_{wp}	**0.5447**	0.8405	**0.6610**
E_a	Words Only : S_w	0.4581	**0.9655**	0.6214
	Picture Only : S_p	0.5099	0.8922	0.6489
	Our Method : S_{wp}	**0.7665**	0.6244	**0.6882**
E_b	Words Only : S_w	0.5305	0.6780	0.5953
	Picture Only : S_p	0.2993	**1.0000**	0.4607
	Our Method : S_{wp}	**0.9242**	0.8026	**0.8592**
E_c	Words Only : S_w	0.7143	0.9868	0.8287
	Picture Only : S_p	0.6726	**1.0000**	0.8042

Table 3. Set 4–3 : Precision and recall ratios with maximum F-measure

Exp.	Method	Precision	Recall	F-measure
E_a	$RS(t_i)$	0.6790	0.6875	0.6832
	Baseline	0.6364	0.0875	0.1538
E_b	$RS(t_i)$	0.5294	0.6545	0.5854
	Baseline	1.0000	0.1091	0.1967
E_c	$RS(t_i)$	0.6491	0.9024	0.7551
	Baseline	0.7500	0.0732	0.1333

Table 4. Set 3–3 : Precision and recall ratios with maximum F-measure

Exp.	Method	Precision	Recall	F-measure
E_a	$RS(t_i)$	0.6210	0.7938	0.6968
	Baseline	0.9090	0.1031	0.1852
E_b	$RS(t_i)$	0.4949	0.8167	0.6164
	Baseline	1.0000	0.1000	0.1818
E_c	$RS(t_i)$	0.6491	0.8810	0.7475
	Baseline	0.7500	0.0714	0.1304

Based on the experimental results in Section 5.2, we let $X = 2$ and $\mu_t = 10$ to compute the context relatedness, i.e., $Rx(t_i|X = 2, \mu_t = 10)$.

We selected two sets of relevant results, Set 4–3 and Set 3–3, as follows.

Set 4–3 the manually assigned relatedness and sharability scores of relevant tweets were more than 3 and 2, respectively.

Set 3–3 the manually assigned relatedness and sharability scores of relevant tweets were more than 2 and 2, respectively.

Table 3 and 4 summarize the results. Our method improved the F-measure by an average of four times and precision ratio by an average of nine times from those of the baseline.

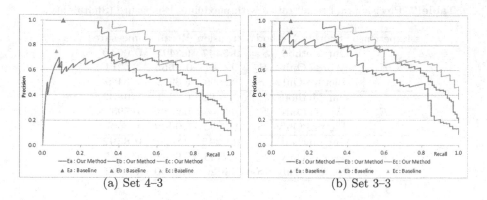

(a) Set 4–3   (b) Set 3–3

Fig. 3. Precision-recall curve of combination of relatedness and sharability

Figure 3(a) and 3(b) are the precision-recall curves of Sets 4–3 and 3–3, respectively. Compared to the baseline, our method greatly improved the recall ratio while maintaining the precision ratio in both sets.

The precision of the baseline is better. One of the considerable reasons is that each tweet obtained by the baseline contains one or more spot (names) which user visited and then has high relatedness to the sightseeing experience. However, its recall is poor and may miss some valuable information. It is to say, our method can find more sightseeing tweets although they do not contain keyword directly related to the sightseeing experience.

Focusing on sightseeing experience E_a, the precision ratio of Set 3–3 was higher than that of Set 4–3 when the recall ratio was lower than 0.7. This is because user U_a posted many tweets about his sightseeing before and after the sightseeing day. This revealed that our method can discover tweets on a sightseeing experience, even though they were posted before or after the sightseeing period. This is why we need collect candidate tweets before and after rather than during the sightseeing period.

6   Conclusion

In this paper, we proposed a novel method of organizing and ranking fragmented tweets related to a certain sightseeing experience to search and share valuable user experiences posted on Twitter. The experimental results revealed that the performance of the proposed method is better than that of keyword-based methods. Especially, taking into consideration the effect of context tweets further improved performance. The experimental results also validate that content sharability is useful in discovering more informative tweets than just considering the relatedness of tweets and user experiences.

Future work includes further evaluation, general improvements in relatedness and content sharability computations, and an application system for organizing and sharing sightseeing experiences.

Acknowledgement. This work is partly supported by KAKENHI(No.25700033) and SCAT Reseach Funding.

References

1. Fujisaka, T., Lee, R., Sumiya, K.: Discovery of User Behavior Patterns from Geo-tagged Micro-blogs. In: ICUIMC, pp. 246–255 (2010)
2. Wu, S., Hofman, J.M., Mason, W.A., Watts, D.J.: Who Says What to Whom on Twitter. In: WWW, pp. 705–714 (2011)
3. Castillo, C., Mendoza, M., Poblete, B.: Information Credibility on Twitter. In: WWW, pp. 675–684 (2011)
4. Kurashima, T., Fujimura, K., Okuda, H.: Discovering Association Rules on Experiences from Large-Scale Blog Entries. In: Boughanem, M., Berrut, C., Mothe, J., Soule-Dupuy, C. (eds.) ECIR 2009. LNCS, vol. 5478, pp. 546–553. Springer, Heidelberg (2009)
5. Ushiama, T., Watanabe, T.: An Automatic Indexing Approach for Private Photo Searching Based on E-mail Archive. In: Gabrys, B., Howlett, R.J., Jain, L.C. (eds.) KES 2006. LNCS (LNAI), vol. 4252, pp. 1111–1118. Springer, Heidelberg (2006)
6. Yin, Z., Cao, L., Han, J., Zhai, C., Huang, T.: Geographical Topic Discovery and Comparison. In: WWW, pp. 247–256 (2011)
7. Hong, L., Ahmed, A., Gurumurthy, S., Smola, A., Tsioutsiouliklis, K.: Discovering Geographical Topics in the Twitter Stream. In: WWW, pp. 769–778 (2012)
8. Arimitsu, J., Ma, Q., Yoshikawa, M.: A User Experience-oriented Microblog Retrieval Method. In: DEIM (2011) (in Japanese)
9. Sadiq, S.W., Zhou, X.F., Deng, K.: Research and Practice in Data Quality. In: APWeb, pp. 41–42 (2008)
10. Agichtein, E., Castillo, C., Donato, D., Gionis, A., Mishne, G.: Finding High-Quality Content in Social Media. In: WSDM, pp. 183–194 (2008)
11. Iwaki, Y., Jatowt, A., Tanaka, K.: How to Calculate the Persuasiveness of Regional Pictures on the Web. In: DEIM (2010) (in Japanese)
12. Hasegawa, K., Ma, Q., Yoshikawa, M.: Trip Tweets Search by Considering Spatio-temporal Continuity of User Behavior. In: Liddle, S.W., Schewe, K.-D., Tjoa, A.M., Zhou, X. (eds.) DEXA 2012, Part II. LNCS, vol. 7447, pp. 141–155. Springer, Heidelberg (2012)
13. Toda, H., Kitagawa, H., Fujimura, K., Kataoka, R.: Topic Structure Mining Using Temporal Co-occurrence. In: ICUIMC, pp. 236–241 (2008)
14. Cui, C., Kitagawa, H.: Topic Activation Analysis for Document Streams Based on Document Arrival Rate and Relevance. In: SAC, pp. 1089–1095 (2005)

Efficient Diverse Rank of Hot-Topics-Discussion on Social Network

Tao Zhu[1], Yumin Lin[2], Ji Cheng[1], and Xiaoling Wang[1]

[1] Shanghai Key Laboratory of Trustworthy Computing,
Institute for Data Science and Engineering, Software Engineering Institute,
East China Normal University, Shanghai, China
[2] Guangxi Key Laboratory of Trusted Software,
Guilin University of Electronic Technology, Guangxi, China
{tzhujs,ymlinbh,chengjirobin}@gmail.com, xlwang@sei.ecnu.edu.cn

Abstract. Social network has become an important way to exchange information, which allows users to post tweets on hot topics and share opinions with others. Typically, tweets on an identical topic are ranked according to either the freshness or the popularity. However, neither of them is ideal, due to the redundancy of tweets' content under the same topic. To address this problem, we intend to diversify the search results in this paper. We propose an effective measure of diversity and a ranking algorithm based on LSH(Locality Sensitive Hash). The proposed solution also works well on the MapReduce framework. Our method can efficiently rank tweets according to their popularity and diversity. We carry out extensive experiments on real-world datasets to verify the effectiveness and efficiency of our methods.

1 Introduction

Recently, online social media services attract many users and become a popular way to share personal opinions. Microblogging, such as Twitter[1] and Sina Weibo[2] is one of the most popular social media services. These platforms group tweets into different topics using hastag[3] and provide the Hot-Topic-Discussion service which allows users to express personal opinions and read about other people's views about a specific hot topic. Ranking is necessary when displaying tweets to users because of the large quantity of tweets under one topic. Conventionally, tweets are ranked according to time stamps. This approach is widely used by social networking applications such as Twitter and Sina Weibo. Another popular ranking strategy is based on the frequency of forwarding and commenting rate of the tweet. However, both ranking methods might result in redundancy on the content of tweets. Hence, we need a more effective strategy to rank the tweets under the same hot topic.

---

[1] http://www.twitter.com
[2] http://www.weibo.com
[3] A hashtag is a word or a phrase prefixed with the symbol #.

F. Li et al. (Eds.): WAIM 2014, LNCS 8485, pp. 522–534, 2014.

Furthermore, Hot-Topic-Discussion draws a lot of attention and receives many queries at the beginning of its lifetime. That is to say, the number of tweets on a topic grows quickly and massive queries about the tweets are simultaneously emerging during the discussion's lifetime. Additionally, users constantly post new messages, repost and comment the existing ones. Thus ranking efficiency and the update cost of the ranking are very important.

In this work, we present an efficient diverse ranking algorithm for the tweets under a hot topic which is eligible for offline ranking and online updating. To summarize, we make the following contributions in this paper:

- Proposing a novel measure to the increase diversity of the top tweets effectively.
- Designing an efficient ranking and updating algorithm based on locality sensitive hashing [1].
- Implementing the ranking algorithm on MapReduce framework to enhance the scalability for the large tweet set.
- Conducting a series of experiments on real-life dataset to verify the efficiency and effectiveness of the proposed algorithm.

The rest of the paper is organized as follows: Related work is given in section 2. We give a formal statement about the diverse ranking problem in section 3, and introduce the LSH-based algorithms for ranking and updating operations in section 4. Besides, we discuss the implementation on the MapReduce framework. In section 5, experiments on real datasets are conducted to verify the efficiency and effectiveness of our approach. And we conclude this work in section 6.

2   Related Work

Information Retrieval(IR) is concerned with finding relevant documents that satisfy user information needs [2] . IR systems rank documents mainly according to the relevance to the query. It has an assumption that the usefulness of a result to the user is independent of the usefulness of the others [3]. For cases where there is a large volume of potentially relevant documents, highly redundant with each other or containing partially or fully duplicative information in extreme, pure relevance ranking is not enough. Hence, beyond pure relevance for document ranking, diversity is ought to be combined to reduce redundancy.

Recently, diversity has gained great attention as it is a way to improve the quality of query results [4–6]. Corbonell and Goldstein [4] introduced the concept of diversity in text retrieval and summarization as well as the maximal marginal relevance, an aggregation of relevance and diversity. Quantum Probability Ranking Principle [6] and the Modern Portfolio Theory[5] introduced similar criterions. They designed different diversity measures and utilized different approaches to aggregate documents' relevance and diversity. However, all of them have a quadratic time complexity and update operation is unfeasible under these measures.

Result diversification has been considered in different applications[7, 8]. In [7], Sofiane Abbar considered the problem of recommending a set of diversified related articles where relevance and diversity are considered independently. First, the most similar documents are retrieved. Second, a diverse subset which maximize the minimum pair-wise distance is constructed by using a 2-approximate greedy algorithm. In [8], location visualization with geo-tagged photos is discussed. The author both took the content similarity and geographic distance to model the redundancy between each objects, and used a clustering-based approach to generate a representative subset. Besides, they designed a hybrid index to achieve query efficiency.

In this work, we consider the problem of diverse ranking for the tweets set on a hot topic. And our focus is the efficiency of ranking and feasibility of update operation for the tweets collection.

3 Problem Definition

This section is to define the diverse ranking problem studied in this work. A tweet is represented as a quadri-tuple $< tid, text, rf, cf >$ where tid is an unique identifier, $text$ is the content, rf is retweeting frequency and cf is comment frequency. $T = \{t_1, t_2, \ldots, t_n\}$ is a set of tweets on a topic O.

To highlight the importance of different tweets under the same topic, we first adopt *popularity* in [10] to measure how popular a tweet is. Intuitively, the tweets retweeted and commented frequently are the popular ones. Hence, we explore retweeting frequency and comment frequency to evaluate the popularity of a tweet formally in Definition 1.

Definition 1 (Popularity). *The popularity of a tweet t_i is a sum of its retweeting frequency and comment frequency.*

$$P(t_i) = \frac{rf_i + cf_i}{Z} \tag{1}$$

Here $Z = \max_{t_j \in T}(rf_j + cf_j)$ is a normalization factor.

In addition, we propose *diversity* to highlight the content novelty that a tweet would contribute to the whole ranking sequence, as some tweets might share similar content. We observe that a tweet is of less diversity if it is similar to a more popular one. Thus, the higher the similarity of a tweet, the lower the diversity of the tweet is. We define diversity formally in Definition 2.

Definition 2 (Diversity). *The diversity of a tweet t_i is negative correlated with the maximum similarity with another more popular one in T.*

$$D(t) = - \max_{t' \in C(t)} Sim(t', t) \tag{2}$$

Here $C(t) = \{t' | t' \in T \wedge P(t') > P(t)\}$ and $Sim()$ is a similarity metric over the content of tweets.

In particular, the similarity metric used here is the Jaccard similarity [11] over K-shinglings [12] set.

Based on the popularity and diversity defined, the diverse ranking is to put the popular and diverse one in the front of the tweet sequence, which is formally defined as follows:

Definition 3 (Diverse Ranking Problem). *The diverse ranking problem is to construct a tweet sequence* $(t'_1, t'_2, \ldots, t'_n)$ *for the tweet collection* T *about topic* O, *according to*

$$\lambda * P(t) + (1 - \lambda) * D(t)$$

Here, λ is a parameter for controlling the tradeoff between popularity and diversity. It is desirable that tweets are ranked in the order of popularity when $\lambda = 1$, whereas diversity is mostly wanted when $\lambda = 0$.

To answer user queries with short response time, it is indispensable to solve the above problem efficiently. Clearly, the diversity ranking can be done at the beginning of the topic discussion, with each tweet's popularity and diversity measured. After that, the ranking order is periodically updated along with the evolvement of a topic discussion. Thus, we can always answer user queries by selecting the top tweets efficiently.

Based on the above discussion, the problems we tackled are listed as follows:

- How to efficiently, in *offline* manner, rank each tweet when the topic discussion began with a set of tweets collected initially?
- How to efficiently, in *online* manner, update the ranking order periodically?

4 Approach

In this section, we introduce how to rank tweets efficiently and update the ranking periodically.

Firstly, we demonstrate a naive method to analyze the bottleneck of the ranking method.

Given a collection of tweets on a particular topic, we can rank all tweets as follows:

step 1 Measuring the popularity of each tweet, and sorting them in descending order by popularity.

step 2 For each tweet, computing content similarity with all other more popular tweets, setting its diversity score to the negative of the maximum similarity.

step 3 Sorting all tweets in order of the aggregation of popularity and diversity.

Assuming the size of tweet collection is n, such approach is associated with n popularity computation, $\frac{n*(n-1)}{2}$ similarity computation and two sort operations. Thus, the most time-consuming part would be the pair-wise similarity computations. If the collection size is very large, such approach is inefficient.

4.1   Lrank: LSH-Based Rank Algorithm

Here we utilize LSH to reduce similarity computation into a small set of tweets pairs. We start the description of the LSH-based rank algorithm with an overview of the LSH algorithm[1].

LSH adopts Minhash functions which can preserve Jaccard similarity so that similar objects have a high probability of colliding in the same bucket. A minhash function $mh_\pi(S)$ outputs an element of set S, which firstly appeares in an element permutation π. Minhash has the property that the probability of two tweets sharing the same minhash value equals with the their Jaccard similarity.

Essentially, given a tweet, the LSH attempts to find all candidate tweets which are similar to it. It is done as follows: firstly, LSH creates b hash functions $g_1, \ldots, g_b$, and their corresponding hash tables $A_1, \ldots, A_b$. Here, each function g_i is a combination of d minhash functions mh_j (i.e. $g_i = mh_1, \ldots, mh_d$). Then each tweets t is stored in bucket $g_i(t)$ of A_i. For a tweet t, we have $t^* \in A_1(g_1(t)) \cup \cdots \cup A_b(g_b(t))$. Here t^* is an approximate nearest neighbor[1] of t. By using LSH, the time complexity of the nearest neighbor query is $O(n^e)$ where $e < 1$[13].

Let us recall the computation of $D(t)$. Formally, given a tweet t and let $C = \{t' | P(t') > P(t) \wedge t' \in U\}$, the key here is to find the nearest neighbor for t in set C. Thus, we can use LSH to accelerate the computation of $D(t)$ by finding an approximate nearest neighbor.

At the beginning of diverse ranking, we compute tweets' popularity(line 2) and partition them into LSH buckets(line 4) as the following pseudo-code.

Algorithm 1. Preprocessing

Input: hash functions $g_1, \ldots, g_b$,
    tweets collection $U = \{t_1, t_2, \ldots, t_n\}$
Output: LSH table $A_1, A_2, \ldots, A_b$
  1. **for each** $t \in U$ **do**
  2. $P(t) = (t.rf + t.cf)/Z$;
  3.   **for** $i = 1$ TO b **do**
  4. $A_i(g_i(t)) \leftarrow A_i(g_i(t)) \cup t$;
  5.   **end for**
  6. **end for**
  7. **return** $A_1, A_2, \ldots, A_b$;

Based on the LSH tables, algorithm 2 only computes the similarity between the tweets pairs in the same bucket. At the beginning, the diversity scores of all tweets are initialize to 0(line 1). Next we check all similar tweets pair (t_1, t_2) in LSH buckets(line 2-3). Without loss of generality, let $P(t_1) > P(t_2)$, then $D(t_2)$ is at most $-Sim(t_1, t_2)$(line 4-9).

The time complexity of algorithm 2 is $O(Mn^{(1+e)} + n \log(n))$ where $e < 1$ and M is the complexity of similarity computation.

Proof *Let us assume there are l buckets with $a_1, a_2, \ldots, a_l$ ($\sum_{i=1}^{l} a_i = b * n$) tweets in each one. Then there are totally $\sum_{i=1}^{l} \frac{a_i * (a_i - 1)}{2}$ similarity computations in algorithm 2.*

Algorithm 2. LSH-based Rank (*lRank*)

Input: tweets set $U = \{t_1, t_2, \ldots, t_n\}$,
     LSH table $A_1, \ldots, A_b$
Output: ranked list $RL = (t_1', t_2', \ldots, t_n')$
  1. set all $R(t_i) = 0$ $(i = 1, 2, \ldots, |U|)$;
  2. **for** $i = 1$ TO b **do**
  3.   **for each** tweets pair (t_1, t_2) in A_i **do**
  4. $r = -Sim(t_1, t_2)$;
  5.   **if** $P(t_1) > P(t_2)$ **then**
  6. $D(t_2) = \min(r, D(t_2))$;
  7.   **else**
  8. $D(t_1) = \min(r, D(t_1))$;
  9.   **end if**
 10.   **end for**
 11. **end for**
 12. sort all tweets based on $\lambda * P(t) + (1 - \lambda) * D(t)$;
 13. set RL as the sorted list;
 14. **return** RL;

On the other hand, querying the nearest neighbor for t' on the LSH table is associated with $\sum_{i=1}^{b} |g_i(t')|$ similarity computations. Therefore, the total times of similarity computations for querying all tweets' nearest neighbor is: $\sum_{j=1}^{n} \sum_{i=1}^{b} |g_i(t_j)| = \sum_{i=1}^{l} a_i * a_i$.

The time complexity of nearest neigbor query on LSH table is $O(Mn^e)$ $(e < 1)$ according to [1]. Therefore, the total complexity of similarity computation in algorithm 2 is $O(Mn^{(1+e)})$ $(e < 1)$. The sort operation has a complexity of $n \log(n)$. Totally, algorithm 2 has a complexity of $O(Mn^{(1+e)} + n \log(n))$. $\square$

Table 1. Shinglings for the Example 1

Shinglings of tweets				
tweets	t_1	t_2	t_3	t_4
shinglings	$\{a, c, d, e\}$	$\{b, c, e, f\}$	$\{a, d\}$	$\{e, f\}$
popularity	0.8	0.6	0.4	0.2
hash1	a	e	a	e
hash2	c	c	d	f

Example 1. *Considering 4 tweets (t_1, t_2, t_3, t_4) whose shinglings are given in Table 1, the LSH partition them into different buckets, $\{\{t_1, t_3\}, \{t_2, t_4\}\}$ under hash function 1, and $\{\{t_1, t_2\}, \{t_3\}, \{t_4\}\}$ under function 2. We sequentially scan all the LSH buckets. In the first bucket $\{t_1, t_3\}$, $Sim(t_1, t_3) = \frac{1}{2}$ and $P(t_1) > P(t_3)$, then $D(t_3) = \min\{0, -\frac{1}{2}\} = -\frac{1}{2}$. After processing the second bucket, we get $D(t_4) = -\frac{1}{2}$. From the third bucket, we get $D(t_2) = -\frac{1}{3}$. For the rest, there is only one tweet and no similarity computation is needed. At last, we get $D(t_1) = 0$, $D(t_2) = -\frac{1}{3}$, $D(t_3) = -\frac{1}{2}$, $D(t_4) = -\frac{1}{2}$.*

4.2   Implementation on MapReduce

When the tweets set is very large, the ranking phase would contain a large number of similarity computations even after being pruned by LSH. A single-machine-implementation is not applicable for extra large datasets and does not have scalability. Thus, we implement the ranking algorithm under the MapReduce framework, which consists of two stages, as in figure 1. The first stage scans the whole tweets set, uses LSH to partition tweets into different buckets and produces a list of similar-tid pairs. The second stage makes similarity comparisons for all tids pairs and evaluates diversity for each tweet.

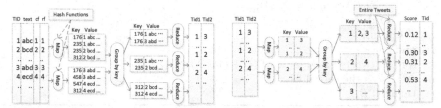

(a) Stage 1 use LSH to partition tweets into different buckets and generate tids-pair

(b) Stage 2 compute diversity for each tweet and rank them in order of an aggregation of diversity and popularity

Fig. 1. Example data flow of the algorithm on MapReduce

Stage 1: Tid-Pair Generation The first stage scans the whole tweets set and computes LSH signatures for each tweet in the mapper. For a tweet t, the mapper will generate b key-value pairs, $(g_1(t), t), (g_2(t), t), \ldots, (g_b(t), t)$, here g_i is hash functions used in LSH.

Each reducer collects tweets sharing the same hash values and then uses a nested loop approach to output tids-pairs, like (tid_1, tid_2) with $P(t_1) < P(t_2)$. Tids-pair (tid_1, tid_2) means one similarity comparison is needed in order to compute the diversity of tid_1.

Stage 2: Diverse Ranking In the second stage, we compute the diversity score of each tweet and rank them by the aggregation of popularity and diversity. Here we distribute the original tweets records to all nodes and store them in the main memory using distributed cache. Since the length of tweet is restricted in 140 characters, the whole tweet file is small and can be stored in the main memory.

In the mapper, we simply output the tid-pairs with the tid of the less popular tweet as key and the more popular one's tid as value.

Each reducer collects all other tweets ids that the one denoted by key is necessarily compared with. We make similarity computations between the tweet with tid *key* and all others tweets whose tid is in the value list. And, the diversity of tweet with tid *key* is easily evaluated by calculating the maximal similarity.

Finally, the reducer outputs $(\lambda * P(t_i) + (1 - \lambda) * D(t_i), t_i)$, which is the final ranking score for tweets.

4.3   Update Algorithm

During the lifetime of a Hot-Topic-Discussion, tweets are frequently updated(i.e. retweet, comment). Single update has micro impact on the diverse ranking order. A bundle of updates will make the ranking change a lot. Therefore we need to re-estimate the popularity and diversity periodically in order to maintain a proper diversed ranking.

If a unpopular tweet receives much attention, its popularity would increase. Thus, this would change its diverse ranking score. In such case, we should re-evaluate the diversity of other tweets. Besides, we also need to re-evaluate the diversity of other tweets according to the following constraints: (1), they were more popular but now are less than the updated one; (2), they take the updated one as a new nearest neighbor which is more popular. The second constraint limits its update-needed tweets into the LSH buckets that the updated one is associated with.

Here we introduce algorithm 3 to re-evaluate a tweet's popularity and diversity. First, we update its popularity and reset its diversity. In line 3-11, we check the relationship between t and other tweets similar with it. If t is less popular, we should update t's diversity(line 7); otherwise, update the other one's(line 9).

Algorithm 3. Update (*update*)

Input: a tweet t

  LSH hash function $(g_1, g_2, \ldots, g_b)$

Output:

1. $P_l = P(t)$;
2. update the popularity of t, $P(t) = t.rf + t.cf$;
3. $P_u = P(t)$;
4. $D(t) = 0$;
5. **for** $i = 1$ TO b **do**
6.   **for each** $t' \in A_i(g_i(t))/t$ **do**
7. $r = -Sim(t, t')$;
8.   **if** $P(t') > P_u$ **then**
9. $D(t) = \min(r, D(t))$;
10.   **else if** $P(t') > P_l$ **then**
11. $D(t') = \min(r, D(t'))$;
12.   **end if**
13.   **end for**
14. **end for**

The complexity of our update algorithm is $O(n^e)$ ($e < 1$). Both update and nearest neighbor query of an input tweet need to scan the related hash buckets. Therefore, algorithm 3 has the same time complexity with nearest neighbor query[13].

Example 2. *Reconsider the example 1. If t_2 received lots of attention and became more popular, the popularity of t_2 is updated. After then, we have $P(t_2) > P(t_1)$. The diversity is re-computed as follows: firstly, we set $D(t_2) = 0$ and pick out all buckets which contain t_2. In this example, they are $\{t_2, t_4\}$ and $\{t_1, t_2\}$. Secondly scan these buckets and compute similarities between t_2 and $\{t_1, t_4\}$.*

Here $P(t_1) < P(t_2)$ and $Sim(t_1, t_2) = \frac{1}{3}$, then $D(t_1)$ is updated to $-\frac{1}{3}$. Considering t_4, because $P(t_2)$ is always bigger than $P(t_4)$, we do not need to re-calculate $D(t_4)$. We get $D(t_1) = -\frac{1}{3}$, $D(t_2) = 0$, $D(t_3) = -\frac{1}{2}$, $D(t_4) = -\frac{1}{2}$.

In addition, we consider the problem of adding a new tweet into the collection. The minor difference between adding and updating is whether we need to insert the new tweet into the LSH tables.

In this section, we discuss the algorithm to solve the diverse ranking problem. We utilize LSH to prune the similarity comparison between the dissimilar tweet pairs. Furthermore, we discuss the MapReduce implementation of our algorithm to enhance the capability of processing large dataset.

5   Experiment

In this section, a series of experiments on real dataset are conducted to verify the effectiveness and efficiency of the proposed algorithms.

5.1   Datasets and Setup

The dataset is crawled from Sina Weibo, which contains three hot topics: Mo Yan[4], Food Safety[5], Medicine[6]. Some statistics are shown in Table 2.

Table 2. Statistical information about the experimenting Weibo datasets

category	Mo Yan	Medicine	Food Safety
Number of tweets	376346	204335	4927

Besides, we use the DBLP [7] dataset to verify the efficiency of our algorithm for the case of processing large datasets under the MapReduce framework. We extract paper records containing title, authors list and journal name.

Experiments are conducted on a desktop machine equipped with Intel(R) Core(TM) i5-3470 CPU @ 3.20GHz, 8.00GB memory, 64-bit Windows 7. We run MapReduce experiments on a 9-node cluster. Each node has four cores, 8GB of RAM. Empirically, LSH's band number is fixed at 16 with 2 min-hash functions used in each one. Totally, 32 hash functions are used in LSH to partition tweets into different buckets.

We compare three ranking methods including *sRank, mRank, lRank* and investigate these algorithms in two aspects: effectiveness and efficiency.

- sRank: the simple ranking method used in real application only based popularity.
- mRank: the diversity ranking approach based on maximal marginal relevance defined in [4].
- lRank: the ranking algorithm proposed in our paper.

---

[4] The discussion about Mo Yan received the Nobel Prize in Literature on 11 October, 2012.

[5] The discussion about the quality of Dumex milk powder.

[6] The discussion about the Poison capsule event exposed on 18 April, 2013.

[7] http://dblp.uni-trier.de/xml/dblp.xml.gz

For effectiveness, we compare the diversity of top results in each ranking list. Here a collection of tweets is considered to be more diverse if it contains more entities. Users may feel tweet set more diverse if each tweet includes more different entities. Hence, the number of entities is an indicator of its diversity. The Standford NER[8] tool is used to extract entities. We use the following metric to measure the diversity of different tweets list L.

$$div_{entity}(L) = |\bigcup_{t_i \in L} NER(t_i)| \qquad (3)$$

For efficiency, we mainly compare the time cost of each algorithm.

5.2   Efficiency

In the first experiment, we study the efficiency of lRank. Here we generate collections of size 500, 1000,..., 4500 for the Food Safety dataset; 1000, 2000, ..., 10000 for the Mo Yan and the Medicine dataset. λ is set to 0.7. It is necessary to mention that λ does not affect the efficiency of rank algorithm (except for $\lambda = 1$, which does not consider the diversity).

The following figures show the performance of each algorithms. We can see that lRank is at least one magnitude faster than mRank. With the increase of the collection size, the gap between lRank and mRank becomes larger.

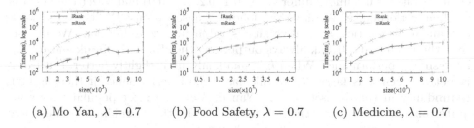

(a) Mo Yan, $\lambda = 0.7$ (b) Food Safety, $\lambda = 0.7$ (c) Medicine, $\lambda = 0.7$

Fig. 2. Time cost under different collection size

Next, we demonstrate the efficiency of update operation. To initialize update operation, tweet collections and the corresponding LSH tables are loaded in the main memory. To simulate the update scenarios, we choose a tweet and increase its popularity randomly. Figure 3(a) reveals that our algorithm has the capacity of processing updates efficiently.

To evaluate the efficiency of our algorithm on MapReduce, we vary the dataset size from 0.1 million to 1.5 million. In figure 3(b), we show the running time on different dataset size. The bar consists of Stage 1 and Stage 2 represents the running time of lRank under the MapReduce framework, while the other bar represents lRank running time on a single machine. As we can see, when the dataset size is 0.1 million, both implementations have similar performance. As the size increases to 0.5 million, running on MapReduce is much more efficient

---

[8] http://nlp.stanford.edu/software/CRF-NER.shtml

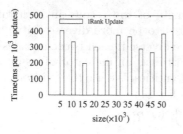   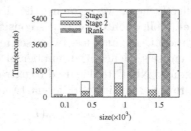

(a) capacity of update opeartion   (b) time cost of parallel algorithm

Fig. 3. Capacity to process update operation and time cost of the mapreduce algorithm

than running on a single machine. The time cost of single-machine version on processing 1(1.5) million-sized dataset is 13890s(21445s), while the MapReduce version of lRank is 6-7 times faster than the single-machine lRank.

5.3   Effectiveness

In the second experiment, we evaluate the effectiveness of our approach. Table 3 shows how many entities are there in the top-k tweets. For mRank and lRank, we evaluate the results under different λ (0.1, 0.2, . . . , 0.9) and select the best one as their performance. We have the following observations: 1) The top-k tweets contain more entities when diversity is taken into consideration. We can see that the mRank and lRank always outperform the sRank. 2) lRank and mRank have similar performance. When k is small, mRank is slightly better. But lRank is better when k is large. 3) sRank has a good performance if there is little redundancy in the dataset. In Mo Yan dataset, the most popular tweets are quite different from each other. Hence, all of the three ranking methods show the same performance among the top-20 tweets.

Table 3. The number of entities in the first 100 tweets, using bold font to highlight the best result

Mo Yan						Food Safety						Medicine				
top-k	5	10	20	60	100	top-k	5	10	20	60	100	top-k	5	10	20	60	100
sRank	35	**51**	**79**	150	201	sRank	13	24	60	111	140	sRank	18	28	41	94	147
mRank	35	**51**	**79**	**157**	229	mRank	**20**	31	61	**125**	176	mRank	**22**	**33**	54	135	202
lRank	35	**51**	**79**	156	**239**	lRank	13	**34**	**63**	121	**197**	lRank	18	30	**58**	**139**	**211**

Figure 4 shows the cumulative distribution of entities over the whole ranking list. As we can see, there are more entities contained in lRank and mRank than sRank in all the three datasets since sRank does not consider diversity. At the tail of the curves, neither mRank nor lRank increases since the last left tweets are duplicated with the ones selected before.

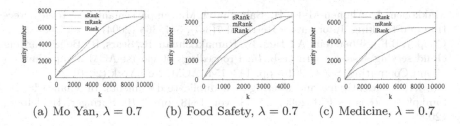

(a) Mo Yan, $\lambda = 0.7$ (b) Food Safety, $\lambda = 0.7$ (c) Medicine, $\lambda = 0.7$

Fig. 4. The cumulative number of entities in each ranking list for different datasets

6   Conclusion

With social network services, people can easily discuss hot topics with each other, which results in a large volume of tweets. In this paper, we discuss the diverse ranking for tweet set under the same hot topic. The improved ranking list is proved to be more informative. Besides, we propose a LSH-based approximate ranking algorithm, which is at least one magnitude faster than MMR. Furthermore, we implement our ranking algorithm on MapReduce framework to handle extra large datasets and enhance scalability. Experiments on real-world dataset verify the efficiency and effectiveness of our approach.

Acknowledgments. This work was supported by the 973 project(No. 2010CB32-8106), NSFC grant (No. 61170085 and 61033007), Shanghai Knowledge Service Platform Project (No. ZF1213) and the general project of Guangxi Provincial Department of Education (No. 2013YB095) .

References

1. Indyk, P., Motwani, R.: Approximate nearest neighbors: Towards removing the curse of dimensionality. In: STOC, pp. 604–613 (1998)
2. Manning, C.D., Raghavan, P., Schütze, H.: Introduction to Information Retrieval. Cambridge University Press, New York (2008)
3. Chen, H., Karger, D.R.: Less is more: probabilistic models for retrieving fewer relevant documents. In: SIGIR, pp. 429–436 (2006)
4. Carbonell, J.G., Goldstein, J.: The use of mmr, diversity-based reranking for re-ordering documents and producing summaries. In: SIGIR, pp. 335–336 (1998)
5. Wang, J., Zhu, J.: Portfolio theory of information retrieval. In: SIGIR, pp. 115–122 (2009)
6. Zuccon, G., Azzopardi, L.: Using the quantum probability ranking principle to rank interdependent documents. In: Gurrin, C., He, Y., Kazai, G., Kruschwitz, U., Little, S., Roelleke, T., Rüger, S., van Rijsbergen, K. (eds.) ECIR 2010. LNCS, vol. 5993, pp. 357–369. Springer, Heidelberg (2010)
7. Abbar, S., Amer-Yahia, S., Indyk, P., Mahabadi, S.: Real-time recommendation of diverse related articles. In: WWW, pp. 1–12 (2013)
8. Liu, J., Huang, Z., Cheng, H., Chen, Y., Shen, H.T., Zhang, Y.: Presenting diverse location views with real-time near-duplicate photo elimination. In: ICDE, pp. 505–516 (2013)

9. Makkonen, J., Ahonen-Myka, H., Salmenkivi, M.: Simple semantics in topic detection and tracking. Information Retrieval 7(3-4), 347–368 (2004)

10. Cooper, B.F., Silberstein, A., Tam, E., Ramakrishnan, R., Sears, R.: Benchmarking cloud serving systems with ycsb. In: Proceedings of the 1st ACM Symposium on Cloud Computing, SoCC 2010, pp. 143–154. ACM, New York (2010)

11. Broder, A.: Identifying and filtering near-duplicate documents. In: Giancarlo, R., Sankoff, D. (eds.) CPM 2000. LNCS, vol. 1848, pp. 1–10. Springer, Heidelberg (2000)

12. Manber, U.: Finding similar files in a large file system. In: USENIX Winter, pp. 1–10 (1994)

13. Datar, M., Immorlica, N., Indyk, P., Mirrokni, V.S.: Locality-sensitive hashing scheme based on p-stable distributions. In: Symposium on Computational Geometry, pp. 253–262 (2004)

Encoding Document Semantic into Binary Codes Space

Zheng Yu[1], Xiang Zhao[2], and Liping Wang[3]

[1] East China Normal University
zyu.0910@gmail.com
[2] National University of Defense Technology
xiangzhao@nudt.edu.cn
[3] East China Normal University
lipingwang@sei.ecnu.edu.cn

Abstract. We develop a deep neural network model to encode document semantic into compact binary codes with the elegant property that semantically similar documents have similar embedding codes. The deep learning model is constructed with three stacked auto-encoders. The input of the lowest auto-encoder is the representation of word-count vector of a document, while the learned hidden features of the deepest auto-encoder are thresholded to be binary codes to represent the document semantic. Retrieving similar document is very efficient by simply returning the documents whose codes have small Hamming distances to that of the query document. We illustrate the effectiveness of our model on two public real datasets – 20NewsGroup and Wikipedia, and the experiments demonstrate that the compact binary codes sufficiently embed the semantic of documents and bring improvement in retrieval accuracy.

1 Introduction

Most of the widely-used methods for documents retrieval utilize word-level information. A representative algorithm is TF-IDF [8,9], it directly compares the word-count vectors of documents to measure their similarity, which only considers individual weighted words; Latent Semantic Analysis (LSA) [4] evaluates the similarity based on the latent semantic between the words of the documents, which takes into consideration semantic content. PLSA [6] and LDA [3] are two probabilistic models that characterize each document with a set of topics(words). Recently, a RBMs-based model [7], which considers more semantic information, is designed to do semantic hashing on documents.

In this paper, we propose a new *deep learning* model that uses a 4-layer stacked *autoencoders* [2] with specific *non-linear* transformation functions to do semantic hashing [7] on documents. With representing documents by word-count vectors, our model takes the vectors as input and further learns compact binary codes to represent the semantic of documents. Further more, the codes learned by our model has an elegant property that semantically similar documents have similar codes, which facilitates the comparison and retrieval. We adopt pre-training and fine-tuning procedures to train our model.

F. Li et al. (Eds.): WAIM 2014, LNCS 8485, pp. 535–539, 2014.
© Springer International Publishing Switzerland 2014

The experimental results show that compared with some traditional methods, our model achieves better performance in terms of document retrieval, and the learned compact binary codes successfully encode the semantic of documents such that they can brilliantly represent the meanings of documents.

2   Model Architecture and Training Techniques

In this section, we describe our model's architecture and the techniques used to do pre-training and fine-tuning. We first preprocess all documents through removing stop words and performing word stemming, then each document is represented as a normalized 2,000-dimensional word-count vector, which is fed to our semantic hashing model.

The architecture of our stacked autoencoders model is 2,000-500-300-128. We gradually decrease the layer's dimension to make the model learn more and more abstract features from the input, and by treating the final 128 dimensional binary codes as memory addresses, we can find semantically similar documents in a time that is independent of the size of document collection [7].

In pre-training process, we greedily train each layer (except the input layer) of the stacked autoencoders model as an individual autoencoder through reconstructing its input as a target. We use sigmoid function as the non-liner activation function for the encoder and decoder parts of the first auto-encoder. The sigmoid function is a bounded differentiable function defined for all real values. It maps all input values to the interval of $(0,1)$, which is much similar to the range of our normalized input values. Thus, this function can appropriately reconstruct the input vectors.

For any node j on the hidden layer of the first autoencoder, the learned feature h_j is:

$$h_j = \alpha(b_j + \sum_k w_{jk}v_k),$$

where $\alpha(x) = 1/(1 + e^{-x})$, b_j is the bias on node j, w_{jk} is the weight between node k (on previous layer) and node j, and v_k is the output value of node k. Then for any node i out the output layer, the value computed on it is:

$$y_i = \beta(b_i + \sum_j w_{ij}h_j),$$

where β is the same sigmoid function as α. The objective of the autoencoder is to use output $\mathbf{y}$ reconstruct input $\mathbf{v}$ by minimizing the loss function J w.r.t all training data S (consists of m examples):

$$J = \frac{1}{m} \sum_{v \in S} \frac{1}{2} ||\mathbf{y} - \mathbf{v}||^2$$

We apply *stochastic gradient descent* (SGD) [1] method to update parameters (W, b) to optimize the autoencoder.

Using the same method, we further pre-train the second and third autoencoders, and this makes our model find a good region of the parameter space, starting from this region, we then fine-tune the parameters to produce a much better model [5].

We add a prediction layer on the top of the network to do further supervised fine-tuning. For the data which is labeled with category or other classification information, the predication layer aims to predicate the input data's label and we use softmax as the activation function. For any node i on the predication layer, the computed value is:

$$y_i = \frac{exp(a_i)}{\sum_{r=1}^{n} exp(a_r)}.$$

Where a is the sum of all values coming into node i. The output vector $\mathbf{y}$ can be regarded as a probability distribution over the labels w.r.t that input data. The target $\mathbf{t}$ is a binary vector, wherein only one element, which corresponds to the true category, is set to 1. We use cross-entropy error as loss function J:

$$J = -\frac{1}{m} \sum_{v \in S} \sum_{i}^{n} t_i \log y_i,$$

Using *gradient descent* method, we iteratively train the model until we find a good local minima of the loss function.

3   Experimental Results

We first do information retrieval task on the 20Newsgroups data set. Fig. 1 shows the precision-recall curves achieved by our model and the TF-IDF, LSA and RMBs-based model. We can see that, our binary codes (auto-encoder based 128d-codes) get the best accuracy at all recall levels. While LSA 128d-codes get the worst result, for RBMs based method and TF-IDF method, they perform better than LSA and achieve similar retrieval accuracy.

We visualize the 128-dimensional codes of all documents from 6 categories obtained by our model (AE-based model) and the RMBs-based model. Fig. 2 and Fig. 3 show that both codes preserve the category structure of documents, but obviously, our learned codes are much better in capturing the structure information so that it significantly cluster the documents into different categories.

We then encode 0.54M Wikipedia articles into the binary codes space. Due to the lack of specific label information of Wikipedia articles, we use the prediction layer to reconstruct the input vector as the objective. Table 1 shows the 5 nearest neighbors of 5 randomly chosen queries (i.e. Wikipedia items). The semantic of the neighbor items are very closely related to the queries' semantic, and this illustrates that our binary codes indeed capture the semantic information of Wikipedia articles and facilitate the fast retrieval of similar Wikipedia items.

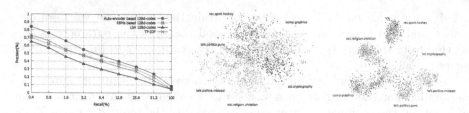

Fig. 1. precision-recall curve   **Fig. 2.** RMBs-based 2D codes   **Fig. 3.** AE-based 2D codes

Table 1. Top-5 nearest neighbors of 5 random Wikipedia items

China	Microsoft	Chinese language	robin	strawberry
India	Adobe systems	Standard Chinese	cardinal	grape
Russia	BearingPoint	Written Hokkien	finch	peach
Brazil	Commodore International	Taiwanese Hokkien	bald eagle	raspberry
Japan	Compaq	Mandarin Chinese	heron	blueberry
Thailand	Netscape	Four Corner Method	cormorant	banana

4   Conclusion and Future Work

We proposed a new deep learning model using stacked autoencoders to encode document semantic into binary codes space, where the codes have a property that semantically similar documents have similar codes, and thus, we can simply measure the similarity according to the Hamming distance between two documents' codes. Future work would analyze this model on short text understanding and also concentrate on the mechanism to enrich short text using a probabilistic knowledge, known as Probase [10].

Acknowledgement. The first author is supported by NSFC61232006.

References

1. Anderson, J.A., Davis, J.: An introduction to neural networks. MIT Press (1995)
2. Bengio, Y.: Learning deep architectures for AI. Foundations and Trends in Machine Learning 2(1), 1–127 (2009)
3. Blei, D.M., Ng, A.Y., Jordan, M.I., Lafferty, J.: Latent dirichlet allocation. Journal of Machine Learning Research 3 (2003)
4. Deerwester, S., Dumais, S.T., Furnas, G.W., Landauer, T.K., Harshman, R.: Indexing by latent semantic analysis. Journal of the American Society for Information Science 41(6), 391–407 (1990)
5. Hinton, G.E., Salakhutdinov, R.: Reducing the dimensionality of data with neural networks. Science 313, 504–507 (2006)
6. Hofmann, T.: Probabilistic latent semantic analysis. In: Proc. of Uncertainty in Artificial Intelligence, UAI 1999, pp. 289–296 (1999)

7. Salakhutdinov, R., Hinton, G.E.: Semantic hashing. Int. J. Approx. Reasoning 50(7), 969–978 (2009)
8. Salton, G.: Developments in automatic text retrieval. Science 253(5023), 974–980 (1991)
9. Salton, G., Buckley, C.: Term-weighting approaches in automatic text retrieval. In: Information Processing and Management, pp. 513–523 (1988)
10. Wu, W., Li, H., Wang, H., Zhu, K.Q.: Probase: a probabilistic taxonomy for text understanding. In: SIGMOD Conference, pp. 481–492 (2012)

LSG: A Unified Multi-dimensional Latent Semantic Graph for Personal Information Retrieval

Yang Huangfu[1,2], Kuien Liu[1], Wen Zhang[1,3], Peng Zhou[1],
Yanjun Wu[1], Qing Wang[1], and Jia Zhu[4]

[1]Institute of Software, Chinese Academy of Sciences, Beijing, 100190, China
[2]University of Chinese Academy of Sciences, Beijing, 100190, China
[3]State Key Laboratory of Software Engineering of Wuhan University, Wuhan, 430072, China
{huangfuyang,kuien,zhangwen,zhoupeng}@nfs.iscas.ac.cn,
yanjun@iscas.ac.cn, wq@itechs.iscas.ac.cn
[4]School of Computer Science, South China Normal University, Guangzhou, 510631, China
Jia@intelligentforecast.com

Abstract. Traditional desktop search engines can merely support keyword-based search as they don't utilize any other information, such as contextual/semantic information, which has been commonly used in internet search. We observe that *a user usually operates some files to complete a task related to a certain topic and organizes these files in some directories.* Inspired by the observation, we propose an approach that considers three relations among personal files to improve desktop search, namely Topic, Task and Location. Each relation is derived from topics of files, user activities log and hierarchy of file system respectively. The heart of our approach is Latent Semantic Graph (LSG), which can measure the three relations with associated score. Based on LSG, we develop a personalized ranking schema to improve traditional keyword-based desktop search and design a novel recommendation algorithm to expand search results semantically. Experiments reveal that the performance of proposed approach is superior to that of traditional keyword-based desktop search.

Keywords: Latent Semantic Discovery, Graph Model, Information Retrieval.

1 Introduction

Personal Information Retrieval, also known as Desktop Search, aims to search personal data stored in the local disks. In recent years, with the explosion of personal data, desktop search has become a hot topic. In order to pinpoint the resources (files), rich meaningful information is needed to be introduced to retrieval model. In Web search, linkage structure has been extensively studied to improve search performance, such as PageRank and HITS. Nevertheless, there is no direct and explicit association structure in local disks. Intuitively, it seems that personal resources are independent with each other. In fact, implicit associations among personal resources exist extensively. These associations can be further used to improve traditional keyword-based search. We observe that users usually operate PC in a common pattern: Operating

F. Li et al. (Eds.): WAIM 2014, LNCS 8485, pp. 540–552, 2014.

some resources to finish a specific task related to a certain topic, and organizing these resources in some directories. For instance, in order to write a paper about desktop search, I read the references stored in D:/research/literature, write and store this ".docx" in D:/research/paper. This observation inspires us that *topic, user behaviors* and *directory structure* are quite useful information for locating resources. We also carried out a well-designed user investigation of 20 skilled researchers. They were asked to select 4 most useful information items for locating local resources in their mind. Fig.1 illustrates the investigation result that strongly supports our observation. Motivated by the upper facts, we propose an approach that exploiting the three kinds of information to improve traditional desktop search. As shown in Fig.2, we denote the three implicit information as {Task, Topic, Location} Relations respectively. The heart of our approach is Latent Semantic Graph (LSG), which is used to measure the three relations with associated score. Based on LSG, we develop a personalized rank schema to improve tradition keyword-based desktop search and design a creative semantic recommendation algorithm to expand the query results.

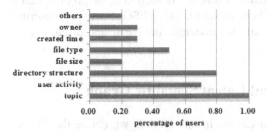
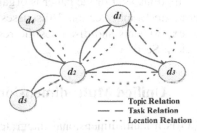

Fig. 1. Result of User Investigation. The user ratio selecting each information item.

Fig. 2. Overview of LSG with {Topic, Task, Location} Relations between Resources {d_1, d_2, d_3, d_4, d_5}

Related Work. A number of researches have tried to introduce extra information into Personal Information Retrieval. Chau et al. [8] exploited the explicit metadata to build the contextual cues and designed navigation-style search. They didn't build any ranking algorithms that exploit these links. Other researchers [2, 9] proposed methods to discover the contextual information from file system hierarchy. Peney et al. [9] mapped the authority and hub nodes to resources and directories respectively and then ranked results with authority score. However, this method can't distinguish resources in the same directory. Recently, many researches [7, 10] put their views on user behaviors. Chen et al. [7] built different relations with respect to different user behaviors and assigned the relations of different type with fixed weights. In this paper, we design novel algorithms to measure the relations. Sawyer et al. [3] applied physical-social interactions to build resource associations, but their approach was limited to emails context. Deng et al. [4] built contextual hierarchies for explicit metadata, but they ignored the implicit information like topic, user behavior pattern etc. In summary, none of above work sufficiently utilized the 3 relations mentioned and presented proper methods to measure these relations. Kim et al. [5] combined many similarity metrics to create association, which is close to our work. However, their approach was strongly dependent on resource tags contributed by users actively. Our approach can measure implicit relations without extra user involvement.

Contribution. The main contribution of this paper includes:

- We design novel methods to measure the {Topic, Task, Location} Relations and propose LSG to integrate them into a unified score. We measure the Topic Relation using a LDA-based method, Task Relation using a Surfing Graph and Location Relation from three aspects as Depth, Transfer Length and Depth Difference of resource organization.
- Based on LSG, we design a new personalized ranking schema, which can reflect user preference and improve keyword-based desktop search results.
- Based on LSG, we design a graph-based recommendation algorithm for query result expansion. We recommend 5 relevant documents for each query result to help user recover memory cues and facilitate search.

The remainder of the paper is organized as follows. In section 2, we propose Latent Semantic Graph. Section 3 discusses how to search using LSG. Section 4 describes the experiments and discusses the results. We conclude the paper and discuss future work in Section 5.

2   Unified Multi-dimensional Latent Semantic Graph

LSG is a multi-dimensional integrated Graph. In this section, we define the {Topic, Task, Location} Relations and present methods to score these relations to propose LSG.

- *Topic Relation* is used to measure the topic similarity between two resources (e.g., d_1, d_2). We denote this relation as $Score_{topic}(d_1, d_2)$.
- *Task Relation* is used to measure the collaborative strength between two resources in history. We denote this relation as $Score_{task}(d_1, d_2)$.
- *Location Relation* is used to measure the relation hidden in the two resources' directory organization. We denote this relation as $Score_{location}(d_1, d_2)$.

2.1   Scoring Topic Relation

In our context, the size of personal resources is varying and large, which results in that term vectors are sparse and high-dimensional. Therefore, cosine similarity between term vectors is not a good way to measure the similarity between resources. In this paper we employ a LDA-based method to measure the similarity of two extractable resources. LDA is a hierarchical Bayesian model which allows us to model a text document as a mixture of topics. After LDA modeling, each resource can be mapped to a latent topic distribution of N dimensions (100 by default). To measure the topic similarity between two resources, we calculate the similarity between the distributions of topics associated with each document, which can effectively resolve the problem due to data sparsity. We use the KL divergence [1], a non-symmetric measure of the difference between two probability distributions X and Y, to measure the distance between two resources. Given distributions X and Y, the KL divergence between them is formalized as: $D_{kl}(X||Y) = \sum_{n=1}^{N} p(x = n) \log \frac{p(x=n)}{p(y=n)}$.

Considering the non-symmetry of KL divergence, namely $D_{kl}(X||Y) \neq D_{kl}(Y||X)$, we use Jensen-Shannon divergence, a symmetric variant of the KL divergence , instead in this paper: $D_{js}(X||Y) = [D_{kl}(X||M) + D_{kl}(Y||M)]/2$, Where $M = \frac{1}{2}(X + Y)$. Smaller JS value means the two resources are more relevant. We use formula 1 to model the Topic Relation between two resources d_1 and d_2 and obtain a normalized score. The obtained JS value can be viewed as the distance between two resources. The curve of formula 1 is steep near 0 and become flat with the growth of JS value. This characteristic ensures that we can easily distinguish the low similar pairs from the high similar ones.

$$Score_{topic}(d_1, d_2) = 1/e^{D_{js}(d_1||d_2)} \tag{1}$$

2.2 Scoring Task Relation

In order to discover the collaborative relation between different resources, we employ a system-level API to monitor and record resource access activities. The recorded information about an access activity is a 4-tuple {*Name, Start, End, Duration*} (see Fig.3(a)). Here, we define a resource being accessed once the resource gets focus. Users tend to access different resources to complete a purpose. Therefore, we define a *Task* as a set of resources for the same purpose. The order of access log in a task reflects the surfing traces of users and collaborative relationships between adjacent resources. Before introducing the task identification algorithm, we first define the concepts used in the algorithm:

Valid Duration (s): The duration of a log must be larger than a given Valid Duration threshold (10s by default). This setting is based on a common assumption that users usually focus on a resource for a relatively long time while finishing a task. The invalid one will be remove from the sequence and split the sequence naturally.
Task Interval (s): A new task starts while the interval between the adjacent logs is larger than a given Task Interval threshold (600s by default).
Task Similarity: We merge adjacent tasks T1 and T2 together and view them as the same task if the similarity between T1 and T2 is larger than a given threshold (0.5 by default). We define the task similarity as

$$\text{Sim}(T1, T2) = \frac{|Number\ of\ common\ resources\ in\ two\ task|}{|Total\ number\ of\ resources\ in\ the\ shorter\ task|} \tag{2}$$

Surfing Graph (SG): We propose a Surfing Graph (SG) to denote each task identi-fied. A surfing graph G is an undirected graph. The node set V_G corresponds to all resources in the task. There is an edge $e(u, v) \in E_G$ for $u, v \in V_G$ iff there is an *access transfer from u to adjacent v. The weight of edge is the transfer frequency happened on this edge regardless of direction.*

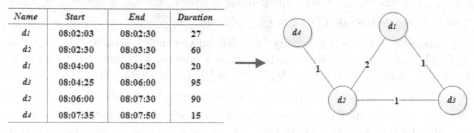

Name	Start	End	Duration
d_1	08:02:03	08:02:30	27
d_2	08:02:30	08:03:30	60
d_1	08:04:00	08:04:20	20
d_3	08:04:25	08:06:00	95
d_2	08:06:00	08:07:30	90
d_4	08:07:35	08:07:50	15

(a) An Example of Task including 4 resources {d_1, d_2, d_3, d_4}   (b) Surfing Graph

Fig. 3. An Example of Surfing Graph

Taking Fig.3 for example, we have a task of 4 resources (e.g., d_1, d_2, d_3, d_4). It is obvious that the surfing trace in this task is $d_1 \rightarrow d_2 \rightarrow d_1 \rightarrow d_3 \rightarrow d_2 \rightarrow d_4$. Thus, the transfer pairs are {$(d_1, d_2), (d_2, d_1), (d_1, d_3), (d_3, d_2), (d_2, d_4)$}. Each pair corresponds to an edge in Surfing Graph. (d_1, d_2) and (d_2, d_1) correspond to the same edge. Therefore, the weight of each edge is measured by the number of transfer pairs corresponding to it.

Algorithm 1. Construction of Task Relation Graph

Input: A user activity log sequence R, Valid Duration VD, Task Interval TI, Task Similarity TS

Output: Task Relation Graph $TARG(V, E)$, where V is vertex set and E is edge set.
1:   $V \leftarrow \emptyset, E \leftarrow \emptyset, TaskList_{new} \leftarrow \emptyset$
2:   $TaskList_{raw} \leftarrow$ **rawTaskList**(R, VD, TI) //split R according to VD and TI
3:   $t \leftarrow TaskList_{raw}[0]$
4:   **for**$(i = 1; i < TaskList_{raw}.size; i++)$ **do**
5:   $t' \leftarrow TaskList_{raw}[i]$
6:   **if** t', t are splited by VD && $sim(t', t) > TS$ **then**   //merge similar raw tasks
7:     $t \leftarrow t + t'$
8:   **else**   add t to $TaskList_{new}$ if $t.size > 2, t \leftarrow t'$ //remove invalid task
9:   **foreach** $t \in TaskList_{new}$ **do**
10:   $SG(V_{sg}, E_{sg}) \leftarrow$ **generateSurfingGraph**(t)
11:   $V \leftarrow V \cup V_{sg}, E \leftarrow E \cup E_{sg}$
12:   **return** $TARG(V, E)$

Algorithm 1 describes the process of task identification (step 1 to 8) and SG merging (step 9 to 12). The identification consists of 3 major steps: (1) split the accessing sequence into clusters according to *Valid Duration* and *Task Interval*; (2) merge similar adjacent clusters (raw tasks) split by *Valid Duration*; (3) remove invalid tasks (size < 3). After identification, we convert all identified tasks to Surfing Graphs and then combine them together as a Task Relation Graph (TARG) where weight of the same edges appearing in the different SG will be added. The TARG reflects the Task Relation between different resources through the weight of edge connecting them. In this paper, the weight of edge is defined as **Transfer Frequency (TF)**. Given the TF of two resources d_1 and d_2, we use Eq.(3) to model Task Relation.

For a certain resource X, we are intent to distinguish those low TF resources interacting with it. A small increment on low TF will result in a big difference on task relation strength. When the TF is high enough, the difference on relation strength with the increment of TF will become small. So those high TF resources can be nearly viewed as the same category, namely highly relevant ones to X. In our model, we use TF^2 to enlarge the difference between small TFs and *log* to shrink them towards steep interval. Therefore, the model can reveal the difference between small TFs.

$$Score_{task}(d_1, d_2) = \frac{\log_4(\text{TF}(d_1,d_2)^2+1)}{1+\log_4(\text{TF}(d_1,d_2)^2+1)} \tag{3}$$

2.3   Scoring Location Relation

Empirically, users tend to arrange their information according to a certain implicit relationship among them, such as related content, same target, etc. In this section, we model the Location Relation between resources from 3 aspects mentioned in [12].

Aspect (a): *Depth*. Users usually organize resources into subdirectories. The resources organized in the subdirectories have closer relationship among each other than those in the parent directories. Recursively, the deeper the two resources stored in the hierarchy of file system, the tighter the relationship between them is. We use the average depth of two resources to denote this aspect: $E(d_1, d_2) = (|d_1| + |d_2|)/2$, where |n| is depth of n.

Aspect (b): *Transfer Length*. Users usually group related resources together. In file system, every resource has a route to every other, which is equivalent to the route between the directories containing the two resources. We refer to the length of the shortest path between two directories as their degree-of-association or transfer-length. Shorter transfer-length means two resources are more related. Thus, if transfer-length is 0, that means the two resources are organized in the same directory. We denote this aspect as equation: $D(d_1, d_2) = (|d_1| - |\lambda|) + (|d_2| - |\lambda|)$, where λ is their lowest common ancestor-or-self directory.

Aspect (c): *Depth Difference*. Users usually group related directories nearby, which means the depths of these directories are close. Under prerequisite of large *Depth* and short *Transfer Length*, if the depths of directories containing the given two resources are closer, the two resources are more relevant. We denote this aspect as equation: $M(d_1, d_2) = ||d_1| - |d_2||$.

By integrating the above aspects, we now obtain the following formula which models Location Relation of 2 resources and returns a normalized score. According to the importance of each aspect, the function penalizes *Transfer Length* and *Depth Difference* harshly and *Depth* leniently. In fact, users won't organize their **valuable** resources at deep level, generally 4-5 level at most. So the influence of *Depth* is limited.

$$Score_{location}(d_1, d_2) = \frac{\sqrt{E(d_1,d_2)}}{1+D(d_1,d_2)^2+M(d_1,d_2)^2+\sqrt{E(d_1,d_2)}}$$ (4)

2.4   Score Aggregation with Linear Regression

After measuring the three relations, we can obtain 3 Relation Graphs. We name them as Topic Relation Graph, Task Relation Graph and Location Relation Graph. The final **Latent Semantic Graph (LSG)** we proposed is a mixture graph to integrate the above 3 Relation Graphs. In this paper, the final association score, namely, the weight of edge in LSG is simply formulized as a linear combination:

$$Score_{LSG}(d_1, d_2)$$

$$= \rho Score_{topic}(d_1, d_2) + \varphi Score_{task}(d_1, d_2) + \omega Score_{location}(d_1, d_2)$$ (5)

Therefore, the parameters ρ, φ, ω can be estimated by a linear regression. We conduct a training set including K (K=200, 40 pairs per person) resource pairs $<d_1, d_2>$ selected from data set provided by participants of our experiments. We ask participants to rate those pairs selected from their own data set with a grade between 0 and 1. Suppose the training set is T = $\{(x_1,y_1),(x_2,y_2),\cdots,(x_K,y_K)\}$, where x_i is triple including 3 relation scores and y_i is relation score rated by user. So we estimate the parameters by solve the following optimization problem.

$$\min_{W} \frac{1}{2}\sum\nolimits_{i=1}^{K}(y_i-w^Tx_i)^2, \quad W = (\rho,\varphi,\omega), \text{ s.t. } \rho + \varphi + \omega = 1$$ (6)

With the limitation of space, we omit some detail about learning process. After a 5-fold cross validation, we average the parameters of 5 models and obtain the approximate parameters as (0.3, 0.6, 0.1) in our context. With the estimated parameters, we combine the three relation graphs and generate the Latent Semantic Graph (LSG).

3   Searching with LSG

When user submits keywords, the content-based search module firstly finds results from index and then feeds them to LSG module. LSG module resorts these results with a personalized ranking schema and gets recommended resources for each result.

Fig. 4. Process of Searching with LSG

Personalized Ranking. In case of Web search, personalized PageRank has been widely used. The original PageRank can be expressed as the solution: $\vec{R}_k = d * M^T \times \vec{R}_{k-1} + (1-d) * \vec{E}$. Here, we design a LSG-based Personalized Ranking Schema to rank the results by rebuilding the Transition Matrix M and personalization vector $\vec{E}$. Generally, user would like to jump to the node which is highly relevant to the current one. Let N_j be the set of resources which the resource j links to in LSG. Then when computing the transfer probability from j to any one in N_j, e.g. resource i, we use the following formula:

$$P(j \rightarrow i) = rM_{ji} = \frac{Score_{LSG}(j,i)}{\sum_{k \in N_j} Score_{LSG}(j,k)} \tag{7}$$

To rebuild $\vec{E}$, let L be the set of resources recorded in user activity log, o_i be the number of occurrences of resource i in the user activity log and N_{index} be the number of indexed resources. If resource i belongs to L, e.g., $o_i \neq 0$, we set $e_i = o_i/\sum_{j \subset L} o_j$, otherwise, $e_i = 1/N_{index}$. The final query score is a combination of TF-IDF score and personalized ranking score. Suppose q is the given query and p is the resource in our disk, then final ranking score: $Score_q(p) = Score_{cosine}(q,p) * Score_{PR}(p)$, where $Score_{cosine}(q,p)$ is cosine similarity between q and p, $Score_{PR}(p)$ is the importance score of p calculated by personalized ranking schema.

Semantic Recommending. Based on LSG, graph-based recommendation methods can be introduced to compute the resource association from a global perspective instead of local pairwise computation of neighborhood. We transplant a graph-based recommendation algorithm IPF [6] into our scenario to recommend semantically-relevant resources for each query result. We name the transplanted IPF as *TIPF* in this paper. Given a query result u and a unknown resource i, there are many propagation paths from u to i in LSG. For u, the recommending score of i is sum of weights of all paths. The path weight can be viewed as the visited probability of i from the resource u. Suppose $P\{d_0, d_1, \ldots, d_n\}$ is the path from the query result u ($d_0 = u$) to resource i ($d_n = i$) , the path weight is defined as: $\psi(p) = \prod_{k=0}^{n-1} P(d_k \rightarrow d_{k+1})$, where $P(d_k \rightarrow d_{k+1})$ is a propagation function defined in Eq.(7). Similarly with IPF, we only consider the short paths (distance < 3) to measure the visited probability because the long path contributes little and is prone to bring in noise. Consequently, suppose $\Gamma(u,i)$ is the set of short paths from u to i, then the recommending score of i to u is defined as:

$$Score_{recommend}(u,i) = \sum_{p \in \Gamma(u,i)} \psi(p) \tag{8}$$

The *TIPF* is implemented by Bread-First-Search on LSG. Finally, we sort the candidate resources according to $Score_{recommend}(u,i)$ and then return the top 5 resources to u. Fig. 8 illustrates an example of semantic recommendation.

4   Experiments

In order to evaluate the effectiveness of LSG in enhancing result ranking and semantic recommendation for query result expansion, we develop a prototype, called IDSE[1],
to implement the searching process with LSG and construct a small scale user study.
Because the experiment on personal data is sensitive, the scale of experiment is hard
to be large. Referring to the experimental scale mentioned in references (e.g, 3 in [5],
5 in [7] and 6 in [10] etc.), we invited five **volunteers** in our institution to participate
in our experiments.

Table 1. An Overview of Data set

User	U1	U2	U3	U4	U5
Data Size (GB)	10.27	3.08	2.38	15.26	0.81
Accessed Resources	926	830	193	508	234
Extractable Resources	2059	4765	1817	3311	349

Data Set. To gather the activities log, we trace participants' behaviors on resources
for two months. The average data set contains *40346* resources in *6783* directories.
The average directory depth is *10* (max *23*). In addition, there are totally *12301* ex-
tractable resources and *2691* resources are recorded in the user log.

Metrics. We take the traditional metrics $\text{Precision} = \frac{|\{retrieved\ relevant\ documents\}|}{|\{retrieved\ documents\}|}$
and $\text{Recall} = \frac{|\{retrieved\ relevant\ documents\}|}{|\{relevant\ documents\}|}$ to evaluate the IR performance.

4.1   Experimental Results

Ranking Performance Evaluation

Comparison. We compare IDSE with two of the state-of-art [11] Desktop Search
tools: Copernic Desktop Search [13] and Google Desktop Search (GDS) [14]. In addition, we provide two versions of prototype: IDSE-based (content-based ranking only)
and IDSE_Imecho (a similar method in [7], which uses fixed relation weight).
Setup. Each participant is asked to design 10 search queries related to their activities
and then send each of queries to the 5 tools respectively. For each query, each participants rate the top 10 results for each tools using grades $\{0, 1, 2, 3, 4, 5\}^2$ where 0 for
an irrelevant result and 5 for a highly relevant one. At each rank, the average precision and recall can be calculated by using the grades rated by our participants. For
each tool, we can calculate the average precision and recall of 50 queries at any rank
k. We use the pooling technology [7] to generate the relevant resources set.

Fig. 5(a) and Fig. 5(b) depict the average precision and recall levels of the 5 tools
at each rank from one to ten respectively. Intuitively, the prototype IDSE with LSG

---

[1]   We share the project of IDSE at GitHub: https://github.com/HarryHuang1990/
idse.git
[2]   The relevant grade will be normalized to [0, 1] when calculating the measure.

outperforms others on both metrics at every rank level. This indicates that IDSE finds more relevant resources and ranks them higher than content-based search tools (GDS, Copernic and IDSE-based). The interpretation of this result is that the GDS, Copernic and IDSE-based are only concerned about hitting key-words and rank their results using TF-IDF models, regardless of any global importance measures for resources. Whereas with semantic links modeled in LSG, IDSE can exactly measure importance for every resource and push resources of interest towards the top of list.

IDSE also outperforms IDSE_Imecho on both metrics. This indicates that exact measurement of relation is better for desktop search than using fixed weight and our approach is effective. IDSE-based is implemented by using an optimized TF-IDF model. Therefore it can find more relevant resources than GDS and Copernic. But it may not rank the high relevant resources towards the top and cause a flat precision curve from top 6 to 10 (depicted in Fig. 5(a)). After adding the LSG, those high relevant but low ranked resources will be ranked high. This results in a visible improvement in terms of precision and recall.

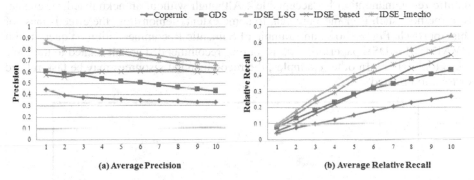

(a) Average Precision   (b) Average Relative Recall

Fig. 5. Comparison of Average Precision and Average Relative Recall from Top 1 to 10

Effectiveness of Semantic Recommendation

Recommendation for each query result is a creative attempt. It can help user to recall the memory cues among resources. The performance of the recommendation mainly relies on users' judgment. We ask each volunteer to rate the each recommended resource for top 5 results returned by IDSE using the same grade method as above. The relevant grade reflects the user satisfaction to the recommended one. Finally, we average the all satisfaction grades in 10 queries to measure a user's satisfaction to our method.

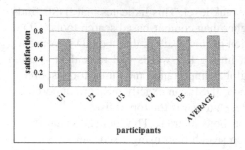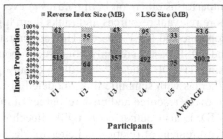

Fig. 6. Average User Satisfaction to Semantic Recommendation for Query Result Expansion.

Fig. 7. Performance on Indexing Space

Fig. 6 illustrates the average assessment result with a five score system where [0, 0.2] for "Very Poor", (0.2, 0.4] for "Poor", (0.4, 0.6] for "Not too Poor", (0.6, 0.8] for "Acceptable", and (0.8, 1] for "Perfect". As we see, all participants feel that our semantic recommendation is "acceptable". Although without an acknowledged evaluation method, such user measure can, to some extent, demonstrate the effectiveness of this approach. Fig. 8 shows an example of Semantic Recommendation on query result "WAIM2014-IDSE paper.docx". 5 resources recommended are all related to the topic "desktop search". In other examples, the recommended resources may be task-related or location-related.

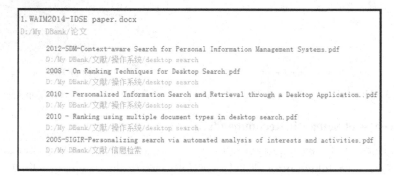

Fig. 8. An Example of Semantic Recommendation for Query Result Expansion

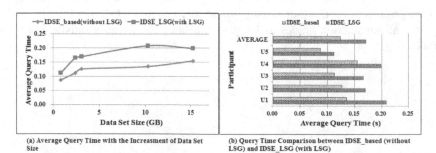

(a) Average Query Time with the Increasement of Data Set Size

(b) Query Time Comparison between IDSE_based (without LSG) and IDSE_LSG (with LSG)

Fig. 9. Performance on Query Time

Scalability

Querying Performance. In Fig. 9(b), the average query time of IDSE across all queries (50 queries) is 0.17s, which is much less than one second. Query time averages are calculated across 5 runs of each query. Fig. 9(a) shows that with the data set size increasing, the average query time appears to grow slowly, especially on big data sets. Although, the average time of search with LSG is a little longer than content-search alone (0.12 on average), it is acceptable. Even though compared with GDS, the participants agree that the performance is also acceptable.

Indexing Space. The IDSE's index consists of reverse index and LSG. Fig. 7 shows the space required by the LSG and reverse index for each data set and on average. This result indicates that the LSG is not a barrier to search. On average, the LSG size is less than 0.85% of the size of the user's data size.

5   Conclusion and Future Work

In this paper, we attempt to exploit {Topic, Task, Location} Relations to improve traditional desktop search performance. We design novel methods to measure these 3 relations properly using information derived from resource latent topics, user access activities and directory hierarchy respectively and then propose LSG, a multi-dimension integrated graph, to integrate the 3 relations using a linear combination. With LSG, a personalized ranking schema is designed to enhance the full-text keyword desktop search and the recommending algorithm TIPF is adapted to help user extend and associate the results returned by the keyword search. Experiments reveal the prototype embedding LSG is superior to the traditional keyword-search desktop search tools. In future work, we are intent to explore Web-behavior-based topic semantic identification for those unextractable resources.

Acknowledgements. We would like to thank Dr. Xianpei Han for providing valuable suggestions. This work is supported by the NSF of China (Nos. 71101138, 61202064 and 91324008), Beijing NSF (No. 4122087) and the Strategic Priority Research Program of the Chinese Academy of Sciences (No. XDA06010600).

References

1. Cover, T., Thomas, J.: Elements of information theory. John Wiley & Sons (2012)
2. Wang, W., Peery, C., Marian, A., Nguyen, T.: Efficient Multidimensional Fuzzy Search for Personal Information Management Systems. TKDE 24(9), 1584–1597 (2012)
3. Sawyer, B., Quek, F., Wong, W.C.: Using physical-social interactions to support information re-finding. In: CHI, pp. 885–910 (2012)
4. Deng, T., Zhao, L., Feng, L., Xue, W.: Information re-finding by context: a brain memory inspired approach. In: CIKM, pp. 1553–1558 (2011)
5. Kim, J., Croft, W.B., Smith, D., Bakalov, A.: Evaluating an associative browsing model for personal information. In: CIKM, pp. 647–652 (2011)
6. Xiang, L., Yuan, Q., Zhao, S., Chen, L.: Temporal recommendation on graphs via long- and short-term preference fusion. In: SIGKDD, pp. 723–732 (2010)

7. Chen, J., Guo, H., Wu, W., Wang, W.: iMecho: an associative memory based desktop search system. In: CIKM, pp. 731–740 (2009)
8. Chau, D., Myers, B., Faulring, A.: What to do when search fails: finding information by association. In: CHI, pp. 999–1008 (2008)
9. Penev, A., Gebski, M., Wong, R.K.: Topic distillation in desktop search. In: Bressan, S., Küng, J., Wagner, R. (eds.) DEXA 2006. LNCS, vol. 4080, pp. 478–488. Springer, Heidelberg (2006)
10. Soules, C., Ganger, G.: Connections: using context to enhance file search. SOSP 39(5), 119–132 (2005)
11. Noda, T., Helwig, S.: Benchmark study of desktop search tools.University of Wisconsin-Madison E-Business Consortium (2005)
12. Ravasio, P., Schär, S.G., Krueger, H.: In pursuit of desktop evolution: User problems and practices with modern desktop systems. TOCHI 11(2), 156–180 (2004)
13. Copernic desktop search, http://www.copernic.com/en/products/desktop-search/
14. Google desktop search, http://desktop.google.com/

HRank: A Path Based Ranking Method in Heterogeneous Information Network

Yitong Li[1], Chuan Shi[1,*], Philip S. Yu[2], and Qing Chen[3]

[1] Beijing University of Posts and Telecommunications, Beijing, China 100876
[2] University of Illinois at Chicago, IL, USA
[3] China Mobile Communications Corporation, Beijing, China

Abstract. Recently, there is a surge of interests on heterogeneous information network analysis. Although evaluating the importance of objects has been well studied in homogeneous networks, it is not yet exploited in heterogeneous networks. In this paper, we study the ranking problem in heterogeneous networks and propose the HRank method to evaluate the importance of multiple types of objects and meta paths. A constrained meta path is proposed to subtly capture the rich semantics in heterogeneous networks. Since the importance of objects depends upon the meta paths in heterogeneous networks, HRank develops a path based random walk process. Furthermore, HRank can simultaneously determine the importance of objects and meta paths through applying the tensor analysis. Experiments on three real datasets show that HRank can effectively evaluate the importance of objects and paths together. Moreover, the constrained meta path shows its potential on mining subtle semantics by obtaining more accurate ranking results.

Keywords: Heterogeneous information network, Rank, Random walk, Tensor analysis.

1 Introduction

It is an important research problem to evaluate object importance or popularity, which can be used in many data mining tasks. Many methods have been developed to evaluate object importance, such as PageRank [9], HITS [1], and SimRank [3]. In these literatures, objects ranking is done in a homogeneous network in which objects or relations are same-typed. However, in many real network data, there are many different types of objects and relations, which can be organized as heterogeneous network. Formally, Heterogeneous Information Networks (HIN) are the logical networks involving multiple types of objects as well as multiple types of links denoting different relations [2]. It is clear that heterogeneous information networks are ubiquitous and form a critical component of modern information infrastructure [2].

Fig. 1(a) shows a HIN example in bibliographic data and Fig. 1(b) illustrates its network schema. In this example, it contains four types of objects: papers (P), authors (A), labels (L, categories of papers) and conferences (C), and links

---

* Corresponding author.

F. Li et al. (Eds.): WAIM 2014, LNCS 8485, pp. 553–565, 2014.
© Springer International Publishing Switzerland 2014

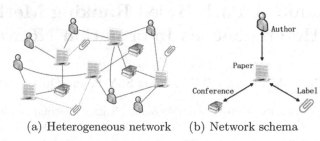

(a) Heterogeneous network   (b) Network schema

Fig. 1. A heterogeneous information network example on bibliographic data. (a) shows heterogeneous objects and their relations. (b) shows the network schema.

connecting them. The link types are defined by the relations between two object types. For example, links exist between authors and papers denoting the writing or written-by relations. In this network, several interesting, yet seldom exploited, ranking problems can be proposed.

- One may pay attention to the importance of multiple types of objects simultaneously, and ask the following questions:

  Q. 1.1 Which are the most influential authors and reputable conferences?

  Q. 1.2 Which are the most influential authors and reputable conferences in data mining field?
- Furthermore, one may wonder which factor mostly affects the importance of objects. So he may ask the questions like this:

  Q. 2 Who are the most influential authors and which factor makes the author most influential?

The ranking analysis in HIN faces the following research challenges. (1) There are different types of objects and links in HIN. If we simply treat all objects equally and apply the random walk as PageRank does in homogeneous network, the ranking result will mix different types of objects together. (2) Different types of objects and links in heterogeneous networks carry different semantic meanings. The random walk along different meta paths has different semantics, which may lead to different ranking results. Here the meta path [4] means a sequence of relations between object types. So a desirable ranking method in HIN should be path-dependent. The study of related work can be seen on [10], which is the extension of this paper.

In this paper, we study the ranking problem in HIN and propose a novel ranking method, HRank, to evaluate the importance of multiple types of objects and meta paths in HIN. For *Q. 1*, a meta path based random walk model is proposed to evaluate the importance of single or multiple types of objects. Although meta path has been widely used to capture the semantics in HIN [6,4], it coarsely depicts object relations. By employing the meta path, we can only answer the *Q. 1.1*. In order to overcome the shortcoming existing in meta path, we propose the *constrained meta path* concept, which can effectively describe subtle semantics. The constrained meta path sets constraint conditions on meta path. Adopting the constrained meta path, we can further answer the *Q. 1.2*. Moreover, in HIN, based on different paths, the objects have different ranking values. The comprehensive importance of objects should consider all kinds of factors (the factors can be embodied by constrained meta paths), which have different contribution to the importance of objects. In order to evaluate the importance of objects and meta paths simultaneously (i.e., answer $Q.$ 2), we further propose a co-ranking method which organizes the relation matrices of objects on different constrained meta paths as a tensor. A random walk process is designed on this tensor to co-rank importance of objects and paths simultaneously. That is, random walkers surf in the tensor, where the stationary visiting probability of objects and meta paths is considered as the HRank score of objects and paths.

2   Preliminary

In this section, we describe notations used in this paper and present some preliminary knowledge.

In heterogeneous information networks, there are multiple object types and relation types. We use the network schema $S = (\mathcal{A}, \mathcal{R})$ to depict the object types and relations existing among object types, where $\mathcal{A} = \{A\}$ is a set of object types and $\mathcal{R} = \{R\}$ is a set of relations. A relation R existing from type S to type T is denoted as $S \xrightarrow{R} T$. Fig. 1(b) shows a network schema of bibliographic information network.

Different from homogeneous networks, two objects in a heterogeneous network can be connected via different paths and these paths have different meanings. These paths are called meta paths which can be defined as follows.

Definition 1 *Meta path [4]. A meta path $\mathcal{P}$ is a path defined on a schema $S =$ $(\mathcal{A}, \mathcal{R})$, and is denoted in the form of $A_1 \xrightarrow{R_1} A_2 \xrightarrow{R_2} \ldots \xrightarrow{R_l} A_{l+1}$ (abbreviated as $A_1 A_2 \ldots A_{l+1}$), which defines a composite relation $R = R_1 \circ R_2 \circ \ldots \circ R_l$ between type A_1 and A_{l+1}, where $\circ$ denotes the composition operator on relations.*

It is obvious that semantics underneath these paths are different. The *"Author-Paper-Author"* (*APA*) path means authors collaborating on the same papers, while the *"Author-Paper-Conference-Paper-Author"* (*APCPA*) path means the authors' papers publishing on the same conferences. Based on different meta paths, there are different relation networks, which may result in different importance of objects. However, meta path fails to capture some subtle semantics. Taking Fig. 1(b) as an example, the *APA* cannot reveal the co-author relations in Data Mining (DM) field. In order to overcome the shortcomings in meta path, we propose the concept of constrained meta path, defined as follows.

Definition 2 *Constrained meta path. A constrained meta path is a meta path based on a certain constraint which is denoted as $CP = \mathcal{P}|\mathcal{C}$. $\mathcal{P} = (A_1 A_2 \ldots A_l)$ is a meta path, while $\mathcal{C}$ represents the constraint on the objects in the meta path.*

Note that the $\mathcal{C}$ can be one or multiple constraint conditions on objects. Taking Fig. 1(b) as an example, the constrained meta path $APA|P.L = $ *"DM"* represents the co-author relations of authors in data mining field through constraining the label of papers with DM. Similarly, the constrained meta path $APCPA|P.L = $ *"DM"*$\&\&C = $ *"CIKM"* represents the co-author relations of authors in CIKM conference and the papers of authors are in data mining field. Obviously, compared to meta path, the constrained meta path conveys richer semantics by subdividing meta paths under distinct conditions.

For a relation $A \xrightarrow{R} B$, we can obtain its constrained transition probability matrix as follows.

Definition 3 *Constrained transition probability matrix.* W_{AB} *is an adjacent matrix between type A and B on relation $A \xrightarrow{R} B$. U_{AB} is the normalized matrix of W_{AB} along the row vector, which is the transition probability matrix of $A \xrightarrow{R} B$. Suppose there is a constraint C on object type A. The constrained transition probability matrix U'_{AB} of constrained relation $R|C$ is $U'_{AB} = M_C U_{AB}$, where M_C is the constraint matrix generated by the constraint condition C.*

The constraint matrix M_C is usually a diagonal matrix whose dimension is the number of objects in object type A. The element in the diagonal is 1 if the corresponding object satisfies the constraint, else the element is 0. Similarly, we can confine the constraint on object type B or both types.

Given a network following a network schema $S = (\mathcal{A}, \mathcal{R})$, we can define the constrained meta path based reachable probability matrix as follows.

Definition 4 *Constrained meta path based reachable probability matrix.* *For a constrained meta path $CP = (A_1 A_2 \cdots A_{l+1} | C)$, the constrained meta path based reachable probability matrix is defined as $PM_{CP} = U'_{A_1 A_2} U'_{A_2 A_3} \cdots U'_{A_l A_{l+1}}$. $PM_{CP}(i,j)$ represents the probability of object $i \in A_1$ reaching object $j \in A_{l+1}$ under the constrained meta path CP.*

When there is a constraint on the objects, we only consider the objects that satisfy the constraint. For simplicity, we use the reachable probability matrix and the M_P to represent the constrained meta path based reachable probability matrix in the following section.

3   The HRank Method

In order to answer the two ranking problems proposed in Section 1, we design two versions of HRank, respectively.

3.1   Ranking Based on Constrained Meta Paths

For the question $Q.\ 1$, we propose the HRank-CMP method based on a constrained meta path $\mathcal{P} = (A_1 A_2 \ldots A_l | C)$.

HRank-CMP is based on a random walk process that random walkers wander between A_1 and A_l along the path. The ranks of A_1 and A_l can be seen as the visiting probability of walkers, which are defined as follows:

$$R(A_l | \mathcal{P}^{-1}) = \alpha R(A_1 | \mathcal{P}) M_{\mathcal{P}} + (1 - \alpha) E_{A_l}$$
$$R(A_1 | \mathcal{P}) = \alpha R(A_l | \mathcal{P}^{-1}) M_{\mathcal{P}^{-1}} + (1 - \alpha) E_{A_1} \tag{1}$$

where $M_{\mathcal{P}}$ and $M_{\mathcal{P}^{-1}}$ are the reachable probability matrix of path $\mathcal{P}$ and $\mathcal{P}^{-1}$. E_{A_1} and E_{A_l} are the restart probability of A_1 and A_l. Note that the path $\mathcal{P}$ is either symmetric ($\mathcal{P} = \mathcal{P}^{-1}$) or asymmetric ($\mathcal{P} \neq \mathcal{P}^{-1}$).

3.2   Co-ranking for Objects and Relations in HIN

There are many constrained meta paths in heterogeneous networks. It is an important issue to automatically determine the importance of paths [4,5], since it is usually hard for us to identify which relation is more important in real applications. To solve this problem (i.e., *Q. 2*), we propose the HRank-CO to co-rank the importance of objects and relations. The basic idea is based on an intuition that important objects are connected to many other objects through a number of important relations and important relations connect many important objects. So we organize the multiple relation networks with a tensor and a random walk process is designed on this tensor. The method not only can comprehensively evaluate the importance of objects by considering all constrained meta paths, but also can rank the contribution of different constrained meta paths.

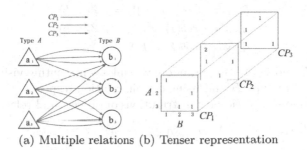

(a) Multiple relations (b) Tenser representation

Fig. 2. An example of multi-relations of objects generated by multiple paths. (a) is the graph representation. (b) is the corresponding tensor representation.

In Fig. 2(a), we show an example of multiple relations among objects. There are three objects of type A, three objects of type B and three types of relations among them. These relations are generated by three constrained meta paths with type A as the source type and type B as the target type. To describe the multiple relations among objects, we use the representation of tensor which is a multidimensional array. We call $X = (x_{i,j,k})$ a 3rd order tensor, where $x_{i,j,k} \in R$, for $i = 1, \cdots, m$, $j = 1, \cdots, l$, $k = 1, \cdots, n$. m and n are the number of objects of type A and type B, respectively, and there are l types of relations among them. $x_{i,j,k}$ represents the times that object i is related to object k through the jth constrained meta path. For example, Fig. 2(b) is a three-way array, where each two dimensional slice represents an adjacency matrix for a single relation. So the data can be represented as a tensor of size $3 \times 3 \times 3$. In the multi-relational network, we define the transition probability tensor to present the transition probability among objects and relations.

Definition 5 *Transition probability tensor. In a multi-relational network, X is the tensor representing the network. F is the normalized tensor of X along the column vector. R is the normalized tensor of X along the tube vector. T is the normalized tensor of X along the row vector. F, R, and T are called the transition probability tensor which can be denoted as follows:*

$$f_{i,j,k} = \frac{x_{i,j,k}}{\sum_{i=1}^{m} x_{i,j,k}} \qquad i = 1, 2, \ldots, m$$

$$r_{i,j,k} = \frac{x_{i,j,k}}{\sum_{j=1}^{l} x_{i,j,k}} \qquad j = 1, 2, \ldots, l \qquad (2)$$

$$t_{i,j,k} = \frac{x_{i,j,k}}{\sum_{k=1}^{n} x_{i,j,k}} \qquad k = 1, 2, \ldots, n$$

$f_{i,j,k}$ can be interpreted as the probability of object i (of type A) being the visiting object when relation j is used and the current object being visited is object k (of type B), $r_{i,j,k}$ represents the probability of using relation j given that object k is visited from object i, and $t_{i,j,k}$ can be interpreted as the probability of object k being visited, given that object i is currently the visiting object and relation j is used. The meaning of these three tensors can be defined formally as follows:

$$f_{i,j,k} = Prob(X_t = i | Y_t = j, Z_t = k)$$
$$r_{i,j,k} = Prob(Y_t = j | X_t = i, Z_t = k) \qquad (3)$$
$$t_{i,j,k} = Prob(Z_t = k | X_t = i, Y_t = j)$$

in which X_t, Z_t and Y_t are three random variables representing visiting at certain object of type A or type B and using certain relation respectively at the time t.

Now, we define the stationary distributions of objects and relations as follows

$$x = (x_1, x_2, \cdots, x_m)^T, y = (y_1, y_2, \cdots, y_l)^T, z = (z_1, z_2, \cdots, z_n)^T \qquad (4)$$

in which

$$x_i = \lim_{t \to \infty} Prob(X_t = i), y_j = \lim_{t \to \infty} Prob(Y_t = j), z_k = \lim_{t \to \infty} Prob(Z_t = k). \qquad (5)$$

From the above equations, we can get:

$$Prob(X_t = i) = \sum_{j=1}^{l} \sum_{k=1}^{n} f_{i,j,k} \times Prob(Y_t = j, Z_t = k)$$

$$Prob(Y_t = j) = \sum_{i=1}^{m} \sum_{k=1}^{n} r_{i,j,k} \times Prob(X_t = i, Z_t = k) \qquad (6)$$

$$Prob(Z_t = k) = \sum_{i=1}^{m} \sum_{j=1}^{l} t_{i,j,k} \times Prob(X_t = i, Y_t = j)$$

where $Prob(Y_t = j, Z_t = k)$ is the joint probability distribution of Y_t and Z_t, $Prob(X_t = i, Z_t = k)$ is the joint probability distribution of X_t and Z_t, and $Prob(X_t = i, Y_t = j)$ is the joint probability distribution of X_t and Y_t. To obtain x_i, y_j and z_k, we assume that X_t, Y_t and Z_t are all independent from each other which can be denoted as below:

$$Prob(X_t = i, Y_t = j) = Prob(X_t = i)Prob(Y_t = j)$$
$$Prob(X_t = i, Z_t = k) = Prob(X_t = i)Prob(Z_t = k) \qquad (7)$$
$$Prob(Y_t = j, Z_t = k) = Prob(Y_t = j)Prob(Z_t = k)$$

Consequently, combining the equations with the assumptions above, we get

$$x_i = \sum_{j=1}^{l} \sum_{k=1}^{n} f_{i,j,k} y_j z_k, \qquad i = 1, 2, \ldots, m,$$

$$y_j = \sum_{i=1}^{m} \sum_{k=1}^{n} r_{i,j,k} x_i z_k, \qquad j = 1, 2, \ldots, l, \qquad (8)$$

$$z_k = \sum_{i=1}^{m} \sum_{j=1}^{l} t_{i,j,k} x_i y_j, \qquad k = 1, 2, \ldots, n.$$

The equations above can be written in a tensor format:

$$x = Fyz, \qquad y = Rxz, \qquad z = Txy \qquad (9)$$

with $\qquad \sum_{i=1}^{m} x_i = 1, \sum_{j=1}^{l} y_j = 1,$ and $\sum_{k=1}^{n} z_k = 1.$

According to the analysis above, we can design the following algorithm to co-rank the importance of objects and relations.

Algorithm 1. HRank-CO Algorithm
___
Input: Three tensors F, T and R, three initial probability distributions x_0, y_0 and z_0 and the tolerance ϵ.
Output: Three stationary probability distributions x, y and z.
  Procedure:
  Set $t = 1$;
  repeat
    Compute $x_t = F y_{t-1} z_{t-1}$;
    Compute $y_t = R x_t z_{t-1}$;
    Compute $z_t = T x_t y_t$;
  until $||x_t - x_{t-1}|| + ||y_t - y_{t-1}|| + ||z_t - z_{t-1}|| < \epsilon$

4　Experiments

In this section, we do experiments to validate the effectiveness of two versions of HRank on three real datasets, respectively.

4.1　Datasets

We use three heterogeneous information networks for our experiments. They are summarized as follows:

DBLP Dataset [6,4]: The DBLP dataset is a sub-network collected from DBLP website [1] involving major conferences in two research areas: database (DB) and information retrieval (IR), which naturally form two labels. The dataset contains 9682 authors, 20 conferences and 22185 papers which are all labeled with one of the two research areas. The network schema is shown in Fig. 3(a).

---

[1] http://www.informatik.uni-trier.de/~ley/db/

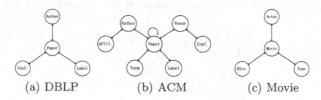

(a) DBLP   (b) ACM   (c) Movie

Fig. 3. The network schema of three heterogeneous datasets. (a) DBLP bibliographic dataset. (b) ACM bibliographic dataset. (c) IMDB movie dataset.

ACM Dataset [6]: The ACM dataset was downloaded from ACM digital library [2] in June 2010. The ACM dataset comes from 14 representative computer science conferences. These conferences include 196 corresponding venue proceedings. The dataset has 12499 papers, 17431 authors, 1903 terms and 1804 author affiliations. The network also includes 73 labels of these papers in ACM category. The network schema of ACM dataset is shown in Fig. 3(b).

IMDB Dataset [7]: We crawled movie information from The Internet Movie Database [3] to construct the network. The related objects include movies, actors, directors and movie types, which are organized as a star schema shown in Fig. 3(c). Movie information includes 5324 actors, 1591 movies, 551 directors and 112 movie types.

4.2 Ranking of Heterogeneous Objects

Here, the experiments validate the effectiveness of HRank-CMP on constrained meta paths.

Experiment Study on Constrained Meta Paths. The experiments are done on the DBLP dataset. We evaluate the importance of authors and conferences simultaneously based on the meta path APC, which means authors publish papers on conferences. Two constrained meta paths ($APC|P.L$ = "DB" and $APC|P.L$ = "IR") are also included, which means authors publish DB(IR)-field papers on conferences. We employ HRank-CMP to rank the importance of authors and conferences based on these three paths. As the baseline methods, we use PageRank and the degree of authors and conferences (called Degree method). We directly run PageRank on the whole DBLP network by ignoring the heterogeneity of objects. Since the results of PageRank mix all types of objects, we select the author and conference type from the ranking list as the final results.

The top ten authors and conferences returned by these five methods are shown in Tables 1 and 2, respectively. As shown in Table 1, the ranking results of these methods on authors all are reasonable, however, the constrained meta paths can find the most influential authors in a certain field. For example, the top three

---

[2] http://dl.acm.org/
[3] www.imdb.com/

Table 1. Top ten authors of different methods on DBLP dataset. The number in the parenthesis of the fifth column means the rank of authors in the whole ranking list returned by PageRank.

Rank	APC	APC|P.L = "DB"	APC|P.L = "IR"	PageRank	Degree
1	Gerhard Weikum	Surajit Chaudhuri	W. Bruce Croft	W. Bruce Croft(23)	Philip S. Yu
2	Katsumi Tanaka	H. Garcia-Molina	Bert R. Boyce	Gerhard Weikum(24)	Gerhard Weikum
3	Philip S. Yu	H. V. Jagadish	Carol L. Barry	Philip S. Yu(25)	Divesh Srivastava
4	H. Garcia-Molina	Jeffrey F. Naughton	James Allan	Jiawei Han(26)	Jiawei Han
5	W. Bruce Croft	Michael Stonebraker	ChengXiang Zhai	H. Garcia-Molina(27)	H. Garcia-Molina
6	Jiawei Han	Divesh Srivastava	Mark Sanderson	Divesh Srivastava(28)	W. Bruce Croft
7	Divesh Srivastava	Gerhard Weikum	Maarten de Rijke	Surajit Chaudhuri(29)	Surajit Chaudhuri
8	Hans-Peter Kriegel	Jiawei Han	Katsumi Tanaka	H. V. Jagadish(30)	H. V. Jagadish
9	Divyakant Agrawal	Christos Faloutsos	Iadh Ounis	Jeffrey F. Naughton(31)	Jeffrey F. Naughton
10	Jeffrey Xu Yu	Philip S. Yu	Joemon M. Jose	Rakesh Agrawal(32)	Rakesh Agrawal

authors of $APC|P.L = "DB"$ are Surajit Chaudhuri, Hector Garcia-Molina and H. V. Jagadish, and all of them are very influential researchers in the database field. Similarly, as we can see in Table 2, HRank with constrained meta paths can clearly find the important conferences in DB and IR fields, while other methods mingle these conferences. For example, the most important conferences in the DB field are ICDE, VLDB and SIGMOD, while the most important conferences in the IR field are SIGIR, WWW and CIKM. Observing Tables 1 and 2, we can also find the mutual effect of authors and conferences.

Table 2. Top ten conferences of different methods on DBLP dataset. The number in the parenthesis of the fifth column means the rank of conferences in the whole ranking list returned by PageRank.

Rank	APC	APC|P.L = "DB"	APC|P.L = "IR"	PageRank	Degree
1	CIKM	ICDE	SIGIR	ICDE(3)	ICDE
2	ICDE	VLDB	WWW	SIGIR(4)	SIGIR
3	WWW	SIGMOD	CIKM	VLDB(5)	VLDB
4	VLDB	PODS	JASIST	CIKM(6)	SIGMOD
5	SIGMOD	DASFAA	WISE	SIGMOD(7)	CIKM
6	SIGIR	EDBT	ECIR	JASIST(8)	JASIST
7	DASFAA	ICDT	APWeb	WWW(9)	WWW
8	JASIST	MDM	WSDM	DASFAA(10)	PODS
9	WISE	WebDB	JCIS	PODS(11)	DASFAA
10	EDBT	SSTD	IJKM	JCIS(12)	EDBT

Quantitative Comparison Experiments. Based on the results returned by five methods, we can obtain five candidate ranking lists of authors in DBLP dataset. To evaluate the results quantitatively, we use the author ranks from Microsoft Academic Search [4] as ground truth. Specifically, we crawled two standard ranking lists of authors in two academic fields: DB and IR. Then we use the *Distance* criterion [8] to compare the difference between our candidate ranking lists and the standard ranking lists. The criterion not only measures the number of mismatches between these two lists, but also considers the position of these mismatches. The smaller *Distance* means the smaller difference (i.e., better performance). Fig. 4 shows the differences of author ranking lists. We can observe that HRank with constrained meta paths achieve the best performances

---

[4] http://academic.research.microsoft.com/

on their corresponding field, while they have the worst performances on other fields. In addition, compared to that of PageRank and Degree, the mediocre performances of HRank with meta path APC further demonstrate the importance of constrained meta path to capture the subtle semantics contained in heterogeneous networks.

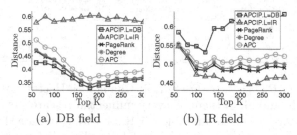

(a) DB field   (b) IR field

Fig. 4. The Distances between the candidate author ranking lists and the standard ranking lists on different fields on DBLP dataset

4.3   Co-ranking of Objects and Paths

Experiment Study on Co-ranking on Symmetric Constrained Meta Paths. In this experiment, we will validate the effectiveness of HRank-CO to rank objects and symmetric constrained meta paths simultaneously. The experiment is done on ACM dataset. First we construct a $(2, 1)$th order tensor X based on 73 constrained meta paths (i.e., $APA|P.L = L_j, j = 1 \cdots 73$). When the ith and the kth authors co-publish a paper together, of which the label is the jth label, we add one to the entries $x_{i,j,k}$ and $x_{k,j,i}$ of X. By considering all the publications, $x_{i,j,k}$ (or $x_{k,j,i}$) refers to the number of collaborations by the ith and the kth author under the jth paper label. In addition, we do not consider any self-collaboration, i.e., $x_{i,j,i} = 0$ for all $1 \leq i \leq 17431$ and $1 \leq j \leq 73$. The size of X is $17431 \times 73 \times 17431$ and the percentage of nonzero entries is $4.126 \times 10^{-4}\%$. In this dataset, we will evaluate the importance of authors through the co-author relations, meanwhile we will analyze the importance of paths.

Table 3. Top 10 authors and constrained meta paths (note that only the constraint (L_j) of the paths ($APA|P.L = L_j, j = 1 \ldots 73$) are shown in the table)

Rank	Authors	Constrained meta paths
1	Jiawei Han	H.3 (Information Storage and Retrieval)
2	Philip Yu	H.2 (Database Management)
3	Christos Faloutsos	C.2 (Computer-Communication Networks)
4	Ravi Kumar	I.2 (Artificial Intelligence)
5	Wei-Ying Ma	F.2 (Analysis of Algorithms and Problem Complexity)
6	Zheng Chen	D.4 (Operating Systems)
7	Hector Garcia-Molina	H.4 (Information Systems Applications)
8	Hans-Peter Kriegel	G.2 (Discrete Mathematics)
9	Gerhard Weikum	I.5 (Pattern Recognition)
10	D. R. Karger	H.5 (Information Interfaces and Presentation)

Table 3 shows the top ten authors (left) and paths (right) based on their HRank values. We can find that the top ten authors are all influential researchers in DM/IR fields, which conforms to our common senses. Similarly, the most important paths are related to DM/IR fields, such as $APA|P.L$ = "$H.3$" (Information Storage and Retrieval) and $APA|P.L$ = "$H.2$" (Database Management). Although the conferences in ACM dataset are from multiple fields, there are more papers from the DM/DB fields, which makes the authors and paths in DM/DB fields ranked higher. We can also find that the influence of authors and paths can be promoted by each other. In order to observe this point more clearly, we show the number of co-authors of the top ten authors based on the top ten paths in Table 4. We can observe that there are more collaborations for top authors in influential fields. For example, although Zheng Chen (rank 6) has more number of co-authors than Jiawei Han (rank 1), the collaborations of Jiawei Han focus on ranked higher fields (i.e., H.3 and H.2), so Jiawei Han has higher HRank score. Similarly, the top paths contain many collaborations of influential authors.

Table 4. The number that the top ten authors collaborate with others via the top ten constrained meta paths (note that only the constraint (L_j) of the paths $(APA|P.L = L_j, j = 1\ldots73)$ are shown in the first row of the table).

Ranked A/CP	1 (H.3)	2 (H.2)	3 (C.2)	4 (I.2)	5 (F.2)	6 (D.4)	7 (H.4)	8 (G.2)	9 (I.5)	10 (H.5)
1 (Jiawei Han)	51	176	0	0	0	0	9	2	2	0
2 (Philip Yu)	51	94	0	0	9	0	3	0	13	0
3 (C. Faloutsos)	17	107	0	5	9	0	3	4	2	0
4 (Ravi Kumar)	73	27	0	3	13	0	18	5	0	0
5 (Wei-Ying Ma)	132	26	0	9	0	0	2	0	30	10
6 (Zheng Chen)	172	9	0	9	0	0	22	0	38	9
7 (H. Garcia-Molina)	23	65	3	0	0	0	1	0	0	4
8 (H. Kriegel)	19	28	3	0	0	0	6	0	7	4
9 (G. Weikum)	82	14	0	4	0	0	8	0	4	0
10 (D. R. Karger)	11	5	13	0	7	4	1	7	0	7

Experiment Study on Co-ranking on Asymmetric Constrained Meta Paths. The experiments on the Movie dataset aim to show the effectiveness of HRank-CO to rank heterogeneous objects and asymmetric constrained meta paths simultaneously. In this case, we construct a 3rd order tensor X based on the constrained meta paths $AMD|M.T = T_j, j = 1\cdots112$. That is, the tensor represents the actor-director collaboration relations on different types of movies. When the ith actor and the kth director cooperate in a movie of the jth type, we add one to the entries $x_{i,j,k}$ of X. By considering all the cooperations, $x_{i,j,k}$ refers to the number of collaborations by the ith actor and the kth director under the jth type of movie. The size of X is $5324 \times 112 \times 551$ and the percentage of nonzero entries is $7.827 \times 10^{-4}\%$.

Table 5 shows the top ten actors, directors and constrained meta paths (i.e., movie type). Basically, the results comply with our common senses. The top ten actors are well known, such as Eddie Murphy, Harrison Ford. Similarly, these directors are also famous in filmdom due to their works. These movie types obtained are the most popular movie subjects as well. In addition, we observe the mutual enhancements of the importance of objects and meta paths again. As we know, Eddie Murphy and Drew Barrymore (rank 1, 4 in actors) are famous

Table 5. Top 10 actors, directors and meta paths on IMDB dataset (note that only the constraint (T_j) of the paths $(AMD|M.T = T_j, j = 1 \ldots 112)$ are shown in the table)

Rank	Actor	Director	Constrained meta path
1	Eddie Murphy	Tim Burton	Comedy
2	Harrison Ford	Zack Snyder	Drama
3	Bruce Willis	Marc Forster	Thriller
4	Drew Barrymore	David Fincher	Action
5	Nicole Kidman	Michael Bay	Adventure
6	Nicolas Cage	Ridley Scott	Romance
7	Hugh Jackman	Richard Donner	Crime
8	Robert De Niro	Steven Spielberg	Sci-Fi
9	Brad Pitt	Robert Zemeckis	Animation
10	Christopher Walken	Stephen Sommers	Fantasy

comedy and drama (rank 1, 2 in paths) actors. Higher ranked directors also prefer popular movie subjects.

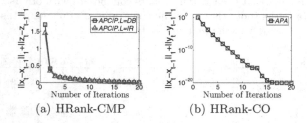

(a) HRank-CMP   (b) HRank-CO

Fig. 5. The difference between two successive calculated probability vectors against iterations based on the two versions of HRank

4.4 Convergence Experiments

In Fig. 5, we show the convergence of HRank on the previous experiments. The results illustrate that the two versions of HRank both quickly converge after no more than 20 iterations. In addition, we can also observe that HRank has different convergence speed in these two conditions. HRank-CMP almost converges on 9 iterations (see Fig. 5(a)). However, HRank-CO for co-ranking converges on 16 iterations (see Fig. 5(b)). We think it is reasonable, since it is more difficult to converge for more objects in HRank-CO. The time and space complexity is analyzed, and three fast computation strategies are designed to fasten the matrix multiplication process in [10].

5   Conclusions

In this paper, we first study the ranking problem in heterogeneous information network and propose the HRank method, which is a path based random walk method. In this method, we introduce the constrained meta path concept to capture the more subtle and refined semantics contained in HIN. In addition, we further put forward a method to co-rank the paths and objects, since the paths effect the importance of objects. Experiments validate the effectiveness and efficiency of HRank on three real datasets.

Acknowledgments. This work is supported by the National Basic Research Program of China (2013CB329603). It is also supported by the National Natural Science Foundation of China (No. 61375058, 61074128, 71231002) and Ministry of Education of China and China Mobile Research Fund (MCM20123021).

References

1. Kleinberg, J.M.: Authoritative Sources in a Hyperlinked Environment. Journal of the ACM 46(5), 604–632 (1999)
2. Han, J.: Mining Heterogeneous Information Networks by Exploring the Power of Links. Discovery Science (2009)
3. Jeh, G., Widom, J.: Simrank: a Measure of Structural-Context Similarity. In: 8th ACM SIGKDD International Conference on Knowledge Discovery and Data Mining, pp. 538–543. ACM (2002)
4. Sun, Y., Han, J., Yan, X., Yu, P.S., Wu, T.: PathSim: Meta Path-Based Top-K Similarity Search in Heterogeneous Information Networks. In: VLDB, pp. 992–1003 (2011)
5. Sun, Y., Norick, B., Han, J., Yan, X., Yu, P.S., Yu, X.: Integrating Meta Path Selection with User-Guided Object Clustering in Heterogeneous Information Networks. In: KDD, pp. 1348–1356 (2012)
6. Shi, C., Kong, X., Yu, P.S., Xie, S., Wu, B.: Relevance Search in Heterogeneous Networks. In: 15th EDBT, pp. 180–191. ACM (2012)
7. Shi, C., Zhou, C., Kong, X., Yu, P.S., Liu, G., Wang, B.: HeteRecom: A Semantic-Based Recommendation System in Heterogeneous Networks. In: KDD, pp. 1552–1555 (2012)
8. Nie, Z., Zhang, Y., Wen, J.R., Ma, W.Y.: Object-level Ranking: Bringing Order to Web Objects. In: WWW, pp. 422–433 (2005)
9. Page, L., Brin, S., Motwani, R., Winograd, T.: The PageRank Citation Ranking: Bringing Order to the Web. Technical report, Stanford University Database Group (1998)
10. http://arxiv.org/abs/1403.7315

Action-Scene Model for Recognizing Human Actions from Background in Realistic Videos[*]

Wen Qu[1], Yifei Zhang[1,2], Shi Feng[1,2], Daling Wang[1,2], and Ge Yu[1,2]

[1]School of Information Science and Engineering, Northeastern University
[2]Key Laboratory of Medical Image Computing (Northeastern University),
Ministry of Education, Shenyang 110819, China
quwen@research.neu.edu.cn,
{zhangyifei,fengshi,wangdaling,yuge}@mail.neu.edu.cn

Abstract. Using single information from person region fails to distinguish similar actions in realistic videos due to occlusions and variation of person. In this paper, we explore the problem of modeling action-scene context from the background regions of the realistic videos. The contextual cues of actions and scenes are formulated in a graphical model representation. A novel Action-Scene Model is proposed to mine the contextual cues with little prior knowledge. The proposed approach can infer actions from background regions directly and is a complement to the existing methods. In order to fuse the contextual cues effectively with other components, a context weight is introduced to measure the contributions of context based on the proposed model. We present experimental results on a realistic video dataset. The experiment results validate the effectiveness of Action-Scene Model in identifying the actions from background regions. And the learned contextual cues can achieve better performance than the existing methods especially for scene-dependent action categories.

Keywords: human action recognition, background context cue, realistic video processing.

1   Introduction

Recognizing human action from realistic videos is the foundation of many practical applications such as on-line video management, video retrieval, robotics, and surveillance. Although a large amount of impressive results have been achieved on datasets recorded in controlled environments, such as KTH [20], WEIZMEN [17], much less progress has been made on realistic videos. This is caused by clutter backgrounds, lots of camera motions and variations of viewpoints in the realistic videos. The realistic

---

[*] Project supported by the National Basic Research 973 Program of China under Grant No. 2011CB302200-G, the Key Program of National Natural Science Foundation of China under Grant No. 61033007, the National Natural Science Foundation of China under Grant Nos. 61370074，61100026, and the Fundamental Research Funds for the Central Universities of China under Grant Nos. N120404007.

F. Li et al. (Eds.): WAIM 2014, LNCS 8485, pp. 566–577, 2014.

videos are usually collected from real-world sources, such as movies [12], TV series, wearable camera videos or websites. Figure 1 depicts some sample frames from these two kinds of datasets. It is obvious that the realistic videos include more complex background settings. Furthermore, the human bodies are occluded or seen from varied viewpoints. Therefore it is essential to explore contextual cues from background regions for human action recognition.

Controlled Video Dataset   *Realistic Video Dataset*

Fig. 1. Example frames from three datasets: KTH (first column), WEIZMEN (second column) and YouTube dataset (third, fourth and fifth columns)

In real-world videos, actions usually happen under particular environments. The scenes in background regions are a rich source providing the contextual cues for actions identification. Many existing methods use scene detectors to detect scene in videos, which are time-consuming in training stage and inflexible to different datasets. Besides, the performance of detectors is prone to affect the recognition precision.

In this paper, we are interested in the problem of learning contextual relationship between actions and scenes without scene detectors. We propose an Action-Scene Model to automatically model the relationship between actions, scenes and background features. The approach only needs to set the number of scenes without knowing the categories. Furthermore, the contextual cues have different contributions for each action class. For example, basketball shooting mostly happens in basketball count. However, walking with a dog happens in diverse scenarios (e.g. grass, road, snowfield, etc.). In this situation the action-scene context has few contributions to recognize walking. To solve this problem, in this paper, we propose a factor to weight the contextual cues automatically when attempting to integrate the contextual cues with other methods.

In detail, we first segment videos into person regions and background regions, and the features of the background regions are extracted. Then the contextual cues of actions, scenes are modeled using a graphical model called Action-Scene Model. Finally, the proposed model gives the probability of action happening in videos according to the background features. The contextual cues can either infer actions from videos directly or integrate with existing methods as a complement. Context weights are computed when contextual cues are integrated with other methods. Figure 2 demonstrates the entire framework of our proposed method.

The main contributions of our work are: 1) Modeling the contextual cues between actions and scenes with less prior knowledge about the scene. Most previous work needs to set scene categories manually. 2) Introducing a context weight to measure the contribution of scene for action recognition automatically. In existing literatures, the context information is added in recognition stage with an empirical weight, which ignores the distinction of action-scene dependency.

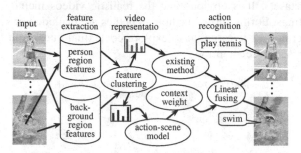

Fig. 2. Overview of the framework

The rest of this paper is organized as follows. Section 2 reviews related work. Section 3 describes feature extraction and video representation. We detail Action-Scene Model in Section 4 and explain the integration with other methods in Section 5. Section 6 presents experimental results, followed by conclusion in Section 7.

2 Related Work

Human action recognition from videos has been an active field for long time. Early work was focus on videos with fixed background [1,20]. However, this scenario is too ideal to apply in real life. As more attention paid to complex scenarios, some recent approaches [9,12,13,16] tried to deal with "wild" videos.

Laptev [12] addressed the problem of automatically annotating actions in movies using multi-channel non-linear SVMs. Liu [13] used both motion and static features to recognize actions. Features integrating additional non-local cues were used in [21]. Ikizler-Cinbis [10] combined the features of object, scene and actions in a multiple instance learning framework. However, these efforts formulated object, scene and action as multiple feature channels without modeling the relationship between them.

Modeling object and scene context has also been explored in recent work. Marszalek [15] exploited the context using movie scripts and developed a joint scene-action SVM-based classifier. Images of typical scene were collected to train several scene detectors. Han [9] proposed a MKGPC method to select and weight multiple features automatically. Jiang [11] modeled the action context from both labeled and unlabeled data, which was focused on action retrieval with a limited number of examples. Object recognition and text mining were used to improve video activity recognition in [16], which mined context from text corpus.

Most of the previous work considering object and scene context [9,10,11,15,16] relies on the pre-trained object and scene detectors. The detector-based method has three limitations: 1) Hard to select a general category of scene and object suitable for all datasets. 2) A lack of training data for every object and scene. It is time-consuming to collect annotated data for each category. 3) The relationship between action and scene, i.e. $P(action|scene)$, learned from texts (movie scripts [15], natural-language text [16]) leads to a semantic gap with visual features.

Unlike previous literatures, the proposed approach in this paper distinguishes from the detector-based method in learning context cues without providing scene types and no pre-trained scene detectors. Inspired by the recent work of action recognition in static images [22] and unsupervised learning method in [17], we use a graphical model to represent action and scene in videos. The action-scene method can capture the underlying action-scene context directly from visual features.

3   Visual Feature Extraction and Video Representation

Learning context between action and scene in videos requires information of scenes. For simplicity, we segment videos into person regions and non-person regions. The non-person regions are viewed as background regions, and these regions provide features of scenes. In this section, we will present the feature extraction from background regions (Section 3.2) and bag-of-feature video representation (Section 3.3).

3.1   Person Detection

The person detection is implemented based on Felzenszwalb's object detector [7] and mean shift tracking [3]. We first use the person detector to find candidate person regions from each frame. The model used in person detector is trained on the data for the PASCAL Visual Object Classes Challenge 2009 [6]. Since there are many false alarms, a sliding window based method is used to discard the false alarms following [11]. And the miss detections are filled with tracking from previous frames using mean shift algorithm.

3.2   Background Region Features Extraction

In order to capture the color, shape and local features of scene, we extract color histogram, Gist descriptor [18] and static interest points from frames that are sampled every 20 frames in each video.

Background Color Feature. The background region is divided into 30×30 (pixels) blocks. Then the color histogram for each block is computed. The RGB color space is discretized into a 512-bins color histogram (3 channel and 8 bins per channel).

Background Shape Feature. The dominant spatial structure is also helpful to describe the scene. One such global image feature has been widely used is the Gist descriptor. Hence, we extract the Gist descriptor for every sampled frame with the original parameter settings in [18].

Background Static Features. Static interest point extracts discriminative points in an image, and it is a useful local feature in object recognition. We first apply 2D Harris detector to obtain interest points. Then 128-dimensional SIFT descriptor [14] for each point is computed.

3.3   Bag-of-Features Video Representation

Bag-of-feature model based video representation has proved to be effective for action recognition [12,13,16,21]. The bag-of-feature model represents a video clip as a vector over a "visual codebook". The "visual codebook" is constructed by clustering visual features detected from videos. The center of a cluster is defined to be a visual word in the visual codebook. Thus, each detected feature in a video clip can be assigned to the unique visual word having the nearest distance. The video representation is an n-dimensional vector (n is the number of clusters), and the ith bin of the vector is the number of features assigned to the ith visual word in the video.

We constructed three codebooks for color, shape and static features respectively. Then these vectors are concatenated to form a visual codebook with 1200 visual words. Finally, a video clip can be represented with a 1200-dimensional vector based on the codebooks.

4   Learning of Action-Scene Relationship

In this section, we introduce how to learn context information between actions and scenes in videos using the Action-Scene Model.

Given a set of video clips, action and scene categories, observed visual words, a video clip is directly associated with actions, and each action has relation with several scenes. Each scene is associated with visual words in the visual codebook (introduced in Section 3.3).

According to the relation between videos, actions, scenes and visual words, we naturally deduce a generative probabilistic model to learn the contextual cues between them. The graphical model corresponding to the generative process of Action-Scene Model is shown in Figure 3. In the figure, nodes are random variables. Shaded node is observed variable, and unshaded ones are unobserved variables. An arrow indicates a conditional dependency between variables. Plates indicate repeated sampling with the number of repetitions given by the variable in the bottom.

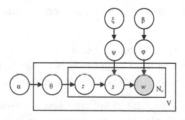

Fig. 3. Action-Scene graphical model. z is action, s is scene and w is visual word in videos. α, β and ξ are hyper parameter of a Dirichlet distribution.

In the Action-Scene Model, each video is seen as a mixture of actions, and each action is a probability distribution over scenes. Furthermore, each scene is associated with a distribution over visual words. Our model is similar with LDA model [2] in the generating process. The LDA model views each video clip as a distribution of actions, and each action is a distribution of visual word [17]. The generating process does not model the relationship between actions and scenes. On the contrary, the Action-Scene Model discovers not only which action happens in a video, but also which scene is associated with the action.

Suppose we have a collection of M video clips, each video v is represented as a set of visual words. Each visual word belongs to the visual codebook including W unique visual words. Table 1 lists the notations used in this paper. Assuming there are K action categories, S scene categories, we can describe the process generating each video v as:

1) For each clip v, there is a distribution over K action categories $\theta_v \sim Dirichlet(\alpha)$.

2) For each visual word vi of the clip v:

   a) Choose an action category z_{vi} according to distribution $Multinomial$ (θ_v).

   b) Choose a scene s_{vi} from $P(s_{vi}|z_{vi}, \psi_z)$, a multinomial probability conditioned on the action z_{vi} and ψ_z. Here, ψ_z is a multinomial distribution of action over scene z_{vi} with $\psi_z \sim Dirichlet(\beta)$.

   c) Choose a visual word w_{vi} from $P(w_{vi}|s_{vi}, \varphi_s)$, a multinomial probability conditioned on the scene s_{vi} and φ_s. And φ_s is the distribution of scene over visual words with $\varphi_s \sim Dirichlet(\xi)$.

Table 1. Notations used in this paper

SYMBOL	DESCRIPTION
S	Number of scenes
K	Number of action classes
M	Number of video clips in dataset
W	Number of unique visual words in dataset
Nv	Number of visual words in video v
vi	The ith visual word in video v
v	A video clip in dataset
z_{vi}	The action associated with the ith visual word in the video clip v
s_{vi}	The scene associated with the ith visual word in the video clip v

Under this generative process, the probability of video corpus having visual word set w, conditioned on α, β and ξ is:

$$P(w,z,s,\theta,\varphi,\psi|\alpha,\beta,\xi) = \prod_{i=1}^{N_v} P(w_{vi}|s_{vi},\varphi_{s_{vi}})p(\varphi_{s_{vi}}|\beta)$$
$$\prod_{i=1}^{N_v} P(s_{vi}|z_{vi},\psi_{z_{vi}})P(\psi_{z_{vi}}|\xi)$$
$$\prod_{v=1}^{V}\prod_{i=1}^{N_v} P(z_{vi}|\theta_v)P(\theta_v|\alpha)$$

(1)

The Action-Scene Model includes the following unknown parameters: the action distribution of video clip θ, the action-scene distribution ψ, the scene distribution φ and the assignments of individual words to action z and scene s. In this paper, we use Gibbs sampling [19] to evaluate the posterior distribution on z and scene s. Then, θ, φ and ψ can be inferred from the evaluated z and s. The parameters α, β and ξ are initialized with fixed values like LDA model (α=50/K, β=0.3, ξ=0.01). For each visual word, the equation needed for the Gibbs sampler is:

$$P(z_{vi} = x, s_{vi} = j | w_{vi} = m, s_{-vi}, z_{-vi}, w_{-vi}, \alpha, \beta, \xi) \tag{2}$$

$$\propto \frac{P(z, s, w | \alpha, \beta, \xi)}{P(z_{-vi}, s_{-vi}, w_{-vi} | \alpha, \beta, \xi)}$$

$$\propto \frac{C^{WS}_{mj,-vi} + \beta}{\sum_m C^{WS}_{m'j,-vi} + W\beta} \cdot \frac{C^{SZ}_{jx,-vi} + \xi}{\sum_{j'} C^{SZ}_{j'x,-vi} + S\xi} \cdot \frac{C^{ZV}_{xv,-vi} + \alpha}{\sum_x C^{ZV}_{x'v,-vi} + K\alpha}$$

where $z_{vi}=x$, $s_{vi}=j$ represent the ith visual word in the video clip is assigned to action x and scene j respectively. $w_{vi}=m$ represents the observation that the ith visual word is the mth visual word in the codebook, and z_{-vi}, s_{-vi} represent all action and scene assignment not including the ith visual word. C^{WS}, C^{SZ} and C^{ZV} are three count matrices. $C^{ZV}_{xv,-vi}$ is the number of times the visual words from video v assigned to action x not including the ith visual word, and $\sum_{x'} C^{ZV}_{x'v,-vi}$ is the number of visual words in video v. $C^{WS}_{mj,-vi}$ is the number of times a visual word m been assigned to scene j not including the current instance, and $\sum_{m'} C^{WS}_{m'j,-vi}$ is the total number of visual words assigned to scene j not including the current instance. $C^{SZ}_{jx,-vi}$ is the number of times the scene j is assigned to action x not including the ith visual word, and $\sum_{j'} C^{SZ}_{j'x,-vi}$ is the number of times scenes are assigned to action z excluding current instance.

Given z, s, w, α, β and ξ, it is straightforward to compute the posterior distributions on θ, φ and ψ. Using the fact that the Dirichlet is conjugate to the multinomial, we can estimate θ, φ and ψ with:

$$\theta_{xv} = \frac{C^{ZV}_{xv} + \alpha}{\sum C^{ZV}_{x'v} + K\alpha} \qquad \varphi_{mj} = \frac{C^{WS}_{mj} + \beta}{\sum C^{WS}_{m'j} + W\beta} \qquad \psi_{jx} = \frac{C^{SZ}_{jx} + \xi}{\sum C^{SZ}_{j'x} + S\xi} \tag{3}$$

where C^{WS}, C^{SZ} and C^{ZV} are counted from the assignment of s and z.

Algorithm 1 describes the process of recognizing action for a new video. In the training stage, the Gibbs sampling is applied over all the visual words in the training set. After a few iterations ($NN=300$ in our experiment), the count matrix C^{WS}, C^{SZ}, C^{ZV} are saved. When classifying a new video clip, we run the Gibbs sampling only on the visual words in the new clip. The sampling process gives the assignments of visual words to action and scene quickly from $P(z_{vi}, s_{vi} | w, z_{-vi}, s_{-vi}, \alpha, \beta, \xi)$. And the count matrices are updated after the assignment. Then the action distribution of this video θ_v is estimated based on the count matrix C^{ZV} using Equation (3).

The video clip is classified with a category index having the largest probability in θ_v. The index is assigned with actual action class label using ground truth labels in the training dataset. Each index corresponds to the most popular action class label within videos belonging to this index.

5   Integrated Action Recognition

The learned context in Section 4 could fuse with other action classification component easily. We explain how to combine the results from Action-Scene Model with existing method in subsection 5.1. In order to avoid negative effect of context, we introduce a factor measuring the contributions of contextual cues in subsection 5.2.

Algorithm 1. Action Recognition with Action-Scene Model

Input: Training set S with V videos, New video v_t with N_t visual words

Output: Prediction label l_t for v_t

1:   Initialize the action and scene assignments z_{vi}, s_{vi}

2: **for** iteration=1:NN

3:      **for** v=1 to V

4:          **for** i=1 to N_v

5:                draw z_{vi}, s_{vi} from $P(z_{vi}, s_{vi}|\, w, z_{-vi}, s_{-vi}, \alpha, \beta, \xi)$

6:                update C^{WS}, C^{SZ}, C^{ZV}

10: Compute the posterior estimates of θ with (3)

11: Draw action category index I for training videos from θ

12: Initialize the action and scene assignment z_{ti}, s_{ti}

13: **for** iteration=1:NN

15:      **for** i=1:N_t

16:          draw z_{ti}, s_{ti} from $P(z_{ti}, s_{ti}|\, w, z_{-ti}, s_{-ti}, \alpha, \beta, \xi)$

17:         update C^{WS}, C^{SZ}, C^{ZV}

20:   Compute the posterior estimates of θ_v with (3)

21:   Draw action category index I_t of v_t from θ_v

22:   Assign l_t with the most popular label in S having same index with I_t

---

5.1   Linear Combination with Other Components

For a video clip v, we denote that F_p, F_b is the input features to the existing component and Action-Scene Model respectively. The classification score integrating the contextual cues is defined as:

$$g(v,c) = g(F_p,c) + f_c \cdot g_{AS}(F_b,c) \tag{4}$$

where f_c is a vector of context weight. The first component is any classifier providing the prediction score of assigning label c to v. The second one is the prediction score of assigning label c to the video using Action-Scene Model.

5.2   Action Oriented Context

Since not every action class occurs in the special condition, it is necessary to leverage the importance of context to different action classes. Therefore, we compute a factor to estimate the strength of relationship between actions and scenes. Suppose variables $AC=\{\, c_1, c_2, ..., c_m\}$ and $SE=\{s_1, s_2, ..., s_n\}$ represent actions and scenes respectively. Then the information about AC captured by SE can be measured by mutual information $I(AC;SE)$:

$$I(AC;SE) = \sum_{c \in AC} \sum_{s \in SE} p(c,s) \log_2 \left(\frac{p(c,s)}{p(c)p(s)} \right) \tag{5}$$

For action class c, we compute f_c to indicate how much c related to scenes. The definition of f_c is:

$$f_c = \sum_{i=1}^{|SE|} p(c,s_i)\log_2 \frac{p(c,s_i)}{p(c)p(s_i)} \tag{6}$$

where f_c is computed based on the count matrix C^{SZ} learned from action-scene model. Larger f_c means the action is more closely related with scenes.

6   Experimental Results

We evaluate our approach on the challenging YouTube dataset provided by Liu et al [13]. The dataset was chosen for following two reasons: All the video clips are collected from YouTube website, which makes them "realistic" and challenging. Besides, each action was recorded under several scenes. The dataset is suitable for studying the effects of scene to action recognition. Following the evaluation methodology of [10,13], we apply leave-one-out cross validation over the 25 subsets of YouTube dataset in the experiments.

YouTube dataset is a large dataset that consists of 1168 videos. The dataset covers 11 action classes: basketball shooting, biking, diving, golf swing, horse riding, soccer juggling, swing, tennis swing, trampoline jumping, volleyball spiking, and walking with a dog. Each kind of action is performed under several different scenes and divided into 25 subsets.

6.1   Parameter Setting

The hidden parameter θ, φ and ψ were estimated based on training set. Fixing the number of actions $K=11$ for all models, we test the effect of S on recognition accuracy in the Action-Scene Model, as illustrated in Figure 4 (a). The actual number of scene categories in dataset is about 17. It shows that the recognition accuracy is the highest when S closes to the actual value. We also look for insights on the effect of β on recognition accuracy in Figure 4 (b). The experiment achieves the best result when $\alpha=50/K$, $\beta = 0.3$ and $\zeta = 0.01$.

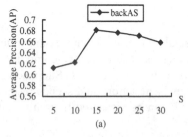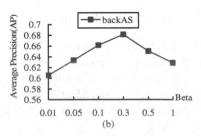

Fig. 4. (a)Average precision vs. the number of scene categories S for YouTube dataset. (b)Average precision vs. value of β for YouTube dataset. Experiments show that beta in range [0.1, 0.5] gets better results for recognition task.

6.2   Performance Evaluation

In order to validate whether the added variable in Action-Scene Model benefits action identification, we compare our Action-Scene Model with LDA model using same features, evaluation methodology and iteration times. In LDA model, the parameters is set as $\alpha=50/K$, $\beta = 0.01$.

Table 2 shows the recognition precision of different methods. The rows "backAS" and "backLDA" refer to the result from Action-Scene Model and LDA model respectively. It shows that the proposed method achieves an improvement over all the action categories. The average improvement in precision is 9.85%. The improvement is especially remarkable for the actions that have strong relation with scenes: basketball (17.73%), horse ride (18.19%) and trampoline jump (10.92%). Figure 5 is a more detail graph of improvement on each action category. It demonstrates that the contextual cues are informative and a good complementary for identifying the actions in realistic videos.

Table 2. Performance from scene features of different method

Method	Average	Basketball	Bike	Dive	Golf	Horse	Soccer	Swing	Tennis	Jump	Spike	Walk
Gist	53.20	38.38	60.69	69.00	61.00	66.00	9.00	42.00	61.00	54.00	81.00	43.09
Color	49.28	33.33	44.83	86.00	65.00	43.00	22.00	27.00	47.00	57.00	73.00	43.90
o+s	**70.67**	47.47	**73.79**	**91.00**	**90.00**	**73.00**	35.00	64.00	75.00	83.00	**89.00**	**56.10**
backLDA	58.28	66.67	29.66	61.54	51.41	49.49	89.74	72.99	80.24	72.27	35.34	31.71
backAS	**68.13**	**84.40**	42.76	66.03	59.86	67.68	**93.59**	**81.02**	**83.83**	**83.19**	43.97	43.09

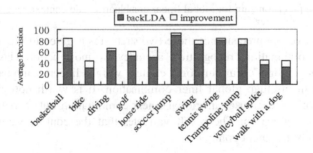

Fig. 5. Average precision of using LDA to recognizes actions from background. The yellow parts are the improvements of action-scene model over using LDA model.

In Table 2, we also compare our results with other methods based on scene features on the YouTube dataset. The Gist and color features of scene are used as single feature channels for recognition in [10], which corresponds to "Gist" and "Color" in Table 2. The result of "backAS" achieves a higher average precision than both of them. Further, "o+s" is the results of combine object feature (e.g. basketball, bike, etc.) and scene feature together [10]. Although lower in average precision, the result of "backAS" is comparable to "o+s" in several action classes (basketball, soccer juggle, swing and tennis swing), which demonstrates the effectiveness of background context modeling. Although "o+s" [10] achieved higher performance when object has explicit meaning for actions, it needs complex pre-processing for object extraction.

6.3   Context Weighting Measurement

We also test whether the contextual cues could be a complement to existing components. The combination of SVM classifier [19] and person region features is the most widely used component in action recognition. We integrate the learned contextual cues with this component under two settings: global linear combination and context weight linear combination. In the global linear combination, two components fused following the formula (4), where f_c is one for every action category. The context weight linear combination use the f_c computed as the formula (6), and the weight are normalized into the range [1, 2].

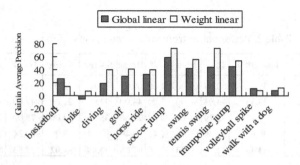

Fig. 6. Average precision gain for global linear combination and context weight combination

The gain in average precision of these two strategies is given in Fig. 6. After using global linear combination, recognition accuracy of most action classes has been improved besides "bike". Furthermore, context weight linear combination achieves larger improvements than global liner combination. It is worth noting that average precision for "bike" decreases in global linear combination but increases in context weight linear combination. The result validates that the context weight effectively captures the contribution of context.

7   Conclusion

In this paper, we have investigated recognizing actions from background regions of realistic videos. An Action-Scene Model is proposed to model and learn the relationship between actions and scenes without scene detectors. To automatically measure the dependency of action on scene, we compute a context weight based on the model. The learned contextual cues are an effective complement to the existing classifier with the context weight. Experimental results validate that Action-Scene Model effectively exploits contextual cues in realistic videos and the combination of the contextual cues indeed improve the recognition precision.

References

1. Blank, M., Gorelick, L., Shechtman, E., Irani, M., Basri, R.: Actions as space-time shapes. In: ICCV, pp. 1395–1402 (2005)
2. Blei, D., Ng, A., Jordan, M.: Latent dirichlet allocation. Journal of Machine Learning Research (3), 993–1022 (2003)
3. Comaniciu, D., Ramesh, V., Meer, P.: Real-time tracking of non-rigid objects using mean shift. In: CVPR, pp. 142–149 (2000)
4. Dalal, N., Triggs, B.: Histograms of oriented gradients for human detection. In: CVPR, pp. 886–893 (2005)
5. Dollar, P., Rabaud, V., Cottrell, G., Belongie, S.: Behavior recognition via sparse spatio-temporal features. In: 2nd Joint IEEE International Workshop on Visual Surveillance and Performance Evaluation of Tracking and Surveillance, pp. 65–72 (2005)
6. Everingham, M., Van Gool, L., Williams, C.K.I., Winn, J., Zisserman, A.: The pascal visual object classes (voc) challenge. IJCV 88(2), 303–338 (2010)
7. Felzenszwalb, P.F., Girshick, R.B., McAllester, D., Ramanan, D.: Object detection with discriminatively trained partbased models. IEEE Trans. Pattern Anal. 32(8), 1627–1645 (2010)
8. Griffiths, T., Steyvers, M.: Find scientific topics. Proceedings of the National Academy of Sciences 101(suppl. 1), 5228–5235 (2004)
9. Han, D., Bo, L., Sminchisescu, C.: Selection and context for action recognition. In: ICCV, pp. 1933–1940 (2009)
10. Ikizler-Cinbis, N., Sclaroff, S.: Object, scene and actions: Combining multiple features for human action recognition. In: Daniilidis, K., Maragos, P., Paragios, N. (eds.) ECCV 2010, Part I. LNCS, vol. 6311, pp. 494–507. Springer, Heidelberg (2010)
11. Jiang, Y., Li, Z., Chang, S.: Modeling scene and object contexts for human action retrieval with few examples. IEEE Transactions on Circuits and Systems for Video Technology 21(5), 674–681 (2011)
12. Laptev, I., Marszalek, M., Schmid, C., Rozenfeld, B.: Learning realistic human actions from movies. In: CVPR, pp. 1–8 (2008)
13. Liu, J., Luo, J., Shah, M.: Recognizing realistic actions from videos in the wild. In: CVPR, pp. 1996–2003 (2009)
14. Lowe, D.: Distinctive image features form scale-invariant keypoints. International Journal of Computer Vision 60(2), 91–110 (2004)
15. Marszalek, M., Laptev, I., Schmid, C.: Actions in context. In: CVPR, pp. 2929–2936 (2009)
16. Motwani, T., Mooney, R.: Improving video activity recognition using object recognition and text mining. In: ECAI (2012)
17. Niebles, J., Wang, H., Fei-Fei, L.: Unsupervised learning of human action categories using spatial-temporal words. IJCV 79(3), 299–318 (2008)
18. Oliv, A., Torralba, A.: Modeling the shape of the scene: a holistic representation of the spatial envelope. IJCV 42(3), 142–175 (2001)
19. Schölkopf, B., Smola, A.: Learning with Kernels: Support Vector Machines, Regularization, Optimization and Beyond. MIT Press (2002)
20. Schuldt, C., Laptev, I., Caputo, B.: Recognizing human actions: a local SVM approach. In: ICPR, pp. 32–36 (2004)
21. Ullah, M., Parizi, S., Laptev, I.: Improving bag of features action recognition with non-local cues. In: BMVC, pp. 1–11 (2010)
22. Yao, B., Fei-Fei, L.: Modeling mutual context of object and human pose in human-object interaction activities. In: CVPR, pp. 17–24 (2010)

Automatically Learning and Specifying Association Relations between Words[*]

Jun Zhang[1], Qing Li[2], Xiangfeng Luo[1], and Xiao Wei[1]

[1] School of Computer Engineering and Science, Shanghai University, Shanghai, China
{zhangjun_shu,luoxf,xwei}@shu.edu.cn
[2] Department of Computer Science, City University of Hong Kong, Hong Kong, China
itqli@cityu.edu.hk

Abstract. One of the most fundamental works for providing better Web services is the discovery of inter-word relations. However, the state of the art is either to acquire specific relations (e.g., causality) by involving much human efforts, or incapable of specifying relations in detail when no human effort is needed. In this paper, we propose a novel mechanism based on linguistics and cognitive psychology to automatically learn and specify association relations between words. The proposed mechanism, termed as ALSAR, includes two major processes: the first is to learn association relations from the perspective of verb valency grammar in linguistics, and the second is to further lable/specify the association relations with the help of related verbs. The resultant mechanism (i.e., ALSAR) is able to provide semantic descriptors which make inter-word relations more explicit without involving any human labeling. Furthermore, ALSAR incurs a very low complexity, and experimental evaluations on Chinese news articles crawled from Baidu News demonstrate good performance of ALSAR.

Keywords: ALSAR, Specify association relation, Information retrieval, Verb valency grammar, Cognitive psychology.

1   Introduction

Searching for interested and/or useful information has become an indispensable part of our daily activities. Current search engine technology, however, still falls short in providing people with adequate level of support. Consider the following example case. When a Ph.D. student, who is very interested in Dr. Wyatt's talk in an academic colloquium held the day before, wants to find some of Wyatt's papers, he may find that he has already forgotten the speaker's name. The only thing he remembers is that

* Research work reported in this paper was partly supported by the key basic research program of Shanghai under grant no. 09JC1406200, the National Science Foundation of China under grant nos. 91024012, 61071110, and 90612010, the National High-tech R&D Program of China under grant no. 2009AA012201, and by the Shanghai Leading Academic Discipline Project under grant no. J50103.

F. Li et al. (Eds.): WAIM 2014, LNCS 8485, pp. 578–589, 2014.

the speaker was a student of Prof. Roger and his name began with 'W'. To help recall the name of Wyatt quickly, the student may adopt the traditional information retrieval way, say, by typing the words 'Prof. Roger', 'student' and 'w' into the search bar and then the system returns a list of related pages which contain these words, as shown in Figure 1(a). However, these pages usually are not close to the user's need as there are many noises in them. Thus, such word matching based search engines may cost users a lot of time to obtain the information they really want.

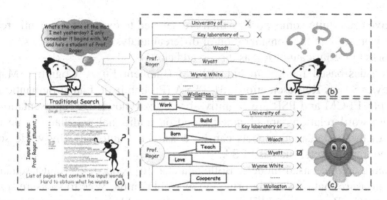

Fig. 1. Targeted Web information obtainable from (a) a traditional search engine, (b) relation-based search engine; and (c) association relation-based search engine

An alternative way for searching is based on the relations between words. Such a kind of search engines, denoted as relation-based search engines, can provide the information associated with the input words more effectively. For example, when 'Prof. Roger' is typed, the search engine may return its related terms, such as university, laboratory etc., as shown in Figure 1(b). With the help of a relation-based search engine, it is possible for users to achieve their goals more accurately. Yet only using association relations may still be inadequate to fulfill the user search requirements. Consider the 'Wyatt' example again, the student may only remember the beginning character 'W' of the name while the information around 'Prof. Roger' contains four similar terms with 'W', thus it may still require the student to check all the four terms. However, if all these association relations are well specified, such a situation can be avoided. In particular, as shown in Figure 1(c), when all the relations around 'Prof. Roger', such as "Prof. Roger was born in Waadt" etc., are well specified, it becomes much easier for the student to find out the destined name (e.g., Wyatt). In general, it can be shown that through the association relations and their specifications users will be able to reach their searching goals more quickly and accurately. Hereafter, we call this kind of search engines as association relation-based search engines.

Based on the theories in linguistics and cognitive psychology, this paper proposes a novel mechanism for automatically learning and specifying association relations between words. This mechanism, which we term as ALSAR, can obtain accurate association relations from the perspective of verb valency grammar [1], and from specifying these association relations with suitable verbs.

The remainder of this paper is organized as follows. In Section 2, we briefly discuss some related work. Section 3 gives the problem definition of our work. Section 4 presents the detailed processes of learning and specifying association relations between words. Evaluations are carried out in section 5. Finally, we conclude our work in Section 6.

2   Related Work

There have been quite some works which are devoted to extracting semantic relations between words. With the consideration of different analysis methodologies, we divide these works into four categories, as discussed below.

1) Statistics-based models include EFCM (Element Fuzzy Cognitive Map) [2], PSR (Power Series Representation) [3], PLSA (Probabilistic Latent Semantic Analysis) [4], etc. EFCM and PSR are cognitive-based text knowledge representation models, where word relations are obtained based on traditional association rule mining [5]. All of these statistical models can be constructed automatically while some of them need training sets, such as PLSA.

2) Search engine-based models: Bollegala et.al [6] proposes a robust semantic similarity measure that uses the information available on the Web to measure similarity between words. Bunescu et.al [7] proposes a supervised learning approach for relation extraction, which is based on a handful of training examples and the Web corpus. These search engine-based models are supposed to acquire more semantics than statistics-based models but they are largely dependent on the results of search engines.

3) Syntax-based models: SRL (Semantic Role Labeling) [8] is a broadly employed text mining technique, as it allows the addition of structured semantic information to plain text. Coppola et.al [9] proposes a general framework for syntax-based relation mining and achieves high accuracy by experimenting with support vector machine as a supervised approach. Oren Etzioni et.al [10] proposes the Open Information Extraction (Open IE) paradigm to develop unlexicalized, domain-independent extractors on English corpus. These models are capable of obtaining rich semantics from texts but they depend on a generic syntactic/structural organization.

4) Expert knowledge-based models include MindNet [11], WordNet [12] and RDF (Resource Description Framework) [13]. This kind of works is excellent in acquiring semantic relations between words and is useful for some special areas. However, all of these works are based on a corpus defined by humans, which costs a lot of human efforts and time.

With respect to the features of the above four models, we know that relations obtained through expert knowledge-based models possess the most abundant semantics while the ones extracted by statistics-based models have the poorest semantics. However, among all of the four, only statistics-based models have an automatic and independent extraction process, which means that the other three have limitations when applied into a dynamic and large scale environment. Meanwhile, while expert knowledge-based models have a good ability of acquiring relations with rich semantics, those relations' semantics are constrained to a pre-established area (for example,

"Parent Of", "Is A", "Subject-Predicate" and so on). Compared to the unlimited kinds of relations in real-world, the amount of relation types in expert knowledge-based models seems too small. For our work, how to automatically and independently discover relations with abundant semantic types is a fundamental and most important task.

3 Problem Definition

In this section, we formulate the problem of automatically learning and specifying association relations. For each document d_i in a document space $DS=\{ d_1, d_2, d_3, \ldots , d_n \}$, a latent relation space, $LRS(d_i)$, which represents all of the possible specific relations in d_i, is defined as a six-tuple:

$$LRS(d_i) = (N, V, E_N, E_{NV}, \eta, \delta) \tag{1}$$

where N is the set of nouns in document d_i, V is the set of verbs, E_N is a set of directed edges denoting association relations between nouns, E_{NV} is a set of directed labeled edges that further explain specific meanings of association relations through verbs, $\eta : N \times N \rightarrow P(E_N)$ is a function that maps the relations between nouns to their associated values ($P(E_N)$ being the power set of E_N), and $\delta : E_N \times V \rightarrow P(E_{NV})$ is a function that maps association relations which are specified with verbs, to their values ($P(E_{NV})$ being the power set of E_{NV}).

At the beginning of our proposed approach, each document d_i is represented in the form of lrs_i, abbreviated for $LRS(d_i)$, whose original state is defined as follows: N_i and V_i include all the nouns and verbs appearing in d_i; E_N contains all the possible relations between the nouns in N_i; and E_{NV} uses all the possible verbs to specify the relations in E_N. However, all the values in the power sets of both E_N and E_{NV} are initialized as zero, which means, $\forall n_1, n_2 \in N$, we have $\eta(n_1, n_2) = 0$, and $\forall en_i \in E_N$, $\forall v_j \in V$, we have $\delta(en_i, v_j) = 0$.

Based on the theories in statistics, linguistics and cognitive psychology, we aim to produce good denotations to the functions of η and δ, denoted as η_{ALSAR} and δ_{ALSAR}, respectively, whose purpose is to find suitable values for identifying whether there exists a relation between two nouns, or justifying whether a verb is fit to specify a relation. Our proposed ALSAR is responsible for producing η_{ALSAR} and δ_{ALSAR}. More specifically, given an original input document d_i, ALSAR processes d_i under the functions of η_{ALSAR} and δ_{ALSAR}, and produces the specific association relation set as output, subjected to the following condition:

$$(\forall n_1, n_2 \in N)(\forall v \in V)(\forall en \in E_N)(\eta(n_1, n_2) \geq \theta \wedge \delta(en, v) \geq \sigma)$$

where θ and σ are some predefined thresholds. The following section gives detailed discussions on how to learn and specify association relations between words.

4   Learning and Specifying Association Relations

As mentioned early, we make use of verb valency grammar [1] to automatically learn the association relations. In linguistics, verb valency refers to the number of arguments controlled by a verbal predicate. Consider, for example, the following two sentences:

> **Sentence 1:**   "*Michael*[1] **wins** the *football game*[2]."
>
> **Sentence 2:**   "*Jackie*[1] **gives** *Sara*[2] a *flower*[3]."

The two sentences show that the nouns depending on "*win*" include "*Michael*" and "*football game*", while the nouns depending on "*give*" include "*Jackie*", "*Sara*" and "*flower*". Thus, the valences of "*win*" and "*give*" are 2 and 3, respectively. Different from the theories in universal grammar proposed by Chomsky [14], verb valency grammar reflects the most basic dependence relations between nouns and verbs.

Consequently, on the basis of verb valency grammar, the process of learning and specifying association relations is conducted to three steps, which are discussed in the following subsections.

4.1   Acquiring Verb Dependent Sets Based on Cognitive Psychology

We first present a detailed discussion on how to automatically acquire verb dependent sets (*VDS*) for each sentence. Although verb valency grammar only describes the fundament dependent relations between nouns and verbs, it is still a challenge for machines to identify whether a noun is dependent on the verb around it. For example, consider a sentence as follows,

Sentence 3:   "In order to **please** *Mary*[1], *William*[2] **went** out and **bought** a bunch of *roses*[3] in a *flower shop*[4] on *Friday*[5]."

There are five noun phrases and three verbs in this sentence. Taking the verb "*bought*" for instance, it is easy for us to understand that the four words, "*William*", "*roses*", "*flower shop*" and "*Friday*", are controlled by "*bought*". However, it is extremely hard for a machine to understand that, as it may regard all of the five noun phrases as the dependents of the verb.

When referring to verb valency grammar, we find the valences of almost all verbs distribute in the range of 0 to 4. Many verbs have the valences of 1, 2 and 3, while a few verbs take four arguments. There is hardly any verb possessing a valence over 5. According to Miller's Law [15] which is a classical theory in cognitive psychology, the reason for the boundary of valency 4 can be attributed to the limited capacity of working memory. Miller's Law pointed out [15] that the number of objects an average human can hold in working memory is seven, plus or minus two. Also, the span is around seven for digits, around six for letters and around five for words [15]. Thus, it is obviously hard for authors to give more than four dependents to a verb, since the total amount of both the verb and its dependents will exceed the capacity of authors' working memory. As a result, based on Miller's Law in cognitive psychology, we define the Max Valency Number (*MVN*) to be 4 for each verb, i.e., *MVN*=4.

However, it is still hard for machines to select suitable nouns to a verb's dependent set, even though its *MVN* is set as 4. In particular, it is non-trivial to identify which kind of nouns is the most possible arguments controlled by a verb. In linguistics, there are many arguments for a verb, such as subject, direct object, indirect object, time adverbial, place adverbial and etc. It is a difficult job for machines to understand that. Thus, through observations on abundant sentences, we conclude below three features that machines can perceive:

1) the nouns or noun phrases that are led by prepositions are always set as time adverbials, place adverbials or some other adverbials;

2) subjects or objects of a verb are always around the verb, i.e., the distance between the subjects/objects and the verb is small;

3) there is always only one verb among the nouns or noun phrases which belong to the same verb's dependent set unless there are conjunctions around that verb.

Note that the above three features show the different importance of the nouns or noun phrases to the verb they are around. For example, to the same verb, the nouns led by prepositions are less important than the ones that are not led by prepositions. Consider again the verb *"bought"* in *Sentence 3*, the nouns *"William"* and *"roses"* have more contributions to *"bought"* than the nouns *"flower shop"* and *"Friday"*, the latter are led by prepositions and represent the place and time, respectively. According to the above three features and Miller's Law, it is possible for machines to automatically generate dependent set for each verb without involving with any human effort. Verb dependent sets (*vds*) will greatly facilitates efficient acquisitions of association relations between words, since it allows confining and reducing the search space to a large extent, as to be discussed next.

4.2 Acquiring Association Relations between Words

On the basis of verb dependent sets, we now present an efficient method to acquire association relations. Compared with traditional selections of transaction set (such as set of sentences or sliding windows), verb dependent set provides a much stronger binding among nouns or noun phrases, which is very useful to the subsequent association relation mining. In this paper, our   proposed algorithm removes the threshold of support as it is affected by the size of transactions. Meanwhile, to alleviate the effects resulting from removing this threshold, we include a check strategy before picking noun pairs from verb dependent sets. Through removing support threshold and adding the check strategy, it becomes possible for our proposed association relation acquiring algorithm to be insensitive to the lengths of texts.

The specific problem the check strategy needs to tackle is as follows. Given a verb dependent set, select the most possible noun pairs that tend to associate with each other according to the characteristics of a language. It should be noted that the setting of the check strategy is language dependent, and in this paper, we mainly focus on the characteristics in Chinese. Herein, the check strategy we use is described as follows:

Through observations on Chinese news, we find that those nouns appearing in both sides of a verb tend to be associated with each other while those nouns appearing in the same side of a verb are likely to have a lower association power among them. This phenomenon can also be explained in English. Consider the following sentence:

Sentence 4:"Both <u>*Jackie*</u>[1] and <u>*William*</u>[2] **bought** *Mary*[3] a bunch of <u>*roses*</u>[4]."

Through Section 4.1, the dependent set of the verb "*bought*" can be obtained, which is {*Jackie, William, Mary, rose*}. It is apparent that the relation between "*Jackie*" and "*Mary*" or between "*William*" and "*rose*" is much stronger than the relation between "*Jackie*" and "*William*" or between "*Mary*" and "*rose*". Consequently, we only consider four relations rather than six relations ($C_4^2 = 6$), which largely reduces the complexity of the subsequent learning process.

Based on *vds* and the above check strategy, we treat *vds* as transaction set and then use Apriori [5] to further mine association relations, which compose of the learnt association relation set, denoted as *lars*.

Compared with traditional association rule mining (such as Apriori [5] and FP-Tree [16]) whose candidate space increases exponentially with the number of words, our algorithm obtains a much smaller candidate space, thereby the complexity of learning association relations is largely reduced.

4.3   Specifying Association Relations between Words

So far there have been many works devoted to specifying the relations between words, including WordNet[12], RDF[13] etc. Relations including "Parent Of", "Is A', "Subject-Predicate" and "Cause-Effect" considered by them though useful for NLP applications in some special domains (such as the medical area), are not necessarily appropriate for general-purpose information retrieval tasks universally. For example, the news event "*melamine milk powder*" in China, the noun "*melamine*" was not available in WordNet's lexical database at that time, let alone the relation between "*milk powder*" and "*melamine*". Even if it had contained both of them, it would be hard for WordNet to use "*contain*" or "*include*", which is a lot closer than "*Part Of*" as per human's understanding, in describing the relation between "*milk powder*" and "*melamine*". Additionally, all of the relations in WordNet or RDF are defined by experts and thus incur a lot of human efforts.

This subsection describes a method for specifically and automatically explaining the relations from the perspective of verbs. Based on the sets of verb dependents and association relations learnt earlier, our basic idea of specifying relations is to use the verbs in verb dependent sets to describe the relations in an association relation set. There are several advantages of using verbs to express the meanings between words: 1) verbs are able to show lots of relation types; 2) verbs are closer to users' general demands than those pre-defined relation types; 3) verbs are possible to catch up with the rapid information updates on the Web; and 4) these verb descriptions are possible to be automatically derived without any human effort. These advantages of using verbs to specify relations demonstrate a good application prospect. Detailed discussions are given as follows.

Give a document d_i, its verb dependent set vds_i, and learnt association relation set $lars_i$ can be obtained (ref. Section 4.1 and 4.2). The objective of our association relation specifying algorithm is to generate a specified association relation set, $sars_i$. It should be noticed that $sars_i$ as a mapping from $lars_i$ to the verb set in d_i is defined as $sars_i = \{((n_1, n_2), v_j) \mid (n_1, n_2) \in lars_i \wedge v_j \in V\}$. The process of acquiring $sars_i$ can be divided into two steps: 1) find verb candidate set for each association relation, and 2) select proper verbs having large contributions from each candidate set.

Based on vds_i and $lars_i$ obtained in Sections 4.1 and 4.2, we are able to use statistical method on these two sets to acquire verb candidate set for each association relation. The way we use here is that, given an association relation $ar_k=(n_1, n_2)$, collect the verb, whose dependent set contains both n_1 and n_2, into ar_k's verb candidate set (vcs). After the process of acquiring vcs, we select proper verbs from it according to the co-occurrence of the association relation and the verbal descriptor.

Overall, relations obtained by applying algorithms in Section 4-5 (which, altogether, constitute our ALSAR mechanism) are very useful for machines to express the semantics in texts, thereby providing better Web services to users of diverse applications including information retrieval, semantic web, personalization, news recommendation, etc. Evaluations on ALSAR are carried out and reported in the next section.

5   Evaluations

5.1   Evaluation Methodology

Instead of trying to compare all kinds of the related algorithms reviewed in Section 2, which would be impossible due to limited accessibility and space available, we mainly focus on comparing our proposed ALSAR with the traditional association rule mining algorithm [5]. To evaluation the quality or accuracy of the learnt relations, we plan to apply the relations into a specific application, namely, classification. The experimental process is conducted in three steps:

1) We use words and their relations to represent each document in a given document set,. Herein, the words are first extracted from the nouns of the top-twenty degrees (which include both in-degrees and out-degrees) in the network of relations between nouns. Weights of these words are defined as follows:

$$w_{ik} = Degree_{ik} \bigg/ \sum_{l=1}^{s} Degree_{il} \qquad (2)$$

where w_{ik} is the weight of the kth word in document d_i, $Degree_{ik}$ is the degree of the kth word in d_i, and s is the number of words in d_i.

2) Assuming there are m topics in the data set, we randomly select a document for each topic and then mix all of the topics together. For each document d_i in the mixed document set, we compute the similarity between d_i and each of the selected documents based on the representation of step 1). The similarity we use here is cosine similarity, defined as:

$$CosineSimilarity\left(d_i, d_j\right) = \frac{\sum_{k=1}^{s'}\left(wf_{ik} \times wf_{jk}\right)}{\sqrt{\sum_{k=1}^{s'}\left(wf_{ik}\right)^2} \times \sqrt{\sum_{k=1}^{s'}\left(wf_{jk}\right)^2}} \tag{3}$$

where wf_{ik} and wf_{jk} are, respectively, the kth feature in documents d_i and d_j, s' is the number of features. Herein, features are comprised of two kinds: words, and unspecified or specified association relations. Weights of words are as defined in eq.(4) while weights of association relations are referred to their confidences acquired in Section 4.2. It should be noted that the same association relations with different verbal descriptors are denoted as different features in this similarity computations. Based on eq.(5), the similarities between each pair of the selected documents belonging to different topics can be obtained. Through these, we allocate document d_i under the topic to which the document most similar to d_i belongs. We repeat this process until all of the documents are allocated.

3) Based on the above classification, we estimate the results and further evaluate the performance by using precision, recall and F-measure.

Through these measures we are able to estimate the classification results for each topic and then, according to the average scores, to evaluate the total effect of each algorithm to be compared with.

5.2   Data Set

Our research group has crawled and accumulated, since 2009, more than 25,000 topics comprised of more than 150,000 Chinese news webpages from Baidu News (see details from http://wkf.shu.edu.cn).

Rather than applying all of the 150,000 Chinese news webpages into our classification evaluation — a time costing job, we randomly select a total of 4321 Chinese news webpages under one hundred topics as our data set. (For each topic, there are about fifty webpages.) Based on these Chinese news webpages and the evaluation methodology discussed in Section 5.1, the following evaluations are carried out.

5.3   On Learning Association Relations

Based on the methodology discussed in Section 5.1, Figure 2 gives a comparison between the traditional association rule mining (*TARM*) algorithm [5] and our proposed *ALSAR*. Herein, *WR_TARM* refers to those classification results that are based on both of the words and association relations extracted by the traditional association rule mining algorithm; meanwhile, *WR_ALSAR* denotes the classification results based on both words and association relations learnt by the algorithm of *ALSAR*. It should be noted that all of the words are extracted based on association relations so as to avoid the influences brought by other words extraction algorithms, such as *TFIDF*.. In order to reduce the influences brought by the random selection of cores, all of the classification tasks have been conducted for a hundred times. Thus, the precisions, recalls and f-measures appearing in figures are all referred to their average values.

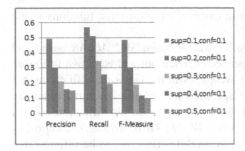 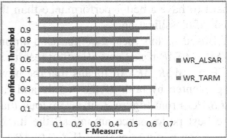

(a) Classification results based on WR_TARM under different support thresholds

(b) Comparisons between WR_TARM-based and WR_ALSAR-based classification results under different confidence thresholds

Fig. 2. Comparisons between *TARM* and our proposed *ALSAR*

From Figure 2(a), we can see that for *WR_TARM* under the same confidence threshold, all of the three metrics (precision, recall and f-measure) decrease very fast with the increase of support thresholds. It indicates that the performance of the traditional method is greatly affected by the settings of the support threshold. This is understandable because the supports of association relations obtained by the traditional method are directly affected by the length of text, yet the lengths of Chinese news webpages in our data set vary very much. So when the support threshold increases, few association relations will be left in those large-sized news webpages, which result in the bad performance as shown in Figure 2(a). Indeed, the traditional association rule mining algorithms are probable to obtain unstable performance when applied into an environment containing various lengths of texts. For our proposed *ALSAR*, the argument of support has been replaced by a newly proposed check strategy, according to the characteristics of a language. This makes our proposed *ALSAR* to be more feasible and appliable to a practical environment in which the documents vary diversely in length.

Figure 2(b) gives a comparison between *WR_TARM*-based and *WR_ALSAR*-based classification results. It can be seen that under different confidence thresholds, all the classification results vary very little. Different from support, it indicates that confidences of association relations are influenced very little by the different lengths of texts. Nevertheless, from Figure 2(b), it is also apparent that in each confidence threshold, *WR_ALSAR* obtains distinctively better performance than *WR_TARM*. In other words, the association relations obtained by our proposed *ALSAR* are better than the ones obtained by the traditional association rule mining algorithm.

5.4 On Specifying Association Relations

In this section, we further verify the effect of the association relations specified by *ALSAR*. The basic idea is to compare the differences between the association relations that have been specified with the ones that have not been specified. If specified association relations make more contributions to the understanding of a text, we regard them to have a better performance than those unspecified association relations in the task of classification.

Based on this idea, a comparison among *W_ALSAR*-based, *WR_ALSAR*-based and *WSR_ALSAR*-based classification results has been carried out and is shown in Figure 3. Here, *WSR_ARSAR* means that the news webpages in the task of classification are represented by both of the words and the specified association relations mined by *ALSAR*. From Figure 3, it is obvious that *WSR_ALSAR*-based classification obtains the best performance among all the three, while *W_ALSAR*-based classification has the worst performance and *WR_ALSAR* is in the middle. Thus, it can be inferred that using verbs to describe the association relations is actually useful for machines to have a better understanding of the texts.

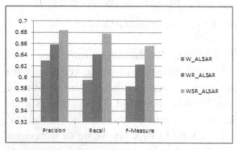

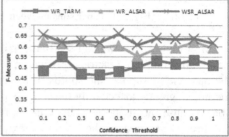

Fig. 3. Comparisons of W_ALSAR-based, WR_ALSAR-based and WSR_ALSAR-based classification results under the condition of confidence=0.1

Fig. 4. Comparisons of WR_TARM-based, WR_ALSAR-based and WSR_ALSAR-based classification results under different confidence thresholds

In addition, Figure 4 gives a comparison among *WR_TARM*-based, *WR_ALSAR*-based and *WSR_ALSAR*-based classification results under different confidence thresholds. It can be seen that, among all of the confidence thresholds, F-Measures obtained by *WSR_ALSAR* always stay above the ones obtained by *WR_ALSAR*, and likewise, *WR_ALSAR* stays above *WR_TARM* in terms of F-Measures. Consequently, it demonstrates that the association relations specified by verbs are really useful for machines to strengthen the ability of text understanding.

6   Conclusions

As the Web enters big data era, it is important to provide an effective support for upper level Web applications, such as information retrieval, knowledge representation, news recommendation etc. Based on verb valency grammar from linguistics and Miller's Law in cognitive psychology, we have developed an automatic association relation learning and specifying mechanism, *ALSAR,* which incorporates two main processes: learning association relations, and specifying the relations with verbal descriptors. Different from the tradition association rule mining, we apply verb valency grammar and Miller's Law into the processes of *ALSAR,* so as to ensure that the relations learnt are well specified, and can be established automatically without human efforts. Experiments on webpages crawled from Baidu News demonstrate better performance of *ALSAR* in comparison with the traditional association rule mining.

For our future study, we plan to construct reasoning rules based on the specified association relations. We also plan to evaluate our mechanism on more data sets, including UCI KDD archive [17] and TREC corpus [18].

References

1. Dixon, R., Aikhenvald, A.: Changing valency: case studies in transitivity. Cambridge University Press, Cambridge (2000)
2. Luo, X.F., Wei, X., Zhang, J.: Guided game-based learning using fuzzy cognitive maps. IEEE Transactions on Learning Technologies 3(4), 344–357 (2010)
3. Luo, X.F., Cai, C.L., Hu, Q.L.: Text knowledge representation model based on human concept learning. In: The 9th International Conference on Cognitive Informatics, pp. 383–390 (2010)
4. Hofmann, T.: Unsupervised Learning by Probabilistic Latent Semantic Analysis. Machine Learning 42(1-2), 177–196 (2001)
5. Agrawal, R., Srikant, R.: Fast algorithms for mining association rules. IBM Research Report RJ9839 (1994)
6. Bollegala, D., Matsuo, Y., Ishizuka, M.: Measuring semantic similarity between words using web search engines. In: Proceedings of the 7th International Conference on World Wide Web, pp. 757–766 (2007)
7. Bunescu, R., Mooney, R.: Learning to extract relations from the Web using minimal supervision. In: Proceedings of the 45th Annual Meeting of the Association of Computational Linguistics, pp. 576–583 (2007)
8. Carreras, X., M'arquez, L.: Introduction to the CoNLL-2005 Shared Task: Semantic Role Labeling. In: Proceedings of CoNLL 2005, pp. 152–164 (2005)
9. Coppola, B., Moschitti, A., Pighin, D.: Generalized framework for syntax-based relation mining. In: The 8th IEEE International Conference on Data Mining, pp. 153–162 (2008)
10. Etzioni, O., Fader, A., Christensen, J., Soderland, S., Mausam: Open information extraction: the second generation. In: Proceedings of the Twenty-Second International Joint Conference on Artificial Intelligence (IJCA 2011), vol. 1, pp. 3–10 (2011)
11. Stephen, D., William, B., Lucy, V.: MindNet: acquiring and structuring semantic information from text. In: Proceedings of the 17th International Conference on Computational Linguistics, pp. 1098–1102 (1998)
12. Fellbaum, C.: WordNet. In: Theory and Applications of Ontology: Computer Applications, pp. 231–243 (2010)
13. Resource Description Framework (RDF) Model and Syntax Specification, W3C Recommendation (1999), http://www.w3.org/TR/1999/REC-rdf-syntax-19990222
14. Chomsky, N.: The minimalist program. MIT Press, Cambridge (1995)
15. Miller, G.: The magical number seven, plus or minus two: some limits on our capacity for processing information. Psychological Review 63(2), 81–97 (1956)
16. Han, J.W., Pei, J., Yin, Y.W.: Mining frequent patterns without candidate generation. In: Proceedings of the 2000 ACM SIGMOD International Conference on Management of Data (SIGMOD 2000), pp. 1–12 (2000)
17. http://kdd.ics.uci.edu/summary.data.type.html
18. http://trec.nist.gov/data.html

A Chinese Question Answering System
for Specific Domain

Tanche Li, Yu Hao, Xiaoyan Zhu, and Xian Zhang

Tsinghua National Laboratory of Intelligent Technology and Systems (LITS)
Department of Computer Science and Technology
Tsinghua University, Beijing, 100084, China
Bing Data Mining Team, Microsoft Corp. Bellevue WA, 98005, USA
litc11@mails.tsinghua.edu.cn, {haoyu,zxy-dcs}@tsinghua.edu.cn,
szhang@microsoft.com

Abstract. We built a domain-specific Chinese interactive Question Answering(QA) system which has already been available as a public service via phone text messaging. The system utilizes Topic Forest [1] as the dialog management model to be capable of keeping track of users' interests. The Question Answering component is a hybrid approach which consists of both a community Question Answering engine [2] and a new knowledge-based QA engine. In the new QA engine, we constructed a semantic pattern matching model to automatically translate question topics and targets generated by the natural language understanding unit to SPARQL [3] queries, which eventually build the final answer. Through experimental data collected from real user survey and case study, our system shows promising results with the comparison over the practical QA system *Siri* as the industrial standard in terms of both accuracy and user satisfaction rates.

1 Introduction

With the increasing development of mobile internet, more and more people found smartphone a handy choice to acquire knowledge and assistance in their daily life. Among many such mobile apps, *Siri* is one of the best example that can accept and respond with natural language sentences, such as "今天天气怎么样? (What's the weather like today)", "打电话给A(Call A)". The power of *Siri* actually reflects the state-of-the-art results of many typical research problems. Apparently, Question Answering (QA) is one of them to answer questions like the former question above. To address the increasing needs to provide more accurate and professional response to user's requests in specific areas, recently, a number of similar applications, such as *Sogou Voice Assistant* and *Xunfei Yudian*, have been launched by other companies, but more focusing on answering domain-specific natural language questions. However, the performance of those applications are still far from satisfactory. Our system is another effort to provide accurate question answering results for specific domains.

After many years of research especially inspired by Text Retrieval Conference(TREC) and the Message Understanding Conferences(MUCs) [4], some QA

F. Li et al. (Eds.): WAIM 2014, LNCS 8485, pp. 590–601, 2014.
© Springer International Publishing Switzerland 2014

systems have achieved good performance [5, 6]. However, mainstream Question Answering systems only focus on English language. Natural language QA systems in Chinese are almost a blank due to many reasons including the complexity and diversity of Chinese language, lack of reliable knowledge base for the backbone of QA engine. So Chinese users have to search the answers in specific search engine, such as *Soso Music*, or in general search engine, which can return direct answers, but only effective when searching by keywords in the traditional way in search engine, thus lack of accuracy and understanding of user's intentions, thus can only return general-purpose contents and leave the burden to find out answers to the users.

In this paper, we provided a solution to address the problems above and materialized it as a publicly available Chinese interactive Question Answering system in musical domain. The solution is a combination of many practical techniques which includes: 1) natural language understanding, 2) knowledge base construction, management and query, 3) dialog system techniques. Section 2 outlines the architecture of the system; Section 3 describes the natural language understanding process; Section 4 describes the algorithm of dialog management; Section 5 introduces the knowledge base and Section 6 presented the experiments in comparison with *Siri* system.

2   System Overview

There are four major components in our system:

- Dialog Management
- Natural Language Understanding
- Knowledge Base
- Answer Constructor

Our system architecture is built around these components as shown in Fig.1. Dialog Management is the central coordinator component [7], which receives natural language questions from user, discomposes the question into subproblems to dispatch to other components to solve, and progressively interacts with

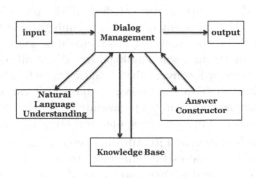

Fig. 1. System Architecture

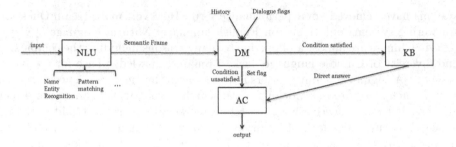

Fig. 2. System Flowchart

users to understand user's intention and provides direct answers. Natural Language Understanding unit (NLU) transforms questions into semantic frames via named entity recognition and patterns. Knowledge Base unit converts the semantic frame into SPARQL sentences and retrieves direct knowledge items to the question. Answer Constructor unit constructs answers in natural language using results from Knowledge Base unit.

The whole system is target driven, that is, the system tries to understand the specific targets of the user's intention, and may interact with user in natural language until the target is clarified. For example, the target in the musical QA system could be a song, a singer, or an album, etc.

To provide answers, the system must be backed with a domain-specific knowledge base. The knowledge base also provides source of knowledge for Natural Language understanding units. For instance, Named Entity recognition, pattern extraction and matching all heavily depends on known entities.

Fig.2 shows the process of our system in detail. The system firstly calls the NLU unit upon the arrival of a question, which identifies the name entity, and then transforms it into predefined categories, e.g. "周杰伦" → "singer". Then the pattern matching module will be invoked to retrieve the specific target of the question and converts this natural language question to an internal semantic frame. The Dialog Management unit hence checks the semantic frame to judge whether specific conditions are satisfied. If the condition are not satisfied, for example, some component is missing or ambiguity is found, the Dialog Management unit will let the Answer Constructor unit ask the user to provide missing information or try to resolve ambiguities. When all conditions are met, the Knowledge Base unit will convert the semantic frame to a SPARQL sentence and get the direct answer from the knowledge base. If it can not get the direct answer, it will invoke the community QA module to get the answer. Finally, the Answer Constructor unit will give users a natural language answer.

With the pluggable system architecture, our system has proved to have strong extensibility and has been successfully migrated to three domains including the original musical field, medical science and China Mobile service. Just for consistence, we will stick with musical domain as example throughout this paper.

3   Natural Language Understanding

3.1   Named Entity Recognition

Entity database is constructed to support dictionary-based Named Entity Recognition. All entity lists are extracted from structured data supported by Tencent. As the system starts, all the entity lists are loaded in the form of Tries indexed by entity aliases. When a question comes, all the possible results of Named Entity recognition are saved in order to reduce the chance of false negatives in entities.

In the musical QA system, we constructed 9 entity lists in total, including album, area, chart, drama, language, singer, song, style and type. Below is an example of Named Entity Recognition. Note that in our entity lists, there is a song named "吴克群" and a singer named "吴克群" as well. The result of Named Entity recognition for question "吴克群是谁唱的(Who sings the song 吴克群)" is shown as below.

> 吴克群是谁唱的
> *song*是谁唱的
> *singer*是谁唱的

3.2   Semantic Pattern Matching

Main aim of natural language understanding component is to find the target of the question and convert the natural language question to a semantic frame. The semantic frame is a collection of slots, and each slot is a piece of structured information extracted from the question, which might be required to find the answer by our Question Answering system. We use semantic patterns to extract semantic frame from the question. Each semantic pattern consists of three components:

- Target : Information slot type to be extracted using the pattern;
- Pattern : Template to be aligned with the question text to extract the target, which is usually a sequence of certain words and target placeholders.
- Priority : The confidence of the semantic pattern

Firstly, the question is matched with all possible patterns and the best matches(with longest match) are chosen. Then the matched semantic patterns are sorted by their corresponding priority and the sorted list are sent to the knowledge base component. For example, suppose we have a semantic pattern:

pattern	target	priority
song是谁唱的	singer	10

The semantic frame of "吴克群是谁唱的" built by pattern matching will be: {target : singer; song : 吴克群}. As there is no pattern like "singer是谁唱的", though the Named Entity recognition results may suggest that "吴克群" can also be a singer, this frame is filtered by pattern matching.

3.3   Semantic Pattern Generation

The semantic patterns are generated automatically as described in F Bu's publication [8]. The targets in the automatically generated patterns are tagged manually. In this way, it is easy for our system to be migrated to other domains.

We recognize the named entity of the questions first. Then we replace each word with an untyped placeholder after POS tagging. Finally we count the appearance of patterns over all questions. The target placeholders of these patterns are labeled manually and count of occurrences are taken as the priority of the pattern.

In our musical QA system, 3 million questions collected from community question answering dataset are used to generate the semantic patterns.

4   Dialog Management

Dialog Management maintains the information context of the conversation between human and computer. The following conversation is an example that the first question-answer pair set up the context for the next question, which implicitly uses the previous answer as the subject.

Q1: 七里香是谁唱的(Who sings the song "Common Jamsine Orange")
A1: 周杰伦(Jay Chou)
Q2: 他有什么专辑(What albums does he have)
A2: 《七里香》、《叶惠美》等("Common Jamsine Orange", "Ye Huimei", etc)

To resolve the problems mentioned above, we build Topic Forest [1] with different topic trees. Each tree has several leaf nodes represented the information of conditions the topic needs. Leaf nodes are connected by *AND* and *OR* in order to judge whether the conditions have satisfied or not.

Meanwhile we build Shared Information Index(SII) [1] to stored the history information. The content of SII can be stored while the topic tree changed.

There are four cases of dialog in our system according to the difference of semantic frame. Target means the aim of the question and the condition means the constrains.

– Both target and condition
  Regard the input as a new question and save the information to SII after getting the direct answer.

– Only target
  Merge the SII into the current topic tree and judge whether the conditions
  satisfy with the relation between leaf nodes.
– Only condition
  Inherit the last target and change to new topic tree.
– None
  Enter the chatting mode.

We show a example of topic tree in Fig.3. The topic tree has inner nodes
and leaf nodes. The inner node has two values, *AND* and *OR*, represented the
relation of its children should satisfy. PP refers to the imported inner nodes.
From the topic tree in Fig.3, a query *(album* || *singer* || *lrc)* is generated.

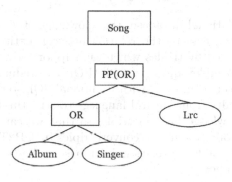

Fig. 3. Topic Tree

We chose topic forest as our dialog management since it is easy to inherit
targets. We only need to maintain a current topic tree. Upon moving to the next
question, if it doesn't have some target, the current topic tree just inherit the
context from the previous sentence.
    Another advantage of topic forest is that it is easy to manage and judge
conditions. We put the conditions into the trees and judge by the relation of leaf
nodes.

5   Knowledge Base

5.1   Knowledge Base Generation

In our system, knowledge is relations and facts between objects in the domain of
interest. Each piece of relation or fact is represented by a triple in our knowledge
base, which consists of two entities and a relation, i.e $t = (s, p, o)$. We regard
the name entity as a subject entity s and object entity o and the target in the
system as a Relation p. The following table is an example of our structured data.
    The Table 1 shows 5 such triples:

Table 1. Structured data example

ID	Name	Singer	Public Time	Language
1	七里香	周杰伦	2008	中文

$$(song, id, 1)$$
$$(song, name, 七里香)$$
$$(song, name, 周杰伦)$$
$$(song, publish\_time, 2008)$$
$$(song, language, 中文)$$

Compare with relational database, the knowledge stored as relation triples provides more direct access to the relation between entities. Meanwhile, the knowledge base is stored by triples which can support advanced query engines like SPARQL[1]. For complex questions, SPARQL can reduce the complexity of the statement effectively compared with traditional SQL so that it decreases the transformation difficulty from natural language to structural queries.

We build our knowledge base in triples so that we can use its advantages mentioned above. In our system, we convert triples by DB2Triples[2]. DB2Triples can associate multiple tables in relational database and generate triples to avoid problems mentioned above.

5.2 Knowledge Base Query

As mentioned above, the semantic frame consists of a target t and a group of conditions C. We define the subject entity of the question as e, and the answer as x.

The target t is a relation in knowledge base and we will search answer x in knowledge base. So our goal is to find the relation triple (e, t, x) in the knowledge base, which is the first constraint of SPARQL.

$C = \{c\}$ defines the condition that the subject entity should meet. A condition in semantic frame is defined as $c = (p, o), p \in R(elation), o \in E(ntity)$. In this condition, the subject entity should meet triples (e, p, o). C can be converted to a group of constraint in triples in this way.

Examples The semantic frame of question "日本摇滚女歌手有什么歌(List some songs of a female Japanese rock singer)" is as follows:

{target=song, gender=女(female), nationality=日本(Japan), style=摇滚(rock)}

---

[1] http://www.w3.org/TR/rdf-sparql-query/

[2] https://github.com/antidot/db2triples

This frame can be converted to 4 constrains:

$$(\ e, song, x)$$
$$(\ e, male, 女)$$
$$(\ e, nationality, 日本)$$
$$(\ e, style, 摇滚)$$

5.3 cQA module

Our system has a cQA(Community Question Answering) module [2] to get the answer if direct answer cannot be retrieved from knowledge base. The cQA module is built by millions of QA pairs extracted from online QA communities, like *Baidu Knows* and *Soso Wenwen*. *Baidu Knows* is a Chinese language collaborative Web-based Collective intelligence by question and answer provided by the Chinese search engine Baidu. *Soso Wenwen* is another community question answering provided by Tencent. When a question comes, we find the similar question in all QA pairs and get a question list. Then we get an answer list from the question list. Finally we rerank the answer list and get the best answer.

6 Experiment

6.1 Baseline system and Our system

Due to the lack of publicly available Chinese QA systems, we chose *Siri* as the baseline system for comparison, which is capable of Chinese Question Answering. *Siri* is an intelligent personal assistant and knowledge navigator in Apple Inc.'s iOS. *Siri* has both speech and text interfaces.

Our system is presented on WeChat. WeChat is a mobile text and voice messaging communication service developed by Tencent in China. We tested our question answering system in musical domain, which is publicly available. User can test our system by chatting with " 音乐问答" in Official Account of WeChat. However, our system interacted with user using text only.

6.2 Dataset

We randomly selected 200 questions from an internet musical QA forum. These natural language questions cover all the targets of our system, including singer, song, album, publish time, style, lyric, height, theme song and so on.

We evaluated the systems in two parts. One part is to compare the accuracy of the system. We manually got the right answer of the question from internet as the golden standard answer, and tested both systems using 200 questions mentioned above, including *Siri*, our system, our system without cQA module and only cQA module.

In the other part, we asked 11 testers to evaluate *Siri* and our system. We provided a question list and the testers were freely choose their way to interact with the system. After that, they filled in a system questionnaire rating their experience comparing *Siri* and our system.

6.3   Results

Testers were asked to identify answers that they thought were right. The result is shown in Table 2. Can't Answer means that the system responds the user that it cannot answer the question, while Error means that the system returns a wrong answer.

Table 2. The distribution of the answers

	Right	Can't Answer	Error
Siri	42%	38%	20%
Our system without cQA	53%	43%	4%
Only cQA	34%	60%	6%
Our system	61%	29%	10%

The results show that our system has a significant improvement than the baseline system.

For instance, there is a question:

狮子座是谁的歌
Whose song is "Leo"?

Our system got the right answer and responded user "曾轶可(Yico Zeng)", while *Siri* cannot gives a correct answer and played the latest song in iPhone instead. It shows that our system has more flexible semantic patterns than the *Siri*.

For another question:

勇敢的心是谁的专辑
Whose album is "Brave Heart"?

Our system searched from the QA engine and got the right answer "汪峰". But *Siri* responded a link of the movie "Brave Heart" in Wikipedia. Apparently *Siri* failed to take use of the word "album" to correctly identify "Brave heard" as a music album instead of a movie in Named Entity Recognition.

The system questionnaires were filled out by users after the experiment. These consisted of questions concerning each system to be rated on a five point scale (1-Worst to 5-Best). It evaluates the various attributes of the systems, including ease of use, response time, understanding and so on. Mean and Median for each system are shown in Table 3. Results show that although our system is not performing significantly better than the baseline system (SQ1,SQ5), users seem to find it more understanding and has more accurate answers than the baseline.

Table 3. System questionnaires responses(Si=Siri, O=our system)

Question	Si Mean	Si Median	O Mean	O Median
SQ1 - Ease of use	4.3	5	3.8	4
SQ2 - Response time	4.5	5	4.5	5
SQ3 - Understanding	3	3	4.6	5
SQ4 - Interesting	4	4	3.9	4
SQ5 - GUI	4.5	5	3.6	4
SQ6 - Answer accuracy	3.2	3	4.3	4

6.4   Analysis

There are 78 questions that our system can not respond the right answers. Our system answered 20 of 200 questions incorrectly. 5 questions are due to the missing of target in our data. These questions' targets aim at the targets that our structured data does not have. For instance, user wants to know

最炫民族风是谁作曲的
The composer of the song "The most dazzling folk style"

There is not a relation for *Composer of a song* in our knowledge base so we can not get the direct answer. We search the answer from community QA module and get the wrong answer. Other missing targets of our data include *Famous songs of singers* and *Part song*. Apparently this type of errors could be improved by training such relations into the sysytem.

8 questions failures are due to the missing of the data in the knowledge base. As our system is not a real-time update system, we do not have some newest data, like "狐狸叫(The fox)". If the user ask the question

狐狸叫是谁唱的
Who sings the song "The fox"

We could not get the answer from knowledge base and respond the wrong answer.

3 questions are due to the errors in our data. Some data in our knowledge base is crawled from internet, such as *Baidu Baike* and *Hudong Baike*. We can not guarantee the accuracy of these data so that there are some errors in our knowledge base. For instance, the theme song of the drama "仙剑奇侠传(Chinese Paladin)" is "杀破狼(Sha po lang)" in fact while our knowledge base responds "陈忠义(Chen Zhongyi)" which is a singer instead.

58 of 200 questions can not be answered by our system. After analyzing all of these 58 questions, we found that 24 questions are due to the missing of target in our data and 27 questions are due to the missing knowledge piece of our data.

Without the cQA module, the distribution of the answers change to 53% right, 43% can't answer and 4% error. The distribution is 34%, 60%, 6% if we only use cQA module. The result shows the importance of cQA module as it can make our system answer more questions, nearly 8%, that can not be answered formerly. It is a positive component for system although it raises the error rate meanwhile. The result shows the cQA module effectively raises the recall though reduces the precision of our system to some extent.

The result in Table 3 shows that users think our system more understanding and dependable. *Siri* mostly answer the question by searching on the internet. However, some users complained that they want to our system tell them the source of the answer to help them judge the confidence, either from the knowledge base or from the cQA module. On the contrary, Users found our system lack of speech recognition so that they had to spend more time on typing. Users also think our GUI a little simple as it only responds a sentence and it reduces their experience. All of the disadvantages mentioned above need further work.

The results in Table 2 and Table 3 show that our system performs better in musical domain than *Siri*, though this is intuitively as expected since it is hard for the open domain system to be optimized towards specific fields.

7   Future Work

We plan to make pattern generation in NLU unit automatically so that we can generate patterns rapidly when we change a domain. We will take advantage of a large amount of QA pairs in community QA and generate pattern by semi-supervised algorithm of machine learning. We will build a specific domain parser if possible [9, 10]. It can enhance the universality and scalability of our system.

8   Conclusion

We presented a domain-specific Chinese interactive Question Answering (QA) system. The system has a dialog management model using Topic Forest to keep track of user's interests. The system can understand natural language questions and transform them into semantic frame using semantic pattern matching model. We used knowledge base to store the structured data and also have a community QA module to answer questions by finding most similar question. We present this system on WeChat and users can chat with it easily on mobile phone.

We evaluated the system against baseline system(*Siri*) with a group of 11 users. The results show our system has a significant improvement than the baseline system in terms of both accuracy and user satisfaction rates. Users found our system more understanding and have more accurate answer. They believed these could be important features in vertical domain question answering. The results were statistically significant. Although there are still some disadvantages in our system, like the GUI and lack of speech recognition, users prefer our system to the baseline overall.

Acknowledgement. This work was partly supported by the following grants from: the National Basic Research Program (973 Program) under grant No. 2012CB316301, the National Science Foundation of China project under grant No. 61332007, Noah's Ark Lab of Huawei Technologies, and the Tsinghua University Initiative Scientific Research Program (with No. 20121088071).

References

1. Wu, X., Zheng, F., Xu, M.: Topic forest: A plan-based dialog management structure. In: IEEE International Conference on Acoustics, Speech, and Signal Processing, Proceedings(ICASSP 2001), vol. 1, pp. 617–620. IEEE (2001)
2. Zheng, Z., Tang, Y., Long, C., Bu, F., Zhu, X.: Question answering system based on community qa. In: The Workshop Programme 23
3. Prud' Hommeaux, E., Seaborne, A., et al.: Sparql query language for rdf. W3C recommendation 15 (2008)
4. Grishman, R., Sundheim, B.: Message understanding conference-6: A brief history. In: COLING 1996, pp. 466–471 (1996)
5. Ittycheriah, A., Roukos, S.: Ibm's statistical question answering system-trec-11. Technical report, DTIC Document (2006)
6. Dang, H.T., Kelly, D., Lin, J.J.: Overview of the trec 2007 question answering track. In: TREC, vol. 7, p. 63 (2007)
7. Galibert, O., Illouz, G., Rosset, S.: Ritel: an open-domain, human-computer dialog system. In: Interspeech, pp. 909–912. Citeseer (2005)
8. Bu, F., Zhu, X., Hao, Y., Zhu, X.: Function-based question classification for general qa. In: Proceedings of the 2010 Conference on Empirical Methods in Natural Language Processing, pp. 1119–1128. Association for Computational Linguistics (2010)
9. Berant, J., Chou, A., Frostig, R., Liang, P.: Semantic parsing on freebase from question-answer pairs. In: Proceedings of EMNLP (2013)
10. Cai, Q., Yates, A.: Large-scale semantic parsing via schema matching and lexicon extension. In: Proceedings of the Annual Meeting of the Association for Computational Linguistics (2013)

Top-k Spatio-textual Similarity Search

Sitong Liu[1], Yaping Chu[1], Huiqi Hu[1], Jianhua Feng[1], and Xuan Zhu[2]

[1] Department of Computer Science and Technology, Tsinghua University
[2] Samsung R&D institute, Beijing, China
{liu-st10,cyp12,hhq11}@mails.tsinghua.edu.cn
fengjh@tsinghua.edu.cn, xuan.zhu@samsung.com

Abstract. Location-based services have attracted significant attention for the ubiquitous smartphones equipped with GPS systems. These services (e.g., Twitter, Foursquare) generate large amounts of spatio-textual data which contain both geographical location and textual description. In this paper, we study a prevalent top-k spatio-textual similarity search problem: Given a set of objects and a user query, find k most relevant objects considering both spatial location and textual description. We make the following contributions: (1) We propose a TA-based framework and devise efficient algorithms to incrementally visit the objects with current highest spatial or textual similarity. (2) We explore a hybrid partition pattern by integrating spatial and textual pruning power. We further propose a partition-based algorithm which can significantly improve the performance. (3) We have conducted extensive experiments on real and synthetic datasets. Experimental results show that our methods outperform state-of-the-art algorithms and achieve high performance.

1 Introduction

With the popularity of global position systems (GPS) in smartphones, location-based services (LBS) have recently attracted significant attention from both academic and industrial communities. Many location-based services generate large amounts of spatio-textual data which contain both geographical location and textual description. For example, Foursquare[1] extracts GPS location and detailed description while uploading users' check-ins of the Point-of-Interest (POI).

In this paper, we explore a top-k spatio-textual similarity search problem, which, given a set of spatio-textual objects and a user query, finds k most relevant objects considering both spatial location and textual description. It has many real applications in existing LBS systems. One example is tag suggestion. Twitter[2] allows users to add geographical tags while posing tweets. Sometimes, users may have difficulty in typing these tags precisely. Thus, to improve user experience, we can utilize the spatial and textual information of a tweet to find similar check-ins in Foursquare and provide the corresponding POIs as a suggestion. Another example is location-aware message delivery. People may be more

---

[1] https://foursquare.com/
[2] https://twitter.com/

F. Li et al. (Eds.): WAIM 2014, LNCS 8485, pp. 602–614, 2014.

concerned about news happening around them. In social networks (e.g, Filckr[3], Twitter), we can extract users' preferences from their profiles and delivery relevant local messages (e.g., images, tweets) by performing a top-k spatio-textual similarity search.

The most relevant study on this problem is IRtree [1] which embeds an inverted index in each node of Rtree. Though this method can solve our problem, it is rather inefficient and may generate large amounts of candidates. The reason is that it pays too much attention to the pruning power of spatial component and fails to explore the feature of textual similarity. To address this limitation and improve the performance, we first propose a TA-based framework which builds spatial and textual index separately. We devise efficient algorithms to incrementally find the object with current highest spatial or textual similarity. We take these objects as candidates and verify them to get the final results. To seamlessly integrate the spatial and textual pruning power, we further propose a hybrid partition pattern and extend it to support our TA-based algorithm.

We summarize our main contributions as follows: (1) We study a prevalent problem called top-k spatio-textual similarity search. We propose a TA-based framework and devise efficient algorithms to incrementally visit the objects with current highest spatial or textual similarity. (2) We explore a hybrid partition pattern by integrating spatial and textual pruning power. We further propose a partition-based algorithm which can significantly improve the performance. (3) We have conducted extensive experiments on real and synthetic datasets. Experimental results show that our methods outperform state-of-the-art algorithms and achieve high performance.

The rest of this paper is organized as follows. We formulate our problem in Section 2 and propose a TA-based framework in Section 3. Section 4 devises an effective hybrid partition pattern and proposes a partition based TA algorithm. Experimental results are provided in Section 5. We review related work in Section 6 and make a conclusion in Section 7.

2   Preliminaries

We first formulate the problem of top-k spatio-textual similarity search in Section 2.1, and then show prevalent algorithm in Section 2.2.

2.1   Problem Statement

Consider a collection of objects $\mathcal{R} = \{r_1, r_2, \ldots, r_{|\mathcal{R}|}\}$. Each object $r \in \mathcal{R}$ includes a spatial location $\mathcal{L}_r$ and textual description $\mathcal{T}_r$, denoted by $r = \{\mathcal{L}_r, \mathcal{T}_r\}$. We use the coordinate of an object to describe its spatial location, denoted by $\mathcal{L}_r = [\mathcal{L}_r.x, \mathcal{L}_r.y]$. And we use a set of tokens to capture the textual description, denoted by $\mathcal{T}_r = \{t_1, t_2, \ldots, t_{|\mathcal{T}_r|}\}$, which describes an object (e.g., Gym, Hotel, Restaurant) or users' interests (e.g., Yoga, Pilates, Jogging). Since tokens may have different importance, we assign each token t_i with a weight $w(t_i)$ (e.g., inverse document frequency idf). A query $q = \{\mathcal{L}_q, \mathcal{T}_q, k, \alpha\}$ includes a spatial

---

[3] https://www.flickr.com/

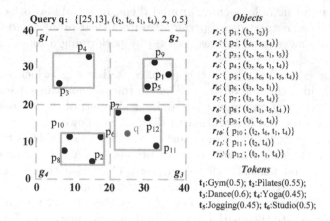

Fig. 1. An example of spatio-textual objects

location $\mathcal{L}_q$, a textual description $\mathcal{T}_q$, a top-k parameter k and a parameter α to balance between the spatial and textual components. Given a query q, our goal is to find k most similar objects in $\mathcal{R}$. Similar to [1], we use a linear interpolation function to measure the spatio-textual similarity.

Definition 1 (Spatio-Textual Similarity). *Given an object r and a query q, the spatio-textual similarity between r and q is defined as:*

$$Sim(r, q) = \alpha Sim_S(r, q) + (1 - \alpha)Sim_T(r, q)$$

where $Sim_S(r, q)$ is the spatial similarity and $Sim_T(r, q)$ is the textual similarity.

In this paper, we use the normalized Euclidean distance and weighted Jaccard to quantify the spatial distance and the textual similarity. Our method can be extended to other spatial and textual similarity functions.

Definition 2 (Spatial Similarity). *Given an object r and a query q, the spatial similarity between r and q is defined as:*

$$Sim_S(r, q) = 1 - \frac{\mathcal{D}_{Euclidean}(\mathcal{L}_r, \mathcal{L}_q)}{\mathcal{D}_{Max}}$$

where $\mathcal{D}_{Max}$ is the maximum distance in the location space.

Definition 3 (Textual Similarity). *Given an object r and a query q, the textual similarity between r and q is defined as:*

$$Sim_T(r, q) = \frac{\sum_{t \in \mathcal{T}_r \cap \mathcal{T}_q} w(t)}{\sum_{t \in \mathcal{T}_r \cup \mathcal{T}_q} w(t)}$$

where $w(t)$ is the weight of token t.

Definition 4 (Top-k Spatio-Textual Similarity Search). *Given a collection of objects $\mathcal{R}$ and a query $q = \{\mathcal{L}_q, \mathcal{T}_q, k, \alpha\}$, a top-k spatio-textual similarity search finds a subset of objects $\mathcal{S} = \{r_1, r_2, \ldots, r_k\}$ which satisfy: (1) $\mathcal{S} \subseteq \mathcal{R}$, (2) $|\mathcal{S}| = k$, and (3) $\forall r_i \in \mathcal{S}, \forall r_j \in \mathcal{R} - \mathcal{S}$, we have $Sim(r_i, q) \geq Sim(r_j, q)$.*

Example 1. Consider the twelve objects in Figure 1. Suppose query $q = \{[25, 13], (t_2, t_6, t_1, t_4), 2, 0.5\}$. For object r_1 where $p_1 = (36, 28)$, its spatial similarity $Sim_S(r_1, q) = 1 - \frac{\sqrt{(36-25)^2+(28-13)^2}}{\sqrt{40^2}} = 0.67$ and its textual similarity $Sim_T(r_1, q) = \frac{w(t_2)}{w(t_3)+w(t_2)+w(t_6)+w(t_1)+w(t_4)} = \frac{0.55}{0.6+0.55+0.5+0.5+0.45} = 0.21$. Thus, the spatio-textual similarity $Sim(r_1, q) = 0.5 \times 0.67 + (1 - 0.5) \times 0.21 = 0.44$. Similarly, for object r_{11}, $Sim_S(r_{11}, q) = 0.83$ and $Sim_T(r_{11}, q) = 0.5$. We then have $Sim(r_{11}, q) = 0.67$. Thus, r_{11} is more similar than r_1 since $0.67 > 0.44$. By calculating the similarity of all the objects, we take the two objects with highest similarity $\{r_{10}/0.86, r_{12}/0.80\}$ as the results.

2.2 Baseline Method

To the best of our knowledge, IRtree [1] is the state-of-the-art method to solve this problem. To build an IRtree, we first construct a R-tree according to the spatial coordinates of all the objects. Then for each node of R-tree, we build an inverted index to map tokens to the children nodes containing these tokens. When coming a query, we traverse IR-tree from the root. At each node, we use the inverted index to estimate the spatio-textual similarity for all its children. The node with largest estimated value will be visited next until we find k objects.

This method may be inefficient and needs to visit significate numbers of objects for the reason that: (1) When visiting a node in IRtree, it needs to calculate the spatio-textual similarity for all its children nodes. However, some children may share few tokens with the query and cannot be the candidates. (2) The token set in each node is the union of tokens from all its subtrees. Thus, especially for nodes in upper level, IRtree cannot efficiently estimate their textual similarity and may waste lots of time in traversing useless nodes. For example, suppose we want to find a gym which offers both yoga and pilates course. Though node n contains both token "yoga" and "pilates" and will be considered first. It is quite possible that the "yoga" and "pilates" courses are actually provided by two different gyms and cannot be the result.

3 Threshold Algorithm Based Approach

In this section, we first propose a threshold algorithm (TA) based framework in Section 3.1 and then devise efficient incremental similarity search algorithms for spatial and textual components in Section 3.2.

3.1 Threshold Algorithm Based Framework

Inspired by the TA algorithm [2], we propose a TA-based framework to efficiently find similar objects for a query. The basic idea is that, if object r is a top-k result

---

Algorithm 1: TA-based Top-k Similarity Search $(\mathcal{R}, q, \mathcal{I}_S, \mathcal{I}_T)$

Input: $\mathcal{R}$: An object set
$\quad\quad\quad$ q: A query formalized as $\{\mathcal{L}_q, \mathcal{T}_q, k, \alpha\}$
$\quad\quad\quad$ $\mathcal{I}_S$: The spatial index; $\mathcal{I}_T$: The textual index
Output: Q: k most similar objects

```
1  begin
2  |   Q ← an empty priority queue;
3  |   θ_ta ← 1, θ_q ← 0;
4  |   while true do
5  |   |   if θ_q ≥ θ_ta then
6  |   |   |   return Q;
7  |   |   object r_s ←BestSpaSearch(L_q, I_S);
8  |   |   if |Q| < k or Sim(r_s, q) > θ_q then
9  |   |   |   add object r_s to queue Q;
10 |   |   |   update threshold θ_q;
11 |   |   object r_t ←BestTextSearch(T_q, I_T);
12 |   |   if |Q| < k or Sim(r_t, q) > θ_q then
13 |   |   |   add object r_t to queue Q;
14 |   |   |   update threshold θ_q;
15 |   |   θ_ta ← αSim_S(r_s, q) + (1 − α)Sim_T(r_t, q);
16 end
```

Fig. 2. Threshold Algorithm Based Framework

of query q, then either their spatial components or textual components should be similar enough. Thus, by building spatial and textual index separately, we can quickly find objects with large spatial similarity or textual similarity. We take these objects as candidates and then verify them to generate the final results.

As shown in Algorithm 1, we build a spatial index $\mathcal{I}_S$ and a textual index $\mathcal{I}_T$ separately (The detail of building $\mathcal{I}_S$ and $\mathcal{I}_T$ is shown in Section 3.2). We use a priority queue Q to dynamically keep k objects with current highest spatio-textual similarity (Line 2). θ_q is the lowest value in Q. At each loop, we incrementally find object r_s with current highest spatial similarity (Line 7). If its spatio-textual similarity $Sim(r_s, q)$ is larger than θ_q, we add r_s to Q and update θ_q (Line 8-10). Similarly, we incrementally find object r_t with current highest textual similarity and update Q with $Sim(r_t, q)$ (Line 11-14). Similar to the TA algorithm, we maintain a threshold θ_{ta} to indicate the maximum spatio-textual similarity of unvisited objects. We update θ_{ta} using the spatial similarity of r_s and the textual similarity of r_t at the end of each loop (Line 15). Once $\theta_q \geq \theta_{ta}$, none of the unvisited objects may get higher similarity than θ_q. We then return the k objects in Q as results (Line 6). Otherwise, we repeat the above process. The correctness is proved in Lemma 1. For constraint of space, we omit the proof.

Example 2. Consider the objects in Figure 1. Given a query $q = \{[25, 13],$ $(t_2, t_6, t_1, t_4), 2, 0.5\}$, if we sort objects according to the spatial similarity, the order will be $\{r_7/0.88, r_6/0.87, r_{12}/0.85, \ldots, r_4/0.60\}$. If we sort them according to the textual similarity, the order will be $\{r_{10}/1.00, r_{12}/0.75, r_3/0.63, \ldots, r_7/0.10\}$. At the first loop, we visit r_7 with current highest spatial similarity and r_{10} with highest textual similarity. We then use their spatio-textual similarity to update Q. Thus, $Q = \{r_7/0.49, r_{10}/0.86\}$ and $\theta_q = 0.49$. Since $Sim_S(r_7, q) = 0.88$ and

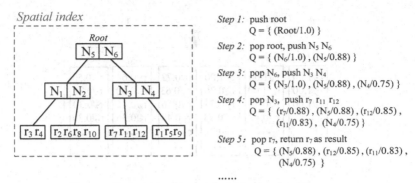

Fig. 3. Incremental spatial similarity search

$Sim_T(r_{10}, q) = 1.00$, $\theta_{ta} = 0.5 \times 0.88 + 0.5 \times 1.00 = 0.94$. Then $\theta_q < \theta_{ta}$, we continue the next loop and visit objects r_6 and r_{12}. Then $Q = \{r_{12}/0.80, r_{10}/0.86\}$ and $\theta_q = 0.80$, $\theta_{ta} = 0.81$. We continue the next loop since $0.80 < 0.81$. We use $r_{12}/0.80, r_3/0.62$ to update Q and $\theta_q = 0.80$, $\theta_{ta} = 0.74$. Since $0.80 > 0.72$, we stop the loop and take $\{r_{10}, r_{12}\}$ in Q as results.

Lemma 1. *Given a query q, the TA-based framework can correctly find k objects with the highest spatio-textual similarity.*

3.2 Incremental Spatial/Textual Similarity Search

Notice that in Algorithm 1, we need an efficient method to incrementally find the object with current highest spatial or textual similarity (Line 7, Line 11).

Incremental Spatial Similarity Search: Many works [3,4] have studied the nearest neighbor search in spatial data. In this paper, we use R-tree [5] as an example to index the spatial components of all the objects and develop [3] to support our incremental search requirement. We traverse R-tree in a top-down pattern and use a priority queue Q to keep the spatial similarity for all the visited nodes. At first, if Q is empty, we initiate it with the root node. Then at each step, we greedily select the node with highest similarity from Q: (1) If it is a non-leaf node, we estimate the upper bound of spatial similarity for all its children and add them to Q; (2) If it is a leaf node, we calculate the real spatial similarity for all the objects within it and add them to Q; (3) If it is an object, we return it as the result.

Example 3. Consider the objects and query q in Figure 1. All the objects are organized as a R-tree in Figure 3. Suppose Q only contains the root node at the moment. We pop the root node and add its children N_5, N_6 to Q. Notice that N_5 is a non-leaf node, we calculate the distance between its minimum bounding rectangle $M_{N_5} = \{(7,4),(18,33)\}$ and the query location $\mathcal{L}_q = [25,13]$, i.e., $\mathcal{D}(M_{N_5}, \mathcal{L}_q) = \sqrt{(25-18)^2 + 0^2} = 7.0$. Then we have $Sim_S(M_{N_5}, \mathcal{L}_q) = 1 - \frac{\mathcal{D}(M_{N_5}, \mathcal{L}_q)}{\mathcal{D}_{Max}} = 1 - \frac{7.0}{\sqrt{40^2}} = 0.88$. Similarly, $Sim_S(M_{N_6}, \mathcal{L}_q) = 1.0$. Thus, $(N_5/0.88)$, $(N_6/1.0)$ are pushed into Q. The following steps are shown in Figure 3. At step 5, the top element in Q is an object, we then return r_7 as the result.

$$T_q = \{\ t_2 \quad t_6 \quad t_1 \quad t_4\ \}$$

t_3	t_2	t_6	t_1	t_5	t_4
r$_7$/1.00	r$_3$/1.00	r$_2$/1.00	r$_8$/0.72	r$_2$/0.64	r$_{11}$/0.50
r$_5$/1.00	r$_{10}$/1.00	r$_5$/0.76	r$_{12}$/0.63	r$_7$/0.60	r$_2$/0.32
r$_4$/1.00	r$_6$/0.58	r$_3$/0.73	r$_5$/0.56	r$_8$/0.46	r$_7$/0.30
...	r$_1$/0.48	r$_{10}$/0.73	r$_3$/0.48	r$_5$/0.36	r$_{12}$/0.30
	...	r$_4$/0.71	r$_{10}$/0.48	...	...
		...	...		

Fig. 4. Incremental textual similarity search

Incremental Textual Similarity Search: Given an object r and a query q, according to Definition 3, we can deduce that $Sim_T(r, q) = \frac{\sum_{t \in \mathcal{T}_r \cap \mathcal{T}_q} w(t)}{\sum_{t \in \mathcal{T}_r \cup \mathcal{T}_q} w(t)} = \frac{\sum_{t \in \mathcal{T}_r \cap \mathcal{T}_q} w(t)}{\sum_{t \in \mathcal{T}_r} w(t) + \sum_{t \in \mathcal{T}_q} w(t) - \sum_{t \in \mathcal{T}_r \cap \mathcal{T}_q} w(t)}$. Since $\sum_{t \in \mathcal{T}_q} w(t)$ is fixed when the query comes, $Sim_T(r, q)$ only depends on $\sum_{t \in \mathcal{T}_r} w(t)$ and $\sum_{t \in \mathcal{T}_r \cap \mathcal{T}_q} w(t)$. Unfortunately, calculating $\mathcal{T}_r \cap \mathcal{T}_q$ for every object is quite an expensive operation especially when $|\mathcal{T}_r|$ or $|\mathcal{T}_q|$ is large. A natural idea is to generate a signature for each token set $\mathcal{T}_r$ and use this signature to estimate the textual similarity of r. The object with highest estimated value is most likely to be the result and will be considered first. Inspired by the prefix filter technique [6], we sort tokens according to the inverse document frequency (idf) and take the first token in $\mathcal{T}_r \cap \mathcal{T}_q$ as the signature, denoted by $\mathcal{T}_r^q$. We can deduce the upper bound of $Sim_T(r, q)$ as Lemma 2. However, $\mathcal{T}_r \cap \mathcal{T}_q$ varies with different query q, then each token in $\mathcal{T}_r$ has the possibility to be $\mathcal{T}_r^q$. Thus, when building the inverted index, for each token t in $\mathcal{T}_r$, we estimate its maximum textual similarity as Lemma 2, denoted by $\mathcal{U}_r^t$, and add $(r, \mathcal{U}_r^t)$ to the corresponding inverted list of t. After inserting all the objects, we sort each inverted list in the descending order of $\mathcal{U}_r^t$. Notice that building index is offline and will not be included in the search time.

In the search stage, we use two priority queues Q_t and Q_{ub} to keep the textual similarity and the estimated similarity of the visited objects. θ_t and θ_{ub} are the maximum value corresponding to Q_t and Q_{ub}. We greedily visit object with highest value in Q_{ub} and verify it by calculating its real similarity. Since objects within each inverted list have been previously sorted according to the estimated textual similarity, we only need to consider the objects in each inverted list successively. When coming a query q, if Q_{ub} is empty, we use $\mathcal{T}_q$ to probe the inverted index and use the first object of the corresponding inverted lists to initiate Q_{ub}. We continuously pop objects from Q_{ub} and use their real textual similarity to update Q_t. When an object is popped from Q_{ub}, we add its successor (the next object from the same inverted list) to Q_{ub}. Once $\theta_t \geq \theta_{ub}$, none of the unvisited objects can get higher similarity than θ_t. We then return the top object in Q_t as the result.

Lemma 2. *Given an object r and a query q, suppose $\mathcal{T}_r = \{t_1, t_2, \ldots, t_{|\mathcal{T}_r|}\}$ and the first token in $\mathcal{T}_r \cap \mathcal{T}_q$ is t_i, then $Sim_T(r, q) \leq \frac{\sum_{t \in \mathcal{T}_r} w(t) - \sum_{k=1}^{k=i-1} w(t_k)}{\sum_{t \in \mathcal{T}_r} w(t)}$.*

Example 4. Consider the objects and query $\mathcal{T}_q = \{t_2, t_6, t_1, t_4\}$ in Figure 1. We sort all the tokens in the descending order of idf and build the inverted index as Figure 4. Suppose Q_{ub} is empty at this moment. We then use $\mathcal{T}_q$ to find the corresponding inverted lists (denoted as $l_{t_2}, l_{t_6}, l_{t_1}, l_{t_4}$) and initiate Q_{ub} with the top object of these inverted lists, i.e., $Q_{ub} = \{(r_3/1.0), (r_2/1.0), (r_8/0.72), (r_{11}/0.5)\}$. At the first loop, we pop $(r_3/1.0)$ from Q_{ub} and add its real textual similarity $(r_3/0.63)$ to Q_t. Meanwhile, its successor $(r_{10}/1.0)$ from l_{t_2} is added to Q_{ub}. Thus, $Q_t = \{(r_3/0.63)\}$ and $Q_{ub} = \{(r_2/1.0), (r_{10}/1.0), (r_8/0.72), (r_{11}/0.5)\}$. Then we have $\theta_t = 0.63$ and $\theta_{ub} = 1.0$. Since $0.63 < 1.0$, we continue to process $(r_2/1.0)$ on top of Q_{ub}. Then $Q_t = \{(r_3/0.63), (r_2/0.38)\}$ and $Q_{ub} = \{(r_{10}/1.0), (r_8/0.72), (r_6/0.58), (r_{11}/0.5)\}$. After processing $(r_{10}/1.0)$, we have $\theta_t = 1.0$ and $\theta_{ub} = 0.76$. Since $1.0 > 0.76$, we return r_{10} as the result.

4   Partition Based Threshold Algorithm

Observe that in Algorithm 1, though we can quickly find the object with highest spatial or textual similarity, it may still not be the final result since the other component is not similar enough. Take r_7 in Figure 1 as an example, though its location is the nearest to the query point, it only shares one token with q. To avoid visiting such objects, a natural idea is to partition them into buckets. Objects within each bucket are close with each other and share similar tokens. Then we can prune objects with low spatial or textual similarity by group. In this section, we discuss how to seamlessly integrate the spatial and textual component to improve the pruning power. We first devise a hybrid partition pattern and then extend TA-based algorithm in Section 3.

To estimate the position of an object, a natural idea is to partition the space into grids and associate each object with the grid containing it. Those grids near the query point are most likely to contain the results. Meanwhile, for each token t in object r, we can estimate an upper bound of textual similarity as Lemma 2, denoted by $\mathcal{U}_r^t$. Those objects whose intersection with $\mathcal{T}_q$ contains larger $\mathcal{U}_r^t$ are most likely to be the results. Based on this idea, we partition objects into buckets according to associative grids and the textual upper bound. Each bucket is marked using a triple $[g, t, \mathcal{U}_g^t]$ where g is the grid ID, t is the token ID and $\mathcal{U}_g^t$ is the minimum textual upper bound of all the objects within this bucket, i.e., $\mathcal{U}_g^t = \min\{\mathcal{U}_r^t | \forall r \in [g, t, \mathcal{U}_g^t]\}$. For example, we partition the spatial space into 2×2 grids as Figure 1, denoted by $g_1 \sim g_4$. Take (g_4, t_4) as an example, three objects $\{(r_2, 0.32), (r_8, 0.30), (r_{12}, 0.23)\}$ fall in grid g_4 and contain t_4. We then divide them into two buckets: $\mathcal{B}_1 = \{(r_2, 0.32), (r_8, 0.30)\}$ and $\mathcal{B}_2 = \{(r_{12}, 0.23)\}$. For $\mathcal{B}_1$, since the lowest value is 0.3, we mark it as $[g_4, t_4, 0.30]$. Similarly, we mark $\mathcal{B}_2$ as $[g_4, t_4, 0.23]$. When coming a query, we estimate a spatio-textual similarity for each bucket and greedily visit the best one. We find top-k similar objects for each visited bucket and combine these results in the end. Two main challenges are under consideration: (1) How to determine the visiting order of buckets. (2) How to find k most similar objects in each bucket.

The Visiting Order of Buckets: To estimate the spatial distance between query q and a bucket $[g, t, \mathcal{U}_g^t]$, we take the center of grid g as a reference point, denoted by g_c. For each object r in $[g, t, \mathcal{U}_g^t]$, we use the distance between $\mathcal{L}_r$ and g_c to represent its position, denoted by $\mathcal{D}(\mathcal{L}_r, g_c)$. Then each bucket $[g, t, \mathcal{U}_g^t]$ corresponds to a spanning of $\mathcal{D}(\mathcal{L}_r, g_c)$, denoted by $[\mathcal{D}_{[g,t,\mathcal{U}_g^t]}^{Min}, \mathcal{D}_{[g,t,\mathcal{U}_g^t]}^{Max}]$ where $\mathcal{D}_{[g,t,\mathcal{U}_g^t]}^{Min} = \min\{\mathcal{D}(\mathcal{L}_r, g_c)|\forall r \in [g, t, \mathcal{U}_g^t]\}$ and $\mathcal{D}_{[g,t,\mathcal{U}_g^t]}^{Max} = \max\{\mathcal{D}(\mathcal{L}_r, g_c)|\forall r \in [g, t, \mathcal{U}_g^t]\}$. For query q, according to the triangle inequality, the minimum distance between $[g, t, \mathcal{U}_g^t]$ and q is $\max\{\mathcal{D}(\mathcal{L}_q, g_c) - \mathcal{D}_{[g,t,\mathcal{U}_g^t]}^{Max}, 0\}$. Meanwhile, since all the objects in $[g, t, \mathcal{U}_g^t]$ have higher $\mathcal{U}_r^t$ than $\mathcal{U}_g^t$, we can use $\mathcal{U}_g^t$ as a lower bound to estimate the textual similarity between $[g, t, \mathcal{U}_g^t]$ and q. For an object r containing token t, if $Sim_T(\mathcal{T}_r, \mathcal{T}_q) > \mathcal{U}_g^t$, then r must appear in $[g, t, \mathcal{U}_g^t]$. Thus, for each bucket, we calculate the spatio-textual similarity between $[g, t, \mathcal{U}_g^t]$ and q as $\alpha(1 - \frac{\max\{\mathcal{D}(\mathcal{L}_q, g_c) - \mathcal{D}_{[g,t,\mathcal{U}_g^t]}^{Max}, 0\}}{\mathcal{D}_{Max}}) + (1 - \alpha)\mathcal{U}_g^t$ and greedily visit the bucket with the highest similarity. For concise of present, we denote the similarity between q and $[g, t, \mathcal{U}_g^t]$ by $Sim(q, [g, t, \mathcal{U}_g^t])$. Notice that we don't need to search top-k objects for every bucket, once $Sim(q, [g, t, \mathcal{U}_g^t])$ is no better than the similarity of current top-k objects, we can stop the searching process and return the k best objects as results.

Top-k Similarity Search within Each Bucket: We extend the TA-based algorithm mentioned in Section 3 to find k objects with highest spatio-textual similarity within each bucket. Similar to Algorithm 1, we keep the spatial and textual index separately. Consider bucket $[g, t, \mathcal{U}_g^t]$ with center point g_c, for the spatial components, we sort all the objects r in $[g, t, \mathcal{U}_g^t]$ in the ascending order of the distance $\mathcal{D}(\mathcal{L}_r, g_c)$. Meanwhile, for the textual components, we keep a copy of objects and sort them in the the descending order of $\mathcal{U}_r^t$. Similar to the TA-based algorithm in Section 3, for each object r, we use $\mathcal{D}(\mathcal{L}_r, g_c)$ and $\mathcal{U}_r^t$ to estimate an upper bound of spatial and textual similarity. Those objects with highest spatial or textual upper bound will be taken as candidates and examined first. Similar to Algorithm 1, we use a priority queue Q to keep the spatio-textual similarity for all the visited objects. The top object in Q has current highest spatio-textual similarity θ_q. At each time, we find the next object r_s (or r_t) with highest spatial (or textual) upper bound. Suppose query q falls outside of grid g, we then take the object farthest from g_c as r_c. According to the triangle inequality, we have $\mathcal{D}(\mathcal{L}_{r_c}, \mathcal{L}_q) \geq \mathcal{D}(\mathcal{L}_q, g_c) - \mathcal{D}(\mathcal{L}_{r_c}, g_c)$. Then $1 - \frac{\mathcal{D}(\mathcal{L}_q, g_c) - \mathcal{D}(\mathcal{L}_{r_c}, g_c)}{\mathcal{D}_{Max}}$ is a maximum spatial similarity of r_c. Meanwhile, we take the object with current highest $\mathcal{U}_r^t$ as r_t. We then use the spatio-textual similarity of r_s and r_t to update Q and θ_q. We set $\theta_{ta} = \alpha(1 - \frac{\mathcal{D}(\mathcal{L}_q, g_c) - \mathcal{D}(\mathcal{L}_{r_c}, g_c)}{\mathcal{D}_{Max}}) + (1 - \alpha)\mathcal{U}_r^t$. Once $\theta_q \geq \theta_{ta}$, we can stop searching top-k objects in current bucket.

Table 1. Dataset Statistics

	# of objects	# of tokens	# of average tokens	Data size	Index size
USA	1 Million	383481	16.94	140.8M	40.28M
Twitter	7 Million	2194073	10.72	529M	147.6M

5 Experimental Study

5.1 Experimental Settings

Datasets: To evaluate our proposed techniques, we conducted extensive experiments on two datasets : USA and Twitter. Table 1 summarizes these two datasets. The Twitter dataset is a real dataset. We crawled 7 million tweets with location and textual information from Twitter[4]. The USA dataset is a synthetic dataset which randomly combines the Points of Interests (POIs) in US and the publications in DBLP.

Experimental Environment: All the algorithms were implemented in C++ and run on a Linux machine with an Intel(R) Xeon(R) CPU E5-2650 @ 2.00GHz and 48GB memory. The algorithms were complied using GCC 4.8.2.

Parameter Setting: Unless stated explicitly, parameters were set as follows by default: $\alpha = 0.5$, $k = 10$. We sorted all the tokens according to IDF.

5.2 Evaluating Different Methods

In this section, we evaluated the performance of three methods: TA-based algorithm (TA-based), partition-based algorithm (Par-TA) and the state-of-art algorithm (IRtree). We randomly selected 10000 objects from USA and Twitter to generate the query sets. Experimental results show that our methods outperform state-of-art algorithm and achieve high performance on all the evaluations.

Evaluation on k: To evaluate the effect of parameter k, we fixed α to 0.5 and varied k from 1 to 50. The result is shown in Figure 5(a) and Figure 5(d). We can see that Par-TA was 2-10 times faster than TA-based and was 21-40 times faster than IRtree. For example, in Twitter, when $k = 20$, IRtree took 3580s, TA-based took 709s while Par-TA only took 163s. Notice that when k increased, IRtree increased obviously while TA-based almost kept a straight line. The reason is that IRtree organizes all the objects as a hierarchical structure and the token set embedded in each node is the union of all the tokens from its subtrees. Thus, especially for nodes in the upper level, IRtree cannot give a precise estimation of textual similarity and will waste lots of time visiting useless nodes. TA-based continuously finds objects with current highest spatial or textual similarity. With the increase of k, TA-based is more likely to find results at an early time.

---

[4] http://twitter.com

Evaluation on the α**:** To evaluate the effect of parameter α, We fixed k to 10 and varied α from 0.1 to 0.9. The result is shown in Figure 5(b) and Figure 5(e). We can see that Par-TA was 3-7 times faster than TA-based and was 3-37 times faster than IRtree. For example, in USA, when $\alpha = 0.3$, IRtree took 29012s, TA-based took 10193s while Par-TA only took 1759s. Notice that IRtree decreased drastically with the increase of α while TA-based and Par-TA almost kept a straight line. The reason is that when α is small, the spatial component only takes a small proportion in the similarity function so that the textual pruning is more important. However, IRtree mainly focuses on the pruning of spatial similarity and fails to give a precise estimation of textual similarity. Thus, IRtree is inefficient and visits lots of useless objects. When α grew larger, the spatial pruning became more and more important so that IRtree achieved good performance. When $\alpha = 0.9$, IRtree even performed almost as fast as Par-TA.

Evaluation on the Number of Query Tokens: To evaluate the performance under different length of query tokens, we generated 10 query sets. Each query set contained 10000 queries and each query had $1 \sim 10$ tokens. The results are shown in Figure 5(c) and Figure 5(f). We can see that Par-TA and TA-based outperformed IRtree. When the length of query tokens increased, all three methods need more time to finish the search query.

Evaluation on Scalability: We evaluated the time and index scalability of our methods. Take Twitter as an example. We varied the object size from 1-7 million. In Figure 5(g), we can see that Par-TA and TA-based almost achieved a linear scalability while IRtree increased drastically. In Figure 5(i), we compared the memory size of different methods. To present their differences clearly, we did not include the memory size of loading data. We can see that IRtree took the most memory since it embedded an inverted index in each node of R-tree. By building the spatial and textual index separately, TA-based slightly reduced the index size. Instead of building an R-tree, Par-TA kept the distance to center point for each grid. That further reduced the index size.

6   Related Works

Top-k Spatial Similarity Search: Many works [2,3,4,7] have studied the problem of top-k spatial similarity search. There are two kinds of solutions: (1) [3,4] indexed objects using a hierarchical structure R-tree [5]. They traversed R-tree from the root and greedily visited current best node. (2) [2,7] regarded the spatial components as a multi-dimensional space. They iteratively selected the best object in each dimension and verified them to get final results.

Textual Similarity Search/Join: Many works [6,8,9,10] have studied on string similarity. [8] proposed a prefix filtering principle to generate signature for each token set and effectively prune those dissimilar objects. Xiao et al [9] proposed a method to solve the top-k set similarity join problem: Given two datasets $\mathcal{R}$ and $\mathcal{S}$, find k most similar pairs from all the pairs of $\mathcal{R}$ and $\mathcal{S}$.

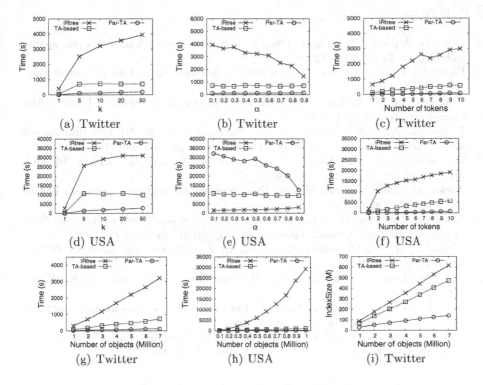

Fig. 5. Evaluation on different methods (10000 queries)

Spatial Keyword Search: There are many studies on spatial keyword search recently [1,11]. Most of them focus on integrating inverted index and R-tree to support spatial keyword search. For example, [11] combined spatial and textual index in different orders.

7   Conclusion

In this paper, we study a prevalent research problem called top-k spatio-textual similarity search. We devise a TA-based framework and explore two efficient methods to incrementally find the objects with current highest spatial or textual similarity. To achieve higher performance, We further develop a partition-based method which can seamlessly integrate the spatial and textual components and prune large number of dissimilar objects. Experimental results show that our methods outperform existing solutions and scale well.

Acknowledgement.   This work was partly supported by NSF of China (61272090), Tsinghua-Samsung Joint Laboratory, "NExT Research Center" funded by MDA, Singapore (WBS:R-252-300-001-490), and FDCT/106/2012/A3.

References

1. Cong, G., Jensen, C.S., Wu, D.: Efficient retrieval of the top-k most relevant spatial web objects. In: PVLDB (2009)
2. Fagin, R., Lotem, A., Naor, M.: Optimal aggregation algorithms for middleware. In: PODS (2001)
3. Roussopoulos, N., Kelley, S., Vincent, F.: Nearest neighbor queries. In: SIGMOD Conference, pp. 71–79 (1995)
4. Katayama, N., Satoh, S.: The sr-tree: An index structure for high-dimensional nearest neighbor queries. In: SIGMOD Conference, pp. 369–380 (1997)
5. Guttman, A.: R-trees: A dynamic index structure for spatial searching. In: SIGMOD Conference, pp. 47–57 (1984)
6. Arasu, A., Ganti, V., Kaushik, R.: Efficient exact set-similarity joins. In: VLDB, pp. 918–929 (2006)
7. Ilyas, I.F., Beskales, G., Soliman, M.A.: A survey of top-k query processing techniques in relational database systems. ACM Comput. Surv. 40(4) (2008)
8. Bayardo, R.J., Ma, Y., Srikant, R.: Scaling up all pairs similarity search. In: WWW, pp. 131–140 (2007)
9. Xiao, C., Wang, W., Lin, X., Shang, H.: Top-k set similarity joins. In: ICDE, pp. 916–927 (2009)
10. Xiao, C., Wang, W., Lin, X., Yu, J.X.: Efficient similarity joins for near duplicate detection. In: WWW, pp. 131–140 (2008)
11. Zhou, Y., Xie, X., Wang, C., Gong, Y., Ma, W.Y.: Hybrid index structures for location-based web search. In: CIKM, pp. 155–162 (2005)

A Query Suggestion Interface
with Features of Queries and Search Results

Shuhei Shogen, Takehiro Yamamoto, and Katsumi Tanaka

Department of Social Informatics, Graduate School of Informatics, Kyoto University,
Yoshida-Honmachi, Sakyo, Kyoto 606-8501, Japan
{shogen,tyamamot,tanaka}@dl.kuis.kyoto-u.ac.jp

Abstract. This paper proposes a query auto-completion interface that displays additional information on query suggestions and their search results along with the original query suggestions. Users of conventional query auto-completion interfaces cannot determine whether they can obtain relevant information from the search results of suggested queries until they actually issue them and check their search results. To address this problem, we propose a query auto-completion interface that incorporates useful information, such as words in the search results and images of the suggested queries, into the original query suggestions.

Keywords: Information retrieval, query suggestion, user interface.

1   Introduction

Query suggestions represent one of the most popular techniques to help users create queries in information retrieval [1, 2]. Many commercial search engines like Google, Bing, and Yahoo! offer interfaces called *query auto-completion interfaces* to display query suggestions to users . They display a list of suggested queries in response to the query that users input into the search box. Query auto-completion interfaces (called *query suggestion interfaces* after this) help users create appropriate queries.

However, we think conventional query suggestion interfaces still have a problem in that it is difficult for users to determine whether they can obtain the relevant information by issuing a target suggested query *until they actually issue the query and check its search results.* Users with this problem cannot recognize useful information to determine whether they can obtain relevant information apart from the query suggestions themselves. If users could check useful information *before* they issued a query, the information would help them make appropriate decisions as to whether they should issue the query.

We propose a method of incorporating such information extracted from query suggestions into the conventional query suggestion interface. We define "a feature of a query suggestion" as an item of useful information for assessing whether users can satisfy their information needs by issuing the suggested query.

F. Li et al. (Eds.): WAIM 2014, LNCS 8485, pp. 615–619, 2014.

We categorized features that we incorporated into a query suggestion interface into three types: (1) search results for the query (*search result features*), (2) the meaning of the query (*meaning features*), and (3) the scope of the topic that the query specified (*topic features*). We also propose methods of obtaining such features. Incorporating the features into query suggestions helps users assess which query suggestion they should issue and it hence helps their search tasks to become more efficient and effective.

2   Method of Extracting Features for Suggested Queries

2.1   Search Result Features

Raw Search Results. The purpose of incorporating the search result feature of a suggested query into a query suggestion interface is to help it express the search results of the query accurately. Naively, by displaying the titles, URLs, and snippets of the top search results of a suggested query as they are, users can easily expect its search results before issuing it. This method displays top-k_1 search results as they are.

Feature Words of Search Results. The interface proposed above displays raw search results. However, if we display too much information to users as features, they would feel high cognitive load to the interface, which might affect search performances of users. Thus we propose a method of obtaining feature words extracted from the Web search results of a suggested query as search result features.

The method of extracting feature words of search results for suggested query q_i works as follows:

1. We obtain top-k_2 search results $R_i = \{r_{i1}, r_{i2}, ..., r_{ik_2}\}$ by issuing query q_i.
2. We extract titles and snippets of R_i and obtain the set of words appear in them. We denote the obtained words as $W_i = \{w_{i1}, w_{i2}, ...\}$.
3. For each word w_{ij} in W_i, we calculate the score of word according to the following formula:

$$f_{sr}(w_{ij}) = \frac{tf_{sr}(w_{ij})}{\sum_{w_{ij} \in W_i} tf_{sr}(w_{ij})} * \log \frac{n}{df_{sr}(w_{ij})}$$

where $tf_{sr}(w_{ij})$ represents the number of frequency of word w_{ij} in R_i, n represents the number of suggested queries, and $df_{sr}(w_{ij})$ represents the number of suggested queries that contain word w_{ij} in their top-k_2 search results.
4. We extract the top-k_1 words according to the value of $f_{sr}(w_{ij})$, as the search result feature for query q_i.

2.2   Meaning Features by Using Feature Images

The purpose of incorporating the meaning feature of a suggested query into a query suggestion interface is to help it express the meaning of the query accurately. We propose a method of obtaining feature *images* extracted from the image search results of a suggested query as meaning features.

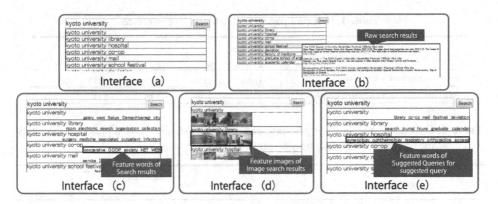

Fig. 1. Interfaces used in user study

The method of extracting images of suggested query q_i works as follows:

1. We obtain top-k_2 *image* search results $A_i = \{a_{i1}, a_{i2}, \ldots, a_{ik_2}\}$ by issuing query q_i .
2. For each image a_{ij} in A_i, we calculate the score of a_{ij} according to the following formula:

$$f_{\text{mean}}(a_{ij}) = \frac{\text{tf}_{\text{mean}}(a_{ij})}{\sum_{a_{ij} \in A_i} \text{tf}_{\text{mean}}(a_{ij})} * \log \frac{n}{\text{df}_{\text{mean}}(a_{ij})}$$

$$\text{tf}_{\text{mean}}(a_{ij}) = k_2 - \text{rank}(a_{ij})$$

where $\text{rank}(a_{ij})$ represent the rank of a_{ij} in the image search of query q_i, and $\text{df}_{\text{mean}}(a_{ij})$ represents the number of suggested queries that contain image a_{ij} in their top-k_2 image search results.

3. We extract the top-k_1 images according to the value of $f_{\text{mean}}(a_{ij})$, as the meaning feature for query q_i.

2.3 Topic Features by Using Suggested Queries for Suggested Query

The purpose of incorporating the topic feature of a suggested query into a query suggestion interface is to help users clearly understand the scope of the topic of the search results. To help users determine they should issue more generalized/specialized queries, we propose a method of obtaining query suggestions of query suggestions and extracting feature words from them as topic features.

Give a suggested query q_i, the method first obtains the set of suggested queries Q_i for q_i. It then extracts queries in Q_i that specialize query q_i as the topic feature of q_i. For example, for the query suggestion "Kyoto University," if the method obtains the suggested query "Kyoto University library" it extracts "library" as a topic feature of "Kyoto University."

Table 1. Relevant pages obtained from five interfaces

	Interface				
	a	b	c	d	e
Relevant pages for a search task	**11.60**	11.20	10.30	10.35	10.35
Relevant pages for a suggested query	2.17	2.39	**2.58**	2.14	2.16

Table 2. Contents and answers to post-task and exit questionnaires (five represented "strongly agree" while one represented "strongly disagree")

	QID	Question	Interface				
			a	b	c	d	e
Post-Task Questionnaire	Q1	Additional information was useful in the search task to choose a suggested query.	-	**3.30**	2.90	2.25	2.95
Exit Questionnaire	Q2	I was confident about issuing a suggested query to satisfy my information needs with the interface.	2.80	**4.20**	3.80	3.20	3.90

3   User Study

As shown in Figure 1, we implemented five interfaces: (a) conventional query suggestions, (b) raw search results as search results features (Section 2.1), (c) feature words as search results features (Section 2.1), (d) feature images as meaning features (Section 2.2), and (e) query suggestions as topic features (Section 2.3). From the preliminary experiment, we set $k_1 = 5$ (the number of features to be displayed) and $k_2 = 10$ (the number of results used for extracting features) in the study.

We asked 10 subjects, 9 males and a female who were in their 20s, to participate in the study. They performed two tasks for each interface (a)–(e), i.e., a total of 10 tasks in the appropriate order. After the subjects had performed each task, they were asked to answer a post-task questionnaire. Also, when they had finished all assigned tasks, the subjects were asked to answer an exit questionnaire.

Search Efficiency. Table 1 shows the results of the study related to search efficiency obtained from the log data. The "relevant pages for a search task" in the table represent the average number of Web pages that they determined to be relevant to the topic of the task. The "relevant pages for a suggested query" represent the average number of relevant pages of query suggestions that the subjects issued.

From the results for the "relevant pages for a search task", we can see that the subjects could find the most relevant pages when they used interface (a). This suggests that the other interfaces required the subjects to spend longer looking at the additional information and thus they took more time to find relevant pages. The "relevant pages for a suggested query", on the other hand, indicate that they could find more relevant pages when they used interface (c). The results imply that by displaying the content of search results as feature words in advance, we prevented them from issuing unnecessary query suggestions.

Subjective Preferences. Table 2 summarizes the participants' subjective preferences obtained from the results for the post-task (Q1) and exit (Q2) questionnaires. With regard to Q1, they awarded interface (d) the lowest value compared with the other interfaces. They also considered the additional information in interface (d) to be useless for choosing a suggested query. With regard to Q2, on the other hand, they awarded interfaces (b)–(e) much higher values than interface (a). The especially high values were remarkable for interface (b). It seemed easy for them to assess the relevance of the search results for a query as additional information was part of the search results. The values were relatively better in interfaces (c) and (e). The reason for this seems to be that the search result feature and the topic feature become clues to assess whether the search results of a query are relevant to their information needs.

4   Conclusions

We proposed methods of obtaining information useful for assessing whether users could obtain information they required from suggested queries and of incorporating such information into query suggestion interfaces as additional information. We proposed three features: search result features, meaning features, and topic features and the methods for extracting these features. In the future, we plan to devise methods of changing the kinds of features according to queries issued by users.

Acknowledgement. This work was supported in part by KAKENHI (#24240013).

References

1. Kato, M.P., Sakai, T., Tanaka, K.: When do people use query suggestion? a query suggestion log analysis. Information Retrieval, pp. 1–22 (2013)
2. Cao, H., Jiang, D., Pei, J., He, Q., Liao, Z., Chen, E., Li, H.: Context-aware query suggestion by mining click-through and session data. In: Proc. of Proc. of KDD, pp. 875–883 (2008)

Finding Photo Sets of Events by Minimizing Misrecognition from Neighbor Events

Bei Liu, Makoto P. Kato, and Katsumi Tanaka

Department of Social Informatics, Kyoto University, Kyoto 606-8501, Japan
{liubei,kato,tanaka}@dl.kuis.kyoto-u.ac.jp

Abstract. This paper presents a method to find a photo set of a given event by avoiding misrecognitions from neighbor events. In contrast with a photo summarization problem that seeks photos with high coverage for a photo collection of an event, our goal is to find photos with high *perceptual quality*, which is defined as the accuracy and quickness to recognize an certain event. We take an approach to proactively avoid misrecognitions of a photo set to improve its perceptual quality, as a photo set that cannot be misrecognized as neighbor events is expected to be one with high perceptual quality. We discuss the types of misrecognitions, namely, sub-event, super-event and sibling-event misrecognition, and then propose three criteria corresponding to each of them. By maximizing combination of three criteria, we generate a photo set that minimizes misrecognition against neighbor events. We empirically demonstrated that our proposed approach can generate photo sets with high perceptual quality in comparison with a baseline method.

Keywords: Photo Set, Event, Misrecognition.

1 Introduction

Nowadays, a large number of photos are flooding our local memory and social photo sharing websites as a result of camera and smart phone's popularization. People often group photos by events. Large storage space brings a torrent of photos for an event. A cover with several photos helps users recognize and recall the event more quickly and much better. What is more, SNS users are more likely to view one event with a few photos shared by their friends rather than all photos. Thus, how to choose photos to form a photo set of an event and improve *perceptual quality*, which is defined as the accuracy and quickness to recognize an event, becomes a significant problem. With higher perceptual quality, the user can recognize and find a desired event from a huge photo collection more precisely and quickly.

The aim of our research is to improve the perceptual quality of a photo set. Although our current work is similar to the problem of photo summarization, the purpose of our work is different. It follows that our approach also differs from those used in photo summarization. For example, Sinha et al.[1] tackled the problem of summarizing personal photos from life events. However, such a

F. Li et al. (Eds.): WAIM 2014, LNCS 8485, pp. 620–632, 2014.

photo set does not necessarily have high perceptual quality, as their approach does not consider other similar events as which a resulting photo set can be misrecognized. Suppose that a user summarizes photos taken in his/her trip to Japan. By the reason of many temples existing in Japan, the user might take many photos of temples. However, one can easily misrecognize the generated summary as "trip to China" since the summary would include many Asian temples, which are hardly distinguishable by foreigners and even people in those two countries. From the viewpoint of perceptual quality, the photo set should have been generated by taking into account neighbor events (e.g. "trip to China") and minimizing possible misrecognitions (e.g. by means of a photo about Japanese castle). Therefore, we analyze relationship between events, and put forward a method of preventing users from misrecognizing a photo set as neighbor events.

We propose three criteria for avoiding possible misrecognitions, namely, sub-event coverage, super-event coverage and membership entropy over sibling events. An algorithm that approximately optimize these criteria is used to generate a photo set with high perceptual quality. Our experimental results showed that our method can achieve high perceptual quality in comparison with a baseline method, while the quality highly depends on queries and the size of photo sets to be generated.

Three contributions of this paper are summarized as follows:
1. We proposed the problem of finding a photo set with high perceptual quality for an event, which is different from the problem of photo summarization,
2. We proposed a method of finding a photo set with high perceptual quality by minimizing possible misrecognitions, and
3. We implemented our approach and conducted experiments by using photos crawled from Flickr. The result was evaluated and compared with a baseline method based on VisualRank.

In the rest of the paper, related work is discussed in Section 2. Section 3 defines the problem investigated in this research. Our approach is presented in Section 4, and followed by experimental explanation and results in Section 5. Finally, we draw a conclusion and discuss future work in Section 6.

2   Related Work

Event representation has been explored in several different study fields for years. As in computer science, there are researches focusing on how to express events in natural language [2]. In multimedia domain, event detection has been popular for both videos and photos. Shen and Fan [3] addressed a challenge of the multimedia field, event detection, by proposing a method for high-level and low-level feature fusion based on collective classification. In recent years, a lot of researches have studied on events that can be predicted from photos. Chen and Roy [4] proposed an approach of detecting Flickr photos depicting events by analyzing users tags and meta data with the photos. A method of mining most informative features for the event recognition from photo collections is proposed in [5]. However, all these researches addressed a problem of detecting events and clustering photos based on events. They do not consider the characteristic of events and event representation. In our research, we focus more on how to select photos for representing an event without any misrecognition.

Many of researches on photo summarization focused on personal photo collections. Platt, Czerwinski, and Field [6] presented an overview of users photo collections generated by an image clustering algorithm, then selected representative images by the most distinctive images. Unsupervised approaches were proposed in [7] for event clustering based on similarity of time and image content. An event-clustering algorithm was developed to automatically segment photos into events and sub-events for albuming based on data/time information and color content of the photos [8]. All these researches focused only on photos of a certain event, while in our research, photos related to neighbors of an event are taken into consideration and they help to improve the perceptual quality of the result photo set.

3   Preliminaries

Many current researches focus on concrete events, and most of them are used in the context of news or sports. While in this research, we focus more on generic events, which are usual and related to our everyday life. More importantly, everyone can easily take photos of these events. By referring to some existing research about general event [9], events are defined as *ones involves human activities and can be specified with time and location*. An event can be represented with some keywords, e.g. "travel Japan" and "hiking summer", where terms "travel" and "hiking" represent activities, while "Japan" specifies the location and "summer" specifies the time.

Note that any term can imply an event in the context of photo, because every term can reflect what a user is doing while taking that photo. For example, term "lavender" is a plant and cannot be regarded as an activity. However, a photo including lavender indicates the activity of viewing lavender. Therefore, every term can represent an event when it is expressed by photos in this research.

The problem we tackle is defined as follows: given an event-related query $e \in E$, a photo collection P, and size of an output photo set n, return a photo set $S_e \subset P_e$ of size n, where E represents all possible events, and P_e is photos related to e from P. The photo collection P is a large set of photos that were taken by different users, while P_e can be ones in a personal photo collection for an application of photo summarization or ones taken for event e by many users. Currently, we use latter ones. Users can decide an appropriate size of output depending on their application, such as thumbnails and cover photos for a personal photo library or SNS posts.

4   Approach

As we mentioned in the introduction part, the main goal of our research is to find a photo set with high perceptual quality. To this end, we propose a method of

(a) Sub-Event Misrecognition (b) Super-Event Misrecognition (c) Sibling-Event Misrecognition

Fig. 1. Three types of misrecognitions from a photo set to an event

minimizing misrecognitions from neighbor events and achieving high perceptual quality accordingly. There are three types of misrecognitions when a user looks at a photo set and tries to tell an event from it, and they are mainly caused by three types of neighbor events.

Sub-event misrecognition means that a user misrecognizes a photo set of an event as its sub-event. Event A is a sub-event of event B if A is included in B. As shown in Figure 1(a), event of "travel" includes several sub-events, such as "transport" (actually it is "travel transport", but we removed "travel" for simplification), "sightseeing" and so on. Three photos used in Figure 1(a) are all related to "transport" which only covers a single sub-event of "travel". This photo set of size three is easily to be taken as "travel transport" by mistake.

Super-event misrecognition is a situation where a user misrecognizes a photo set of an event as its super-event. Super-event is an inverse meaning of sub-event. As seen in the example given by Figure 1(b), super-events of "travel in Kyoto" are "Kyoto" (all the events that can happen in Kyoto) and "travel". If a photo set only includes photos of Kyoto, people would regard the photo set as "Kyoto" rather than "travel in Kyoto".

Sibling-event misrecognition indicates a case where a photo set is misrecognized as an event's sibling-event. Event A is a sibling event of event B if A is a sub-event of B's super-event and A share similar contents with B. For example, we can see from Figure 1(c) that a photo set for "conference party" is also suitable for "birthday party" which is a sibling event.

4.1 Neighbor Events Generation

In order to minimize misrecognition from neighbor events, we need to generate neighbor events for a given event. Every event e consists of a set of keywords, $K_e \subset K$, where K represents all possible keywords. Suppose an event is associated with a set of tags: $T_e \subset T$, where T represents all the tags. Note that noisy tags, which are not related to content of a photo, such as "nikon" and "photo" have been removed in advance.

There are three types of similarity between two tags: visual similarity, context similarity and semantic similarity. *Visual similarity* of two tags $t1$ and $t2$, VisualSim($t1, t2$), can be obtained by Euclidean Distance between visual features of them[10]. Visual feature of each tag is obtained from global visual features, such as color (RGB and HSV), and local feature with 1000-D bag of visual words. Current research efforts in content-based image retrieval (CBIR) have proved that these common used features could measure visual contents similarity between images quite effectively [11]. *Context similarity*, ContextSim($t1, t2$), measures whether two tags are similar by considering their neighbor tags [12]. For example, "cloud" and "sky" are often tagged with the same set of tags, such as "blue, water, tree", so they are very similar based on tag context. By *semantic similarity*, SemanticSim($t1, t2$), we mean whether two tags has the same meaning and we implement it by using path similarity of wordnet[13].

Generate Sub-Events. We can specify sub-event by adding one keyword to the original event's keyword set and ones that can be easily get confused with the event will be selected. The added tags are expected to be important in the event, and to be visually similar to the event. We first compute tf(term frequency in a photo collection of an event)-idf (inverted photo frequency) of all tags attached to event-related photos P_e. A tag is denoted by t and tf-idf(t, e) represents its importance for e. Then we calculate visual similarity between e and each tag, which is defined by VisualSim(t, e). Relevance of a tag t to event e is harmonic mean of its tf-idf and visual similarity value: SubRel(t, e) = $\frac{2\text{tfi-df}(t,e)\text{VisualSim}(t,e)}{\text{tfi-df}(t,e)+\text{VisualSim}(t,e)}$. MMR (Maximal Marginal Relevance) is utilized to reduce the redundancy in resulting sub-events. We use context similarity between tags to be diversity metric. Let $W_e \subset T_e$ be the current sub-event tag set. The strategy is to find tags which have high relevance with e and is most diverse to the current set:

$$\text{MMRSub}(t, e) = \underset{t_i \in T_e \setminus W_e}{\text{argmax}} \left[\theta \text{SubRel}(t_i, e) - (1 - \theta) \max_{t_j \in W_e} \text{ContexSim}(t_i, t_j) \right]$$

After finding top ten sub-event tags with maximal marginal relevance, we generate sub-events by adding each sub-event tag to the original event's keyword set K_e, i.e. Sub(e) = $\{v | K_v = K_e \cup \{w\} \wedge w \in W_e\}$.

Generating Super-Events. We simply generate super-events by removing each keyword from the original event's keyword set K_e, as subsets with one smaller size of keywords usually represent a more general concept than the original keyword sets. Thus, Sup(e) = $\{u | K_u \subset K_e \wedge |K_u| = |K_e| - 1\}$. For example, super-events of event "travel Japan" are "travel" and "Japan".

Generating Sibling-Events. An event's sibling events are sub-events of its super-events. However, since we want to find sibling-events that are likely to lead to misrecognition, we change the relevance and diversity metrics when utilizing MMR. For each super-event c of an event e, the absent word k is one that generalizes the event, i.e. $k = K_e - K_u$ ($u \in$ Sup(e)). For example, "Japan" from "travel Japan" makes super-event "travel" location non-specified. Sibling-events of "travel Japan" under "travel" are sub-events of "travel" which can specify a location and the added tag should be similar to "Japan" in the context of remaining tags. In addition, the added tag should have short distance with the absent word in regard to visual features. As a result, context similarity and visual similarity between tags are combined for relevance computation, i.e. SibRel(t, k) = $\frac{2\text{ContextSim}(t,k)\text{VisualSim}(t,k)}{\text{ContextSim}(t,k)+\text{VisualSim}(t,k)}$.

Semantic similarity is used for diversity, because tags with the same meaning should be avoided. Similarly, let $W_u \subset T_u$ be the current sibling-event tag set under one of the super-events u. The target function of MMR algorithm is:

$$\text{MMRSib}(t, u, e) = \underset{t_i \in T_u \setminus W_u}{\text{argmax}} \left[\phi \text{SibRel}(t_i, k) - (1 - \phi) \max_{t_j \in W_u} \text{SemanticSim}(t_i, t_j) \right].$$

With top ten tags with maximal marginal relevance W_u, we obtain sibling events under each super-event u, i.e. $\text{Sib}(e, u) = \{v | K_v = K_u \cup \{w\} \wedge w \in W_u\}$.

4.2 Sub-event Coverage

A photo set which covers only a single or a few sub-events may cause sub-event misrecognitions. So an ideal photo set should cover as many sub-events as possible. By using the example in Figure 1(a), a photo set that contains "transport", "food", and "landscape" would be better to present the event "travel" than one that only contains "transport".

The sub-event coverage can be measured by borrowing an idea in search result diversification, which aims to retrieve search results that cover as many topics as possible in response to a given query [14]. The approach used in search result diversification is to estimate the probability that all the topics are covered with at least one search result, and to find a set of search results that maximizes this probability. Analogously, we estimate the probability that all the sub-events are covered with at least one photo, and to find a set of photos that maximizes this probability. Thus, the sub-event coverage $\text{SubCov}(S, e)$ is defined as follows:

$$\text{SubCov}(S, e) = \sum_{v \in \text{Sub}(e)} P(v|e)(1 - \prod_{s \in S}(1 - P(c = 1|s, v))), \tag{1}$$

where e is a given event, S is a photo set, $\text{Sub}(e)$ refers to sub-events of e, $P(v|e)$ is the probability that e contains sub-event v as well, and $P(c = 1|s, v)$ is the probability that photo s covers sub-event v. We assume a unique distribution for $P(v|e)$ due to the lack of prior knowledge for this probability, i.e. $P(v|e) = \frac{1}{|\text{Sub}(e)|}$. An intuitive interpretation of this formula is that $\text{SubCov}(S, e)$ becomes high if at least one of the photos in a photo set S has high probability $P(c = 1|s, v)$ for all the sub-events of e.

Below, we discuss a method of estimating the probability $P(c = 1|s, v)$. A basic assumption here is that photo s is likely to cover sub-event v if s is similar to photos of v. We use k-nearest neighbor distance k-NND(s, v), which is the average distance of k-nearest neighbor photos of photo s in photo set P_v, to measure the similarity between a photo and photos of an event. In addition to its simplicity, the computation of the k-nearest neighbor distance is efficient, since k-nearest neighbor search has been extensively studied in the literature. We obtain the following formula by taking the inverted distance of k-NND(s, v) with an exponential function: $P(c = 1|p, e) = \exp(-\lambda \cdot \text{k-NND}(s, v))$, where λ is a parameter that controls the shape of this distribution.

In summary, the sub-event coverage $\text{SubCov}(S, e)$ measures how likely a photo set can prevent sub-event misrecognitions. A photo set with high sub-event coverage is expected to avoid sub-event misrecognitions, and consequently achieve high perceptual quality.

4.3   Super-Event Coverage

A photo set that covers only one or few super-events may cause super-event misrecognitions. To cover all super-events $\text{Sup}(e)$ of event e, there should be at least one photo related to each super-event in the photo set S. Thus, we use the average of sub-event coverage for all its super-events to get super-event coverage:

$$\text{SupCov}(S, e) = \frac{1}{|\text{Sup}(e)|} \prod_{v \in \text{Sup}(e)} \text{SubCov}(S, v). \tag{2}$$

Only if all the super-events have high sub-event coverage, super-event coverage becomes high. For example, a photo set with high super-event coverage for event "travel in Kyoto" would be one that has high sub-event coverage for both events "travel" and "Kyoto".

4.4   Membership Entropy over Sibling-Events

A photo similar to just a few sibling-event can cause sibling-event misrecognitions. Thus, any photo in the output photo set should be similar to sibling-events as evenly as possible or not similar to any sibling event under their shared super-event. In the example of "conference party" in Figure 1(c), any photo in a photo set should evenly represent all the sibling-events of "conference party", or be dissimilar to any of the sibling-events such as "birthday party". Formally, photo s in photo set S for event e should evenly cover sibling-events $\text{Sib}(e, u)$ of event e in terms of super-event u. This idea can be implemented by using an entropy over sibling-events as proposed in work on query suggestion [15]. Letting $P_e(v|s)$ be the probability that photo s covers sibling-event v of e, the membership entropy of s over sibling-events $v \in \text{Sib}(e, u)$ under super-event u of e is defined as $H^s_{e,u} = -\sum_{v \in \text{Sib}(e,u)} P_e(v|s) \log P_e(v|s)$, where the probability $P_e(v|s)$ is defined by using k-nearest neighbor distance $k\text{-NND}(s, v)$ in the previous section:

$$P_e(v|s) = \frac{\exp(-\lambda \cdot k\text{-NND}(s, v))}{\sum_{v' \in \text{Sib}(e,u)} \exp(-\lambda \cdot k\text{-NND}(s, v'))}. \tag{3}$$

Here, we used a similar idea to the probability $P(c = 1|s, v)$ used in the previous section, i.e. a photo is likely to cover an event if the photo is similar to photos of the event. Note the difference is that the probability $P_e(v|s)$ follows a multinomial distribution, while the probability $P(c = 1|s, v)$ follows a binomial distribution.

The criterion of entropy for super-events of e with respect to sibling-events for a photo is formulated as follows:

$$\text{SibEnt}(s, e) = \frac{1}{|\text{Sup}(e)|} \sum_{u \in \text{Sup}(e)} \hat{H}^s_{e,u}, \tag{4}$$

where $\text{Sup}(e)$ is a set of super-events of event e, and $\hat{H}^s_{e,u}$ is the 0-1 normalized version of the entropy $H^s_{e,u}$ (i.e. $\hat{H}^s_{e,u} = H^s_{e,u} / \log |\text{Sib}(e, u)|$).

Having calculated the membership entropy for each photo in a photo set, we aggregate these scores by using the following formula:

$$\text{MemEnt}(S, e) = 1 - \prod_{s \in S} (1 - \text{SibEnt}(s, e)). \tag{5}$$

Although it is possible to take the average of $\text{SibEnt}(s, e)$ for each photo in a photo set S, we selected this formula so that $\text{MemEnt}(S, e)$ is *submodular* for efficient photo set generation as explained later.

4.5   Photo Set Generation

By combining three criteria we introduced above, sub-event coverage, super-event coverage and membership entropy over sibling events, we get an objective function $f(S, e)$ to be maximized:

$$f(S, e) = \alpha \text{SubCov}(S, e) + \beta \text{SupCov}(S, e) + \gamma \text{MemEnt}(S, e), \tag{6}$$

where α, β, and γ are parameters that determine which criteria to be emphasized.

Our objective now can be reformulated as a problem of finding a photo set of size n for a given event e that maximizes the objective function $f(S, e)$. Unfortunately, it is a NP-hard problem to find an optimal photo set. When photos can belong to multiple photo sets, there may not exist a single ordering of photo sets such that the objective function of $f(S, e)$ is maximized for all possible S. The reason is that a set of photos optimal for $f(S', e)$, where $|S'| = n - 1$, need not be subset of optimal of $f(S, e)$, where $|S| = n$.

As the set function $f(S, e)$ is monotonic and submodular (see Appendix), we can apply the greedy algorithm and guarantee the result returns $(1 - 1/e)$-approximation of the maximum[16], which often gives a good approximation to the optimum. We start with an empty photo set S, and iteratively add a photo $p \in P_e$ to S that maximizes $f(S \cup \{p\}, e)$ until the size of the photo set $|S|$ reaches n.

5   Experiments

We conducted experiments to demonstrate the effectiveness of our proposed method by using photos crawled from Flickr. Due to two phases in our approach that includes generating neighbor events and a photo set, they were implemented independently. All data in our experiment was collected from photo sharing social website Flickr, which contained photos taken by different users. We first collected about 1.3 million photos with 500 seed queries from the website to form photo collection P. Photos containing event's keywords are used as P_e.

5.1   Generating Neighbor Events

In this experiment, we generated neighbor events of 25 events with our proposed method and a simple baseline method. In our method, parameters are set according to our preliminary experiments: $\theta = 0.7$, $\phi = 0.7$. As a baseline, we used tf-idf to rank tags and generated sub-events by adding top-ranked tags to

Table 1. Neighbor Events of "travel Japan spring"

Method	Neighbor Event	Result
Our Method	Sub-Event	'travel japan spring park', 'travel japan spring cherry', 'travel japan spring tree', 'travel japan spring tokyo'
	Super-Event	'travel japan', 'travel spring', 'japan spring'
	Sibling-Event	'travel japan garden', 'travel spring taiwan', 'japan spring museum', 'japan spring architecture'
tf-idf	Sub-Event	'travel japan spring flower', 'travel japan spring tokyo', 'travel japan spring home', 'travel japan spring sky'
	Super-Event	'travel japan', 'travel spring', 'japan spring'
	Sibling-Event	'travel japan tokyo', 'travel spring flower', 'japan spring museum', 'japan spring architecture'

a target event. Super-events were produced the same way as in our method, while sibling-events were generated by using sub-events extracted by the simple baseline method.

Table 1 shows some examples generated with our method and baseline method. We can see that neighbor events generated by our methods are closer to reality. Especially our proposed method could generate better sibling-events probably because we considered the absent term of each super-event. On the other hand, baseline method using tf-idf generated many duplicate sibling-events probably due to the lack of considering similarity between tags.

5.2   Generating Photo Set

Setup. VisualRank (VR)[17] was used for a baseline method, because it can generate representative photos for a given event due to the fact that it gives high score to photos that are similar to many other photos in a collection. We simply used photos with the highest VR scores as an output photo set.

We used 25 events as inputs and applied five methods to generate photo sets of five types of sizes for each event. The five methods we tested are VisualRank (VR), sub-event coverage only (Sub), super-event coverage only (Sup), membership entropy over sibling events only (Sib), and combination of three criteria (ALL) which maximizes value of f in Equation (6). Having conducted preliminary experiments, three parameters in the objective function were set empirically as follows: $\alpha = 0.26$, $\beta = 0.42$, $\gamma = 0.32$. We used five types of the size of photo sets, i.e. $n = 1, 3, 5, 10, 20$. The parameter k of k-nearest neighbor was set to 10, and the parameter λ was set to 25 in this experiment. In total, there are 625 photo set for (event, method, size of photo set) combinations.

To evaluate the perceptual quality of each photo set, we utilized Lancers[1], a crowd sourcing service in Japan. Five users were assigned to each photo set. As we mentioned earlier, the perceptual quality of a photo set is better weighted by the accuracy and quickness of user's perception of an event by looking at a photo set. According to our request, they were required to answer what events

---

[1] http://www.lancers.jp/

Fig. 2. Example Results of event "walk dog"

they recognized from a photo set in a few seconds, which could guarantee the quickness of perception in a relative high level. We then quantized the accuracy of perception by judging the agreement between the original event keywords and their labels. This comparison was conducted by one of the authors, since this task was neither difficult nor subjective. The agreement was measured at a three-point scale: mismatch (0), partial match (1), and match (2). Match is given if the label has the same meaning as an event used, while the assessor gave partial match to labels that partially overlap keywords of the event.

Results. Figure 2 give photo sets of event "walk dog" generated by "Sib", "ALL" and "VR" method. As we can see, "VR" gives very similar photos, while fails to cover possible sub-events, such as "walk dog leash". It also lacks differentiation from sibling-event events like "raise dog" and "walk stroll". "Sib" successfully avoid misrecognition from these sibling-events.

Figure 3 shows an overall comparison of results from five methods. The horizontal axis is the number of photos in the photo set, and vertical axis is average score we get for average perceptual quality of each size. The overall trend is clear: with more photos in the photo set, the perceptual quality is improved. However, too many photos, such as 10 or 20 photos cannot help with it significantly. A two-way ANOVA shows a significant difference in both the type of methods ($F(4, 600) = 3.27$, $p < 0.05$) and size of photo sets ($F(4, 600) = 68.3$, $p < 0.01$). Significant interaction was not found: $F(16, 600) = 0.217$.

Three criteria (Sub, Sup and Sib) show different performances on various sizes. With three or five photos, sub-event coverage is the best probably due to its coverage of good representative contents over an event. Nevertheless, sup-event coverage is not as good as other two criteria. Membership entropy over sibling-events exhibits best performance with one photo. This may attribute to the fact that it maximizes an event's difference from sibling-events that can cause misunderstanding and users can easily figure out the right event without hesitation. Unfortunately, we find that the combination of three criteria does not reach our expectation. This failure may be the result of following reasons.

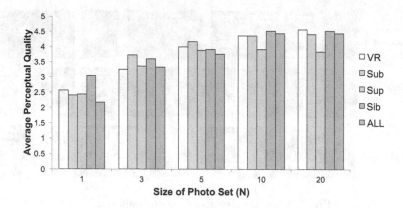

Fig. 3. Perceptual Quality of five Methods

The parameters of every criterion's weight are fixed. While according to our findings, they should vary based on size of photo set since performance of each criterion differs in sizes. Meanwhile, performance of each criterion differs with various events, which indicates the necessity of adjusting parameters according to events. We can also find that baseline method outperforms well when there are too many photos (20 photos). The reason may be that it will include diverse photos when the size increases too much.

6   Conclusion and Future Work

We propose to use a photo set to present an event and utilize event coverage and membership entropy to minimize misrecognition from neighbor events. Three types of misrecognition are summarized, and respectively, three criteria are used to deal with each type. To generate the final photo set, a function combining three criteria is given to be maximized. We compare the performance of VisuaRank and our methods. Results from our experiment prove that our proposed approach can improve perceptual quality in different sizes of photo set.

Apparently, there are still a lot of work needed to do in this research. As our definition of event consists of activity, time and place, it is better to make them different when considering relationship between events. Currently, we focus on only the selection of photos to form a photo set, composition of photos and aesthetic factors should be taken into consideration in our future research.

Acknowledgments. This work was supported in part by the following projects: Grants-in-Aid for Scientific Research (No. 24240013) from MEXT of Japan, and Microsoft Research CORE Project.

References

1. Sinha, P., Jain, R.: Extractive summarization of personal photos from life events. In: Proc. of ICME, pp. 1–6 (2011)
2. Pustejovsky, J., Castano, J.M., Ingria, R., Sauri, R., Gaizauskas, R.J., Setzer, A., Katz, G., Radev, D.R.: Timeml: Robust specification of event and temporal expressions in text. In: Proc. of IWCS (2003)
3. Shen, Y., Fan, J.: Leveraging loosely-tagged images and inter-object correlations for tag recommendation. In: Proc. of MM, pp. 5–14 (2010)
4. Chen, L., Roy, A.: Event detection from flickr data through wavelet-based spatial analysis. In: Proc. of CIKM, pp. 523–532 (2009)
5. Imran, N., Liu, J., Luo, J., Shah, M.: Event recognition from photo collections via pagerank. In: Proc. of MM, pp. 621–624 (2009)
6. Platt, J.C., Czerwinski, M., Field, B.A.: Phototoc: Automatic clustering for browsing personal photographs. Technical Report MSR-TR-2002-17, Microsoft Research (2002)
7. Cooper, M., Foote, J., Girgensohn, A., Wilcox, L.: Temporal event clustering for digital photo collections. ACM TOMCCAP 1(3), 269–288 (2005)
8. Loui, A.C., Savakis, A.: Automated event clustering and quality screening of consumer pictures for digital albuming. IEEE TOMM 5(3), 390–402 (2003)
9. Zacks, J.M., Speer, N.K., Swallow, K.M., Braver, T.S., Reynolds, J.R.: Event perception: a mind-brain perspective. Psychological Bulletin 133(2), 273–293 (2007)
10. Silberer, C., Ferrari, V., Lapata, M.: Models of semantic representation with visual attributes, pp. 572–582 (2013)
11. Makadia, A., Pavlovic, V., Kumar, S.: A new baseline for image annotation. In: Forsyth, D., Torr, P., Zisserman, A. (eds.) ECCV 2008, Part III. LNCS, vol. 5304, pp. 316–329. Springer, Heidelberg (2008)
12. Jeh, G., Widom, J.: Simrank: a measure of structural-context similarity. In: Proc. of KDD, pp. 538–543 (2002)
13. Pedersen, T., Patwardhan, S., Michelizzi, J.: Wordnet: Similarity: measuring the relatedness of concepts. In: Proc. of HLT-NAACL, pp. 38–41 (2004)
14. Agrawal, R., Gollapudi, S., Halverson, A., Ieong, S.: Diversifying search results. In: Proc. of WSDM, pp. 5–14 (2009)
15. Kato, M.P., Sakai, T., Tanaka, K.: Structured query suggestion for specialization and parallel movement: effect on search behaviors. In: Proc. of WWW, pp. 389–398 (2012)
16. Nemhauser, G.L., Wolsey, L.A., Fisher, M.L.: An analysis of approximations for maximizing submodular set functions I. Mathematical Programming 14(1), 265–294 (1978)
17. Jing, Y., Baluja, S.: Visualrank: Applying pagerank to large-scale image search. IEEE TPAMI 30(11), 1877–1890 (2008)
18. Krause, A., Guestrin, C.: Near-optimal observation selection using submodular functions. In: Proc. of AAAI, pp. 1650–1654 (2007)

7   Appendix: Proof for Section 4

According to [16], submodularity can be defined as below.

Definition 1 (Submodularity). *Given a finite ground set N, a set function $2^N \to \mathbb{R}$ is submodular if and only if for all sets $S, T \subset N$ such that $S \subset T$, and $d \in N \setminus T$, $f(S \cup \{d\}) - f(S) \geqslant f(T \cup \{d\}) - f(T)$.*

Lemma 1. *$f(S, e)$ is a monotonic and submodular function.*

Proof. More photos in a photo set returns higher value, which indicates the function is increasing monotonic. For $S_1 \subset S_2$, we can have $f(S_1, e) \leqslant f(S_2, e)$.

The class of submodular functions is closed under non-negative linear combinations[18]. Three criteria in the function are all non-negative, so we just need to prove the submodularity of each function individually.

According to [14], SubCov(S, e) is submodular. In addition, SupCov(S, e) is also submodular as it is a non-negative linear combination of SubCov(S, e). In this case, we just need to have MemEnt(S, e) submodular which can be also proved.

Let S, T be two arbitrary sets of photos related by $S \subset T$. Let m be a photo not in T. S' denotes $S \cup \{m\}$, and T' for $T \cup \{m\}$ similarly.

$$\text{MemEnt}(S', e) - \text{MemEnt}(S, e)$$
$$= 1 - \prod_{s \in S}(1 - \text{SibEnt}(s, e)) \cdot (1 - \text{SibEnt}(m, e)) - (1 - \prod_{s \in S}(1 - \text{SibEnt}(s, e)))$$
$$= \prod_{s \in S}(1 - \text{SibEnt}(s, e)) \cdot \text{SibEnt}(m, e)$$

Similarly, we can establish that:

$$\text{MemEnt}(T', e) - \text{MemEnt}(T, e) = \prod_{s \in T}(1 - \text{SibEnt}(s, e)) \cdot \text{SibEnt}(m, e)$$
$$= \prod_{s \in T \setminus S}(1 - \text{SibEnt}(s, e)) \cdot \prod_{s \in S}(1 - \text{SibEnt}(s, e)) \cdot \text{SibEnt}(m, e)$$

For all s, SibEnt(s, e) is the average value of normalized entropies, which means SibEnt(s, e) has value between 0 and 1. Thus, $\prod_{s \in T \setminus S}(1 - \text{SibEnt}(s, e)) \leqslant 1$. Therefore, we conclude that

$$\text{MemEnt}(S', e) - \text{MemEnt}(S, e) \geqslant \text{MemEnt}(T', e) - \text{MemEnt}(T, e)$$

As a result, the function MemEnt(S, e) is submodular and thus objective function $f(S, e)$ is also submodular.

Online Community Transition Detection

Biying Tan[1], Feida Zhu[1], Qiang Qu[2], and Siyuan Liu[3]

[1] School of Information Systems
Singapore Management University
[2] Department of Computer Science
Aarhus University
[3] Heinz College
Carnegie Mellon University

Abstract. Mining user behavior patterns in social networks is of great importance in user behavior analysis, targeted marketing, churn prediction and other applications. However, less effort has been made to study the evolution of user behavior in social communities. In particular, users join and leave communities over time. How to automatically detect the online community transitions of individual users is a research problem of immense practical value yet with great technical challenges. In this paper, we propose an algorithm based on the Minimum Description Length (MDL) principle to trace the evolution of community transition of individual users, adaptive to the noisy behavior. Experiments on real data sets demonstrate the efficiency and effectiveness of our proposed method.

1 Introduction

Recent years have witnessed the growth of community detection as one of the major directions in social network mining. A community can be defined as a group of users sharing some common properties. As users migrate from a community to another, communities could form and dissolve. As a result, social networks in real life are highly dynamic with evolving communities. In this paper, we are particularly interested in the discovery of community transition of individual users in social networks. This study is important for understanding user behavior patterns, which can be used to support many real-life applications. For example, in targeted marketing, users' migration to another online community often foretells the emergence of new interests and, accordingly, new opportunities for marketing [1, 2]. Similar applications can also be found in churn prediction in which the detection of community transition within a sliding window could indicate a user's potential service switch. Furthermore, the discovery of impending migration is useful in social identity linkage across social networks [3].

Unfortunately, most of the existing methods on community mining assume an underlying social network which is static [4–9]. Though there are some recent studies to investigate the dynamics of social networks [10–13] and model dynamic communities based on community structure, most of them do not examine user interaction, which is in fact one of the most important dynamics of a

F. Li et al. (Eds.): WAIM 2014, LNCS 8485, pp. 633–644, 2014.

social network [14]. For instance, in a social network like Twitter, interactions of tweeting and retweeting form conversations, which propagate information within and across communities. In this paper, we define a sequence of individual interaction networks snapshots to model dynamic communities.

To the best of our knowledge, very few studies have been devoted to the problem of discovering community transitions. The problem is challenging in that (I) The transition between communities is, in general, an irregular and infrequent event for an individual user, which adds an extra degree of difficulty for detection due to the absence of regular patterns; and (II) A single observed interaction may not substantiate a user's immediate inclination. For example, the chat with an insurance agent does not always indicate that the user is inclined to buy insurance.

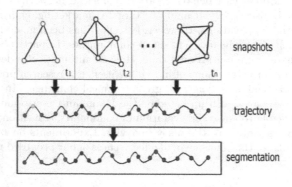

Fig. 1. The proposed framework

In summary, we make the following contributions in this work. We formalize the problem of community transition discovery for social network users. A trajectory is built to represent users' community evolution. We propose an automatic trajectory segmentation method based on the principle of Minimum Description Length (MDL) [15], which frees our solution from user-defined parameters such as the number of segments or the threshold to report a transition. Furthermore, our method can detect transitions in a noisy environment by replacing the noise with the underlying real observation. The framework of our method is illustrated in Figure 1. The experiments demonstrate the efficiency and effectiveness of our approach in discovering the community transition on real datasets.

The remainder of the paper is organized as follows. In Section 2, we define the underlying interaction network, community evolutionary trajectory and trajectory segmentation. Then we design an encoding scheme to determine a good segmentation of the community evolutionary trajectory in Section 3. Section 4 presents the corresponding algorithms, and we report the experimental results in Section 5. Section 6 reviews the related work and Section 7 concludes the paper.

2 Problem Statement

In this section, we present the preliminaries and formulate the problem.

Consider a graph, G, consisting of a set, $V(G)$, of $|V(G)|$ nodes ($|V(G)| \geq 2$) and a set, $E(G)$, of edges linking pairs of nodes, $e(u_i, u_j) \in E(G)$ for $u_i, u_j \in V(G)$. We term a graph G as an ego network if a node $u_0 \in V(G)$ satisfying for any $e(u_0, u_i) \in E(G)$, there exists $u_i \in V(G)$. Given a set $\Sigma = \{\varsigma_1, \varsigma_2, \ldots, \varsigma_k\}$ of labels, a labeling function, $L(V) \mapsto \Sigma$, maps nodes in G to labels in Σ.

Given such an underlying ego network G, we define a_{ij} as an interaction between two users u_i and u_j, which is attached with a weight representing the frequency of interactions between the two users. Such interactions, for example, can be emails or replies.

Consider an ego network G which is evolving over time, we use $G^{(t)}$ to represent G at t-timestamp. Given a network snapshot $G^{(t)}$, we use a $|V^{(t)}| \times |V^{(t)}|$ matrix to represent an interaction network $I^{(t)} = G^{(t)}$, where an element $a_{ij}^{(t)}$ is a weight of the edge $e^{(t)}(u_i, u_j) \in E(G^{(t)})$ at timestamp t, and each user $u_i \in V(G^{(t)})$ is associated with a set of labels $\Sigma_{u_i}^{(t)}$.

In this paper, community is defined as the subgroup of users sharing the same properties, then we group the users with a same label into a community. Given a sequence of interaction network snapshots, i.e., $I = \{I^{(1)}, I^{(2)}, \ldots, I^{(t)}\}$, the $n^{(i)}$ communities detected at i-th snapshot are denoted by $C(i) = \{C_1^{(i)}, C_2^{(i)}, \ldots, C_{n_i}^{(i)}\}$, where $C_j^{(i)} \in C(i)$ is a subgraph of $I^{(i)}$. The members $V(C_j^{(i)})$ within $C_j^{(i)}$ share a same label, such as school, location, gender, thus we have the label of $C_j^{(i)}$, $L(C_j^{(i)}) = \bigcap L(V(C_j^{(i)}))$, $L(C_j^{(i)}) \neq L(C_{j'}^{(i)})$ for $j \neq j'$. As a user that interacts more often with a community has more to do with that community and therefore we introduce the notion of dominant community which denotes the community with maximal interactions.

Definition 1 (Dominant Community). *The community which has the maximal interactions with user u_0 at i-timestamp is defined as the dominant community $D^{(i)}$. We define the community activity to measure the frequency of interactions, which is computed as the ratio of intra-community interactions to the total number of interactions at i-timestamp, i.e.,*

$$CA(C_j^{(i)}) = \frac{\sum_{u_m, u_n \in V(C_j^{(i)})} a_{mn}^{(i)}}{\sum_{u_m, u_n \in I^{(i)}} a_{mn}^{(i)}}, \tag{1}$$

where the minimum is 0 indicating no interaction within this community, while the maximum is 1 representing no interaction outside this community.

Tracking dominant communities along the sequence of interaction network snapshots enables to discover valuable knowledge about the behavior evolution regarding to communities in social network. To motivate this, we construct a community evolutionary trajectory using the sequence of dominant communities.

Definition 2 *(Community Evolutionary Trajectory).* *A community evolutionary trajectory of length t for user u_0 is a sequence consisting of t dominant communities, i.e.,*

$$\mathcal{T} := \{L_1, L_2, ..., L_t\}, \tag{2}$$

where each trajectory element L_i is the label of the dominant community $D^{(i)}$.

There are two reasons to construct the evolutionary trajectory using dominant communities: 1) The communities within which the users interact frequently are more important, which reveals stronger correlation between users and communities. 2) Community evolution trajectory provides an efficient method to study the community evolution. Traditional algorithms with regard to community evolution problems are of high computational complexity. Tracking how communities evolve over time based on sequence reduces the complexity by mapping complex graph structure into a community label, which is more efficient, especially in large social networks.

In order to track the evolution of communities transitions, we partition the community trajectory into a finite set of segments, and each segment has the maximal discriminating power. The goodness measurement of the segmentation will be made precisely in the next section, which formulates our cost objective function.

Definition 3 *(Community Trajectory Partitions).* *For a trajectory $\mathcal{T}$ of length t, L_i with $0 < i \le t$ denotes its i-element. A trajectory segment is denoted by $L_{i...j}$ for $0 < i \le j \le t$. The set of consecutive communities which are assigned into the p-th segment $1 \le p \le n$ is denoted by $\mathcal{T}_p$. The partitions do not overlap, in the sense that $\mathcal{T}_p \bigcap \mathcal{T}_{p'} = \oslash$ for $p \ne p'$. Given two consecutive communities $\mathcal{T}_1 = L_{i...j}, \mathcal{T}_2 = L_{j+1...k}$, the labels of two segments are denoted by $L(\mathcal{T}_1) = L_i \bigcup ...L_j, L(\mathcal{T}_2) = L_{j+1} \bigcup ...L_k$, respectively. We say that a community transition has occurred if $\frac{L(\mathcal{T}_1) \bigcap (\mathcal{T}_2)}{L(\mathcal{T}_1) \bigcup (\mathcal{T}_2)} < \epsilon$, and we call $j + 1$ a change point.*

Based on the above definition, the partition of a sequence can be regarded as a 1-dimensional classificaiton problem, and the best partition is a set of homogeneous segments, within which there exists the communities with the same label. We treat a community evolutionary trajectory as a sequence, an encoding scheme is proposed to describe the sequence. As a perfect segment is the repetitions of the same character, we can use fewer characters to describe the sequence, which leads to a smaller encoding cost. According to the Minimal Description Length (MDL) principle, the best encoding scheme is the one which leads to the minimal encoding cost. The problem of finding a good partition can be converted to the problem of determining a good encoding method. Thus, we define the problem as follows:

The goal: Given a community evolutionary trajectory $\mathcal{T}$, find the best partition $P(\mathcal{T})$ by identifying a set of change points, which leads to the minimal encoding cost.

3   Methodology

To achieve the goal that finding the best partition with minimal encoding cost, a lossless encoding scheme is introduced. Consider a lossless compressor that output s' from s, s' can be seen as another description of s. If s' is shorter than s, then s' is a better description than s, which gives a smaller encoding cost. Given a community evolutionary trajectory, our target is to identify a set of change points that leads to the optimal partition. The problem can be rephrased as to pick up the unreasonable change points. The unreasonable change points mainly consist of the noisy ones. For example, given a trajectory $T = NNNUNNNAUUUUUUU$, the first occurring "U" and "A" can be considered as such noisy change points, and the optimal partition should be $\{\{NNNUNNNA\}, \{UUUUUUU\}\}$. Under the MDL principle, we design a cost objective function with two parts, the first one is the cost to encode the optimal segmentation, and the second is the cost to identify the unreasonable change points.

3.1   Segment Encoding

We start by converting the trajectory into a sequence whose elements are numerical characters. Let Num be the function that maps each community label to a numerical character $L_i \mapsto Char(j)$. For example, given a trajectory $T - NNNNNSNNUUUU$, and Num maps each label to a character as follows: $(S, 0), (N, 1), (U, 2)$, so that we have the corresponding sequence $S = Num(T) = 111110112222$. After converting the trajectory, we introduce how to encode the trajectory segments.

As our goal is to decompose the trajectory into best segments, each segment can be seen as a repetition of some characters. For example, given an input sequence: $S = 111110112222$, our encoding schema partitions this sequence into five consecutive segments: $P(S) = \{\{11111\}, \{0\}, \{11\}, \{2222\}\}$. A segment can be encoded into two fields, encoding the repeating character symbol into one field, and the length of the segment in the other:

- $Char^{S_p}(j)$: the repeating character of segment S_p,
- $Len(S_p)$: the length of segment S_p.

Note that, $Char^{S_p}(j)$ and $Len(S_p)$ is constant, which has no effect on the final segmentation. To facilitate the encoding, we define the entropy for a sequence under a partition S_p as

$$E(S) = -\sum_{x \in S} p(x) \log p(x), \tag{3}$$

where x is the element in S, $p(x)$ is the probability of x that appears in S. $E(S)$ measures how homogeneous the sequence is. The minimum of $E(S)$ is 0 when all the elements are the same, and the maximum is $\log |S|$ when the sequence is a pure random sequence. Thus, we obtain the segment encoding cost as the follows:

Definition 4 *(Segments Encoding Cost)*

$$C^{S_p} = \sum_{S_P} m + E(S), \tag{4}$$

where m is the number of bits needed to encode a segment. The cost is a sum of encoding cost of all the segments.

3.2   Identifying the Unreasonable Change Points

Given a trajectory with a set of change points, the cost of identifying the change point is heavily related to the occurrence frequency of the corresponding community. For instance, given a trajectory $T = NNNUNNNAUUUUUUU$, obviously, the cost of identifying the time that "A" appears as an unreasonable change point is much less than the one to identify the first time "U" appears, as we need to consider eight candidate "U" in order to identify. Thus it leads to more cost to substitute the community which occurs frequently in the trajectory. It is also sound to interpret this with our problem statement, more penalty should be given to treat a frequently occurring community as a noise. The encoding scheme for a substitution is encoded in a way similar as a segment encoding, which includes:

- $Sub^{S_p}(k, k')$: the substitution of character k' for k,
- $Pos^{S_p}(k)$: the position of character k in the sequence.

Similarly, the description complexity for substitution $\mathcal{T}$ is constant, which is equal to the encoding cost of a segment. Thus, we obtain the substitution encoding cost as the follows:

Definition 5

$$C^{S_b} = \sum_{S_b} m, \tag{5}$$

where m is the number of bits needed to encode a substitution. The cost is a sum of encoding cost of all the substitutions.

Given a community evolutionary sequence S, our goal is to partition it into a number of segments, and compress each segment which leads to a minimal encoding cost. The total encoding cost is the sum of the segments encoding cost and the cost of all the substitutions. We formulate the total encoding cost in the following:

Definition 6 *(Total Encoding Cost)*

$$C = \sum_{S_p} C^{S_p} + \sum_{S_b} C^{S_b}, \tag{6}$$

where C^{S_p} is the encoding cost of p-th partition, and C^{S_b} is the encoding cost of b-th substitution.

In the next section, we present a search algorithm to find the optimal solution, based on the cost objective function.

4   Algorithm

In this section, based on the encoding scheme and cost function introduced in section 3, an algorithm is introduced to find the optimal segmentation in the presence of noise. Algorithm 1 is to find all the change points existing in the trajectory. The change point is a point in the trajectory that is different from the one located in the immediate left/right. The algorithm we apply to find the optimal segmentation is basically to substitute some unreasonable change points in order to find the optimal partition under the measurement of the cost function. Algorithm 2 presents the approximate algorithm to achieve the target. Basically, we use Depth-First Search (DFS) to search for the optimal partition from the answer space. Procedure OP presents the search logic with one node in the search tree. Given the current partition, OP constructs the change point sets for both two directions, PL and PR. Then OP tries to substitute each of them with the character which makes it distinguished, i.e., the substitution merges one change point into the segment left or right to it. After the substitution, OP computes the encoding cost of the new partition. If the encoding cost is smaller than the value of the original partition passed to OP, OP is recursively called with the new partition. Global value minCost and minString are used to store the answer. They are kept updating during the whole search procedure, with the minimal cost and its corresponding partition.

Algorithm 1. Find potential change points (left)

1: **procedure** FCPL(S, P)
2:     $P \leftarrow \oslash$
3:     $cur \leftarrow S_0$
4:     **for** $i \leftarrow 1, |S|$ **do**
5:         **if** $cur \neq S_i$ **then**
6:             $cur = S_i$
7:             $P \leftarrow P \bigcup S_i$
8:         **end if**
9:     **end for**
10: **end procedure**

Algorithm 2. Find optimal partition

1: $optimalStr = S$
2. $minCost = Cost(S)$
3: **procedure** OP(S)
4:     $PL \leftarrow$ FCPL(S)
5:     $PR \leftarrow$ FCPR(S)
6:     $min = Cost(S)$
7:     **for** each p in $PL(PR)$ **do**
8:         $p' \leftarrow$ the point locates in the immediate left(right) of p
9:         $S' \leftarrow Substitute(S, p, p')$
10:         $cost = Cost(S')$
11:         **if** $cost \geqslant min$ **then return**
12:         **else**
13:             **if** $cost < minCost$ **then**
14:                 $optimalStr = S'$
15:                 $minCost = cost$
16:             **end if**
17:             OP(S')
18:         **end if**
19:     **end for**
20: **end procedure**

5   Experimental Results

In this section, we will evaluate our method on real, large social network datasets. We perform the experiments on a PC with a Intel(R) Core(TM) i5-2300 2.80GHz CPU and 12GB RAM and the algorithm is implemented with JAVA.

Table 1. The statistics of Renren dataset

	Max	Avg
Number of Communities/User	10	2
Number of Timestamps/User	1105	580

5.1   Dataset

We conduct our study on the Renren Network dataset[1]. Renren Network is a leading social networking service in China, which is similar as Facebook. As Renren is one of the largest real name based social network in China, and Renren user provides a list of education background in their profiles, thus we can observe the evolution of their communities through the observation of interactions with different groups of friends along with the evolution of users' identities. The dataset consists of more than 16 million replies to 6,437 users' status posts, and the profile information of 690,926 unique users are involved in the dataset.

The interaction network snapshot $I^{(t)}$ is constructed from the observed reply messages, each edge in the network corresponding to a reply between two users. We use a window size of one day to construct the interaction network snapshots, and then generate the community evolutionary trajectory based on the network snapshots. The education and occupation information are chosen as the labels of users, and each label is associated with a time period constraint. The users with the same label are grouped into a community in each network snapshot, the community with the maximal community activity is selected as the element of the trajectory. The statistics of Renren dataset is shown in Table 1.

5.2   Effectiveness

In this subsection, we evaluate the effectiveness of our method for finding optimal segmentation in the following two experiments.

In the first experiment, we match the reasonable change points identified by our method with the ground truth provided by the labels in users' profiles. More specifically, we compute the smallest difference (day) between the results with any change point in the ground truth. The variance of the results is reported in Figure 2. We can observe that most dots locate in the left side, indicating our method has a high accuracy. Many dots gather in the lower left corner, which demonstrates our method can detect the transition happened in the latest two

---

[1] www.renren.com

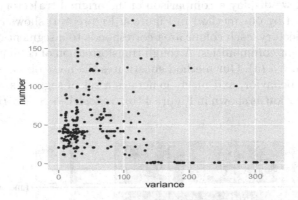

Fig. 2. Variance distribution of the results

months. The dots in the right part present the cases of low accuracy, which could be due to the delayed update or data missing of their profiles for some users. The other reason is due to the short sequence generated from inactive users.

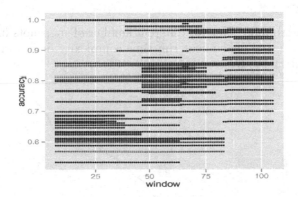

Fig. 3. Comparison of accuracy with different tolerance windows

Secondly, we estimate our method under different tolerance window. The tolerance window is created by moving the change point on the left and right with the distance of the tolerance value. Then we determine if any change point in ground truth falls within that window. The comparison of the accuracy with different tolerance from 7 to 105 days is listed in Figure 3. We can observe an obvious increase of accuracy along with increase of the tolerance window. The results show that our method can achieve 60%~70% accuracy when the window is smaller than 35 days, and the accuracy can reach to 75%~87% when the size of tolerance window is between 47 and 62 days. The results showed in Figure 3 is consist with what we observe in Figure 2.

Furthermore, we display a comparison of the original trajectory and the trajectory generated by our method in Figure 4. Figure 4(a) shows an example of the original trajectory, each colour area corresponds to a segment of consecutive identical dominant communities. A rough transformation of communities can be observed in Figure 4 (a). Our method substitutes the noise observation with the underlying real behavior, and nicely generates a set of segments representing a community evolution as shown in Figure 4 (b): $SchoolA \rightarrow SchoolB \rightarrow SchoolC$.

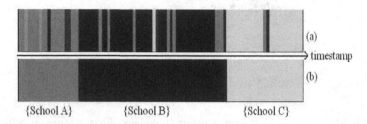

Fig. 4. Comparison of the original trajectory (a) with the trajectory (b) generated by our method

5.3 Scalability

We illustrate the running time for different number of snapshots in Figure 5. As shown in Figure 5, the near linear run time complexity demonstrates the high scalability of our method.

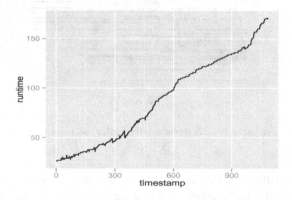

Fig. 5. Runtime (ms) vs number of timestamps (day)

6 Related Work

Several papers have studied on the community evolution in social networks. In general, we consider three broad class of temporal analysis on online communities: 1) properties or phenomena of evolving community analysis, 2) stage of dynamic community identification, 3) dynamic community detection and the community changes identification. In the first class work, Leskovec et al. [16] discovered the shrinking diameter phenomena on time-evolving networks. In the second class work, Palla et al. [17] quantifies the events in community evolution: growth, merging, birth, contraction, splitting and death. In the last class work, Tang et al. [11] proposed a clustering algorithm to mine the evolution of communities in social network with multiple entities. Aggarawal and Yu [18] proposed a method to find community changes in dynamic graphs, which requires user-specified parameters. Sun. et al. [19] and Ferlez at al. [20] proposed a parameter-free framework to discover communities and community changes in dynamic networks, relying on the MDL principle. Overall, all the above studies have a drawback that they cannot determine the evolution of users across communities individually, they all track the evolution of the whole network.

7   Conclusions and Future Work

In this paper, we propose a parameter-free method to discover the community transition for individual users in dynamic networks. We start by constructing a trajectory to represent the evolution of communities, then a trajectory segmentation approach is proposed to discover the best partition, based on the MDL principle. Our method is automatic, which requires no user-specified parameters, like the threshold of a transition and the number of segments. Furthermore, our method is adaptive to distinguish between the noise and underlying true behavior. The experiments on real dataset show that our method has a high accuracy to find the community transitions, and our method is fast and scalable

Our paper is a preliminary study on the detection of community transition, which still needs a lot of improvements. First, we should improve the method by incorporating all the communities evolved in the user's interactions, and detecting long-term transition and short-term transition. Second, we should enlarge the experiment by evaluating our method on different datasets, and comparing some state-of-art methods.

Acknowledgement. This research has been supported by the Singapore National Research Foundation under its International Research Centre @ Singapore Funding Initiative and administered by the IDM Programme Office, Media Development Authority (MDA) and the Pinnacle Lab at Singapore Management University.

References

1. Dasgupta, K., Singh, R., Viswanathan, B., Chakraborty, D., Mukherjea, S., Nanavati, A.A., Joshi, A.: Social ties and their relevance to churn in mobile telecom networks. In: Proceedings of the 11th International Conference on Extending Database Technology: Advances in Database Technology, pp. 668–677. ACM (2008)
2. Richter, Y., Yom-Tov, E., Slonim, N.: Predicting customer churn in mobile networks through analysis of social groups. In: SDM, pp. 732–741 (2010)

3. Liu, S., Wang, S., Zhu, F., Zhang, J., Krishnan, R.: Hydra: Large-scale social identity linkage via heterogeneous behavior modeling

4. Duch, J., Arenas, A.: Community detection in complex networks using extremal optimization. Physical Review E 72(2), 27104 (2005)

5. Fortunato, S., Barthelemy, M.: Resolution limit in community detection. Proceedings of the National Academy of Sciences 104(1), 36–41 (2007)

6. Lancichinetti, A., Fortunato, S.: Community detection algorithms: a comparative analysis. Physical Review E 80(5), 56117 (2009)

7. Fortunato, S.: Community detection in graphs. Physics Reports 486(3), 75–174 (2010)

8. Kovács, I.A., Palotai, R., Szalay, M.S., Csermely, P.: Community landscapes: an integrative approach to determine overlapping network module hierarchy, identify key nodes and predict network dynamics. PloS One 5(9), e12528 (2010)

9. Leskovec, J., Lang, K.J., Mahoney, M.: Empirical comparison of algorithms for network community detection. In: Proceedings of the 19th International Conference on World Wide Web, pp. 631–640. ACM (2010)

10. Kim, M.-S., Han, J.: A particle-and-density based evolutionary clustering method for dynamic networks. Proceedings of the VLDB Endowment 2(1), 622–633 (2009)

11. Tang, L., Liu, H., Zhang, J., Nazeri, Z.: Community evolution in dynamic multimode networks. In: Proceedings of the 14th ACM SIGKDD International Conference on Knowledge Discovery and Data Mining, pp. 677–685. ACM (2008)

12. Tantipathananandh, C., Berger-Wolf, T., Kempe, D.: A framework for community identification in dynamic social networks. In: Proceedings of the 13th ACM SIGKDD International Conference on Knowledge Discovery and Data Mining, pp. 717–726. ACM (2007)

13. Yang, T., Chi, Y., Zhu, S., Gong, Y., Jin, R.: Detecting communities and their evolutions in dynamic social networks a bayesian approach. Machine Learning 82(2), 157–189 (2011)

14. Wilson, C., Boe, B., Sala, A., Puttaswamy, K.P., Zhao, B.Y.: User interactions in social networks and their implications. In: Proceedings of the 4th ACM European Conference on Computer Systems, pp. 205–218. ACM (2009)

15. Barron, A., Rissanen, J., Yu, B.: The minimum description length principle in coding and modeling. IEEE Transactions on Information Theory 44(6), 2743–2760 (1998)

16. Leskovec, J., Kleinberg, J., Faloutsos, C.: Graphs over time: densification laws, shrinking diameters and possible explanations. In: Proceedings of the Eleventh ACM SIGKDD International Conference on Knowledge Discovery in Data Mining, pp. 177–187. ACM (2005)

17. Palla, G., Barabási, A.L., Vicsek, T.: Quantifying social group evolution. Nature 446(7136), 664–667 (2007)

18. Aggarwal, C.C., Yu, P.: Online analysis of community evolution in data streams. In: Proceedings of the SIAM International Conference on Data Mining (SDM 2005), pp. 56–67 (2005)

19. Sun, J., Faloutsos, C., Papadimitriou, S., Yu, P.S.: Graphscope: parameter-free mining of large time-evolving graphs. In: Proceedings of the 13th ACM SIGKDD International Conference on Knowledge Discovery and Data Mining, pp. 687–696. ACM (2007)

20. Ferlez, J., Faloutsos, C., Leskovec, J., Mladenic, D., Grobelnik, M.: Monitoring network evolution using mdl. In: IEEE 24th International Conference on Data Engineering, ICDE 2008, pp. 1328–1330. IEEE (2008)

PPML: Penalized Partial Least Squares Discriminant Analysis for Multi-Label Learning

Zongjie Ma[1], Huawen Liu[1,2,*], Kaile Su[1], and Zhonglong Zheng[1]

[1] Department of Computer Science, Zhejiang Normal University, China
hwliu@zjnu.edu.cn
[2] NCMIS, Academy of Mathematics and Systems Science, CAS, China

Abstract. Multi-label learning has attracted widespread attention in machine learning, and many multi-label learning algorithms have been witnessed. However, two main challenging issues remain: the high dimension of data and the label correlation. In this paper, a new classification method, called penalized partial least squares discriminant analysis for multi-label learning (PPML), is proposed. It aims at performing dimension reduction and capturing the label correlations simultaneously. Specifically, PPML first identifies a latent space for the variable and label space via partial least squares discriminant analysis (PLS-DA). To tackle with the problem of high dimensionality in solving PLS-DA, a ridge penalization is exerted on the optimization problem. After that, the latent space is used to construct learning model. The experimental results on the standard public data sets indicate that PPML has better performance than the state-of-the-art approaches.

Keywords: multi-label learning, partial least squares discriminant analysis, ridge regression, dimension reduction

1 Introduction

Multi-label classification has attracted widespread attention in machine learning, and has been applied to many fields, such as gene function [1], semantic annotation of images [2], and so on [3]. Currently, many learning approaches for multi-label data have been developed. They can be roughly divided into two groups: problem transformation and algorithm adaptation [4]. The former transforms the multi-label problems into a set of single-label problems, which can be solved with the traditional classification algorithms. Binary Relevance (BR) and Label Powerset (LP) [4] are two typical problem transformation approaches. As BR and LP can not work well if the label set has large numbers, Random k-labelsets (RAkEL) [5] employs LP to learn a corresponding classifier by breaking the original set of labels into a number of small random subsets. Nevertheless, RAkEL shows low efficiency when the label set is large and sparse. Algorithm adaptation exploits the traditional learning algorithms to handle the

---

* Corresponding author.

F. Li et al. (Eds.): WAIM 2014, LNCS 8485, pp. 645–656, 2014.
© Springer International Publishing Switzerland 2014

multi-label problems. For instance, Multi-Label k-Nearest Neighbor (ML-kNN) [6] is based on the k-Nearest Neighbor (kNN) algorithm, which is a lazy learning approach and ignores the label correlations. Backpropagation for Multi-label Learning (BP-MLL) [7] is a neural network approach based on the popular backpropagation algorithm. However, with the increase of the sample dimensions, the efficiency of the learning algorithms becomes a serious problem.

The so-called "curse of dimensionality" resulted from the high dimensionality of data brings enormous challenges to multi-label classification [8]. The existing dimensionality reduction methods are not very appropriate for the multi-label classification problems. Principal component analysis (PCA) [9], which is a classical dimensionality reduction approach and widely used in practice, ignores the label information. Linear discriminant analysis (LDA) [10], a popular supervised dimensionality reduction algorithm, does not consider the dependence between the variables and labels. Canonical correlation analysis (CCA) [11] maximizes the correlation between two blocks of variables, but produces a generalized eigenvalue problem with higher computational expense.

The challenge has been attempted to solve in multi-label classification by researchers during the past few years. For example, Huang and Zhou [12] encode the local influences of label correlations in a LOC code in order to exploit label correlations locally. In [13], a new criterion named PRO LOSS is proposed, which concerns the prediction on all labels and the rankings of only relevant labels. In order to high-dimensional data, Liang Sun et al. [14] structure an equivalent least-squares formulation for CCA under a mild condition. A semi-supervised framework is obtained in [15], which performs optimization for dimension reduction and multi-label inference. Bayesian network structure is used to encode the conditional dependencies of both the labels and feature sets in [16]; Nevertheless, this high-order approach may lead to high model complexities.

In this work, we present a new general framework for multi-label classification, capturing the label correlations and dimensionality reduction simultaneously. The proposed framework is named as penalized partial least squares discriminant analysis for multi-label classification (PPML). It aims at reducing the dimension of muliti-label data and obtaining the latent variables between the variable and label spaces. Specifically, we adopt the technique of partial least squares discriminant analysis (PLS-DA) [17] to discover the latent variables between the variable and label spaces of data, so as to model the correlation between them. To tackle with the problem of "large p, small n", a ridge penalization [18] is further performed on the object optimization function of PLS-DA. After obtaining the latent space, we build a discriminant model and use to predict the label sets for new instances in terms of the regression coefficient of the latent variables.

The rest of the paper is divided into the following sections: Section 2 briefly reviews the state-of-the-arts of multi-label learning. Our proposed method, PPML, is presented in Section 3. Section 4 is the part of experimental comparative study. Finally, we conclude the paper.

2   Related Work

In this section, we briefly review the representative multi-label learning methods. More details can be found in good papers (e.g., [4]).

Generally, multi-label learning algorithms can be divided into two categories: problem transformation and algorithm adaptation. The problem transformation methods firstly transform the multi-label problems into a set of corresponding single label ones, which can be solved by the traditional classification approaches. The characteristic of this kind of methods is fitting the multi-label data to learning algorithms. Binary relevance (BR) is a typical example of this kind methods [4]. It builds a binary classifier for each label occurring in data, where the instances are considered as positive if they contain the label and negative otherwise. Note that BR does not take the correlation of labels into consideration.

CLR [19] and MLStacking [20] exploit the pairwise correlation between the labels to construct classification models. Specifically, they first take a pair of labels as a new label at each time. The instances involve only one of labels in the pair are considered as positive or negative, depending on the containing label. This kind of learning algorithms belong to the second-order transformation methods [21]. Contrastively, RAkEL [5] is a high-order transformation method, where the correlation between multiple labels is involved. However, they have relatively high complexity, and can not handle the high-dimensional data effectively. Other problem transform algorithms, such as Pruned Problem Transformation (PPT) [22] and Classifier Chains (CC) [23] have similar situations.

The second learning methods, i.e., algorithm adaptation, cope with the multi-label learning problems by extending the traditional learning algorithms directly, so as to adapt to the multi-label data. In other words, this kind of methods is fitting learning algorithms to data. As a representative example, ML-kNN extends the traditional kNN learning algorithm, so that it can handle the multi-label data appropriately. ML-DT [24] is another first-order adapting method, where decision tree has been revised according to the properties of multi-label data. Ranking Support Vector Machine (Rank-SVM) [25] is a second-order method, which tries to find maximum margins within the multi-label data. A high-order method named LEAD [16] adopts Bayes learning to deal with the multi-label data. It encodes the conditional dependencies of the variables and labels simultaneously.

How to exploit the correlation of labels and variables of multi-label data is still an open issue for multi-label learning. Although there are some multi-label learning algorithms, exploiting the label dependencies for the multi-label data, they have relatively high complexity, resulting in low robust to high dimensional data. This paper presents a novel approach named PPML. It takes both the correlation of variables and labels and dimension reduction into account simultaneously.

3 PPML Methodology

In this section, we first briefly give the formal concepts of multi-label learning and partial least squares discriminant analysis, and then propose a new learning framework for the multi-label data.

3.1 Multi-Label Classification

Without loss of generality, let $X \in R^{n \times m}$ and $Y \in R^{n \times d}$ be the variable and the label spaces, respectively, where n is the number of instances, m is dimensions of instances and d is the number of class labels involving in instances. In multi-label learning, each instance $x_i \in X (i = 1, 2, ..., n)$ is a vector of variables, and it corresponds to a possible multi-label set $y_i \in Y$, where y_i equals to 1 if the corresponding instance x is tagged with the i-th class label, otherwise $y_i = 0$. Given a data set $D = \{(x_i, y_i) \mid 1 \le i \le n\}$ consisting of n multi-label instance, the purpose of multi-label learning is to build a classifier $g \colon X \to 2^Y$ from D, and then use this model to predict the labels of unseen instances. From this definition, one may observe that the output of multi-label classification model is a subset of labels, not a single label. This is the distinguished difference of multi-label learning to the traditional ones.

3.2 Partial Least Squares

Partial least squares regression (PLS regression) is a statistical method that finds a linear regression model by projecting the predicted variables and the independent variables to a new space. It shows some similar properties to principal components regression, which tries to find hyper-planes of minimum variance between the predicted and independent variables. PLS regression is particularly suited to the case of high dimensionality, where there is multi-collinearity among the variables. In this case, standard regression will always fail.

PLS tries to locate the fundamental relations between two matrices, i.e. a latent variable approach to modeling the covariance structures in these two spaces. Assume that $X \in R^{n \times m}$ and $Y \in R^{n \times d}$ denote the independent and predicted variables. Usually, they can be decomposed by the common latent components $T \in R^{n \times k}$ as follows:

$$X = TP^T + E \tag{1}$$
$$Y = TQ^T + F \tag{2}$$

where $T = (t_1, t_2, ..., t_k) \in R^{n \times k}$ denote the score (latent) vectors of X and Y, i.e., the latent space. $P = (p_1, p_2, ..., p_k) \in R^{m \times k}$ and $Q = (q_1, q_2, ..., q_k) \in R^{d \times k}$ are the loading vectors of X and Y, respectively. $E \in R^{n \times m}$ and $F \in R^{n \times d}$ are the residual matrixes.

According to the linear transformation, we make an assumption that the latent components T is a linear tramsformation of X as follows:

$$T = XW \tag{3}$$

where $W \in R^{m \times k}$ is the weight matrix. After T is constructed, Q^T in Eq.(2) can be obtained by solving the least squares problem, i.e.,

$$Q^T = (T^T T)^{\dagger} T^T Y \tag{4}$$

where $(T^T T)^{\dagger}$ is the Moore-Penrose inverse of $T^T T$. Substituting (3) into (2), for Y we have its regression form:

$$Y = XB + F \tag{5}$$

where the regression coefficient B is $B = WQ^T = W(T^T T)^{\dagger} T^T Y$.

From the definitions above, we know that a PLS model tries to find the multidimensional direction in the X space that explains the maximum multidimensional variance direction in the Y space. Depending on the tasks, PLS have many variants. For example, partial least squares discriminant analysis (PLS-DA) is widely used when the Y is categorial.

Given a new instance X_{new}, its categories or labels can be predicted in terms of the PLS model (i.e., Eq.(5)) as follows:

$$\hat{Y} = X_{new}B = X_{new}(WQ^T) \tag{6}$$

PLS-DA is an efficient dimension reduction tool for handling the multi-label data. However it is not a specifical method for the cases of "large p, small n" and feature selection. Besides, the results obtained by PLS-DA are often difficult to be interpreted. Just for this reason, we impose a ridge penalization on the PLS-DA model to alleviate this problem.

3.3   PLS-DA with Ridge Penalization

For the high dimensionality of multi-label data, obtaining the PLS-DA model in a straightforward way becomes unfeasible, especially when the variables are highly collinear.

According to the Eq.(1) and (2), one may observe that PLS aims at locating the common latent variables T of X and Y, such that their covariance is maximal. That is to say, the object function of PLS-DA can be represented in a equivalent form as follows:

$$\begin{aligned} argmax_{p,q} \ cov(Xp, Yq) \\ s.t. \quad \|p\| = 1, \|q\| = 1 \end{aligned} \tag{7}$$

where $cov(Xp, Yq)$ is the covariance of Xp and Yq, p and q are the loading vectors of X and Y respectively.

The optimization problem of Eq.(7) can be solved through the following Lagrange function:

$$L(p, q) = p^T X^T Y q - \frac{\beta}{2}(p^T p - 1) - \frac{\theta}{2}(q^T q - 1) \tag{8}$$

where β and θ are Lagrange multipliers. After differentiating Eq.(8) with respect to p and q respectively, and let them equal to zero, we have the following equivalent problems:

$$X^T Y Y^T X p = \lambda p \qquad (9)$$
$$Y^T X X^T Y q = \lambda q \qquad (10)$$

Thus, the optimization problem is now transformed into the problem of solving eigenvalues and eigenvectors. Note that the vectors p and q are the eigenvectors of $X^T Y Y^T X$ and $Y^T X X^T Y$, respectively. An intuitive way of obtaining the eigenvectors p and q is to perform the technique of singular value decomposition (SVD) on $X^T Y$:

$$X^T Y = P \Sigma Q^T \qquad (11)$$

where the eigenvectors $P \in R^{m \times r}$ and $Q^T \in R^{r \times d}$ are orthogonal, $\Sigma \in R^{r \times r}$ is the diagonal matrix consisting of the singular values. (p, q) is a pair of eigenvectors corresponding to the eigenvalue in Σ.

It should be mentioned that Eq.(11) may be ill-posed as the dimensionality of data is larger than the number of data. Thus, it should be penalized for the consideration of numerical computing and practical applications. Here we exert a l_2-norm penalty on Eq.(11). This is also known as ridge regularization, which is a method for solving badly conditioned linear regression problems. The benefits of ridge regularization are most striking in the presence of multi-collinearity and penalize the size of the regression coefficients, resulting in shrinking the regression coefficients toward zero [26], [27]. Sparse property is more prefer because it can yield easily interpretable results. Moreover, with the increase of the number of labels, the label space Y is usually sparse. Thus, it is necessary to shrinking the loading vectors of Y.

After applying l_2-norm penalty, the loading vector q of Y can be obtained by solving the constraint optimization problem as follows:

$$q^{ridge} = argmin_q \left\| q^T M - p \right\|_2^2 + \lambda \left\| q \right\|_2^2 \qquad (12)$$

where $M = Y^T X$, λ is the regularization parameter for the loading vector q. For q, when λ is enough small, the weight coefficients of some variables compressed to zero by comparing with a threshold. Let $L(q)$ be the Lagrange function of Eq.(12), we have

$$\begin{aligned} L(q) &= \left\| q^T M - p \right\|_2^2 + \lambda \left\| q \right\|_2^2 \\ &= q^T M M^T q - 2 q^T M p^T + p p^T + \lambda q^T q \end{aligned} \qquad (13)$$

After taking the derivative of Eq.(13) with respect to q and setting it to zero, we can obtain q^{ridge} as

$$q^{ridge} = [M M^T + \lambda I]^\dagger M p^T \qquad (14)$$

Substituting M in Eq.(14) with Eq.(11), we further have

$$\begin{aligned} q^{ridge} &= P(\Sigma^2 + \lambda I)^\dagger \Sigma Q^T p^T \\ &= \sum_{i=1}^d \frac{\sigma_i^2}{\sigma_i^2 + \lambda} P_i (Q_i^T p^T) \\ &= \sum_{i=1}^d f(\sigma_i) P_i (Q_i^T p^T) \end{aligned} \qquad (15)$$

Alg. 1 The framework of PPML for the multi-label data

Input:
  X, Y: The training data with the variable and label spaces
  $\bar{X}$: The new instances
  λ: The regularization parameter
  h: The number of iterations
  θ: The threshold value for prediction
Output:
  $\bar{Y}$: the predicted labels of $\bar{X}$
Training:
  Initialize T, P, Q and W as $T=[\ ]$,$P=[\ ]$,$Q=[\ ]$,$W=[\ ]$
  For $i=1$ **to** h
    Obtain the initial values of p and q according to Eq.(11)
    Repeat
      Obtain p according to q, and normalize it
      Obtain q according to Eq.(14), and normalize it
      $u = Yq$
      Compute w as $w = X^T u$, and normalize it
      $t = Xw$
    Until convergence
    Update X,Y as $X = X - tp^T$; $Y = Y - tq^T$
    Update T, P, Q and W as:
      $T \leftarrow [T, t], P \leftarrow [P, p], Q \leftarrow [Q, q], W \leftarrow [W, w]$
  End For
Predicting:
  Compute the real-valued outputs space of $\bar{X}$ according to Eq.(6)
    $O = \bar{X}WQ^T$
  Compute the final predictive labels space $\bar{Y}$:
    $\bar{Y}_{ij} = \begin{cases} 1, & O_{ij} \geq \theta \\ 0, & O_{ij} < \theta \end{cases}$

where $f(\sigma_i)$ is the shrinkage factors.

Based on the analysis above, we propose a new multi-label learning framework called PPML (Penalized Partial least squares discriminant analysis for Multi-label Learning). As the name indicates, our method employs PLS-DA with ridge regularization to handle the classification problem of high-dimensional multi-label data. Algorithm 1 presents the framework of PPML in detail. It exploits Nonlinear Iterative Partial Least Squares (NIPALS) [17] to obtain p and q. Alternatively, PPML can also be implemented with other forms like PLS-SB [28] and SIMPLS [29].

PPML works in a straightforward way and can be easily understood. It mainly consists of two stages, i.e., model training and result predicting. Specifically, in the training stage, two loops are nested. The major purpose of the inner loop is to get the loading vectors p and q of the variable and label spaces respectively, while the outer loop aims at yielding all loading vectors (P and Q), the latent components T and the coefficients W, so as to build the PPML learning model with PLS-DA. In the predicting stage, the prediction value of a new instance is a real-value vector in terms of (6). Later the output will be transformed into a vector of $\{0, 1\}$ by comparing with a given threshold θ, which is often empirically set as 0.5.

Table 1. General information of the experimental data sets

Datasets	Inst.	Var.	Labels	L.Card.	L.Dens.
Medical	978	1449	45	1.245	0.028
Arts	5000	462	26	1.636	0.063
Entertainment	5000	640	21	1.420	0.068
Health	5000	612	32	1.662	0.052
Recreation	5000	606	22	1.423	0.065
Reference	5000	793	33	1.169	0.036
Science	5000	743	40	1.451	0.036

4   Experiments

4.1   Data Sets

In our experiments, seven public data sets from the real-world applications were adopted. They are *Medical, Arts, Entertainment, Health, Recreation, Reference,* and *Science.* The *Medical* data set was used in the Medical Natural Language Processing Challenge[1] in 2007. In this data set, each instance contains brief free-text summary of a patient symptom history. The last six benchmark data sets were collected from Yahoo. They cover different domains in web page categorization.

Table 1 summaries the general information of the benchmark data sets used in experiments, where *Inst.* and *Var.* denote the number of instances and the dimensionality of data for each data set respectively. In addition, *L.Card.*, representing label cardinality, is the average number of labels per instance, while *L.Dens.*, standing for label density, is the fraction of the cardinality according to the number of labels.

4.2   Comparison of Algorithms

To demonstrate the effectiveness of our algorithm, nine multi-label learning algorithms have been adopted in comparing with PPML. They are BP-MLL [7], BRkNN [30], IBLR_ML [31], LP [4], MLkNN [6], PPT [22], CC [23], MLStacking(MLS) [20], and MAHR [33]. They are representatives of the state-of-the-art multi-label learning algorithms, and stand for different learning manners. They can deal with the multi-label problems and have relatively better performance and efficiency.

The performance of the learning algorithms heavily relies on their parameters. In our experiments, default value was assigned for each parameter as did in the MULAN software package[2]. MULAN [32] is an open source Java library for multi-label learning. It brings many popular multi-label learning algorithms together. For the MAHR classifier, its parameters was set as recommended by

---

[1] http://www.computationalmedicine.org/challenge/
[2] http://mlkd.csd.auth.gr/multilabel.html

the authors in the literature [33], that is, the number of boosting rounds was two times of variables for each data set.

4.3 Evaluation Metrics

Multi-label classification needs more complex evaluation metrics than traditional classification. In order to roundly evaluate the performance of PPML and other algorithms, we took four commonly used evaluation metrics. They are *Ranking Loss*, *One-Error*, *Coverage* and *Average Precision* [4].

Ranking Loss (RL) indicates the mis-ordered degree of couples of labels, where an irrelevant label has higher rank than a relevant one. *One-Error* (OE) estimates how many times the top-ranked label is irrelevant to the true class labels for each instance. *Coverage* (Cov) obtains the number of the steps that are needed, on average, to move down the ranked list of labels, in order to cover the whole relevant labels of the instance. *Average Precision* (AP) evaluates the average fraction of true labels ranked above a particular label.

Since *Ranking Loss*, *One-Error* and *Coverage* evaluate the loss of the prediction results, the smaller the metric values, the better the performance of learning algorithms. On the contrary, for *Average Precision*, the larger value indicates the better performance.

4.4 Experimental Results

In the experiments, 10-folds cross-validation was performed on each combination of classifier and data set. The experiments were carried out under the platform of MULAN. Table 2 shows the comparison results of classification performance of classifiers in terms of four evaluation metrics, where the mean value of each algorithm was ecorded on each data set.

From the experimental results in Table 2, one can notice that PPML is promising. It has better performance than others in most cases. For example, PPML achieved the best performance on five over seven data sets at the aspect of *Ranking Loss*. Even on the *Health* and *Science* data sets, the performance of PPML is just slightly worse than the corresponding best one, not the worst one.

Similar situations also present on the *One-Error* and *Coverage* metrics, where PPML outperformed other popular multi-label learning algorithms on six and five over seven benchmark data sets respectively. The one-error of PPML on *Medical* is 13.61%, which is slight higher than that of MAHR.

For the measure of *Average Precision*, PPML is the best in comparing with other classifiers. The performance of PPML is predominant and significantly better than the rest learning algorithms over all of the seven benchmark data sets. For example, on the *Arts* and *Recreation* data sets, the average precisions of PPML are 60.59% and 62.14% respectively, while the highest precisions of other classifiers are 50.92% and 52.01%, achieved by MLS and MAHR respectively.

Table 2. Experimental results of classifiers on four evaluation metrics, where ↓ means the smaller, the better, and ↑ means the larger, the better. Bold value shows the winner on each dataset.

	PPML	BPMLL	BRkNN	IBLR_ML	LP	MLkNN	PPT	CC	MLS	MAHR
				Ranking Loss(↓)						
Medical	**0.0175**	0.4321	0.0596	0.0890	0.1277	0.0573	0.1067	0.0990	0.0910	0.0410
Arts	**0.1379**	0.4545	0.2331	0.1602	0.3946	0.1575	0.2951	0.2512	0.1610	0.1886
Entertainment	**0.1165**	0.3763	0.3858	0.1395	0.4838	0.1364	0.3353	0.2364	0.1399	0.1630
Health	0.0660	0.2469	0.1953	0.0604	0.4310	0.0723	0.2155	0.1316	**0.0611**	0.0656
Recreation	**0.1464**	0.5634	0.2475	0.1811	0.4381	0.1804	0.3265	0.2605	0.1797	0.2513
Reference	**0.0826**	0.2694	0.2472	0.0918	0.4545	0.0889	0.2896	0.1675	0.0905	0.1206
Science	0.1159	0.4564	0.2023	0.1145	0.4480	0.1177	0.3729	0.2311	**0.1139**	0.2182
				One-Error(↓)						
Medical	0.1361	0.9674	0.2347	0.2918	0.1572	0.2388	0.1949	0.2123	0.3081	**0.1123**
Arts	**0.4788**	0.9844	0.8410	0.6124	0.7194	0.6338	0.6612	0.6674	0.6142	0.5801
Entertainment	**0.4368**	0.9582	0.7342	0.6044	0.6294	0.6360	0.5360	0.5356	0.6116	0.4542
Health	**0.2664**	0.9936	0.7228	0.4094	0.5288	0.4598	0.4314	0.3816	0.4138	0.3080
Recreation	**0.4726**	0.9778	0.7670	0.6466	0.6648	0.6664	0.6158	0.6232	0.6462	0.5470
Reference	**0.3838**	0.9644	0.8964	0.4908	0.5864	0.4838	0.5206	0.5476	0.4928	0.4160
Science	**0.5238**	0.9885	0.6244	0.5874	0.7244	0.5940	0.7096	0.6568	0.5876	0.5660
				Coverage(↓)						
Medical	1.3174	20.3123	3.9756	5.3643	7.9194	3.7633	6.4061	6.1643	5.5163	2.9185
Arts	5.3223	13.0170	7.7288	5.6838	12.1808	5.5830	9.6152	8.6368	5.6956	6.6663
Entertainment	3.2930	8.3088	9.0084	3.6808	10.9336	3.5900	7.9446	6.0716	3.6912	4.4364
Health	3.7082	9.1972	8.6658	**3.2216**	16.2388	3.6586	9.1798	6.6592	3.2486	3.6703
Recreation	4.1814	12.9760	6.2270	4.7388	10.7832	4.7090	8.2528	6.8510	4.7050	6.5649
Reference	3.3472	9.1452	8.9134	3.5550	15.6426	3.4244	10.1608	6.1446	3.4960	4.7160
Science	6.1550	19.6508	10.1670	5.9508	20.0474	5.9376	17.0720	11.7990	**5.9304**	10.9840
				Average Precision(↑)						
Medical	**0.8871**	0.1124	0.7958	0.7497	0.7923	0.8039	0.7967	0.8029	0.7421	0.8791
Arts	**0.6059**	0.1344	0.3752	0.5098	0.3554	0.5032	0.4294	0.4517	0.5092	0.5055
Entertainment	**0.6599**	0.1860	0.4044	0.5445	0.3959	0.5341	0.5169	0.5503	0.5408	0.6196
Health	**0.7707**	0.1817	0.3712	0.6896	0.4719	0.6485	0.6200	0.6805	0.6858	0.7504
Recreation	**0.6214**	0.1210	0.3767	0.5016	0.3925	0.4899	0.4731	0.4953	0.5022	0.5201
Reference	**0.7004**	0.1613	0.3080	0.6064	0.4260	0.6110	0.5342	0.5647	0.6080	0.6441
Science	**0.5809**	0.0849	0.4698	0.5305	0.2938	0.5228	0.3556	0.4382	0.5306	0.4844

5   Conclusions

In this paper, a new multi-label learning framework, called PPML, is proposed to deal with the multi-label problems. It mainly exploits partial least squares discriminant analysis (PLS-DA) to achieve the purpose of performing dimension reduction and capturing the label correlations simultaneously. To cope with the multi-collinearity problem resulted from the high dimensionality of data, a ridge regularization penalty is further exerted on the object optimization function of PLS-DA. The experimental results are encouraging and show that PPML is promising in comparison with the other state-of-the-art algorithms.

In the future, we will make an attempt to find other more efficient methods and combine them with PLS-DA to tackle with the problems of multi-label learning.

Acknowledgements. The authors are grateful to the anonymous referees for their valuable comments and suggestions. This work is partially supported by the National NSF of China (61100119, 61170108, 61170109, 61272130, and 61272468), the NSF of Zhejiang province (LY14F020012), Postdoctoral Science Foundation of

China (2013M530072), and the Open Project Program of the National Laboratory of Pattern Recognition (NLPR) (201204214).

References

1. Barutcuoglu, Z., Schapire, R., Troyanskaya, O.: Hierarchical multi-label prediction of gene function. Bioinformatics 22(7), 830–836 (2006)
2. Yang, S., Kim, S., Ro, Y.: Semantic home photo categorization. IEEE Transactions on Circuits and Systems for Video Technology 17, 324–335 (2007)
3. Read, J.: Scalable multi-label classification, PhD thesis, University of Waikato, Hamilton, New Zealand (2010)
4. Tsoumakas, G., Katakis, I., Vlahavas, I.: Mining multi-label data. In: Maimon, O., Rokach, L. (eds.) Data Mining and Knowledge Discovery Handbook, 2nd edn., Spring (2010)
5. Tsoumakas, G., Katakis, I., Vlahavas, I.: Randomk-Labelsets for Multi-Label Classification. IEEE Transactions on Knowledge and Data Engineering 23(7), 1079–1089 (2011)
6. Zhang, M.-L., Zhou, Z.-H.: ML-kNN: a lazy learning approach to multi-label learning. Pattern Recognition 40(7), 2038–2048 (2007)
7. Zhang, M.-L., Zhou, Z.-H.: Multi-label Neural Network with Applications to Functional Genomics and Text Categorization. IEEE Transactions on Knowledge and Data Engineering 18(10), 1338–1351 (2006)
8. Bellman, R.: Adaptive Control Processes: A Guided Tour. Princeton University Press, Princeton (1961)
9. Jolliffe, I.: Principal Component Analysis. Springer, New York (1986)
10. Fisher, R.: The use of multiple measurements in taxonomic problems. Annals of Eugenics 7, 179–188 (1936)
11. Hotelling, H.: Relations between two sets of variables. Biometrika 28, 312–377 (1936)
12. Huang, S.-J., Zhou, Z.-H.: Multi-label learning by exploiting label correlations locally. In: Proceedings of the 26th AAAI Conference on Artificial Intelligence (AAAI 2012), Toronto, Canada, pp. 949–955. AAAI Press (2012)
13. Xu, M., Li, Y.-F., Zhou, Z.-H.: Multi-Label Learning with PRO Loss. In: Proceedings of the 27th AAAI Conference on Artificial Intelligence (AAAI 2013), Bellevue, WA (2013)
14. Sun, L., Ji, S.-W., Ye, J.-P.: Canonical Correlation Analysis for Multilabel Classification: A Least-Squares Formulation. Extensions, and Analysis, Pattern Analysis and IEEE Transactions on Machine Intelligence 33(1), 194 (2011)
15. Qian, B., Davidson, I.: Semi-supervised dimension reduction for multi-label classification. In: Proceedings of the 24th AAAI Conference on Artificial Intelligence (2010)
16. Zhang, M.-L., Zhang, K.: Multi-label learning by exploiting label dependency. In: Proceedings of the 16th ACM SIGKDD International Conference on Knowledge Discovery and Data Mining (KDD 2010), Washington, D.C., pp. 999–1007 (2010)
17. Wold, H.: Path Models with Latent Variables: The NIPALS Approach. In: Blalock, H.M., Aganbegian, A., Borodkin, F.M., Boudon, R., Capecchi, V. (eds.) Quantitative Sociology: International Perspectives on Mathematical and Statistical Modeling, pp. 307–357. Academic Press, New York (1975)

18. Hoerl, A., Kennard, R.: Ridge regression: Biased estimation for nonorthogonal problems. Technometrics 12(1), 55–67 (1970)
19. Frnkranz, J., Hllermeier, E., Menca, E.L., Brinker, K.: Multilabel classification via calibrated label ranking. Machine Learning 23(2), 133–153 (2008)
20. Tsoumakas, G., Dimou, A., Spyromitros, E., Mezaris, V., Kompatsiaris, I., Vlahavas, I.: Correlation-based pruning of stacked binary relevance models for multi-label learning. In: Proceedings of the Workshop on Learning from Multi-Label Data (MLD 2009), pp. 101–116 (2009)
21. Zhang, M.-L., Zhou, Z.-H.: A review on multi-label learning algorithms. In: IEEE Transactions on Knowledge and Data Engineering (2013) doI:10.1109/TKDE.2013.39
22. Read, J.: A pruned problem transformation method for multi-label classification. In: Proc. 2008 New Zealand Computer Science Research Student Conference (NZC-SRS 2008), pp. 143–150 (2008)
23. Read, J., Pfahringer, B., Holmes, G., Frank, E.: Classifier chains for multi-label classification. In: ECML/PKDD 2009, pp. 254–269 (2009)
24. Clare, A.J., King, R.D.: Knowledge Discovery in Multi-label Phenotype Data. In: Siebes, A., De Raedt, L. (eds.) PKDD 2001. LNCS (LNAI), vol. 2168, pp. 42–53. Springer, Heidelberg (2001)
25. Elisseeff, A., Weston, J.: A kernel method for multi-labelled classification. In: Dietterich, G., Becker, S., Ghahramani, Z. (eds.) Advances in Neural Information Processing Systems, vol. 14, pp. 681–687. MIT Press, Cambridge (2002)
26. Mairal, J., Bach, F., Ponce, J., Sapiro, G.: Online learning for matrix factorization and sparse coding. Journal of Machine Learning Research 11(1), 19–60 (2010)
27. Jenatton, R., Mairal, J., Obozinski, G., Bach, F.: Proximal methods for sparse hierarchical dictionary learning. In: Proceedings of the International Conference on Machine Learning, ICML (2010)
28. Sampson, P., Streissguth, A., Barr, H., Bookstein, F.: eurobehavioral effects of prenatal alcohol: Part II. Partial Least Squares Analysis, Neurotoxicology and Teratology 11(5), 477–491 (1989)
29. De Jong, S.: SIMPLS: an alternative approach to partial least squares regression. Chemometrics and Intelligent Laboratory Systems 18(3), 251–263 (1993)
30. Spyromitros, E., Tsoumakas, G., Vlahavas, I.P.: An Empirical Study of Lazy Multilabel Classification Algorithms. In: Darzentas, J., Vouros, G.A., Vosinakis, S., Arnellos, A. (eds.) SETN 2008. LNCS (LNAI), vol. 5138, pp. 401–406. Springer, Heidelberg (2008)
31. Cheng, W., Hüllermeier, E.: Combining instance-based learning and logistic regression for multilabel classification. Machine Learning 76(2-3), 211–225 (2009)
32. Tsoumakas, G., Spyromitros-Xioufis, E., Vilcek, J., Vlahavas, I.: Mulan: A Java Library for Multi-Label Learning. Journal of Machine Learning Research 12, 2411–2414 (2011)
33. Huang, S.-J., Yu, Y., Zhou, Z.-H.: Multi-label hypothesis reuse. In: Proceedings of the 18th ACM SIGKDD Conference on Knowledge Discovery and Data Mining, Beijing, China, pp. 525–533 (2012)

Authorship Attribution with Very Few Labeled Data: A Co-training Approach

Mengdi Fan, Tieyun Qian*, Li Chen, Bin Liu, Ming Zhong, and Guoliang He

State Key Laboratory of Software Engineering,
Wuhan University, Wuhan, China
fanmengdi8336@163.com, {qty,binliu,glhe}@whu.edu.cn,
ccnuchenli@163.com, mike.clark.whu@gmail.com

Abstract. Authorship attribution refers to the task of identifying the authors of a set of documents. Early studies in this area either used book length texts or assumed that there were a large number of training documents. The focus of modern authorship attribution has been shifted to the analysis on small online texts. This is realistic since in the real life it is hard to collect the training texts. However, the small size of training data makes the authorship attribution much more difficult. In this paper, we present a novel co-training method to iteratively recognize a few unlabeled data to augment the training set. Specifically, each document is first partitioned into two distinct views, i.e., lexical and syntactic view. And then, a two view semi-supervised method, co-training, is adopted to exploit the large amount of unlabeled documents. Our experiment results based on real data show that the proposed method can effectively exploit unlabeled data to improve the classification performance.

Keywords: authorship attribution, very few labeled data, co training.

1 Introduction

Authorship attribution (AA) is a traditional problem and has been studied by many researchers [11,35,8,12,6]. The early work on AA focused on analyzing Shakespeare's plays and Bronte Sisters' novels. Later on, it was used to identify other literary works such as American and English literature and news articles. More recently, AA was applied to online texts such as emails [37], blogs [21], forum posts [31] and reviews [17]. The problem of AA is useful in many applications such as forensic investigations, detection of copyright infringement and internet plagiarism.

Existing approaches on authorship attribution are mainly based on supervised classification [39,34,6,17,29]. The major weakness of this method is that, for each author a large number of his/her articles are necessary to be used as the training data. For example, book length texts were used to classify the authorship of Bronte Sisters' novels [19]. However, in the real life, it is actually hard to collect sufficient labeled data. For instance, in most cases of forensics, only one small text is available for a specific author, which are not enough to serve as the training data. The small number of labeled documents makes it extremely challenging for supervised learning to train an accurate classifier.

---

* Corresponding author.

F. Li et al. (Eds.): WAIM 2014, LNCS 8485, pp. 657–668, 2014.

In this paper, we consider the problem of authorship attribution with very few labeled data. Little work has been done in this area. A similar problem was once attempted in [22]. However, the number of training samples in each class is greater than 115 and 129 for the small and the large data set, respectively, which is still very large. In contrast, we consider a much more difficult problem, where the size of training samples in our setting is extremely small, i.e., 10 samples per author for training. Luyckx and Daelemans also evaluated the effect of training set size [24], but their algorithm was not particularly designed for coping with few training data.

We propose a new framework to address the authorship classification problem with limited training data using a co-training framework in this paper. Following the basic idea of co-training, i.e., utilizing two sufficient and redundant views on the data, we partition the documents into two natural parts of lexical and syntactic structures and build two classifiers separately on two views. The predictions of each classifier on un-labeled examples are used to augment the training set of the other. This process repeats until a termination condition is satisfied, and the enlarged labeled set is finally used to train a classifier and make predictions on the test data. By exploiting the redundancy in the human languages, we tackle the problem of very few training data. Experiments on a real world data set show that the proposed co-training framework can effectively incorporate the unlabeled data to help improve classification performance by a large margin.

The rest of this paper is organized as follows. Section 2 reviews related work. Section 3 presents our co-training method for authorship attribution. Section 4 provides experimental results. Finally, Section 5 concludes the paper.

2   Related Works

Authorship attribution has received a great deal of attention in recent years. A variety of approaches have been developed for this problem. Existing methods can be categorized into two main themes. One focuses on finding appropriate features for quantifying the authors' writing style, and the other focuses on developing efficient and effective techniques to perform the classification task.

There is a body of literature examining the effects of different features. The use of function words could date back about half a century ago [26]. Since then, various features have been proposed for modeling writing styles. The features that have been investigated include function words [1,2], length features [7,8], richness features [11,19], punctuation frequencies [8], character n-grams [9,12], word n-grams [28], POS n-grams [7,13], and rewrite rules [11].

There are also a number of works that study the use of machine learning methods in attribution. An early study used Bayesian statistical analysis [26], but later work focused exclusively on classification, including discriminant analysis [35], PCA [14], neural networks [8,39], multi-layer perceptrons [8], clustering [28], decision trees [36,38], and SVM [5,7,19,12]. Among them, SVM is regarded as one of the best approaches for solving this problem [23,17].

The main problem in traditional research is the unrealistic size of the training set. Basically, a size of about 10,000 words per author is regarded to be a reasonable size

[2,4,7]. Even when no long documents available, hundreds of short texts can be selected [10,13] such that the total amount of words is large enough for training. A recent work [24] introduced the problem of AA with limited data. Instead of presenting an effective approach to deal with the problem, the authors only investigated the effect of limited data in authorship attribution. Several ensemble-based methods were also introduced [33,21] with the basic idea of feature set subspacing. While the subspacing technique is appropriate for high dimensional feature space and sparse data, it is not developed for handling the problem with very few labeled data. Co-training, on the other hand, is a representative learning mechanism which combines both labeled and unlabeled data under a two view setting [3,27]. Although the co-training paradigm has been successfully employed in many areas, the problem of authorship attribution was rarely addressed using a co-training framework. Kourtis and Stamatatos once introduced a variant of the self-training method for AA [22]. However, the number of labeled documents is still very large in that work, i.e., about 115 and 129 documents per author on average. Moreover, the self-training method in [22] uses two classifiers on one view. In contrast, we adopt a two-view co-training framework and an extremely small number of documents (10 documents per author) are used as the initial training data.

3   A Co-training Algorithm for Authorship Attribution

The problem of authorship attribution can be defined as follows: Let $A = \{a_1, ..., a_k\}$ be a set of k authors (classes) and $D = \{D_1, ..., D_k\}$ be k sets of documents with D_i being the document set of author $a_i \in A$. Each (training or testing) document is represented as a feature vector. Each feature represents a piece of information about the document, e.g., a word or a syntactic tag. A model or classifier is then built from the training data and applied to the test data to determine the author a of each test document d, where a is from A ($a \in A$).

One of the main challenges for modern authorship attribution is the small number of training samples. To address this problem, we propose a co-training approach which integrates two views into one framework. Co-training is semi-supervised method. It begins with a small set of labeled data, and enlarges the labeled data set by adding unlabeled data. The key aspect of co-training algorithm is the property of two views. It has been shown that the redundancy in two views contributes more information than the single view in practice and theory [3,25]. Since human languages naturally consists of two parts: a lexicon and a grammar, we apply the co-training framework to authorship attribution with limited annotated documents.

3.1   The Overall Framework

In the context of authorship attribution, each document has two views of features: features about lexical structure and features about syntactic structures. Hence two classifiers can be co-trained using these two views. The overall framework is shown in Algorithm 1.

Algorithm 1. Co Training on Lexical and Syntactic Views (CTLSV)

Input: A small set of labeled documents $L = \{l_1, ..., l_r\}$, a large set of unlabeled documents $U = \{u_1, ..., u_s\}$, a set of test documents $T = \{t_1, ..., t_t\}$
Parameters: the number of iterations k, the size of selected unlabeled documents u, the number of top predicted unlabeled documents p
Output: t_i's class assignment ($t_i \in T$)
Steps:

1. Extract lexical and syntactic views $L_l, L_s, U_l, U_s, T_l, T_s$ for L, U, and T.
2. Loop for k iterations:
   - Randomly select u unlabeled documents U' from U;
   - Learn the first view classifier C_l from L based on lexical features L_l;
   - Use C_l to label documents from U' based on U_l;
   - Create a document subset U_{tl} by choosing p most confidently predicted documents from U';
   - Learn the second view classifier C_s from L based on syntactic features L_s;
   - Use C_s to label documents from U' based on U_s;
   - Create a document subset U_{ts} by choosing p most confidently predicted documents from U';
   - $U = U - U'$, $L_l = L_l \bigcup U_{ts}$, $L_s = L_s \bigcup U_{tl}$;
3. Learn the first view classifier C_l from L based on lexical features L_l;
4. Use C_l to label t_i in T based on T_l;
5. Learn the second view classifier C_s from L based on syntactic features L_s;
6. Use C_s to label t_i in T based on T_s;

In Algo. 1, step 1 extracts two types of views from the labeled, unlabeled, and test data, respectively. Step 2 iteratively co-trains two classifiers by adding the most accurately predicted data from the other view into labeled set. The algorithm first randomly selects a small set of u documents. Although we can directly select from the large unlabeled set U, it is shown [3] that a smaller pool can force the classifier C_l and C_s to select instances that are more representative of the underlying distribution that generates U. Hence we set he parameter u to 100, which is about 1/80 percent of the whole unlabeled set. It then iterates for the following steps. First, use the lexical and syntactic view on current labeled set to train a classifier C_l and C_s, respectively. Second, allow each of these two classifiers to examine the unlabeled set U' and select p samples it most confidently labels as positive. The examples selected by the lexical classifier C_l are added to the document set of syntactic view L_s with the label assigned, and those selected by the syntactic classifier C_s are added to the document set L_l of lexical view. Finally, the u documents are removed from the unlabeled pool U'. Steps 3-6 are used to assign the test document to a category (author) using the classifier learned from the first and second view in the augmented labeled date, respectively.

Once we get the prediction values from the different classifiers, some additional algorithms can be added to decide the final author attribution for each test document t_k. One simple general method is voting. Unfortunately, this method is not appropriate for this task as we only have two classifiers. There are also other methods, which can depend on what output value the classifier produces. Here we present two strategies. These methods require the classifier to produce a predicted score, which can reflect the positive and negative certainty. Many classification algorithms produce such a score, e.g., SVM, logistic regression, and naïve Bayesian. Here we use SVM as an example. For each test case, SVM outputs a positive or negative score which can be interpreted as the certainty that a test case is positive or negative.

The two methods are given below:

1. **ScoreSum:** The learned model/classifier is first applied to classify all test cases in T. Then for each test document t_k, this method sums up all scores of positive classifications. It then assigns t_kf to the author with the highest scores.
2. **ScoreMax:** This method also works similarly except that it finds the maximum classification score). And then the decision is made similarly.

3.2 Lexical Features

The lexical features are used to compose a lexical view for a document. It is straightforward to view a text article as a bag-of-words, like that has been widely used in topic-based text classification. We represent each article by a vector of word frequencies. The vocabulary size for word unigram in our experiment is *195274*. We do neither word stemming nor stop word removal as in text categorization. This is because some of the stop words are actually function words which have been demonstrated discriminative for authorship identification. In addition, stemming can be also harmful to information extraction of an author.

3.3 Syntactic Features

The syntactic features are used to compose a syntactic view for a document. Existing studies have shown the usefulness of syntactic information in supervised authorship classification [7,13]. In this paper, we use four typical content-independent structures including n-grams of POS tags ($n = 1..3$) and rewrite rules.

The syntactic features are extracted from the parsed syntactic trees. For example, the tree for sentence "This is the best book in the set" is as follows:

```
(ROOT
  (S
    (NP (DT This))
    (VP (VBZ is)
      (NP
        (NP (DT the) (JJS best) (NN book))
        (PP (IN in)
          (NP (DT the) (NN set)))))
    (. .)))
```

This tree contains 15 POS 1-grams:

```
S NP DT VP VBZ NP NP
DT JJS NN PP IN NP DT NN,
```

14 POS 2-grams:

```
S|NP NP|DT DT|VP VP|VBZ ... DT|NN,
```

13 POS 3-grams:

```
S|NP|DT NP|DT|VP DT|VP|VBZ ... NP|DT|NN,
```

and 7 rewrite rules:

```
S->NP+VP NP->DT NP->DT+JJS+NN NP->DT+NN
PP->IN+NP NP->NP+PP VP->VBZ+NP
```

Each POS n-gram or rewrite rule is encoded like a single pseudo-word and assigned a unique number id. The vocabulary sizes for POS 1-grams, POS 2-grams, POS 3-grams, and rewrites in our experiment are *63, 1917, 21950*, and *19240*, respectively. These four types of syntactic structures are merged into a single vector. Hence the syntactic view of a document is represented as a vector with 43140 components.

4   Evaluation

In this section, we evaluate the proposed approach. We first introduce the experiment setup, and then present the results using different parameter settings. Finally we compare our results with four types of baselines. All our experiments use the $SVM^{multiclass}$ classifier [15] with default parameter settings.

4.1   Experiment Setup

We conduct experiments on the IMDB data set [30]. This data set has 62,000 reviews by 62 users (1,000 reviews per user). It is publicly available upon the request to authors. We randomly select 10 authors for experiments. We do not use a large number of authors in our experiments because when the number of authors increases, the performance of supervised classification deteriorates quickly even with many training instances [34,21]. Thus it is very difficult, if not impossible, to evaluate the effects of co-training algorithm. For each author, we further split his/her documents into the labeled, unlabeled, and test set, 1% of one author's documents, i.e., 10 documents per author, are used for training, 79% are used as unlabeled data, and the rest 20% for testing. We extract and compute the lexical features directly from the raw data, and we use the Stanford PCFG parser [18] to generate the grammar structure of sentences in each document for extracting syntactic features. We normalize each feature's value to [0, 1] interval by dividing by the maximum value of this feature in the training set.

We report classification accuracy as the evaluation metric.

4.2   Baseline Methods

We first implement the semi-supervised learning approach presented in [22]. This approach self-trains two classifiers from the character 3-gram view using CNG and SVM classifiers. Our results show that the performance of CNG+SVM is very poor. CNG is a profile-based method which builds the classification model based on the dissimilarity of the profile of the text from each of the profiles of the candidate authors. Its accuracy is only 5.80% with the original 10 training documents. And this directly leads to the failure of the whole self-training framework. Thus we do not list the results for this method due to the space limitation.

Now we give four other baseline methods.

- SLR (Single Lexical Representation of documents): This is a widely used method for authorship attribution (AA) [28,20]. Each document is represented as a feature vector consisting of the frequency of word unigrams.
- SSR (Single Syntactic Representation of documents): This is another popular approach in AA [7,13,17]. Each document is represented as a feature vector consisting of the frequency of syntactic tokens.
- CLSR (Combined Lexical and Syntactic Representation of documents): This method represents each document as a combined feature vector consisting of the frequency of syntactic and lexical tokens [16].
- SSLF (SubSpacing on Lexical Features of documents): This is an ensemble method based on feature subspacing [32,21]. The word unigrams in a document are partitioned into several disjoint parts, and a supervised learning method is used to learn a classifier on each part. Then the classifiers are applied to the testing documents separately. Finally a ScoreMax or ScoreSum strategy is adopted to determine the class of each document.

4.3   Experimental Results

In this section, we exploit the parameter sensitivity and then compare co-training to other baseline methods.

Effects of the Number of Iterations k

We first evaluate the effects of the number of iterations k. Figure 1 depicts the plots of accuracy versus parameter k using the classifier by two single views as well as the two combining strategies.

We have the following observations from Figure 1.

- The classification performance increases with a larger number of iterations. This indicates that the learning system benefits more from a larger labeled data set. It achieves the best results when the number of iterations reaches 40. After that, there is a small drop for the lexical classifier and the curve for syntax-based classifier goes steadily. Since a larger number of iterations brings about more overhead, there is a tradeoff between the accuracy and the parameter k. In many cases, an iteration number k = 10 is already good enough. We use this as the default parameter setting in the following experiments unless explicitly stated.

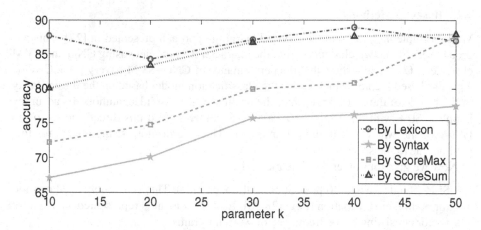

Fig. 1. Evaluations on different iteration numbers k. Other algorithm parameters are set to u=100, and p=30.

- The lexicon-based classifier performs the best among four approaches. It even outperforms two combined methods. This strongly demonstrates that the lexicon-based classifier can be helped a lot by co-training on the syntax view. On the other hand, the syntax-based classifier is helped less from co-training. The reason can be that there are less syntactic tokens and hence it is less capable of learning an accurate classifier.

Effects of the Number of the Top Predicted Unlabeled Documents p

We evaluate the effects of number of the top predicted unlabeled documents. Figure 2 shows the results. In Fig. 2, we see an upward trend in general, which is similar to that in Figure 1, showing that more labeled examples result in better performance. We also notice that the lexicon-based classifier gets the best result when selecting 30 top-predicted samples. This can be due to the fact that the labeled set contains more noises when more documents are added. Note that the classifier can not always predict an unlabeled data correctly. And thus it brings down the performance. All our experiments below use the model trained with p = 30.

Comparing with the Baselines

Among the four baselines, SLR and SSR are one-view based classifier, and the CLSR and SSLF are combined method. We first compare the results of our co-training method under lexical and syntactic view with those of SLR and SSR. The results are summarized in Table 1, where the number n in *CTLSVn* denotes the number of iterations.

From Table 1, it is clear that the proposed co-training framework outperform the traditional supervised training by a large margin on both the lexicon-based classifier and the syntax based classifier. For example, after co-training 10 iterations, our CTLSV

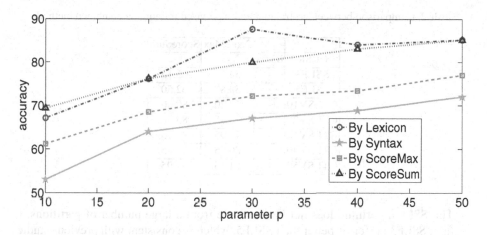

Fig. 2. Evaluations on different p. Other algorithm parameters are set to u=100, and k=10.

Table 1. Comparison between co-training and supervised training on single view

	Lexicon-based Classifier	Syntax-based Classifier
Supervised training	SLR: 63.85	SSR: 50.05
CTLSV10	85.70	67.10
CTLSV20	84.30	70.10
CTLSV30	87.15	75.75
CTLSV40	88.95	76.25
CTLSV50	87.00	77.55

method using a lexicon based classifier reaches an accuracy of 85.70%, significant better than that of SLR, which is only 63.85%. Similarly, the performance by syntax- based classifier increases from 50.05 to 75.75 after 30 iterations, showing a 51.35% lead.

Next, we compare our results on the combined classifier with CLSR and SSLF. The results are shown in Table 2.

In Table 2, the number n in *CTLSVn* denotes the number of iterations. And the number m in $SSLFm$ denotes the number of feature subset, which means that the unigrams are randomly partitioned into m (3 or 5) splits. Classifiers are trained and tested on each split of the labeled and test data using SVM, and then the final decisions are made using the ScoreSum and ScoreMax strategies as those used in our framework. For CLSR, we combine the lexicon and syntax feature vector into one long vector, and directly apply SVM to the labeled and test data. Note that we are unable to use this combination approach in our CTLSV because we co-train classifiers on two different views, that is to say, the instances selected and added to one labeled set may be not same as those added to the other labeled set. Hence the final labeled set of L_s is different from L_l.

We highlight three important points from Table 2.

– The ScoreMax strategy performs worse than ScoreSum. This indicates that the decision made from a number of scores is much more reliable than that made from only one score.

Table 2. Comparison between co-training and supervised training on combined methods

		ScoreMax	ScoreSum
CLSR	63.05	-	-
SSLF3	-	54.50	63.15
SSLF5	-	50.80	62.50
CTLSV10	-	72.20	80.00
CTLSV20	-	74.75	83.40
CTLSV30	-	79.95	86.75
CTLSV40	-	80.85	87.65
CTLSV50	-	81.45	87.95

- The SSLF algorithm does not benefit more from a large number of partitions. In fact, SSLF3 performs better than SSLF5, which is consistent with previous studies [32].
- The proposed CTLSV framework performs the best among all methods. The performance of CLSR is very close to that that of SSLF using ScoreMax, but both of them are significantly worse than that of CTLSV. For example, with 40 iterations, our accuracy value is 87.75 while the accuracy for SSLF3 and CLSR is only 63.15% and 63.05%, respectively.

5   Conclusion

In this paper, we investigate the problem of authorship attribution with very few labeled data. We present a novel co-training framework which utilizes the two natural views in human languages, i.e., the lexical and syntactic views. By iteratively learning a small number of samples from unlabeled data, the proposed approach effectively augments the labeled set and increases the performance. Experimental results on a real data set show that our method can significantly improve the classification accuracy. It outperforms the state-of-the-art methods by a large margin.

Our current study focuses on co-training using two views and evaluates on a small number of authors. In the future, we plan to extend our work by integrating more views and experimenting with more authors. In addition, our current experiment only involves one data set. Further experiments are required to determine the general behavior of co-training framework for authorship attribution.

Acknowledgements. The work described in this paper has been supported in part by the NSFC projects (61272275, 61232002, 61202036, 61272110, and U1135005), and the 111 project(B07037).

References

1. Argamon, S., Levitan, S.: Measuring the usefulness of function words for authorship attribution. In: Literary and Linguistic Computing pp. 1–3 (2004)
2. Argamon, S., Whitelaw, C., Chase, P., Hota, S.R., Garg, N., Levitan, S.: Stylistic text classification using functional lexical features: Research articles. J. Am. Soc. Inf. Sci. Technol. 58, 802–822 (2007)
3. Blum, A., Mitchell, T.: Combining labeled and unlabeled data with co-training. In: Proceedings of the 11th Annual Conference on Computational Learning Theory, pp. 92–100 (1998)
4. Burrows, J.: All the way through: Testing for authorship in different frequency data. Literary and Linguistic Computing 22, 27–47 (2007)
5. Diederich, J., Kindermann, J., Leopold, E., Paass, G., Informationstechnik, G.F., Augustin, D.S.: Authorship attribution with support vector machines. Applied Intelligence 19, 109–123 (2000)
6. Escalante, H.J., Solorio, T., Montes-y Gómez, M.: Local histograms of character n-grams for authorship attribution. In: Proceedings of the 49th Annual Meeting of the Association for Computational Linguistics: Human Language Technologies, vol. 1, pp. 288–298 (2011)
7. Gamon, M.: Linguistic correlates of style: authorship classification with deep linguistic analysis features. In: Proceedings of the 20th International Conference on Computational Linguistics (2004)
8. Graham, N., Hirst, G., Marthi, B.: Segmenting documents by stylistic character. Natural Language Engineering 11, 397–415 (2005)
9. Grieve, J.: Quantitative authorship attribution: An evaluation of techniques. Literary and Linguistic Computing 22, 251–270 (2007)
10. van Halteren, H.: Author verification by linguistic profiling: An exploration of the parameter space. ACM Transactions on Speech and Language Processing 4, 1–17 (2007)
11. van Halteren, H., Tweedie, F., Baayen, H.: Outside the cave of shadows: using syntactic annotation to enhance authorship attribution. Literary and Linguistic Computing 11, 121–132 (1996)
12. Hedegaard, S., Simonsen, J.G.: Lost in translation: authorship attribution using frame semantics. In: Proceedings of the 49th Annual Meeting of the Association for Computational Linguistics, vol. 2, pp. 65–70. Human Language Technologies (2011)
13. Hirst, G., Feiguina, O.: Bigrams of syntactic labels for authorship discrimination of short texts. Literary and Linguistic Computing 22, 405–417 (2007)
14. Hoover, D.L.: Statistical stylistics and authorship attribution: an empirical investigation. Literary and Linguistic Computing 16, 421–424 (2001)
15. Joachims, T.: (2007), http://www.cs.cornell.edu/people/tj/svm_light/old/svmmulticlass_v2.12.html
16. Kaster, A., Siersdorfer, S., Weikum, G.: Combining text and linguistic document representations for authorship attribution. In: SIGIR Workshop: Stylistic Analysis of Text for Information Access (STYLE), pp. 27–35 (2005)
17. Kim, S., Kim, H., Weninger, T., Han, J., Kim, H.D.: Authorship classification: a discriminative syntactic tree mining approach. In: Proceedings of the 34th International ACM SIGIR Conference on Research and Development in Information Retrieval, pp. 455–464 (2011)
18. Klein, D., Manning, C.D.: Accurate unlexicalized parsing. In: Proceedings of the 41st Meeting of the Association for Computational Linguistics, pp. 423–430 (2003)
19. Koppel, M., Schler, J.: Authorship verification as a one-class classification problem. In: Proceedings of the Twenty-First International Conference on Machine Learning (2004)
20. Koppel, M., Schler, J., Argamon, S.: Computational methods in authorship attribution. J. Am. Soc. Inf. Sci. Technol. 60(1), 9–26 (2009)

21. Koppel, M., Schler, J., Argamon, S.: Authorship attribution in the wild. Lang. Resources & Evaluation 45, 83–94 (2011)
22. Kourtis, I., Stamatatos, E.: Author identification using semi-supervised learning. In: Notebook for PAN at CLEF 2011 (2011)
23. Li, J., Zheng, R., Chen, H.: From fingerprint to writeprint. Communications of the ACM 49, 76–82 (2006)
24. Luyckx, K., Daelemans, W.: Authorship attribution and verification with many authors and limited data. In: Proceedings of the 22nd International Conference on Computational Linguistics, pp. 513–520 (2008)
25. Maria-Florina, B., Avrim Blum, K.Y.: Co-training and expansion: Towards bridging theory and practice. In: Advances in Neural Information Processing Systems (2004)
26. Mosteller, F.W.: Inference and disputed authorship: The Federalist. Addison-Wesley (1964)
27. Nigam, K., Analyzing, G.R.: Analyzing the effectiveness and applicability of co-training. In: Proceedings of the 9th International Conference on Information and Knowledge Management, pp. 86–93 (2000)
28. Sanderson, C., Guenter, S.: Short text authorship attribution via sequence kernels, markov chains and author unmasking: an investigation. In: Proceedings of the 2006 Conference on Empirical Methods in Natural Language Processing, pp. 482–491 (2006)
29. Seroussi, Y., Bohnert, F., Zukerman, I.: Authorship attribution with author-aware topic models. In: Proc. of The 50th Annual Meeting of the Association for Computational Linguistics (ACL), pp. 264–269 (2012)
30. Seroussi, Y., Zukerman, I., Bohnert, F.: Collaborative inference of sentiments from texts. In: De Bra, P., Kobsa, A., Chin, D. (eds.) UMAP 2010. LNCS, vol. 6075, pp. 195–206. Springer, Heidelberg (2010)
31. Solorio, T., Pillay, S., Raghavan, S., Montes Y Gómez, M.: Modality specific meta features for authorship attribution in web forum posts. In: Proceedings of the 5th International Joint Conference on Natural Language Processing, pp. 156–164 (2011)
32. Stamatatos, E.: Ensemble-based author identification using character n-grams. In: Proc. of the 3rd Int. Workshop on Textbased Information Retrieval, pp. 41–46 (2003)
33. Stamatatos, E.: Author identification using imbalanced and limited training texts. In: Proc. of the 4th International Workshop on Text-based Information Retrieval, pp. 237–241 (2007)
34. Stamatatos, E.: A survey of modern authorship attribution methods. Journal of The American Society for Information Science and Technology 60, 538–556 (2009)
35. Stamatatos, E., Kokkinakis, G., Fakotakis, N.: Automatic text categorization in terms of genre and author. Comput. Linguist. 26, 471–495 (2000)
36. Uzuner, Ö., Katz, B.: A comparative study of language models for book and author recognition. In: Proceedings of the 2nd International Joint Conference on Natural Language Processing, pp. 969–980 (2005)
37. de Vel, O., Anderson, A., Corney, M., Mohay, G.: Mining email content for author identification forensics. Sigmod Record 30, 55–64 (2001)
38. Zhao, Y., Zobel, J.: Effective and scalable authorship attribution using function words. In: Lee, G.G., Yamada, A., Meng, H., Myaeng, S.-H. (eds.) AIRS 2005. LNCS, vol. 3689, pp. 174–189. Springer, Heidelberg (2005)
39. Zheng, R., Li, J., Chen, H., Huang, Z.: A framework for authorship identification of online messages: Writing-style features and classification techniques. Journal of the American Society of Information Science and Technology 57, 378–393 (2006)

Edges Protection in Multiple Releases
of Social Network Data

Liangwen Yu[1,2], Yonggang Wang[3], Zhengang Wu[1,2],
Jiawei Zhu[1,2], Jianbin Hu[1,2,*], and Zhong Chen[1,2]

[1] Institute of Software, School of EECS, Peking University, China
[2] MoE Key Lab of High Confidence Software Technologies (PKU)
[3] National Computer Emergency Response Team and Coordination Center, China
{yulw,wuzg,zhujw,hjbin,chen}@infosec.pku.edu.cn, wyg@cert.org.cn

Abstract. With the increasing popularity of online social networks, such as twitter and weibo, privacy preserving publishing of social network data has raised serious concerns. Previous works only consider a single static release of social network data, which are not inadequate for analyzing the evolution of social networks. In this paper, we focus on the problem of preserving edges when edges are deleted or added in multiple releases of social network data. To achieve this objective, we propose the *Dynamic Safety Condition*, which effectively constrains nodes partition to ensure sparsity of edges between any two group. Using this condition, we devise the heuristic algorithm DEP, which anonymizes a sequential graphs to satisfy the privacy objective. Finally, we verify the effectiveness of the algorithm through experiments.

Keywords: social network, privacy preserving, data publishing, edges protection.

1 Introduction

Nowadays, online social network sites, such as facebook, twitter and weibo, have received dramatic interest, more and more people join in various social networks. People use online social networks to share data, which produce lots of social network data. Using data mining methods to analyze social network data, you can get lots of meaningful results: to optimize the search engine in the social networks, to improve the existing social networks, to study the characteristics of social networks. If this data is directly exposed to researchers, it will cause the privacy disclosure, which leads us to study how to effectively anonymize so as to protect sensitive information in social networks while maximizing the social network's utility analysis.

A social network can be modeled as a graph in which each node represents an individual, and the connections between individuals are summarized by the edges. Most of prior privacy protection techniques in social network data publishing focus primarily on static social networks. However, one single snapshot is

---

* Corresponding author.

F. Li et al. (Eds.): WAIM 2014, LNCS 8485, pp. 669–680, 2014.
© Springer International Publishing Switzerland 2014

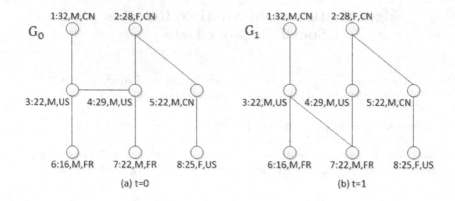

Fig. 1. Snapshots of a dynamic social network at $t = 0$ and $t = 1$

not useful for analyzing the evolution of social networks, such as how the popularity of individuals changes or how a disease spreads over time. To support such analysis, privacy issues in multiple releases of social network data urgently need to be addressed.

In this paper, we consider how to protect edges between individuals in multiple releases of social network data. For example, figure 1 shows two snapshots of a dynamic social network at time $t = 0$ and $t = 1$. The graph G_0 evolves into G_1 after adding an edge and deleting an edge. Our focus is on the privacy of edges between individuals in dynamic social networks, so we do not study the privacy of nodes' attributes.

Bhagat et al.[1] proposes edges protection for anonymizing a single graph. It masks the mapping via grouping the nodes of the graph. This technique retains the entire graph structure but perturbs the mapping from labels to nodes. Bhagat et al.[2] further use link prediction algorithms to model the evolution, however, the prediction graph could partly predict the newly added edges. Based on [1], Wang et al.[3] design a constraint in the grouping procedure, taking into account the effect of time to avoid revealing privacy with multiple releases. Both Bhagat et al.[2] and Wang et al.[3] assume edges and nodes are only added to the graph, not deleted. In this paper, graph model is extended to allow edges deletion, which is close to the evolution of social networks. In this work, we make the following contributions:

1. We solve the problem of edges deletion in dynamic edges protection.
2. We propose the *Dynamic Safety Condition*, which effectively constrains nodes partition to ensure sparsity of edges between any two group.
3. We devise the heuristic algorithm DEP, which anonymizes a sequential graphs to satisfy the privacy objective.
4. We evaluate our approach on two real datasets.

This paper is organized as follows. Section 2 gives the problem definition. Section 3 introduces the anonymized method. Section 4 reports experimental results. Section 5 presents the related work. Section 6 concludes the paper.

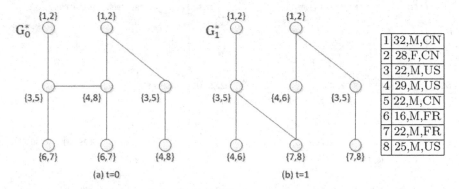

Fig. 2. The released graphs G_0^*, G_1^*

2 Problem Definition

In this paper, a time-stamped social network is modeled as an undirected simple graph $G_t(V_t, E_t, L_t, \mathcal{L})$, where V_t is the set of nodes corresponding to the individuals at time t, E_t is the set of edges that represent the relationships between individuals at time t, L_t is the set of labels at time t and a function $\mathcal{L} : V_t \to L_t$ assigns each node a label. A label has an identity(such as user id or name), and a set of properties(such as age, gender and country). Let $\mathcal{G} =< G_0, G_1, ..., G_T >$ be the sequence of graphs representing the network observed at timestamps $t = 0, 1, ..., T$ respectively. We mainly solve the problem of edges deletion in the evolution of social networks. The operation of node deletion could also be treated as retaining the node and deleting all edges related to the node in our paper, so the quantity of nodes is always increased. Thus, we have $V_{t-1} \subseteq V_t$ and $L_{t-1} \subseteq L_t$ where $t = 1$ to T. Let G_i^* denotes the anonymized graph of G_i at time $t = i$.

Given $\mathcal{G}$ and a constant k, the protection objectives guaranteed by this paper are:

1. For any edge $e \in E_t$, without background knowledge the probability that an attacker identifies a node v_x involved in e is at most $\frac{1}{k}$;
2. For any two nodes $v_x, v_y \in V_t$, without background knowledge the probability that an attacker identifies that there is an edge between them is at most $\frac{1}{k}$.

To protect edges, [1] focuses on masking the mapping via grouping the nodes of the graph. They propose a *Safety Condition* that each node must interact with at most one node in any group and no edges exist in a group. [2] defines the *Edge Identification* which measures the likelihood of identifying an interaction. Both *Safety Condition* and *Edge Identification* ensure sparsity of interactions between nodes of any two groups, but *Safety Condition* is more restrictive than *Edge Identification*.

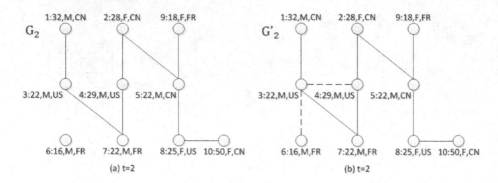

Fig. 3. The original graph G_2 and supplementary graph G'_2

Definition 1 (Safety Condition). *A grouping of nodes in graph G, satisfies the Safety Condition if*

$$\forall(v,w),(v,z) \in E : w \in g \wedge z \in g \Rightarrow w = z;$$

$$\forall(v,w) \in E : v \in g \wedge w \in g \Rightarrow v = w.$$

Definition 2 (Edge Identification). *Given a pair of groups of nodes, $g_x \subset V$ and $g_y \subset V$, their Edge Identification $EI(g_x, g_y)$ is the ratio of the number of edges between the two groups to the maximum number of such edges*

$$EI(g_x, g_y) = \frac{|(g_x \times g_y) \cap E|}{|g_x| \cdot |g_y|}$$

In figure 2, G_0^* and G_1^* are the released graphs of original graphs in figure 1. They are respectively generated using the method in [1] with $k = 2$. Intersecting the candidate sets in G_1^* with those in G_0^*, the nodes in $\{4, 6, 7, 8\}$ could be easily identified in the released graphs. Then, the adversary could conclude that there is a relationship between node 4 and node 7 in both G_0^* and G_1^*, which obviously violates privacy purposes.

3   The Anonymized Method

[3] applies the algorithm DMRA(Decreasing Multiple Releases Anonymization) in anonymizing dynamic social network data. This approach first anonymizes G_T, and then gradually produces $G_{T-1}, G_{T-2}, ..., G_0$. The same anonymized procedure is used in our paper. To protect deleted edges, we introduce the supplementary graph G'_T, where $V'_T = V_T$ and $E'_T = \cup_{t=1}^{T}(E_{t-1} \setminus E_t) \cup E_T$. So the supplementary graph G'_T contains all the deleted edges in the evolution of social networks. For example, figure 3(a) shows the following original graph G_2 at time $t = 2$, and figure 3(b) shows the corresponding supplementary graph G'_2. Therefore, G'_2 includes two deleted edges which are represented by dash lines in figure3(b).

3.1 Dynamic Safety Condition

Definition 3 (Dynamic Safety Condition). *A grouping of nodes V_t for the graph G_t, satisfies the dynamic Safety Condition if*

$$\forall v, w \in V_t : if v \in g \land w \in g \Rightarrow \exists Set \in \{V_0\} \cup (\cup_{i=1}^{t}\{V_i \setminus V_{i-1}\}), \{v,w\} \subseteq Set;$$

$$\forall (v, w) \in E_t : if v \in g \land w \in g \Rightarrow v = w;$$

$$\forall group\ g_x, g_y, m\ is\ the\ number\ of\ edges\ between\ g_x\ and\ g_y, m \leqslant \frac{|g_x| \cdot |g_y|}{k}.$$

Based on [3], we give the definition of *Dynamic Safety Condition*. The first condition requires that the nodes in one group are created at the same time. The second condition constrains edges not exist in one group. The third condition limits the number of edges between any two groups.

Theorem 1. *If nodes partition satisfies Dynamic Safety Condition, the anonymized graphs can dynamically protect edges against attacks without background knowledge.*

Proof. The first condition in *Dynamic Safety Condition* requires that the nodes in one group are created at the same time. Using this condition in our anonymized method, we constrain the size of groups in anonymized graphs is no less than k. Thus, without background knowledge the probability that an attacker can identify a node is at most $\frac{1}{k}$. The second condition constrains edges not exist in one group. For each edge in our anonymized graphs, there are at least k candidate labels for the two endpoints of it. Thus, without background knowledge the probability that an attacker can identify a node involved in an edge is at most $\frac{1}{k}$. The third condition constrains the number of edges m between any two groups is no more than $\frac{|g_x||g_y|}{k}$. *Edge Identification* is the probability an attacker can attach to a particular pair of users between any two groups. With respect to nodes partition, *Edge Identification* equals to $\frac{|(g_x \times g_y) \cap E|}{|g_x| \cdot |g_y|} = \frac{m}{|g_x| \cdot |g_y|} \leqslant \frac{\frac{|g_x| \cdot |g_y|}{k}}{|g_x| \cdot |g_y|} = \frac{1}{k}$. After anonymizaiton, without background knowledge the probability that an attacker can identify that there is an edge between two nodes is at most $\frac{1}{k}$. Therefore, *Dynamic Safety Condition* can make our anonymized graphs satisfy privacy objectives.

3.2 The DEP Alogrithm

The algorithm DEP has a similar structure as the algorithm in [3]. The input of DEP are a sequential original graphs $\mathcal{G} = <G_0, G_1, ..., G_T>$, a supplementary graph G'_T and a parameter k. The output of DEP are a set of groups with size at least k and a sequential anonymized graphs $G_0^*, G_1^*, ..., G_T^*$. This algorithm first puts all the nodes of G'_T into a *NodeList* and sorts the *NodeList* based on the properties of nodes' label to enable nodes in a group have the similar values of the properties. Each time, we select a *SeedNode* from the *NodeList* and remove it from the list, and insert the *SeedNode* into the first group that has fewer than k nodes, if performing this insertion would not violate the *Dynamic Safety Condition*(line 4-11). If no group can be found which satisfies this condition, or all groups existed have at least k nodes, then a new group containing only *SeedNode* is created(line 12-14). For any group g which cannot reach size k, we first delete this group from the group set. Then we take each node in g in turn, and insert it into the first group which size is no less than k under *Dynamic Safety Condition*(line 15-21). In some cases, there are no suitable groups for the insertion operation. We add $(k - |g|)$ noise nodes into g and randomly assign

---

Algorithm 1. Dynamical Edges Protection (DEP) Algorithm

Input: A sequential original graphs $\mathcal{G} = < G_0, G_1, ..., G_T >; G'_T(V'_T, E'_T, L'_T, \mathcal{L}), V'_T = V_T, E'_T = \cup_{t=1}^{T}(E_{t-1} \setminus E_t) \cup E_T, L'_T = L_T$;a parameter k.

Output: A set of groups $S = \{g_1, g_2, ..., g_m\}, g_i \cap g_j = \emptyset, i, j = 1, 2, ..., m, i \neq j, |g_i| \geq k, i = 1, 2, ..., m$;a sequential anonymized graphs $< G_0^*, G_1^*, ..., G_T^* >$.

1: $S = \emptyset$;
2: $NodeList = \emptyset$;
3: put all the nodes of G'_T into the $NodeList$ and sort the $NodeList$ based on the properties of nodes' label;
4: **while** $NodeList.size() > 0$ **do**
5:    $SeedNode = NodeList.head()$ and remove it from $NodeList$;
6:    $flag = true$;
7:    **for** each group $g \in S$ with $|g| < k$ **do**
8:       **if** inserting $SeedNode$ into g doesn't violate dynamic safety condition **then**
9:          $g = g \cup \{SeedNode\}$;
10:         $flag = false$;
11:         break;
12:    **if** $flag$ **then**
13:       group $g' = $ new group($SeedNode$);
14:       $S = S \cup \{g'\}$;
15: **for** each group $g \in S$ with $|g| < k$ **do**
16:    $S = S - \{g\}$;
17:    **for** each node $u \in g$ **do**
18:       **for** each group $g' \in S$ with $|g'| \geq k$ **do**
19:          **if** inserting u into g' doesn't violate dynamic safety condition **then**
20:             $g = g - \{u\}; g' = g' \cup \{u\}$;
21:          break;
22:    **if** $|g| > 0$ **then**
23:       insert $(k - |g|)$ noise nodes into g;
24:       randomly assign labels to noise nodes following the label distribution in G'_T;
25:       $S = S \cup \{g\}$;
26: assign the corresponding label list to each node according to its group in G'_T;
27: delete all edges $\in E'_T \setminus E_T$ from G'_T to generate G_T^*;
28: **for** $t = T$ to 1 **do**
29:    delete all nodes $\in V_t \setminus V_{t-1}$, all edges $\in E_t \setminus E_{t-1}$ and add all edges $E_{t-1} \setminus E_t$ from G_t^* to generate G_{t-1}^*

---

labels to noise nodes following the label distribution in G'_T, then return g back to group set(line 23-26). After grouping, we assign the corresponding label list to each node according to its group in G'_T. Finally, we delete all edges in $E'_T \setminus E_T$ from G'_T to generate G^*_T, then we delete all nodes in $V_t \setminus V_{t-1}$, all edges in $E_t \setminus E_{t-1}$ and add all edges $E_{t-1} \setminus E_t$ from G^*_t from G^*_t to generate G^*_{t-1} for $t = T$ to 1.

3.3   The Running Example

Figure 4 shows the running example of anonymizing a sequential original graphs $\mathcal{G} = < G_0, G_1, G_2 >$ with $k=2$. Firstly, we establish the supplementary graph G'_2 based on $\mathcal{G}$, and group the nodes in V'_2. Figure 4(a) shows the procedure of groups partition. Observing the *Dynamic Safety Condition*, V'_2 is partitioned into $\{1,2\}, \{3,5\}, \{4,6\}, \{7,8\}$,and $\{9,10\}$. Secondly, we assign the corresponding label list to each node according to its group in G'_2. Thirdly, we delete all edges in $E'_2 \setminus E_2$ from G'_2 to generate G^*_2. Last, we delete all nodes in $V_2 \setminus V_1$, all edges in $E_2 \setminus E_1$ and add edges in $E_1 \setminus E_2$ from G^*_2 to generate G^*_1, and also generate G^*_0 in the same way .

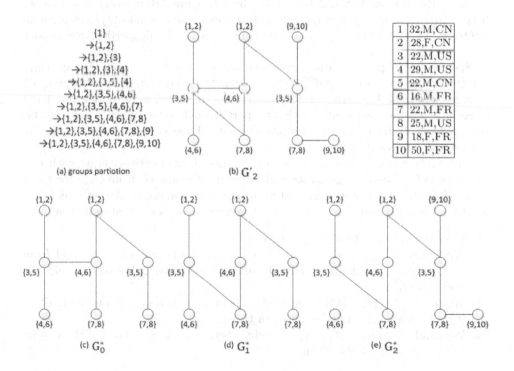

Fig. 4. The running example

4   Experimental Results

In this section, we conduct aggregate network queries on two datasets. For each query, its relative error is $\frac{|n-n'|}{n}$. Where n and n' are the results of the query on the original and anonymized graphs, respectively. In this paper, we test two kinds of queries based on [1, 3, 4]:

1. One hop query: The number of nodes pairs represented by a 4-tuple (v_1, c_1, v_2, c_2), where v_1 has the characteristic c_1 (the characteristic could be one or more properties of label), and v_2 has the characteristic c_2.
2. Two hops query: The number of nodes tuples represented by a 6-tuple $(v_1, c_1, v_2, c_2, v_3, c_3)$, where v_1 has the characteristic c_1, v_2 has the characteristic c_2, and v_3 has the characteristic c_3.

ca-CondMat: This dataset shows a Condense Matter collaboration network which is built from the scientific collaborations between authors' papers from January 1993 to April 2003(available at http://snap.stanford.edu/data/ca-CondMat.htm). It contains 23,133 nodes and 186,936 edges. An undirected edge is created in the graph if two authors co-authored a paper. We use the Adult dataset from the UC Irvine Machine Learning Repository(available at http://archive.ics.uci.edu/ml/machine-learning-databases/adult/) to assign a label to each node in the graph, and only consider a set of 3 properties: age, gender and country.

Speed Dating: This dataset is from a publicly available Speed Dating study conducted by Fisman et al.[5](available at http://andrewgelman.com/2008/01/the_speeddating_1/). After sanitizing the data, it contains 552 nodes and 4,184 edges. Each node is associated with 3 properties, such as age, gender and country, chosen from 21 properties in the original data. Each edge represents a dating between two participants.

To simulate the evolution of social networks, we respectively start with two datasets as G_0, Next, we generate each subsequent graph G_t from G_{t-1} for $t = 1$ to 2. This simulation uses four evolution parameters a, b, c, d. We set a as 5%, b as 10%, c as 20% and d as 10%. At each iteration, G_t is evolved in four steps:

1. We randomly delete $a|E_{t-1}|$ edges in G_{t-1}.
2. We create $b|V_{t-1}|$ new nodes. For each new node, we assign a label from Adult dataset to it and randomly makes a connection between it and the nodes in G_{t-1}.
3. We randomly create $c|V_{t-1}|$ new edges between new nodes and nodes in G_{t-1} if there is no existing edge between them.
4. We randomly create $d|E_{t-1}|$ new edges between nodes in G_{t-1} if there is no existing edge between them.

We conduct 100 queries, and use the average relative query error to measure the utility of the anonymized graphs. We also consider a efficient sort order in [1] over the properties of nodes' label : Country, Gender, and Age(CGA). In order to perform queries on the anonymized graphs with the label list, we use Sampling

Consistent Graphs in [1] which are consistent with the anonymized graphs, and analyze the sampled graphs. As the anonymized methods in [2, 3] do not consider the situation of edges deletion, we couldn't compare our results with them. Figure 5 and figure 6 show the average relative error of one hop queries and two hops queries respectively. It is easy to observe that average relative error increases as k rises. This is because the increasing of k leads to more alternative labels for the nodes in the group. The result on one hop queries is better than that of two hops queries, as two hops queries use 2 conjunctions. Last, sorting the properties of nodes' label is necessarily needed in our anonymized method.

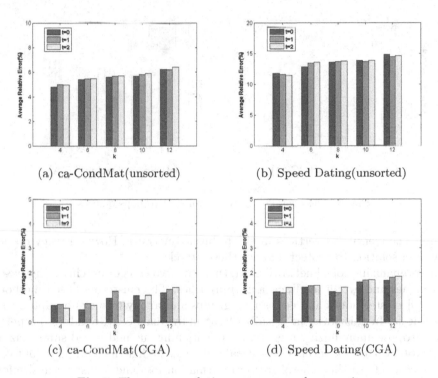

(a) ca-CondMat(unsorted)   (b) Speed Dating(unsorted)

(c) ca-CondMat(CGA)   (d) Speed Dating(CGA)

Fig. 5. The average relative error on one hop queries

5   Related Work

The problem of privacy protection in social networks is first proposed in [6], where the authors demonstrate that the naive anonymization strategy which replaces all identifiers of individuals with randomized integers is not sufficient by both active and passive attacks. In active attacks, an adversary maliciously plants a subgraph in the network before it is published and uses the knowledge of the planted subgraph to re-identify nodes and edges in the published network. In passive attacks, an adversary simply uses a small uniquely identifiable subgraph

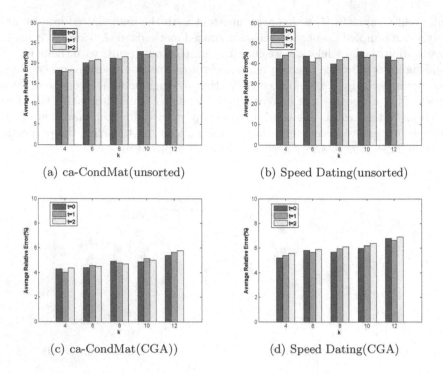

(a) ca-CondMat(unsorted)   (b) Speed Dating(unsorted)

(c) ca-CondMat(CGA))   (d) Speed Dating(CGA)

Fig. 6. The average relative error on two hops queries

to infer the identity of vertices in the published network. However, they do not provide a solution to protect against these attacks.

In anonymizing social network data, there are two categories: clustering-based approaches and graph modification approaches. The clustering-based methods [1, 7–10] cluster nodes and edges into groups and anonymize a subgraph into a super-node. In this way, the details about individuals can be hidden properly. [7, 9] propose anonymizing a graph by partitioning the nodes and summarizing the graph at partition level. The critical difference is that the method in [9] takes into account both the generalization information loss and the structural information loss during the clustering procedure. Zheleva and Getoor[8] focuses on the case where there are multiple types of edges but only one type of nodes, and applies clustering-based method in protecting relationships disclosure. Cormode et al.[10] focuses on the problem of anonymizing bipartite graphs. Based on [10], Bhagat et al.[1] further constructs a model of the rich interaction graph, and proposes three approaches in protecting users' rich interaction.

The graph modification methods[11–19] anonymize a graph by modifying(such as, adding and/or deleting) edges and nodes in a graph. Hay et al.[15] proposes an approach that obeys a rule of random edge additions and deletions in anonymizing the graph, this method can effectively resist some kinds of attacks but suffers a significant cost in utility. Edge randomization techniques are further explored in [16], whose goal is to preserve the spectral properties of the graph. While the network utility is much improved, the effect on anonymity is not quantified. Liu et al.[17] perturbs the weights of some edges to retain the shortest path and the approximate cost of the path between some pairs of nodes in the original network. Liu and Terzi[14] first introduces the k-anonymity model from the relational data to the social network data, and proposes k-degree anonymity to protect each individual in a group consisting of at least k nodes of the same degree. Zhou and Pei[13] proposes the stronger model that each individual in a group consisting of at least k nodes of the same degree and sharing 1-neighbourhood isomorphism. Zou et al.[18] proposes a k-automorphism model that each individual in a group consisting of at least k nodes without any structural difference. Cheng et al.[19] designs a k-isomorphism model that the anonymous graph consists of k disjoint isomorphic subgraphs to protect nodes and relationships. [11, 12] based on different models apply l-diversity in protecting nodes re-identification and nodes' sensitive attribute.

6 Conclusions

In this paper, we protect edges in anonymizing multiple releases of social network data. We present the *Dynamic Safety Condition* and the heuristic algorithms DEP. The *Dynamic Safety Condition* constrains nodes partition to ensure sparsity of edges between any two group in the evolution of social networks. The DEP algorithm anonymizes a sequential graphs to satisfy the privacy objective. We prove the effectiveness of the proposed anonymized method through extensive experiments.

Acknowledgment. This work is partially supported by the HGJ National Significant Science and Technology Projects under Grant No. 2012ZX01039-004-009, Key Lab of Information Network Security, Ministry of Public Security under Grant No.C11606, the National Natural Science Foundation of China under Grant No. 61170263.

References

1. Bhagat, S., Cormode, G., Krishnamurthy, B., Srivastava, D.: Class-based graph anonymization for social network data. Proceedings of the VLDB Endowment 2(1), 766–777 (2009)
2. Bhagat, S., Cormode, G., Srivastava, D., Krishnamurthy, B.: Prediction promotes privacy in dynamic social networks. In: Proceedings of the 3rd Conference on On-line Social Networks, p. 6. USENIX Association (2010)
3. Wang, C.-J.L., Wang, E.T., Chen, A.L.: Anonymization for multiple released social network graphs, pp. 99–110 (2013)
4. Yuan, M., Chen, L., Yu, P.S.: Personalized privacy protection in social networks. Proceedings of the VLDB Endowment 4(2), 141–150 (2010)
5. Fisman, R., Iyengar, S., Kamenica, E., Simonson, I.: Gender differences in mate selection: Evidence from a speed dating experiment. The Quarterly Journal of Economics 121(2), 673–697 (2006)

6. Backstrom, L., Dwork, C., Kleinberg, J.: Wherefore art thou r3579x?: anonymized social networks, hidden patterns, and structural steganography. In: Proceedings of the 16th International Conference on World Wide Web, pp. 181–190. ACM (2007)

7. Hay, M., Miklau, G., Jensen, D., Towsley, D., Weis, P.: Resisting structural re-identification in anonymized social networks. Proceedings of the VLDB Endowment 1(1), 102–114 (2008)

8. Zheleva, E., Getoor, L.: Preserving the privacy of sensitive relationships in graph data. In: Bonchi, F., Malin, B., Saygın, Y. (eds.) PInKDD 2007. LNCS, vol. 4890, pp. 153–171. Springer, Heidelberg (2008)

9. Campan, A., Truta, T.M.: A clustering approach for data and structural anonymity in social networks. In: Privacy, Security, and Trust in KDD Workshop, PinKDD 2008 (2008)

10. Cormode, G., Srivastava, D., Yu, T., Zhang, Q.: Anonymizing bipartite graph data using safe groupings. Proceedings of the VLDB Endowment 1(1), 833–844 (2008)

11. Tai, C., Yang, D., Yu, P., Chen, M.: Structural diversity for privacy in publishing social networks. In: Proc. of SDM (2011)

12. Zhou, B., Pei, J.: The k-anonymity and l-diversity approaches for privacy preservation in social networks against neighborhood attacks. Knowledge and Information Systems 28(1), 47–77 (2011)

13. Zhou, B.: Preserving privacy in social networks against neighborhood attacks. In: IEEE 24th International Conference on Data Engineering, ICDE 2008, pp. 506–515. IEEE (2008)

14. Liu, K., Terzi, E.: Towards identity anonymization on graphs. In: Proceedings of the 2008 ACM SIGMOD International Conference on Management of Data, pp. 93–106. ACM (2008)

15. Hay, M., Miklau, G., Jensen, D., Weis, P., Srivastava, S.: Anonymizing social networks. Computer Science Department Faculty Publication Series, p. 180 (2007)

16. Ying, X., Wu, X.: Randomizing social networks: a spectrum preserving approach. In: Proceedings of the 2008 SIAM International Conference on Data Mining (SDM 2008), pp. 739–750. Citeseer (2008)

17. Liu, L., Wang, J., Liu, J., Zhang, J.: Privacy preservation in social networks with sensitive edge weights. In: 2009 SIAM International Conference on Data Mining (SDM 2009), pp. 954–965. Sparks, Nevada (2009)

18. Zou, L., Chen, L., Özsu, M.: K-automorphism: A general framework for privacy preserving network publication. Proceedings of the VLDB Endowment 2(1), 946–957 (2009)

19. Cheng, J., Fu, A., Liu, J.: K-isomorphism: privacy preserving network publication against structural attacks. In: Proceedings of the 2010 International Conference on Management of Data, pp. 459–470. ACM (2010)

Trust Prediction with Temporal Dynamics

Guoyong Cai[1], Jiliang Tang[2], and Yiming Wen[1]

[1] Guanxi Key Lab of Trusted Software,
Guilin University of Electronic Technology, Guilin 541004, PRC
[2] Computer Science and Engineering, Arizona State University, Tempe 85281, USA

Abstract. In open society-based applications, inferring unknown trust relations attracts increasing attention in recent years. Most existing work assumes
that trust relations are static. In this paper, we incorporate temporal dynamics in
trust prediction by modeling the dynamics of user preferences in two principled
ways. Initial experiments on a real-world data set are conducted and the results
demonstrate the effectiveness of the proposed models.

Keywords: trust prediction, trust dynamism, trust network.

1   Introduction

Most existing work on online trust assumes a static trust model [1-2]. The assumptions are usually unacceptable for long-running society-based online application systems. As time goes on, a user's interests may change, a product's attraction or freshness may decay, etc. Therefore we have to consider temporal dynamics for a more
accurate trust prediction. We study - (1) how to model temporal dynamics and (2)
how to incorporate temporal dynamics for trust prediction.

2   Problem Statement

Let $u = \{u_1, u_2, ..., u_n\}$ be the set of users, $t = \{t_1, t_2, ..., t_m\}$ be the set of timestamps.
Assume that $\mathcal{G} = \{ G_1, G_2, ..., G_m\}$, where $G_i \in \mathrm{R}^{n \times n}$ is the matrix representation of
trust relations at t_i, where $G_i(j,k) = 1$ if u_j establishes a trust relation with u_k, $G_i(j,k) = 0$
otherwise. Our task is to predict the value of G_{m+1} given a time series of an observed
data matrix $\{G_1, G_2, ..., G_m\}$ for the past time periods $[t_1, t_m]$.

3   Trust Prediction Models

In [3], we propose a trust prediction framework based on low-rank matrix factorization. Here we will use this trust model as the basic model while incorporating temporal dynamics.

F. Li et al. (Eds.): WAIM 2014, LNCS 8485, pp. 681–686, 2014.

3.1   Temporal Weight Matrix Factorization (*TWMF*)

Assume that g_{ij}^t is the timestamp when u_i trusts u_j. The influence of trust relation between u_i to u_j at g_{ij}^t to whether u_i trusts others at the current time t is related to the distance between g_{ij}^t and t. That is the prediction error $\| G(i,j) - U(i,:)VU^T(j,:) \|_2^2$ is related to the time distance. A less error value is related to a lower decaying value, and a larger error is related to a larger decaying value. Since different users may change differently, they should have different decaying ratios. Therefore, we introduce η_i as a personalized decaying ratio for u_i. Then an exponential time function $e^{-\eta_i(m-g_{ij}^t)}$ is adopted to weight the trust prediction error as $e^{-\eta_i(m-g_{ij}^t)} \| G(i,j) - U(i,:)VU^T(j,:) \|_2^2$. Thus our first framework (*TWMF*) is to solve the following minimization problem.

$$\min_{U,V,\eta_i} \sum_{i=1}^n \sum_{j=1}^n e^{-\eta_i}(m-g_{ij}^t) \| G(i,j) - U(i,:)VU^T(j,:) \|_2^2 + \alpha \sum_{i=1}^n \| \eta_i \|_2^2 \quad (1)$$

$$+ \beta \| U \|_F^2 + \gamma \| V \|_F^2 \qquad s.t. \ U,V \geq 0, \ \eta_i \geq 0, \ \forall i \in [1,n]$$

An alternative optimization method is adopt to solve Eq.(1) shown as following algorithm 1.

Algorithm 1. *TWMF* for Trust Prediction

Input: $\{G_1, G_2, ..., G_m\}$, α, β and γ. **Output**: $\hat{G}$

1: $G = \sum_{t=1}^m G_t$; 2: Initialize V randomly; 3: Initialize η_i randomly; 4: Initialize $U(i,:)$ randomly

5: **while** Not convergent or not reach the maximal iteration

6: Update $\eta_i \leftarrow \eta_i \sqrt{\dfrac{\sum_{j=1}^n c_{ij} e^{-\eta_i(m-g_{ij}^t)}}{2\alpha\eta_i + \sum_{j=1}^n c_{ij} e^{-\eta_i(m-g_{ij}^t)}}}$; 7: Update $U(i,k) \leftarrow U(i,k) \sqrt{\dfrac{a_i^T(k)}{[U(i,:)A_i](k)}}$

8: $V(i,k) \leftarrow V(i,k) \sqrt{\dfrac{B(i,k)}{C(i,k)}}$

9: **end while**

10: $\hat{G} = UVU^T$

where $c_{ij} = \| G(i,j) - U(i,:)VU^T(j,:) \|_2^2$, $a_i = \sum_{j=1}^n b_{ij}^t G(i,j)VU^T(j,:) + \sum_{j=1}^n b_{ji}^t G(j,i)V^T U^T(j,:)$,

$A_i = \sum_{j=1}^n b_{ij}^t VU^T(j,:)U(j,:)V^T + \sum_{j=1}^n b_{ji}^t V^T U(j,:)^T U(j,:)V + \beta I$,

$B = \sum_{i=1}^n \sum_{j=1}^n b_{ij}^t G(i,j)U^T(i,:)U(j,:)$, $C = \sum_{i=1}^n \sum_{j=1}^n b_{ij}^t U^T(i,:)U(j,:)V^T U^T(i,:)U(j,:) + \gamma V$ · For a pair of users $\langle u_i, u_j \rangle$ without a trust relation, the likelihood of them establishing trust relation is indicated by $\hat{G}(i, j)$

3.2   Temporal Smoothness Matrix Factorization (*TSMF*)

An alternative way to model temporal dynamics for trust prediction is to model the evolution of user preferences. Let $U=\{U_1,U_2,...,U_m\}$ be the set of user preference matrices of u, where $U_t \in R^{n \times d}$ is the user preference matrix at the t-th timestamp. Let $V=\{V_1,V_2,...,V_m\}$ be set of correlation matrices, where $V_t \in R^{d \times d}$ captures the correlations of U_t at the t-th timestamp. *TSMF* for trust prediction is found by solving the following equation 2.

$$\min_{U,V} \quad \sum_{t=1}^{m} \| G_t - U_t V_t U_t^T \|_F^2 + \alpha \sum_{t=1}^{m} \| V_t \|_F^2 + \beta \sum_{t=1}^{m} \| U_t \|_F^2 + \gamma \sum_{t=2}^{m} \| U_t - U_{t-1} \|_F^2$$

$$s.t. \quad U_t \geq 0, \ V_t \geq 0, \ t \in [1, m]$$

(2)

Where $\gamma \sum_{t=2}^{m} \| U_t - U_{t-1} \|_F^2$ is the temporal smoothness term and γ is introduced to control the contribution from the temporal regularization. The rationale behind temporal regularization is that user preferences tend to evolve gradually; hence, a user preference matrix between consecutive timestamps should be as smooth as possible. Similar to the *TWMF* framework, we adopt an alternative optimization method for Eq.(2). And the detail is omitted for space limited.

4   Experiments

We collect a dataset from Epinions[1] to evaluate our proposed models. This dataset spans the length of 12 years, ranging from Jul 05, 1999 to May 08, 2011. We split the whole dataset into 10 timestamps, i.e., $T = \{t_1, ..., t_{10}\}$, where t_{10} contains data after Jan 11, 2010 and t_1 to t_9 each contain data for one year. We choose trust relations from t_i to t_{i+l-1} as old trust relations O with the time window size l, and trust relations in $t+l$ as new trust relations to predict N. Let A be the set of user pairs excluding pairs in O. We follow the common metric in [4] to evaluate the performance of trust prediction. In this paper, we choose the time window size as 4 and then vary i from 1 to 6.

---

[1] http://www.epinions.com/

4.1   Comparison of Different Trust Predictors

We choose the following trust predictors as our baseline methods:

— **TP**: The trust relations are inferred through trust propagation by four atomic propagations: direct propagation, co-citation, transpose trust, and trust coupling[5].

— **NMF**: This predictor performs a non-negative matrix factorization in the whole trust network $G = \sum_{i=1}^{m} G_i$ as in paper [4]. It assumes that the trust network is static and ignores the temporal dynamics.

— **TF**: With timestamps, trust relations can be represented as a 3-D tensor. TF performs tensor factorization on the tenor representation $A \in R^{n \times n \times m}$, where $A(:,:,i) = G_i \quad i \in [1, m]$

— **DW-NMF**: This predictor first combines trust relations with a decay weight $G = \sum_{i=1} e^{-\eta(t-i)} G_i$ and then perform matrix factorization on **G**.

The parameters of all baseline methods are selected via cross-validation. For TSMF, we set $\gamma = 1$. α and β are empirically set to 0.1. The comparison results are demonstrated in Fig. 1 (left) and we make the following observations.

- Methods based on matrix factorization and tenor factorization obtains better performance than trust propagation.
- Compared to NMF, DW-NMF and TF gain significantly performance improvement. Both DW-NMF and TF consider temporal dynamics, while NMF assumes that trust relations are static. These results show the importance of temporal dynamics in trust prediction. TWMF and TSMF outperform all baseline methods.
- TWMF always gives a little precise prediction than TSMF. Because the change speed varies among different users, by automatically learning a personalized decay ratio η_i, varying change speed of different users can be more accurately captured in TWMF. TSMF cannot capture personalized preference evolution.
- In general, as time goes on, the baseline methods decrease the prediction accuracy gradually; while our TWMF and TSMF keep quite stable prediction accuracy.

4.2   Impact of Time Smoothing Parameter

To investigate the impact of time smoothing parameter γ on the framework TSMF, we vary the values of γ as $\{1e-3, 1e-2, 0.1, 0.5, 0.7, 1, 10, 100\}$ and see how the changes of γ effect the performance of TSMF. The results are shown in Fig.1 (right), and we observe the following:

- When γ changes from 1e-3 to 1e-2, the performance improves considerably. When γ is very small, *TSMF* learns U_i mainly from G_i, which is too sparse to learn an accurate user preference matrix.
- When γ changes from 0.1 to 10, the performance first increases, reaches its peak value, and then degrades. In certain regions, the performance of TSMF is very stable.
- When γ changes from 10 to 100, the performance decreases drastically. When γ is large, the smoothing regularization will dominate the learning process, also resulting in an inaccurate estimation of user preference matrices.

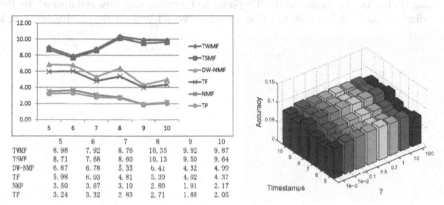

	5	6	7	8	9	10
TWMF	8.98	7.92	8.76	10.35	9.92	9.87
TSMF	8.71	7.68	8.60	10.13	9.50	9.64
DW-NMF	6.87	6.78	5.33	6.41	4.32	4.99
TF	5.98	6.03	4.81	5.39	4.02	4.37
NMF	3.50	3.67	3.10	2.80	1.91	2.17
TP	3.24	3.32	2.83	2.71	1.88	2.05

Fig. 1. (left) Performance Comparison for Different Trust Predictors in Epinions Note that the y-axis denotes prediction accuracy*100, and the x-axis denotes the time by unit of year (as time window is set to 4, the first predicted year starts from 5).   Fig.1 (right) The Impact of Time Smoothing Parameter γ on *TSMF*.

5   Conclusions

In this paper, we provide two ways to model temporal information based on matrix factorization and propose two unsupervised trust prediction models based on temporal weight matrix factorization and time smoothing matrix factorization, respectively. The experimental results on a real-world dataset show that our proposed models significantly outperform the state-of-the-art trust predictors. Further experiments are conducted to gain a deep understanding of the importance of temporal dynamics in trust prediction.

Acknowledgements. This work was supported by the NSFC grant (No.61063039), Project of Guangxi Key Lab of Trusted Software (No.kxrj201202).

References

1. Jøsang, A., Ismail, R., Boyd, C.: A survey of trust and reputation systems for online service provision. Decis. Support Syst. 43(2), 618–644 (2007)
2. Tavakolifard, M., Almeroth, K.C.: A taxonomy to express open challenges in trust and reputation systems. JCM 7(7), 538–551 (2012)
3. Tang, J., Gao, H., Hu, X., Liu, H.: Exploiting homophily effect for trust prediction. In: Proceedings of the Sixth ACM International Conference on Web Search and Data Mining, pp. 53–62. ACM (2013)
4. Liben-Nowell, D., Kleinberg, J.: The link-prediction problem for social networks. Journal of the American Society for Information Science and Technology 58(7), 1019–1031 (2007)
5. Guha, R., Kumar, R., Raghavan, P., Tomkins, A.: Propagation of trust and distrust. In: Proceedings of the 13th International Conference on World Wide Web, pp. 403–412. ACM (2004)

Trust Inference Path Search Combining Community Detection and Ant Colony Optimization

Yao Ma, Hongwei Lu, Zaobin Gan*, and Yizhu Zhao

Huazhong University of Science and Technology, Wuhan, China
mayaobox@qq.com, {luhw,zgan,missbamboo}@hust.edu.cn

Abstract. Finding trust inference paths for unfamiliar users in online social networks is a fundamental work of trust evaluation. Most existing trust inference path search approaches apply classical brute-force graph search algorithms, which leads to high computation costs. To solve this issue, we propose a trust inference path search approach combining community detection and ant colony optimization. First, the singular value decomposition signs method is utilized to process the trust relationship matrix in order to discovery the trust communities. Then, by taking the communities as different colonies, we use the ant colony optimization to find the optimal trust inference path along which the witness has the maximum deduced referral belief. The released pheromones in previous trust inference path searches help subsequent searches to reuse previous experience and save path search costs. Comparative experiments show that the proposed trust inference path search approach outperforms the existing ones on path search efficiency and trust inference accuracy.

Keywords: Trust inference path search, community detection, ant colony optimization, social network analysis.

1   Introduction

Popularized Online Social Network (OSN) applications, especially the Social Network Sites (SNSs), provide people with great convenience for information sharing, collaborating and interacting. Trust in online social networks plays an important role for users to make trusted decisions when facing unfamiliar co-partners or environments. For example, the "Web of Trust" in Epinions.com builds a community of trusted members for users and makes personalized recommendations. It is beneficial for the buyers to evaluate the trustworthiness of the unfamiliar recommenders, sellers or service providers before making purchase decisions. Given a pair of users who have no interaction experience, trust transitivity based trust inference can deduce the trust opinion between them by applying trust discounting and consensus operations to the trust propagation

---

* Corresponding author.

F. Li et al. (Eds.): WAIM 2014, LNCS 8485, pp. 687–698, 2014.
© Springer International Publishing Switzerland 2014

paths [7]. In the large-scale OSN, there are a great number of such paths and how to efficiently find appropriate trust inference paths emerges as a question.

Most existing trust inference path search approaches ignore the structure characteristics of the trust network, which makes the path search blind and cost expensive. Ant Colony Optimization (ACO)[1], inspired by the pheromone trail laying and following behavior of some ant species, is a metaheuristic for solving hard combinatorial optimization problems. Similar to the ants that find the shortest paths connecting to the food, users in the ONS also want to find the most reliable trust inference path connecting to the target participants. However, the users flock with shared interests, preferences or opinions etc. and they compose the different communities. The structure of social networks attracts much attention and the research of community detection in social networks derives [3]. The ideal communities in trust networks should be like this: the members in the same community trust each other and the distrust relationships do not appear in one community. We try to detect the communities in the trust network by trust and distrust relationships and cluster users as different colonies for ant colony optimization. Since users in one community have similar trusting and distrusting preferences, the clustering can help the path search have a clear sense of direction and tend to find trustworthy recommenders. Moreover, existing trust inference path search approaches do not accumulate and reuse search experience, so they cannot reduce the path search costs even for repeated path search requests. It also inspires us to utilize the ACO to solve this issue.

The main contribution of this paper includes a trust community detection method and an ACO based Trust Inference Path Search algorithm (ACO-TIPS). The proposed trust community detection method utilizes Singular Value Decomposition (SVD) signs method [2] to process the trust relationship matrix to detect communities in the trust network and label them as colonies. For a given pair of source and target users, the ACO utilizes the pheromones of colonies close to the source user as experience information and the distances between the candidates and the target user in the singular vector space as heuristic information to find the optimal or near-optimal trust inference path solutions.

2   Related Work

Classical brute-force graph search based approaches are the mainstream in the field of the trust inference path search. Jøsang et al. [7] used the Depth First Search algorithm to find all the possible paths connecting the source and target participants in the trust network. Similarly, Hang et al. [5] proposed CertProp with three search strategies (*shortest, fixed, selection*) to find the paths for trust evaluation. The trust inference requires to find the best path connecting to each witness. Although the search algorithm is not detailed, the Depth First Search algorithm is obviously preferred to find all the possible paths. TidalTrust [4] utilizes a modified Breadth First Search to first find the trust inference path with the minimum depth and continue to find any other paths at the minimum depth. The trust inference paths with the maximum strength will be used in the calculation for inferring trust. In [13], the Breadth First Search algorithm is also used to find the trust propagation paths within the minimum depth for further trust evaluations. Ma et al. [11] proposed a bidirectional path search approach based on Dijkstra's algorithm to find the trust inference path with the minimum deduced uncertainty.

There are also stochastic trust inference path search approaches. TrustWalker [6] performs random walks on the trust network to solve the recommendation issues for cold start users. Repeated random walks take into account both the trust values of the neighbors and the similarities between the target item and the items rated by the neighbors. Thus, it makes a good combination of trust based and collaborative filtering based recommendation. Liu et al. [10] modeled the optimal social trust path selection as the classical Multi-Constrained Optimal Path (MCOP) selection problem and proposed the Heuristic Social Context-Aware trust Network discovery algorithm (H-SCAN) based on the K-Best-First Search. This method shows better performance than Time-To-Live Breadth First Search, Random Walk Search and High Degree Search.

The brute-force search based approaches are computation costly and the path search experience cannot be accumulated and reused for all the methods mentioned above. So, given a pair of source and target participants, repeated requests for the trust path between them will lead to repeated path searches at the same or similar computation cost, unless the previous search results are saved. Obviously, it is infeasible to save such paths for the dynamic large-scale trust networks.

3 Proposed Trust Inference Path Search Approach

3.1 Trust Communities Detection

The trust network can be formally described by a directed graph $G = < V, E >$, where V represents the set of participants and E represents the set of trust relationships. Binary trust relationships (i.e. trust and distrust relationships) are considered in this paper. So, $\forall e_{v_1 \to v_2} \in E, \exists | s_{e_{v_1 \to v_2}} \in \{1, -1\}$ and $v_1, v_2 \in V$. Here $s_{e_{v_1 \to v_2}}$ is the sign of the trust relationship, $s_{e_{v_1 \to v_2}} = 1$ means v_1 trusts v_2, $s_{e_{v_1 \to v_2}} = -1$ means v_1 distrusts v_2.

The trust relationship matrix noted by $T_{|V| \times |V|} = (t_{ij})$ is a $|V| \times |V|$ sparse matrix, where $t_{ij} = s_{e_{v_i \to v_j}}$ and $1 \leq i, j \leq |V|$. This matrix is different with the adjacent matrix because it contains the -1 elements. So it not only describes how participants connect with others but also shows their opinions on trust worthiness. By decomposing this matrix with truncated SVD, we can cluster the participants by how they trust and distrust others and/or how they are trusted and distrusted by others with less dimensions. The decomposed trust relationship matrix with rank k can be represented by:

$$T'_{|V| \times |V|} = U_{|V| \times k} S_{k \times k} V^T_{|V| \times k} \tag{1}$$

Here T' is the best possible rank k approximation to T and $k < rank(T)$. The value of k can be chosen by plotting the descending ordered singular values of T and finding the turning point of the line. Thus, the entries of S are the k dominant singular values and the rows of U and V can be regarded as the coordinates of the participants in the k dimensional spaces.

The *SVD signs* [2] is a clustering method which makes the singular value deposition of the adjacent matrix of the undirected graph and uses the sign patterns of the singular vectors to cluster the entries. In this paper, we apply this method to process the trust relationship matrix T so as to detect trust communities in the trust network. Since the matrix T is asymmetric, clustering methods by rows of U or V have different meanings. If the rows of U that have the same sign patterns on the k dimensions are classified into one cluster, this may lead to up to 2^k clusters. It clusters the participants by how they trust and distrust others. Similarly, the sign patterns of the rows of V are also applicable and this clusters the participants by how they are trusted and distrusted by others.

For examples shown in Fig.1, trust and distrust relationships are distinguished as solid and dotted arrows. The left example shows that A and B both trust V_1, V_3, V_n and distrust V_2, V_{n-1}, and they are probably classified into the same cluster by using rows of U. In the example on the right side of Fig.1, A and B are both trusted by V_1, V_3, V_n and distrusted by V_2, V_{n-1}. Thus, they are probably classified into the same cluster by using rows of V. For simplicity, we only use the sign patterns of the rows of U and ignore the rows of V. Given a pair of participants who have the similar trusted and distrusted participants, they may be classified into the same cluster, or to say, colony.

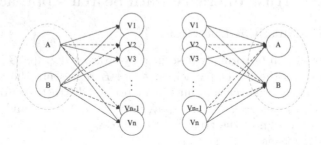

Fig. 1. The meanings of clusterings by rows of U or V

The instance in Fig.2 illustrates that, by building the trust relationship matrix and set $k = 2$, the ten vertices in the trust network are classified into three clusters according to the sign patterns $((-, +), (+, -)$ and $(-, -))$ of the rows of $U_{10 \times 2}$ (on the right side of Fig.2). In this clustering result, the trust relationships lie between vertices in the same cluster and the distrust relationships lie between vertices in different clusters, which satisfies the expectation of ideal trust community. Thus, the SVD sign based trust community detection method is feasible.

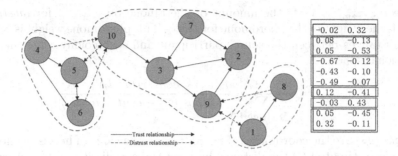

-0.02	0.32
0.08	-0.13
0.05	-0.53
-0.67	-0.12
-0.43	-0.10
-0.49	-0.07
0.12	-0.41
-0.03	0.43
0.05	-0.45
0.32	-0.11

Fig. 2. SVD sign based trust community detection

3.2 ACO with Multiple Colonies

Following the subjective logic proposed by Jøsang et al. [8], we infer the trust relationship of a pair of unfamiliar participants by applying the trust inference path (e.g. $v_1 \rightarrow v_2 \rightarrow \ldots \rightarrow v_n$) with trust discounting operators $\omega_n^{1:\ldots:n-1} = \omega_2^1 \otimes \omega_3^2 \otimes \ldots \otimes \omega_n^{n-1}$, where ω is the *subjective opinion* and $\otimes$ denotes the *trust discounting operator*. The Uncertainty Favoring Discounting operator (noted as $\otimes_1$) and Opposite Belief Favoring Discounting (noted as $\otimes_2$) operator are also introduced in [8]. Here, only the last hop of the path is on *functional trust* and the former ones are on *referral trust*. Modified ACO is utilized in this paper to find appropriate trust inference paths.

Transition possibilities of ants. With each participant labeled by a unique colony identifier, we can obtain the coordinates of the colony centers in the k dimensional singular vector space. Given a pair of source and target participants v_S and v_T, the corresponding colony labels $label(v_S)$ and $label(v_T)$ (belong to $\{label_i | 1 \le i \le |\{label_i\}|\}$), and the coordinates of the colony centers $center(label_i)$. For each round of ACO, numbers of ants perform random walks from the source vertex. Each ant chooses its next hop by computing the transition possibilities for the successors of the current vertex. Without loss of generality, given the current vertex v_A and its successors $suc(v_A)$, the transition possibility for the successor v_B ($v_B \in suc(v_A)$) can be obtained by the following equation

$$p_{v_A \rightarrow v_B} = \frac{\tau_{v_A \rightarrow v_B}^{\alpha} \cdot \eta_{v_A \rightarrow v_B}^{\beta}}{\sum\limits_{v_i \in suc(v_A)} \tau_{v_A \rightarrow v_i}^{\alpha} \cdot \eta_{v_A \rightarrow v_i}^{\beta}} \qquad (2)$$

where $\tau_{v_A \rightarrow v_B}$ is the total amount of pheromone on $e_{v_A \rightarrow v_B}$ and $\eta_{v_A \rightarrow v_B}$ is the value of heuristic information. Parameters α and β determine the relative influence of the pheromone trails and the heuristic information. Since we consider multiple colonies, the pheromones of similar colonies are also utilized.

$$\tau_{v_A \rightarrow v_B} = \sum\limits_{1 \le i \le |\{label_i\}|} \psi_{(v_A \rightarrow v_B, label_i)} \cdot w_{label_i} \qquad (3)$$

Here $\psi_{(v_A \to v_B, label_i)}$ is the amount of pheromone on $e_{v_A \to v_B}$ for $label_i$ and w_{label_i} is the weight of the pheromone for $label_i$. This pheromone weight is related to the distances between the source participant and each colony center in the k dimensional space.

$$w_{label_i} = \frac{1/\|pos(v_S) - center(label_i)\|}{\sum\limits_{1 \le j \le |\{label_j\}|} 1/\|pos(v_S) - center(label_j)\|} \tag{4}$$

where $pos(v_S)$ are the coordinates of v_S. Moreover, the value of heuristic information is also estimated by the distance between the coordinates of the successors and the target participant.

$$\eta_{v_A \to v_B} = 1/\|pos(v_B) - pos(v_T)\| \tag{5}$$

After the computation of transition possibilities for the successors, a random hop can be determined and the current vertex is updated. If the current vertex is the target participant, the ant stops and the path is recorded for further path selection.

Selection of the optimal trust inference path. At the end of each round, all the found paths are compared by computing the deduced referral trust about the last recommender (i.e. the witness) as the quality of the path. Given a connecting path $path_i$ denoted as $[v_{(i,1)}, v_{(i,2)}, \ldots, v_{(i,n_i)}]$, where n_i is the number of vertices along $path_i$, $v_{(i,1)} = v_S$, $v_{(i,n_i)} = v_T$ and $1 \le i \le |\{path_i\}|$, the inferred v_S's opinion about the witness $v_{(i,n_i-1)}$ on referral trust would be $w_{(i,n_i-1)}^{(i,1):\ldots:(i,n_i-2)}$. The first element in the round bracket denotes the index of the path and the second element denotes the index of the vertex along this path.

For each round of the ACO search, the optimal trust inference path among all the connecting paths found in this round is the path $path_k$ with the maximum indirect referral belief $b_{(k,n_k-1)}^{(k,1):\ldots:(k,n_k-2)}$. The optimization problem can be formally described as to find $path_k$ that

$$b_{(k,n_k-1)}^{(k,1):\ldots:(k,n_k-2)} = \max_{path_i}\{b_{(i,n_i-1)}^{(i,1):\ldots:(i,n_i-2)}\} \tag{6}$$

where $1 \le k \le |\{path_i\}|$ and $b_{(i,n_i-1)}^{(i,1):\ldots:(i,n_i-2)}$ can be obtained by $\omega_{(i,n_i-1)}^{(i,1):\ldots:(i,n_i-2)} = \omega_{(i,2)}^{(i,1)} \otimes \omega_{(i,3)}^{(i,2)} \otimes \ldots \otimes \omega_{(i,n_i-1)}^{(i,n_i-2)}$ with all hops on referral trust.

After the determination of the optimal trust inference path, we update the pheromones for colony $label(v_S)$ along this path by the following equation and ignore the rest found paths.

$$\psi_{(v_A \to v_B, labelv_S)} = \rho \cdot \psi_{(v_A \to v_B, labelv_S)} + \frac{e}{1 - b_{(k,n_k-1)}^{(k,1):\ldots:(k,n_k-2)}} \tag{7}$$

where $e_{v_A \to v_B}$ is an arbitrary edge of $path_k$, ρ is the evaporation rate of pheromones and e is the reinforcement factor to enhance the pheromones of the best path since the start of the algorithm.

3.3   ACO Based Trust Inference Path Search Algorithm

The overall trust inference path search algorithm first clusters the users by the trust community detection method. Then, by taking the clustered users as different colonies, multiple rounds of ACO are performed to find the optimal or near-optimal trust inference path. The ACO-TIPS algorithm can improve the performance of each round of search gradually by utilizing, releasing and updating pheromones. In order to simplify the trust inference path search problem, only one optimal trust propagation path is found for each search without considering the fusions of multiple paths. The detailed algorithm is described in the Algorithm 1.

Algorithm 1 ACO based Trust inference path search algorithm

Require: Trust network $G = (V, E)$, source and target participants v_S, v_T.
Ensure: $path = [v_S, \ldots, v_T]$ {if the path does not exist, it returns null}.

1. Get the trust relationship matrix $T_{|V| \times |V|}$ and its singular values.
2. Determine k, make the truncated SVD of $T_{|V| \times |V|}$ and cluster the participants by the SVD sign method.
3. **while** search round $round \leq$ the maximum $round_{max}$ **do**
4.   Set the current vertex $v_c(ant_i)$ to v_S for each ant_i.
5.   **while** current path depth $depth \leq 7$ **do**
6.   **for** $ant_i \in \{ant_i | ant_i.state == active\}$ **do**
7.   **if** $v_c(ant_i)$ has no successors **then**
8. $ant_i.state \leftarrow inactive$
9.   **end if**
10.   Compute the transitive possibilities $p_{v_c(ant_i) \to v_s(ant_i)}$ by Eq.(2) where $v_s(ant_i)$ belongs to the successors of $v_c(ant_i)$.
11.   Scale the possibilities of the edges without pheromones by the exploring factor θ and the ones with pheromones by $1 - \theta$.
12.   Choose one successor $v_s(ant_i)$ randomly and replace the current vertex $v_c(ant_i) \leftarrow v_s(ant_i)$.
13.   **if** $v_c(ant_i) == v_T$ **then**
14. $ant_i.state \leftarrow inactive$, and $ant_i.found \leftarrow success$
15.   **end if**
16.   **end for**
17. $depth \leftarrow depth + 1$
18.   **end while**
19.   Assemble the paths $\{path_i\}$ passed by ant_i where $ant_i.found == success$ and determine the optimal $path_k$ by Eq.(6).
20.   **if** $b_{(k,n_k-1)}^{(k,1):\ldots:(k,n_k-2)} > b_{best}(v_S, v_T)$ **then**
21. $b_{best}(v_S, v_T) \leftarrow b_{(k,n_k-1)}^{(k,1):\ldots:(k,n_k-2)}$, and $e \leftarrow e_{strength}$
22.   **end if**
23.   Update $\psi_{(v_A \to v_B, label_{v_S})}$ by Eq.(7) where $e_{v_A \to v_B}$ belongs to the edges of $path_k$.
24. $round = round + 1$
25. **end while**
26. **return** $path_k$

In this algorithm, we address the *exploit-vs-explore* dilemma by introducing an exploring factor θ $(0 < \theta < 1)$. When computing the the transitive possibilities, the edges without pheromones share the possibility that equals to θ and the ones with pheromones share the possibility that equals to $1 - \theta$. This mechanism is disabled when all the edges from the current vertex do or do not have pheromones. It protects the algorithm from the premature convergence at the initial rounds of searches and makes the subsequent searches able to find better paths. Moreover, the found path with the maximum deduced referral belief since the start of algorithm is rewarded by a strengthened reinforcement factor $e_{strength}$. This can help the pheromones of the global best found path avoid to be submerged. Generally, the number of ants starts with a great number and then decreases gradually. After several rounds of trust inference path searches, it can find the trust inference path with high quality by one round of search with a small number of ants.

If we denote the number of ants as m and the maximum path depth as d (in this paper $d = 7$), in the worst case, the times of vertex scan would be $m \cdot (d-1)$ in one round and the time complexity for one round of search is $O(m)$.

4   Experiments and Analysis

In this section, experiments are carried out on the Epinions data set to compare the performance of the proposed ACO-TIPS approach with the representative TidalTrust[4], CertProp(Sel.)[5] and H-SCAN[10] approaches on the path search efficiency and the applicability to the trust inference.

4.1   Data Set Description

Epinions is a consumer reviews web site that helps people make informed buying decisions by valuable consumer insight and personalized recommendations. The extended Epinions data set released by [12] is available at trustlet.org which describes the trust and distrust relationships among users and their ratings on other user's articles. The sampling method based on random walk introduced in [9] is utilized to scale down the original data set and the data set for experiments contains 33036 users who issued 84141 trust and distrust statements.

4.2   Methodology and Metrics

First, the subjective opinion between arbitrary pair of users should be obtained. Given a pair of users (v_A and v_B), we count the number of the source user's ratings on target user's articles as the number of observations (noted as $n_{rating} \geq 0$) and get the mean rating (noted as $m_{rating} \in [1,5]$). Then, $\omega_B^A = (b_B^A, d_B^A, u_B^A, a_B^A)$ can be obtained by:

$$\begin{cases} b_B^A = (m_{rating} - 1) \cdot (1 - u_B^A)/4 \\ d_B^A = (5 - m_{rating}) \cdot (1 - u_B^A)/4 \\ u_B^A = 2/(2 + n_{rating}) \\ a_B^A = 0.5 \end{cases} \tag{8}$$

Before the trust inference path search, the users are clustered by the trust community detection method with the left singular vector. The trust relationship matrix is reordered so that the rows corresponding to the users within the same cluster are together and the same reordering is also applied to the columns. The reordered trust relationship matrix is plotted in Fig.3. The blue dots represent trust relationships and the red ones represent distrust relationships. The transverse lines are the borders of clusters. In this figure, we can find that there are clusters where the users mainly trust each other in the same cluster and overall distrust the users in some other clusters.

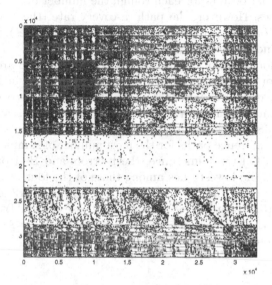

Fig. 3. Reordered trust relationship matrix after clustering

In order to validate the performance of ACO-TIPS in terms of trust inference path search efficiency and trust inference accuracy, we use a standard Leave-one-out evaluation technique with 500 randomly chosen sample user pairs. Before the trust inference path search starts, the original trust statement is masked. All the comparative path search approaches are performed to find the trust inference path connecting the source user and the target user. When the path searches terminate, the number of the found paths and the scanned vertices per path are recorded as the metrics for the trust inference path search efficiency. The approach with higher path discovery rate and lower scanned vertices per path shows better path search performance. Then, the trust inference is performed by applying the found trust inference paths with trust discount operators $\otimes_1$ and $\otimes_2$ respectively. The deduced subjective opinion is compared with the original subjective opinion obtained by the ground truth in terms of P-error and B-error introduced in [5] as the metrics for trust inference accuracy. The lower errors in trust inference reflect better applicability of the path to trust inference.

In order to give a computation bound in the experiments, the maximum number of scanned vertices is set to 5000 and the maximum depth of path is set to

7 according to the small world theory. For TidalTrust, CertProp(Sel.) and H-SCAN, repeated searches for the same user pair yield the same or similar results and costs, and thus they are performed only once. The proposed ACO-TIPS is performed in 40 rounds with decreased number of ants. We choose the mean performance of the last 5 rounds to make comparisons.

4.3   Results and Analysis

Experiment results of the ACO-TIPS are illustrated in Fig.4a and Fig.4b. As we decrease the number of ants for each round, the number of scanned vertices per path also decreases. However, the path discovery rate (the number of searches that find at least one path divided by the number of samples), the P-errors and B-errors for $\otimes_1$ and $\otimes_2$ are all floating at a stable level.

The mean performance of the last 5 rounds of ACO-TIPS are compared with the performance of TidalTrust, CertProp(Sel.) and H-SCAN in Table.1. On trust inference path search, ACO-TIPS reaches the highest discovery rate 63.56% (9.59% higher than TidalTrust's 58%) and the lowest mean scanned vertices per path is 735.538 (38.9% less than TidalTrust's 1204). It means that ACO-TIPS can find the trust inference paths for the most number of samples with the lowest mean search cost. On trust inference, the P-errors and B-errors of ACO-TIPS for $\otimes_1$ and $\otimes_2$ are the lowest ones among those of the four trust inference path search approaches (P-error1, B-error1, P-error2 and B-error2 of ACO-TIPS are 27.25%, 21.61%, 15.55% and 21.88% lower than those of TidalTrust respectively). This implies that the trust inference with the path found by ACO-TIPS reaches higher trust inference accuracy.

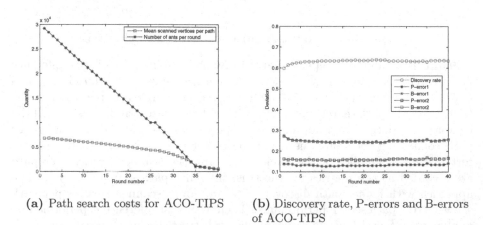

(a) Path search costs for ACO-TIPS   (b) Discovery rate, P-errors and B-errors of ACO-TIPS

Fig. 4. Experimental results of ACO-TIPS

Thanks to the pheromones released by the previous rounds of searches in ACO-TIPS, the subsequent rounds of searches can easily discover the trust inference paths with much less costs than previous ones. It implies that ACO-TIPS

Table 1. Performance comparisons of TidalTrust, CertProp(Sel.), H-SCAN and ACO-TIPS

Term	TidalTrust	CertProp(Sel.)	H-SCAN	ACO-TIPS
Discovery Rate	0.58	0.452	0.478	**0.6356**
Scanned vertices per path	1204	4963.02	793.2	**735.54**
P-error1	0.1868	0.3323	0.2109	**0.1359**
B-error1	0.3225	0.5696	0.3799	**0.2528**
P-error2	0.1936	0.3784	0.2392	**0.1635**
B-error2	0.3209	0.567	0.3785	**0.2507**

can accumulate and reuse former path search experience to speed up the path search. In ACO-TIPS, only the optimal path of each round is qualified to add pheromones. Furthermore, the optimal path since the start of the algorithm even gets additional pheromones. Therefore, ACO-TIPS tends to find better and better trust inference path after rounds of searches and make the trust inference more and more accurate until it finds the optimal path.

5   Conclusions and Future Work

In this paper, we propose a trust inference path search approach combining the trust community detection and the ACO with multiple colonies. The SVD sign clustering method is applied to detect trust communities according to how users trust and distrust others. Then, the trust communities are regarded as colonies in the ACO. By selecting the found path with the maximum deduced referral belief, the ACO-TIPS can accumulate and reuse path search experience to find the optimal or near-optimal trust inference path efficiently. This optimal path can connect the source user to the target user with the most trustworthy witness, which leads to accurate trust inference results.

Since the trust network in the experiments is static, the so-called "exploit" or "explore" period is transient. In fact, the evaporation of the pheromones and stochastic routing can make the ACO-TIPS applicable to the dynamic trust network, which needs to be further validated with suitable experiment environments. Also, the idea of accumulating and reusing experience in trust inference path search can be applied to the trust inference based recommendations for better accuracy and lower costs.

Acknowledgments. This research is funded by the National Nature Science Foundation of China (No. 61272406) and the Fundamental Research Funds for the Central Universities (HUST:2013TS101).

References

1. Dorigo, M., Stttzle, T.: Ant colony optimization: Overview and recent advances. In: Gendreau, M., Potvin, J.Y. (eds.) Handbook of Metaheuristics. International Series in Operations Research & Management Science, vol. 146, pp. 227–263. Springer, US (2010)
2. Douglas, E.P.: Computing and applying trust in web-based social networks. Master's thesis, College of Charleston (2008)
3. Fortunato, S.: Community detection in graphs. Physics Reports 486(3), 75–174 (2010)
4. Golbeck, J.A.: Computing and applying trust in web-based social networks. Ph.D. thesis, University of Maryland College Park (2005)
5. Hang, C., Wang, Y., Singh, M.P.: Operators for propagating trust and their evaluation in social networks. In: Proc. of the 8th Intl. Conf. on Autonomous Agents and Multiagent Systems, pp. 1025–1032 (2009)
6. Jamali, M., Ester, M.: Trustwalker: a random walk model for combining trust-based and item-based recommendation. In: Proc. of the 15th ACM SIGKDD Intl. Conf. on Knowledge Discovery and Data Mining, pp. 397–406. ACM (2009)
7. Jøsang, A., Hayward, R., Pope, S.: Trust network analysis with subjective logic. In: Proc. of the 29th Australasian Computer Science Conf., pp. 85–94. Australian Computer Society, Inc. (2006)
8. Jøsang, A., Marsh, S., Pope, S.: Exploring different types of trust propagation. In: Stølen, K., Winsborough, W.H., Martinelli, F., Massacci, F., et al. (eds.) iTrust 2006. LNCS, vol. 3986, pp. 179–192. Springer, Heidelberg (2006)
9. Leskovec, J., Faloutsos, C.: Sampling from large graphs. In: Proc. of the 12th ACM SIGKDD Intl. Conf. on Knowledge Discovery and Data Mining, pp. 631–636. ACM (2006)
10. Liu, G., Wang, Y., Orgun, M.A., Liu, H.: Discovering trust networks for the selection of trustworthy service providers in complex contextual social networks. In: Proc. of IEEE 19th Intl. Conf. on Web Services (ICWS 2012), pp. 384–391. IEEE (2012)
11. Ma, Y., Lu, H., Gan, Z.: Trust inference path search with minimum uncertainty for e-commerce. In: Proc. of the 10th Conf. on Web Information Systems and Applications, pp. 133–137. IEEE (2013)
12. Massa, P., Avesani, P.: Trust-aware bootstrapping of recommender systems. In: Proc. of ECAI Workshop on Recommender Systems, pp. 29–33. IOS (2006)
13. Shekarpour, S., Katebi, S.: Modeling and evaluation of trust with an extension in semantic web. Web Semantics: Science, Services and Agents on the World Wide Web 8(1), 26–36 (2010)

ExNa: An Efficient Search Pattern for Search Engines*

Xiao Wei[1], Xiangfeng Luo[1,*], Qing Li[2], and Jun Zhang[1]

[1] School of Computer Engineering and Science, Shanghai University, Shanghai, China
{xwei,luoxf,zhangjun_shu}@shu.edu.cn
[2] Department of Computer Science, City University of Hong Kong, Hong Kong
qing.li@cityu.edu.hk

Abstract. As the Web enters Big Data age, users and search engines may find it more and more difficult to effectively use and manage such big data. To solve this problem, this paper presents a novel search pattern named ExNa by defining its model and basic operations in detail, which can help to design new type of search engines. To validate the ExNa search pattern, we develop a prototype news search engine named KNOWLE, which shows that KNOWLE equipped with ExNa can improve the efficiency of search system.

Keywords: Search pattern, Index structure, Interaction process, User profile.

1   Introduction

As the Web enters Big Data age, users and search engines may find it more and more difficult to effectively use and manage such big data. New types of search engines are emerging to solve the problem [1] and different search engines serve users in different modes. A search engine mainly consists of three parts: index generation, index, and search service. Each part is contributed to the efficiency of the entire search engine system. We adopt and adapt the term Search Pattern to depict the components of a search system that are directly related to the search process.

2   ExNa Search Pattern

Definition 1. Search Pattern (SP) A search pattern, *SP*, is a model to describe the components of a search system that are directly related to the search process. It is denoted as a triple $SP = \langle RI, UP, IM \rangle$, where *RI* is the index of resources, *UP* denotes user profiles, and *IM* is the interaction mechanism between users and the index of resources.

Based on the definition of SP, we propose a new search pattern named ExNa, which is introduced from the following three aspects: Index Structure, User profile, and Interactive Mechanism.

---

* Research work reported in this paper was jointly supported by a Strategic Grant from City University of Hong Kong (Project No. 7002912), the National Science Foundation of China under grant no. 61071110, and the key project of shanghai municipal education commission under grant no. 13ZZ064.
* Corresponding author.

F. Li et al. (Eds.): WAIM 2014, LNCS 8485, pp. 699–702, 2014.
© Springer International Publishing Switzerland 2014

2.1   The Multi-layered Semantic Link Network Index Structure

ExNa requires a multi-layered network index structure, which should have rich semantic relations, high efficient storage, and free-styled interaction with users.

Definition 2: A Multi-layered Semantic Link Network Index Structure, MNIS, is defined as $MNIS = \langle G, S_L, S_V, R, M \rangle$, where $G = \{G_i \mid G_i = \{item_j\}\}$ is the set of all the index items in the system, $G_i = \{item_j\}$ is the set of index items which belongs to the ith layer, $item_j$ is the jth index item in G_i ; $S_L = \{SR_i\}$ is the set of all kinds of semantic relations among the index items in G, $SR_i = \{SLN_j\}$ is the set of different types of semantic relations at the ith layer, with each type of semantic relation being organized in a semantic link network and stored by an adjacent

matrix $SLN_j = \begin{pmatrix} sw_{11} & \cdots & sw_{1m} \\ \vdots & sw_{ij} & \vdots \\ sw_{m1} & \cdots & sw_{mn} \end{pmatrix}$, sw_{ij} is the weight of the semantic relation between index

$term_i$ and $term_j$; $S_v = \{< term_i, term_j > \mid term_i \in G_i; term_j \in G_{i+1}\}$ is the set of semantic mappings between two adjacent layers in the vertical direction; $R = \{res_j\}$ is the set of indexed resources, res_j is the jth indexed resource; $M = \{< term_i, res_j > \mid term_i \in G, res_j \in R\}$ is the set of mappings from the index items to resources.

The MNIS provides the basic framework to index web resources into a semantic-rich form, while the final realizations of MNIS may be different according to the selections of different semantic granularities and dimensions.

2.2   User Profile

A user profile records a user's interests and is used to provide personalized services. ExNa supports multiple levels of user profiles to meet the needs of different search engines. Here we define a user profile supporting four levels as $UP = \langle \mathbb{I}, \mathbb{R}, \mathbb{M} \rangle$, where $\mathbb{I}$ records the interests at different semantic granularities and it is defined as $\mathbb{I} = \{I_i \mid i \in N \wedge 0 < i \triangleleft |G|\}$, G is the semantic granularity, $|G|$ is the number of granularities, I_i records the interests at the ith semantic granularity G_i and is defined as $I_i = \{< node_j, in > \mid node_j \in G_i\}$, $node$ is the basic unit in different semantic granularities, in is the interest degree of $node_j$; $\mathbb{R}$ records the semantic relations among user's interests in different semantic granularities and is defined as $\mathbb{R} = \{R_i \mid i \in N \wedge 0 < i \triangleleft |G|\}$, R_i records the semantic relations in granularity G_i and is defined as $R_i = \{< node_k, node_j, \omega > \mid node_k \in I_i, node_j \in I_i\}$, the triple $< node_k, node_j, \omega >$ denotes the semantic relation between $node_i$ and $node_j$, with ω being the weight of the semantic relation; $\mathbb{M} = \{big\ granularity,\ small\ granularity,\ top\text{-}down,\ bottom\text{-}up,\ up\text{-}and\text{-}down,\ zigzag,...\}$, $\mathbb{M}$ is the set of search modes which records how a user uses a search engine. Everyone has his/her fixed search pattern when he/she uses the search engine for a long time. These patterns are both the modes of using search engine and the mode of acquiring information, which can be summarized from the interaction processes. The four levels of UP can be gotten when some constraints are added on UP, which are listed as follows.

1) UP1: UP1's constraints are $|\mathbb{I}|=1, \mathbb{R}=\phi, \mathbb{M}=\phi$, which means UP1 only records users' independent interests in one semantic granularity.

2) UP2: UP2's constraints are $|\mathbb{I}|>1, \mathbb{R}=\phi, \mathbb{M}=\phi$, which means UP2 records users' independent interests in different semantic granularities and forms the mutil-layers of user interests.

3) UP3: UP3's constraints are $|\mathbb{I}|>1, |\mathbb{R}|>1, \mathbb{M}=\phi$, which means UP3 records users' interests in different semantic granularities, and the interests of each granularity are organized in a semantic link network. UP3 is the mutil-layered network structure of interests.

4) UP4: UP4's constraints are $|\mathbb{I}|>1, |\mathbb{R}|>1, \mathbb{M}\neq\phi$, which means UP4 adding the search modes on UP3.

2.3   Interaction Mechanism

Interaction mechanism provides free-styled interaction with MNIS for user to acquire their desired information efficiently. A user can change his interests in different semantic granularities, different semantic dimensions, and different nodes in semantic link network. During the interaction, the static interests can be recorded directly into $UP_1 \cdots UP_3$, while the dynamic interests can be summarized as search modes to be updated to UP_4. ExNa provides the following interaction mechanisms (IMs):

(1) Narrow $\mathbb{N}^-$: $\mathbb{N}^-$ is an operation to switch the focus from a coarse semantic granularity to a finer-grained one in MNIS, defined as $\mathbb{N}^-(r_0)=\{r_c | r_c \text{ is the child of } r_0\}\oplus UP_2$, where r_0 is the currently focused index item, $\{r_c\}$ is the set of children nodes of r_0 at the lower layer. $\oplus$ is a filter operator to sort r_0 by UP_2 and select the top-n from the sorted results.

(2) Broaden $\mathbb{N}^+$: $\mathbb{N}^+$ is the inverse operation of $\mathbb{N}^-$, which means increasing the granularity and shift the focus to a higher level. Generally, the multi-granularity index is organized as a tree structure, in which each node has only one parent node. So $\mathbb{N}^+$ is defined without the filter operation $\oplus$ as follows: $\mathbb{N}^+(r_0)=\{r_p | r_p \text{ is the parent of } r_0\}$.

(3) Expand $\mathbb{E}$: $\mathbb{E}$ is a semantic expand operation to support user travelling in a semantic link network at a certain semantic granularity. The semantic expand in the jth dimension, SLN_j , from node R_0 is defined as $\mathbb{E}(r_0, SLN_j)=\{r_n | (\exists r_0 \to r_n \in SLN_j) \wedge (sw_{r_0 \to r_n} > \alpha)\}\oplus UP_3$, where $\{r_n\}$ is the set of neighbors of r_0 in SLN_j. All the weights of the semantic relations from r_0 to r_n must be larger than a threshold α. The meaning of $\oplus$ is the same to the one in $\mathbb{N}^+$, $\mathbb{E}(R_0, SLN_j)$ is the final result after getting filtered by UP_3.

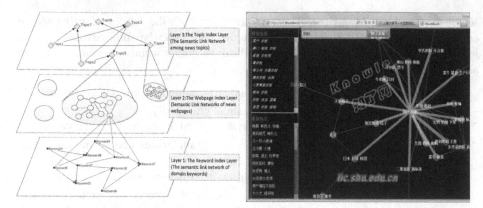

Fig. 1. The Multi-layered Semantic Link Network Index Structure of KNOWLE

Fig. 2. The prototype of ExNa: KNOWLE (http://wkf.shu.edu.cn/aln/aln.html)

(4) Pattern Switch $\mathbb{X}$ **:** This operator is to change the search mode from Narrow to Expand or from Expand to Narrow. The pattern switch happens when the search mode changes actively or passively. Active change is done by user, while passive change is done by the search engine according to the mode $\mathbb{M}$ in UP_4. $\mathbb{X}$ is defined

as $\mathbb{X} = \begin{cases} N^+/N^- \Rightarrow E \\ E \Rightarrow N^+/N^- \end{cases}$, where $N^+/N^- \Rightarrow E$ and $E \Rightarrow N^+/N^-$ mean the interactive direction changes from vertical to horizontal and from horizontal to vertical respectively.

3 Prototype Search Engine of ExNa

To validate the proposed search pattern ExNa, we develop a prototype search engine using ExNa named KNOWLE [2], which is a Chinese news search system. KNOWLE gathers web news from major news websites in China, clusters them into news topics, and organizes them into the multi-layered Semantic Link Network Index Structure, as shown in Fig.1. KNOWLE provides a free-styled interaction mechanism for users to acquire their desired information efficiently and the interface of KNOWLE is shown in Fig.2. The experiments on KNOWLE show that ExNa is an efficient search pattern to improve the efficiency of search engine.

References

1. ICSTI Insight: Next Generation Search, http://www.icsti.org/IMG/pdf/insight_2010_july.pdf
2. Wei, X., Luo, X., Li, Q., Zhang, J.: KNOWLE: Searching News in the Search Pattern of Knowledge Flow. In: Lin, X., Manolopoulos, Y., Srivastava, D., Huang, G. (eds.) WISE 2013, Part II. LNCS, vol. 8181, pp. 527–530. Springer, Heidelberg (2013)

A Correlation-Based Semantic Model
for Text Search*

Jing Sun, Bin Wang, and Xiaochun Yang

College of Information Science and Engineering,
Northeastern University, Liaoning, 110819, China
jsunneu@gmail.com, {binwang,yangxc}@mail.neu.edu.cn

Abstract. With the exponential growth of texts on the Internet, text
search is considered a crucial problem in many fields. Most of the traditional text search approaches are based on "bag of words" text representation based on frequency statics. However, these approaches ignore the
semantic correlation of words in the text. So this may lead to inaccurate
ranking of the search results. In this paper, we propose a new Wikipedia-
based similar text search approach that the words in the texts and query
text could be semantic correlated in Wikipedia. We propose a new text
representation model and a new text similarity metric. Finally, the experiments on the real dataset demonstrate the high precision, recall and
efficiency of our approach.

Keywords: text search, Wikipedia, semantic correlation.

1 Introduction

In recent years, all kinds of information are in rapid expansion and the texts on
the Internet have grown exponentially. The text similarity computing and text
search have become important information retrieval technologies.

Text search includes two main parts: text representation model and text similarity metric. Currently, there are many common text representation models
[4,1,5]. However, all these methods do not consider the semantic correlation of
the words in the text.

There are also many text similarity metrics [3,6]. Traditional text similarity
metrics use BOW (bag of words) method which gets the words from text and
statistics the frequency of each word appears in the text (such as TF-IDF). The
vector space model (VSM) [4] is generally used to represent text, and then the
Cosine similarity is applied to calculate the similarity between texts. However,
traditional BOW-based text search methods have some disadvantages: first, the
traditional methods consider that the words in the text are independent, ignore

---

* The work is partially supported by the National Natural Science Foundation of China
(Nos. 61322208, 61272178, 61129002), the Doctoral Fund of Ministry of Education of
China (No. 20110042110028), and the Fundamental Research Funds for the Central
Universities (No. N120504001, N110804002).

F. Li et al. (Eds.): WAIM 2014, LNCS 8485, pp. 703–706, 2014.

the correlation among the words; second, if two texts use two different sets of words to describe the same topic, traditional methods would get a small similarity, which cannot indicate the real text similarity.

The problem is that given a collection of texts D, a query text q, and a knowledge base, find a good semantic text search model, including semantic-based text representation and text semantic similarity metric, which can represent the texts semantically and search the texts matching with the query accurately.

In this paper, we propose a new similar text search method based on Wikipedia, which considers the semantic correlation among the words in the text.

2   A Correlation-Based Semantic Model

This paper presents a new correlation-based semantic text similarity model, which considers not only the word frequency in the text vector, but also considers the semantic correlation among words.

2.1   Framework of Correlation-Based Semantic Text Search

The framework of our method for leveraging semantic correlation among the words in the texts by Wikipedia to improve text search is presented in Fig. 1.

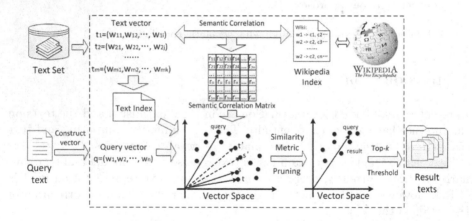

Fig. 1. The framework of correlation-based semantic text search

Firstly, we transform the texts to vector space representation (we use 2-dimension to illustrate). As the words in the texts could be semantic connected in Wikipedia, we use Wikipedia to compute the semantic correlation among the words in the texts and the query text. Then we can get a semantic correlation matrix which records the semantic correlations between the text and the query text. Using the semantic correlation-based text Similarity metric, we could compute the similarity among text vectors. Finally, we can get semantic-based text search results.

2.2   Semantic Correlation among Words Exploiting Wikipedia

Given two words w_1 and w_2, let W_1 and W_2 be the sets of words that w_1 and can w_2 connect to in Wikipedia, respectively. Let $Category(w_1)$ and $Category(w_2)$ be the sets of categories that w_1 and w_2 belong to in Wikipedia. We use the following formula to compute the semantic correlation between w_1 and w_2.

$$correlation(w_1, w_2) = (\frac{|W_1 \cap W_2|}{|W_1 \cup W_2|} + \frac{|Category(w_1) \cap Category(w_2)|}{|Category(w_1) \cup Category(w_2)|})/2. \quad (1)$$

As each text is a collection of words, the semantic correlation of two texts can be represented by a semantic correlation matrix M. M is a m×m matrix. We use r_{ji} to indicate the semantic correlation between w_i and w_j, m to indicate the total number of words in these two text.

2.3   Correlation-Based Semantic Text Similarity Model

In this section, we introduce our new correlation-based semantic text similarity model (CSM). Given two texts s and t, a semantic correlation metric M, the semantic correlation-based text similarity metric is define as follows,

$$sim(s, t) = sccos(s, t, M) = \frac{sMt}{|s||t|} \quad (2)$$

The sM is equivalent to s multiplying a weight matrix M. The weight matrix M associates with both s and t.

3   Experiments

We have conducted experiments on NYTimes news articles real dataset which includes 300,000 documents, and the number of words in the vocabulary is 102,660. We compared our CSM approach with the TF-IDF-based Cosine (COS) and Wikepedia enhances method (WEM)[2]. We manually constructed five queries $(Q_1, Q_2, ..., Q_5)$.

Fig. 2 shows the effectiveness of precision and recall when k=10.

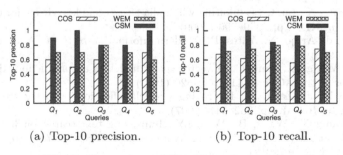

(a) Top-10 precision.   (b) Top-10 recall.

Fig. 2. Comparison of Top-10 precison and recall

For these five queries, WEM is a little better than Cosine method both on top-k precision and recall. But WEM sometimes has poor precision, it is because WEM introduces the concepts in the Wikipedia, this may bring noise and ambiguous information. Our CSM approach performs best. Our approach CSM approach can achieve more than 0.8 top-10 precisions for these five queries. We could get both high top-10 precision and top-10 recall.

Fig. 3 demonstrates the precision and recall with different k. As shown in

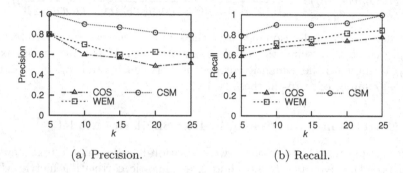

(a) Precision.   (b) Recall.

Fig. 3. Precision, recall vary k

Fig. 3(a), with the change of k, our CSM approach is better than COS and WEM. We can see that with different k, CSM can get high precision and recall. Our CSM approach can achieve more than 0.8 precisions and recalls.

We can also get similar results on the precision and recall with different *threshold* query.

4   Conclusion

In this paper, we study semantic-based similar text search. We propose a novel semantic correlation-based text search approach, which considers the Wikipedia-based semantic correlation among the words in the texts and the query text.

References

1. Hotho, A., Staab, S., Stummme, G.: Wordnet inproves text doucument clustering. In: SIGIR, pp. 143–152 (2003)
2. Hu, X., Zhang, X., Lu, C.: Exploiting wikipedia as external knowledge for document clustering. In: KDD, pp. 389–396 (2009)
3. Ribeiro, B., de Arajo, N., Yates, B.: Modern information retrieval. Addison-Wesley Longman (1999)
4. Salton, G., Wong, A., Yang, C.: A vector space model for automatic indexing. Communications of the ACM 18(11), 613–620 (1957)
5. Wang, P., Hu, J., Zeng, H.: Improving text classification by using encyclopedia knowledge. In: ICDM, pp. 332–341 (2007)
6. Zhu, H., Yang, X., Wang, B., Wang, Y.: Improving text search on hybrid data. In: Bao, Z., et al. (eds.) WAIM 2012 Workshops. LNCS, vol. 7419, pp. 192–203. Springer, Heidelberg (2012)

Finding Relatedness between Research Papers
Using Similarity and Dissimilarity Scores

Qamar Mahmood, Muhammad Abdul Qadir, and Muhammad Tanvir Afzal

Center for Distributed and Semantic Computing
Mohammad Ali Jinnah University, Islamabad, Pakistan
{qamar,aqadir,mafzal}@jinnah.edu.pk

Abstract. Identification of relatedness between research papers is very important in recommender systems, information retrieval, and document categorizing etc. The state-of-the-art approaches compute relatedness using similarity functions by exploiting terms from content and metadata, however, dissimilarity between research papers is not used to compute the relatedness. This research proposes an approach for finding relatedness based on similarity as well as dissimilarity based characteristics of document's content. The relatedness between documents has been computed by combining and normalizing similarity and dissimilarity. It was found that results of our experiment are encouragingly comparable to other content based similarity measuring techniques.

Keywords: Content Based Similarity, Document Relatedness, Information Retrieval, Synonym Based Similarity.

1   Introduction and Related Work

The problem which we need to solve is that we want to identify related documents by using similarity as well as dissimilarity factors from a given set of documents. Similarity and dissimilarity factors are related to contents of research papers: such as terms involved in text (title and abstract) of paper and synonyms of these terms. Similarity can be defined on basis of concept discussed by Lin [1], according to which similarity between two documents can be represented on basis of three factors: commonality, difference and identity. Aslam et al. [2] have worked on pair wise object similarity using this information theoretic approach of similarity.

Problem of detecting related documents is helpful in finding the documents which are similar to a document specified in a query provided by some user in an information retrieval system such as a citation indexer. Content based similarity techniques focus on textual contents of documents and are further categorized as word based [3] and sentence based [4, 5, 7] similarities. Different methods used for content based similarity are Euclidean distance, cosine similarity, relative entropy, Jaccard coefficient and Pearson correlation coefficient which are used by Huang [3]. Lee et al. [6] evaluated different content based similarity measures with human way of finding similarity.

F. Li et al. (Eds.): WAIM 2014, LNCS 8485, pp. 707–710, 2014.
© Springer International Publishing Switzerland 2014

Our proposed technique is different from content based similarity techniques in a way that it uses similarity as well as dissimilarity factor for finding relatedness between the documents.

Key components of our proposed approach to find the relatedness between documents using pair wise similarity are described below:

1. Contents of research papers such as title and abstract of research papers were used.
2. Keywords/terms from contents of research papers were extracted.
3. Terms/keywords with exact match were found from two papers in comparison.
4. Synonyms were found for each keyword/term extracted from research papers in comparison.
5. Set of synonyms which were common to two research papers in comparison found.
6. Difference sets of synonyms for two research papers in comparison were found.
7. All these measures were normalized to compute a relatedness measure between two research papers in comparison.

We have found that results of our technique were encouraging when compared to those of previously available content based similarity measuring techniques.

2   Proposed System Architecture

Our proposed technique's system architecture is shown in Fig. 1, according to which PaperA and PaperB represent two documents for which we need to compute document relatedness. Term extractor is used to extract terms from titles and abstracts of PaperA and PaperB. These terms are named as TermSetA and TermSetB.

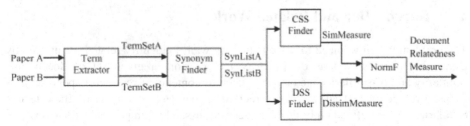

Fig. 1. Architecture Diagram of Proposed System

Synonym Finder is finding synonyms for each of terms in TermSetA and TermSetB by using a publicly available synonyms dataset, list of these synonyms are labeled as SynListA and SynListB respectively. CSS (Common Synonym Set) Finder and DSS (Difference Synonym Set) Finder is taking both SynListA and SynListB as input and provides SimMeasure (Similarity Measure) and DissimMeasure (Dissimilarity Measure) as output. These measures are passed to a normalization function NormF which provides document relatedness measure between PaperA and PaperB.

3 Results and Discussion

We have applied our proposed technique on research papers available in data set of DBLP. Results of proposed system were compared with document relatedness measures computed by human rators. These human rators were provided with a pre defined document relatedness rating criteria to provide a document relatedness measure in form of numeric values between 0 and 1, as shown in Table 1.

Table 1. Document relatedness rating criteria recommended for human rators

	Not Closely Similar	Below Average	Near Average	Above Average	Highly Above Average	Near the Identity
Recommended Measure Score	0 to 0.25	0.25 to 0.35	0.35 to 0.5	0.5 to 0.75	0.75 to 0.9	0.9 to 1.0

Results computed by human rators for research papers used in our experiment were compared with results of our proposed technique. Fig. 2 shows these comparisons which were very encouraging for proposed technique.

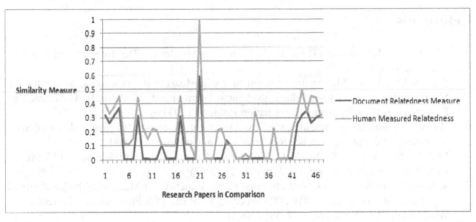

Fig. 2. Results of proposed system as compared with human rators assessments of relatedness measures

As is obvious from Fig. 2 that document relatedness computed by using proposed technique had some differences from the measures computed by human rators. Table 2 provided information about comparison of different content based similarity techniques [3] with each other as well as with our proposed technique. It was found that our proposed technique was producing results which were comparable with other techniques, specifically KLD (Kullback-Leibler Divergence).

Table 2. Comparison of proposed technique with other content based similarity measures

	Euclidean	Cosine	Jaccard	Pearson	KLD	Proposed Technique
Accuracy	0.43	0.55	0.67	0.49	0.7	0.72
Precision	0.65	0.63	0.75	0.77	0.66	0.78
Recall	0.45	0.23	0.35	0.28	0.44	0.39

4   Conclusions and Future Work

We have proposed a new approach for finding relatedness between the document by using similarity and dissimilarity factors in a combined way which represents novelty of our idea. We have found that our technique was encouragingly comparable to other previously available content based similarity measuring techniques. There was some limitation in our technique that it was based on word level analysis of text of research papers in comparison, not doing anything with sentence structure.

In future, we are planning to work on identification of relationships between words occurring in a sentence or paragraph using co-occurrence relationships to improve the similarity measure accuracy of our technique.

References

1. Lin, D.: An Information-Theoretic Definition of Similarity. The International Machine Learning Society (1998)
2. Aslam, J.A., Frost, M.: An Information theoretic Measure for Document Similarity. In: SIGIR 2003: Proceedings of the 26th Annual International ACM SIGIR Conference on Research and Development in Information Retrieval (2003)
3. Huang, A.: Similarity Measures for Text Document Clustering. In: Proceedings of Sixth New Zealand Computer Science Research Student Conference (2008)
4. Metzler, D., Dumais, S., Meek, C.: Similarity Measures for Short Segments of Text. In: Advances in Information Retrieval, pp. 16–27 (2007)
5. Sahami, M., Heilman, T.D.: A Web-based Kernel Function for Measuring the Similarity of Short Text Snippets. In: WWW 2006 Proceedings of the 15th International Conference on World Wide Web Pages, pp. 377–386 (2006)
6. Lee, M.D., Pincombe, B.M., Welsh, M.B.: An empirical evaluation of models of text document similarity. In: Annual Conference, Cognitive Science Society, Stresa (2005)
7. Yih, W., Meek, C.: Improving Similarity Measures for Short Segments of Text. AAAI (Association for Advancement of Artificial Intelligence) (2007)

Context-Dependent Sentiment Classification
Using Antonym Pairs and Double Expansion

Zhifei Zhang[1], Duoqian Miao[1], and Bo Yuan[2]

[1] Department of Computer Science and Technology, Tongji University,
Shanghai 201804, P.R. China
[2] Department of Chinese Language and Literature, Tsinghua University,
Beijing 100084, P.R. China
zhifei.zzh@gmail.com, dqmiao@tongji.edu.cn,
yuanb10@mails.tsinghua.edu.cn

Abstract. Sentiment classification at word-level plays a fundamental role in many sentiment-related tasks. Context-free sentiment classification circumvents the essential factors in language use and is error prone due to oversimplification in its assumption. Context-dependent sentiment classification poses new challenges to researchers. We propose a novel approach to automatically determine the contextual polarity with the help of antonym pairs. First, neighboring nouns are extracted as context information within a predefined distance of sentiment words. Secondly, the polar posterior probabilities of sentiment words are derived based on Bayes' theorem. Finally, the polarity of one sentiment word with the context of one neighboring noun is determined by Bidirectional Rule and Unidirectional Rule. In addition, we define a new similarity measure, which combines semantic distance with edit distance, for double expansion, i.e., Context Expansion and Target Expansion. The experimental results on two real-world data sets validate the effectiveness of our approach.

Keywords: sentiment classification, context-dependent, antonym pairs, word similarity.

1 Introduction

For the past decade, the studies of sentiment classification in natural language processing, due to its extensive applications [1], have been increasingly drawing attention from the researchers around the world. The aim of sentiment classification is to identify the subjectivity of texts and recognize the polarity ("positive" or "negative").

Sentiment classification mainly falls into three levels: document-level, sentence-level and word-level, among which word-level sentiment classification serves as the basis for the other two. The polarity of some words remains stable despite the context variation while the polarity of others depends on the context. The former can be referred to as context-free words and the latter as context-dependent words. Sentiment classification of context-dependent words is more challenging in that it requires deeper and more thorough understanding of natural language incorporating many syntactic, semantic and pragmatic factors. For example, the word "high" conveys a negative polarity in "high

F. Li et al. (Eds.): WAIM 2014, LNCS 8485, pp. 711–722, 2014.

cost", but indicates a positive polarity in "high quality". In fact, such context-dependent sentiment words can not be discarded in sentiment classification [2].

It is worth mentioning that antonyms can help inferring the polarity in natural language. For example, "low" is the antonym of "high", then we can infer its polarity in "low cost" and "low quality" as the opposite as "high". Antonym pairs like ("high","low") and ("big","small") are helpful in sentiment detection [3]. However, no previous work has attempted to utilize such antonym pairs for context-dependent sentiment classification.

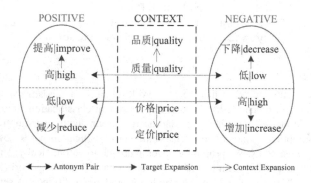

Fig. 1. A toy model of context-dependent sentiment classification

Figure 1 portrays a toy model of context-dependent sentiment classification with antonym pairs. It is often the case that two words in an antonym pair share the same context yet opposite polarity, as in the example of "高(低)质量|high(low) quality". To recognize the polarity of context-dependent sentiment words, we begin with choosing a seed set of antonym pairs such as (" 高|high","低|low") and ("大|big"," 小|small") , which are frequently used in opinionated texts. Next, we improve the scalability through double expansion, i.e., Context Expansion and Target Expansion, which is similar to double propagation [4]. Context Expansion refers to the process of expanding neighboring nouns with synonymy identification [5], e.g., "质量|quality"→"质地|quality" (as shown in Figure 1). Target Expansion refers to the process of expanding seed sentiment words to synonymous adjectives [6], e.g., "大|big"→"巨大|huge", or semantically correlated verbs, e.g., "高|big"→"提高|improve" (as shown in Figure 1).

In the most relevant work [2], the task of disambiguating dynamic sentiment ambiguous adjectives is transformed to sentiment expectation of noun. However, the impact of polarity detection with antonym pairs is ignored. We focus on context-dependent sentiment classification using antonym pairs.

The rest of this paper is organized as follows: In Section 2 we describe the polarity classification method for antonym pairs. Section 3 illustrates the double expansion method in detail. The evaluation results of our approach through sentiment classification experiments are presented in Section 4. Finally, we conclude and discuss the future work in Section 5.

2   Polarity Classification for Antonym Pairs

2.1   Context Information Extraction

Context information can be more accurately represented by aspects, which can be obtained from domain experts, or automatic methods[7]. However, aspect extraction of sentiment words is beyond the scope of this paper. The task of disambiguating ambiguous adjectives is simplified into sentiment classification of neighboring nouns [2]. Similarly, we extract the context information from neighboring nouns of ambiguous adjectives within a predefined distance.

Given the segmented and labeled sequence of a sentence $(w_1/t_1, w_2/t_2, \cdots, w_n/t_n)$, we assume that w_i is an ambiguous adjective from antonym pairs. Neighboring nouns, denoted as nn, can be matched by templates shown in Table 1.

Table 1. Matching templates for context information extraction

Template	nn	Example
$t_{i-1} = n$	w_{i-1}	价格 高|price is high
$t_{i-2} = n$ and $t_{i-1} \neq c$	w_{i-2}	价格 很 高| price is very high
$t_{i-3} = n$ and $t_{i-2} \neq c$ and $t_{i-1} \neq c$	w_{i-3}	价格 不是 很 高| price is not very high
$t_{i+1} = n$	w_{i+1}	高 质量|high quality
$t_{i+2} = n$ and $t_{i+1} \neq c$	w_{i+2}	高 的 成本| high cost
$t_{i+3} = n$ and $t_{i+2} \neq c$ and $t_{i+1} \neq c$	w_{i+3}	高 的 服务 质量| high service quality

Note: "n" indicates a noun, "c" indicates a conjunction.

2.2   Polar Posterior Probability

Definition 1 (**Antonym Pair**). *An antonym pair is formalized as a tuple $pair = (u, v)$, where u and v are ambiguous and antonymous adjectives. Two antonymous adjectives with the same context generally have the opposite polarities, i.e.,*

$$Polarity(u|tn) = -Polarity(v|tn) \qquad (1)$$

In this paper, we only discuss eight antonym pairs listed in Table 2. They are all one-character words and frequently used in opinionated texts [2].

Table 2. Eight antonym pairs

u	v	u	v	u	v	u	v
高|high	低|low	大|big	小|small	多|many	少|few	快|fast	慢|slow
深|deep	浅|shallow	长|long	短|short	轻|light	重|heavy	厚|thick	薄|thin

The collocations of ambiguous adjectives and neighboring nouns are saved in a polarity decision table.

Definition 2 (Polarity Decision Table). *A polarity decision table is formalized as a quad $PDT = (U, C \cup D, V, f)$, where,*

U: a finite nonempty set of objects, e.g., $\{e_1, e_2, \cdots, e_{12}\}$ in Table 3;
C: a finite nonempty set of condition attributes, $C = \{nn, sw\}$ in this paper, $sw = u$ or $sw = v$;
D: a finite nonempty set of decision attributes, $D = \{label\}$ in this paper, label labels the polarity of w;
V: $V = \cup V_a$, V_a is a nonempty set of values of $a \in C \cup D$. Thus, $V = V_{nn} \cup V_{sw} \cup V_{label}$ in this paper, V_{nn} represents all neighboring nouns, V_{sw} contains 16 words from 8 antonym pairs, and $V_{label} = \{1, -1\}$ or $V_{label} = \{1, 0, -1\}$;
f: $f = \{f_a | f_a : U \to V_a\}$, f_a is an information function that maps an object in U to one value in V_a.

In general, it is hard to annotate the polarity label for each context-dependent word for training. It is not necessary for our method which is on the basis of the following assumption.

Assumption 1. The polarity label of one word in a sentence is consistent with the polarity label of the sentence, i.e., the same if the sentence is affirmative, while the opposite is the case if the word is in the scope of negation.

Definition 3 (Polar Posterior Probability). *A polar posterior probability is a probability that a sentiment word sw is positive or negative given a neighboring noun nn, denoted as $P(sw = 1|nn)$ and $P(sw = -1|nn)$ respectively.*

The polar posterior probabilities for antonym pairs can be computed according to Bayes' theorem [8].

$$P(sw = 1|nn) = \frac{P(nn|sw = 1)P(sw = 1)}{\sum\limits_{l \in V_{label}} P(nn|sw = l)P(sw = l)} \tag{2}$$

$$P(sw = -1|nn) = \frac{P(nn|sw = -1)P(sw = -1)}{\sum\limits_{l \in V_{label}} P(nn|sw = l)P(sw = l)} \tag{3}$$

where,

$$P(sw = l) = \frac{count(sw, label = l) + 1}{|V_{label}| + \sum\limits_{s \in V_{label}} count(sw, label = s)} \tag{4}$$

$$P(nn|sw = l) = \begin{cases} \frac{count(nn, sw, label = l) + 1}{count(sw, label = l) + 1} & count(sw, label = l) \neq 0 \\ 0.001 & count(sw, label = l) = 0 \end{cases} \tag{5}$$

In Eqs. (4) and (5), the function "$count(X)$" returns the number of objects in U that the condition X is met. To eliminate zero probabilities, we use add-one smoothing, which simply adds one to each count.

We can determine the polarity of sentiment words by the following two rules: Bidirectional Rule and Unidirectional Rule.

Bidirectional Rule. If two sentiment words from an antonym pair both have the polar posterior probabilities given the same context, a bidirectional rule is made. The polarity of u from an antonym pair $pair = (u, v)$ given a neighboring noun nn is obtained:

$$Polarity(u|tn) = \begin{cases} 1 & P(u = 1|nn) > P(u = -1|nn) \wedge P(v = 1|nn) < P(v = -1|nn) \\ -1 & P(u = 1|nn) < P(u = -1|nn) \wedge P(v = 1|nn) > P(v = -1|nn) \\ 0 & otherwise \end{cases}$$

(6)

We compute the Z-score statistic with one-tailed test to perform the significant test. The hypothesized value P_0 is set to 0.7 [9]. The statistical confidence level is set to 0.95, whose corresponding Z-score is -1.64. If Z-score is greater than -1.64, the difference of two posterior probabilities is significant.

$$Z(nn, u, l) = \frac{P(u = l|nn) - P(u = -l|nn) - P_0}{\sqrt{\frac{P_0(1-P_0)}{\min\{count(nn,u,label=l),count(nn,u,label=-l)\}+1}}}$$

(7)

Unidirectional Rule. If only one sentiment word from an antonym pair has the polar posterior probabilities given the same context, a unidirectional rule is made.

$$Polarity(u|nn) = \begin{cases} 1 & Z(nn, u, 1) > -1.64 \wedge Z(nn, u, -1) < -1.64 \\ -1 & Z(nn, u, -1) > -1.64 \wedge Z(nn, u, 1) < -1.64 \\ 0 & otherwise \end{cases}$$

(8)

2.3   An Example of Polarity Decision Table

An example of polarity decision table is given in Table 3. The fourth column is the polarity label inferred from the training data. According to *Assumption 1*, if sw is in the scope of negation, *label* is opposite to the polarity of the sentence, otherwise the same. Table 4 lists the results of context-dependent polarity classification, which are satisfying.

Table 3. An example of polarity decision table

U	nn	sw	$label$	U	nn	sw	$label$
e_1	价格|price	高|high	-1	e_7	噪音|noise	大|big	1
e_2	价格|price	高|high	-1	e_8	噪音|noise	大|big	-1
e_3	价格|price	低|low	1	e_9	噪音|noise	大|big	-1
e_4	质量|quality	高|high	1	e_{10}	噪音|noise	小|small	1
e_5	质量|quality	高|high	1	e_{11}	速度|speed	慢|slow	1
e_6	质量|quality	高|high	1	e_{12}	速度|speed	慢|slow	-1

Table 4. Results of context-dependent polarity classification

nn	sw	Polarity	Eq.	Remark
质量|quality	高|high	1	(8)	$Z(nn, sw, 1) = -0.22, Z(nn, sw, -1) = -2.84$
质量|quality	低|low	-1	(1)	
价格|price	高|high	-1	(6)	$P(sw = 1\|nn)(0.25) < P(sw = -1\|nn)(0.75)$
价格|price	低|low	1	(6)	$P(sw = 1\|nn)(1.00) > P(sw = -1\|nn)(0.00)$
噪音|noise	大|big	-1	(6)	$P(sw = 1\|nn)(0.40) < P(sw = -1\|nn)(0.60)$
噪音|noise	小|small	1	(6)	$P(sw = 1\|nn)(1.00) > P(sw = -1\|nn)(0.00)$
速度|speed	慢|slow	0	(8)	$Z(nn, sw, 1) = -2.16, Z(nn, sw, -1) = -2.16$

3 Double Expansion

3.1 Context Expansion

Context expansion is equivalent to finding synonyms of the contextual word. A synonym dictionary can be used directly, but its coverage is limited. Automatically finding synonyms is transformed to semantic similarity measure.

HowNet is widely used in semantic similarity measure for Chinese words. Each word is described by several concepts. The similarity between w_i and w_j is equal to the maximum similarity of all concepts of the two words [10], denoted by $Sim_h(w_i, w_j)$. If w_i or w_j is out of HowNet lexicon, $Sim_h(w_i, w_j)$ equals 0. We utilize a modified similarity measure from the perspective of edit distance [11].

$$Sim_e(w_i, w_j) = \frac{1}{1 + EditCost(w_i, w_j)} \quad (9)$$

where $EditCost(w_i, w_j)$ is the minimum cost of character insertion and deletion operations needed to transform one word to another. The cost of inserting or deleting an character ch is set as in [12], where NEG is a set of negation characters, such as "不\|not" and "无\|none".

$$Cost(ch) = \begin{cases} 1 & \text{if } \text{delete } ch \wedge ch \notin NEG \\ 0.1 & \text{if } \text{insert } ch \wedge ch \notin NEG \\ \infty & \text{if } ch \in NEG \end{cases} \quad (10)$$

In context expansion, we combine the above two similarity measures. We give a higher weight to Sim_h, i.e., $0.5 < \alpha \leq 1$. If Sim of two words is greater than a predefined threshold θ ($\theta = 0.6$ in the following examples), they are considered to be synonymous.

$$Sim(w_i, w_j) = \alpha \cdot Sim_h(w_i, w_j) + (1 - \alpha \cdot Sim_h(w_i, w_j)) \cdot Sim_e(w_i, w_j) \quad (11)$$

Property 1. $0 \leq Sim(w_i, w_j) \leq 1$.

Property 2. If w_i or w_j is out of HowNet lexicon, $Sim(w_i, w_j) = Sim_e(w_i, w_j)$.

Example 1 ("质量\|quality" and " 质地\|quality"). $Sim_h = 0.93$, $Sim_e = 0.48$, let $\alpha = 0.6$, $Sim = 0.77 > 0.6$. Thus, "质量\|quality"→"质地\|quality".

3.2   Target Expansion

Target expansion is to find more adjectives or verbs which are related to the target sentiment words.

Polar Adjective Expansion. Find adjectives, each of which has the high similarity with the target sentiment word. The expansion is the same with context expansion. A higher weight to Sim_h is given, i.e., $0.5 < \alpha \leq 1$.

Example 2 ("大|big" and "巨大|huge"). $Sim_h = 1.00$, $Sim_e = 0.91$, let $\alpha = 0.6$, $Sim = 0.96 > 0.6$. Thus, "大|big"→"巨大|huge".

Polar Verb Expansion. Find verbs, each of which has the same trend as the target sentiment word. The semantic similarity between " 高|high" and "提高|high" (0.13) is less than that between "低|high" and "提高|high" (0.24), which is obviously wrong. Hence, a higher weight to Sim_e is given, i.e., $0 \leq \alpha < 0.5$.

Example 3 ("高|high" and "提高|improve"). $Sim_h = 0.13$, $Sim_e = 0.91$, let $\alpha = 0.4$, $Sim = 0.92 > 0.6$. Thus, "高|high"→"提高|improve".

With the help of polar verb expansion, we can obtain verbs expressing the same trend as the corresponding adjective. We define a verb list for one adjective.

Definition 4 (Verb Set for Adjective). *Given an ambiguous adjective sw, a verb for sw satisfies that the similarity is greater than θ. All such words comprise a verb set for adjective sw, denoted by $VSA(sw)$.*

$$VSA(sw) = \{w|Sim(w, sw) > \theta\} \tag{12}$$

Polar Verb Re-Expansion. $VSA(sw)$ can also provide help for polar verb expansion. If the minimum Sim (computed with the similarity measure in context expansion) between a new word and all words in $VSA(sw)$ is greater than θ, the new word is considered to express the same meaning with sw.

Example 4 ("高|high" and "增加|increase"). $VSA("$ 高|high")={"提高|improve", " 增高|rise"}, $Sim("$ 提高|improve"," 增加|increase")=0.72, $Sim("增高|improve","增加|increase")=0.79$, $\min\{0.72, 0.79\} = 0.72 > 0.6$. Thus, " 高|high"→"增加|increase". But if we compute the similarity directly, $Sim = 0.53 < 0.6$.

4   Experiments

4.1   Data Sets

We conduct experiments on two real-world data sets. One data set is from Task 1 of Chinese Opinion Analysis Evaluation 2012, denoted as COAE [13], and the other is

Table 5. Statistics of two data sets

Data	COAE	SEMEVAL
Positive	598	1202
Negative	1295	1715
Neutral	507	0
Total	2400	2917

from Task 18 of Evaluation Exercises on Semantic Evaluation 2010, denoted as SE-MEVAL [2]. Their statistics are listed in Table 5.

The distribution of sentiment words from 8 antonym pairs is shown in Figure 2. The total of 960 sentiment words appear in 709 sentences on COAE data set, and 4991 sentiment words appear in 2846 sentences on SEMEVAL data set. These words are often used in opinionated texts, especially " 高|high","低|low", "大|big"," 小|small", " 多|many","少|few", and "重|heavy".

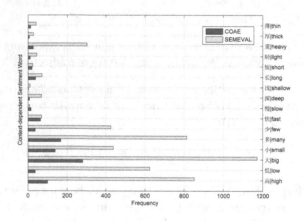

Fig. 2. Distribution of sentiment words from antonym pairs

4.2 Experiment Settings

- Preprocessing: All texts are automatically word-segmented and POS-tagged with ICTCLAS [14]. A sentence is divided into a few sub-sentences with some punctuation marks (,.!?).
- Sentiment Lexicon: We construct a sentiment lexicon with Affective Lexicon Ontology [15] and some common expressions. There are 28567 entries in our sentiment lexicon.
- Sentiment Classification Method:
  Baseline The method proposed by Turney [16] is used as the baseline, which discards the context-dependent words discussed in this paper.

NB Naive Bayes method is directly used to classify [17].

SVM Support Vector Machines method is directly used to classify [17].

Approach-1 Add the step of polarity classification for antonym pairs based on Baseline.

Approach-2 Add the step of context expansion based on Approach-1.

Approach-3 Add the step of target expansion based on Approach-2.

– Evaluation: The evaluation criteria is micro-average F-measure $micro\text{-}F_1$.

4.3 Performance of Sentiment Classification for Antonym Pairs

The comparative results between three baselines and Approach-1 are demonstrated in Figure 3. The improvement on SEMEVAL data set is obvious, because the data set is designed to disambiguate sentiment ambiguous adjectives per se, and almost each sentence contains ambiguous adjectives. After executing polarity classification for antonym pairs, many sentences can be truly labeled the polarity. The performance of Approach-1 on COAE data set is slightly improved due to the small percentage of sentences containing such adjectives. We also find that Approach-1 is better than NB and SVM.

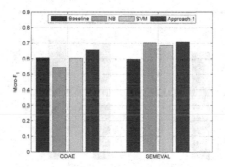

Fig. 3. Performance of sentiment classification for antonym pairs

Table 6. Behavior analysis of selected words

	#Selected Words			#Other Polar Words	
	#Positive	#Negative	#Neutral	#Context-dependent	#Context-free
COAE	264	328	261	79	3128
SEMEVAL	1841	2082	699	110	7723
TOTAL	2105	2410	960	189	10851

Table 6 shows the behavior analysis of the selected 16 words. Among all the selected words, 35% are positive, 40% are negative and the remainder are neutral due to the lacking of contextual information or matching rules. The number of other context-free polar words is nearly twice that of the selected words. The number of the selected words is about three times that of other context-dependent polar words.

4.4 Performance of Double Expansion

There are two important parameters, α and θ, in double expansion. α measures the different weight of Sim_h and Sim_e, θ is a similarity cut-off. According to several experiment results, the optimal parameter settings are given in Table 7. The similarity cut-off of different expansion on a certain data set is the same. The performance of double expansion is shown in Figure 4.

Table 7. Optimal parameters in all expansions

	COAE	SEMEVAL
Context Expansion	$\alpha = 0.7, \theta = 0.6$	$\alpha = 0.7, \theta = 0.9$
Polar Adjective Expansion	$\alpha = 0.7, \theta = 0.6$	$\alpha = 0.7, \theta = 0.9$
Polar Verb Expansion	$\alpha = 0.3, \theta = 0.6$	$\alpha = 0.3, \theta = 0.9$
Polar Verb Re-Expansion	$\alpha = 0.7, \theta = 0.6$	$\alpha = 0.7, \theta = 0.9$

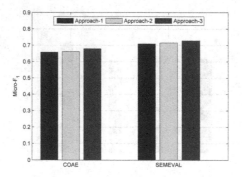

Fig. 4. Performance of double expansion

Because more contextual polarities are discovered, double expansion further improve the performance of sentiment classification. The improvement on SEMEVAL data set is greater than that on COAE data set.

4.5 Annotation Consistence Analysis

As mentioned above, we assume that the polarity annotation of a word is consistent with the sentence. To validate the rationality, Cohen's kappa coefficient [18] is used as a statistical measure of inter-annotator agreement.

$$\kappa = \frac{Pr(a) - Pr(e)}{1 - Pr(e)} \tag{13}$$

where, $Pr(a)$ is the relative observed agreement among annotators, and $Pr(e)$ the hypothetical probability of chance agreement.

Table 8. Contingence table of two annotators

Assumption\Proposed Method	$label = 1$	$label = -1$
$label = 1$	PP	PN
$label = -1$	NP	NN

In this paper, there are two annotators: *Assumption 1* and our proposed method. The contingence table is shown in Table 8:

$Pr(a)$ and $Pr(e)$ in Eq. (13) are computed respectively:

$$Pr(a) = \frac{PP + NN}{PP + PN + NP + NN} \tag{14}$$

$$Pr(e) = \frac{(PP + PN) \times (PP + NP) + (NP + NN) \times (PN + NN)}{(PP + PN + NP + NN) \times (PP + PN + NP + NN)} \tag{15}$$

We can figure out $Pr(a) = 0.86$ and $\kappa = 0.73$. Two annotators are consistent enough, and the assumption is reasonable.

5 Conclusions

In this paper, we propose a novel approach to automatically determine the polarity of context-dependent words with antonym pairs. To the best of our knowledge, this is the first context-dependent sentiment classification scheme which utilizes antonym pairs. According to Bayes' theorem, two polar posterior probabilities are obtained. We also initiate two kinds of rules, i.e., Bidirectional Rule and Unidirectional Rule, and assign the polarity to sentiment words. In addition, we define a new similarity measure which combines semantic distance with edit distance for Context Expansion and Target Expansion. Our approach is effective in improving the overall performance of sentiment classification.

In the future, we would like to dive into more accurate context information extraction which can help to filter noisy neighboring nouns. We consider that antonym pairs deserve further research.

Acknowledgments. This work is partially supported by the National Natural Science Foundation of China (No. 61273304, and No. 61202170), the Specialized Research Fund for the Doctoral Program of Higher Education of China (No. 20130072130004) and the Fundamental Research Funds for the Central Universities.

References

1. Pang, B., Lee, L.: Opinion mining and sentiment analysis. Foundations and Trends in Information Retrieval 2(1-2), 1–135 (2008)
2. Wu, Y., Wen, M.: Disambiguating dynamic sentiment ambiguous adjectives. In: Proceedings of the 23rd International Conference on Computational Linguistics, pp. 1191–1199. ACL, Stroudsburg (2010)

3. Mohammad, S., Dorr, B., Hirst, G.: Computing word-pair antonymy. In: Proceedings of the 2008 Conference on Empirical Methods in Natural Language Processing, pp. 982–991. ACL, Stroudsburg (2008)
4. Qiu, G., Liu, B., Bu, J., Chen, C.: Expanding domain sentiment lexicon through double propagation. In: Proceedings of the 21st International Joint Conference on Artifical Intelligence, pp. 1199–1204. Morgan Kaufmann Publishers Inc., San Francisco (2009)
5. Hu, M., Liu, B.: Opinion extraction and summarization on the web. In: Proceedings of the 21st National Conference on Artificial Intelligence, pp. 1621–1624. AAAI Press, Palo Alto (2006)
6. Lu, Y., Castellanos, M., Dayal, U., Zhai, C.: Automatic construction of a context-aware sentiment lexicon: an optimization approach. In: Proceedings of the 20th International Conference on World Wide Web, pp. 347–356. ACM, New York (2011)
7. Jo, Y., Oh, A.H.: Aspect and sentiment unification model for online review analysis. In: Proceedings of the 4th ACM International Conference on Web Search and Data Mining, pp. 815–824. ACM, New York (2011)
8. Papoulis, A., Pillai, S.U.: Probability, random variables, and stochastic processes. Tata McGraw-Hill Education (2002)
9. Zhang, L., Liu, B.: Identifying noun product features that imply opinions. In: Proceedings of the 49th Annual Meeting of the Association for Computational Linguistics: Human Language Technologies, pp. 575–580. ACL, Stroudsburg (2011)
10. Zhu, Y., Min, J., Zhou, Y., Huang, X., Wu, L.: Semantic orientation computing based on HowNet. Journal of Chinese Information Processing 20(1), 14–20 (2006)
11. Lin, D.: An information-theoretic definition of similarity. In: Proceedings of the 15th International Conference on Machine Learning, pp. 296–304. Morgan Kaufmann Publishers Inc., San Francisco (1998)
12. Che, W., Liu, T., Qin, B., Li, S.: Similar Chinese sentence retrieval based on improved edit-distance. Chinese High Technology Letters 7, 15–19 (2004)
13. Liu, K., Wang, S., Liao, X., Xu, H.: Overview of Chinese opinion analysis evaluation 2012. In: Proceedings of the 4th Chinese Opinion Analysis Evaluation, pp. 1–32 (2012)
14. Zhang, H., Yu, H., Xiong, D., Liu, Q.: HHMM-based Chinese lexical analyzer ICTCLAS. In: Proceedings of the 2nd SIGHAN Workshop on Chinese Language Processing, pp. 184–187. ACL, Stroudsburg (2003)
15. Xu, L., Lin, H., Pan, Y., Ren, H., Chen, J.: Constructing the affective lexicon ontology. Journal of the China Society for Scientific and Technical Information 27(2), 180–185 (2008)
16. Turney, P.D.: Thumbs up or thumbs down?: semantic orientation applied to unsupervised classification of reviews. In: Proceedings of the 40th Annual Meeting on Association for Computational Linguistics, pp. 417–424. ACL, Stroudsburg (2002)
17. Pang, B., Lee, L., Vaithyanathan, S.: Thumbs up?: sentiment classification using machine learning techniques. In: Proceedings of the 2002 Conference on Empirical Methods in Natural Language Processing, pp. 79–86. ACL, Stroudsburg (2002)
18. Cohen, J.: A coefficient of agreement for nominal scales. Educational and Psychological Measurement 20(1), 37–46 (1960)

A Novel Knowledge Network Framework for Financial News Navigation

Lili Zhou[1], Hanchao Wang[1], Lei Zhang[1], Enhong Chen[1], and Jun Chen[2]

[1] University of Science and Technology of China
{zhoulili,hanchaow,stone}@mail.ustc.edu.cn
[2] Xinhua News Agency
cheneh@ustc.edu.cn, chenjun2008@xinhua.org

Abstract. Nowadays, various financial news retrieval platforms are provided to help users, especially for financial professionals and hobbyists to make right decisions. In those platforms, users usually get information by searching the relevant news via keywords or clicking the recommended news with the similar topic in the clicked web page. However, such ways to obtain financial information cannot effectively meet users' further needs. They are eager to obtain the relevant news with different domains in a short time. To address this problem, we propose a novel four-layers-based knowledge network framework for financial news navigation. Experiments on real data sets demonstrate the effectiveness and efficiency of our proposed framework.

Keywords: retrieval platform, knowledge network, financial news navigation.

1 Introduction

Currently, lots of financial news retrieval platforms including general search engines (e.g., baidu) and financial domain vertical websites (e.g., sina) arise to facilitate access to financial information for users, especially for financial professionals and hobbyists. They are eager to get the latest financial information to help them make right decisions. In these platforms, users can search the relevant news by keywords or click the corresponding label in the navigation menu to obtain financial information they need. If wanting to know the ins and outs of the clicked news, users have to continue to seek those related news through the aforementioned two ways. Such a process is time-consuming and laborious. Although the majority of news pages list some recommended news links with the similar topic for extension reading, this does not alleviate the problem. Meanwhile, existing studies on financial news mainly focus on how to organize and present financial news for users more friendly [1] or mining the underlying information in financial news [2]. But All of them do not solve this problem properly.

For this, we propose a novel four-layers-based knowledge network framework for financial news navigation. Specifically, we first crawl large amounts of financial news from many popular financial web portals. Second, we apply topic model

F. Li et al. (Eds.): WAIM 2014, LNCS 8485, pp. 723–727, 2014.

and classification methods to extract knowledge from the collected news corpus. Based on these knowledge sets, we then construct a four-layers-based knowledge subnet for each news report. Finally, we implement and visualize this knowledge network with a popular used library D3.js.

2   The Four-Layers-Based Knowledge Network

Constructing the knowledge network framework consists of two steps, i.e., offline step and online step. Next, we detail the involved techniques and implementations respectively.

2.1   Knowledge Definition and Extraction

In the offline step, we crawl real datasets and extract the knowledge from them. Here we define knowledge as financial information contained in financial news, which can be expressed as an industry label or a topic, etc. Among these knowledge, the topics and industry label of one news report are usually difficult to obtain. Fortunately, Latent Dirichlet Allocation [3] which is a classic topic model can be used to extract the latent topics from vast amounts of financial news effectively. Since the headline often represents the core idea of a news report, we consider both the title and news content and give different weights to them when using LDA model. To recognize the industry label, we choose Support Vector Machine [4] due to its high accuracy and efficiency.

2.2   Knowledge Network Construction and Visualization

In the online step, we construct the dynamic knowledge network composed of numerous knowledge subnets. Fig.1 shows the full view.

Suppose a user opens a news page, then the first layer of the knowledge subnet for this news report shown in Fig.1(a) is presented behind the news content. In Fig.1(a), the center node denotes the clicked news (hereafter we call Main News) and each of the other three nodes connecting to Main News stands for a news set of the common nature. For example, if Main News's industry label is real estate, the node named "industry policy" represents some relevant news whose labels are all real estate control policy. If the user is interested in macro data after reading Main News, he or she can click the corresponding news-cluster node to get detailed information. Fig.1(b) shows the second layer of the knowledge subnet where each blue node denotes a similar topic set with a description of several words and each topic in the set is contained by some news in its parent node. We use a simple clustering method named K-Means [5] to aggregate those similar topics. And click one blue node, the specific topics in this set are displayed shown in Fig.1(c). Furthermore, the user also can find top-n relevant news for one interesting topic by clicking the corresponding topic-node. As shown in Fig.1(d), there are 5 macro economy news reports whose main topics are all "Topic C". We find the top-5 news reports according to the probability value of this topic

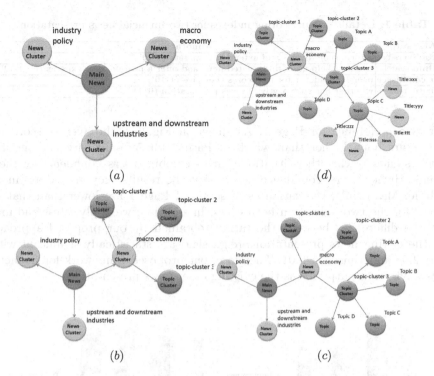

Fig. 1. The structure of a knowledge subnet for one financial news report

in all news reports. Finally, the user can read one of these news by clicking it. Through the whole process, users can conveniently grasp the information with different topics

We visualize each layer of the knowledge network with a popular used library D3.js (http://d3js.org/) which can bring data to life using HTML, SVG and CSS.

3   Experiments

To validate the proposed framework, we implement our proposed framework focusing on real estate according to Section 2. All the implementations in Java are on a Windows 7 PC with Intel 1-core i3 3.10GHz CPU, 4GB of main memory and a 64-bit operating system. On this basis, we design a scoring system for comparing two different kinds of news presentation, that is, the novel news page with the knowledge subnet embedded in denoted as Page 1 and the original news page denoted as Page 2. We offer four measures to users for marking the two different pages. Table 1 presents meanings of these measures. In order to avoid human bias, we deploy this scoring system on the server to let external users grade them after using it.

So far, there are more than two thousand scoring records. We randomly select 600 pieces of data among them for final statistics. Among these users, 399 users

Table 1. Details of four scoring indexes for two financial news presentations

Measure	Annotation
info-diversity	information diversity of extended news sets (1-10')
relInfo-search	whether help users find related news and offer a positive user experience. (1-10')
user-confidence	users' confidence for the scores they give (1-10')
whether-like	whether like such a way of news presentation (0/1)

express a preference for Page 1. As shown in Fig.2(a), the average scores of Page 1 are both higher than which of Page 2 for info-diversity and relInfo-search, similar to the other situation that we combine users' confidence for their ratings. Higher users' confidence means that the results they graded are more credible. Meanwhile, the variances of scores of Page 1 are lower than that of Page 2 for the two indexes in both cases. In addition, we apply z-test and find that the differences between the ratings obtained by our proposed approach and the exiting news presentations (e.g., sina) are statistically significant with $|z| \geq 2.58$ and thus $p \leq 0.01$. Therefore, our proposed framework for financial news navigation outperforms the existing other news presentations.

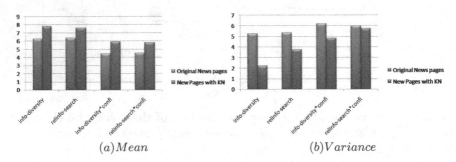

(a)Mean   (b)Variance

Fig. 2. The scoring results: mean and variance

4   Conclusion

In this study, we presented a four-layers-based knowledge network framework for financial news navigation. The experiments focusing on real estate industry demonstrated that the proposed framework can effectively satisfy users' further needs. Furthermore, the idea of using knowledge network to facilitate users' access to financial information can be generally applicable to other domains of news navigation (e.g., sports news). In the future, we would like to incorporate the behavior data such as geographic information and users' browsing history logs into the proposed framework for personalized recommendation.

Acknowledgments. This research was partially supported by grants from the Research Fund for the Doctoral Program of Higher Education of China (Grant No. 20113402110024), the Science and Technology Development of Anhui

Province, China (Grants No. 13Z02008-5 and 1301022064), the International Science & Technology Cooperation Plan of Anhui Province (Grant No. 1303063008), the Nature Science Foundation of Anhui Education Department (Grant No. KJ2012A273).

References

1. Wang, H., Wang, Z.: Mobile financial news mashup development based on yql. In: 2013 Fifth International Conference on Computational and Information Sciences (ICCIS), pp. 1717–1720. IEEE (2013)
2. Alanyali, M., Moat, H.S., Preis, T.: Quantifying the relationship between financial news and the stock market. Scientific Reports 3 (2013)
3. Blei, D.M., Ng, A.Y., Jordan, M.: Latent dirichlet allocation. The Journal of Machine Learning Research 3, 993–1022 (2003)
4. Sebastiani, F.: Machine learning in automated text categorization. ACM Computing Surveys (CSUR) 34(1), 1–47 (2002)
5. Jain, A.K.: Data clustering: 50 years beyond k-means. Pattern Recognition Letters 31(8), 651–666 (2010)

How Do People Communicate through Different Social Connections?

Keke Cai[1], Yu Zhao[2], Jie Tang[2], and Li Zhang[1] and Zhong Su[1]

[1] IBM Research China
[2] Department of Computer Science and Technology, Tsinghua University, China
{caikeke,lizhang,suzhong}@cn.ibm.com, zhaoyu121407@gmail.com,
tangjie@keg.cs.tsinghua.edu.cn

Abstract. This paper presents a comparison study to identify the communication patterns of people through different social connections. Advances in technology have brought many communication channels for people in daily life, like E-mail, blogs/micro-blog and mobile telecommunication etc. Now and in the future it is going to be critical that people use multiple channels of communication to reach others. The understanding of people's choice of communication channels is becoming quite important. In this paper, we specifically selected two of the most significant channels as the objects for comparison. One is online social network, e.g., Twitter as representative of such networks, and another is mobile telecommunication. The corresponding social network is therefore constructed for each communication channel. Based on that, we conduct a series of investigation, including temporal analysis, geographical analysis and topological analysis. Generally, what we have found in this study is that people's communication through different channels shows the differences in various aspects.

1 Introduction

Advances in technology have brought many communication channels for people in daily life, like E-mail, blogs/micro-blog and mobile telecommunication etc. Now and in the future it is going to be critical that people use multiple communication channels to reach each other. One interesting and also fundamental question is: would people exhibit consistent or totally different communication habits in different communication channels? To explore this problem, in this paper we specifically selected two of the most significant channels as the objects for comparison. One is online social network, e.g., Twitter as representative of such networks, and another is mobile telecommunication.

Online social network(OSN), emerged as a new medium, has flourished as never before. Meanwhile, continuous developments in mobile telecommunication technology have improved the way by which people engage with each other and constructed another kind of mobile social network(MSN). No matter online social network or mobile social network, all they provide a platform where individuals with similar interests or commonalities can be connected with one another easily and efficiently. In recent years, the exploration of user social communication

F. Li et al. (Eds.): WAIM 2014, LNCS 8485, pp. 728–739, 2014.

behaviors has captured a lot of attention. However, most of existing work focus on one particular kind of social network separately. In this paper, we conduct a deep analysis to compare the patterns of user's behaviors through online and mobile social connections from various perspectives, including temporal, geographical and topological features. To the best of our knowledge, there is no previous study investigating the behavioral patterns comparisons from multiple perspective. Essentially, this paper tries to answer the questions:

1. Does the communication in different social networks have distinguished temporal properties?
2. Does the geographic location influence the communications in different social network?
3. Are there some significant topological difference between different social network?

This paper is organized as follows. In section 2, we give a detailed description of data collection for mobile and online social network. Then we implement the behavior analysis from three dimensions of time, location and topology in section 3, 4 and 5. Finally, in Section 6 and 7, we summarize related works and come to a conclusion.

2 Data Set

In this paper, the experiments are conducted on two data sets. One is Reality Mining dataset[1][4] , which contains the call logs, SMS logs, GSM cell information and Bluetooth scan collected from 106 subjects for more than 9 months. Another is Twitter data [2], which consists of 103,452 Twitter user information and 211,509,594 tweets for about 9 months.

The Reality Mining Datasetconsists of the tracking data of people's daily activities on the smart phones. It is the first mobile data set with rich personal behavior and interpersonal interactions and contains 162,699 communication records (128,542 call logs and 34,157 short message logs). Specifically, the information utilized in our study includes:

- Date, time, duration, direction and originator/recipient of each call and short message.
- A label indicating whether a call is missed or not.
- The location information, which is in the form of GSM cell information.
- The inferred proximity information. Bluetooth device can detect other Bluetooth devices within a range of 5-10m. Based on it, we can deduce whether two subjects are in close proximity to each other.

For the data set for online social network analysis, we choose Twitter as the source for data collection since it is one of the most popular online social networks

---

[1] http://reality.media.mit.edu/
[2] https://twitter.com

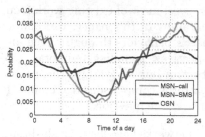 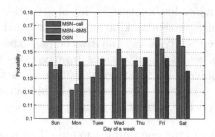

(a) Probability of Communication vs. Time of a Day

(b) Probability of Communication vs. Day of a week

Fig. 1. Communication Periodicity in Mobile and Online Social Network

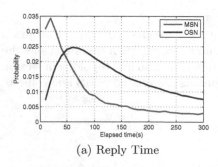 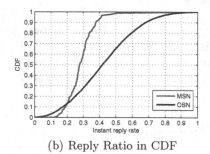

(a) Reply Time

(b) Reply Ratio in CDF

Fig. 2. Instant Reply in Mobile and Online Social Network

and described as "the SMS of the Internet", and record various information of user's interactions on Twitter, including time, frequency, forward and comment etc, as well as user personal profile, like location, fans and followers etc.

3 Temporal Analysis of Communication

In this section, we lay emphasis on various time factors of user interaction in mobile and online social network. For better understanding of these factors, we proceed with a definition of these factors and then a detailed observations and analysis.

3.1 Periodicity of Communication

Periodicity of communication is the tendency of communication happening at regularly-spaced periods of time. To explore the periodicity facts under different social networks, Our study gives the observations of 24-hour and one-week periodicity under different social networks.

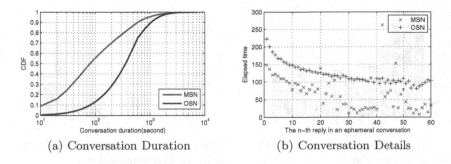

(a) Conversation Duration   (b) Conversation Details

Fig. 3. Conversation in Mobile and Online Social Network

- *24-hour periodicity.* As shown in Figure 1(a), the communications in mobile social network, either call or SMS, exhibit a clear periodicity of lower probability in the morning and higher probability at night. But comparatively, the periodicity in online social network is not quite clear and presents a smoother curve of communication probability through a day.
- *one-week periodicity.* Similarly, the probability of communication in online social network fluctuates less than that in mobile social network during a week. As shown in Figure 1(b), the variance of phone-call and SMS probability during a week is 2.5×10^{-4} and 1.1×10^{-4} respectively, while the variance of communication probability in online social network is 1.8×10^{-5}.

3.2   Instant Reply

Given a time limit θ, if user a sends a message to user b at time t, and user b sends a message back to user a within the time limit θ, the message sent back is defined as an *instant reply*. In this paper, instant reply is evaluated from two aspects more detailedly, including reply time and is reply rate.

- *Reply Time.* Figure 2(a) shows the probability of the elapsed time before message reply. From this figure, it can be seen that more time is taken for response in online social network. Especially, the curve for online social network peaks at about 60s, while the peak in mobile social network is about 45s. It proves that the communication online is more random than that in mobile social network.
- *Reply Rate.* Figure 2(b) gives the Cumulative Distribution Functions(CDF) curves of users with respect to the instant reply rate. We observe that the reply rate for most mobile users is between 0.2 and 0.4. It means that less than half of the messages will be replied. Comparatively, the reply rate varies widely for online users and the maximal reply rate can achieve 0.8.

3.3   Conversation

If user a and user b iteratively send and reply message to each other , all the messages in the discussion thread form a *conversation*. In this section, we inves-

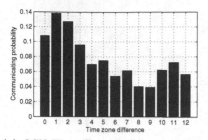
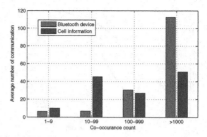

(a) OSN Time Zone Difference vs. Communication Probability

(b) MSN Co-occurrence Count vs. Average Number of Communications

Fig. 4. Geographical Analysis in Mobile Social Network and Online Social Network.

tigate the *conversation* by observing the *conversation duration* and *conversation details*.

- *Conversation Duration.* Figure 3(a) illustrates the CDF curves of conversation duration in different social networks. The first observation is that most conversations last less than 1000 seconds. The second one is there are rare short conversation (shorter than 100 seconds) in online social network, while short conversations occupy about 55% in mobile social network. This observation is consistent with the observations for instant reply. Online users tend to communicate with each other freely and the conversation can last for long time.
- *Details of Conversation.* To capture the micro-level characteristics of conversation, we plot the details of elapsed time in each round of conversation. Illustrated by Figure 3(b), the average time in the beginning of a conversation is longer (150-200s), but as the conversation progresses, the interaction averagely takes less and less time. This phenomenon can be observed in both mobile and online social network. It reveals the process of how a conversation gradually captures people's attention and interests.

4   Geographical Analysis

Location is an important element in social network analysis[11]. It is expected that the communcation through social network can break the geographical restrictions and becomes more freely and easily. In this section, our study will investigate whether or not the communication between people in different social network is limited by geographic constraints. The trend of communication in regard to geographic distance is studied.

- *Online social network.* As we know, the exposure of location information in online social network always depends on user themselves. Generally, it is hard to extract the location information accurately and completely. As

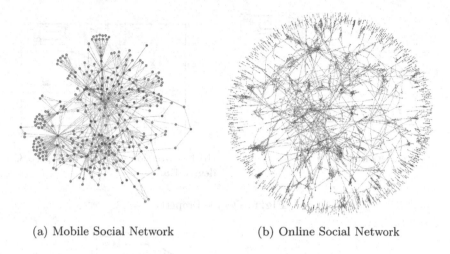

(a) Mobile Social Network   (b) Online Social Network

Fig. 5. Communication Network Structure

a substitution, we consider the time zone as an approximate indicator of location to measure the geographic distance between online users.

Figure 4(a) demonstrates the probability distribution of communication with respect to time zones. It is interesting to find that most communication occurs between users with 1 hour of time zone difference. As the time zone difference increases, the communication probability decreases. However, when the time zone difference reaches more than 4 hours, the fluctuation of communication probability flattens out. The result indicates that the communications between online users is out of distance constraints. Online users would like to make new friends and expand their social circles.

– *Mobile social network.* In this part of study, the geographic co-occurrence is estimated as the average geographic distance between users. Two methods are applied to measure the co-occurrence. One is based on Bluetooth Device, which can record the proximity within 5-10m range. Another is based on GSM, which can cover about 100 meter distance. The communication distribution with respect to the number of co-occurrence is demonstrated in Figure 4(b). It can be seen that the higher co-occurrence of people within close proximity to each other, the more communication they have.

5   Topological Analysis

Except temporal and spatial factor, topological structure hidden behind the communication is also important for the mining of user behavior patterns on social network. Before the detailed study and comparison, let's first take a look at the communication structure of different social network.

Figure 5(a) shows the Reality Mining communication structure. The red nodes represent subjects to be observed in our data and grey ones represent nodes

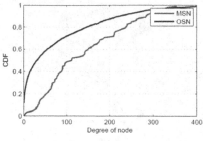
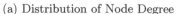

(a) Distribution of Node Degree

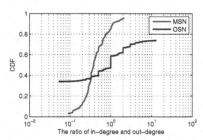

(b) Distribution of Node In-degree/Out-degree Ratio

Fig. 6. Degree Property

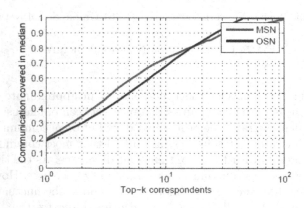

Fig. 7. Communication Covered by Top-k Contacts.

having communication with the red nodes. Here, we ignore all the nodes that have less than one degree for clarity. A linkage will be setup between users with more than 5 times interaction and the red link highlights the connection between two red nodes. Moreover, the width of line represents the correlation coefficient of two nodes. The more width of the link, the more correlation it has. Similarly, Figure 5(b) shows the communication structure of online social network, where the inner nodes constitute a large sub-network and represent the communities with close communication.

Discussion above gives a high-level introduction of the different topological structures in social networks. In the following, we will implement more deep discussion on the granularity of node, node pair and network perspectively.

5.1   Node-level Analysis

- *Degree Property.* Generally, degree is used to describe the association strength of a given node with other nodes. By observing the degree of each node, we plot the CDF curves as shown in Figure 6(a). In this study, the link between

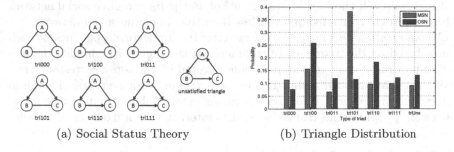

(a) Social Status Theory   (b) Triangle Distribution

Fig. 8. Social Status Theory in different social networks

Table 1. Experiment results of communication detection

nDCG@k		100	500	1000	5000
	Avg-Deg	0.175	0.200	0.244	**0.399**
MSN	CN-Num	0.239	**0.243**	**0.256**	0.256
	CN-Str	**0.315**	0.226	0.254	0.287
	Avg-Deg	0.508	0.598	0.631	0.686
OSN	CN-Num	**0.793**	**0.794**	**0.796**	**0.792**
	CN-Str	0.653	0.699	0.717	0.745

nodes is considered as undirected. It can be observed that over 60% of nodes in online social network have less than 50 degrees, whereas in mobile social network about 80% of nodes have more than 40 degrees.

To more accurately explore the communication between nodes, we further make a study under the directed graph. An arch from A to B will be established if A has a talk to B. Based on the directed graph, we consider the ratio of in-degree to out-degree ($\frac{d_{in}}{d_{out}}$) and get the corresponding CDF as illustrated in Figure 6(b). The result for online social network shows that there are more than 30% of nodes without any in-degree and more than 25% without any out-degree. Besides, about half of nodes has more out-degree than in-degree, 40% of nodes has more in-degree and 10% of nodes has the same in-degree and out-degree. In mobile social network, about 85% of users have more out-degrees, but few have no in-degree or no out-degree. As a summary of the difference, the ratio of degree in mobile social network varies from 0.1 to less than 10, whereas the ratio in online social network varies from less than 0.1 to more than 10.

– *Communication Reciprocity.* Previous studies reported the low level of reciprocity on Twitter [9]. It is also proved by our experiments. The experimental results indicate that 74.25% of linked users is connected one-way, and only 25.75% pairs have reciprocal relationship. Comparatively, reciprocity in mobile social network is originally thought to be higher. But we surprisingly discovered from the experiment result that mobile social network also has this property. It shows that only 21.9% of user pairs in mobile social network has reciprocal relationship, even less than that in online social network.

- *Communication Diversity.* We characterize the top-k correspondents for each user based on the frequency of communications. Figure 7 highlights the median portion of users' communications covered by the top-k correspondents. We find that for both mobile and online social network over 50% of communications are covered by top-5 correspondents. It shows that in any social network people would like to or need to interact with a fixed set of friends.

5.2   Pair-wised Analysis

After the node-level discussion, let's turn to the pair-wised analysis to explore the influence of social connections on communication. A general question is whether the users who share many common neighbors in social network are more likely to be as friends? We transfer this exploration to a prediction problem. The task is to predict the communication link between people in different social network by considering different factors of topological structure.

To do it, we select four months of data and split them into two equal parts. Then, the next step is to detect unprecedented communication links in the second time period by referring the topological characteristics generalized from the first time period. Here, we consider three kinds of topological feature.

- *Average Degree(Ave-Deg).* Average degree is to describe the linkage status of user. One hypothesis is that user with more linkages tends to build new connection with others. Given two users r and s, the average degree is defined as below.

$$AveDeg(r, s) = \frac{2d(r)d(s)}{d(r) + d(s)} \tag{1}$$

  where $d(s)$ means the link degree of user s.

- *Common Neighbor Number(CN-Num).* Common neighbors can always indicate the affinity of users. Intuitively, users with more common neighbors have more chance to talk with each other.

- *Common Neighbor Strength(CN-Str).* Besides the observation of common neighbors, the similarity of connection strength with neighbors is another evidence for affinity. Given two users r and s, the strength similarity is defined as below,

$$CNStr(r, s) = \sum_{v \in N(r,s)} \frac{2m(v, r)m(v, s)}{m(v, r) + m(v, s)} \tag{2}$$

  where $N(r, s)$ is the set of all common neighbors of user r and s, $m(v, r)$ is the communication count between user v and r.

By considering the factors defined above respectively, we can get a ranking list of user pair. The top ones are users predicted to be connected. We use nDCG@k to evaluate the performance. From the results shown in Table 1, we have some interesting findings.

- For online social network, all the three topological features are good evidence for communication prediction. Inversely, the communication probability is hard to be predicted for mobile social network. It proves again that users are easier to make friends through online social network. Furthermore, it shows that the transparence of online social network is actually a promotion for people to connect to friends of friends.
- The performance of CN-Num and CN-Str based prediction is better than that of Ave-Deg based prediction. However, it can be noted that the performance of Ave-Deg has a notable raise with the increase of k.
- CN-Str based approach performs better than CN-Num based approach for mobile social network, but it is on the other side for online social network.

5.3   Network-Based Analysis

Node-level and Pair-wised analysis explain the influence of topological features on communication from micro perspective. In this section, we will discuss it from a macro angle of view to disclose the role of people in their communication.

We first give an introduction of social status theory. According to the theory of Social status[5], an arch from A to B indicates that A has a higher status than B and all triangles in the network should be acyclic.

Based on the theory of social status, Figure 8(a) lists 7 types of triads. For easy understanding, given a triad (A, B, C), we use 1 to denote a directed arch and 0 to denote an undirected edge. Thus label 011 denotes an undirected edge between A and B , and two directed links from C to A and B respectively. To be noticed, one triad defined in this figure violates the social theory, and is labeled as unsatisfied triad.

We make a statistics of all 7 types of triangles on both mobile and online social network. As illustrated in Figure 8(b), mobile and online social network have similar results in violation cases and for both social network, only small part of triangles violate the theory. However, these two social networks have totally different distributions on other triads. In mobile social network, the case of "101" is more distinguished than others, but comparatively, the most outstanding cases in online social network are "100" and "110".

6   Related Work

Various online and social networking services provide different channels for communication. It therefore spurred the researches into the characteristics study for different social networking[2].

Twitter, as a new form of online social network, has attracted much attention in recent years. Characterizing user behavior in Twitter is discussed widely.

Different work explored this problem from different angles. Sue et al. analyzed information spreading and impact of retweet over the entire Twitter sphere[9]. Wu et al. investigated the flow of information among different categories of users[15]. Huberman et al. reported that the number of friends is actually smaller than the number of followers or followings[6].

Mobile social network, as a traditional and typical kind of social network, has been studied a lot from the perspective of communication behavior. Onnela et al. analyzed the structure and tie strengths in call-based mobile communication networks[13]. Banjo et al. developed a theoretical model to illustrate the social effects for cell phone usage in public places[1]. Miklas et al. investigated how to use mobile systems to exploit people's social interactions [12]. Yuan et al. studied the inference of user emotional status from mobile communication network[16]. Wang et al. explored the influential nodes in mobile social network based on communication status[14].

Some communication analysis on other kinds of social network also have been studied, such as, the work from Karagiannis et al. on email social network analysis[8], the work of Leskovec et al. on instant-messaging network mining[10] and the logs-based comparison of search patterns across different platforms from Kamvar et al.[7].

Existing approaches did a lot of work on social communication study. However, they barely explore this problem against different social networks simultaneously. This paper presents a comparison study to identify the communication patterns of people through different social connections. It is a new and interesting topic for research.

7   Conclusion

In this paper, we discussed and presented user communication properties in different social network from three aspects: time, location and topology.

- The observations from temporal analysis included: 1) Communications in mobile social network exhibit a clear periodicity, but no any regularity can be found in online social network; 2) The messages in Mobile social network always gets low reply rate but more quick response, while the opposite case in online social network; 3) During a conversation, the reply time in both social network reduces gradually; 4) The conversation in online social network has longer duration compared with that in mobile social network.
- The observations from geographical analysis showed that compared with mobile social network, the communications in online social network happened without geographical restrictions.
- The observations from topology analysis found that: 1)There is very low level of reciprocity for both mobile and online social network; 2) Over 50% of communications in both mobile and online social network happened with only small set of contact.

References

1. Banjo, O., Hu, Y., Sundar, S.S.: Cell Phone Usage and Social Interaction with Proximate Others: Ringing in a Theoretical Model. In: ICA (2006)
2. Benevenuto, F., Rodrigues, T., Cha, M., Almeida, V.: Characterizing user behavior in online social networks. In: Proceedings of the 9th ACM SIGCOMM Conference on Internet Measurement Conference, IMC 2009, pp. 49–62. ACM, New York
3. Boyd, D., Golder, S., Lotan, G.: Tweet, tweet, retweet: Conversational aspects of retweeting on twitter. In: HICSS, pp. 1–10. IEEE Computer Society (2010)
4. Eagle, N., Pentland, A., Lazer, D.: Inferring Social Network Structure using Mobile Phone Data. PNAS (2007)
5. Guha, R.V., Kumar, R., Raghavan, P., Tomkins, A.: Propagation of trust and distrust. In: WWW, pp. 403–412 (2004)
6. Huberman, B.A., Romero, D.M., Wu, F.: Social networks that matter: Twitter under the microscope. First Monday 14(1) (2009)
7. Kamvar, M., Kellar, M., Patel, R., Xu, Y.: Computers and iphones and mobile phones, oh my!: a logs-based comparison of search users on different devices. In: Quemada, J., Len, G., Maarek, Y.S., Nejdl, W. (eds.) WWW, pp. 801–810. ACM (2009)
8. Karagiannis, T., Vojnovic, M.: Behavioral profiles for advanced email features. In: Quemada, J., Len, G., Maarek, Y.S., Nejdl, W. (eds.) WWW, pp. 711–720. ACM (2009)
9. Kwak, H., Lee, C., Park, H., Moon, S.B.: What is twitter, a social network or a news media? In: WWW, pp. 591–600 (2010)
10. Leskovec, J., Horvitz, E.: Planetary-scale views on a large instant-messaging network. In: WWW, pp. 915–924 (2008)
11. Li, N., Chen, G.: Analysis of a location based social network. In: CSE (4), pp. 263–270. IEEE Computer Society (2009)
12. Miklas, A.G., Gollu, K.K., Chan, K.K.W., Saroiu, S., Gummadi, K.P., de Lara, E.: Exploiting Social Interactions in Mobile Systems. In: Krumm, J., Abowd, G.D., Seneviratne, A., Strang, T. (eds.) UbiComp 2007. LNCS, vol. 4717, pp. 409–428. Springer, Heidelberg (2007)
13. Onnela, J.P., Saramaki, J., Hyvonen, J., Szabo, G., Lazer, D., Kaski, K., Kertesz, J., Barabasi, A.L.: Structure and tie strengths in mobile communication networks. PNAS 104, 7332–7336 (2006)
14. Wang, Y., Cong, G., Song, G., Xie, K.: Community-based greedy algorithm for mining top-k influential nodes in mobile social networks. In: KDD, pp. 1039–1048 (2010)
15. Wu, S., Hofman, J.M., Mason, W.A., Watts, D.J.: Who says what to whom on twitter. In: Srinivasan, S., Ramamritham, K., Kumar, A., Ravindra, M.P., Bertino, E., Kumar, R. (eds.) WWW, pp. 705–714. ACM (2011)
16. Zhang, Y., Tang, J., Sun, J., Chen, Y., Rao, J.: Moodcast: Emotion prediction via dynamic continuous factor graph model. In: ICDM, pp. 1193–1198 (2010)

From Trajectories to Path Network:
An Endpoints-Based GPS Trajectory Partition
and Clustering Framework

Hua Yuan[1,*], Yu Qian[1], Baojun Ma[2], and Qiang Wei[3]

[1] School of Management and Economics, University of Electronic Science
and Technology of China, 610054 Chengdu, China
{yuanhua,qiany}@uestc.edu.cn
[2] School of Economics and Management, Beijing University of Posts and Telecommunications,
100876 Beijing, China
mabaojun@bupt.edu.cn
[3] School of Economics and Management, Tsinghua University, 100084 Beijing, China
weiq@sem.tsinghua.edu.cn

Abstract. In this paper, we aim to mine the interesting locations and the frequent travel sequences in a given geo-spatial region. Along this line, a new partition method is proposed to divide the trajectories into a set of line segments and the geographical-similar endpoints are clustered into groups to detect the fixed territories. Also, a path network is generated to show the linkage relations between these fixed territories. The proposed method can be used to detect frequent movement paths as well as fixed territories from GPS trajectories efficiantly.

Keywords: GPS trajectory, stationary sub-trajectory, clustering, path network.

1 Introduction

With the widespread usage of miniaturized GPS devices, recording the trace data of moving objects become extremely easy and useful work while it provides enormous business opportunity in geography navigation and recommendation system[1]. During the past years, a bunch of research has been performed based on individual location history represented by GPS trajectories. These works include detecting individual locations [2], recognizing user-specific activities[3,4] and predicting traveler's movement [2]. The trajectory pattern mining problem was introduced in [5]. Following this work, some important efforts had been devoted, such as[6,7,8].

Finding some key information, such as characteristic points[9,10] and representative line[10], is the feasible method to help people extracting useful information from GPS trajectory. However, it is hard to get such information because the recorded GPS data are always non-uniformity, sparse, lost and inconsistency with the endpoints in cases that the users turn on and off the GPS-enabled devices casually. In this paper, we present an endpoints-based GPS trajectory partition and clustering framework to mine users' interested locations and frequent travel sequences in a given geo-spatial region.

---

* Corresponding author.

F. Li et al. (Eds.): WAIM 2014, LNCS 8485, pp. 740–743, 2014.

2   The Method

2.1   Stationary Sub-trajectory

Let $TR = \{g_1 g_2 ... g_i g_{i+1} ... g_n\}$ denote the *trajectory* of the object moving from g_1 to g_n. For any sub-trajectory $STR = \{g_s g_{s+1} ...... g_t\} \subseteq TR$, $\overrightarrow{g_s g_{s+1}}$ is the *first move action*. Now, suppose the projection of $g_i, (i = s + 1, ..., t)$ on $\overrightarrow{g_s g_{s+1}}$ is g_i', then the distance between g_i and g_i' is called *position disturbance* (denoted by $d_{g_i} = |g_i g_i'|$).

A sub-trajectory STR is a *stationary sub-trajectory* (SST) if the position disturbance of each GPS point $g_i, s + 1 < i \leq t$ changes not rapidly with respect to the *first move action* $\overrightarrow{g_s g_{s+1}}$ while the moving behavior of g_{t+1} changes rapidly.

2.2   Trajectory Partitioning Method

In fact, the key issue for partitioning a TR into SSTs is to find out all the *characteristic points*[10] in it. In this section, we propose a new trajectory partitioning algorithm which aims at finding the points where the behavior of a trajectory changes rapidly. The main idea is to check the value of $d_{g_i}, i = 2, ..., n$ with respect to the present *first move action* (Algorithm 1).

Algorithm 1. Trajectory Partitioning Algorithm

1: **Input**: Trajectory $TR = \{g_1 g_2 g_3 ...... g_n\}$, threshold d_0;
2: **Output**: A set of characteristic points CP;
3:   $g_1 \rightarrow CP$; $i = 1$;
4: **repeat**
5:     *first move action*$= \overrightarrow{g_i g_{i+1}}$;
6:     **for** $j = i + 2$ to n **do**
7:         **if** $d_{g_j} \geq d_0$ **then**
8:             $g_{j-1} \rightarrow CP$; $i = j - 1$;
9:         **end if**
10:    **end for**
11: **until** $g_n \rightarrow CP$.
12: **return** CP.

2.3   Characteristic Points Clustering and Path Network

Given N trajectories, the clustering processes can be specified as follows: Firstly, we partition each $TR_i = \{g_{1_i}, ..., g_{n_i}\}, i = 1, ..., N$, into m_i SSTs. Then, we introduce the general clustering method to cluster all the points in $CP = \cup_{i=1}^{N} CP_i$ into l clusters: $C_1, ..., C_l$, based on the Euclidean distance of paired points. The element number of each cluster is $|C_i|, i = 1, ..., l$. Finally, we calculate the centroid point c_i of cluster $C_i, (i = 1, ..., l)$ and use these centroid point to represent each cluster.

The edge between two nodes c_i and c_j $(i \neq j)$ can be used to represent all the possible SSTs whose start point in C_i (or C_j) and end point in C_j (or C_i). In another word, it approximates to the common path between area c_i to c_j. We can construct an undirected *path network* by connecting these fixed territories.

3   Experimental Results

3.1   Efficiency and Preciseness

The following experiments are based on the real GPS datasets[1]. The results about efficiency and preciseness are shown in Figure 1. It indicates that: 1) The efficiency of our method ($d_0 = 1$) is more faster than that of the method presented by [10]. 2) The computation speed will become more faster with increasing value of d_0. 3) The curves in Figure 1 are almost straight lines, these are experimental evidences about the truth that both these two methods have computation complexity of $O(n)$.

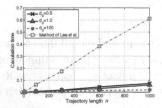

Fig. 1. Comparison of Efficiency   **Fig. 2.** Comparison of Preciseness

Figure 2 shows the results about preciseness, the black dash line is the base of raw trajectory. The pink line is the number of characteristic points generated with the method presented by Lee et al., and the blue line is that with our method. The red line presents the number of characteristic points found by these two methods commonly. We can see that, 1) Both the preciseness of these two methods would become worse along with the increase of trajectory length, n. 2) The method presented by Lee et al. has advantage in preciseness because it keeps more GPS points as characteristic points.

3.2   Path Network

In the following, we choose randomly 8 users' data to conduct the experiments. Their path networks are generated as in Figure 3. Each network is a spatial map of the typical motions for a specific user, and it contains valuable information about: the frequent pathes, the fixed territories, and the movement correlations among representative points.

4   Conclusion

In this work, we try to trim GPS trajectories into *fixed territories* and *frequent path* to generate a spatial map of typical motions, i.e., *path network*, by taking into account of users' historic travel experiences as well as the correlation between locations.

The methods proposed in this work are efficient for mining information hidden in the trajectory data, especially the frequent path, fixed territories and movement intention, which can provide business opportunity in geography information service.

---

[1] http://research.microsoft.com/en-us/projects/urbancomputing/

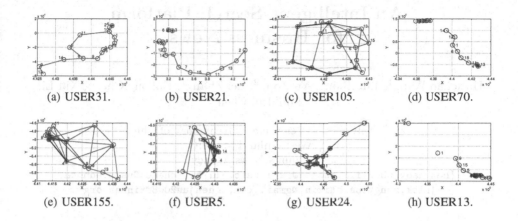

(a) USER31.   (b) USER21.   (c) USER105.   (d) USER70.

(e) USER155.   (f) USER5.   (g) USER24.   (h) USER13.

Fig. 3. Path networks generated for the eight selected users

Acknowledgements. This work is supported by the NSFC(Nos. 71271044/U1233118/71102055), the SRFDP (20100185120024) and the Fundamental Research Funds for the Central Universities (No. 2014RC0601).

References

1. Zheng, Y., Zhou, X. (eds.): Computing with Spatial Trajectories. Springer, Berlin (2011)
2. Ashbrook, D., Starner, T.: Using gps to learn significant locations and predict movement across multiple users. Personal and Ubiquitous Computing 7(5), 275–286 (2003)
3. Patterson, D.J., Liao, L., Fox, D., Kautz, H.: Inferring high-level behavior from low-level sensors. In: Dey, A.K., Schmidt, A., McCarthy, J.F. (eds.) UbiComp 2003. LNCS, vol. 2864, pp. 73–89. Springer, Heidelberg (2003)
4. Zheng, Y., Xie, X.: Learning location correlation from gps trajectories. In: Proceedings of the MDM 2010, pp. 27–32 (2010)
5. Giannotti, F., Nanni, M., Pinelli, F., Pedreschi, D.: Trajectory pattern mining. In: Proceedings of the 13th ACM SIGKDD, pp. 330–339 (2007)
6. Monreale, A., Pinelli, F., Trasarti, R., Giannotti, F.: Wherenext: a location predictor on trajectory pattern mining. In: Proceedings of the KDD 2009, pp. 637–646 (2009)
7. Ye, Y., Zheng, Y., Chen, Y., Feng, J., Xie, X.: Mining individual life pattern based on location history. In: Proceedings of the MDM 2009, pp. 1–10 (2009)
8. Wei, L.-Y., Zheng, Y., Peng, W.-C.: Constructing popular routes from uncertain trajectories. In: Proceedings of the 18th SIGKDD, pp. 195–203 (2012)
9. Lee, J.-G., Han, J., Li, X., Gonzalez, H.: Traclass: trajectory classification using hierarchical region-based and trajectory-based clustering. Proceedings of the VLDB Endowment 1(1), 1081–1094 (2008)
10. Lee, J.-G., Han, J., Whang, K.-Y.: Trajectory clustering: a partition-and-group framework. In: Proceedings of the 2007 ACM SIGMOD (2007)

An Intelligent Search Platform
for Business News

Hanchao Wang[1], Lili Zhou[1], Yu Zong[2], Lei Zhang[1], Enhong Chen[1], Xin Li[1],
and Jun Chen[3]

[1] University of Science and Technology of China
[2] West Anhui University
[3] Xinhua News Agency
{hanchaow,zhoulili,stone}@mail.ustc.edu.cn, cheneh@ustc.edu.cn,
{nick.zongy,ustclixin}@gmail.com, chenjun2008@xinhua.org

Abstract. Living in a data driven world, the business news is very crucial for making economic decisions. To help decision makers obtain related business news quickly, two kinds of providers for business news,
i.e., the search engine (e.g., Google News) and business portals (e.g.,
Reuters), are widely used. Though the keyword-based search engine is
simple and easy to use, it has relatively low precision of the returned
results and cannot directly provide news of particular business domains
such as currency and real estate. In contrary, the portals can provide a
variety of news of specific business domains, but it is difficult for users
to browse since the front page looks so bloated and has many irrelevant
ads. To solve the above problems, in this paper we propose and implement a platform named **I**ntelligent **S**earch **P**latform for **B**usiness **N**ews
(ISPBN). This new platform not only combines the advantages of both
search engine and portals, but also provides further analysis to discover
the hidden relationships of different business news. To be specific, we incorporate automatic classification technology into the search platform to
organize and retrieve business news in different domains. Furthermore,
to fast guide users finding diversified and useful news, we construct a
dynamic knowledge network graph to display the hidden relationships
among news. Finally, we show the performance of our subsystems and
present the final user interface of the proposed search platform.

Keywords: intelligent search, search engine, business news, web mining.

1   Introduction

With the high-speed development of the Internet, the information including all
kinds of business news on the World Wide Web (WWW) is growing at an exponential rate. There is no doubt that the business news contains immense wealth
and is very important to help people make decisions. However, people have to
face the serious problem of information overload. Therefore, how to help users
acquire valuable business news easily and quickly becomes a very vital problem.

F. Li et al. (Eds.): WAIM 2014, LNCS 8485, pp. 744–755, 2014.

Generally, there are two widely used providers for business news, namely the search engine (e.g., Google News [1]) and business portals (e.g., Reuters [2]). Most of us are now using search engine for information retrieval since it is very simple and easy to use. Users only need to input keywords, then they could acquire relevant results. However, too many returned results lead to relatively low precision and recall [1], and users have to spend a large amount of time on finding out useful information. More importantly, some users just care about news related to a particular field (e.g., currency, real estate) which the search engine cannot offer. Although some business portals provide the users with relatively professional and authoritative news of different business domains, there exists two drawbacks as follows. a) The home pages in these portals display all kinds of news related to different fields, which look bloated and huge. Therefore, it may confuse the users who are used to getting news by using search engine. b) These portals just simply display the news, and they cannot find out the hidden relationships of different business news, for example, the news about "housing price" may be related to news of "real estate control policy" or "building material industries".

In order to address the problems mentioned above, in this paper, we propose and implement a platform called Intelligent Search Platform for Business News (ISPBN). In this platform, we design and implement a vertical search engine system which incorporates automatic classification technology to organize and retrieve business news in different domains. In this way, our platform combines the advantages of both search engine and portals. The user can not only acquire news by keyword-based query, but also browse the news of specific fields by its category. To help users quickly understand business stories, we apply Name Entity Recognition (NER) techniques to recognize the entities (e.g., person names, time) in the clicked Web pages. Furthermore, we propose to construct dynamic knowledge network of news. Based on this knowledge network, we can discover the hidden relationships of news in different domains for news navigation. For example, if a user clicks a news story about "housing price", he can get the related news about "control policy" of "real estate" or news about "furniture industry". Our contributions can be summarized as follows:

- We propose and implement an Intelligent Search Platform for Business News (ISPBN). This novel platform combines the advantages of both search engine and business portals to help users acquire business news easily and quickly.
- We build a business thesaurus and implement an interactive management system used for optimizing our models such as classification model.
- We construct a dynamic knowledge network of news, which can be used to discover the hidden relationships of different business news. Based on this knowledge network, users can easily get the useful news they want.

The rest of this paper is organized as follows. We introduce the overview of ISPBN in Section 2 and describe the design of ISPBN in Section 3. We present the experimental results and final user interface in Section 4 and discuss the related work in Section 5. Finally, we conclude our work in Section 6.

---

[1] https://news.google.com/

[2] http://www.reuters.com/

2   Overview of ISPBN

We give an overview of ISPBN and introduce how each part of ISPBN works together in this section. As shown in Fig. 1, the architecture of our platform consists of four parts. 1) A vertical search engine including Intelligent Crawler,

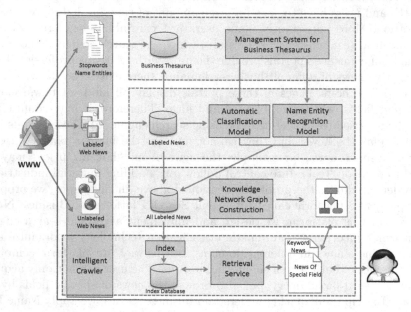

Fig. 1. The architecture of ISPBN

Index and Retrieval Service. 2) A Knowledge Network Graph Construction module. 3) Two models including Automatic Classification Model and Name Entity Recognition Model. 4) A Management System for Business Thesaurus. In the first part, we implement an intelligent crawler based on Nutch [3] and employ index and retrieval technologies based on Solr [4]. The crawler can get three kinds of data: the ordinarily unlabeled business news as the main part of index database, the specially labeled news used for training classification model and extracting feature words of each category, some famous name entities and stopwords. In the second part, we construct the knowledge network graph which can display the hidden relationships of different business news by analyzing all labeled news. In the third part, we integrate the search engine with two models, i.e., automatic classification model based on SVM [2] to provide the users with news of a specific field, and NER model to recognize important name entities of business news. Both of them are finished before indexing. In the fourth part, we design and implement a management system of business thesaurus for the sake of optimizing our models. Our thesaurus consists of stopwords, famous name entities and feature words which are extracted form the specially labeled news.

---

[3] http://nutch.apache.org/
[4] http://lucene.apache.org/solr/

3   Implementation of ISPBN

In this section, we describe the design and implementation of each system in our platform in detail. At first, we present the search engine system combined with automatic classification and NER techniques. Then we introduce how to build and manage the business thesaurus. Finally, we describe how to construct the dynamic knowledge network graph.

3.1   Vertical Search Engine

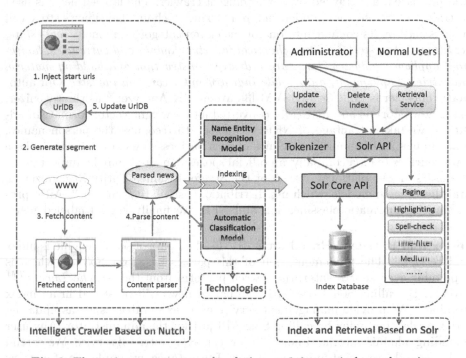

Fig. 2. The main components and techniques of the vertical search engine

The vertical search engine is the foundation of our platform. In this subsection, we not only introduce the vertical search engine, but also describe how to integrate automatic classification and NER technologies which are used for classifying unlabeled news and recognizing important name entities. Fig. 2 presents the main components and techniques of the vertical search engine. From Fig. 2, we can easily know that our vertical search engine is divided into three parts. We introduce each part as follows.

Intelligent Crawler Based on Nutch. As we all know, a search engine is formed with crawler, index and retrieval. The crawler is mainly responsible for acquiring a variety of data (e.g., web page, documents, videos, etc.). In our platform, we implement an intelligent crawler mainly used for crawling all kinds of business news. The implementation of our crawler is based on Nutch which is an open source web crawler software from the *Apache Lucene*^{TM} project. The workflow of our crawler is divided into five steps. The crawler first injects start urls to Url Database (UrlDB). Then it generates segment which contains urls scheduled for fetching. After that, it fetches the content and parses the content by using content parser. Finally, it updates UrlDB with extracted urls. In order to build a vertical search engine in business field, we just crawl portals related to business. We periodically download business news from main business portals of China such as *sina.com* and *163.com*.

Automatic Classification and NER. Two kinds of business news, labeled and unlabeled, are crawled by our intelligent crawler. The labeled news is used to train our classification model. Before indexing, each unlabeled news story will be classified by its content into one of the eleven categories, namely, *real estate, textile industry, steel industry, chemical industry, finance and currencies, household appliance industry, furniture industry, construction and building material industries, employment, automobile industry and decoration industry.* Our automatic classifier is built on LIBSVM [3]. We use IK Analyzer [5] as the tokenizer. Each news story is automatically converted into a weight vector format. In addition, we take advantage of NER technology to recognise the person names, place names and organization names of each business news story, because these name entities (e.g., a company, an official spokesman, etc.) may be important to the users. In our platform, we adopt the Stanford Named Entity Recognizer [4] and the Chinese models which use distributional similarity clusters [6]. After processing by automatic classification and NER, we finally get all labeled news.

Index and Retrieval Based on Solr. The last part of search engine is index and retrieval. The implementation of this part is based on $Solr^{TM}$ which is a popular open source enterprise search platform from the *Apache Lucene*^{TM} project. By calling Solr Core API, the parsed news will be stored in an index database. We provide management service of index database such as updating index or deleting index by calling Solr API just for the administrator. Any user including administrator and normal user, can enjoy retrieval service. We extend and enrich the retrieval functions of Solr. For instance, we implement functions like the results paging, highlighting both of title and content, spell check, time filter, the number of medium statistics, browsing news by category and so on.

3.2   Business Thesaurus

Each special field has its own feature words. When training a classification model, each news story is converted into a weight vector of the words. The weight can be calculated by using some classical and popular methods such as TF and IDF. Words in different fields have different weights. In a special field, we regard the words whose weight is higher than the general words as the feature words of this field. These feature words is vital for training models. Hence, we build a business thesaurus and implement a management system. In this subsection, we

---

[5] http://code.google.com/p/ik-analyzer/
[6] http://nlp.stanford.edu/software/CRF-NER.shtml

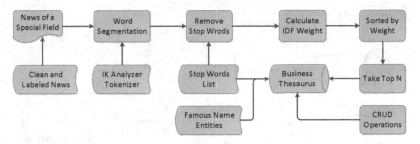

Fig. 3. The process of building and managing business thesaurus

introduce how to generate the feature words of each predefined field such as real estate, and we show how to extend and manage the thesaurus.

Fig. 3 illustrates the whole process of building and managing business thesaurus. Our thesaurus consists of three parts: the feature words of each predefined field, the stop words, the famous name entities (e.g., some person names, some company names, etc.). We take the news which is used for training classifier as the data source to extract feature words for each special field. Each news story must do Chinese word segmentation first. Then we remove stop words through the stop words list and calculate the weight using IDF. After that, we sort words in ascending order by their weight and take top N [7] words. Finally we put these words into the thesaurus database. We put stop words list into the database too, thus we can manage them by manual to optimize our model. In addition we use special crawler to acquire some famous name entities to assist our NER model. We implement the basic CRUD (create, read, update, delete) operations on this thesaurus database and build the management system. Experts could modify some inaccurate words or add new words in manual considering that there are a great number of noise in our corpus. The feedback of experts can be helpful for optimizing our models.

3.3 Dynamic Knowledge Network

News stories of different domains may have hidden relationships. For instance, a news story about housing price may have relationships with news stories of three major categories, namely upstream and downstream industries (e.g., construction and building material industries, furniture industry), industry policy such as real estate control policy, and macro-economy (e.g., stock, currency). If we can extract topics of news and find out the relationships of news according to their topics, we could guide and extend the users' interest. That's the main idea of our network. In this subsection, we mainly introduce how to construct and display the dynamic knowledge network graph. As shown in Fig. 4, the whole construction process consists of two main parts, namely offline topic extraction and online displaying.

---

[7] N is a user specified parameter. In our platform, it is set to 500 and it can be changed according to the size of the corpus.

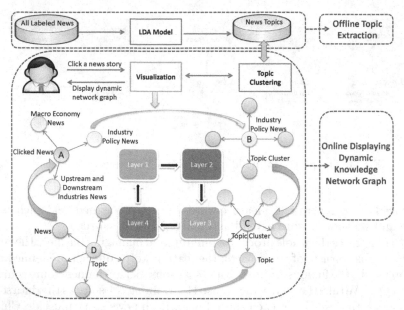

Fig. 4. The construction process of dynamic knowledge network graph. In this graph, (1) Node A represents a news story and is displayed by its title; (2) Node B represents a set of news stories and is displayed by its category; (3) Node C represents a cluster of topics namely a topic cluster and is displayed by a set of words, as we cluster the similar topics of a specific category; (4) Node D represents a specific topic and is displayed by a set of words.

Offline Topic Extraction. If a user clicks a news story, maybe he is interested in the news stories that have the same topic. We try to discover the hidden relationships of news according to their topics by using LDA (Latent Dirichlet Allocation) model [5]. LDA is a three-level hierarchical Bayesian model, in which each document of a collection is modeled as a finite mixture over an underlying set of topics. Each topic is represented by a distribution over words. In our platform, we use JGibbLDA [8] to train LDA model and extract each news story's topics [6]. At the same time, the topics will be stored in a database. We design a table to store a news story's distribution over topics and another table to store a topic's distribution over words.

Online Displaying. When a user clicks a news story, as shown in Fig. 4, our platform will provide him with the hidden relationships through displaying the dynamic network graph. The network graph consists of four layers in due order. Each layer is the extension and supplement of the former layer. In the first layer, the center node denotes the news story that the user clicks. The other three nodes connected to the center node are a set of news stories which are related to the center news story and belong to three major categories including

---

[8] http://jgibblda.sourceforge.net/

macro economy, upstream and downstream industries, and industry policy. For instance, a user may care about housing price and he clicks a news story whose topics are about housing price. The first layer will extend the center news from three views to guide the user's interest. If the user is interested in news of the industry policy such as real estate control policy, he could click the industry policy node. Thus we step into the second layer, the center node is the industry policy and its each child node denotes a cluster of similar topics. It is the further arrangement and extension of the first layer. The user could click a topic cluster node that interests him. Thus we step into the third layer and each new node is a specific topic of the topics cluster. If the user clicks one of the unfolded nodes, the top K relevant news stories come out immediately as its children nodes. At the same time, we step into the fourth layer. The user could click and then browse the relevant news stories in detail. So far, we finish the construction process of dynamic knowledge network graph for the business news.

4   Experimental Results and Final User Interface

In this section, we evaluate the performance of the classifier and the impacts of our management system of thesaurus on the classifier. Then we present the final user interface of each system in the platform.

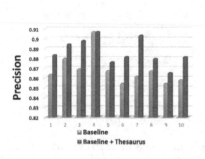

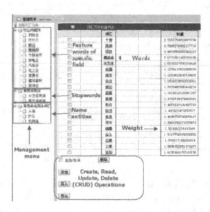

Fig. 5. The precision of the classifier

Fig. 6. The user interface of management system of business thesaurus

Classification Performance and the Management System Evaluation. To evaluate the performance of our classifier, a total of 5500 online news stories are collected by our crawler from five specific portals. These news stories are manually classified into eleven categories as mentioned in subsection 3.1, and each category has 500 news stories. We adopt 10-Cross-validation as the evaluation methodology and the average performance is reported. In order to evaluate the impacts of our management system of thesaurus on the classifier, we do another 10-Cross-validation experiments as a comparison. Fig. 5 shows the result of our experiment. In the baseline method, we only use IK Analyzer to do Chinese Word Segmentation, while in our method we use IK Analyzer [9] combined with our thesaurus to do Chinese Word Segmentation. As shown, almost every set experiment in our method has the higher precision than the baseline method. It indicates that the thesaurus could optimize our model. The baseline has an overall precision of 86.84%, and in our method, the overall precision is improved to 88.73%. The user interface of our management system of our thesaurus is shown in Fig. 6.

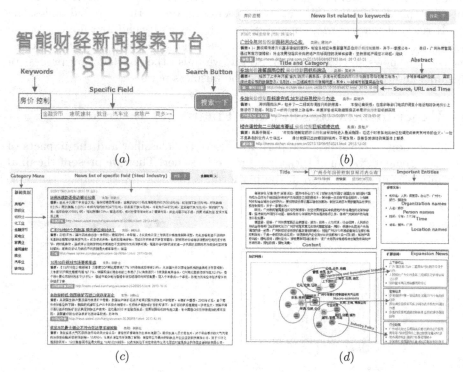

Fig. 7. The user interface of ISPBN: (a) The home page of ISPBN; (b) Get news by keyword-based search; (c) Browse news of specific fields; (d) Display the dynamic knowledge network graph, recommend name entities and extension news.

Final User Interface. Fig. 7 shows the user interface of our platform. From the home page in Fig. 7 (a), the user can not only acquire business news by inputting keywords as shown in Fig. 7(b), but also browse news of specific fields as shown in Fig. 7 (c). What'more, once the user clicks a news story, he will jump to another page which displays the dynamic knowledge network graph and recommends the important name entities and relevant news to the user in order to extend and guide their interest as shown in Fig. 7 (d).

---

[9] http://code.google.com/p/ik-analyzer/

5 Related Work

We present a brief survey of the relevant existing approaches both for intelligent search and web mining. In our platform we employ more web mining techniques aiming to make the search platform more intelligent and human-friendly.

5.1 Intelligent Search

In order to overcome the shortcomings of the traditional keyword based search engine, a great many researchers have focused on intelligent search. One of the hottest research topics is the Semantic Web [7]. There are a large number of related work on Semantic Web based intelligent search [8,9,10,11] and on Agent based intelligent search [12,13,14]. Specifically, [8] proposed a platform called QuestSemantics which provides automated ontology-based metadata creation and resource annotation based on a detailed ontological model of the domain, and [9] used the context-aware metadata which is stored in domain ontology. [10] presented the semantic web based search engine called SWISE which extracted metadata information of the pages and use the power of ontology. [11] investigated the semantic search performance of search engines by comparing with three keyword-based search engines. An agent based intelligent search framework for product information using ontology mapping was proposed in [12], while [13] presented the architecture of an agent-based intelligent search engine system for effective web mining. In this paper, we adopt web mining techniques to make our platform more intelligent. Our platform not only supports keyword-based search service, but also provides other useful service such browsing news by categories and knowledge graph navigation service.

5.2 Web Mining

Web mining is classified into three categories: Web content mining, Web structure mining and Web usage mining [15]. Web content mining aims to discover the useful information from web contents. There are two popular approaches used in web content mining, namely Agent based search approach and database approach [16,12]. There exists many popular web mining techniques for unstructured data such as information extraction techniques [17,18], topic detection and tracking techniques [19,20], summarization techniques [17], classification techniques [17,21], clustering techniques [20] and information visualization techniques [19,22]. For example, [17] proposed a system called Financial Information Digest System including three tasks (i.e., classification, information extraction, and information enquiry). [21] presented a Theme-Based News Retrieval System which incorporated a classification framework based on Support Vector Machines. We also adopted SVM in our platform, however we built a management system of business thesaurus to further improve the accuracy of the classification model. In addition, we proposed to construct a novel knowledge network which can be used to mine the hidden relationships of different business news. [19] also tried to understand the mutual relationships between information flows and social activity. Note that [19] aimed to explain abnormal financial market volatility, while we tried to guide and extend the users' interest.

6   Conclusion and Future Work

In this paper, we proposed and implemented an Intelligent Search Platform for Business News (ISPBN). This platform could satisfy enterprise search requirement because it combines the features of two open source projects of Apache (namely Nutch and Solr). Moreover, we integrated classification techniques into the search engine, so as to make it easy for users to acquire and browse news stories in different domains as well as by using keyword query. In the mean time, we built a management system of business thesaurus which could be used for optimizing models. In addition, to help users quickly understand business stories, we also incorporated name entity recognition techniques into the search platform to recognize the name entities (e.g., person names, organization names) in the clicked Web pages. Furthermore, we proposed to construct dynamic knowledge network graph based on business news, which can be used to find out the hidden relationships of news in different domains for news navigation. Based on this platform, users can easily and quickly get the information they need.

For the future work, we try to build the users' personal profiles and implement a personal recommendation system according to the browsing and searching history log. We hope that our platform could provide the users with more accurate, personalized and intelligent service to help them make decisions.

Acknowledgments. This research was partially supported by grants from the National Science Foundation for Distinguished Young Scholars of China (Grant No. 61325010), the National High Technology Research and Development Program of China (Grant No. 2014AA015203), the Science and Technology Development of Anhui Province, China (Grants No. 13Z02008-5 and 1301022064), the International Science & Technology Cooperation Plan of Anhui Province (Grant No. 1303063008), the National Key Technology Research and Development Program of the Ministry of Science and Technology of China (Grant No. 2012BAH17B03) and the Nature Science Research of Anhui (Grant No. 1208085MF 95).

References

1. Chakrabarti, S.: Data mining for hypertext: A tutorial survey. ACM SIGKDD Explorations Newsletter 1(2), 1–11 (2000)
2. Vapnik, V.: The nature of statistical learning theory. Springer (2000)
3. Chang, C.C., Lin, C.J.: LIBSVM: A library for support vector machines. ACM Transactions on Intelligent Systems and Technology 2, 27:1–27:27 (2011), Software available at http://www.csie.ntu.edu.tw/~cjlin/libsvm
4. Finkel, J.R., Grenager, T., Manning, C.: Incorporating non-local information into information extraction systems by gibbs sampling. In: Proceedings of the 43rd Annual Meeting on Association for Computational Linguistics, pp. 363–370. Association for Computational Linguistics (2005)
5. Blei, D.M., Ng, A.Y., Jordan, M.I.: Latent dirichlet allocation. The Journal of Machine Learning Research 3, 993–1022 (2003)

6. Phan, X.H., Nguyen, L.M., Horiguchi, S.: Learning to classify short and sparse text & web with hidden topics from large-scale data collections. In: Proceedings of the 17th International Conference on World Wide Web, pp. 91–100. ACM (2008)
7. Berners-Lee, T., Hendler, J., Lassila, O., et al.: The semantic web. Scientific American 284(5), 28–37 (2001)
8. Tamma, V.: Semantic web support for intelligent search and retrieval of business knowledge. IEEE Intelligent Systems 25(1), 84–88 (2010)
9. Khattak, A.M., Mustafa, J., Ahmed, N., Latif, K., Khan, S.: Intelligent search in digital documents. In: IEEE/WIC/ACM International Conference on Web Intelligence and Intelligent Agent Technology, WI-IAT 2008, vol. 1, pp. 558–561. IEEE (2008)
10. Shaikh, F., Siddiqui, U.A., Shahzadi, I., Jami, S.I., Shaikh, Z.A.: Swise: Semantic web based intelligent search engine. In: 2010 International Conference on Information and Emerging Technologies (ICIET), pp. 1–5. IEEE (2010)
11. Tumer, D., Shah, M.A., Bitirim, Y.: An empirical evaluation on semantic search performance of keyword-based and semantic search engines: Google, yahoo, msn and hakia. In: Fourth International Conference on Internet Monitoring and Protection, ICIMP 2009, pp. 51–55. IEEE (2009)
12. Inamdar, S., Shinde, G.: An agent based intelligent search engine system for web mining. Research, Reflections and Innovations in Integrating ICT in Education (2008)
13. Kim, W., Choi, D.W., Park, S.: Agent based intelligent search framework for product information using ontology mapping. Journal of Intelligent Information Systems 30(3), 227–247 (2008)
14. Hai-long, C.: Design and realization of intelligent search engine based on multi-agents [j]. Journal of Harbin University of Commerce (Natural Sciences Edition) 2, 010 (2009)
15. Al-Azmi, A.A.R.: Data, text, and web mining for business intelligence: A survey. International Journal of Data Mining & Knowledge Management Process 3(2) (2013)
16. Srividya, M., Anandhi, D., Ahmed, M.I.: Web mining and its categories-a survey. International Journal of Engineering and Computer Science, IJECS 2(4), 1338–1345 (2013)
17. Lam, W., Ho, K.S.: Fids: an intelligent financial web news articles digest system. IEEE Transactions on Systems, Man and Cybernetics, Part A: Systems and Humans 31(6), 753–762 (2001)
18. Domenech, J.: An intelligent system for retrieving economic information from corporate websites. In: Proceedings of the 2012 IEEE/WIC/ACM International Joint Conferences on Web Intelligence and Intelligent Agent Technology, vol. 1, pp. 573–578. IEEE Computer Society (2012)
19. Hisano, R., Sornette, D., Mizuno, T., Ohnishi, T., Watanabe, T.: High quality topic extraction from business news explains abnormal financial market volatility. PloS One 8(6), e64846 (2013)
20. Dai, X.Y., Chen, Q.C., Wang, X.L., Xu, J.: Online topic detection and tracking of financial news based on hierarchical clustering. In: 2010 International Conference on Machine Learning and Cybernetics (ICMLC), vol. 6, pp. 3341–3346. IEEE (2010)
21. Maria, N., Silva, M.J.: Theme-based retrieval of web news. In: Suciu, D., Vossen, G. (eds.) WebDB 2000. LNCS, vol. 1997, pp. 26–37. Springer, Heidelberg (2001)
22. Gupta, V., Lehal, G.S.: A survey of text mining techniques and applications. Journal of Emerging Technologies in Web Intelligence 1(1), 60–76 (2009)

Structured Sparse Linear Model
for Social Trust Prediction

Deng Yi, Yin Zhang, Yuqi Wang, and Baogang Wei

College of Computer Science, Zhejiang University, Hangzhou, China
{yideng,zhangyin98,hunter3g601,wbg}@zju.edu.cn

Abstract. Social trust prediction aims at predicting the missing trust relations between online users. In this paper, we propose a novel and scalable structured sparse linear model for social trust prediction from a global neighborhood-based collaborative filtering perspective. We formulate the prediction problem as a set of independent linear regression problems regularized by pairwise elastic net, to automatically learn correlation coefficients between a user and its most similar neighbors. In order to deal with large-scale sparse social trust data, we utilize efficient hashing techniques and stochastic coordinate descent algorithm to cut down the computational cost of training model. The experimental results on three real-world data sets show that our approach can significantly outperform the other tested methods in terms of prediction quality and efficiency.

Keywords: Social Trust Prediction, Structured Sparse Linear Model, Pairwise Elastic Net, Hashing Technique, Stochastic Coordinate Descent.

1 Introduction

Nowadays, many social websites allow users to vote each other to facilitate the trustworthiness evaluation. For example, users can give trust/distrust votes to other users based on their reviews of items on Epinions[1] and Ciao[2]. We can represent these users(nodes) and their trust/distrust votes(directed edges) as a trust graph. This trust graph can also be transferred into a matrix, if user i trusts user j , then the corresponding matrix entry is 1, otherwise 0. Trust prediction is therefore to predict the missing values in this matrix and can be viewed as a special case of the general link prediction in social networks[1].

The challenges of developing trust prediction algorithms are three-fold: (1) *Scalability*, a trust graph often consists of millions of users, (2) *Sparsity*, just extremely sparse trust relations are available, making trust prediction a difficult task. (3) *Skewness*, a small proportion of users often specify most of trust relations while a large proportion of users specify few trust relations[2]. Hence, an ideal method of trust prediction is supposed to efficiently handle large-scale

---

[1] http://www.epinions.com/
[2] http://www.ciao.co.uk/

F. Li et al. (Eds.): WAIM 2014, LNCS 8485, pp. 756–767, 2014.

data sets, while addressing the problem of sparseness and skew distribution in user-specified trust relations.

In this paper, we propose a novel and scalable Structured Sparse Linear Model(SSLM) for social trust prediction. SSLM formulates the prediction problem as a set of independent linear regression problems regularized by pairwise elastic net. Furthermore, we apply efficient hashing techniques and stochastic coordinate descent algorithm to expediting the computation of our model in training phase. Our main contributions are summarized as follows:

- We treat trust prediction problem as a set of linear regression problems. To learn a set of structured sparse correlation coefficients between a user and other similar users, we introduce the pairwise elastic net[3] regularizer into each linear regression problem, which takes advantage of the prior user similarity information in the trust graph and brings both sparsity and structured grouping effect to the coefficients vector.
- We utilize efficient hashing techniques, i.e., one permutation hashing[4] and locality sensitive hashing[5], to efficiently calculate the similarity values between each pair of users, which makes our model much more scalable.
- We propose a scalable stochastic coordinate descent algorithm for solving each linear regression problem regularized by pairwise elastic net, in order to significantly reduce the computational cost of training process. A simulation study demonstrates the efficiency and effectiveness of our stochastic coordinate descent algorithm.
- We conducted experiments on the data sets from three real social websites, i.e., Epinions, Ciao and Slashdot. The experimental results show that our approach can significantly perform better than other tested methods in terms of efficiency and trust prediction quality.

The rest of this paper is organized as follows: Section 2 briefly introduces some related works. In Section 3, we describe our structured sparse liner model for social trust prediction and present how to apply hashing techniques and the efficient stochastic coordinate descent algorithm into our model. Section 4 reports the experimental evaluation and results. Finally, we conclude our work in Section 5.

2   Related Work

In recent years, many researchers have studied social trust prediction problem. Existing trust prediction methods can be roughly divided into the following categories:

The first category is supervised methods[6][7], which first construct features from available sources and then train a binary classifier based on these features by considering existing trust relations as labels. However, online trust relations often follow a power law distribution, which makes the classification problem extremely imbalanced. The performance of these methods is also sensitive to the sampled negative samples[8].

The second category is based on trust propagation. The authors in [9] proposed several atomic propagations such as direct propagation, cocitation propagation, transpose propagation and trust coupling propagation. TidalTrust[10], which is a continuous trust inference algorithm, leverages the path length from the source to sink and various properties of continuous ratings. However, trust propagation based methods strongly depend on existing trust relations among users and they might fail when existing trust relations are sparse[2].

The last category is low-rank approximation methods, which can be done using the singular value decomposition[11] or trace norm minimization[12]. However, there still exists some issues with these methods, i.e., the observed entries are assumed to be sampled uniformly at random, but typical real-world trust graphs are empirical examples of power laws[13]. Hence, the author in [1] proposed a robust matrix completion method to alleviate these problems. Some authors also utilized matrix factorization to seek a low-rank approximation[2][14].

Our approach is a global neighborhood-based approach that models the correlation structure among nodes in trust graphs based on a structured sparse linear model. Our model is derived from the previous work [15] for recommendation scenario by introducing the elastic net regularizer capable of automatic selecting groups of correlated variables, in order to accomplish high-quality social trust predictions.

3   Structured Sparse Linear Model for Trust Prediction

In this section, we first introduce our structured sparse linear model for social trust prediction. Then, we present how to use hashing techniques to significantly improve the efficiency and scalability of our method.

3.1   The Proposed Model

Previous studies in sociology and our life experience suggest that the persons in the same social circle often exhibit similar behavior and tastes[16]. This phenomenon signals that we can infer the trust value for an individual user from his or her social neighbors. As shown in Figure 1, the trust value from someone (e.g., Rose or Tom) to Jack can be expressed as a linear combination of trust values which this user has given to Jack's social neighbors, assuming that the aggregation coefficients between Jack and his neighbors can be learnt in advance.

More formally, Given a user-user trust relation matrix $\mathbf{A}$ of size $M \times M$, whose entry a_{ij} is 1 if user i has trusted user j, otherwise the value is 0, M is the number of users. The missing trust value from user i to user j can be calculated as a sparse aggregation of trust values which user i has given to user j's social neighbors, and can be formulated as:

$$a_{ij} = \hat{\mathbf{a}}_\mathbf{i}^\mathbf{T} \mathbf{w_j}, \tag{1}$$

where $\hat{\mathbf{a}}_\mathbf{i}^\mathbf{T} \in \mathbb{R}^K$ is a row vector of trust values which user i has given to user j's nearest social neighbors, K is the number of user j's social neighbors, $\mathbf{w_j} \in \mathbb{R}^K$ is a sparse column vector of aggregation coefficients.

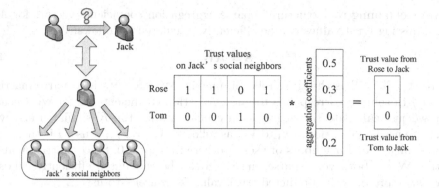

Fig. 1. An Illustration of Our Motivation

In order to effectively predict all the missing trust values, we need to learn a set of sparse aggregation coefficients vectors $(\mathbf{w_j}, j = 1, 2, ..., M)$ from the trust relation matrix $\mathbf{A}$. We then propose the following regularized optimization problem to learn a structured sparse coefficients vector $\mathbf{w_j}$ for user j:

$$\min_{\mathbf{w_j}} \|\mathbf{a_j} - \hat{\mathbf{A}}\mathbf{w_j}\|_2^2 + \eta |\mathbf{w_j}|^{\mathbf{T}}\mathbf{P}|\mathbf{w_j}|, \qquad (2)$$

where $\mathbf{a_j} \in \mathbb{R}^M$ is the j-th column of matrix $\mathbf{A}$, $\mathbf{w_j} \in \mathbb{R}^K$ is the coefficients vector of user j, and the constant η is the regularization parameter. Matrix $\hat{\mathbf{A}} \in \mathbb{R}^{M \times K}$ consists of K columns chosen from matrix $\mathbf{A}$ which indicate K most similar social neighbors of user j and are ranked in descending order of their similarity values. Note that each user is always trusted to himself or herself, so j-th trust value in $\mathbf{a_j}$ should not be utilized to solve the coefficients vector $\mathbf{w_j}$. Fortunately, we can easily set the j-th row in $\hat{\mathbf{A}}$ to 0 to avoid it happening.

The second item of Eq.(2), $|\mathbf{w_j}|^{\mathbf{T}}\mathbf{P}|\mathbf{w_j}|$, is the pairwise elastic net[3] regularizer term, where $\mathbf{P}$ is a symmetric and positive semidefinite(PSD) matrix with nonnegative entries. In this paper, we consider matrix $\mathbf{P}$ as $\mathbf{I} + \mathbf{1}\mathbf{1}^{\mathbf{T}} - \mathbf{R}$, then this regularizer term can be expressed as:

$$|\mathbf{w_j}|^{T}\mathbf{P}|\mathbf{w_j}| = |\mathbf{w_j}|^{T}(\mathbf{I} + \mathbf{1}\mathbf{1}^{\mathbf{T}} - \mathbf{R})|\mathbf{w_j}| = \|\mathbf{w_j}\|_2^2 + \|\mathbf{w_j}\|_1^2 - |\mathbf{w_j}|^{T}\mathbf{R}|\mathbf{w_j}|, \qquad (3)$$

where $\mathbf{I}$ is an identity matrix, $\mathbf{1}\mathbf{1}^{T}$ is a matrix of all ones entries and $\mathbf{R} \in \mathbb{R}^{K \times K}$ is a similarity matrix whose entry r_{ij} is the vector similarity between user i and user j in matrix $\hat{\mathbf{A}}$. When learning a coefficients vector $\mathbf{w_j}$, enforcing this regularizer term will simultaneously incur sparsity and structured grouping effect of the learnt coefficients vector $\mathbf{w_j}$, which encodes intrinsic and delicate relations between other users and user j.

If the number of user is M, we need to solve M optimization problems shown in Eq.(2) to obtain M sparse aggregation coefficients vectors. Because these problems are independent to each other, they can be solved in parallel. Furthermore, we will present a stochastic coordinate descent algorithm in Section 3.3 to solve each optimization problem significantly faster than the traditional way.

After obtaining the structured sparse aggregation coefficients vectors for M users, missing trust values can be efficiently calculated as follows:

$$\mathbf{S} = \mathbf{AW}, \tag{4}$$

where $\mathbf{A} \in \mathbb{R}^{M \times M}$ is the trust relation matrix, $\mathbf{W} \in \mathbb{R}^{M \times M}$ is a sparse matrix whose j-th column corresponds to an aggregation coefficients vector $\mathbf{w_j}$. Recall that we just calculate the aggregation coefficients for top-K similar users, we therefore need to transform $\mathbf{w_j}$ to $\mathbf{W_j}$ as follows: $W_{ij} = w_{kj}$, if user i is the k-th most similar social neighbors of user j, otherwise $W_{ij} = 0$. Since matrix $\mathbf{A}$ and matrix $\mathbf{W}$ are both very sparse, matrix $\mathbf{S}$ can be calculated efficiently, whose entry s_{ij} represents the predicted trust value from user i to user j.

Note that matrix $\mathbf{P}$ in Eq.(3) is not always PSD, so we can use a shrinkage parameter as described in [3] to ensure $\mathbf{P}$ to be PSD, that is

$$\hat{\mathbf{P}} = \theta\mathbf{I} + (1-\theta)\mathbf{P} = \mathbf{I} + (1-\theta)\mathbf{1}\mathbf{1}^T - (1-\theta)\mathbf{R}, \tag{5}$$

where $\frac{\gamma}{1+\gamma} \leq \theta \leq 1$, $\gamma = -min(0, \lambda_{min}(\mathbf{P}))$, $\lambda_{min}(\mathbf{P})$ is the minimum eigenvalue of matrix $\mathbf{P}$. Through the above transformation, matrix $\mathbf{P}$ is sure to be PSD and our formulation is a convex optimization problem.

3.2   Incorporating Hashing Techniques

In our proposed model, generating two matrices $\hat{\mathbf{A}}$ and $\mathbf{R}$ are two important steps before solving each coefficients vector $\mathbf{w_j}$. Both steps need to calculate the similarities between each pair of users. In this paper, we just use the binary trust relation information to help computing the similarity values. Intuitively, if two users are trusted by almost the same set of users, they are very likely to be social trusted neighbors.

Although there are many similarity measures, i.e., Jaccard coefficient, cosine similarity and Pearson correlation coefficient, we decide to calculate the similarity values by the Jaccard coefficient because of binary trust relation vectors. However, calculating the Jaccard coefficients between millions of users in real-world applications still costs a huge amount of time and memory space. Hence, we utilize hashing techniques that have been widely used for efficiently estimating similarity in massive data to expedite calculating the Jaccard coefficients.

Figure 2 shows the workflow of our approach. We first apply the one permutation hashing[4] which is an effective and efficient MinHash technique to each column of the trust relation matrix $\mathbf{A}$ (i.e.,$\mathbf{a_1}, \mathbf{a_2}, ..., \mathbf{a_M}$) and generate short signatures (i.e.,$\mathbf{s_1}, \mathbf{s_2}, ..., \mathbf{s_M}$) for them. In our experiments, we choose the length of signature as $|s| \approx |a|/30$, and the average error between the original Jaccard coefficient and the estimated one is less than 0.01. Hence, we can directly calculate the Jaccard coefficient on these short signatures to generate matrix $\mathbf{R}$, which saves a lot of time and memory cost.

Moreover, in order to efficiently find the similar users to generate matrix $\hat{\mathbf{A}}$, we can further utilize the locality sensitive hashing(LSH) to hash these signatures into several buckets. After we hash these signatures for several times, similar

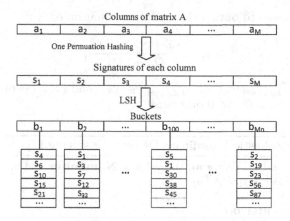

Fig. 2. The Workflow of Incorporating Hashing Techniques. M is number of columns, n is the number of LSH hash functions.

users are more likely to be hashed into the same bucket than dissimilar ones and most of the dissimilar users will never be hashed to the same bucket. Hence, we only need to check a small number of users in some buckets. For example, to find out the users most similar to a specific user j (i.e., a_1), we only need to check several buckets (i.e., b_2, b_{100}) which have the signature s_j (i.e., s_1) and consider users in these buckets as candidate similar users. We then sort them in descending order of the Jaccard coefficients between user j's signature (i.e., s_1) and candidates (i.e., s_3, s_7, s_5, s_{30}, etc.).

3.3   Stochastic Coordinate Descent Algorithm

The traditional way to solve the optimization problem in Eq.(2) is the coordinate descent algorithm. Generally speaking, the coordinate descent algorithm holds all elements in $\mathbf{w_j}$ fixed except for the i-th element w_{ij} and then solves $\frac{\partial L}{\partial w_{ij}} = 0$ to get w_{ij} accordingly. Cycling through each element in $\mathbf{w_j}$ iteratively until $\mathbf{w_j}$ converges, we could get the final optimal solution. From Eq.(2), we have

$$
\begin{aligned}
L(\mathbf{w_j}) &= \|\mathbf{a_j} - \hat{\mathbf{A}}\mathbf{w_j}\|_2^2 + \eta |\mathbf{w_j}|^T \mathbf{P} |\mathbf{w_j}| \\
&= \mathbf{a_j}^T \mathbf{a_j} - 2\mathbf{q}^T \mathbf{w_j} + \mathbf{w_j}^T \mathbf{Q}\mathbf{w_j} + \eta \sum_i^K \sum_k^K P_{ik} |w_{ij} w_{kj}|,
\end{aligned}
\tag{6}
$$

where $\mathbf{Q} = \hat{\mathbf{A}}^T \hat{\mathbf{A}}$, $\mathbf{q} = \hat{\mathbf{A}}^T \mathbf{a_j}$ and w_{ij} represents the i-th element in $\mathbf{w_j}$. We can calculate the partial derivative as follows:

$$
\frac{\partial L}{\partial w_{ij}} = -2q_i + 2\mathbf{Q_i}^T \mathbf{w_j} + 2\eta sgn(w_{ij}) \sum_{k=1}^K P_{ik} |w_{kj}|.
\tag{7}
$$

We then set $\frac{\partial L}{\partial w_{ij}} = 0$ to obtain the updating equation of w_{ij}:

$$w_{ij} = \frac{sgn(c)(|c| - b)_+}{Q_{ii} + \eta P_{ii}}, \tag{8}$$

where $c = q_i - \sum_{k \neq i}^{K} Q_{ik} w_{kj}$, $b = \eta \sum_{k \neq i}^{K} P_{ik}|w_{kj}|$ and $(z)_+$ denotes the positive part, which is z if $z > 0$ and 0 otherwise.

Algorithm 1. Stochastic Coordinate Descent for solving Eq.(2)

Input: Matrix $\hat{\mathbf{A}}$ and $\mathbf{P}$, Vector $\mathbf{a_j}$, the value of $\mathbf{N}$, **numiter** and η.
1. Compute $\mathbf{Q} = \hat{\mathbf{A}}^{\mathbf{T}}\hat{\mathbf{A}}, \mathbf{q} = \hat{\mathbf{A}}^{\mathbf{T}}\mathbf{a_j}$;
2. Initialize $\mathbf{w_j}$ with 0.
3. **for** t=1,2,...,**numiter do**
4.   Sample i uniformly at random from set $\{1, 2, ..., K\}$;
5.   Update w_{ij} according to Eq.(8);
6. **end for**
Output: Optimal solution $\mathbf{w_j^o}$.

In each iteration, a single element of $\mathbf{w_j}$ is updated by Eq.(8). The computational cost of calculating each $\mathbf{w_j}$ is $O(K^2 t)$, where t is the number of iterations. When the number of users is M, the computational cost of solving our model is $O(MK^2 t)$, which is too expensive to deal with large-scale social trust predictions. Hence, we propose a *stochastic* coordinate descent algorithm to efficiently solve each problem in Eq.(2). Inspired by [17], we suggest to choose one element uniformly at random from the coordinate set, instead of cycling through all elements. This simple modification leads to a significant reduction of iterations while achieving the expected accuracy.

Algorithm 1 shows our stochastic coordinate descent algorithm. The algorithm initializes $\mathbf{w_j}$ to be all 0. At each iteration, we pick an element i uniformly at random from the coordinate set $\{1, 2, ..., K\}$ to update w_{ij} by Eq.(8). After T iterations, our stochastic algorithm will converge to the optimal solution $\mathbf{w_j^o}$. The computational cost of calculating each $\mathbf{w_j}$ is therefore reduced to $O(KT)$. In most cases, T is much smaller than Kt. Hence, our stochastic coordinate descent algorithm significantly cuts down the computational time in the training phase.

3.4 A Simulation Study

We conducted a simulation study to show the effectiveness and efficiency of our stochastic coordinate descent algorithm. We simulate data from the linear model: $\mathbf{y} = \mathbf{X}\beta_* + \varepsilon$, where $\mathbf{X} \in \mathbb{R}^{n \times p}$ is the design matrix, $\mathbf{y} \in \mathbb{R}^n$ is the response vector, $\beta_* \in \mathbb{R}^p$ are the unknown weights and $\varepsilon \in \mathbb{R}^n$ is a zero-mean i.i.d. Gaussian noise vector. n and p are the number of observations and predictors respectively. We can find an estimate $\hat{\beta}$ of β_* by solving the same optimization problem in Eq.(2):

$$\min_{\hat{\beta}} \|\mathbf{y} - \mathbf{X}\hat{\beta}\|_2^2 + \eta |\hat{\beta}|^T \mathbf{P}|\hat{\beta}| \tag{9}$$

The data set is simulated as follows: We set $p = 5000$. The rows of the design matrix $\mathbf{X}$ are i.i.d. draws from the Gaussian distribution $\mathcal{N}(0, \Sigma)$, where the correlation matrix $\Sigma \in \mathbb{R}^{p \times p}$ has entries $\Sigma_{ij} = 0.5^{|i-j|}$. β_* is a sparse vector which has 1000 non-zeros entries drawn independently from the standard uniform distribution. The pair-wise similarity matrix $\mathbf{R}$ is constructed as $R_{ij} = |\mathbf{X}_i^{\mathbf{T}} \mathbf{X}_j|$, where $\mathbf{X}_i$ denotes the i-th column of $\mathbf{X}$. We train our model on a training set of $n = 2000$. Parameters are set with a validation set of $n = 1000$. Then, we test the performance on a testing set with $n = 4000$.

Table 1 shows the results of the traditional and stochastic coordinate descent algorithm. We find that our stochastic algorithm not only achieves a comparable mean-squared error(MSE) on estimating $\mathbf{y}$ and β_* against the traditional one, but also beats the traditional algorithm in terms of efficiency. We also show the convergence of our stochastic algorithm in Figure 3. The MSE of estimating β_* goes down as the number of iterations increases. Hence, our stochastic algorithm is empirically proven to converge.

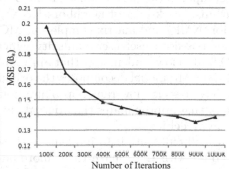

Table 1. The Simulation Results. CD indicates the traditional coordinate descent algorithm and SCD indicates the stochastic one.

Methods	MSE($\mathbf{y}$)	MSE(β_*)	time(s)
CD	0.1378	0.1327	49.06
SCD	0.1407	0.1353	17.15

Fig. 3. The Convergence of Our Stochastic Algorithm

4   Experimental Evaluation

In this section, we describe some experiments conducted to evaluate the performance of our Structured Sparse Linear Model(SSLM) and compare it with other popular methods.

4.1   Data Sets

We utilized three real-world trust graph data sets to evaluate the performance of different methods, i.e., Epinions[3], Slashdot[4] and Ciao[5]. Epinions and Ciao are both product review sites where users can give trust/distrust votes based on their opinions on others' reviews on products. Slashdot allows users to tag each other as friends or foes based on their submitted technology news. Note that these

---

[3] http://snap.stanford.edu/data/soc-Epinions1.html
[4] http://snap.stanford.edu/data/soc-Slashdot0811.html
[5] http://www.public.asu.edu/~jtang20/datasetcode/truststudy.htm

original data sets only record trust relations, and most of distrust relations are unknown. Hence, we only utilized trust relations in our experiments. For each data set, we filtered out the users who were trusted by less than 5 users to obtain a data set that has sufficient trust relations. The statistics of our data sets are summarized in Table 2.

Table 2. Statistics for the Experimental Data Sets

Data set	#Users	#Trust Relations	Density
Epinions	7,655	300,363	0.513%
Ciao	5,174	99,146	0.370%
Slashdot	8,948	436,208	0.544%

4.2 Comparison Methods and Details

We compared the performance of our SSLM with that of other four popular algorithms, i.e., k-nearest neighbors(KNN), SVD[11], non-negative matrix factorization(NMF) and matrix completion(MC)[12]. KNN predicts the missing trust values based on user-user similarity which is calculated by Jaccard coefficient. SVD finds a low-rank matrix that approximates the original trust relation matrix and the NMF model decomposes the original matrix into two non-negative factor matrices to seek a low-rank approximation. The MC model also seeks a low-rank matrix that approximates the original one, but the rank is automatically chosen by the model itself.

We utilized Matlab to implement these algorithms and optimized the Matlab codes with C to make them more efficient. We also employed the Parallel Computing Toolbox in Matlab to solve our model in parallel. All the experiments were conducted on a workstation equipped with two 4-core Intel Xeon E5620(2.40GHz) CPUs and 16GB RAM.

4.3 Evaluation Metrics

Since in many trust-related online applications, what we need is to obtain each user's most trusted users, rather than predict the missing trust values. Hence, we create a testing set by randomly selecting one of its trusted users for each user as the testing example. The remaining trust relations are treated as the training set used to train the examined trust prediction models. Afterward, missing trust values are generated by each model. We can obtain a list of most trusted users for each user in descending order of predicted trust values. In order to measure the prediction quality, we define the prediction accuracy(PA) as follows:

$$PA = \frac{\sum_{i=1}^{M} |L_i \cap T_i|}{M},\tag{10}$$

where L_i is a list of user i's most trusted users, T_i is the trusted user of user i in the testing set and M is the total number of users. In our experiment, we set the length of each list $|L_i| = 10$. PA measures the ability of each model to correctly predicate a list of trusted users for each user.

Table 3. The prediction accuracy(PA) on Different Data Sets

Data Sets	KNN	SVD	NMF	MC	SSLM(100)	SSLM(300)	SSLM(500)
Epinions	0.1568	0.1410	0.1285	0.1916	0.1932	0.1969	0.1975
Ciao	0.0707	0.0798	0.0700	0.0862	0.0876	0.0893	0.0901
Slashdot	0.0540	0.0586	0.0516	0.0745	0.0748	0.0762	0.0768

4.4 Experimental Results

For all the above methods, we tuned the best parameters to made them achieve the best performance. We only report the best performance results.

Trust Prediction Results. The prediction accuracy(PA) of different methods on three data sets are presented in Table 3. We selected different K values in SSLM and denoted them as SSLM(K) ($K = 100, 300, 500$) to study their performance. The results in Table 3 show that SSLM can generate better PA than other methods over all the data sets. We also find that the PA of SSLM improves as the K value increases. That is because a larger K value means a larger size of neighborhood, the more relations between user j and its neighbors can be learned in the aggregation coefficients vector $\mathbf{w_j}$, which gives rise to the better prediction quality. Another observation is that the MC model has a comparable performance with SSLM(100). However, it often takes several hours(e.g., about 31 hours on the Epinions data set) to train the MC model, while our SSLM(100) only takes several minutes(e g , about 3 minutes on the Epinions data set) for training the model. Moreover, SSLM(300) and SSLM(500) can achieve better PA than the MC model on all data sets.

Efficiency Evaluation. We utilized the Epinions data set as an example to illustrate the efficiency of our hashing based procedure. Table 4 presents the time cost of computing similarities between all pairs of users through two different methods. The first method is the original Jaccard coefficient and the second one is our hashing-based procedure. All methods were implemented in C to make sure that any time difference in performance is due to methods themselves. In Table 4, we find that our hashing-based procedure just takes about 1/6 time cost against the original Jaccard coefficient, which makes our SSLM model efficiently handle large-scale data sets.

Table 4. Time of Different Similarity Computation Procedures

Method	Time(s)
Jaccard	9,041
Hashing	1,446

Table 5. Time(ms) of Our Approach with Different Neighborhood Sizes and Optimization Algorithms

Methods	CD	SCD
SSLM(100)	0.708	0.586
SSLM(300)	6.09	4.08
SSLM(500)	17.39	10.61

We also evaluated the efficiency of the traditional coordinate descent(CD) and stochastic coordinate descent(SCD) algorithm on the Epinions data set. We report the average time of each iteration of these two algorithms in Table 5. Both algorithms were written in C and executed serially to objectively compare their efficiencies. For three neighborhood sizes, Table 5 shows that the larger the K value is, the faster our stochastic coordinate algorithm runs in each iteration. With respect to SSLM(500), our stochastic approach cuts off the 39% average execution time of each iteration, compared to the traditional approach. Hence, our stochastic coordinate descent algorithm can significantly boost the efficiency of our proposed SSLM approach.

5   Conclusions and Future Work

In this paper, we propose a novel and scalable Structured Sparse Linear Model (SSLM) for social trust prediction. SSLM formulates the trust prediction problem as a set of independent regularized linear regression problems. Scalable hashing techniques and stochastic coordinate descent algorithm are also incorporated into our model to reduce the computational cost of system training. The experiments on simulated and real-world data sets show that our approach outperforms other examined approaches in terms of prediction ability and scalability. In the future, we intend to extend our model to utilize additional user information, e.g., users' profiles and item rating information, and derive more scalable stochastic coordinate descent algorithms.

Acknowledgments. This work was supported by Zhejiang Provincial Natural Science Foundation of China (No. LQ13F020001), the Chinese Knowledge Center for Engineering Sciences and Technology Project, 973 Program (No.2012CB316400), the Special Funds for Key Program of National Science and Technology (No. 2010ZX01042-002-003), the Program for Key Cultural Innovative Research Team of Zhejiang Province, the Fundamental Research Funds for the Central Universities and the Opening Project of State Key Laboratory of Digital Publishing Technology.

References

1. Huang, J., Nie, F., Huang, H., Lei, Y., Ding, C.H.Q.: Social trust prediction using rank-k matrix recovery. In: IJCAI (2013)
2. Tang, J., Gao, H., Hu, X., Liu, H.: Exploiting homophily effect for trust prediction. In: Proceedings of the Sixth ACM International Conference on Web Search and Data Mining, WSDM 2013, pp. 53–62. ACM, New York (2013)
3. Lorbert, A., Eis, D., Kostina, V., Blei, D.M., Ramadge, P.J.: Exploiting covariate similarity in sparse regression via the pairwise elastic net. In: Proceedings of the Thirteenth International Conference on Artificial Intelligence and Statistics (AISTATS 2010), vol. 9, pp. 477–484 (2010)

4. Li, P., Owen, A.B., Zhang, C.H.: One permutation hashing. In: NIPS, pp. 3122–3130 (2012)
5. Indyk, P., Motwani, R.: Approximate nearest neighbors: Towards removing the curse of dimensionality. In: STOC, pp. 604–613 (1998)
6. Nguyen, V.A., Lim, E.P., Jiang, J., Sun, A.: To trust or not to trust? Predicting online trusts using trust antecedent framework. In: Ninth IEEE International Conference on Data Mining, ICDM 2009, pp. 896–901 (2009)
7. Liu, H., Lim, E.P., Lauw, H.W., Le, M.-T., Sun, A., Srivastava, J., Kim, Y.A.: Predicting trusts among users of online communities: An epinions case study. In: Proceedings of the 9th ACM Conference on Electronic Commerce, EC 2008, pp. 310–319. ACM, New York (2008)
8. Wang, D., Pedreschi, D., Song, C., Giannotti, F., Barabasi, A.L.: Human mobility, social ties, and link prediction. In: Proceedings of the 17th ACM SIGKDD International Conference on Knowledge Discovery and Data Mining, pp. 1100–1108 (2011)
9. Guha, R., Kumar, R., Raghavan, P., Tomkins, A.: Propagation of trust and distrust. In: Proceedings of the 13th International Conference on World Wide Web, WWW 2004, pp. 403–412. ACM (2004)
10. Golbeck, J.: Generating predictive movie recommendations from trust in social networks. In: Stølen, K., Winsborough, W.H., Martinelli, F., Massacci, F. (eds.) iTrust 2006. LNCS, vol. 3986, pp. 93–104. Springer, Heidelberg (2006)
11. Billsus, D., Pazzani, M.J.: Learning collaborative information filters. In: Proceedings of the Fifteenth International Conference on Machine Learning, ICML 1998, pp. 46–54 (1998)
12. Cai, J.F., Candès, E.J., Shen, Z.: A singular value thresholding algorithm for matrix completion. SIAM J. on Optimization 20(4), 1956–1982 (2010)
13. Meka, R., Jain, P., Dhillon, I.S.: Matrix completion from power-law distributed samples. In: Advances in Neural Information Processing Systems, vol. 22, pp. 1258–1266 (2009)
14. Huang, J., Nie, F., Huang, H., Tu, Y.C.: Trust prediction via aggregating heterogeneous social networks. In: Proceedings of the 21st ACM International Conference on Information and Knowledge Management, CIKM 2012, pp. 1774–1778 (2012)
15. Ning, X., Karypis, G.: Sparse linear methods for top-n recommender systems. In: Proceedings of the 2011 IEEE 11th International Conference on Data Mining, ICDM 2011, pp. 497–506. IEEE Computer Society, Washington, DC (2011)
16. McPherson, M., Lovin, L.S., Cook, J.M.: Birds of a feather: Homophily in social networks. Annual Review of Sociology 27, 415–444 (2001)
17. Shalev-Shwartz, S., Tewari, A.: Stochastic methods for l1 regularized loss minimization. In: Danyluk, A.P., Bottou, L., Littman, M.L. (eds.) ICML. ACM International Conference Proceeding Series, vol. 382, p. 117. ACM (2009)

Finding Time-Dependent Hot Path
from GPS Trajectories*

Yijiao Chen, Kaixi Yang, Hao Hu, Zhangqing Shan,
Renchu Song, and Weiwei Sun**

School of Computer Science, Fudan University, Shanghai, China
{chenyijiao,yangkaixi,huhao,shanzhangqing,songrenchu,wwsun}@fudan.edu.cn

Abstract. Finding hot path is important in many scenarios like trip
planning, traffic management and animal movement studies. However,
in practice, the hotness of paths may change over time, e.g., for one
path, it is hotter in morning rush hour than in the midnight. This paper
studies how to find time-dependent hot path.

Given two locations, a departure time and a travel time limit, our task
is to get a hot path highly fitting the real-time physical world within
a user-specified travel time limit. We first analyze the change of edge
hotness in different time ranges by learning historical GPS trajectories.
Then, with the time-dependent hotness information, we propose an effective algorithm to answer the hot path query mentioned above. Extensive
experiments on a real dataset show that our methods outperform the
baseline approaches in terms of both effectiveness and efficiency.

1 Introduction

Hot path (or popular path) is a path that is frequently passed through by people
or vehicles. It is useful in many scenarios, e.g., trip planning, traffic management
and animal movement studies. Many works [1–5] have been proposed to find hot
path. However, most of them do not aim to finding time-dependent hot paths. In
practice, the hotness of a path may change over time. This paper studies how to
answer Time-Dependent Hot Path(TDHP) query. We begin by two motivating
examples and then introduce the TDHP problem.

Example 1. Lily is on vacation. She arrives at a city for the first time and wants
to find an attractive path which has been chosen by many travellers. If it is a
sunny morning, travellers may prefer paths with natural landscapes, but if it is
an evening, travellers may prefer paths with fantastic architectures shining at
night. We can see that hot path is related to time of a day.

Example 2. Assume a taxi to be idle. To meet potential passengers as soon as
possible, the taxi driver would like to drive on roads that are frequently passed

---

* This research is supported in part by the National Natural Science Foundation of
  China (NSFC) under grant 61073001.
** Correponding author.

F. Li et al. (Eds.): WAIM 2014, LNCS 8485, pp. 768–780, 2014.

through by taxis or pedestrians. During morning rush hours he/she could choose roads around residential area because many people go to work by taxi, while in the evening he/she prefers roads around entertainment area since there would be a significant demand of going home for passengers.

The two examples above illustrate the necessity of finding time-dependent hot path. Furthermore, we visualize the distribution of real trajectory points on map. Figure 1 draws Beijing taxi trajectory points in respective time ranges(denoted by tr). We can see that the points in 1(a) and 1(b) are both quite dense but cover different edges, but that in 1(c) is relatively sparse. Therefore, it is not appropriate to answer a hot path query without the time factor.

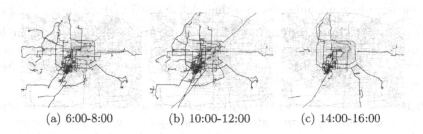

(a) 6:00-8:00   (b) 10:00-12:00   (c) 14:00-16:00

Fig. 1. An visual verification of time-dependent hotness

Given a source and a destination, the TDHP is to find an optimal path (with the maximum hotness) with a departure time and a user-specified travel time limit.

To our best knowledge, this is the first work that addresses the time-dependence of hot path problem. However, it is a non-trival work and we have identified two challenges. First, there is no available edge hotness data for path finding. We need to add time-dependent hotness information to the road network. Second, finding this hot path is difficult. As hotness changes over time, every time one edge is visited in path finding process, the time when moving objects arrive this edge should be known for hotness calculation. This is so time-consuming that some appropriate fast algorithm must be applied in this problem.

To answer the query, we first employ a clustering method towards all timestamps of trajectory points to construct hotness function for each edge. Then, we solve the TDHP problem by an A*-like algorithm. We also design an appropriate estimation function and give two strategies to improve the performance.

The contributions are as follows.

- We formally define the TDHP problem to find hot path with time-dependence and travel time limit.
- We propose a method to construct hotness function from a real trajectory dataset.
- We propose a new searching strategy for TDHP and we use two optimizing methods to improve the performance.

- We design innovative evaluation methods and perform extensive experiments on a real GPS trajectory dataset to show the effectiveness and efficiency of our algorithm.

The rest of the paper is organized as follows. We formally define our problem in Section 2. We present hotness function construction method in Section 3. The search strategy for hot path finding is introduced in Section 4. Section 5 includes our experimental study. The related work is discussed in Section 6. Finally Section 7 concludes the paper.

2   Problem Definition

We first introduce preliminaries and then define the TDHP problem.

Definition 1 (Time-Dependent Hotness Graph). *A time-dependent hotness graph is a directed graph $G_T(V, E, H)$ (or G_T for short): $V = \{v_i\}$ is a set of nodes; $E \subseteq V \times V$ is a set of edges; $H = \{h_{i,j}(t)\}$ is a set of hotness functions for each $e_{i,j}$ where t is a time variable in a day.*

Definition 2 (Trajectory). *A trajectory traj is a sequence of GPS points. It can be transformed by map-matching into a sequence of neighbour nodes with timestamps: $p_1 \to p_2 \to \ldots \to p_n$. p_i can also be described as (v_i, t_i) where t_i records the time arriving at v_i.*

Definition 3 (Edge Hotness). *Given a trajectory dataset $Traj$, the hotness of an edge $e_{i,j}$ at time t is defined as:*

$$\mathcal{H}_{i,i+1}(t) = \frac{1}{2\Delta t} |\{traj | traj \in Traj, p_i \to p_{i+1} \in traj, and\ p_i.t \in [t - \Delta t, t + \Delta t]\}| \tag{1}$$

where $[t - \Delta t, t + \Delta t]$ is a nearby time range(denoted by TR) around t and $\{traj\}$ are trajectories that pass through $e_{i,j}$ during this time range.

Due to different hotnesses in different times, it is more appropriate to count those time-nearby trajectories than all trajectories in a day. To divide the count by TR's length is because the hotness is meaningful in unit time. In addition, for simplicity, we assume that edge hotness keeps fixed while an object is traveling on the edge.

Definition 4 (Path). *A path p is a sequence of neighbour nodes, i.e., $p : v_1 \to v_2 \to \ldots \to v_n$, and we define E_p as all edges passed through by p.*

Definition 5 (Path Hotness). *Given a departure time t_d, the hotness of a path p is defined as the average hotness of E_p.*

$$\mathcal{H}_p(t_d) = \frac{1}{|E_p|} \sum_{e_{i,i+1} \in E_p} \mathcal{H}_{i,i+1}(t_i) \tag{2}$$

where t_i is the arrival time for $e_{i,i+1}$. We denode the travel time for $e_{i,i+1}$ by $e_{i,i+1}.tt$ and hence $t_i = t_{i-1} + e_{i-1,i}.tt$.

Problem 1 (TDHP Problem). Given a time-dependent hotness graph $G_T(V, E, H)$ and a query with a source v_s, a destination v_e, a departure time t_d and a travel time limit tl, the Time-Dependent Hot Path(TDHP) problem is to find a path with maximal hotness:

$$\mathcal{H}_{p*}(t_d) = \max_p \mathcal{H}_p(t_d), s.t. \sum_{e_{i,j} \in E_p} e_{i,j}.tt \leq tl \tag{3}$$

It seems not practical that moving objects spend all the time on the way and they never stop during driving. Considering the stops are uncertain and too many constraints will bring more unnecessary complication, we assume that moving objects never stop in our problem.

3   Hotness Function Construction

To solve problem 1, we need edge hotness function $h_{i,j}(t)$ and an efficient algorithm. These issues are discussed next.

3.1   Hotness Function Analysis

The $\mathcal{H}$ is a continuous function and provides a statistic solution for constructing $h_{i,j}(t)$. However, a continuous hotness function would bring serious efficiency problems. First, it is space-consuming if we store hotness information for all possible time stamps. Second, it will waste much running time if the hotness is calculated during the on-line path finding process.

In this paper, we propose a TimeParti algorithm to discrete the continuous hotness function $\mathcal{H}$. Generally speaking, edge hotness does not change sharply during a short tr. So we can discrete the continuous $\mathcal{H}$ by partitioning a day into appropriate trs and then calculate a hotness value for each tr. Therefore, nearby times having similar hotness can be merged to share the same hotness.

3.2   TimeParti Algorithm

We begin by presenting the desired properties of partitioned TRs, and then introduce our TimeParti algorithm.

Definition 6 (Timestamp Set). *For an edge $e_{i,j}$, a timestamp set ts_{t_1,t_2} records timestamps in a given time range $[t_1, t_2)$, where for each $t \in ts_{t_1,t_2}$, $\exists traj \in Traj$ that arrives $e_{i,j}$ at t.*

Given an initial ts which covers $[t_{start}, t_{end})$, assume that the time range partition process finally generates n time ranges $\{[t_k, t_{k+1})|1 \leq k < n\}$. And the hotness function is correspondingly constructed as follows:

$$h_{i,j}(t) = \frac{1}{t_{k+1} - t_k} |ts_{t_k, t_{k+1}}|, t \in [t_k, t_{k+1}) \tag{4}$$

Indeed, Equation 4 is an approximation of $\mathcal{H}$. It counts trajectories passing a given edge in a certain tr, while the $\mathcal{H}$ counts trajectories in a tr around a certain time. The notion of time range partition is practical, because we only need to store a small number of hotness values to answer for any time in $[t_{start}, t_{end})$.

There are two objectives in the time range partition. First, the hotness for any two timestamps $t_1, t_2 \in tr$ should be similar. Second, the number of trs should be bounded. Too many trs would be space-consuming. In this paper, we measure the quality of time range partition for an edge $e_{i,j}$ by a classic standard metric SSE(Sum of Squares for Error):

$$SSE(e_{i,j}) = \sum_{k=1}^{n} SSE([t_k, t_{k+1}))\tag{5}$$

where $SSE([t_k, t_{k+1}))$ is the SSE of an tr that is defined as:

$$SSE([t_k, t_{k+1})) = \frac{1}{|ts_{t_k,t_{k+1}}|} \sum_{t \in ts_{t_k,t_{k+1}}} \left(t - \bar{t}\right)^2 \tag{6}$$

where $\bar{t}$ is the average time in $ts_{t_k,t_{k+1}}$.

TimeParti algorithm is a modified bisecting K-means clustering method. We choose bisecting K-means rather than the famous K-means. This is because the number of time ranges is not fiexed for different edges. In TimeParti, we set a SSE bound(denoted by $SSEbd$) to stop the partition when the SSE of all existing time ranges are smaller than $SSEbd$. We use an example in Figure 2 to describe TimeParti. For a certain edge, 2(a)we initialize a ts_1 including all timestamps for trajectories that arrive $e_{i,j}$, 2(b)partition ts_1 into ts_2 and ts_3, 2(c)select a ts with the biggest SSE(assume $ts=ts_3$), and partition it into ts_4 and ts_5, 2(d) next select ts_2 and partition it into ts_6 and ts_7. Then the partition can be stopped, because the SSE of ts_4, ts_5, ts_6 and ts_7 are all below $SSEbd$. If there exists $|ts|=1$ and $t \in ts$, then we enlarge the corresponding time range by $[t - \delta, t + \delta]$(here we set δ=30min for example).

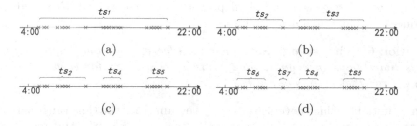

Fig. 2. An example of TimeParti algorithm

4 Hot Path Finding Algorithm

This section introduces TDHP algorithm first, and then proposes techniques for optimization.

4.1 TDHP Algorithm

We novelly extend the A* algorithm by estimating path hotness. A* is an algorithm for goal-directed search that keeps a priority queue PQ of alternative nodes ordered by an evaluation function. In TDHP, we redefine the evaluation function by an estimated path hotness function $f(x)$ for node x. $f(x)$ determines the order in which the search visits nodes in PQ. Differing from traditional A*, in TDHP, the bigger $f(x)$, node x is more likely to be extended.

$f(x)$ is sum of two functions: the past path hotness function(denoted by $g(x)$) and a future path hotness function(denoted by $h(x)$). For $g(x)$, instead of using $\mathcal{H}_p$, we give a slightly rough version using $h_{i,j}(t)$ which has been constructed in

---

Algorithm 1: The TDHP Algorithm

1 **Input:** a time-dependent hotness graph G_T, a query: v_s, v_e, t_d, and tl
2 **Output:** The time-dependent hottest path TDHP
3 **Procedure:**
  1: $PQ \leftarrow \emptyset$; $CloseSet \leftarrow \emptyset$;
  2: $PQ.\mathrm{enqueue}(v_s)$;
  3: **while** $PQ \neq null$ **do**
  4:     $v \leftarrow \mathrm{dequeue}(PQ)$;
  5:     **if** $v = v_e$ **then**
  6:         **return** TDHP;
  7:     **end if**
  8:     $CloseSet.\mathrm{add}(v)$;
  9:     **for** $each\ u \in v.adjacentNodes$ **do**
 10:         $t \leftarrow (v,u).tt + FT(u \Rightarrow v_e)$;
 11:         **if** $tt(v_s \rightarrow v) + t > tl$ **or** $u \in CloseSet$ **then**
 12:             **continue**;
 13:         **end if**
 14:         $f(u) \leftarrow g(u) + h(u)$;
 15:         **if** $u \notin PQ$ **then**
 16:             $PQ.\mathrm{add}(u)$;
 17:         **else**
 18:             **if** $f(u) > f_{old}(u)$ **then**
 19:                 $update(PQ, u)$;
 20:             **end if**
 21:         **end if**
 22:     **end for**
 23: **end while**

---

the previous section, to simplify the calculation. Assume p^- to be a path from v_s to x, we have

$$g(x) = \frac{1}{|E_{p^-}|} \sum_{e_{i,i+1} \in E_{p^-}} h_{i,i+1}(t_i) \tag{7}$$

$h(x)$ is a heuristic estimate of the hotness from x to v_e. Assume E_R to be a set of edges that could possibly passed through in tl, then $h(x)$ can be defined as:

$$h(x) = \frac{1}{|E_R|} \sum_{e_{i,j} \in E_R} \overline{h}_{i,j}(t)$$

where $\overline{h}_{i,j}(t_i)$ is an estimated edge hotness based on $h_{i,j}(t_i)$. We discuss $\overline{h}_{i,j}(t_i)$ in next subsection.

The pseudocode for this algorithm is given in Algorithm 1. A max priority queue PQ is utilized to determine the node with the maximum $f(x)$ value. In the *while* loop, we extract v with the maximum $f(x)$ from PQ. Then for each neighbour node u: we use a fastest-travel time pruning strategy in line 10-11 and update PQ with $f(x)$ in line 14-19. Once the destination node is extended, the algorithm will stop and return the result hot path in line 6.

4.2   Details on Optimization

We now discuss the pruning strategy and the calculation of $h(x)$ in algorithm 1.

Pruning Strategy. With a travel time limit, possible nodes between any two locations can be restricted in an area(denoted by A). Follow Algorithm 1 and assume that u is a node that is to be extended from v, let $tt(v_s \rightarrow u)$ be the total travel time from v_s to u, and let $FT(u \Rightarrow v_e)$ be the fastest travel time from u to v_e, then u would be pruned if the following condition is satisfied:

$$tt(v_s \rightarrow u) + FT(u \Rightarrow v_e) > tl$$

This strategy can be applied to selection of E_R for $h(x)$ with little change. For $\forall e_{i,j} \in E$, $e_{i,j}$ would be pruned if the following condition is satisfied:

$$tt(v_s \rightarrow u) + FT(u \Rightarrow v_i) + e_{i,j}.tt + FT(v_j \Rightarrow v_e) > t$$

The left part infers that the fastest travel time of a path that starts from v_s, passes through u and $e_{i,j}$ and finally arrives at v_e.

To obtain the travel time for each edge, we add up all actual travel times from $Traj$ and calculate average as the edge's travel time. For those edges with no trajectories passing through, we first calculate an average speed $\overline{v}$ from $Traj$, and then estimate the travel time by dividing the edge length with $\overline{v}$.

Hotness Function Filtering. The quality of $h(x)$ is influenced not only by selection of E_R but also by $\overline{h}_{i,j}(t)$ for each edge. It would be difficult to calculate $h_{i,j}(t)$ because we have no idea about in which time range the object will arrive

$e_{i,j}$. $\overline{h}_{i,j}(t)$ is a filtered consequence of $h_{i,j}(t)$ and we use it to estimate the hotness when an object is passing $e_{i,j}$. Next we propose the filtering method and how this $\overline{h}_{i,j}(t)$ is calculated.

The filtering method is based on two observations: for $e_{i,j}$, first, the time arriving at v_i can not be earlier than some time t_{early} because it is limited by the graph topological structure. Second, the time leaving v_j can not be later than some other time t_{late} because of the travel time limit. Then we can filter the $h_{i,j}(t)$ by $[t_{early}, t_{late}]$ and average the left hotness for $h(x)$.

The time range can be easily computed in the following way: $t_{early} = t_s + t(s \Rightarrow x) + FT(x \Rightarrow v_i)$ and $t_{late} = t_s + tl - FT(v_j \Rightarrow e)$. After that, for $\forall tr \in TR$, tr would be filtered if either of the following two conditions is satisfied:

$$t_{k+1} < t_{early}, t_k > t_{late}$$

Assume TR_{left} as the left time ranges after hotness function filtering, then $\overline{h}$ can be calculated by averaging all hotness corresponding to TR_{left}. The pseudocode for this algorithm is given in Algorithm 2. The pruning strategy is used in line 3 and the hotness function filtering is used in line 7.

Finally, we present the time complexity analysis. The calculation of $g(x)$ can make use of the parent node of x, so its time complexity is $O(1)$. Assume that TR_{max} is the time range set of an edge with biggest $|TR|$. Hence, $h(x)$ costs $O(|E||TR_{max}|)$. Therefore, the time complexity of TDHP is $O(|N|log|N| + |E|^2|TR_{max}|)$.

5 Experiments

In this section, we test the efficiency and effectiveness of both TimeParti and TDHP algorithms.

Algorithm 2: The future path hotness function $h(x)$

---

1 **Input:** a time-dependent hotness graph G_T, a query and node u
2 **Output:** a future hotness
3 **Procedure:**
  1: $h_{value} \leftarrow 0$, $count \leftarrow 0$;
  2: **for** each $ei, j \in E$ **do**
  3:    **if** $tt(s \rightarrow u) + FT(u \Rightarrow v_i) + e_{i,j}.tt + FT(v_j \Rightarrow v_e) > tl$ **then**
  4:        **continue**;
  5:    **end if**
  6:    **for** $\forall tr \in TR$ **do**
  7:        **if** $tr.ft > t_{late}$ or $tr.lt < t_{early}$ **then**
  8:           **continue**;
  9:        **end if**
 10:      $h_{value} \leftarrow h_{value} + h_{i,j}$;
 11:      $count++$;
 12:    **end for**
 13: **end for**
 14: **return** $h_{value}/count$;

5.1 Dataset

We use a graph dataset and a real-world trajectory dataset in our experiments. The graph dataset is a Beijing road network from OpenStreetMap, including 2011 nodes and 5118 edges. The trajectory dataset we use for hotness function construction is offered by [6, 7]. It contains real GPS trajectories of 10,357 taxis during the period of Feb. 2 to Feb. 8, 2008 in Beijing. The total number of points in this dataset is about 15 million and the total distance of the trajectories reaches to 9 million kilometers. In the preprocessing, we did map matching and transformed each trajectory into an edge sequence with time stamps. Table 1 depicts the time distribution of all trajectory points.

Table 1. Statistics on trajectory time stamps

0:00-4:00	4:00-8:00	8:00-12:00	12:00-16:00	16:00-20:00	20:00-24:00
2.20%	35.21%	38.89%	18.47%	3.70%	1.51%

5.2 Parameter Tuning on TimeParti

The TimeParti algorithm is sensitive to the tuning parameter $SSEbd$. Figure 3(a) shows how the clustering error is affected by $SSEbd$. The average edge SSE increases as $SSEbd$ increases. Figure 3(b) gives a relation between $SSEbd$ and the query time of TDHP. As we can see, a larger $SSEbd$ will make the hot path finding more efficient, because less time ranges are produced and traversed in h(x). Observing the query time slightly decreases when $SSEbd$ increases to 1.0, we set $SSEbd$=1.0.

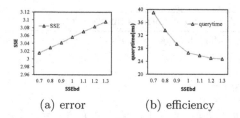

(a) error   (b) efficiency

Fig. 3. SSEbd tuning

5.3 Evaluation on TDHP

In the following, we illustrate the search results of TDHP algorithm and compare the results with a baseline approach(RICK) in [6] in three ways. We show in the experiments that TDHP performs better in time-dependent hotness environment.

Metrics and Baseline. We use two criterions: path hotness and trajectory similarity to evaluate the effectiveness of TDHP. The path hotness defined in Definition 7 makes a quantization on the hotness of a path. Given a ground-truth hottest path from the trajectory dataset, the trajectory similarity infers how close the result path is to the ground-truth. Different from [6], we select from the dataset the ground-truth with the highest hotness score defined in Section 5 for a given query. In addtion, we compare the results with the existing approach(RICK) in [6]. In [6], given a sequence of locations, they find a hot path connecting any two contiguous locations. In this experiment, the parameters of RICK are set as $|q| = 2$ and $k = 1$.

Querydata Preperation. We set three time ranges and design three query sets respectively. The query source-destination pairs are the same in three sets but the departure times are in corresponding time ranges. Taking the trajectory points' time distribution in Table 1 into consideration, we set the time ranges as 4:00-8:00, 8:00-12:00 and 12:00-16:00. In addition, we also run experiments on a dataset ignoring the time-dependence, denoted as average time. For each query q, we collect all trajectories travelling from the $q.s$ to $q.e$ during a time interval around $q.t$. Then these trajectories are ranked by their hotness. We can assume that the greater the hotness is, the respective trajectory is more likely to be a hot path found by an experienced and professional driver.

Performance on Path Hotness. Figure 4 verifies the relation between the distance of query nodes and the path hotness. To study the path hotness difference, we compare TDHP with 3 competitors:(1)RICK,(2)the hotness of ground-truth hottest path and (3)average hotness of all trajectories that from v_s to v_e. The x-coordinate refers to the length of shortest path(SP) between query node pairs. We can see in Figure 4 that the path hotness(in Definition 7) decreases as the distance increases. It is because the increasing distance leads to more edges. Assume the number of trajectories on the path to be fixed, then path hotness would get smaller with a bigger denominator. The TDHP performs better than RICK because the evaluation function in TDHP only counts those edge hotnesses in satisfied time ranges.

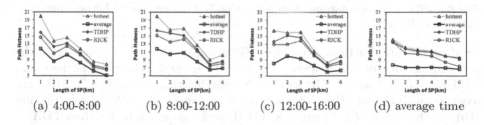

   (a) 4:00-8:00   (b) 8:00-12:00   (c) 12:00-16:00   (d) average time

Fig. 4. Path hotness comparison

Performance on Trajectory Similarity. Figure 5 verifies the error of TDHP. To evaluate the difference between a path and the corresponding ground-truth hottest trajectory, we apply the length-Normalized Dynamic Time Warping distance(NDTW)[6]. For each query, we get top-k hottest trajectories as ground-truth and average the k NDTWs as the error of the path finding algorithms. Figure 5 shows that the NDTW of TDHP slightly grows as the length of SP

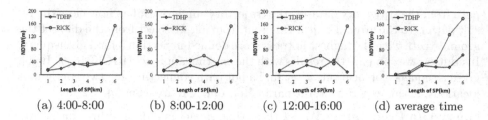

(a) 4:00-8:00   (b) 8:00-12:00   (c) 12:00-16:00   (d) average time

Fig. 5. Trajectory error comparison

increases. We can also infer that the hot path of RICK is hot but is quite different from drivers' experience.

Efficiency. We investigate the query time of TDHP and show the results in Figure 6. In the path finding, we improve the efficiency of the path generation by a travel time pruning strategy. As shown in Figure 6(a), TDHP outperforms RICK a lot. The reason is that every time the $h(x)$ is called, it is only the time ranges in $h_{i,j}(t)$ for $\forall\, e_{i,j} \in E_R$ that would be checked to calculate the $\overline{h}_{i,j}(t)$. But in RICK, for $\forall\, e_{i,j} \in E_R$, every trajectory that passes $e_{i,j}$ would be checked to be included into a set of existing trajectories(details in [6]). Figure6(b) compares the times that time ranges are visited in TDHP with the times that trajectories are visited in RICK. As we can see, the query time is in direct proportion to the visiting number.

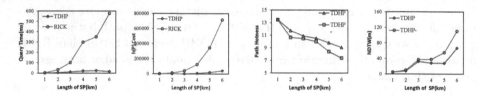

Fig. 6. Efficiency comparison between TDHP and RICK

Fig. 7. Performance of hotness function filtering

Evaluation on Hotness Function Filtering In the end, we investigate the performance of hotness function filtering in Figure 7. The path hotness and NDTW of TDHP and TDHP without hotness filtering(denoted by TDHP-) are recorded in 7(a) and 7(b) respectively. As we can see in both Figure 7(a) and 7(b), as the length of SP increases, TDHP performs much better than TDHP-. This is because by calculating edge's earliest arrival time and latest leaving time, TDHP filters a few time ranges, and limits the possible arrival time of each edge in time ranges that are left after filtering.

6   Related Work

Our TDHP problem is highly related to hot route finding, time-dependent hotness, trajectory pattern mining and time-dependent fastest path.

Hot Route Finding. It is [1] that first put forward the end-to-end hot route problem and assumed there was no background graph information. It used a modified DBSCAN algorithm to cluster intersections and a markov model to calculate transfer probabilities between intersections, and finally discovered the most popular route. Given a location sequence and a time span, the work [2] presented a framework to get top-k hottest paths passing through the location sequence within a time span. P.Kumar[3] provided users with a path strategy that fitted most people's driving behaviour. However, all the works above ignored the hotness time-dependence.

The work [4] proposed a method to on-line achieve hot motion paths that are close to the query location. A FlowScan algorithm presented by [5] clustered road segments. Both works got a set of hot roads, instead of continual paths.

Time-Dependent Hotness. The study [8] analysed a trajectory dataset and it showed the distribution of trajectory points changed over time. The work [9] visualized this phenomenon of a small part of the map. They just gave a subjective explaination but did not systematically quantize the edge hotness. Several studies [10–12] discovered period moving trends of objects that share alike routes. To better make use of these trends, the study [13] clustered trajectory points at fixed intervals, hence converted GPS trajectories into cluster sequences. But these works lack of a straightforward definition and expression of hotness,thus their efficiency of algorithm is not good.

Trajectory Pattern Mining. Several works [10–12] focused on mining trajectory patterns. By tagging these patterns in the graph, some hot paths could be found. However, these popular paths were discrete and scattered over the whole graph. Given a certain source-destination pair, there may be no consecutive paths that could pass through the two locations. Therefore, retrieving hot paths from trajectory patterns is not a good solution for the problem in this paper.

Time-Dependent Fastest Path. Given a graph with edge travel time changing over time, time-dependent fastest path aims at finding a path of which the total travel time is minimized. The work [14] studied the online algorithm for fastest path in time-dependent graph and presented a technique speeding up path calculation. However, the problem addressed in this paper is fairly different from this kind of problem, for it is the hotness rather than the travel time that is dynamic. Besides, the fastest path only demanded that the travel time of path should be minimized. But TDHP is to get a path with maximized average hotness of edges as long as the travel time would not surpass a specified value.

7   Conclusion

In this paper, we studied the problem of time-dependent hot path finding between two given locations by using users' daily moving trajectories. We first converted GPS points to road segment sequences with timestamps by a map-matching algorithm. Then we proposed a novel method to construct time-dependent hotness function for each edge. Based on the hotness function, we proposed an A*-like algorithm to find the optimal TDHP. We conducted extensive studies and confirmed that our algorithms outperformed than the baseline algorithm in terms of both effectiveness and efficiency.

References

1. Chen, Z., Shen, H.T., Zhou, X.: Discovering popular routes from trajectories. In: ICDE, pp. 900–911 (2011)
2. Wei, L.-Y., Zheng, Y., Peng, W.-C.: Constructing popular routes from uncertain trajectories. In: KDD, pp. 195–203 (2012)
3. Kumar, P., Singh, V., Reddy, D.: Advanced traveler information system for Hyderabad city. IEEE Transactions on Intelligent Transportation Systems 6(1), 26–37 (2005)
4. Sacharidis, D., Patroumpas, K., Terrovitis, M., Kantere, V., Potamias, M., Mouratidis, K., Sellis, T.K.: On-line discovery of hot motion paths. In: EDBT, pp. 392–403 (2008)
5. Li, X., Han, J., Lee, J.-G., Gonzalez, H.: Traffic density-based discovery of hot routes in road networks. In: Papadias, D., Zhang, D., Kollios, G. (eds.) SSTD 2007. LNCS, vol. 4605, pp. 441–459. Springer, Heidelberg (2007)
6. Yuan, J., Zheng, Y., Xie, X., Sun, G.: Driving with knowledge from the physical world. In: KDD, pp. 316–324 (2011)
7. Yuan, J., Zheng, Y., Zhang, C., Xie, W., Xie, X., Sun, G., Huang, Y.: T-drive: driving directions based on taxi trajectories. In: GIS, pp. 99–108 (2010)
8. Guo, D., Liu, S., Jin, H.: A graph-based approach to vehicle trajectory analysis. J. Location Based Services 4(3&4), 183–199 (2010)
9. Li, Q., Zeng, Z., Zhang, T., Li, J., Wu, Z.: Path-finding through flexible hierarchical road networks: An experiential approach using taxi trajectory data. Int. J. Applied Earth Observation and Geoinformation 13(1), 110–119 (2011)
10. Giannotti, F., Nanni, M., Pedreschi, D., Pinelli, F., Renso, C., Rinzivillo, S., Trasarti, R.: Mobility data mining: discovering movement patterns from trajectory data. In: Computational Transportation Science, pp. 7–10 (2010)
11. Jeung, H., Yiu, M.L., Zhou, X., Jensen, C.S., Shen Discovery, H.T., Shen, H.T.: Discovery of convoys in trajectory databases. PVLDB 1(1), 1068–1080 (2008)
12. Kalnis, P., Mamoulis, N., Bakiras, S.: On discovering moving clusters in spatiotemporal data. In: Medeiros, C.B., Egenhofer, M., Bertino, E. (eds.) SSTD 2005. LNCS, vol. 3633, pp. 364–381. Springer, Heidelberg (2005)
13. Li, Z., Ding, B., Han, J., Kays, R., Nye, P.: Mining periodic behaviors for moving objects. In: KDD, pp. 1099–1108 (2010)
14. Ding, B., Yu, J.X., Qin, L.: Finding time-dependent shortest paths over large graphs. In: EDBT, pp. 205–216 (2008)

A Two-Phase Model for Retweet Number Prediction

Gang Liu[1], Chuan Shi[1,*], Qing Chen[2], Bin Wu[1], and Jiayin Qi[1]

[1] Beijing University of Posts and Telecommunications, Beijing, China
{liugang519,shichuan,wubin}@bupt.edu.cn, qijiayin@139.com
[2] China Mobile Communications Corporation
chenqing@chinamobile.com

Abstract. With the surge of social media, micro-blog has become a popular information share tool, in which retweeting is a basic way to share and spread information. It is important to predict the retweet number for influence measure and precision market. Contemporary methods usually consider it as a classification or regression problem directly, which can be regarded as one-phase models. However, they cannot accurately predict the number of retweet. In this paper, we propose a two-phase model to predict how many times a tweet can be retweeted in Sina Weibo. That is, the model first classifies tweets into several categories, and then does regression on each category. Extensive experiments on real Sina Weibo dataset show that our model is a general framework to achieve better performances than traditional one-phase prediction model without complex feature extraction.

Keywords: Social media, Sina Weibo, Retweet, Classification, Regression.

1 Introduction

Recently, there is a surge of social media. Many social network services have emerged, among which micro-blog service is a platform of sharing, spreading and acquiring message based on users' relationship. People can post messages of up to 140 characters through Web or smart phones to share information timely. Micro-blog services gain worldwide popularity. As the most popular micro-blog service, Twitter [1] had about 500 million registered users in 2012 and these users post about 340 million messages every day. In China, Sina Weibo [2] had 503 million registered users before March 2013.

In micro-blog network, retweet is the main way to spread messages. When a user posts a message, this message will be pushed to the user's followers. When followers see this message, he/she can choose to retweet the message, so the message will be pushed to his/her followers. By retweet, messages can be continued to spread in the micro-blog network. Therefore, the times of retweet (i.e., retweet number) can be as an important indicator of the message's influence. Predicting the retweet number of a message in micro-blog network (i.e., tweet) has practical significance in evaluating

---

* Corresponding author.

F. Li et al. (Eds.): WAIM 2014, LNCS 8485, pp. 781–792, 2014.
© Springer International Publishing Switzerland 2014

the influence and the value of a tweet. What's more, it contributes to controlling the spread of illegal information like rumors.

There has been some studies [3,4,5,6,7,8,9,10] on the retweet prediction in micro-blog network. Many of them consider the problem as a two-classification problem, which predicts whether a message will be retweeted or not. Some studies [9] also treat it as a multi-classification problem. However, it is difficult to determine the threshold of multiple classifiers. There are a few works predicting the retweet number directly [10]. All these work can be considered as one-phase model as shown in Fig. 1(a). Due to the complexity of retweet behavior, it is hard to accurately predict the retweet number with these one-phase models. Moreover, most of work focuses on English micro-blog services like Twitter and few studies are on Chinese micro-blog services.

In this paper, we first analyze the characteristics of retweet in real Sina weibo data-set and point out it is not rational to directly do regression on training data due to the power law distribution of retweet number. Then we propose a two-phase model to predict retweet number. Fig. 1(b) illustrates the basic idea of our model. In the first phase, the model classifies tweets into one of multiple categories, where the classification threshold can be automatically determined by the 80/20 rule. In the second phase, the model does regression on each category to predict the retweet number. The two-phase model has the following advantages: (1) It is a general framework, which can employ any classifier or regression model in it; (2) It is a simple but effective method without complex feature extraction. Extensive experiments on Sina Weibo data validate the above benefits through achieving better performances than one-phase models under different model settings.

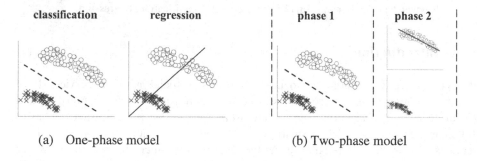

(a)   One-phase model   (b) Two-phase model

Fig. 1. Different methods on retweeting behavior

This paper is organized as follows. In Section 2, we describe related work about micro-blog network and retweet behavior analysis. In Section 3, we describe our dataset and analyze the characteristics of retweet behavior. Section 4 illustrates our two-phase model for retweet number prediction. Section 5 provides numbers of experiments to validate our model. Finally, we summarize our findings and conclude with Section 6.

2   Related Work

Since micro-blog has become a popular social network service, a lot of studies have been done to explore its traits. Some studies are about micro-blog network structure and user characteristics. Java et al. [11] introduce the basic functions and features of Twitter and give a preliminary analysis of its social networking features, the results indicate that Twitter showed some power law distribution and small world properties.

As retweet behavior is the key mechanism for information diffusion in micro-blog network, many studies focus on retweet behavior. Boyd et al. [12] took a detailed analysis of retweet in Twitter to explore how and why people retweet a tweet. Suh et al. [13] examined a number of features that might affect retweet, and they found that the number of followers and followees seem to affect retweet behavior, while the number of past tweets does not affect a tweet to be retweeted. Zhang Yang et al. [8] analyzed the importance of different features and investigate the feasibility of applying classification method and proposed a feature-weighted model. Their model can predict a major fraction of tweets(nearly 86%). Bandari et al. [9] proposed a model to predict popularity of new articles. They classified articles to three classes based on their retweet number 1-20, 20-100 and 100-2400. The model can predict ranges of popularity on Twitter with an overall 84% accuracy. Most of studies above focus on Twitter and few studies are on direct retweet number.

There are also a few work on retweet number prediction of Chinese micro-blog. Li Ying-le et al. [10] proposed a prediction model based on SVM algorithm with five features: user influence, user activity, interest similarity, the importance of micro-blog content and users' closeness. The experiment with Sina Weibo data shows a good result that the predict accuracy is up to 86.63%. However, the features they extracted are very complex and expensive and thus the model is not suitable for large-scale data.

3   Data and Features

This section describes the dataset and features extracted from tweets. Then we analyze the characteristic of retweet behavior.

3.1   Dataset Preprocess

We use Sina Weibo API to collect tweets for three months from April 2013 to July 2013 and finally get 54M tweets and 142K different users. In real micro-blog network, retweet number of a tweet will change with time. For example, when a tweet is just posted, the retweet number may be 0. After an hour, the retweet number may increase. Because we want to build a prediction model to predict the final retweet number of a tweet, we filtered data and got 49M tweets which exist in micro-blog network more than 30 days. Since retweet number of a tweet will trend to be stable with the time passed by, we consider that the retweet number will be stable after 30 days.

3.2   Feature Description

We extracted 28 features from the tweet and the tweet's creator. Most of them can be directly crawled from Sina Weibo API, and these features have been proved to be effective in some papers [8],[13]. Moreover, most of the features are basic information from dataset and they don't require complex computation. Table 1 and Table 2 show the details. In Table 1, we also compute 4 features which describe the tweet's creator's influence.

Table 1. Feature about the tweet creator

Feature	Explanation	Feature	Explanation
GD	Gender of the tweet's creator	VR	The tweet's creator is a verified user
NL	The length of the nickname of the tweet's creator	VT	The verified type of the tweet's creator
FON	The number of the followers who follow tweet's creator	ED	The number of days since the tweet's creator registered
FRN	The number of the friends who are followed by tweet's creator	MSPD	SN/ED
BFN	The number of the friends who and tweet's creator follow each other	MFPD	FON/ED
FAN	The number of the favorites which the tweet's creator has	MFPS	FON/SN
SN	The number of tweets of tweet's creator post	MAFPS	(FON-FRN)/SN

Table 2. Feature about the tweet

Feature	Explanation	Feature	Explanation
HI	The tweet has hashtag in text	TL	The length of the text
HC	The number of the hashtag in text	TM	The month of the tweet which was created
AI	The tweet has @ in text	TD	The day of the tweet which was created
AC	The number of @ in text	TH	The hour of the tweet which was created
HI	The tweet has link in text	TW	The week of the tweet which was created
HC	The number of link in text	HOI	The tweet was created on holiday
PI	The tweet has pictures	EH	The hours since the tweet was created

3.3   Characteristic of Retweet Behavior

We take a basic statistics analysis on retweet behavior of users.

For retweet number, we count tweets of different retweet number and calculate its cumulative distribution. Fig. 2(a) shows the log distribution of total tweets over all data, demonstrating a long tail shape. We can see the cumulative distribution of different retweet number in Fig. 2(b). Apparently, the retweet number is extremely unbalanced.

For example, 66.4% of all tweets' retweet number is less than 1 and 90% of that is no more than 30. The tweets that retweet number more than 1000 is less than 1%.

For users, we compute the mean retweet number of each user (i.e., MRN). Fig. 2(c) shows the cumulative distribution of different mean retweet number of users. From Fig. 2(c), we can see that 63.1% of all users 's mean retweet number is less than 1 and users with retweet number less than 30 is 90%. While, less than 0.8% users' mean retweet number is more than 1000. Fig. 2(d) is also a cumulative distribution. We sort the mean retweet number according to descending order. From Fig. 2(d), the top 10% tweets come from 0.78% of users. 7.8% of all users post top 90% tweets. Obviously, only a small part of users can post tweets with high retweet number.

In all, we can get two observations: (1) In the micro-blog network, the retweet number and number of tweets comply with the power-law distribution, and only a small part of tweets' retweet number is high. (2) For users in micro-blog network, only a small percentage of them have the potential to post a tweet with high retweet number.

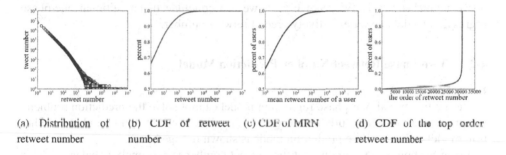

(a) Distribution of retweet number   (b) CDF of retweet number   (c) CDF of MRN   (d) CDF of the top order retweet number

Fig. 2. The characteristic of retweet behavior

4   Two-Phase Prediction Model

In this Section, we first discuss the disadvantage of one-phase models to predict retweet number, and then describe our two-phase retweet number prediction model.

4.1   One-Phase Prediction Model

In order to predict the retweeting number, a direct solution is to do regression on training data including features and its retweet number. This solution is called one-phase prediction model in this paper. Here, we select four one-phase regression models: LeastMedSq[14], LinearRegression, M5P[15] and MultilayerPerceptron to predict retweet number. We random select 1 million tweets from the whole dataset, because our dataset is too large to train a model in time. 80% of the data is training set and the rest is test set. We draw the prediction result of four one-phase models in Fig. 3, which shows the relation of prediction values and real values.

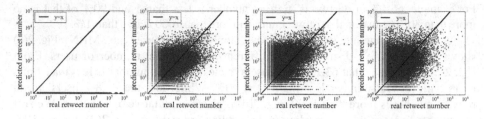

(a) LeastMedSq   (b) LinearRegression   (c) M5P   (d) MultilayerPerceptron

Fig. 3. One-phase prediction model scatter diagram

In Fig. 3(a), we can see LeastMedSq model predicts all tweet with a retweet number 0 or 1. Because most of tweets in training set have retweet number less than 1. Apparently, this model cannot solve the prediction problem. In Fig. 3(b)(c)(d), the prediction results are chaos. Even for most of tweets with retweet number 0 or 1, these models predict high retweet numbers. These three models also cannot predict retweet number precisely. From Fig. 3, we can conclude that traditional one-phase regression model cannot effectively predict retweet number.

4.2   Two-Phase Retweet Number Prediction Model

Basic Idea

Since the traditional one-phase regression model cannot solve the prediction problem, we propose a two-phase prediction model. The different between one-phase prediction model and two-phase prediction mode is shown in Fig. 1.

From Section 3.3, we know that the retweet number is extremely unbalanced. This characteristic may lead some models to predict small values, which can get less error. And comparing with predicting whether a tweet can be retweeted, predicting the retweet number is more complex. There are many features affecting the retweet number of a tweet. For example, the tweet is about special events, like earthquake or the tweet is created at a special time, like someone's birthday. What's more, the user may pay money to some people to get a high retweet number, like marketing and so on. Because of the unbalances distribution and the essential complexity, the one-phase can hardly predict very well.

To reduce the influence of the two problems above, we propose two-phase model. We know that only a small number of user can post tweets with high retweet number from Section 3.3. If we can classify tweets into some classes based on retweet number, the influence of the two problems will be reduced in each class. So the regression models which build in each class will get a better prediction. In our two-phase model, we build a multi-classification model on dataset in the first-phase. And in the second-phase, we build a regression model for each class. With "Divide and Conquer" strategy, our two-phase model can solve the unbalanced and complex problem in a way.

Table 3. Notations

Symbol	Description	Symbol	Description
T	The training set	L	The set of threshold values
x_i	the feature vector of the ith tweet in T	N	The set of class number which includes $1,2,3\ldots,\|L\|+1$
X	The set of feature vector in T	c_i	The class number of the ith tweet in T
r_i	The retweet number of the ith tweet in T	C	The set of class number in T
R	The set of retweet number in T	$\|S\|$	The size of set S

Algorithm 1. Two-phase retweet number prediction model training algorithm

Input: X, R, L, N
Output: Two-phase retweet prediction model
1   **for** $i = 1$ to $|T|$ **do**
2   compute the class number c_i of the ith tweet based on r_i and thresholds in L
3   **end for**
4   train classification model CLF   based on X and C
5   classify X and R into $/N/$ different parts based on different class number in C
6   **for** $i = 1$ to $|N|$ **do**
7   train regression model REG-i in part-i
8   **end for**
9   **return** (CLF,REG-1,REG-2,...,REG-$|N|$)

Algorithm 2. Two-phase retweet number prediction model predicting algorithm

Input: models CLF,REG-1,REG-2,...,REG-$|N|$, the feature of a tweet x
Output: the prediction retweet number r_p
1   use classification model CLF to classify x, return the class number c
2   use the regression model REG-c to predict the r_p based on x
3   **return** r_p

Classification Phase and Regression Phase

In first-phase of our model is the classification part. This part is important to our model because if the classification model doesn't have a good precision, our model cannot work very well. But there are two problems here: (1) How to select thresholds for classifying? (2) How many classes should we make?

For the first problem, we should not ignore the unbalanced distribution of the dataset. Considering the proportion of the tweets with different retweet number, we recommend to use the 80/20 Rule to choose thresholds. For example, if we want to classify tweets into 2 classes, the threshold will be retweet number of the tweet at the 80% position with a sorted datasets on retweet number. If we want to classify tweets into 3 classes, the positions will be 80% and 96%. Obviously, there are many other methods to select thresholds to classify tweets. In Section 5, we discuss other methods with experiments. For the second problem, in Section 5, we also discuss the different influence on our model with different number of classes.

From Section 4.1, we have known training regression models directly on the whole dataset cannot get a good model. To prove the effectiveness of our two-phase model, we will still choose the same 4 regressions in Section 4.1 on experiments.

Algorithm Framework

Before describing the training algorithm, some notations are introduced in Table 3. In our two-phase model framework, if we classify tweets into n classes, we should train (n+1) models. The training algorithm and predicting algorithm are described in Algorithm 1 and 2.

5   Experiments

In this section, we conduct a series of experiments on Sina Weibo dataset. Since our dataset is too large, we random select 1 million tweets on experiments. First, we conduct experiments to select classification model. Then, we compare the prediction results of one-phase model and two-phase model. Last, we discuss some parameters of our model on experiments.

5.1   Effectiveness Experiments

Classification Model Comparison

Firstly, we compare the performances of different classifiers. We use the 80/20 Rule to classify tweets into three classes, the thresholds in our dataset is 5 and 118. So, the range of retweet number in three classes are 0-5, 6-118, 119-MAX. We choose 4 classification models, RandomForest (RF), Logistic (LO), DecisionTree (DT) and NativeBayes (NB) to experiment. We random select 60%, 70%, 80% and 90% of the dataset as training set and the rest as test set.

Table 4. The results of different classification models

M	60%			70%			80%			90%		
	P	R	F1	P	R	F1	P	R	F1	P	R	F1
RF	85.5%	86.8%	86.1%	85.7%	87.0%	86.3%	86.0%	87.2%	86.6%	86.0%	87.2%	86.6%
LO	77.5%	81.5%	79.4%	77.6%	81.5%	79.5%	77.7%	81.6%	79.6%	77.8%	81.7%	79.7%
DT	84.0%	85.1%	84.5%	84.3%	85.3%	84.8%	84.6%	85.6%	85.1%	84.7%	85.7%	85.2%
NB	75.5%	79.7%	77.5%	75.6%	79.7%	77.6%	75.6%	79.8%	77.6%	75.7%	80.0%	77.8%

In Table 4, we compute precision (P), recall (R) and F1 value to evaluate each classification model. The result of RandomForest model is the best. So, we choose RandomForest model as our first phase model in the following experiments.

Two-phase Model vs One-phase Model
In this part, we compare results of one-phase model and two-phase model. We still use the same regression models in Section 4: LeastMedSq (LMS), LinearRegression (LR), M5P and MultilayerPerceptron (MP). We use MAE and RAE to evaluate the results. They are defined as follows:

$$\text{MAE} = \frac{1}{n}\sum_{i=1}^{n}|p_i - r_i| \tag{1}$$

$$\text{RAE} = \frac{\sum_{i=1}^{n}|p_i-r_i|}{\sum_{i=1}^{n}|r_i-r_m|} \tag{2}$$

where p_i is the prediction retweet number of the ith tweet in test set and r_i is the real retweet number. r_m is the mean retweet number in training set.

In order to fully test this model, we still random select 60%, 70%, 80% and 90% of the dataset as training set and the rest as test set to do experiments. From Table 5, we can see that two-phase model get a better prediction for each regression model. In the four two-phase models, the combination of RandomForest model and LeastMedSq model gets the best results.

Table 5. Comparation of one-phase and two-pahse models

Method	60%		70%		80%		90%	
	MAE	RAE	MAE	RAE	MAE	RAE	MAE	RAE
RF+LR	66.77	69.65%	65.76	69.29%	69.44	70.08%	73.89	71.28%
LR	104.15	108.64%	102.63	108.14%	101.89	102.83%	106.97	103.19%
RF+MP	65.81	68.65%	63.01	66.40%	62.37	62.95%	76.85	74.13%
MP	161.46	168.43%	132.02	139.11%	123.59	124.73%	388.33	374.61%
RF+LMS	**47.89**	**49.96%**	**47.54**	**50.09%**	**53.39**	**53.88%**	**58.22**	**56.16%**
LMS	51.59	53.81%	51.35	54.11%	57.42	57.95%	62.27	60.08%
RF+M5P	67.03	69.92%	63.88	67.31%	60.6	61.16%	71.66	69.13%
M5P	89.66	93.52%	84.13	88.65%	92.41	93.26%	106.97	103.19%

Most of time, we want to predict the approximate range of the retweet number of a tweet, rather than a specific value. So, we define the range of a specific number. Supposing a, b and n are positive integers, a<b and $10^a < n < 10^b$, the range of n is:

$$\text{range}(n) = \left[n - \frac{10^b-10^a}{m}, n + \frac{10^b-10^a}{m}\right] \tag{3}$$

where m is a parameter to control the radius.

Supposing n_p is the prediction retweet number and n_r is the real retweet number, when n_p satisfies

$$n_p \in [n_r - \frac{10^{\lceil log_{10}(n_r)\rceil} - 10^{\lfloor log_{10}(n_r)\rfloor}}{m}, n_{r+} \frac{10^{\lceil log_{10}(n_r)\rceil} - 10^{\lfloor log_{10}(n_r)\rfloor}}{m}] \qquad (4)$$

the prediction is right, otherwise is wrong. Then prediction accuracy is defined as follows.

$$Acc = \frac{the\ number\ of\ right\ predictons}{the\ number\ of\ all\ predictions} \qquad (5)$$

Because of the unbalanced distribution of the retweet number, most of the retweet number is 0 and 1. If a model predicts all tweets with a small number, it will get high prediction accuracy. To avoid this phenomenon, we random select an extremely tough test set in the rest of the 1 million training set. The test set has 100k tweets with retweet number 0-100, 100k tweets with retweet number 101-1000 and 100k tweets with retweet number more than 1001. In the following experiments, we use this set as test set. We still random select 60%, 70%, 80% and 90% of the 1 million tweets to be training set. From Fig. 4, we can see that two-phase models have a better performance than one-phase models.

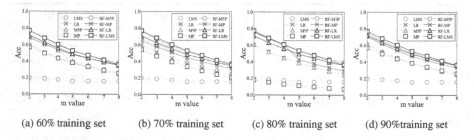

(a) 60% training set   (b) 70% training set   (c) 80% training set   (d) 90%training set

Fig. 4. Comparison of one-phase model and two-phase model

5.2   Parameters Study

In previous experiments, we classify tweets into three classes based on the 80/20 Rule. Here, we test the influence of different class number on our model. We classify tweets into 2-6 classes with thresholds 5, 118, 766, 4143 and 16866. The details are in Table 6. In Fig. 5, with the increased number of classes, the model gets better prediction. But when the number is more than 4, the precision become stable.

In previous experiments, we extract feature from the tweet's creator and the tweet. To find out which kind of feature has more contribution to prediction, we use each of the two kind of feature alone to train the model. From Fig. 6, we can see that the tweet's creator(user) has more influence on prediction. This result tells us that when we want to evaluate the influence of a tweet, we should pay more attention to who posts the tweet rather than the tweet itself.

Table 6. Thresholds of different number of classes

	5	118	766	4143	16866
2-class	√	×	×	×	×
3-class	√	√	×	×	×
4-class	√	√	√	×	×
5-class	√	√	√	√	×
6-class	√	√	√	√	√

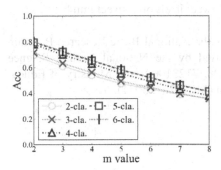

Fig. 5. Effect of different classes

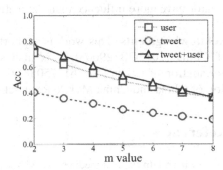

Fig. 6. Effect of different features

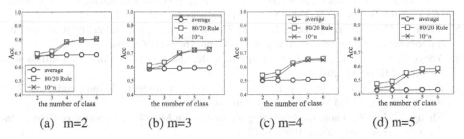

(a) m=2   (b) m=3   (c) m=4   (d) m=5

Fig. 7. Effect of different methods of classification

In Section 4, we say that there are other methods besides the 80/20 Rule to classify tweets. Here we discuss other methods based on the retweet number rather than proportion. Obviously, choosing 10^n as thresholds is an easy way. If we want 2 classes, the threshold is 10; 3 classes, the thresholds are 10, 100 and so on. Another method can be average allocation on retweet number. First, we choose the maximum retweet number as MAX_R. Then if we want 2 classes, the threshold is MAX_R/2; 3 classes, the thresholds are MAX_R/3, 2MAX_R/3 and so on. But in dataset of tweets, there are always some tweets that their retweet number like explosion. To avoid tweets like this, we filter retweet number that has less than 10 tweets heuristically, and then choose the MAX_R. We conduct lots of experiments to compare the three methods. From Fig. 7, we can see that no matter under what circumstances, the method of the 80/20 Rule gets the best prediction.

6   Conclusion

Retweet behavior is the key mechanism for information diffusion in micro-blog network. Retweet number, which denotes how many times that a tweet can be retweeted, is good measurement of both influence in diffusion and value in market of a tweet. To predict retweet number of a tweet in Sina Weibo, we build a two-phase retweet number prediction model. Experiments conducted on real dataset in Sina Weibo show that our two-phase retweet number model has better performance than traditional one-phase prediction model. In the experiments, we also find that the features of tweet's creator have more influence than the feature of tweet itself on retweet number.

Acknowledgments. This work is supported by the National Basic Research Program of China (2013CB329603). It is also supported by the National Natural Science Foundation of China (No. 61375058, 61074128, 71231002) and Ministry of Education of China and China Mobile Research Fund (MCM20123021).

References

1. Twitter, http://en.wikipedia.org/wiki/Twitter
2. Sina Weibo, http://en.wikipedia.org/wiki/Sina_Weibo
3. Petrovic, S., Osborne, M., Lavrenko, V.: RT to Win! Predicting Message Propagation in Twitter. In: ICWSM (2011)
4. Ma, H., Qian, W., Xia, F., et al.: Towards modeling popularity of microblogs. J. Frontiers of Computer Science 7(2), 171–184 (2013)
5. Yang, Z., Guo, J., Cai, K., Tang, J., Li, J., Zhang, L., Su, Z.: Understanding retweeting behaviors in social networks. In: Proc. of CIKM, pp. 1633–1636 (2010)
6. Yang, J., Counts, S.: Predicting the speed, scale, and range of information diffusion in twitter. In: ICWSM (2010)
7. Peng, H., Zhu, J., Piao, D., Yan, R., Zhang, Y.: Retweet Modeling Using Conditional Random Fields. In: ICDM Workshops, pp. 336–343 (2011)
8. Zhang, Y., Lu, R., Yang, Q.: Predicting Retweeting in Microblogs. Journal of Chinese Information Processing 26(4), 109–114 (2012)
9. Bandari, R., Asur, S., Huberman, B.: The pulse of news in social media: forecasting popularity. In: ICWSM (2012)
10. Li, Y., Yu, H., Liu, L.: Predict algorithm of micro-blog retweet scale based on SVM. Application Research of Computers 30(9), 2594–2597 (2013)
11. Java, A., Song, X., Finin, T., et al.: Why we twitter: understanding microblogging usage and communities. In: WebKDD, pp. 56–65 (2007)
12. Boyd, D., Golder, S., Lotan, G.: Tweet, tweet, retweet: Conversational aspects of retweeting on Twitter. In: 43rd Hawaii International Conf. on System Sciences, pp. 1–10 (2010)
13. Suh, B., Hong, L., Pirolli, P., Chi, E.H.: Want to be Retweeted? Large Scale Analytics on Factors Impacting Retweet in Twitter Network. In: SocialCom/PASSAT, pp. 177–184 (2010)
14. Rousseeuw, P.J., Leroy, A.M.: Robust regression and outlier detection (1987)
15. Wang, Y., Witten, I.H.: Induction of model trees for predicting continuous classes. In: ECML, pp. 128–137 (1997)

Finding Vacant Taxis Using Large Scale GPS Traces*

Zhen Qiu[1,3], Hongyan Li[1,3,**], Shenda Hong[1,3], Yiyong Lin[1,3], Nana Fan[2,3],
Gaoyan Ou[2,3], Tengjiao Wang[2,3], and Lilue Fan[1,3]

[1] Key Laboratory of Machine Perception(Peking University),
Ministry of Education, China
[2] Key Laboratory of High Confidence Software Technologies(Peking University),
Ministry of Education, China
[3] School of Electronics Engineering and Computer Science,
Peking University, Beijing, China
lihy@cis.pku.edu.cn

Abstract. In modern cities, more and more vehicles, such as taxis, have
been equipped with GPS devices for localization and navigation. The
GPS-equipped taxis can be viewed as pervasive sensors and the large
scale traces allow us to reveal many hidden "facts" about the city dynamics. In this paper, we aim to estimate the wait time and probability
of taking a vacant taxi according to time and position. Further more, we
provide recommendations for passengers who want to take a vacant taxi.
To achieve these objectives, firstly we preprocess the large scale taxi GPS
traces data set to generate the Map Grid Based(MGB) index. Secondly,
with the MGB index, we apply the nonhomogeneous Poisson process
corrected by the conditions of road and weather(NPPCRW) method to
perform estimation and recommendation. We build our system based
on a large scale real-world GPS traces data set generated from more
than 12000 taxis in Beijing over a 110 days period. Then we validate the
system with extensive evaluations including in-the-field user studies.

Keywords: GPS traces, large scale, vacant taxi, estimation, recommendation.

1 Introduction

As an indispensable part of urban public transportation, taxi plays a significant
role in people's daily life [1]. However, more and more urban people suffer from
waiting too long for taxis. The reasons come from the following aspects. Firstly,
there exists imbalance between supply and demand. Secondly, passengers lack
the information of the vacant taxis. Based on the past experience, a skillful
passenger can choose an effective position. However, many people gain little skill

---

* This work was supported by Natural Science Foundation of China (No.60973002 and
No.61170003), the National High Technology Research and Development Program
of China (Grant No. 2012AA011002), and MOE-CMCC Research Fund.
** Corresponding author.

F. Li et al. (Eds.): WAIM 2014, LNCS 8485, pp. 793–804, 2014.
© Springer International Publishing Switzerland 2014

about how long to wait and where is better for waiting. Thus, it is imperative to provide some information for passengers to find vacant taxis, such as wait time, probability of taking a vacant taxi and useful positions.

Nowadays, GPS sensors are equipped in the taxis for dispatching and ensuring security in many big cities, like New York, Singapore and Beijing. The initial functions of GPS devices are localization, navigation, and scheduling. As the usage of the GPS sensors and wireless communication units has increased a lot, a large amount of taxi GPS traces data have been accumulated. Besides the applications like "location-based services", some fine-new functions are emerging with the collected GPS trace data set, ranging from urban design[2] to traffic prediction[3]. Based on these fresh functions, researchers have carried out many interesting studies including human mobility patterns discovering[4][5], traffic conditions detecting [6][7] and taxi mobility intelligence mining [8][9][10].

Our study is built upon the GPS traces of taxis in Beijing. The data set is collected from more than 12,000 taxis which make up approximately one-fifth of the total in Beijing. Since the data set is large scale, we firstly apply MapReduce framework[11] to preprocess the data set and generate the MGB index. Then we propose NPPCRW method to estimate the wait time and probability of taking a vacant taxi at a given time and position. Finally we recommend a preferable position for passengers.

The major contributions of our work include:

1. We divide the map by grids, each position will be mapped into the corresponding grid. We also propose an approach to generate MGB index based on the map grid. With the large scale data set, we apply MapReduce framework to finish the data preprocess work.
2. We propose NPPCRW method to estimate the wait time and probability of taking a vacant taxi at a given time and position. We also provide a recommendation on where to wait a vacant taxi for passengers based on NPPCRW method.
3. We conduct several experiments to evaluate our methods on real data set, compared with state-of-the-art methods, our method achieves a higher estimation accuracy with less time and provide a more useful recommendation.

The rest of the paper is organized as follows. Section 2 introduces the related work. In section 3, we give an overview of our work. Section 4 describes the data sets and data preprocessing. Section 5 presents our model and describes how to estimate the wait time and probability of taking a vacant taxi. Section 6 provides a recommendation for passengers. Section 7 shows our experiments and evaluations on our approach. Section 8 concludes the research and gives directions for future studies.

2   Related Work

Taxi traces have become a hotspot research in recent years. Existing work can be grouped into three categories: 1)Improving Taxi Dispatching, 2)Providing Recommendations For Taxi Drivers and 3)Providing Recommendations For Passengers.

2.1 Improving Taxi Dispatching

Taxi dispatching systems have attracted more and more attentions of researchers with the popularization of GPS sensors and the development of intelligent transportation systems[12]. Yang et al. [13] build a model to analyze some influencing factors on customer waiting time. Yamamoto et al. [14] propose a fuzzy clustering and adaptive routing approach to improve dispatching system. The goal of their studies is to improve centralized dispatching. Compared with them, our research pays more attentions to estimate the waiting time for a vacant taxi at a given time and position.

2.2 Providing Recommendations for Taxi Drivers

Yuan et al. [15] mine smarting driving directions for drivers based on the past GPS traces data set collected from a large number of taxis. Ge et al. [16] develop a recommender system for taxi drivers. Their system can recommend a sequence of pick-up points or parking positions for maximizing the taxi driver's profit. Moreover, Li et al. [17] use L1-Norm SVM to discover the most recognizable features to distinguish the performance of the taxis. However, all these studies concerned about how to serve the taxi drivers. Our goal is to provide services for passengers.

2.3 Providing Recommendations for Passengers

Phithakkitnukoon et al. [18] propose a model to predict the number of vacant taxis at a given position and to provide the information for both passengers and taxi drivers. They employ a method based on the Naïve Bayesian classifier and obtain the prior probability distribution from the historical data. However, their data size is relatively small containing the traces of only 150 taxis, which might not be enough to present the general rules with the weak statistical significance. Yuan et al. [19] propose an approach to make recommendation for passengers based on the knowledge of 1) passengers mobility patterns and 2) taxi drivers picking-up/dropping-off behaviors learned from the taxis GPS traces. However, their work ignores the conditions of road and weather. In addition, their recommendation is based on road-segment level. Actually, different positions in the same road have different probabilities of taking a vacant taxi. Thus their method lacks of enough accuracy.

3 Overview

In this section, we first give some preliminary definitions which are related to our work. Then we show the framework of our system.

3.1   Preliminary Definitions

Definition 1. *(Map Grid): The map is partitioned into square grids, and each grid G_i can also be divided up into nine small square grids, i.e., G_i : $g_{i1}, g_{i2}, \ldots, g_{i9}$. We map each position into G_i by latitude-longitude at first, and then find the accurate small grid $g_{ik}(1 \le k \le 9)$ in where the position is.*

Definition 2. *(Taxi State): The working taxi has three states: occupied, cruising, parking. O represents a taxi occupied by a passenger, C represents a taxi traveling without a passenger, P represents a taxi waiting for a passenger.*

3.2   System Framework

The framework of our system is illustrated in Fig.1. Firstly, we apply MapReduce framework to preprocess the large scale GPS traces data set. Then we generate data and index based on map grid. Next, we filter the data, which is queried from grid data using MGB index, by the conditions of road and weather, then we propose NPPCRW method to estimate the wait time and probability of taking a vacant taxi. Finally, we provide a recommendation for passengers.

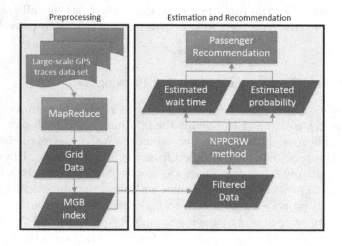

Fig. 1. System framework

4   Data Set Preprocessing

In this section, we first describe our three data sets: taxis GPS traces data set, road networks data set and weather data set. Then we introduce how to generate the MGB index by utilizing these data sets.

4.1   Date Set Description

The large scale real-world taxi GPS data set is collected from more than 12,000 taxis in Beijing, over a period of 110 days in the year of 2010. In this data set, each taxi is equipped with a GPS device with a sampling-rate of about one record per minute. Table 1 lists the field for each GPS entry, along with a sampling entry. The road networks data set contains features of each road, such as width, length, direction, etc. Moreover, we crawl the weather data set of 110 days from Internet, the features of weather can be regarded as temperature, rainy, sunny, etc.

Table 1. Fields for a GPS entry with samples

TaxiID	State	Time	Longitude	Latitude	Speed	Direction
431498	0	20121101095636	116.4243011	40.0727348	43	88
129521	2	20121101095634	116.1922989	39.9446297	0	258

4.2 Mapping Position to Map Grid

According to Definition 1, we divide the map into grids including $G_1, \ldots, G_9$, as shown in Fig.2. Each position can be mapped to the map grid based on its latitude-longitude. If G_5(the middlemost big grid) is the chosen grid, we continue to turn G_5 into sudoku with area threshold. If the position is in g_{55}, we only consider the traces information of G_5. If it is in g_{52}, we should consider both G_5 and G_2. In the same way, g_{54} corresponds to G_5 and G_4, g_{56} corresponds to G_5 and G_6, g_{58} corresponds to G_5 and G_8. If the position is in g_{51} we should take account of G_1, G_2, G_4 and G_5. Similarly, g_{53} corresponds to G_3, G_2, G_6 and G_5, g_{57} corresponds to G_7, G_4, G_8 and G_5, g_{59} corresponds to G_9, G_6, G_8 and G_5.

4.3 Generating MGB Index

We design a hash function to map a position(longitude, latitude) to the corresponding grid G_t:

$$key = \lfloor longitude \times a \rfloor \times b + \lfloor latitude \times c \rfloor \tag{1}$$

The difference range of both longitude and latitude of Beijing is 1 degree, which represents one hundred kilometers. Thus, we regard 0.001 degree(one hundred meters) as the reasonable range of grid, and set $a = c = 1000, b = 100000$ in order to assure the key is unique.

The taxi GPS data set is large scale, single node becomes less practical with limited computing and storage resources. We apply distributed process by the framework of MapReduce. In the Map function, we generate the key by longitude-latitude, and regard the raw data as the value. Then the MapReduce library groups together all values associated with the same key.

In order to acquire exact GPS traces data quickly, we should obtain index using files which are generated by the Map function. It is better to sort the records in the index file by the key, thus the prediction model can utilize binary search method to query traces information more quickly.

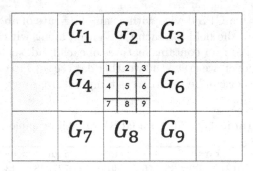

Fig. 2. Map the position to the grid

5   The Prediction Method

NPPCRW method first filters the obtained data based on weather conditions, then adopts NPP(non-homogeneous Poisson process) to simulate the vacant taxis' activities, finally corrects the simulations based on road conditions. We define the rate parameter of NPP as a time-dependent function $\lambda(t)$.

5.1   Estimation of Waiting Time

We denote a taxi arriving time as T_{taxi}. Thus at time t and position p, the wait time can be defined as

$$T_w^{t,p} = f_{road}(T_{taxi}) \tag{2}$$

where f_{road} is defined as the correct function based on road conditions.

According to Definition 2, the vacant taxi contains two states: P and C, we do not need wait time for P taxi. The arriving event of C taxi conforms to NPP. Therefore, we have:

$$T_{taxi} = Pr(P|t,p) * 0 + Pr(C|t,p) * T_{next} \tag{3}$$

We partition one day into K small time units, the length of each unit is τ(where Δt can be divisible by τ, e.g., $\tau = 5$ minutes, $\Delta t = 15$ minutes). $D(p_0, \delta)$ represents the corresponding grids area according to subsection 4.2. $\#_{k,j}(C), \#_{k,j}(P)$ and $\#_{k,j}(O)$ denote the number of cruising, parking and occupied taxis.

$$Pr(C|t,p) = \frac{\sum_{k=\lfloor t-\Delta t/\tau \rfloor}^{\lfloor (t+\Delta t)/\tau \rfloor} \sum_{j \in D(p,\delta)} \#_{k,j}(C)}{\sum_{k=\lfloor t-\Delta t/\tau \rfloor}^{\lfloor (t+\Delta t)/\tau \rfloor} \sum_{j \in D(p,\delta)} (\#_{k,j}(C) + \#_{k,j}(P) + \#_{k,j}(O))} \tag{4}$$

We just need to estimate T_{next}. Let $N(t)$ represents the number of arriving vacant taxis by time t, $\Lambda(t)$ represents the cumulative intensity function of taxi arriving rate. According to the NPP. We have following properties:

$$\Lambda(t) = \int_0^t \lambda(\tau)d\tau \tag{5}$$

$$Pr\{N(t) = k\} = Pr\{N(t+s) - N(s) = k\} = e^{-\Lambda(t)}\frac{\Lambda(t)^k}{k!} \tag{6}$$

The cumulative distribution function (CDF) of the next vacant taxi arriving within passengers' endurance limitation wait time T_e is:

$$Pr\{T_{next} < T_e\} = 1 - Pr\{T_{next} > T_e\} = 1 - Pr\{N(t) = 0\} = 1 - e^{-\Lambda(t)} \tag{7}$$

Thus the probability density function (PDF) of T_{next} is:

$$P(T_{next}) = \frac{dPr\{T_{next} < t\}}{dt} = \lambda(t)e^{-\Lambda(t)} \tag{8}$$

where $\lambda(t)$ can be simplified into piecewise constant function, we assume the rate function has a cycle of 24 hours. Thus, the time interval t_i is one hour, we have

$$\widehat{T}_{next} = E(T_{next}) = \int_0^{+\infty} t\lambda(t)e^{-\Lambda(t)}dt = \int_0^{+\infty} t\lambda e^{-\Lambda(t)}dt = \frac{1}{\lambda} \tag{9}$$

We employ the maximum likelihood estimation(MLE) to estimate λ. We define the number of vacant taxis at position j, time k of ith day as $N_{k,j}^i(C)$. The likelihood function is

$$L(\lambda) = \prod_{i=1}^{n} Pr\{N(t) = N_{k,j}^i(C)\} \tag{10}$$

Then we can get $\lambda = \frac{\sum_{i=1}^n N_{k,j}^i(C)}{nt_i}$, thus T_{taxi} can be estimated as:

$$T_{taxi} = \frac{\sum_{k=\lfloor t-\Delta t/\tau\rfloor}^{\lfloor(t+\Delta t)/\tau\rfloor} \sum_{j\in D(p,\delta)} \#_{k,j}(C)nt_i}{\sum_{k=\lfloor t-\Delta t/\tau\rfloor}^{\lfloor(t+\Delta t)/\tau\rfloor} \sum_{j\in D(p,\delta)}(\#_{k,j}(C) + \#_{k,j}(P) + \#_{k,j}(O))\sum_{i=1}^n N_{k,j}^i(C)} \tag{11}$$

We can assume the number of vacant taxis is directly proportional to the width(W) of road, because if the road is wider, there are more taxis. We also assume the the number of vacant taxis is inversely proportional to the length(L) of road, because if the road is longer, there are more passengers who want to take a taxi. In addition, we regard direction(D) as a weighted factor. Thus, we can get the following formula:

$$f_{road}(T_{taxi}) = \frac{(\theta_0 + \theta_1 T_{taxi} + \theta_2 D)W}{L} \tag{12}$$

We apply Eq.(11) into Eq.(12) and adopt gradient descend method to estimate all the parameters. Finally, we apply Eq.(12) into Eq.(2) to estimate the wait time of taking a vacant taxi.

5.2   Probability of Taking Vacant Taxis

We define the probability of one vacant taxi arriving within time T_e as P_{taxi}. According to Eq.(7), P_{taxi} can be formulated as:

$$P_{taxi} = Pr\{T_{next} < T_e\} = 1 - e^{-\Lambda(t)} \tag{13}$$

6   Position Recommendation for Passengers

Based on NPPCRW method, we can estimate the wait time and probability of taking a vacant taxi at position p_0 and time t_0. Furthermore, we also could provide a direct recommendation on where to take a taxi. Our recommendation assumes that passengers have their endurance limitation wait time T_e, if the estimated wait time exceeds T_e, we need to provide them another position which is not very far. The goal of our recommendation is to make a balance tradeoff between the wait time and walk distance.

Our recommendation provides candidate places P_c for passengers:

$$P_c = \{p : Dist(p, p_0) < ST_e\} \tag{14}$$

where S is the common walk speed, $Dist$ is a function calculating distance between two positions.

We can define the walk time T_{walk} from p_0 to p is $\frac{Dist(p,p_0)}{S}$. Then we define the cost function of total time from p_0 to p:

$$Cost(p) = T_{walk} + T_w^{t_0 + T_{walk}, p_0} \tag{15}$$

Finally, we choose position p_{best} in set $P_{candidate}$ with minimal $Cost(p)$:

$$p_{best} = \{p \in P_{candidate} : argminCost(P)\} \tag{16}$$

7   Experiments

In this section, we conduct comprehensive experiments to evaluate our method. In the first experiment, we will evaluate our estimation for wait time and probability of taking a vacant taxi. T-Finder method [19] is our compared method. In the second experiment, we will further evaluate our recommendation for passengers, Zheng's method [20] is selected as our compared method.

7.1   Evaluation on Estimation for Wait Time and Probability

We perform the in-the-field study for estimation in some positions of Beijing: p_1, p_2, p_3 and p_4 on Zhongguancun Main Street(shortened as Z), p_5, p_6, p_7 and p_8 on The Summer Palace Road(shortened as S), p_9, p_{10}, p_{11} and p_{12} near Wudaokou Railway Station(shortened as W). For each position, the users are involved in calculating the wait time for the first vacant taxi at several given times. Therefore, the users' records are the real wait time at given positions and times. We also apply our NPPCRW method and T-Finder method to estimate the wait time. Here we use percent error to evaluate the relative accuracy of the estimation. The percent error is defined as:

$$percent\ error = \frac{|\ real\ value - estimate\ value\ |}{real\ value} \times 100\% \qquad (17)$$

Fig.3 shows the percent error of the wait time of NPPCRW method and T-Finder method[19] with 30 test cases. The average percent error of NPPCRW is 29.95% which is lower than T-Finder's 31.58%. The result shows that both of their estimations are reasonable, but NPPCRW is more accurate. We also compare the cost time except for preprocess time of both methods, Fig.4 shows that NPPCRW method costs much less time. The result reflects the MGB index plays an important role in reducing cost time when faced with large scale data set.

In order to evaluate the probability of taking a vacant taxi, we also perform the in-the-field study at $p_1, \ldots, p_{12}$ and some given times. Table 2 gives the overall results of evaluation. *Rank* denotes the real ranking according to the average number of vacant taxis encountered at a certain time (denoted by #).

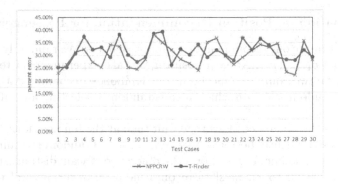

Fig. 3. The percent error of the wait time by estimation

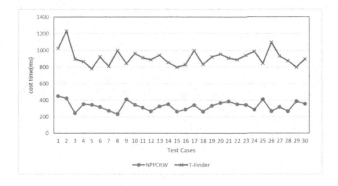

Fig. 4. The cost time of estimation

Table 2. Evaluating probability of taking a vacant taxi

Time & Position	8:30-8:35				9:30-9:35				10:00-10:05			
	p_1	p_2	p_3	p_4	p_5	p_6	p_7	p_8	p_9	p_{10}	p_{11}	p_{12}
#	5.1	3.2	3.0	1.2	1.8	2.2	3.2	8.3	6.2	2.4	3.3	3.2
$Rank$	1	2	3	4	4	2	3	1	1	4	2	3
$Rank_{T-Finder}$	1	3	2	4	2	3	4	1	1	2	3	3
$Rank_{NPPCRW}$	1	2	3	4	4	3	2	1	1	4	3	2

Time & Position	13:30-13:35				14:30-14:35				15:00-15:05			
	p_1	p_2	p_3	p_4	p_5	p_6	p_7	p_8	p_9	p_{10}	p_{11}	p_{12}
#	4.2	2.8	3.2	1.3	2.5	3.2	2.2	5.3	4.2	5.7	1.5	2.4
$Rank$	1	3	2	4	3	2	4	1	2	1	4	3
$Rank_{T-Finder}$	2	1	3	4	2	3	4	1	1	2	3	3
$Rank_{NPPCRW}$	1	2	3	4	3	2	4	1	3	1	4	2

$Rank_{T-Finder}$ and $Rank_{NPPCRW}$ stand for ranking according to probability of finding a vacant taxi based on T-Finder's method[19] and our NPPCRW method. It is obvious that our NPPCRW method is better than the compared method.

7.2   Evaluation on Position Recommendation for Passengers

We further evaluate our recommendation for passengers. We randomly choose ten positions from $p_1, \ldots, p_{12}$. At each position, we take experiments at ten different times. We have two compared methods: one is Zheng's recommendation[20], the other one is random selection which means randomly selecting a position within the range.

Fig.5 shows the difference of average wait time at the ten positions, our recommendation is significantly better than Zheng's[20] and random recommendations in terms of the wait time. Fig. 6 shows the difference of walk distance, our recommendation is similar to Zheng's[20] recommendation in terms of distance. And they are better than random selection. Therefore, our recommendation is more useful for passengers.

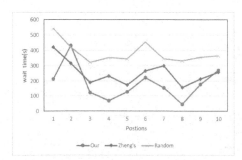

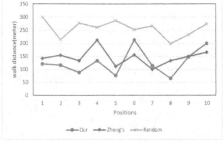

Fig. 5. Difference of average wait time   **Fig. 6.** Difference of walk distance

8   Conclusion

In this paper, we preprocessed the large scale taxi GPS traces data set and generate MGB index, then we proposed NPPCRW method to estimate wait time and probability of taking a vacant taxi on the filtered data which considers conditions of both road and weather. Further more, we provided a recommendation for passengers with a useful position. We evaluated our method by extensive experiments including a series of in-the-field studies. The experimental results show that our estimation and recommendation are more accurate than state-of-the-art methods.

In the future, we attempt to combine some other methods such as machine learning to improve the performance of our estimation and recommendation. We also plan to develop related applications which can serve for passengers.

References

1. Lu, J., Wang, W.: Confirming method of urban taxi quantity. Journal of Traffic and Transportation Engineering 4(1), 92–95 (2004)
2. Zheng, Y., Liu, Y., Yuan, J., Xie, X.: Urban computing with taxicabs. In: Proceedings of the 13th International Conference on Ubiquitous Computing, pp. 89–98. ACM (2011)
3. Castro, P.S., Zhang, D., Li, S.: Urban traffic modelling and prediction using large scale taxi GPS traces. In: Kay, J., Lukowicz, P., Tokuda, H., Olivier, P., Krüger, A. (eds.) Pervasive 2012. LNCS, vol. 7319, pp. 57–72. Springer, Heidelberg (2012)
4. Gonzalez, M.C., Hidalgo, C.A., Barabasi, A.L.: Understanding individual human mobility patterns. Nature 453(7196), 779–782 (2008)
5. Li, X., Pan, G., Wu, Z., Qi, G., Li, S., Zhang, D., Zhang, W., Wang, Z.: Prediction of urban human mobility using large-scale taxi traces and its applications. Frontiers of Computer Science 6(1), 111–121 (2012)
6. Chang, H.W., Tai, Y.C., Hsu, J.Y.J.: Context-aware taxi demand hotspots prediction. International Journal of Business Intelligence and Data Mining 5(1), 3–18 (2010)
7. Reades, J., Calabrese, F., Sevtsuk, A., Ratti, C.: Cellular census: Explorations in urban data collection. IEEE Pervasive Computing 6(3), 30–38 (2007)
8. Wong, K., Wong, S., Bell, M., Yang, H.: Modeling the bilateral micro-searching behavior for urban taxi services using the absorbing markov chain approach. Journal of Advanced Transportation 39(1), 81–104 (2005)
9. Liu, L., Andris, C., Ratti, C.: Uncovering cabdrivers behavior patterns from their digital traces. Computers Environment and Urban Systems 34(6), 541–548 (2010)
10. Yang, H., Fung, C., Wong, K., Wong, S.: Nonlinear pricing of taxi services. Transportation Research Part A: Policy and Practice 44(5), 337–348 (2010)
11. Dean, J., Ghemawat, S.: Mapreduce: simplified data processing on large clusters. Communications of the ACM 51(1), 107–113 (2008)
12. Lee, D.H., Wang, H., Cheu, R.L., Teo, S.H.: Taxi dispatch system based on current demands and real-time traffic conditions. Transportation Research Record: Journal of the Transportation Research Board 1882(1), 193–200 (2004)
13. Yang, H., Yang, T.: Equilibrium properties of taxi markets with search frictions. Transportation Research Part B: Methodological 45(4), 696–713 (2011)

14. Yamamoto, K., Uesugi, K., Watanabe, T.: Adaptive routing of cruising taxis by mutual exchange of pathways. In: Lovrek, I., Howlett, R.J., Jain, L.C. (eds.) KES 2008, Part II. LNCS (LNAI), vol. 5178, pp. 559–566. Springer, Heidelberg (2008)

15. Yuan, J., Zheng, Y., Zhang, C., Xie, W., Xie, X., Sun, G., Huang, Y.: T-drive: driving directions based on taxi trajectories. In: Proceedings of the 18th SIGSPA-TIAL International Conference on Advances in Geographic Information Systems, pp. 99–108. ACM (2010)

16. Ge, Y., Xiong, H., Tuzhilin, A., Xiao, K., Gruteser, M., Pazzani, M.: An energy-efficient mobile recommender system. In: Proceedings of the 16th ACM SIGKDD International Conference on Knowledge Discovery and Data Mining, pp. 899–908. ACM (2010)

17. Li, B., Zhang, D., Sun, L., Chen, C., Li, S., Qi, G., Yang, Q.: Hunting or waiting? Discovering passenger-finding strategies from a large-scale real-world taxi dataset. In: 2011 IEEE International Conference on Pervasive Computing and Communi-cations Workshops (PERCOM Workshops), pp. 63–68. IEEE (2011)

18. Phithakkitnukoon, S., Veloso, M., Bento, C., Biderman, A., Ratti, C.: Taxi-aware map: Identifying and predicting vacant taxis in the city. In: de Ruyter, B., Wichert, R., Keyson, D.V., Markopoulos, P., Streitz, N., Divitini, M., Georgantas, N., Mana Gomez, A. (eds.) AmI 2010. LNCS, vol. 6439, pp. 86–95. Springer, Heidelberg (2010)

19. Yuan, N., Zheng, Y., Zhang, L., Xie, X.: T-finder: A recommender system for finding passengers and vacant taxis (2012)

20. Zheng, X., Liang, X., Xu, K.: Where to wait for a taxi? In: Proceedings of the ACM SIGKDD International Workshop on Urban Computing, pp. 149–156. ACM (2012)

Logo Detection and Recognition
Based on Classification

Yifei Zhang[1,2], MingMing Zhu[1], Daling Wang[1,2], and Shi Feng[1,2]

[1] School of Information Science and Engineering, Northeastern University
[2] Key Laboratory of Medical Image Computing(Northeastern University)
Ministry of Education, Shenyang 110819, P.R. China

Abstract. Online product frauds in the booming e-commerce market have become a major concern for market surveillants and commercial companies. The logo detection plays a crucial role in preventing the increasing online counterfeit trading attempts. In this paper, a novel method based on Random Forest classification with multi-type features is presented to detect the logo regions on arbitrary images and the detected logo regions are further recognized using the visual words with spatial correlated information. Extensive experiments have been conducted on realistic and noise images with different logos. The results show that the proposed method is able to detect the logo regions, and the recognition performance outperforms the well-known Viola-Jones approach for recognizing the arbitrary logos on realistic images.

Keywords: logo detection, Random Forest classification, visual words, logo recognition.

1   Introduction

As more and more people are willing to go shopping online, how to monitor and prevent the increasing online product fraud cases has become a major concern for both market surveillants and commercial companies [1]. Some online trading counterfeits of name brands or well-known products are always produced with misleading advertisements. Although text based searching methods have been used for illegal online trading detection, it is difficult for users to detect the counterfeits without brand-related words in their descriptions. To protect the interests of both dealers and customers, an automatic logo recognition system is required to prevent illegal trading activities and preserve the brand rights, even though there is not any corresponding brand related descriptions.

In logo recognition, one fundamental problem is to detect the logo regions in an image. The logo detection can be considered as a rare event detection and has a high demand of low false position rates. Although the adaptive boosting based Void-Jones approaches have become a commonplace in the general purpose pattern detection application [2,3], it suffers from the low noise-tolerant problem in the training datasets, which also is general in the application of logo image detection [4,5]. Actually, Logos are always made with all kinds of materials such

F. Li et al. (Eds.): WAIM 2014, LNCS 8485, pp. 805–816, 2014.
© Springer International Publishing Switzerland 2014

as fabrics, glasses, leathers and metal. The larger intra-class variations or class noise (mis-labeled) would be created on the logo images' textures, intensities and design details. Therefore, a noise-tolerant classifier is essential for the application of the logo detection and recognition.

As the appearances of the same logo may vary due to different lighting conditions, scaling, and rotation effects during imaging, some logo images may respond poorly to some simple features or single feature. To improve the performance of logo detection and recognition, more methods need to be introduced as view cues to enhance the feature detection on logo images.

In this paper, we propose a noise-tolerant classification method based on Random Forest for logo detection and recognition. Firstly, a fusion scheme of multi-type feature methods is designed to extract features from logo images, and normalize them using the Bag of Words model (BOW). Then, a random forest classification algorithm is used to detect logo regions by defining classification scoring rules, and the robustness of logo detection on realistic noise pictures is further improved using the comprehensive scoring on all decision trees. Lastly, the visual words with space information are used for logo recognition to improve the accuracy of results.

The rest of the paper is organized as follows. Section 2 introduces the related work on logo detection and recognition. Section 3 analyzes the characteristics of the logo images and proposes the algorithm for a fusion of multi-type feature extraction methods. Section 4 describes the Random Forest based classification method for logo detection and further presents the optimization approach for logo recognition. The experimental results on the realistic datasets are provided in section 5, and a conclusion and future work are presented in Section 6.

2   Related Work

There are mainly two approaches for detecting logo regions. The first direction is automatic region matching using local or global characteristics. Joly and Buisson used SIFT features to search images with a special logo by a query expansion strategy [6]. David S. Doermann et al added some limited conditions in the logo recognition process, such as the invariants on the relationship between a circle and a line [7]. Li et al introduced a system architecture for segmentation-free and layout-independent logo detection and recognition in document images [8]. Hassanzadeh and Pourghassem extracted the spatial structure features of images for the logo recognition in noise document images after image segmentation by the method of binarization [9]. Although, the characteristic based methods can lead to higher accuracy for the specific logo recognition, such as the document images, they suffer from the shape and material diversity of arbitrary logo images. Some logos may respond poorly to special features or shapes, and the direct boundary extraction may fail due to boundary shielding or noise images.

Another direction is to detect logo regions based on machine leaning methods. Bayesian belief network was used for document logo detection and recognition in the literature [10], where a logo hypothesis was computed by matching the nodes of a logo's RAG (Region Adjacency Graph). Such a basic framework is efficient for the well-defined topological structures of logos. When processing with complex contours or incomplete boundaries, this method often falls to extract the sub-regions of logos and build an effective RAG for logos. As an application of the pattern detection, although the Viola-Jones can perform a fast and accurate detection in terms of Fale Positive (FP) [11], its performance degrades significantly on noisy training datasets [12]. For most logo images, the training dataset often has larger inter-class variations than the training datasets for other regular objects such as face. Yu and Vrizlynn updated the Viola-Johes classifier on the noisy training logo data by detecting the extreme training situation and adjusting the training constraints [13]. As the noisy training data would be trimmed adaptively in the training processing, more data with the same logo would be needed for training and the accuracy of classifier could be affected on the small dataset or the dataset with more noisy images. what's more, more classifiers need to be built for multi-class logo detection while every classifier will be built only for one logo.

In this paper, a logo detection based on Random Forest learning is proposed for a wide range of the logo images in training and detection. More feature extraction methods are fused to enhance the discriminative feature phases. Then the classifier is designed for multi-class logo detection using Random Forest learning, which can avoid an extreme stage training situation by building more decision trees in the random forest. Finally, considering the fixed structure of logo images, a new method of similarity calculation based on spatial information is proposed for logo recognition.

3   Feature Extraction and Fusion

Different from general images, various logos always have the same or similar features on some aspect. For example, Heineken logo and Star Bucks logo have the same shapes as shown in figure 1(a), and Barilla logo and Pepsi logo have the similar colors as shown in figure 1(b). The similarity of different logos causes that it is difficult to detect logos only using simple or single type of features. Moreover, logo images are smaller than general images, so we can extract and process multi-type features of logo images with not larger computation cost.

(a)   (b)

Fig. 1. Logos with similar shape or color

In this research, considering the inherent attributes of logos and the different effects during image taking, we extract the hybrid features of logo images for the classifier training, including texture features ($F_{texture}$), canny edge features (F_{canny}), shape features (F_{shape}), Histogram of Oriented Gradient (HOG) features [14] (F_{HOG}), color features (F_{color}), Scale Invariant Feature Transform (SIFT) features [15] (F_{SIFT}), Speed Up Robust Feature (SURF) features [16] (F_{SURF}). Then a logo image with class label can be described with F_{logo} as follows:

$$F_{logo} = (< F_{texture}, F_{canny}, F_{shape}, F_{HOG}, F_{color}, F_{SIFT}, \\ F_{SURF} >, logo) \tag{1}$$

where *logo* is a class label of the logo image.

Since the numbers of some features depend on the sizes of images (such as HOG) and the numbers of extracted interest points (such as SURF and SIFT), we need to normalize image features using some rules for building the classifier. In order to get the same number values for the same feature on all logo images, we first normalize logo images using linear interpolation to the same size for feature extraction in this paper. For SURF and SIFT features obtained by detecting the image interest points, their numbers vary with the numbers of the extracted interest points in different images. Here we use BOW to obtain uniform k statistic features of SURF and SIFT on scaling invariability for logo images. In the other word, the BOW model is utilized to normalize the image features by preserving the original feature points of images.

4   Logo Detection and Recognition

4.1   Random Forest Based Classifier Training

From the discussion in section 3, we know that logos' hybrid features need to be extracted for logo detection, and that there are also noise data in logo images. To perform the classification better on the noise dataset with high dimensions, we select Random Forest classification approach for logo detection. Random Forest is a machine learning algorithm combining the integration learning theory and random subspace method, which gets multi decision trees in training processing for multi-class classification [17].

Algorithm 1 gives the approach of the logo classifier building. Supposing we construct n decision trees for the Random Forest classifier, for each one, we firstly need to randomly select N cases from the original dataset D to create the training dataset, and randomly select m features from M image features as the training features, Then we construct decision tree t on the selected dataset using the selected features. Here the decision tree is created by C4.5 algorithm [17].

4.2   Logo Detecting

The logos are usually superimposed on a nook and corner of boards or commodities as well as the two images shown in Figure 2. The detection may fail if based

---

Algorithm 1. RandomForest Classification

Input: D: logo image set with class label,
        N: image number sampled in D,
        M: feature number of every image,
        n: number of decision tree,
        m: selected number of features;
Output: RandomForest classifier with n trees;
Description:
1) $F = T = \Phi$; \\feature and tree sets initializing
2) $for\ i = 1$ to n
3)   { Sample N cases randomly from D;
4) $if\ m << M$
5)   { Selecting m features randomly from M features on each image into F;
6)   Building decision t into T by a best split;
7)   }
8) $endif$
9) }

---

on global features or features of main objects on these images. So we need to detect the logo areas on subspaces of an image. In general object detection application, the sliding window is always used to detect target areas [13]. However, the sizes of logo regions often vary with the objects that they are adhered to in images, thus a wide range of sizes of windows need to be defined and detected on only one image. In this paper, we get subregions for logo detecting using random partition.

    (a)   (b)

Fig. 2. Logos existed as a small part of images

We randomly partition each image into $M \times N$ non-overlapping rectangular patches and perform such partitioning Q times independently. Thus we will obtain a subimage pool with $M \times N \times Q$ patches. Each patch is denoted as $patch_{q(x,y,w,h)}$, where $q \in \{1, 2, \ldots, Q\}$ expresses the q-th partition, x and y are the coordinates of the starting point of the patch in the q-th partition, w and h mean the width and the height of the patch respectively, while x and y are randomly generated in each partition. For a given partition $q \in \{1, 2, \ldots, Q\}$, the $M \times N$ patches are non-overlapping, while the patch $patch_{q_1(x,y,w,h)}$ and $patch_{q_2(x,y,w,h)}(q_1, q_2 \in \{1, 2, \ldots, Q\}$ and $q_1 \neq q_2)$ from different partitions may be overlapped. We apply the classifier created in Section 4.1 to classify every partition patch, and the unoverlapped patches which scores from the classifier are larger than the defined threshold θ are selected into the candidate set for further logo recognition.

The Random Forest classifier is made up of several decision trees. When inputting a test sample p, all decision tree would make a decision and give a class estimate. Here the last score of every partition will be decided by the most frequency of classes outputted by these decision trees. The classification score of p will be computed by:

$$score_p = \frac{max(n_1, n_2, \ldots, n_c)}{N_t} \tag{2}$$

where N_t is the total number of decision trees contained in the Random Forest classifier and $n_1, n_2, \ldots, n_c$ respectively expresses the decision number of class $1, 2, \ldots, c$ given by N_t decision trees for p.

Algorithm 2. Logo detection

Input: an image I, partition times Q;
Output: detected logo patch set P;
Description:
1) $P = \Phi$;
2) $for\ i = 1$ to Q
3) $\{$ Randomly partition I into $M \times N$ patches;
4) $\quad for\ j = 1$ to $M \times N$
5) $\quad \{F_{logo} = (<\boldsymbol{F}_{texture}, \boldsymbol{F}_{canny}, \boldsymbol{F}_{shape}, \boldsymbol{F}_{HOG}, \boldsymbol{F}_{color}, \boldsymbol{F}_{SIFT}, \boldsymbol{F}_{SURF} >, logo)$
6) $\quad\quad score_j = RandomForest(patch_j)$;
7) $\quad\quad if\ score_j > \theta$
8) $\quad\quad\quad for$ each p from P
9) $\quad\quad\quad\quad if\ patch_j \cap p = \Phi$;
10) $\quad\quad\quad\quad\quad P \leftarrow patch_j$;
11) $\quad\quad\quad\quad elseif\ patch_j \cap p \neq \Phi$ and $score_p < score_j$
12) $\quad\quad\quad\quad\quad$ Delete p from P;
13) $\quad\quad\quad endif$
14) $\quad\quad endif$
15) $\quad \}$
16) $\}$

The processing of logo detection based on classification is described in Algorithm 2. Firstly we partition a given image to patches with different sizes and extract the feature vectors on every patch. Then every partition is classified by Random Forest classifier and result in a classification score. When a received score is larger than the threshold θ, we consider there is a high probability that a logo will be presented in the corresponding partition. As an image might include more than one logo, we select the patches with the highest score respective in non-overlapping areas as the candidate logo regions in the image.

4.3   Logo Recognizing Based on Spatial Structure

After logo detecting, each subregion in the candidate set have a class tag from results of classification and such tags may be looked as the results of logo recognition. However, the extracted feature vectors in logo detection are used for global object areas without spatial depended relation. For example, the characters in word "STARBUCKS " are reordered to a different sequence as shown in figure 3(b) and result in a totally disparate lexeme, while the two images in figure 3 have the same visual words in the feature space.

Since product frauds always mislead customers by a little difference in their logos from luxury or well-recognized products, the results of classification in section 4.2 usually fail to detect a nice distinction of two similar images with same visual words like in figure 3. Therefore, we need to add spatial information into visual features for logo recognition according to the intrinsic characteristics of logos.

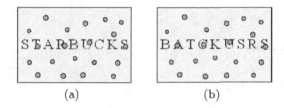

(a)   (b)

Fig. 3. Different images with the same visual words

Inspired by the work of Zhang and Cheng [18], in this paper the object regions are recognized using visual words with spatial correlated information to a specific logo. The geometry-preserving in an image area is expressed by the co-occurrence of visual phrases in relative displacements. Firstly the object region and the logo image are partitioned into the small patches with the same size. Then the visual words are recovered in each patch following the result of clustering on BOW in section 3. The co-occurrence matrixes of visual words are constructed at the relative offsets respectively on the object region, the logo image and between the object region and the logo image. Lastly Formula 3 is used to compute the similarity between the object region and the logo image for logo recognition. The processing steps are briefly described as follows:

1. We construct the visual word tables on the feature space for each patch following BOW after partitioning the logo image and the object region with the fixed size.
2. Given the max offset s, we construct the co-occurrence matrixes of visual words at the relative offsets respectively on the object region, the logo image and between the object region and the logo image, and get a statistic table of visual words at each offset. If the number of visual words at some offset is larger than the certain length len, the visual words included at this offset are composed into a visual phrase $phrase$.

3. For each co-occurrence matrix, we count up the frequency of visual phrases co-occur on the same relative offsets on its responding images.
4. The similarity of the object region and the specific logo are computed for logo recognition.

Supposing r represents a subregion in the candidate set and l is a logo in the logo dataset, we give the definition of the similarity computation as follows:

$$VP\_Sim(r, l) = \frac{co\_vp_{rl}}{\sqrt{co\_vp_r \cdot co\_vp_l}} \tag{3}$$

where co_vp_{rl} denotes the number that the same visual words co-occur in the same relative offset of two images; co_vp_r and co_vp_l respectively represent the numbers that visual words occur in image r and image l. Given a threshold α, if $VP\_Sim(r, l) > \alpha$, the object region r is regarded as the target logo l, otherwise it is misclassified.

5 Experiments

5.1 Experiment Setup

- **Training the classifier.** For the classifier training, we apply Waikato Environment for Knowledge Analysis (Weka) [19] integration environment. The features of images are declared following the Attribute-Relation File Format (ARFF) that Weka supports, where we define all the feature types as the relational classes and multi-type features as the nested structures in Weka.
- **Training Dataset.** We used a recent proposed logo dataset DBMM-contest [20] as the training dataset for training the classifier. The logo dataset is composed of 112 images divided into 10 different classes, while the logo images belonging to the same brand vary with lighting, scaling, and rotation. The distribution of the dataset is shown in Table 1.
- **Testing Dataset.** Since there is no benchmark dataset for the logo detection on the real images, we manually annotated 100 images with a wide range of sizes as the test data, while the test images covered all the logo classes included in the training set, as well as some images don't contain any logos and some contain one or more logos.
- **Evaluation Measure.** We use $F\text{-}Score$ to measure the performance of algorithm. In Formula 4, S presents the number of correct recognized images in one category. C is the number of the recognized images in one category. We denote the number of manual tagged images in one category as R.

$$Precision = \frac{S}{C}, \ Recall = \frac{S}{R}, \ F - Score = \frac{2 \times Precision \times Recall}{Precision + Recall} \tag{4}$$

5.2 Experiment Results

Feature Selection. Firstly we analyze the average performance of all-type logo detection using different combinations of image features. For the features depended on the sizes of images or the numbers of interest points, such as HOG, SURF and SIFT, we normalize their number respectively as k dimensions as described as in section 3. For the other features, we respectively extract 128 HSV color statistic features, 8 texture features based on GLCM [21], 8 canny operators [22] and 7 shape invariant moments [23].

The experiment result is shown in figure 4 and the parameter k is used to limit the dimensions of the features. In the figure, we can see that it has the worst performance when SIFT and SURF features are not used at all for any dimension features. Moreover, SURF is more distinguished feature than the others for the classifier training. Obviously, the more satisfied results can be achieved using SURF+HOG+others features or SIFT+SURF+HOG+others features when about $k = 100$ and the similar performances are got using these two combined features. Considering the complexity of SIFT extracting, we will use 100 dimension SURF+HOG+others features of images for logo detection in the following experiments.

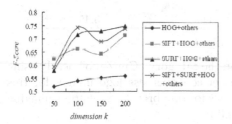 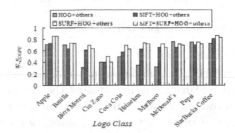

Fig. 4. F-Score of different features and dimensions

Fig. 5. F-Score of logo detection for different categories

Logo Detection. Random partition is used to get subregions for logo detecting and the partition time is aromatically estimated for every detected image by defining the stop condition. Here similarity parameter α is set to 0.7 and the maximum size of candidate is set to be 3. In other words, we think the logo number in one image is not more than 3. Table 2 presents the scores of the top ranked subregions for patches on an image. We can see that the patches No.1, No.5 and No.7 achieve the better classification scores. The patch No.7 will be detected as a candidate region with the largest classification score in non-overlapping areas and the class label "Pepsi "will be returned for it.

Figure 5 shows the average performance of logo detection using the different feature combinations respectively for different categories when $k = 100$. Generally, SURF is the important feature for nearly all logos, while the separate features may be remarkable for only few logos such as HOG for "McDonald's "

Table 1. DBMM Logo Dataset

Class	Logo	Number
Apple		10
Banilla		8
Birra Moretti		6
Cin Zano		3
Coca Cola		10
Heineken		13
Marlboro		14
McDonald's		16
Pepsi		12
StarBucks Coffee		20

Table 2. Patch scores

ID	Patch	Classification score
1		0.7
2		0.38
3		0.3
4		0.2
5		0.65
6		0.4
7		0.85
8		0.23
9		0.33
10		0.27
11		0.45

and "Pepsi ". The poorer results are got on the category "Cin Zano" nearly using all feature combinations, as the training set contains less logo images for this brand. So we need more logo training images for one brand on different imaging conditions to get better classification results.

Logo Recognition. In Table 3, we compare the average performance of logo recognition using spatial correlated information and the classification performance before further recognition. Here the similarity parameter α is set to be 0.7 too. We can see from Table 3 that the further recognition has a better F-$Score$ compared with the classification result in nearly all categories except "Cin Zano ". It is due to the poorer classification result from the insufficiency of training cases.

At last, we compare our results with the N-TEC in the literature [13]. As the N-TEC only construct the classifier for one logo each time, We select the brand "Benilla", "StarBucks"and "Cin Zano" respectively to the training classifier and compare it with our method. The comparing result is shown in Table 4. Generally speaking, our logo method can achieve a better or comparable performance compared with N-TEC. For the dataset with less logo images, such as the brand "Cin Zano" including 3 logo images, the performance of our method is better than N-TEC due to the multi decision trees learning from different feature combinations. The comparable performances are obtained for the brands with more training logo images, but N-TEC need to train more classifiers for different logos.

Table 3. Performance of logo recognition

Class	Detection F-Score	Recognition F-Score
Apple	85.79%	86.31%
Banilla	71.38%	73.73%
Birra Moretti	69.64%	69.97%
Cin Zano	49.74%	49.74%
Coca Cola	69.43%	70.21%
Heineken	70.59%	73.33%
Marlboro	66.87%	67.68%
McDonald's	76.54%	78.11%
Pepsi	71.48%	73.25%
StarBucks Coffee	78.27%	80.97%

Table 4. Performance comparison

Class	Random Forest F-Score	N-TEC F-Score
Banilla	73.73%	64.63%
StarBucks	80.97%	80.21%
Cin Zano	49.74%	38.27%

6 Conclusion and Future Work

In this paper, we explore to use the Random Forest classifier training method to detect logos from real images and the accuracy of the log recognition is further improved by using image spatial visual information. The key idea is that more feature extraction methods are fused to enhance the distinguishable visual phrases and the multi-class classification is designed to avoid an extreme stage training situation. Considering the variety of logo sizes adhered to images, the random partitioning is used to detect target regions instead of the sliding window for saving the cost of computation. The experiment results validate that the performance of logo recognition outperforms the well-known adaptive boosting based Viola-Johes classifier on the real noise images with multi-category logos.

For the logo recognition adapting to different imaging conditions, we will consider a larger training data set to prove the robust of the method in our further work, and further analysis the features working well on logos comparing to other general images.

Acknowledgments. This work is supported by the State Key Development Program for Basic Research of China (Grant No. 2011CB302200-G), State Key Program of National Natural Science of China (Grant No. 61370074, 61100026).

References

1. Otim, S., Grover, V.: E-commerce: a brand name's curse. Electronic Markets 20(2), 147–160 (2010)
2. Zhang, C., Zhang, Z.: Boosting-based face detection and adaptation. Synthesis Lectures on Computer Vision 2(1), 1–140 (2010)
3. Castrillón, M., Déniz, O., Hernández, D., Lorenzo, J.: A comparison of face and facial feature detectors based on the viola–jones general object detection framework. Machine Vision and Applications 22(3), 481–494 (2011)

4. Mita, T., Kaneko, T., Hori, O.: Joint haar-like features for face detection. In: Tenth IEEE International Conference on Computer Vision, ICCV 2005, vol. 2, pp. 1619–1626. IEEE (2005)
5. Zhang, C., Platt, J.C., Viola, P.A.: Multiple instance boosting for object detection. In: Advances in Neural Information Processing Systems, pp. 1417–1424 (2005)
6. Joly, A., Buisson, O.: Logo retrieval with a contrario visual query expansion. In: Proceedings of the 17th ACM International Conference on Multimedia, pp. 581–584. ACM (2009)
7. Doermann, D., Rivlin, E., Weiss, I.: Applying algebraic and differential invariants for logo recognition. Machine Vision and Applications 9(2), 73–86 (1996)
8. Li, Z., Schulte-Austum, M., Neschen, M.: Fast logo detection and recognition in document images. In: 2010 20th International Conference on IEEE Pattern Recognition (ICPR), pp. 2716–2719 (2010)
9. Hassanzadeh, S., Pourghassem, H.: A fast logo recognition algorithm in noisy document images. In: 2011 International Conference on Intelligent Computation and Bio-Medical Instrumentation (ICBMI), pp. 64–67. IEEE (2011)
10. Wang, H.: Document logo detection and recognition using bayesian model. In: Pattern Recognition (ICPR), 2010 20th International Conference on, pp. 1961–1964. IEEE (2010)
11. Viola, P., Jones, M.J.: Robust real-time face detection. International Journal of Computer Vision 57(2), 137–154 (2004)
12. Tuzel, O., Porikli, F., Meer, P.: Pedestrian detection via classification on riemannian manifolds. IEEE Transactions on Pattern Analysis and Machine Intelligence 30(10), 1713–1727 (2008)
13. Chen, Y., Thing, V.L.L.: A noise-tolerant enhanced classification method for logo detection and brand classification. In: Kim, T.-h., Adeli, H., Fang, W.-c., Villalba, J.G., Arnett, K.P., Khan, M.K. (eds.) SecTech 2011. CCIS, vol. 259, pp. 31–42. Springer, Heidelberg (2011)
14. Dalal, N., Triggs, B.: Histograms of oriented gradients for human detection. In: IEEE Computer Society Conference on Computer Vision and Pattern Recognition, CVPR 2005, pp. 886–893. IEEE (2005)
15. Lowe, D.G.: Object recognition from local scale-invariant features. In: The Proceedings of the Seventh IEEE International Conference on Computer vision, vol. 2, pp. 1150–1157. IEEE (1999)
16. Bay, H., Ess, A., Tuytelaars, T., Van Gool, L.: Speeded-up robust features (surf). Computer Vision and Image Understanding 110(3), 346–359 (2008)
17. Prinzie, A., Van den Poel, D.: Random forests for multiclass classification: Random multinomial logit. Expert Systems with Applications 34(3), 1721–1732 (2008)
18. Zhang, Y., Chen, T.: Efficient kernels for identifying unbounded-order spatial features. In: IEEE Conference on Computer Vision and Pattern Recognition, CVPR 2009, pp. 1762–1769. IEEE (2009)
19. Data mining software in java
20. Dbmm12-contest logo image database
21. Haralick, R.M., Shanmugam, K., Dinstein, I.H.: Textural features for image classification. IEEE Transactions onSystems, Man and Cybernetics (6), 610–621 (1973)
22. McIlhagga, W.: The canny edge detector revisited. International Journal of Computer Vision 91(3), 251–261 (2011)
23. Hu, M.K.: Visual pattern recognition by moment invariants. IRE Transactions on Information Theory 8(2), 179–187 (1962)

Business Intelligence by Connecting Real-Time Indoor Location to Sales Records

Jessie Danqing Cai

SAP Asia, 30 Pasir Panjang Road, Singapore 117440
Jessie.cai@sap.com

Abstract. Indoor positioning systems have many technological varieties and application scenarios. Through development of an indoor LBS service for targeted retail use case we have adopted Wi-Fi signal strength fingerprinting considering cost, resolution and scaling-up factors. A mobile app is developed and rolled out in a small scale to connect customer location data with their loyalty status and the retailer's product sales. Through this research attempt we have discovered some less communicated aspects of ILBS including the lack of accuracy, the response time lag, and the privacy concern which is most likely beyond the help of technology advancement. On the up side, valuable business insight can be generated by mashing up of transactional sales data with non-transactional location data.

Keywords: Indoor location, signal fingerprinting, context detection, application scenario, business applications, data privacy, mobile marketing.

1 Introduction

In the last decade, many technologies to detect indoor location have been proposed and adopted including GPS, RFID, UWB, acoustic based, cellular based, WLAN based and many more. A good summary of existing techniques considering both resolution and scale of application can be found in [1]. Retail among all has been the most mature industry where Indoor Location Based Services (ILBS) have flourished. Players in this field range from start-ups to MNCs. The ability to weave consumer location context into browsing, social sharing and buying activities is the common mandate of ILBS, which presumably will have more potential to drive online clicks into increased offline sales since people around the world spend most of their time indoor. It has become an inseparable part of behavior based personalization in mobile marketing & commerce.

To deliver a proof of concept with a local retail partner, we have selected Wi-Fi signal strength triangulation considering constraints such as cost and coverage. Through the POC development we found the accuracy of collected indoor location coordinates to be limited and unstable. According to available references such as [2], changes in settings and customer crowds attenuated the signal sensing. We have therefore applied Bayesian smoothing which was able to reduce data jumping but not quite significantly.

F. Li et al. (Eds.): WAIM 2014, LNCS 8485, pp. 817–823, 2014.

A mobile app was developed for the POC and rolled out in a small scale to connect customer location data with their loyalty status (member or non-member) and the retailer's product sales. Through this research attempt we have discovered certain less communicated aspects of LBS namely the lack of accuracy, the response time lag, and most importantly the privacy concern of users. We have found people to be far less willing to share their locations than expected or assumed in previous reports. Nevertheless, meaningful business intelligence was generated out of this proof of concept during its test implementation in the week of Great Singapore Sales in year 2013.

2   System Setup

In general, GPS, signal strength triangulation via Cell Tower locating, Wi-Fi network, BLE (Bluetooth low energy), scanning of RFID (radio frequency identification), UWB (ultra wideband), mapping of Magnetic Field, and processing of video content are some of the technologies adopted in indoor LBS applications. For those interested in details of these technologies such as locating accuracy and setup effort, [1], [2] and [4] can be referred to. We have selected Wi-Fi signal triangulation considering cost of implementation and the ease of scaling up.

Six Wi-Fi access points and one master router for signal strength triangulation were implemented in our retail business partner's physical store. Relative coordinates were generated and passed to real time location server (RTLS) in the cloud through 3G network, and consumed by our internal server as a data service. Bayesian smoothing was applied to improve the locating accuracy as recommended by literatures such as [5], [6] and [7]. The data captured from the 6 designated Wi-Fi routers (access points) was consisted of following dimensions: (1) time stamp, (2) device mac address, (3) floor level, (4) X coordinates, and (5) Y coordinates. The routers scanned the 2.4 GHz Wi-Fi spectrum and formed a wireless mesh network without transmitting. They communicated with one another through the 5GHz Wi-Fi spectrum. Data generated were passed to cloud server via 3G mobile broadband network every a few seconds. The access points were installed in an area around 900 square meters as shown in following Fig. 1. Note that the placements of routers as blue dots are for illustration only and not exact to the scales.

Before setting up the Wi-Fi infrastructure to do signal fingerprinting, we tried also built-in sensors in smartphone including GPS, compass and accelerometer for location estimation. The results were very far from satisfying because too many assumptions including starting point and walking pace need to be taken care of for the reading from these sensors to make any sense. BLE (Bluetooth low energy) was tested as well during scoping phase. Reports said Apple has been working on the retrieval of micro locations through iBeacon which in essence is BLE locating [10]. It was dropped out of project scope because around then we had found very few people actually enable Bluetooth option of their smartphones for a prolonged period of time.

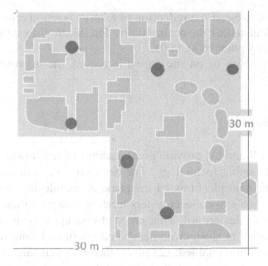

Fig. 1. Area of implementation

3   Bayesian Smoothing

Wi-Fi signal strength fingerprinting assumes linear decay of signal strength with LOG distance hence simple triangulation of signal strength from three access points could identify a relative position in a two-dimensional space. Actual internal setting such as furniture, display shelves, and crowd movement however has been proved to attenuate the signal strength measurement substantially, as reported in many literatures including [7] and [8]. Bayesian inference is able to help by using a predefined motion model to transform on site visitor's trajectory into a polygon to associate it with the link model derived from the indoor layout.

Fig. 2 below shows the testing data collected out of a 4-hour time window out of the retail venue during which one test user carrying the Wi-Fi enabled mobile device had stayed in the yellow circled area. Data series 1 was the original reading collected, and data series 2 was redrawn after applying Bayesian smoothing as described in [7].

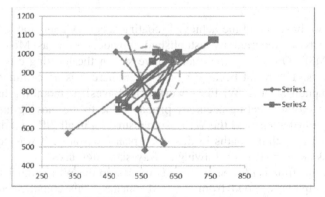

Fig. 2. Effect of Bayesian smoothing

For this dataset alone we can see reduction of relative position error over 25%, where relative error is calculated as the accumulative Euclidean distance error compared to the assumed stationary point (600, 900).

However when the size of on-site crowd increased, the improvement in accuracy became less satisfying.

4   Retail Scenario

Literatures such as [3] and [4] provide good readings of real world application scenarios for indoor positioning technology. For our study we collaborated with a local retail venue hosting multiple branded counters. A mobile app was developed and rolled out in a small scale to connect testers location data with their loyalty status and the retailer's product sales record. This proof of concept system was able to do the following: (1) visualize in store customers' location in real time if they carry smart devices with Wi-Fi option enabled; (2) push marketing and sales notifications to selected group of users who have installed our mobile app and detected to be around the retail store; (3) allow such users of mobile app to navigate in the physical store; (4) connect to user's Facebook account for product and interest suggestions; (5) connect with the retailer's POS system to track the in store conversion rate. This POC basically covers most major use cases of existing indoor LBS applications in retail scenarios, namely: venue navigation, crowd behavior monitoring, vicinity marketing, and in-situ deal recommendation.

One thing to add is the smartphone penetration rate in Singapore is fairly high – close to 150% as reported in mid of 2012 [9]. It is one of the most mobile savvy populations in Asia. People in general are open to try out new apps and share with their friends. After the development was done, we conducted several user tests. The users walked around the open plan retail venue around 900 square meters of size, while actively using their mobile phone with their Wi-Fi option enabled. The findings are summarized in next section.

5   Findings

First of all, we have found the achieved locating accuracy comparable to commercial grade products as reported by established providers such as Motorola Solutions and Cisco MSE. Our own experiments have shown the locating error to be around 5 to 10 meters. The result became worse when there was no continuous and active data consumption through Wi-Fi network. The Bayesian smoothing had not significantly improved the locating accuracy probably due to the moving nature of on-site crowd. The crowd attenuated the Wi-Fi signal strength in all different directions. Also there were many walking paths in this open plan retail area. It was difficult to have effective link node models for drawing of Bayesian inferences.

Secondly, the time lag for access points to pick up the Wi-Fi signal search from individual smartphones varied from 1 to 10 minutes. This complicated the accuracy issue further more and led to larger location bias. The users using the navigation app could not see their own location pins moving along most of the time. When the pin did move, it jumped from place to place. The good news was the time lag and location bias did not impact much on the vicinity marketing scenario. 100% of users staying for longer than 10 minutes received the marketing push notification. For users merely walking by, the chance decreased to around 50%.

Considering both time lag and location bias, the location based crow behavior analytics were found with low credibility in real time. However the accumulated statistics over a certain amount of time such as an hour or a day become more reliable and insightful. Following shows the accumulated number of footfalls from 1st of June to 15th of July 2013. Different line types/colors are used for different branded counters. It is clear that "brand A" represented by Solid Blue Line was quite successful in attracting onsite consumers in the first 2 weeks of June which coincided with the week of Great Singapore Sale, especially when compared to "brand B" represented by Dashed Red Line. The Latter was however able to attract more number of visits for the rest of time window.

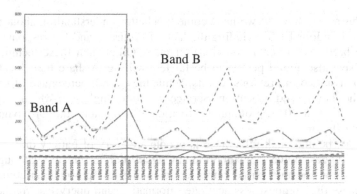

Fig. 3. Accumulated daily footfall data (non-transactional)

When we combined the above footfall data with the number of sales from the 2 mentioned brands to compute the Conversion Rate, which is Number of Deals divided by Number of Onsite Visits, it turned out Conversion Rate of "brand B" had been consistently higher than that of "brand A". It said the marketing effectiveness of "brand A" can still be improved and persuade more on site visitors to actual buying customers.

In addition we have tried to mash up the amount of dwell time with the amount of money spent for selected brands, as illustrated in Figure 4. The correlation coefficient was found slightly larger than 0.5.

Last but not the least we have done a survey with the retail partner to conclude the research. The feedbacks can be summarized as: (1) Mobile ILBS for end consumers are attractive but there is lack of evidence in the tracked conversion rate. (2) The micro location based crowd behavior report is considered less useful than the tracking of conversion rate. Visibility of conversion rate is considered most useful for business users. (3) The business user prefers monthly subscription of above reports instead of owning and viewing of the real time location data.

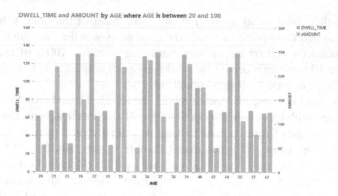

Fig. 4. Mash-up of Dwell Time (non-transactional) with Amount Spent (transactional)

6   Conclusion

Through this research effort we have come to a better understanding about the different aspects of indoor LBS including the lack of accuracy and the response time lag. Both have been elaborated in previous chapters. Other than these technical limitations, we have also found people to be far less willing to share their locations than expected or assumed in previous reports. While the overall percentage of smartphone owners who have used Location Service is quite high – 74% as reported in FactBrowser in year 2013, prolonged activation of it for non-outdoor navigation purposes is actually rare.

One reason behind such reluctance is people are wary of battery drain. GPS chip requires the biggest power consumption among all mobile sensors. Although experts have claimed the location based reminders consume less power than commonly assumed because they request less accurate information and query less frequently than navigation, most smartphone owners prefer to turn off the Location Service option to save power. The same concern is held towards the Bluetooth option and Wi-Fi option. Relatively speaking, more people tend to leave the Wi-Fi option on which is an advantage of Wi-Fi signal strength triangulation.

Another contributing factor is the privacy concern and it is becoming increasingly prominent. Criticisms towards the possibility of revealing uniquely identifiable data in social sharing activities have never stopped. Their influence gradually shows both in country level regulations and individual level conversations. For the mobile phone users we have come across in this research project – from C-level users in their 50s to interns in their early 20s - all responded that they only checked in locations via mobile phones when necessary. Once they learnt Wi-Fi access points being able to pick up the Mac Addresses of their smart devices, the uniform reaction was that they would make sure the Wi-Fi option of their smartphones is turned off when not in use.

Tero Kuittinen [11] has cautioned that privacy concern could kill the mobile LBS market in a slow way. From what we have learnt about the existing technologies to enable LBS and mobile phone users' overall attitude towards location sharing, we tend to agree the limitations of LBS applications are still too many for them to make great market impact or drive noticeable behavior change.

Nevertheless, meaningful business intelligence can be generated by combining real time location footfalls with sales record, such as the inference of Conversion Rate. It brought much wanted clarity of business operation effectives. And this would not be possible without connecting the real time non-transactional data to transactional sales database.

References

1. M. Vossiek, et al (2003), "Wireless local positioning—Concepts, solutions, applications," in Proc. IEEE Wireless Communications Network Conference, Aug. 2003, pp. 219–224.
2. Liu, H., et al (2007), Survey of Wireless Indoor Positioning Techniques and Systems, IEEE Transactions on systems, man and cybernetics – part C: Applications and Reviews, VOL. 37, NO. 6, NOVEMBER 2007
3. Christian Prehofer (2009), Real-world experiences with indoor location based services, Proceeding ICPS '09 Proceedings of the 2009 international conference on Pervasive services, Pages 143-144
4. Zeimpekis, V, et al (2003), Taxonomy of Indoor and Outdoor Positioning Techniques for Mobile Location Services, 2003 ACM 1073-0516/01/0300-0034
5. Liu, Jingbin, et al (2012), A Hybrid Smartphone Indoor Positioning Solution for Mobile LBS, Sensors 2012, 12, 17208-17233
6. Derr, K., Manic, M. (2008), Wireless Indoor Location Estimation Based on Neural Network RSS Signature Recognition (LENSR), 3rd IEEE Conference on Industrial Electronics and Applications, JUNE 2008
7. Madigan, D, et al (2005), Bayesian Indoor Positioning Systems, INFOCOM 2005, 24th Annual Joint Conference of the IEEE Computer and Communications Societies. Proceedings IEEE, Page(s): 1217 - 1227 vol. 2
8. Jones, K, et al (2007), Improving Wireless Positioning with Look-ahead Map-Matching, Mobile and Ubiquitous Systems: Networking & Services, 2007. MobiQuitous 2007. Fourth Annual International Conference on Digital Object Identifier, pp. 1–8.
9. Elizabeth Tan (2012), Singapore hits mobile penetration rates of 148.9%, E27, May 16 2012
10. Richard Padilla (2013), Apple Retail Stores to Integrate iBeacon Systems to Assist with Sales and Services.MacRumores, November 16 2013
11. Tero Kuittinen (2013), Are privacy concerns killing the mobile location-based services market? BGR, September 16 2013.

Search 360 - Question Answering in Business Intelligence

Falk Brauer[1], Bob Massarczyk[1], and Nicolas Kuchmann-Beauger[2,*]

[1] SAP Research & Innovation
firstname.lastname@sap.com
[2] Paris Descartes University
nicolas.beauger@etu.parisdescartes.fr

Abstract. Question Answering (Q&A) from structured data is a technique that may revolutionize enterprise search, especially in the area of Business Intelligence (BI). In order to make BI more accessible to end-users, some efforts have been made in the search for existing reports. However, the problem of converting an end-user's natural language input to a valid structured query in an ad-hoc fashion hasn't been sufficiently solved yet. In this paper we demonstrate a Q&A system for Business Intelligence. It translates of a former research prototype into practice. The main innovation is that it operates on arbitrary multi-dimensional data, recognizes semantics of a user's questions and translates it to a structured query model and generates ad-hoc reports from natural language.

1   Introduction

In the last decades data warehouses became an important information source for decision making and controlling. A lot of progress has been made to support casual end-users by allowing interactive navigation inside complex reports or dashboards (e.g. by interactive filtering or calling OLAP-operations such as drill-down in a user-friendly way). In addition there has been a lot of effort in making reports or dashboards searchable. However, most casual users still have to rely on pre-canned reports that are provided by the IT-department of a company because todays' Business Intelligence (BI) self-service tools still require a lot of technical insights such as an understanding of the data warehouse schema. This is especially cumbersome because data warehouses grew dramatically in size and complexity. A popular use-case for BI is for instance the segmentation of customers to plan marketing campaigns (e.g. to derive the most valuable, middle-aged customers in a certain region). It is not unusual that business users who plan a campaign have to cope with hundreds of key performance indicators (KPIs) and attributes, which they have to combine in an ad-hoc fashion to cluster their customer base. A keyword or even natural language-based interface to formulate their information need would ease this task a lot.

The system presented in this paper supports the whole process of defining and executing a domain or application-specific Question Answering system. It is an industry ready solution that translates research prototypes similar to [1,2,3] into practice and is therefore interesting for both researchers and practitioners. The main difference between the research prototypes described in [1,2,3] and the system presented here is

---

* Work done while being affiliated with SAP Research.

F. Li et al. (Eds.): WAIM 2014, LNCS 8485, pp. 824–827, 2014.

that the research prototypes relied on RDF or other graph-based engines, while the demonstrated system is natively implemented in SQLScript inside a columnar database and thus provides the scalability that is required to cope with real-world data warehouses with a size of several hundred gigabytes and can generate relevant ad-hoc reports in milliseconds.

2   Demo Description

Search 360 operates on data warehouses that have a well-defined schema (here a columnar database with extension for multi-dimensional reporting). Schema elements of a multi-dimensional data warehouse can be distinguished in dimensions, i.e. analytical axis that can be used for data exploration (filtering, level of aggregation, navigation in hierarchies), and measures, i.e. key performance indicators that shall be aggregated along a given set of dimensions. A typical business intelligence query from the tourism industry is shown in Fig. 1 in form of SQL, where *customer* is the dimension, *revenue* is the measure and other dimensions are used as filters, i.e. customers in *Palo Alto* with an age below *20*. Note that the formula to compute the revenue and the join paths are usually maintained in the data warehouse and are hidden from the multidimensional reporting layer above.

SELECT cust.NAME AS customer, sum(inv.DAYS * serv.PRICE) AS revenue

FROM Singapore cust

INNER JOIN Sales sales ON (sales.CUST_ID = cust.CUST_ID)

INNER JOIN Invoice inv ON (inv.INV_ID = sales.INV_ID)

INNER JOIN Service serv ON (inv.SERVICE_ID = serv.SERVICE_ID)

WHERE cust.CITY = 'Palo Alto' AND cust.AGE <= 20

GROUP BY customer ORDER BY revenue LIMIT 5

Fig. 1. Example query for "top 5 customers from Singapore with age below twenty"

The system has currently two user interfaces, one running on a mobile device (left in Fig. 2), enabling the user also to enter his questions via Speech-to-Text technology, and a desktop version (right in Fig. 2). The mobile version allows the user in addition to navigate the result set via speech.

The underlying system work as follows: In a first step the query is parsed. Entities which are part of the schema are recognized leveraging information extraction technologies (tokenization, stemming, lemmatization, etc.). In addition phrases for recognizing natural language patterns, e.g., to identify numerical ranges such as "age below twenty", are identified and variables of this patterns are normalized (e.g., convert "twenty" to the number "20" which can be used in a technical query). In a second step the algorithm leverages structural patterns to generate potential technical queries. Parts of the structural patterns are relations among the schema elements (e.g., that recognized dimensions and measures belong to one fact table, i.e. can be used together in a technical query), constraints that apply to phrases (e.g., that numerical ranges can only be applied to numerical values), etc. In a next step we derive a set of technical queries, which are than ranked by their complexity (the more information captured from the question the higher the score) and the confidence (e.g., if only parts of a schema element's name were matched or spelling mistakes were recognized the score will be lowered). As a result we derive a ranked list of technical queries, where the top-k queries are executed (k depends on the actual client, i.e. mobile or desktop) and are visualized then to be presented to the user.

Fig. 2. User Interfaces of Search 360 on mobile and desktop

During the demonstration we will show different types of queries that are supported and explain the mechanism how the system can be configured for different domains and which parts of the configuration can be automated. In addition we will show the former research prototype and explain the differences and limitations with respect to the here described system.

3   Related Work

In the BI domain, three systems are more closely related than the previous approaches. First, QUASL[1] a RDF/SparQL-based system which offers a lot of flexibility. However, it cannot scale to real-world data warehouses. Second, SODA [2] is a keyword-based search system over data warehouses. It uses some kinds of patterns to map keywords and some operators in the user's query to rules to generate SQL fragments. However, this system does not focus on "using natural language processing to interpret the input" [2]. Our proposal is thus much more powerful, e.g., to provide means for including user context or more complicated natural language patterns and relate them with other background knowledge, which is of utermost importance as stated in [3]. Last, SAFE [4] is an answering system dedicated to mobile devices in the medical domain. It uses patterns, i.e. pre-defined SparQL queries with placeholders for variables. Each pattern has assigned a predefined natural language representation (i.e. a question that the user can understand) and the challenge is to rank these questions according to a keyword input posed by the user. Our approach goes beyond this idea by the ability to describe complex relations (constraints) among the recognized entities in a declarative way and map them into a technical query.

References

1. Kuchmann-Beauger, N., Brauer, F., Aufaure, M.-A.: QUASL: A framework for question answering and its Application to business intelligence. In: Proc. RCIS 2011 (2011)
2. Blunschi, L., Jossen, C., Kossmann, D., Mori, M., Stockinger, K.: Soda: Generating sql for business users. In: Proc. VLDB 2012 (2012)
3. Hearst, M.A.: "Natural" search user interfaces. Commun. ACM 54(11), 60–67 (2011)
4. Orsi, G., Tanca, L., Zimeo, E.: Keyword-based, context-aware selection of natural language query patterns. In: Proc. EDBT/ICDT 2011, pp. 189–200 (2011)

HybridPG: A SSD-Friendly High Performance Hybrid Storage Design for PostgreSQL

Chunling Wang, Jiangtao Wang, Zhiliang Guo, and Xiaofeng Meng

Renmin University of China, Beijing, China
{wangchunling,jiangtaow,guozhiliang,xfmeng}@ruc.edu.cn

Abstract. Flash-based solid state drives (SSDs) provide higher IOPS than disk drives. As a permanent data storage device, SSD exhibits more affordable prices than RAM. So we decide to use SSD as an extended cache for PostgreSQL to improve the performance of OLTP. In this demonstration, we mainly modify the shared buffer part of PostgreSQL and present a hybrid PostgreSQL based on SSD and HDD, called HybridPG. This demonstration can compare the performance between original PostgreSQL and HybridPG in TPC-C benchmark with different system parameters.

1   Introduction

Solid state drives (SSDs) outperform disk drives in throughput and energy consumption for database workloads [1] and people tend to use SSD to store data in order to improve the performance of database. But the price per unit capacity of SSD is still much higher than disk drives. So it is more cost-effective to use SSD as a cache layer between main memory and hard disk for database than totally to replace disk drives by SSD.

For most contemporary SSDs, random writes are obviously slower than sequential writes. Managing data in SSD like DRAM buffer may not get the expected performance and can decline the life of SSD. Therefore, we should turn small random writes to large sequential writes to utilize high sequential bandwidth and internal parallelism of SSD for higher throughput [2]. Now we propose an efficient strategy for using SSD as an extended cache for database and implement a Hybrid PostgreSQL based on SSD and HDD, called HybridPG.

In HybridPG, data pages evicted out from DRAM buffer will be selectively cached in SSD. Data in SSD is organized by data blocks and each data block contains fixed number of data pages. When the size of data pages in SSD goes beyond a tunable threshold, dirty pages will be flushed to disk by a novel replacement algorithm designed with write optimization. If we fail to get a page from the DRAM buffer, we try to look up for it in SSD; if it doesn't locate at SSD either, HDD will be searched. With SSD as an extended cache, HybridPG performs better than original PostgreSQL on TPC-C benchmark.

F. Li et al. (Eds.): WAIM 2014, LNCS 8485, pp. 828–831, 2014.
© Springer International Publishing Switzerland 2014

2   System Architecture

Figure 1 depicts the system architecture of HybridPG. The system consists of three components: 1) Meta data management 2) DRAM data management 3) SSD data management.

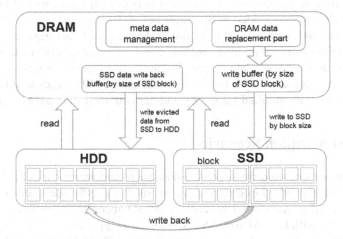

Fig. 1. The architecture of HybridPG

1) **Meta data management.** It includes following four data structures :
   - Page-level mapping table. It is an array to record the block id and offset in data block for data pages as they are organized as blocks in SSD.
   - Hash index. It is actually a hash table for indexing data pages in SSD, which takes page id as hashing key and the corresponding location in SSD as value.
   - Block-level mapping table. It is an array of structures which records the block id and metadata of pages belong to a bock in SSD. Page metadata consists of page id and state (valid, dirty& frequency).
   - Bit map. This data structure is an essential part for space recovery, and each bit of it represents whether the corresponding data block is used (1) or free (0).
2) **DRAM data management.** In HybridPG, we did not change the replacement strategy for shared buffer in PostgreSQL. A flash dirty flag is added for each page kept in the buffer to indicate whether it has been updated since it flushed to SSD. When evicting a page from DRAM, if its flash dirty flag is false and it exists in SSD, there is no need to flush it to SSD again. And we add a write buffer with the same size as data block in SSD to cache victim data pages and flush it to SSD once it is full. Besides, another in-memory write buffer is added for caching data blocks evicted from SSD to disk for reducing writes on HDD and its size is the same to a data block.
3) **SSD data management.** This part mainly manages the strategies of data to flush in and out the extended SSD cache.
      When the write buffer is full, we write it to a free data block in SSD. Though the positions of data block may be random, free space inside a data block is sequential.

By transforming several random data page writes to a sequential data block write, we reduce the write counts and can take advantage of the internal parallelism of SSD.

When the number of data pages in SSD goes beyond a tunable threshold, the replacement algorithm for SSD cache will be triggered to flush several data pages from SSD to disk. According to the previous study, we should consider the frequency of data more than recentness when evicting data pages [3]. Thus, we put forward a replacement algorithm considering the frequency of data blocks.

A time window is maintained to record the frequency of data pages cached in SSD in order to reduce the computing cost and avoid the negative effect incurred by frequency accumulation of data pages. Frequencies of these pages are added to represent the frequency of this block. When the size of SSD comes to a threshold, we evict the block with minimal frequency and flush out dirty data pages in this block.

3   Demonstration Scenario

In the demonstration, we provide some adjustable parameters for users to observe the performance of HybridPG in different conditions. We can decide whether only clean data pages or all pages should be cached in SSD according to different application scenarios, along with DRAM size, SSD size and SSD type.

Here is an example of the demonstration. The TPC-C benchmark database and workload were created by the BenchmarkSQL tool. Size of the data set is 500 warehouses (about 51GB), the client number of connections is 500. Memory of both original PostgreSQL and HybridPG is set to 1GB. We select 32GB of SSD to serve as the extension cache. And the total experiment time is 2 hours. Figure 2 shows the test results of the two systems measured by tpmC (transactions per minute). We can observe that the tpmC of HybridPG is 48 percent higher than it of Original PostgreSQL. It is due to the fact that disk write operations in HybridPG are much reduced, for most of frequently accessed data pages are cached in SSD.

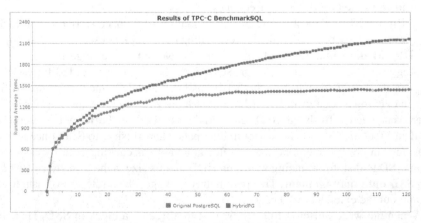

Fig. 2. Original PostgreSQL VS. HybridPG in TPC-C benchmark

Acknowledgements. This research was partially supported by the grants from the Natural Science Foundation of China (No. 61379050, 91224008); the National 863 High-tech Program (No. 2013AA013204); Specialized Research Fund for the Doctoral Program of Higher Education(No. 20130004130001), and the Fundamental Research Funds for the Central Universities, and the Research Funds of Renmin University(No. 11XNL010).

References

1. Lee, S.-W., Moon, B., Park, C.: Advances in Flash Memory SSD Technology for Enterprise Database Applications. In: SIGMOD, pp. 863–870 (June 2009)
2. Chen, F., Lee, R., Zhang, X.: Essential Roles of Exploiting Internal Parallelism of Flash Memory Based Solid State Drives in High-Speed Data Processing. In: HPCA, pp. 266–277 (2011)
3. Zhou, Y., Philbin, J.F., Li, K.: The Multi-Queue Replacement Algorithm for Second Level Buffer Caches. In: Pro of USENIX 2001, Boston, Massachusetts, USA, pp. 91–104 (June 2001)

SmartCare:Location Based e-Care Assistant on Azure

Ji Cheng, Mengdi Wang, Bo Li, and Xiaoling Wang

Shanghai Key Laboratory of Trustworthy Computing,
Institute for Data Science and Engineering, Software Engineering Institute,
East China Normal University, Shanghai, China
chengjirobin@gmail.com

Abstract. Facing the severe population aging trend in the World, the assistance of information technology is necessary when taking care of senior citizens. With the development of human-computer interaction the accessibility of electronic devices are greatly enhanced. This demonstration paper presents a system aiming at assisting guardians to nursing the elders mainly involving cloud computing, location based service and crowdsourcing; includes three components: (1)A phone app which allows guardians to schedule items; with regarding to the location and time the app alerts the elders with text-to-speech engines. (2)A wearable bracelet collects the vital signs of the elders and sends data to the cloud. (3)A real-time data storage and analysis service deployed on Azure, which collects the vital signs and GPS data of the elders then detect abnormal trajectories and vital signs, if detected it alerts the guardian.

1 Introduction

Information Technology is infiltrating our lives, greatly changing and improving the way we live. However senior citizens received little attention in this revolution, due to the myth of the senior citizens are not capable of operating digital devices. As more and more auxiliary accessing technologies like text-to-speech technologies and touch screens are prevailing every handhold device. The accessibility of electronic devices is greatly enhanced and attracting more and more older users. In [1] we can see that the population aging is unprecedented, pervasive and enduring, from 1950 to 2050 the proportion of older persons will soar from 8% to 21% and this trend in China is even worse [2]. It's of great importance to utilize information technology to assist the younger citizens to nurse their older parents.

To cope with this challenge, we propose our system SmartCare to help younger citizens nurse their older parents or community physicians taking care of senior citizens. Due to the reliability, scalability and computation ability of cloud computing service we mainly deployed an online probability graph[3] based user anomaly moving trajectory detection algorithm on the cloud. Conventionally in anomaly trajectory detection, like [4], it considers that for every trajectory with the same source and destination should have the same path. But in our case each person's trajectory is varied even with the same source and destination. The anomaly trajectory of the older person could indicate fraud of various kind.

F. Li et al. (Eds.): WAIM 2014, LNCS 8485, pp. 832–834, 2014.

We also implemented a few other intelligent monitoring mechanisms to help the guardians to take care of the older persons. For instance we build a free medicine information database by crowdsourcing[5].

2 The Framework

The main components of SmartCare are the smart phone app, the wearable bracelet and the data service on Azure. (Fig. 1) shows our system architecture.

The wearable bracelet measures heart rate and body temperature of the wearing person and transfers them to the smart phone via Bluetooth.

The smart phone app mainly utilized text-to-speech and image barcode reading technology and consists of Data Collector, Location Based Alarm and Medication Alarm. The Data Collector collects the vital-signs from the Bracelet send them to the data service on the cloud along with the phone's GPS data. Location Based Alarm is activated when the senior citizen is close to a schedule item set by the guardian. Medication Alarm uses text-to-speech engine to speak repeatedly the name and usage of the medicine to alert older person and reads the barcode on the medicine box to ensure the correct medicine is taken by the camera on the phone, if this process is not succeeded it will alert the guardian. Both logs generated are sent to the cloud and stored.

Windows Azure demonstrated outstanding computation ability and reliability [6] thus we deployed our cloud service on Azure. Our cloud service mainly involves probability graphs, crowdsourcing and location based range query tenancies and mainly consists of storage manager, anomaly detector, diary generator and location analyzer. Storage manager receives the data from the smart phone and stores the user information on the SQL like tables (referred to as SQL), while the GPS data and vital-sign data on the key value pair like table (referred to as table) on Azure. It also collects the medicine information input by users, our frequency and support based algorithm selects common entries and displays to future users to select the proper one. If an item has a high selection frequency it's considered as true and stored. Anomaly detector detects abnormal location and vital-sign of the senior citizens. We use Hidden Markov Model to model this

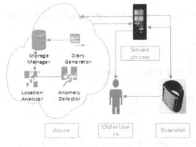

(a) Architecture of SmartCare

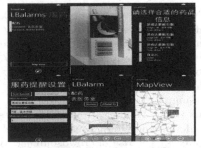

(b) Selected Interfaces of iCare

Fig. 1. The Architecture and Interface of SmartCare

problem. First we find the point of interests and consider the point of interests as states and use weather, date, time, phone call records as features. We trained an HMM for each user according to history trajectories of each user. When user reach a new point of interest, we predict the probability of this new point and if it's beneath the threshold then it is considered as an anomaly. To detect the abnormal vital signs by calculating the cosine similarity between the current data vector and the averaged data vector from the previous hour. If any anomaly is detected the guardian of the older person will be alerted by phone or text message detailing the plausible scenario and the location of the incident. The Dairy Generator generates a daily summary of the senior citizen, according to the logs submitted by phone app and the moving trajectory. Lastly the location Analyzer returns the nearby business of a GPS position to the phone app.

3   Demonstration Scenario

We implemented the phone app on Win-Phone, and built the wearable bracelet based on sensor modules. Selected interfaces of SmartCare are shown in(Fig. 1).

Our system mainly contains the following scenarios: (1) Monitors the moving trajectory and vital signs of the older person by an online detection of trajectory and vital sign anomalies service on the cloud. (2) Enables guardians to schedule the timetable of the senior citizens, and alerts the senior citizens if their current location is near the scheduled item. (3) A dairy is generated regarding of the data collected on the cloud, helping guardians to know the daily life of the senior citizens better. (4) As most senior citizens need long-term medication our system also provides a mechanism to assist and supervise the older person taking correct medicine on time. (5) If any anomalies are detected or any procedure failed our system will notify the guardian for the incident.

Acknowledgments. This work was supported by the 973 project(No. 2010CB328106), NSFC grant (No. 61170085 and 61033007), Program for New Century Excellent Talents in China (No.NCET-10-0388) and Shanghai Knowledge Service Platform Project (No. ZF1213).

References

1. UN Human Development Report (2005), http://www.un.org/esa/population/publications/worldageing19502050/
2. Chinese Population Aging Trends Forecasts, http://www.cncaprc.gov.cn/yanjiu/33.jhtml
3. Kollar, D., Friedman, N.: Probabilistic graphical models: principles and techniques. The MIT Press (2009)
4. Daqing, Z., Li, N., Zhou, Z.-H., Chen, C., Sun, L., Li, S.: iBAT: detecting anomalous taxi trajectories from GPS traces. In: Proceedings of the 13th International Conference on Ubiquitous Computing, pp. 99–108. ACM (2011)
5. Howe, J.: The rise of crowdsourcing. Wired Magazine 14(6), 1–4 (2006)
6. Johnston, S.J., O'Brien, N.S., Lewis, H.G., et al.: Clouds in space: Scientific computing using windows azure. Journal of Cloud Computing 2(1), 1–10 (2013)

Leveraging Focused Locations for Web Search

Xiaoxiang Zhang, Peiquan Jin, Sheng Lin, Shouhong Wan, and Lihua Yue

University of Science and Technology of China, 230027, Hefei, China
jpq@ustc.edu.cn

Abstract. In this paper, we propose to identify and use focused locations to improve Web search. Focused locations refer to the most important ones in a Web page, and the key issue is how to determine them and apply in Web search process. We present a location-assisted search tool called *LAST* (*Location-Assisted Search Tool*) to demonstrate the extraction and use of focused locations in Web search. After an overview of the general features of *LAST*, we discuss the architecture and implementation of *LAST*. And finally, a case study of *LAST*'s demonstration is presented.

Keywords: Web search, Location, Re-ranking.

1   Introduction

Many queries in Web search are related with time [1] and locations [2]. However, a Web page may mention lots of locations in its content, and generally they have different meaning and importance to the page. Therefore, we have to identify the *focused locations* for each Web page and design location-friendly ranking algorithms for search engine. Basically, focused locations in a Web page represent the most appropriate locations embedded explicitly or implicitly in the content.

Location-related queries have been studied by many re-searchers in related fields. Some of them are focused on building a vertical search engine using real geographical locations identified by longitudes and latitudes [3], but they did not pay attention to the focused locations. Some other works are about personalizing Web search using location data [4] or location-aware query processing [5]. However, to our best knowledge, there are very few previous works took into account focused locations.

We present a location-assisted search tool called *LAST* (*Location-Assisted Search Tool*) to demonstrate the extraction and use of the focused locations in Web search. *LAST* is designed to optimize the search results when answering location-related queries, concentrating on focused location extraction and new ranking methods. *LAST* is an optimized meta-search engine. It is built upon Google and Bing and can also support other types of search engines.

2   The Architecture of LAST

Figure 1 shows the architecture of *LAST*. In the **Snippet Processing** module, we use search API or crawlers to process the top-k results returned by Google and Bing.

F. Li et al. (Eds.): WAIM 2014, LNCS 8485, pp. 835–838, 2014.

In the **Focused Locations Extraction** module, we extract the focused locations for every Web page returned from an existing search engine. In this module, the algorithm for focused location extraction proposed in [6, 7] is used. In the **User Locations Extraction** module, we determine user locations on the basis of users' IP addresses. The user locations play an important role in the ranking procedure of the results. User locations are also studied in many areas, e.g., advertisement recommendation. In the **Query Location Extraction** module, we extract the locations included in the query string. Geographic predicates in queries are usually treated as common textual words in traditional search engines. However, in *LAST*, they are processed in a different way from the textual keywords. In our system, we simply determine query locations by looking up a tailor-made gazetteer.

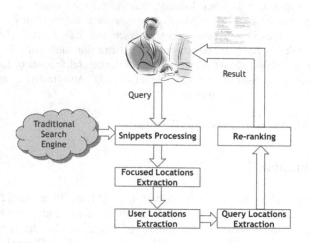

Fig. 1. The architecture of *LAST*

In the **Re-ranking** module, we re-rank the results returned by existing search engines. In order to compare *LAST* with existing search engines, both the original results and the re-ranked results of *LAST* will be shown in the user interface.

Generally, given a query Q, the ranking of a Web page A can be calculated by the following formula:

$$R_n(Q,A) = \alpha \cdot R_0(Q,A) + \beta \cdot rel(A.focus, Q.ip) + \gamma \cdot rel(A.focus, Q.search) \tag{1}$$

Here, α, β, and γ are the parameters for weight adjustment. They are range from 0 to 1, and their sum is 1. $R_0(Q, A)$ is the ranking score of A from a generic search engine. The score is reflected by A's order among all the results returned. $A.focus$ means the focused location in page A, $Q.ip$ is the user location which is extracted from the user's IP address, and $Q.search$ is the query location extracted from the query. $rel(X,Y)$ is the correlation score between two locations (X and Y).

3   Prototype and Demonstration

So far, *LAST* has been built up on the basis of Google and Bing. The interface of *LAST* is shown in Fig.2. the left frame shows the results returned from Google or Bing. In the right frame, the user location, values of parameters as well as some other information are shown on the top. The re-ranking results are shown with a rank-changed mark for each result. Also, we design two buttons (up and down) to get the users' feedback, which will be used to improve the re-ranking algorithms in the future. We also prepared several *<query, user location>* pairs to show the improvement by our re-ranking algorithms. Users can use the given pairs directly as well as input a new query to see the performance of *LAST*. For the given pairs, the original top-300 ranking results have been pre-processed, while for the user-input situation, *LAST* is focused on the original top-50 results due to the time cost.

Fig. 2. The interface of *LAST*

4   Conclusion

In this paper, we proposed to extract and use focused locations in Web search and presented a location-assisted search tool called *LAST*. *LAST* can determine the focused locations for Web pages and used them to re-rank the search results by incorporating different types of locations in Web search.

Acknowledgement. This paper is supported by NSFC (No. 61379037 and No. 71273010), the NSF of Anhui Province (no. 1208085MG117), and the OATF project in USTC.

References

[1] Zhao, X., Jin, P., Yue, L.: Automatic Temporal Expression Normalization with Reference Time Dynamic-Choosing. In: Proc. of COLING, pp. 1498–1506 (2010)

[2] Sanderson, M., Kohler, J.: Analyzing geographic queries. In: Proc. of GIR (2004)

[3] Cao, X., Cong, G., Jensen, C., Ng, J., et al.: SWORS: A System for the Efficient Retrieval of Relevant Spatial Web Objects. PVLDB 5(12), 1914–1917 (2012)

[4] Bennett, P., Radlinski, F., White, R.: Inferring and Using Location Metadata to Personalize Web Search. In: Proc. of SIGIR, pp. 135–144 (2011)

[5] Yu, B., Cai, G.: A Query-Aware Document Ranking Method for Geographic Information Retrieval. In: Proc. of GIR, pp. 49–54 (2007)

[6] Zhang, Q., Jin, P., Lin, S., Yue, L.: Extracting Focused Locations for Web Pages. In: Wang, L., Jiang, J., Lu, J., Hong, L., Liu, B. (eds.) WAIM 2011 Workshops. LNCS, vol. 7142, pp. 76–89. Springer, Heidelberg (2012)

[7] Jin, P., Zhang, X., Zhang, Q., Lin, S., Yue, L.: Ranking Web Pages by Associating Keywords with Locations. In: Wang, J., Xiong, H., Ishikawa, Y., Xu, J., Zhou, J. (eds.) WAIM 2013. LNCS, vol. 7923, pp. 613–618. Springer, Heidelberg (2013)

Author Index

Printed in the United States
By Bookmasters